ALGEBRAIC REVIEW

$$\frac{a}{d}+\frac{b}{d}=\frac{a+b}{d} \qquad \frac{a}{b}+\frac{c}{d}=\frac{ad+bc}{bd} \qquad \frac{a}{b}\cdot\frac{c}{d}=\frac{ac}{bd} \qquad \frac{a/b}{c/d}=\frac{ad}{bc}$$

$$ax+ay=a(x+y) \qquad x^2-y^2=(x+y)(x-y)$$

$$x^2+2xy+y^2=(x+y)^2 \qquad x^2-2xy+y^2=(x-y)^2$$

$$x^2+(a+b)x+ab=(x+a)(x+b)$$

$$acx^2+(ad+bc)x+bd=(ax+b)(cx+d)$$

$$x^3+y^3=(x+y)(x^2-xy+y^2)$$

$$x^3-y^3=(x-y)(x^2+xy+y^2)$$

$ax^2+bx+c=0$ has roots $x=\dfrac{-b\pm\sqrt{b^2-4ac}}{2a}$

$$a^m\cdot a^n=a^{m+n} \qquad (a^m)^n=a^{mn} \qquad (ab)^n=a^nb^n \qquad a^{-n}=1/a^n$$

$$a^{p/q}=\sqrt[q]{a^p}=(\sqrt[q]{a})^p$$

IMPORTANT LAWS AND FORMULAS

Triangle: Let the angles be A, B, and C and let the lengths of the opposite sides be a, b, and c. Then:

1. $A+B+C=180°$
2. Area $=\frac{1}{2}$ (length of side) (length of altitude on that side).
3. **Law of Sines:** $\dfrac{\sin A}{a}=\dfrac{\sin B}{b}=\dfrac{\sin C}{c}$
4. **Law of Cosines:** $c^2=a^2+b^2-2ab\cos C$
   $b^2=a^2+c^2-2ac\cos B$
   $a^2=b^2+c^2-2bc\cos A$

Circle: Let r be the radius and θ be the radian measure of the central angle of a sector. Then:

1. Circumference $(C)=2\pi r$
2. Area $(A)=\pi r^2$
3. Area of sector $=\frac{1}{2}r^2\theta$
4. Length of subtended arc $=r\theta$.

ELEMENTARY TECHNICAL MATHEMATICS

Under the Editorship of Calvin A. Lathan

ELEMENTARY TECHNICAL MATHEMATICS

James F. Connelly
Robert A. Fratangelo

Monroe Community College

Macmillan Publishing Co., Inc.
New York
Collier Macmillan Publishers
London

Printed in the United States of America

Macmillan Publishing Co., Inc.
866 Third Avenue, New York, New York 10022

Collier Macmillan Canada, Ltd.

Library of Congress Cataloging in Publication Data

Connelly, James Francis, (date)
Elementary technical mathematics.

Includes index.
1. Mathematics—1961– I. Fratangelo, Robert A., joint author. II. Title.
QA39.2.C65 512'.1 77-3003
ISBN 0-02-324430-5

Printing: 1 2 3 4 5 6 7 8 Year: 8 9 0 1 2 3 4

To our wives, Ginger and Jo, and our children,
Craig, Colleen, Charleen, and
Paula, Joseph, Debra, and Donna

PREFACE

This text has been written to meet the needs of students in the various technology programs offered in community colleges, junior colleges, and technical institutes. The material is intended to provide the student with a firm understanding of basic mathematical principles. It is our belief that such an understanding is necessary for a study of the applications of mathematics to technology and for the successful completion of more specialized courses.

The material has been written with the student in mind and is intended to be readable in a nontheoretical fashion with emphasis placed on practicability and applicability rather than on mathematical rigor. In this way, the applications of mathematics can be meaningfully illustrated and the student is able to develop more fully the basic skills needed by technicians.

Mathematical concepts, methods, and applications are carefully explained in all instances and motivated wherever possible. The more than 600 examples in the text are presented in detail and graded in degree of difficulty. Each exercise set is also graded in degree of difficulty and reflects the mathematics and applications of the corresponding section. In this way, the most important aspects of each section are reinforced and reemphasized.

Answers are provided for most odd-numbered exercises at the end of the book. Many of the numerical results were arrived at with the aid of a calculator. Since calculators "round off" results, the student may note some discrepancies in the answers. Such variances are generally due to the fact that most calculators have a higher degree of accuracy than the values given in Appendix F, Tables I through V. This should not cause the student any discomfort.

In Appendix E, we discuss some of the uses and applications of the hand-held calculator. At the end of this Appendix is a reference to the problems in the text that may be more easily solved with the aid of a calculator. Properly used, the calculator can greatly increase the efficiency of the student. It is our feeling, however, that the calculator does not offer a cure for arithmetic and algebraic deficiencies. The use of the calculator must complement computational skills and assist in the development of mathematical logic. For these reasons, we have discussed the development and use of many mathematical concepts in detail. This is especially true in the sections dealing with trigonometric functions and their corresponding values, and common and natural logarithms.

The use of the text depends on the instructor, the program, and the needs and background of the students. Since these factors will vary from school to school, we take special note of the following features.

Chapters 1, 2, and 3 are intended to provide the student with the minimum algebraic background necessary to the successful completion of the remainder of the text. The chapters can be taught, in full, used as a review course, or omitted, depending on the needs of the class.

At the end of each chapter, we have included both review questions and review exercises. Such problems provide the student with the opportunity to integrate mathematical ideas and applications on a chapter by chapter basis.

Most students acquire a mastery of mathematics only through practice and use. Thus more than 3000 exercises and 1000 review questions and review exercises are provided in the text so that students will have ample opportunity to develop and sharpen their skills.

Topics dealing with geometry, the metric system, dimensional analysis, approximations, rounding, significant digits, and uses of the hand-held calculator have been placed in the appendixes. In this way, the instructor may introduce any of these topics when necessary and revisit them when possible.

It is indeed a pleasure to acknowledge the many people who have assisted us in the preparation of the text. We wish to thank Professor James E. Anderson of the University of Utah and Calvin A. Lathan of Monroe Community College for reading the manuscript and making many valuable suggestions. We are indebted to Everett Smethurst and Leo Malek, editors at Macmillan, and William Setek, Jr., a colleague, for their concern and encouragement during the preparation and production of this book. To Pamela Dretto, for her faithful and competent typing of the manuscript we express our thanks for a job well done.

Once again, we owe an immeasurable debt of thanks to our wives, Ginger and Jo. Without their patience, understanding, and encouragement, this task might never have been completed.

J. F. C.
R. A. F.

CONTENTS

CHAPTER 1

Real Numbers and Their Properties

CHAPTER 2

Operations with Algebraic Expressions

CHAPTER 3

Equations in One Variable

CHAPTER 4

Functions and Graphs

CHAPTER 5

Inequalities

CHAPTER 6
Systems of Equations

CHAPTER 7
The Trigonometric Functions

CHAPTER 8
Graphs of the Trigonometric Functions

CHAPTER 9
Exponential Functions and Logarithms

CHAPTER 10
Complex Numbers

CHAPTER 1

Real Numbers and Their Properties

SECTION 1

Classification of Numbers

We begin our study of technical mathematics by discussing types of numbers. The first numbers that we shall use are the **positive integers**. They are represented by the symbols 1, 2, 3, and so on. Next are the **negative integers**, represented by the symbols -1, -2, -3, and so on. The combinations of the negative integers, the positive integers, and the number zero, represented by the symbol 0, constitute the numbers called the **integers**.

Any number that can be represented by the division of one integer by another is called a **rational number**. Thus, every integer can be considered to be a rational number, since such numbers may be written as $\frac{1}{1}$, $\frac{3}{1}$, $\frac{-3}{1}$, and so on. Fractions such as $\sqrt{2}/3$ and $\pi/2$ are not rational numbers, since they cannot be expressed as the ratio of two integers. Numbers that cannot be expressed as the ratio of two integers are called **irrational numbers**. The rational and the irrational numbers together form the **real numbers**.

Example 1.1

The number 3, or $\frac{3}{1}$, is considered to be both an integer and a rational number. The number $-\frac{3}{4}$ is a rational number. The number $\sqrt{2}$ is an irrational number. The numbers 3, $-\frac{3}{4}$, and $\sqrt{2}$ are all real numbers.

The real numbers can also be described as **nonterminating decimals**. The decimal $0.3\bar{3}$ means that the symbol 3 repeats forever. The decimal 0.2 can also be expressed as the nonterminating decimal $0.2\bar{0}$, since addition of zeros to the right of the symbol 2 does not alter the value of the decimal 0.2. Thus, both $0.3\bar{3}$ and $0.2\bar{0}$ are real numbers. Since $0.2\bar{0} = \frac{1}{5}$ and $0.3\bar{3} = \frac{1}{3}$, we classify all nonterminating decimals that repeat as **real rational numbers** and all nonterminating decimals that do not repeat as **irrational real numbers**.

Example 1.2

The decimal $0.14\overline{14}$ is a rational real number, because it repeats. The decimal 0.010203 . . . is an irrational real number, because it does not repeat.

We shall use the real numbers exclusively throughout this text, with the exception of a discussion of the **complex numbers**. At that point it will be necessary to discuss the square root of a negative number, which is not part of the real number system.

Exercises for Section 1

Classify each number according to the following categories: positive integer, negative integer, rational number, irrational number, real number, nonreal number. (If a number belongs to more than one category, so indicate.)

1. -3 **2.** $\frac{4}{3}$ **3.** 2.37

4. $1.\overline{14}$ **5.** 0 **6.** $-\frac{19}{6}$

7. 3 **8.** $0.25\bar{0}$ **9.** $\sqrt{4}$

10. $\sqrt{5}$ **11.** $\sqrt{-4}$ **12.** π

13. 0.020202 . . . **14.** 0.020406 . . . **15.** 0.020020002 . . .

SECTION 2

Basic Operations and Laws

The four basic operations with real numbers are addition, subtraction, multiplication, and division. The **addition** of two numbers a and b is denoted as $a + b$. Thus, $3 + 4 = 7$. When the number b is **subtracted** from the number a, the difference is denoted as $a - b$. Thus, $7 - 3 = 4$. The product of two real numbers a and b is a real number c such that $a \times b = c$. Thus, $2 \times 3 = 6$. When the number a is divided by b, we denote it as $a \div b$, or a/b, and we say that a/b equals a real number c if a is equal to the product of b and c. Thus, $8/4 = 2$, since $8 = 4 \times 2$.

We must exercise some care when we consider division. *Division by zero is not defined.* This can be demonstrated from the definition of division above. Suppose we assume that $1/0 = c$. This would imply that $1 = 0 \cdot c$, but this is impossible, since $0 \cdot c = 0$. Thus, we say that $1/0$ is *undefined*. The situation where we have $0 \div 0$, or $0/0$, is also impossible. For if we insisted that $0/0$ equaled the real number c, we would have $0 \cdot c = 0$, which is true for *any* real number but not a definite real number. To make a distinction between this case and the previous case, we call 0/0 **indeterminate**.

Example 2.1

(a) $\dfrac{4+3}{2-2} = \dfrac{7}{0}$ is undefined

(b) $\dfrac{2-2}{2-2} = \dfrac{0}{0}$ is indeterminate

(c) $\dfrac{3-3}{-6} = \dfrac{0}{-6} = 0$

In summary, we have:

1. $\dfrac{a}{b}$ is undefined if $a \neq 0$ and $b = 0$.
2. $\dfrac{a}{b}$ is indeterminate if $a = 0$ and $b = 0$.
3. $\dfrac{a}{b}$ equals zero if $a = 0$ and $b \neq 0$.

Zero is a very special real number which separates the positive real numbers and the negative real numbers into two collections. The negative numbers are preceded by a minus sign (−); the positive numbers are preceded by a plus sign (+) or by no sign at all. The two collections together are called the **signed numbers**, and we formulate the following basic operations concerning them.

Addition

1. To add two numbers with like signs, we add their numerical values and prefix their common sign.
2. To add two numbers with unlike signs, we find the difference between their numerical values and prefix the sign of the number with the greater numerical value.

Example 2.2

(a) $2 + 4 = 6$

(b) The numerical values of -3 and -5 are 3 and 5, respectively. Thus, $(-3) + (-5) = -(3 + 5) = -8$.

(c) $2 + (-4) = -2$

Subtraction

To subtract one number b from another number a, we change the sign of b and add.

Example 2.3

(a) $2 - 6 = 2 + (-6) = -4$
(b) $(-3) - 4 = -3 + (-4) = -7$
(c) $5 - (-7) = 5 + 7 = 12$

Multiplication

1. To multiply two numbers having like signs, multiply their numerical values together and prefix with a plus sign.
2. To multiply two numbers having unlike signs, multiply their numerical values together and prefix with a minus sign.

Example 2.4

$(2)(3) = 6 \qquad (-2)(-4) = 8 \qquad (-5)(3) = -15 \qquad (4)(-8) = -32$

Division

1. To divide two numbers having like signs, divide their numerical values and prefix with a plus sign.
2. To divide two numbers having unlike signs, divide their numerical values and prefix with a minus sign.

Example 2.5

$$\frac{8}{4} = 2 \qquad \frac{-6}{2} = -3 \qquad \frac{12}{-4} = -3 \qquad \frac{-18}{-3} = 6$$

Now that we have established how to add and multiply real numbers, we formulate the following laws for these operations.

Commutative Law for Addition: States that the order of the addition of two numbers does not alter the result.

Thus, $1 + 2 = 2 + 1 = 3$ or, in general, $a + b = b + a$.

Commutative Law for Multiplication: states that the order of the multiplication of two numbers does not alter the result.

Thus, $(4)(6) = (6)(4) = 24$ or, in general, $(a)(b) = (b)(a)$.

Associative Law for Addition: states that the terms of a sum may be grouped in any order without altering the result.

Thus, $1 + 2 + 3 = 1 + (2 + 3) = (1 + 2) + 3 = 6$ or, in general,

$a + b + c = a + (b + c) = (a + b) + c$.

Associative Law for Multiplication: states that the factors of a product may be grouped in any order without altering the result.

Thus, $3 \cdot 4 \cdot 5 = 3 \cdot (4 \cdot 5) = (3 \cdot 4) \cdot 5 = 60$ or, in general,

$abc = a(bc) = (ab)c.$

Distributive Law for Multiplication says that the product of a number a by the sum of b and c is equal to the sum of the products of a and b and a and c.

Thus, $3(2 + 4) = 3 \cdot 2 + 3 \cdot 4 = 18$ or, in general, $a(b + c) = ab + ac.$

Exercises for Section 2

In Exercises 1 through 28, perform the indicated operation (if possible).

1. $13 + 6$ **2.** $4 + (-11)$ **3.** $(-4) + (-6)$

4. $(-8) \cdot 0$ **5.** $(-6) \cdot (-3)$ **6.** $\frac{-6}{-3}$

7. $(+7) - (+9)$ **8.** $\frac{-1}{-1}$ **9.** $(-7) - (-9)$

10. $(-8) - (+10)$ **11.** $9 - 0$ **12.** $0 - (-7)$

13. $\frac{-24}{2}$ **14.** $\frac{24}{-2}$ **15.** $\frac{-24}{-2}$

16. $\frac{3 \cdot 0}{2 - 2}$ **17.** $\frac{0}{0 - 4}$ **18.** $\frac{5 - 5}{5 - 5}$

19. $(-6)(8 + 3)$ **20.** $(-4)(7 - 9)$ **21.** $\frac{4 \cdot 0 \cdot 7}{2}$

22. $\frac{6(-6)}{-3}$ **23.** $\frac{3(-4)(-5)}{6 - 4}$ **24.** $14 - (6 + 2)$

25. $6 + (-5) + (-7)$ **26.** $4(7 - 6)$ **27.** $3(9 - 12)$

28. $\frac{3(-4) - (2)(4)}{3 - 8}$

29. What is the sign of the product of an odd number of negative factors? Of the product of an even number of negative factors?

SECTION 3

Number Line and the Ordering of the Real Numbers

All real numbers may be represented graphically as points on a line. This line is referred to as the **real number line** and is shown in Figure 3.1. The point on the line corresponding to the real number zero is called the **origin**. The positive real numbers are associated with points to the right of the origin, and the negative real numbers are associated with the points on the line to the left of the origin.

Figure 3.1

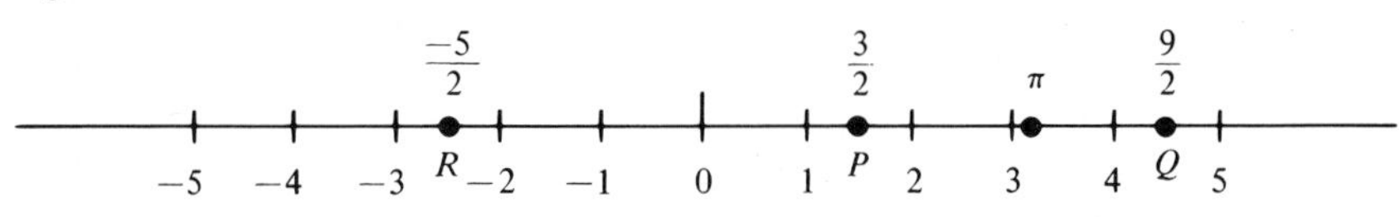

The position of the real numbers on the number line establishes an order to the real numbers. If a point Q lies to the right of another point P on the number line, we say that the number corresponding to Q is **greater than** the number corresponding to P, or that the number corresponding to P is **less than** the number corresponding to Q. The symbols $>$ and $<$ are used to represent "greater than" and "less than," respectively. These symbols are called **inequality signs**.

Example 3.1

In Figure 3.1 we see that the real number $-\frac{5}{2}$ corresponding to the point R is less than the real number $\frac{3}{2}$ corresponding to the point P. Symbolically, we write $-\frac{5}{2} < \frac{3}{2}$ or $\frac{3}{2} > -\frac{5}{2}$. Similarly, from Figure 3.1 we have $\frac{9}{2} > \pi$ or $\pi < \frac{9}{2}$.

Another useful mathematical concept is the **absolute value** of a real number. We say that the absolute value of a real number is the **undirected distance** between the origin and the real number on the number line. The absolute value of a real number a is denoted $|a|$.

Example 3.2

(a) $|6| = 6$, since the distance between 0 and 6 on the number line is 6 units.

(b) $|-6| = 6$, since the distance between 0 and -6 is 6 units.

(c) $|6 - 4|$ or $|4 - 6|$ is equal to 2, since the distance between 6 and 4 is 2 units.

It should be noted that we can define the absolute value of a real number, a, as follows:

$$|a| = \begin{cases} a & \text{when } a > 0 \\ 0 & \text{when } a = 0 \\ -a & \text{when } a < 0 \end{cases}$$

Exercises for Section 3

1. *Place the appropriate inequality symbol (< or >) between each pair of real numbers.*

(a) $4, 6$ (b) $0, -1$ (c) $-\frac{1}{2}, -\frac{3}{2}$

(d) $-1, 3$ (e) $3, -3$ (f) $3.14, \pi$

(g) $-\frac{3}{5}, -\frac{3}{4}$ (h) $0.2, 0.\bar{2}$ (i) $\sqrt{4}, \sqrt{9}$

(j) $|-6|, |-4|$

2. *Find the absolute value of each of the following.*

(a) $|3|$ (b) $|-3|$ (c) $|-8|$
(d) $|0 - 8|$ (e) $|6 - 2|$ (f) $|2 - 6|$
(g) $|6 - (-3)|$ (h) $|-4 - 7|$ (i) $|-\sqrt{4}|$

3. *Find the absolute value of each of the following.*

(a) $|6 \cdot 2|$ (b) $|(-6) \cdot 2|$ (c) $|(-8) \cdot 0|$

(d) $|4(7 - 6)|$ (e) $|4(6 - 7)|$ (f) $\left|\frac{6(-3)}{6}\right|$

(g) $\left|\frac{4 \cdot 0 \cdot 7}{2}\right|$ (h) $|2 - 3(-4)|$ (i) $|2 + 3(-4)|$

SECTION 4

Exponents

When a number a is multiplied by itself n times, the product $\underbrace{a \cdot a \cdot a \cdots a}_{n}$ (n of them) is denoted by the symbol a^n, which can be read "a to the nth power" or "the nth power of a." The number a is called the **base**, and the positive integer n is called the **exponent**.

We now state and illustrate some important laws governing the uses of these exponents for positive integers.

$$a^m \cdot a^n = a^{m+n} \tag{4.1}$$

$$(a^m)^n = a^{mn} \tag{4.2}$$

$$(ab)^n = a^n \cdot b^n \tag{4.3}$$

$$\left(\frac{a}{b}\right)^n = \frac{a^n}{b^n} \qquad b \neq 0 \tag{4.4}$$

$$\frac{a^m}{a^n} = \begin{cases} a^{m-n} & \text{if } m > n, \quad a \neq 0 \\ 1 & \text{if } m = n, \quad a \neq 0 \\ \dfrac{1}{a^{n-m}} & \text{if } m < n, \quad a \neq 0 \end{cases} \tag{4.5}$$

Example 4.1

(a) $3^2 \cdot 3^3 = 3^{2+3} = 3^5$, since $3^2 \cdot 3^3 = (3 \cdot 3)(3 \cdot 3 \cdot 3) = 3^5$
(b) $(2^3)^2 = 2^{3(2)} = 2^6$, since $(2^3)^2 = (2 \cdot 2 \cdot 2)(2 \cdot 2 \cdot 2) = 2^6$
(c) $(3 \cdot 2)^2 = 3^2 \cdot 2^2$, since $(3 \cdot 2)^2 = (3 \cdot 2)(3 \cdot 2) = (3 \cdot 3)(2 \cdot 2) = 3^2 \cdot 2^2$

(d) $\left(\frac{2}{3}\right)^3 = \frac{2^3}{3^3}$, since $\left(\frac{2}{3}\right)^3 = \left(\frac{2}{3}\right)\left(\frac{2}{3}\right)\left(\frac{2}{3}\right) = \frac{2 \cdot 2 \cdot 2}{3 \cdot 3 \cdot 3} = \frac{2^3}{3^3}$

(e) $\frac{4^3}{4^2} = 4^{3-2} = 4^1$, since $\frac{4^3}{4^2} = \frac{4 \cdot 4 \cdot 4}{4 \cdot 4} = 4$

(f) $\frac{5^2}{5^6} = \frac{1}{5^{6-2}} = \frac{1}{5^4}$, since $\frac{5^2}{5^6} = \frac{5 \cdot 5}{5 \cdot 5 \cdot 5 \cdot 5 \cdot 5 \cdot 5} = \frac{1}{5^4}$

(g) $\frac{3^2}{3^2} = 3^0 = 1$, since $\frac{3^2}{3^2} = \frac{3 \cdot 3}{3 \cdot 3} = 1$

If n is a positive integer, we define $a^{-n} = 1/a^n$, $a \neq 0$. Thus, $3^{-2} = 1/3^2$ and $5^{-3} = 1/5^3$. We also say that $a^0 = 1$ if $a \neq 0$. Thus, $3^0 = 1$ and $(-2)^0 = 1$. These two definitions allow us to extend the use of the laws of exponents given by formulas (4.1) through (4.5) to include zero and the negative integers.

Example 4.2

(a) $3^{-2} = \frac{1}{3^2} = \frac{1}{9}$, $\frac{a^2}{a^6} = a^{2-6} = a^{-4} = \frac{1}{a^4}$

(b) $\frac{4^5}{4^5} = 4^{5-5} = 4^0 = 1$, $2^{-3} \cdot 2^{-2} = 2^{-5} = \frac{1}{2^5}$

(c) $(3 \cdot 2)^0 = 3^0 \cdot 2^0 = 1$

The question that now arises is: Can we attach any meaning to such expressions as $2^{1/2}$, $3^{1/3}$, $8^{2/3}$, $a^{3/4}$, and so on? The answer is yes, if we make the following agreement.

If n is a positive integer and if a and b are such that $a^n = b$, then a is called an nth **root** of b. If b is positive, it may have more than one nth root. In such cases we will call the positive nth root the **principal** nth *root* and denote it $\sqrt[n]{b}$ or $b^{1/n}$. The symbol $\sqrt{\ }$ is called the **radical sign,** n is called the **index**, and b is called the **radicand**. If $n = 2$, it is common practice to express $b^{1/2}$ simply as $\sqrt{b}$ (read "square root of b").

Example 4.3

(a) $3^{1/2} = \sqrt{3}$ is read "square root of three"
(b) $3^{1/3} = \sqrt[3]{3}$ is read "cube root of three"
(c) $3^{1/4} = \sqrt[4]{3}$ is read "fourth root of three"

Example 4.4

$4^{1/2} = \sqrt{4} = 2$, since $2^2 = 4$. Note that -2 is a square root of 4, since $(-2)^2 = 4$, but it is *not* the principal second or square root of 4.

Example 4.5

$(-4)^{1/2} = \sqrt{-4}$ is nonreal, since there is no real number whose square will equal -4.

Example 4.6

(a) $-\sqrt{16} = -4$, $2 + \sqrt{4} = 2 + 2 = 4$
(b) $8^{1/3} = \sqrt[3]{8} = 2$ and 2 is the principal third root of 8
(c) $(-8)^{1/3} = \sqrt[3]{-8} = -2$ and -2 is the principal third root of -8
(d) $16^{-1/4} = \dfrac{1}{16^{1/4}} = \dfrac{1}{\sqrt[4]{16}} = \dfrac{1}{2}$

We now can give significance to the real numbers whose exponents are rational numbers. We say that

$$a^{p/q} = (a^{1/q})^p \qquad p \text{ and } q \text{ are positive integers} \tag{4.6}$$

It should be carefully noted that $a^{p/q}$ will not be a real number when a is negative and q is even (unless p is also even).

Example 4.7

(a) $8^{2/3} = (8^{1/3})^2 = 2^2 = 4$
(b) $25^{3/2} = (25^{1/2})^3 = 5^3 = 125$
(c) $(-9)^{3/2} = (-9^{1/2})^3$ is nonreal, since $(-9)^{1/2}$ is nonreal
(d) $(-27)^{2/3} = (-27^{1/3})^2 = (-3)^2 = 9$
(e) $(-1)^{2/3} = (-1^{1/3})^2 = (-1)^2 = 1$

With formula (4.6) we can now extend formulas (4.1) through (4.5) to include exponents that are rational numbers.

Example 4.8

(a) $8^{-2/3} = \dfrac{1}{8^{2/3}} = \dfrac{1}{(8^{1/3})^2} = \dfrac{1}{2^2} = \dfrac{1}{4}$

(b) $\left(\dfrac{1}{16}\right)^{1/4} = \dfrac{1^{1/4}}{16^{1/4}} = \dfrac{1}{2}$

(c) $\dfrac{2^{-1/2}}{2^{1/2}} = 2^{-(1/2)-(1/2)} = 2^{-1} = \dfrac{1}{2}$

(d) $\dfrac{2^0 - 2^{-2}}{2 - 2(2)^{-2}} = \dfrac{1 - \dfrac{1}{2^2}}{2 - \dfrac{2}{2^2}} = \dfrac{1 - \dfrac{1}{4}}{2 - \dfrac{2}{4}} = \dfrac{\dfrac{3}{4}}{\dfrac{6}{4}} = \dfrac{3}{6} = \dfrac{1}{2}$

Summary

1. $a^{1/n} = \sqrt[n]{a}$, where n is an even positive integer and $a > 0$, will be a real positive number.
2. $a^{1/n} = \sqrt[n]{a}$, where n is an odd positive integer and $a < 0$, will be a negative real number.
3. $a^{1/n} = \sqrt[n]{a}$, where n is an odd positive integer and $a > 0$, will be a positive real number.
4. $a^{1/n} = \sqrt[n]{a}$, where n is a positive even integer and $a < 0$, will *never be a real number*.

The exponent laws (4.1) through (4.5) can be used to deal with fractional exponents as long as the $a^{1/n}$ is a real number.

Exercises for Section 4

In Exercises 1 through 43, evaluate (if possible).

1. 2^3

2. $(-3)^2$

3. $2^4 \cdot 2^2$

4. $(3^2)^2$

5. $\left(\dfrac{2}{3}\right)^3$

6. $\left(\dfrac{1}{2}\right)^0$

7. $\dfrac{3^5}{3^2}$

8. $\dfrac{4^6}{4^8}$

9. $(2 \cdot 4)^2$

10. $16^{1/2}$

11. $(-16)^{1/4}$

12. $32^{1/5}$

13. $\dfrac{5^{2/3}}{5^{-1/3}}$

14. $1^{1/4}$

15. $(-1)^{3/2}$

16. $\sqrt{121}$

17. $-\sqrt{121}$

18. $\sqrt{-81}$

19. $\left(\frac{16}{9}\right)^{1/2}$

20. $\left(\frac{16}{9}\right)^{3/2}$

21. $\left(\frac{4}{25}\right)^{-1/2}$

22. $27^{4/3}$

23. $4^{1/4} \cdot 8^{1/2}$

24. $\frac{4^{1/2}}{4^{-1/2}}$

25. $(6^{2/3})^3$

26. $(5^4)^{1/2}$

27. $\left(\frac{1}{4}\right)^{5/2}$

28. $49^{-3/2}$

29. $\frac{3^{1/2} \cdot 3^{1/3}}{3^{-1/2} \cdot 3^{-2/3}}$

30. $8^{2/3} \cdot 16^{-3/4} \cdot 2^0$

31. $\sqrt{32}$

32. $\sqrt{54}$

33. $\sqrt{24}$

34. $(\sqrt[3]{25})^3$

35. $(\sqrt[4]{5})^4$

36. $\sqrt[3]{-125}$

37. $(3^{1/2})^{1/2}$

38. $\sqrt{\sqrt{3}}$

39. $(2^{2/3})^{3/2}$

40. $(3^{3/4})^{4/3}$

41. $\sqrt{16^{1/2}}$

42. $(\sqrt{16})^{1/2}$

43. $\sqrt{\sqrt{16}}$

In Exercises 44 through 48, state the principal nth root for each.

44. $\sqrt[3]{-125}$

45. $\sqrt[4]{81}$

46. $\sqrt[5]{-1}$

47. $\sqrt[6]{1}$

48. $\sqrt[3]{-64}$

SECTION 5

Radicals

In Section 4 we established the fact that $a^{1/n} = \sqrt[n]{a}$ is called a radical. We shall frequently find it necessary to simplify expressions involving such radicals. The laws (4.1) through (4.5) allow us to deal with these expressions. Expressions such as $\sqrt{-4}$, $\sqrt{-16}$, $\sqrt[4]{-1}$, and $\sqrt[6]{-64}$ are nonreal numbers and are *not to be considered* when we are using the laws of exponents.

Example 5.1

(a) $(\sqrt{3})(\sqrt{12}) = \sqrt{(3)(12)} = \sqrt{36} = 6$
(b) $(\sqrt[3]{2})(\sqrt[3]{4}) = \sqrt[3]{8} = 2$
(c) $(\sqrt{6})(\sqrt{5}) = \sqrt{30}$
(d) $\sqrt{18} = \sqrt{(9)(2)} = 3\sqrt{2}$
(e) $\sqrt{125} = \sqrt{(25)(5)} = 5\sqrt{5}$
(f) $\sqrt[3]{54} = \sqrt[3]{(27)(2)} = 3\sqrt[3]{2}$
(g) $(\sqrt{5})^2 = (\sqrt{5})(\sqrt{5}) = \sqrt{25} = 5$

Example 5.2

The expression $\sqrt{8} + \sqrt{50} + \sqrt{98}$ can be simplified as follows: $\sqrt{8} = \sqrt{(4)(2)} = 2\sqrt{2}$, $\sqrt{50} = \sqrt{(25)(2)} = 5\sqrt{2}$, and $\sqrt{98} = \sqrt{(49)(2)} = 7\sqrt{2}$. Thus,

$$\sqrt{8} + \sqrt{50} + \sqrt{98} = 2\sqrt{2} + 5\sqrt{2} + 7\sqrt{2} = 14\sqrt{2}$$

It should be noted that we combined the three terms to obtain $14\sqrt{2}$ because the radicals in each term were **similar** or **like** radicals.

Example 5.3

The expression $\sqrt{2} + \sqrt[3]{2} + \sqrt[4]{2}$ is in its simplest form, since none of the terms have like radicals that can be combined.

Example 5.4

The expression $\sqrt{5}/\sqrt{2}$ can be simplified as follows:

$$\frac{\sqrt{5}}{\sqrt{2}} = \sqrt{\frac{5}{2}} = \sqrt{\frac{5}{2}\cdot\frac{2}{2}} = \sqrt{\frac{10}{4}} = \frac{\sqrt{10}}{2}$$

Note that by multiplying the fraction 5/2 by 2/2 we made the denominator of the fraction a perfect square root.

Example 5.5

$$\frac{\sqrt[3]{3}}{\sqrt[3]{2}} = \sqrt[3]{\frac{3}{2}} = \sqrt[3]{\frac{3}{2}\cdot\frac{2^2}{2^2}} = \sqrt[3]{\frac{12}{2^3}} = \frac{\sqrt[3]{12}}{2}$$

Note that by multiplying the fraction 3/2 by $2^2/2^2$ we made the denominator of the fraction a perfect cube root.

Example 5.6

$$\begin{aligned}\sqrt{3}(\sqrt{6} + \sqrt{8}) &= (\sqrt{3})(\sqrt{6}) + (\sqrt{3})(\sqrt{8}) \\ &= \sqrt{18} + \sqrt{24} \\ &= \sqrt{(9)(2)} + \sqrt{(4)(6)} \\ &= 3\sqrt{2} + 2\sqrt{6}\end{aligned}$$

The terms $3\sqrt{2}$ and $2\sqrt{6}$ cannot be combined since the radicals are not similar.

Example 5.7

(a) $(\sqrt[3]{3})(\sqrt{3}) = 3^{1/3} \cdot 3^{1/2} = 3^{1/3+1/2} = 3^{5/6} = \sqrt[6]{3^5}$

(b) $(\sqrt{2})(\sqrt[3]{3})$ cannot be simplified by using formula (4.1), since the base numbers are different. The product $(\sqrt{2})(\sqrt[3]{3})$ can be simplified as follows:

$$(\sqrt{2})(\sqrt[3]{3}) = 2^{1/2} \cdot 3^{1/3} = 2^{3/6} \cdot 3^{2/6} = (2^3)^{1/6} \cdot (3^2)^{1/6} = (8 \cdot 9)^{1/6} = \sqrt[6]{72}$$

Example 5.8

$$\frac{\sqrt[3]{4}}{\sqrt[4]{4}} = \frac{(4)^{1/3}}{(4)^{1/4}} = 4^{(1/3)-(1/4)} = 4^{1/12} = \sqrt[12]{4}$$

Note that formula (4.5) can be applied, since we have the same base number.

Example 5.9

Since $\sqrt[3]{2}/\sqrt{3} = 2^{1/3}/3^{1/2}$, we cannot apply formula (4.5), but we could apply (4.4) as follows:

$$\frac{2^{1/3}}{3^{1/2}} = \frac{2^{2/6}}{3^{3/6}} = \frac{\sqrt[6]{2^2}}{\sqrt[6]{3^3}} = \sqrt[6]{\frac{2^2}{3^3}} = \sqrt[6]{\frac{2^2 \cdot 3^3}{3^3 \cdot 3^3}} = \sqrt[6]{\frac{(4)(27)}{3^6}} = \frac{\sqrt[6]{108}}{3}$$

Example 5.10

The expression $\sqrt[3]{\sqrt{27}}$ can be simplified by applying formula (4.2) as follows:

$$\sqrt[3]{\sqrt{27}} = (27^{1/2})^{1/3} = 27^{1/6} = \sqrt[6]{27} = \sqrt[6]{3^3} = 3^{3/6} = 3^{1/2} = \sqrt{3}$$

Exercises for Section 5

Simplify and collect like radicals (when possible) for each of the following.

1. $(\sqrt{3})^2$

2. $(\sqrt[3]{4})^3$

3. $(\sqrt[4]{2})^4$

4. $(\sqrt{4})(\sqrt{5})$

5. $(\sqrt{2})(\sqrt{18})$

6. $(\sqrt[3]{3})(\sqrt[3]{9})$

7. $\sqrt[3]{81}$

8. $\sqrt[3]{32}$

9. $\sqrt[3]{108}$

10. $\sqrt{98}$

11. $\sqrt{75}$

12. $\sqrt{72}$

13. $\sqrt{180}$

14. $(\sqrt[3]{3})(\sqrt{3})$

15. $(\sqrt[3]{4})(\sqrt[4]{8})$

16. $3\sqrt{2} + 4\sqrt{2} - 6\sqrt{2}$

17. $6\sqrt[3]{3} - 4\sqrt[3]{3} - 2\sqrt[3]{4}$

18. $4\sqrt{5} + 2\sqrt[3]{3} + 3\sqrt{5} - \sqrt[3]{3}$

19. $\sqrt{32} - \sqrt{8} + \sqrt{18}$

20. $3\sqrt{12} + 2\sqrt{48} - 2\sqrt{75}$

21. $\sqrt{98} + 2\sqrt{18} - 3\sqrt{128}$

22. $2\sqrt{48} - 3\sqrt{12} + 4$

23. $\sqrt{\frac{1}{2}}$

24. $\sqrt{\frac{4}{3}}$

25. $\frac{\sqrt{5}}{\sqrt{3}}$

26. $\frac{\sqrt{10}}{\sqrt{7}}$

27. $\frac{\sqrt{10}}{\sqrt{5}}$

28. $\frac{\sqrt[4]{8}}{\sqrt[3]{4}}$

29. $\frac{\sqrt[3]{5}}{\sqrt[3]{4}}$

30. $\frac{\sqrt{8}}{\sqrt[3]{4}}$

31. $\frac{\sqrt{6}}{\sqrt[3]{2}}$

32. $\sqrt[4]{\sqrt[3]{5}}$

33. $\sqrt{\sqrt[3]{7}}$

34. $\frac{\sqrt[3]{3}}{\sqrt[4]{4}}$

35. $\dfrac{\sqrt{\frac{4}{3}}}{\sqrt[3]{\frac{1}{2}}}$

REVIEW QUESTIONS FOR CHAPTER 1

1. What kinds of numbers constitute the integers?
2. Describe a rational number.
3. What are real numbers?
4. What are the four basic operations with real numbers?
5. Why is division by 0 not defined?
6. What is 0/0 called?
7. Explain the difference between undefined and indeterminate.
8. Describe the makeup of the real number line.
9. What is the absolute value of a real number?
10. In the expression a^n, what do we call a and what do we call n?
11. What is the value of a^0 if $a \neq 0$?
12. Can a number have more than one nth root?
13. Can a number have more than one principal nth root?

14. Under what conditions is $a^{p/q}$ not a real number?

15. Why is $\sqrt{2} + \sqrt{3} + \sqrt{5}$ in simplest form?

REVIEW EXERCISES FOR CHAPTER 1

In Exercises 1 through 10, simplify each expression.

1. $10 + 6$

2. $-2 + (-12)$

3. $\dfrac{-6 - (-8)}{2}$

4. $0 - (-4)$

5. $0 \cdot (-4)$

6. $\dfrac{(2 + 3) \cdot 0}{5}$

7. $(2 - 3) \cdot (4 - 5)$

8. $(3 + 4) \cdot (4 + 5)$

9. $3(5 + 8)$

10. $3 \cdot 5 + 3 \cdot 8$

In Exercises 11 through 20, find the absolute value.

11. $|5|$

12. $|-5|$

13. $|5 - 2|$

14. $|2 - 5|$

15. $|-2(3 - 8)|$

16. $\left|\dfrac{4(-2)}{6}\right|$

17. $\left|\dfrac{4 \cdot 5 \cdot 0}{2}\right|$

18. $|-2 \cdot \sqrt{9}|$

19. $\left|\dfrac{-2(3 + 6)}{5}\right|$

20. $\left|\dfrac{4 - 3 + (-7)}{2 - 6}\right|$

In Exercises 21 through 30, determine if each expression is undefined, indeterminate, or zero.

21. $\dfrac{3 + (-3)}{4 + 1}$

22. $\dfrac{4 \cdot (2 + 3)}{6 - (4 + 2)}$

23. $\dfrac{4 + 0}{4 \cdot 0}$

24. $\dfrac{4 \cdot 0}{4 + 0}$

25. $\dfrac{4 \cdot 0}{4 \cdot 0}$

26. $\dfrac{2 + (3 - 5)}{4 - (2 + 6)}$

27. $\dfrac{2 + (3 - 5)}{2 - (5 - 3)}$

28. $\dfrac{3 \cdot 5}{2 - (7 - 5)}$

29. $\dfrac{[4 + (2 - 3)] \cdot 0}{5 + (2 - 1)}$

30. $\dfrac{6 + (2 \cdot 5)}{-5 + (8 - 3)}$

In Exercises 31 through 50, evaluate each expression when possible.

31. 2^4

32. $-(2)^4$

33. $(-2)^4$

34. $\left(\dfrac{3}{4}\right)^3$

35. $(6 \cdot 9 \cdot 24)^0$

36. $\dfrac{2^5}{2^8}$

37. $\left(\dfrac{9}{16}\right)^{1/2}$

38. $81^{3/4}$

39. $25^{-3/2}$

40. $(36)^{-3/2}$ **41.** $\dfrac{3^{5/4}}{3^{-5/4}}$ **42.** $(7^6)^{1/3}$

43. $(\sqrt[3]{2})^6$ **44.** $\sqrt{49}$ **45.** $\sqrt{48}$

46. $\sqrt[3]{-8}$ **47.** $\sqrt{\sqrt{81}}$ **48.** $\sqrt[8]{1}$

49. $\sqrt[8]{-1}$ **50.** $\sqrt[7]{-1}$

In Exercises 51 through 60, simplify each expression if possible.

51. $\sqrt{24}$ **52.** $\sqrt[3]{24}$ **53.** $\sqrt{50}$

54. $\dfrac{\sqrt{18}}{\sqrt{3}}$ **55.** $\dfrac{\sqrt{12}}{\sqrt{3}}$ **56.** $\sqrt{2} + \sqrt{8} - \sqrt{18}$

57. $\sqrt{12} + \sqrt{27} + 2\sqrt{3}$ **58.** $3\sqrt{8} - 2\sqrt{12} + \sqrt{18}$

59. $2 + 2\sqrt{2} + 3\sqrt{72}$ **60.** $4\sqrt{2} - 2\sqrt{75} + \sqrt{50} + 3\sqrt{48}$

CHAPTER 2

Operations with Algebraic Expressions

SECTION 6

Algebraic Expressions

A single number or letter that represents a number, or a combination of numbers and letters that represent numbers, is called an **algebraic expression**.

Example 6.1

2, $3x$, $x - 4$, $\dfrac{4}{3x}$, $3x^2y^3$, and $x^2 + xy + y^2$ are algebraic expressions.

An algebraic expression that does not contain plus or minus signs or inequality signs is called a **term**. An expression consisting of one term is called a **monomial**. Algebraic expressions containing two terms are called **binomials**. Those containing three terms are called **trinomials**. If an algebraic expression consists of more than one term, it is also called a **multinomial**.

Example 6.2

(a) x, $3x^2$, $4x/5y$, $-2xy^3x^2$ are called monomials, since they are all single terms.

(b) $6x^2 + 2y$ is a binomial, since it consists of the two terms $6x^2$ and $2y$.

(c) $3x^2 + 2xy + 4y$ is a trinomial, since it consists of the three terms $3x^2$, $2xy$, and $4y$.

(d) The expressions $6x^2 + 2y$ and $3x^2 + 2xy + 4y$ are also called multinomials.

Example 6.3

The term $-3xy$ means "-3 times x times y." Thus, if $x = 2$ and $y = 4$, the value of the term $-3xy = -3(2)(4) = -24$.

The factors of a term such as $4xy$ are 4, x, and y. We call 4 the **numerical coefficient** of the term $4xy$. We call $4x$ the *coefficient of* y in the term $4xy$ and we call $4y$ the **coefficient of** x in the term $4xy$. We say that **like terms** are terms that differ only by numerical coefficients.

Example 6.4

The terms $2x$, $-3x$, and $9x$ are like terms, since they only differ by numerical coefficients. The terms $4a^2b$, $12a^2b$, and $-8a^2b$ are also like terms.

Two or more like terms in an algebraic expression may be combined into one term. We say that an algebraic expression with two or more terms is completely simplified when all like terms in the expression have been combined.

Example 6.5

(a) The expressions $8x^2y + 3x^2y - 4x^2y$ can be simplified to the monomial $7x^2y$ by noting that $8x^2y + 3x^2y - 4x^2y = 11x^2y - 4x^2y = 7x^2y$.

(b) The expression $2x + 4x - 3y$ can be simplified to a binomial by combining the like terms $2x$ and $4x$ to obtain $6x - 3y$.

(c) The expression $6 + 2x^2y - 3xy^2$ is in its simplest form, since none of the terms are like terms and thus no combining is possible.

It should be noted that the concept of like terms is consistent with the concept of like radicals presented in the preceding section.

Exercises for Section 6

In Exercises 1 through 5, state the number of terms in each expression.

1. $2 + 3 - 4$

2. $xy + y$

3. $x^2y^3 + 2x + 1$

4. $2x + 2y + 2b + 2c$

5. $\frac{x}{y} + 4z - \frac{y}{x} + 6$

In Exercises 6 through 12, state the coefficient indicated for each term.

6. Coefficient of x in the term $4x$.

7. Coefficient of ab in the term abc^2.

8. Coefficient of z in the term $2xy^2z$.

9. Numerical coefficient of $-3abc$.

10. Numerical coefficient of $x^2y^3z^3$.

11. Numerical coefficient of $-xyz^2$.

12. Numerical coefficient of $ab/2$.

13. Determine which of the following groups of terms contain all like terms.
(a) $3x^2, -3x^2, 12x^2$
(b) $4xz, -2xz, xz, xz/2$
(c) $x^2y, 2x^2y, xy^2$
(d) $y, 2y^2, -y, 4y, 5$
(e) $\sqrt{2}, 3\sqrt{2}, 8\sqrt{2}$
(f) $\sqrt{2}, \sqrt[3]{2}, \sqrt[4]{2}$

In Exercises 14 through 21, evaluate each expression when $a = 1$, $b = -2$, $c = 3$, and $d = 0$.

14. $2a^2 + 3ab$

15. $b^3 - 2ab + bd$

16. $\dfrac{3ab + bc}{a}$

17. $\dfrac{b^2 - c^2}{c}$

18. $\dfrac{a}{b} + \dfrac{d}{c}$

19. $abcd$

20. $a^2 + b^2 + c^2$

21. $a^3 - b^3 - c^3$

22. Simplify the following:
(a) $2x + 3x + 5x$
(b) $2xy + 3xy + x^2y$
(c) $3\sqrt{2} + 4\sqrt{2} - 2\sqrt{2}$
(d) $a^2b + 2ab^2 + 3a^2b + 4ab^2$
(e) $\sqrt{18} + \sqrt{72} + \sqrt{24}$
(f) $3\sqrt{54} - 2\sqrt{24} + \sqrt{96}$
(g) $\sqrt{50} + \sqrt{125} + \sqrt{25}$

23. Write an algebraic expression for each of the following.
(a) The sum of 3, x, y, and z.
(b) x subtracted from y.
(c) The product of x and y added to their difference.
(d) The sum of x and y divided by their product.
(e) Three times a number increased by 5.
(f) A number minus the number increased by 7.

Answer the following:

24. The length of a rectangle is l meters (m) and the width is 4 m less than the length.
(a) Write an expression for the perimeter of the rectangle.
(b) Write an expression for the area.

25. It takes x pounds (lb) of copper and y lb of zinc to pour a casting. How much metal is needed to pour 12 castings?

26. The area of a square is given by s^2, where s is the length of a side. If the length of the side is increased by 5, express the new area algebraically.

27. The union scale for a certain job is p dollars per hour (hr). The scale for a second job is \$5/hr more than half the amount for the first job. Show algebraically how much the second job pays.

SECTION 7

Addition and Subtraction of Algebraic Expressions

When we write algebraic expressions it is often necessary to group certain terms together. Symbols of grouping, notably parentheses ( ), braces { }, or brackets [ ], are often used to show that terms contained within them are considered to be a single quantity.

Example 7.1

The sum of the two algebraic expressions $3x^2 + 2x - y$ and $4x + 2y$ may be written $(3x^2 + 2x - y) + (4x + 2y)$. The subtraction of $4x + 2y$ from $3x^2 + 2x - y$ may be written $(3x^2 + 2x - y) - (4x + 2y)$.

We add two or more algebraic expressions by combining like terms. One way to achieve this is to arrange the expressions in rows with like terms in the same columns and then add the columns.

Example 7.2

To add $3x^2 + 2x - y$ and $4x + 2y$ we write in column fashion, as follows:

$$\begin{array}{r} 3x^2 + 2x - \ \ y \\ + 4x + 2y \\ \hline 3x^2 + 6x + \ \ y \end{array}$$

Thus, the result is $3x^2 + 6x + y$. We could also have written in row fashion, as follows:

$$(3x^2 + 2x - y) + (4x + 2y) = 3x^2 + 6x + y$$

We subtract two algebraic expressions by changing the sign of *every term* in the expression that is being subtracted, and then we *add* this result to the second expression.

Example 7.3

To subtract $x^2 + 2x - 3$ from $3x^2 + 5x - 6$, we can write

$$\begin{aligned} (3x^2 + 5x - 6) - (x^2 + 2x - 3) &= 3x^2 + 5x - 6 - x^2 - 2x + 3 \\ &= 2x^2 + 3x - 3 \end{aligned}$$

Example 7.4

Subtract $x^3 - 2x + 1$ from $4x^3 + 2x^2 - x - 3$ using the column method.

Solution

To subtract $x^3 - 2x + 1$ we first change the sign of each of its terms and then add in column fashion, as follows:

$$\begin{array}{r} 4x^3 + 2x^2 - x - 3 \\ -\ x^3 + 2x - 1 \\ \hline 3x^3 + 2x^2 + x - 4 \end{array}$$

Examples 7.2 and 7.3 suggest the following laws on the removal of symbols of grouping.

Laws for Removal of Grouping Symbols

1. If a + sign or no sign precedes a grouping symbol, the symbol may be removed without affecting the terms within it.
2. If a − sign precedes a grouping symbol, the symbol may be removed if *each sign* of the terms within is changed.

The following examples further illustrate this law.

Example 7.5

To simplify $5x - [3x^2 - (2x^2 + 5y)]$, we proceed as follows:

$$\begin{aligned} 5x - [3x^2 - (2x^2 + 5y)] &= 5x - [3x^2 - 2x^2 - 5y] \\ &= 5x - 3x^2 + 2x^2 + 5y \\ &= 5x - x^2 + 5y \end{aligned}$$

Example 7.6

To simplify $-\{2x - [3 - (4 - 3x)] + 6x\}$, we proceed as follows:

$$\begin{aligned} -\{2x - [3 - (4 - 3x)] + 6x\} &= -\{2x - [3 - 4 + 3x] + 6x\} \\ &= -\{2x - 3 + 4 - 3x + 6x\} \\ &= -\{5x + 1\} = -5x - 1 \end{aligned}$$

Examples 7.5 and 7.6 show us that if more than one grouping symbol is present, we remove the inner ones first.

Exercises for Section 7

In Exercises 1 through 7, find the sum of each algebraic expression in both column and row fashion as shown in Example 7.2.

1. $3x - 2,\ 4x + 5$ **2.** $7 - 3x,\ -8 + x$

3. $x^2 + 2y^2 - 2x + y,\ x - 2y + x^2 - 4y^2$

4. $3y^2 + x^2 - 3x$, $8y^2 - 4x^2$

5. $3xy - 2yz + 6xz$, $4xz + 2yz - 8xy$

6. $x^2 - 2x + 6$, $-2x^2 + 4x - 3$, $x^2 - 2x + 4$

7. $x^3 + 2x^2 + x + 1$, $x^3 + 2x - 6$, $x^2 + 6x - 5$

In Exercises 8 through 23, simplify each algebraic expression.

8. $3a - (2b - 3a + c)$

9. $4y - (2y + 5x + 6)$

10. $(b - 4) + (3 - 4b)$

11. $6 - 3 - (10 + 2y)$

12. $2x - 2y - z + 3y - x$

13. $-(2a + b) - (3b - 2a)$

14. $-(x + y + 1) - (-x + y - 3)$

15. $3 - [2x + y - (3y - x)]$

16. $-\{7b - [a + (b - 2c) + 3c] + 2a\}$

17. $-\{-[-(2x - 3y) + 2]\}$

18. $3 + 2(x + 2y) - 3(2y - x)$

19. $4ab + (a - b) + 2b$

20. $[3 - (r^2 - 8)]$

21. $6 + 4[-6 - (3 + 2)]$

22. $(-gt^2 + 62t + 100) - (-gt^2 + 144t + 160)$

23. $V - [2V - (EI + V)]$

24.–28. Subtract the second polynomial from the first in Exercises 1 through 5.

29. Subtract $4ab$ from $5a^3 + 6ab + 1$ and add the result to $2a^2 - 3ab + 4$.

30. Subtract $x^2 + 2x + 7$ from the sum of $5x^2 - 6x + 2$ and $3x^2 - 3x - 4$.

31. The width w of a rectangle is increased by 2 units and its length l is decreased by 1 unit. Find an expression for the perimeter of this rectangle and simplify it.

32. An equilateral triangle with side x is decreased by 2 units. Find an expression for its perimeter and simplify it.

33. The diagonal of a rectangular prism is given by the formula $d = \sqrt{l^2 + w^2 + h^2}$. Determine d if:
(a) $l = 2$, $w = 4$, $h = 4$.
(b) $l = 4$, $w = 2$, $h = 2$.
(c) $l = 3$, $w = 5$, $h = 4$.

SECTION 8
Multiplication of Algebraic Expressions

To multiply two or more monomials, we use the laws of exponents given in Section 4, the rules for products of signed numbers, and the associative and commutative laws for multiplication in Section 2. We illustrate this in the following example.

Example 8.1

To find the product of the monomials $-2x^2y$, $3xy^3$, and $-4xy^2$, we write $(-2x^2y)(3xy^3)(-4xy^2)$. Now we use the commutative and associative laws to rearrange the expression as $[(-2)(3)(-4)][(x^2)(x)(x)][(y)(y^3)(y^2)]$. Finally, we use the rules of signed numbers and the law of exponents (4.1), to obtain

$$[(-2)(3)(-4)][(x^2)(x)(x)][(y)(y^3)(y^2)] = 24x^4y^6$$

Thus, $(-2x^2y)(3xy^3)(-4xy^2) = 24x^4y^6$.

To multiply a monomial and a multinomial, we multiply each term of the multinomial by the monomial and then combine the results.

Example 8.2

We find the product of $3x^2y^3$ and $2xy - 2x^3 + xy^2$ as follows:

$$\begin{aligned}(3x^2y^3)(2xy - 2x^3 + xy^2) &= (3x^2y^3)(2xy) + (3x^2y^3)(-2x^3) \\ &\quad + (3x^2y^3)(xy^2) \\ &= 6x^3y^4 - 6x^5y^3 + 3x^3y^5\end{aligned}$$

With practice the middle step in both Examples 8.1 and 8.2 may be done mentally.

To multiply a multinomial by another multinomial, we multiply each term of one multinomial by each term of the other and then add the results.

Example 8.3

To multiply $x^2 + x + 1$ by $2x - 1$, we proceed as follows:

$$\begin{array}{rl} & x^2 + x + 1 \qquad \text{(A)} \\ & \underline{2x \quad - 1} \\ \text{[by multiplying (A) by } 2x] & 2x^3 + 2x^2 + 2x \\ \text{[by multiplying (A) by } -1] & \underline{\quad\ - x^2 - x - 1} \\ \text{(by adding)} & 2x^3 + x^2 + x - 1 \\ \text{Thus, } (x^2 + x + 1)(2x - 1) = & 2x^3 + x^2 + x - 1 \end{array}$$

Example 8.4

To multiply $(3x - 2)$ by $(3 - x)$ we could proceed as in Example 7.3 or as follows:

$$\begin{aligned}(3x - 2)(3 - x) &= (3x - 2)(3) + (3x - 2)(-x)\\ &= 9x - 6 - 3x^2 + 2x\\ &= -3x^2 + 11x - 6\end{aligned}$$

Example 8.5

$(x + 2)^2$ means

$$\begin{aligned}(x + 2)(x + 2) &= (x + 2)(x) + (x + 2)(+2)\\ &= x^2 + 2x + 2x + 4\\ &= x^2 + 4x + 4\end{aligned}$$

We note that the resultant product has for its first term the square of x, for its second term twice the product of x and 2, and for its last term the square of 2.

Example 8.5 suggests a shorter method for expanding expressions of the form $(a + b)^2$. The result will be the *sum of the square of a, twice the product of a and b, and the square of b*. Thus,

$$(a + b)^2 = a^2 + 2ab + b^2 \tag{8.1}$$

Example 8.6

The width of a rectangle is w feet (ft) and its length is l ft. If the width of the rectangle is increased by 2 ft and its length is decreased by 1 ft, express the area of the rectangle in terms of w and l.

Solution

If w ft is the original width, the new width can be expressed as $w + 2$ ft. Similarly, the new length can be expressed as $l - 1$ ft. Thus, the area, A, can be expressed as follows:

$$\text{area} = \text{length} \times \text{width}$$

$$A = (l - 1)(w + 2)\ \text{ft}^2$$

or

$$A = (lw - w + 2l - 2)\ \text{ft}^2$$

Exercises for Section 8

In Exercises 1 through 26, find the product indicated.

1. $(-3ab^2)(2a^2b^3)$

2. $(-2x^2y)(2xy)(-5x^3y^4)$

3. $5rs(2r + s)$

4. $x^2y^3(x^2 - y^2)$

5. $(3m - 2)(2m + 1)$

6. $(x + y)(x - y)$

7. $(2x + 1)^2$

8. $(4 - x)^2$

9. $(3a^2b)(a^2 + 2ab + b^2)$

10. $(5mn)(m^3 + 2mn + n)$

11. $(x^2 - 5x + 3)(x + 2)$

12. $(2x^3 + x + 1)(x^2 - 1)$

13. $(x^3 + 1)(x^4 + x^2 + 3)$

14. $(2x + y - 3)(2x + y + 3)$

15. $(x^2 - xy + y^2)(x^2 + xy + y^2)$

16. $(abc + 1)^2$

17. $2x(3x + 1)(x + 1)^2$

18. $(t + s)(4t - 3s)$

19. $(b + 4)(4 + b)$

20. $(a^2 + b^2)^2$

21. $(a^2 + b^2)(a^2 - b^2)$

22. $-(m + n)(m - n)$

23. $(3 - y^2)^2$

24. $3a(2a - b)(6a^2 + 3ab)$

25. $(3x^2y - \frac{1}{2})^2$

26. $(x + \frac{1}{2})(2x + \frac{3}{2})$

27. The width of a rectangle w is increased by 3 units and the length l is increased by 2 units. The area of this rectangle is given by $(w + 3)(l + 2)$. Simplify this expression.

28. A square whose side s is increased by 1 unit has an area expressed as $(s + 1)(s + 1) = (s + 1)^2$. Simplify this expression.

29. Simplify $(a + b + c)^2$. ***Hint:*** Consider $(a + b + c)^2$ as $[(a + b) + c]^2$ and use formula (7.1).

30. Multiply (98)(103) by expressing 98 as $100 - 2$ and 103 as $100 + 3$.

In Exercises 31 through 35, use the method of Exercise 30.

31. 48(31)

32. 37(62)

33. $(19)^2$

34. $(21)^2$

35. 35(15)

SECTION 9

Division of Algebraic Expressions

The quotient of a monomial divided by a monomial can be achieved simply by using formula (4.5), as illustrated in the following examples.

Example 9.1

We divide $12x^3y^4z^2$ by $-4x^5y^2z$ as follows:

$$\frac{12x^3y^4z^2}{-4x^5y^2z} = \left(\frac{12}{-4}\right)\left(\frac{x^3}{x^5}\right)\left(\frac{y^4}{y^2}\right)\left(\frac{z^2}{z}\right)$$

$$= (-3)\left(\frac{1}{x^2}\right)(y^2)(z) = -\frac{3y^2z}{x^2}$$

Example 9.2

We divide $8a^2b^3c$ by $4abcd$ as follows:

$$\frac{8a^2b^3c}{4abcd} = \left(\frac{8}{4}\right)\left(\frac{a^2}{a}\right)\left(\frac{b^3}{b}\right)\left(\frac{c}{c}\right)\left(\frac{1}{d}\right)$$

$$= (2)(a)(b^2)(1)\left(\frac{1}{d}\right) = \frac{2ab^2}{d}$$

To divide a multinomial by a monomial, we divide each term of the multinomial by the monomial and then combine the results.

Example 9.3

We divide $8x^3y - 4x^2y^2 + 6x$ by $2xy$ as follows:

$$\frac{8x^3y - 4x^2y^2 + 6x}{2xy} = \frac{8x^3y}{2xy} + \frac{-4x^2y^2}{2xy} + \frac{6x}{2xy}$$

$$= 4x^2 - 2xy + \frac{3}{y}$$

Algebraic expressions whose terms are of the form ax^n, where n is a nonnegative integer, are called a **polynomial** in x. The greatest value of n in the expression is called the **degree** of the polynomial. Thus, a polynomial containing two or more terms is a special case of a multinomial.

Example 9.4

(a) $x^4 + 3x^2 + 1$ is a fourth-degree polynomial consisting of three terms.

(b) $3x^2$ is a second-degree polynomial consisting of one term.

(c) $x^3 + x^2 + x^{1/2} + 1$ is a multinomial but not a polynomial, since the term $x^{1/2}$ has an exponent which is not a nonnegative integer.

(d) $x^{1/2}$ has for its exponent the number $1/2$.

(e) $x^2 + 1/x$ is a multinomial but not a polynomial, since the term $1/x = x^{-1}$ has for its exponent the number -1.

(f) 6 is a polynomial of degree zero.

Rules for Division of a Polynomial by a Polynomial

1. Arrange the terms of both polynomials in descending powers of x.
2. Divide the first term in the dividend polynomial by the first term in the divisor polynomial. This result will be the first term in the quotient polynomial.

3. Multiply the divisor polynomial by the first term of the quotient polynomial, and then subtract the result from the dividend polynomial to obtain a new dividend polynomial.
4. Now use the new dividend and repeat steps 2 and 3 until the remainder obtained has a degree lower than the degree of the divisor polynomial or is equal to zero.
5. Express the result in the form

$$\frac{\text{dividend polynomial}}{\text{divisor polynomial}} = \text{quotient polynomial} + \frac{\text{remainder polynomial}}{\text{divisor polynomial}}$$

Example 9.5

To divide $4x^4 + 3x^3 + 2x + 1$ by $x^2 + x + 2$, we proceed as follows:

$$\begin{array}{r|l}
 & 4x^2 - x - 7 \\
\hline
x^2 + x + 2 & 4x^4 + 3x^3 \qquad + 2x + 1 \\
 & 4x^4 + 4x^3 + 8x^2 \\
\hline
 & \quad - x^3 - 8x^2 + 2x + 1 \\
 & \quad - x^3 - x^2 - 2x \\
\hline
 & \qquad - 7x^2 + 4x + 1 \\
 & \qquad - 7x^2 - 7x - 14 \\
\hline
 & \qquad\qquad 11x + 15
\end{array}$$

Thus,

$$\frac{4x^4 + 3x^3 + 2x + 1}{x^2 + x + 2} = 4x^2 - x - 7 + \frac{11x + 15}{x^2 + x + 2}$$

If the remainder polynomial in the division process is equal to zero, we say that the divisor polynomial is a **factor** of the dividend polynomial and the quotient is also a factor of the dividend polynomial.

Example 9.6

To divide $x^3 - 3x^2 + 5x - 6$ by $x - 2$, we proceed as follows:

$$\begin{array}{r|l}
 & x^2 - x + 3 \\
\hline
x - 2 & x^3 - 3x^2 + 5x - 6 \\
 & x^3 - 2x^2 \\
\hline
 & \quad - x^2 + 5x - 6 \\
 & \quad - x^2 + 2x \\
\hline
 & \qquad 3x - 6 \\
 & \qquad 3x - 6 \\
\hline
 & \qquad\quad 0
\end{array}$$

Thus,

$$\frac{x^3 - 3x^2 + 5x - 6}{x - 2} = x^2 - x + 3$$

and we say that $x - 2$ is a factor of $x^3 - 3x^2 + 5x - 6$, since the remainder is equal to zero. The quotient polynomial $x^2 - x + 3$ is also a factor of $x^3 - 3x^2 + 5x - 6$, and we can express the result as follows:

$$x^3 - 3x^2 + 5x - 6 = (x - 2)(x^2 - x + 3)$$

Exercises for Section 9

Perform the division as indicated.

1. $\dfrac{9x^6}{3x^2}$

2. $\dfrac{8x^3y}{4xy^3}$

3. $\dfrac{35a^2bc}{-7ab^2}$

4. $\dfrac{18r^2s^3t}{-4rst^2}$

5. $\dfrac{20x^3y^4 - 15x^2y + 5x^4y^2}{5x^2y^2}$

6. $\dfrac{4a^2b^2 - 6a^3b^3c^3 + 7abc}{abc}$

7. $\dfrac{x^6 - 3x^4 + x^3 + 6x^2 + x}{x^3}$

8. $\dfrac{x^5 - x^3 + 3x^2 + 1}{3x^2}$

9. $\dfrac{x^5 - 3x^4 + x^3 + 6x^2 + x}{x}$

10. $\dfrac{v^3 - 3v^2 + v + 2}{2v}$

11. $\dfrac{x^2 + 6x + 5}{x + 5}$

12. $\dfrac{x^2 + 3x + 2}{x + 1}$

13. $\dfrac{x^2 + 2x - 8}{x - 2}$

14. $\dfrac{x^2 + 2x - 8}{x + 4}$

15. $\dfrac{x^2 - x - 20}{x - 5}$

16. $\dfrac{x^2 - x - 20}{x + 4}$

17. $\dfrac{x^2 - x - 19}{x - 5}$

18. $\dfrac{x^2 + 6x + 9}{x + 3}$

19. $\dfrac{x^2 + 6x + 9}{x - 3}$

20. $\dfrac{x^3 + x^2 - x - 1}{x - 1}$

21. $\dfrac{2x^3 + 3x^2 - 18x + 4}{x - 2}$

22. $\dfrac{16x^4 + 4x^3 + 2x^2 - x + 1}{2x + 1}$

23. $\dfrac{x^6 - x^4 + x^3 - x + 1}{x^2 + 1}$

24. $\dfrac{x^4 - 4x^3 - 7x^2 + 22x + 24}{x - 3}$

25. $\dfrac{a^3 - 8}{a - 2}$

26. $\dfrac{y^4 + 16}{y - 2}$

27. $\dfrac{v^3 + 1}{v + 1}$

28. $\dfrac{x^5 + 1}{x + 1}$

29. $\dfrac{x^5 + x^4 + x^3 + x^2 + x + 1}{x^3 + 1}$

30. $\dfrac{x^4 - x^3 - 2x^2 - 3x - 3}{x^2 + x + 1}$

SECTION 10

Factors of Algebraic Expressions

When we consider the product of two positive integers such as $4(5) = 20$, we say that 4 is a factor of 20, because 20 divided by 4 yields 5, another positive integer. We also say that 5 is a factor of 20, since $20 \div 5 = 4$.

We use a similar idea to discuss the factors of an algebraic expression. We say that 3 is a factor $6x + 3$, since $(6x + 3) \div 3$ yields $2x + 1$, which is another algebraic expression. We also say that $x + 1$ is a factor $x^2 - x - 2$, since $(x^2 - x - 2) \div (x + 1)$ yields $x - 2$, another algebraic expression. The factorization process for algebraic expressions is generally restricted to finding the factors of polynomials whose numerical coefficients are integers. Unless otherwise stated, this will always be assumed. Such polynomials will be called **prime** if their only factors are plus or minus themselves or plus or minus 1.

Most of our factoring will be based on our experience with multiplication. Thus, it is necessary to become familiar with some special products that frequently occur. Each of the following general forms can be verified by multiplication.

$$ax + ay = a(x + y) \tag{10.1}$$

$$x^2 - y^2 = (x + y)(x - y) \tag{10.2}$$

$$x^2 + 2xy + y^2 = (x + y)^2 \tag{10.3}$$

$$x^2 - 2xy + y^2 = (x - y)^2 \tag{10.4}$$

$$x^2 + (a + b)x + ab = (x + a)(x + b) \tag{10.5}$$

$$acx^2 + (ad + bc)x + bd = (ax + b)(cx + d) \tag{10.6}$$

$$x^3 + y^3 = (x + y)(x^2 - xy + y^2) \tag{10.7}$$

$$x^3 - y^3 = (x - y)(x^2 + xy + y^2) \tag{10.8}$$

$$\begin{aligned} ac + bc + ad + bd &= (a + b)c + (a + b)d \\ &= (a + b)(c + d) \end{aligned} \tag{10.9}$$

If formulas (10.1) through (10.9) are read from right to left, they can be used to determine the factors of the expression on the left.

Example 10.1

To factor $8x^2 + 6$, we observe that both the numerical coefficients 8 and 6 have 2 as a common factor. Thus, we can apply formula (10.1) to obtain

$$8x^2 + 6 = 2(4x^2) + 2(3) = 2(4x^2 + 3)$$

Example 10.2

To factor $x^2 - 16$, we observe that x is the square of x^2 and 4 is the square of 16. Thus, we can apply formula (10.2) to obtain

$$x^2 - 16 = (x)^2 - (4)^2 = (x + 4)(x - 4)$$

Similarly, since $3x$ is the square of $9x^2$ and 5 is square of 25, we apply formula (10.2) and factor as follows:

$$9x^2 - 25 = (3x)^2 - (5)^2 = (3x + 5)(3x - 5)$$

Example 10.3

To factor $4x^2 + 12x + 9$, we observe that $2x$ is the square of $4x^2$ and that 3 is the square of 9. We can use formula (10.3), since the middle term $12x = 2(2x)(3)$. Thus,

$$4x^2 + 12x + 9 = (2x + 3)^2$$

Example 10.4

To factor $9x^2 - 24x + 16$, we observe that $3x$ is the square of $9x^2$ and -4 is the square of 16. We can use formula (10.4), since the middle term $-24x = 2(3x)(-4)$. Thus,

$$9x^2 - 24x + 16 = (3x - 4)^2$$

Example 10.5

The polynomial $x^2 + 4x + 16$ *cannot be factored*, even though it appears to fit the form of the left side of formula (10.3). We see that x is the square of x^2 and 4 is the square of 16, but the middle term $4x \neq 2(x)(4)$.

The polynomial $x^2 + 4x + 16$ also appears to fit the form of the left side of formula (10.6) but is still *not factorable*, since $ac = 1$, $ad + bc = 4$, and $bd = 16$, but $ad + bc \neq ac + bd$.

Finally, we say that expressions such as $x^2 + 4x + 16$ are *prime* if they cannot be factored.

Example 10.6

To factor $2x^2 + 5x + 3$, we consider formula (10.6) and we observe that the binomial factors we seek must have first terms that are the factors of $2x^2$, and they must have second terms that are the integral factors of 3. We then list the possible factorizations and check them by multiplication. Among the possibilities for factorization are

$$(2x + 1)(x + 3) = 2x^2 + 5x + 3 \qquad \text{(A)}$$

$$(2x - 1)(x - 3) = 2x^2 - 5x + 3 \qquad \text{(B)}$$

We select (A) since the middle term of this result is the same as the middle term of $2x^2 + 5x + 3$. The key to this trial-and-error method is to examine each possibility to see if the sum of the inside and outside products is equal to the middle term of the given trinomial.

A procedure for factoring the types of trinomials given in formulas (10.5) and (10.6) by use of (10.9) is shown in Exercise 51.

Example 10.7

To factor $x^3 + 8$, we observe that x is the cube root of x^3 and that 2 is the cube root of 8. Thus, we can apply formula (10.7) to obtain

$$\begin{aligned} x^3 + 8 = (x)^3 + (2)^3 &= (x + 2)[x^2 - (2)(x) + 2^2] \\ &= (x + 2)(x^2 - 2x + 4) \end{aligned}$$

Formula (10.7) can also be used to factor $8x^3 + 27y^3$ as follows:

$$\begin{aligned} 8x^3 + 27y^3 &= (2x)^3 + (3y)^3 \\ &= (2x + 3y)[(2x)^2 - (2x)(3y) + (3y)^2] \\ &= (2x + 3y)(4x^2 - 6xy + 9y^2) \end{aligned}$$

Example 10.8

To factor $x^6 - 1$ we can use formula (10.8), since

$$\begin{aligned} x^6 - 1 &= (x^2)^3 - (1)^3 \\ &= (x^2 - 1)[(x^2)^2 + (x^2)(1) + (1)^2] \\ &= (x^2 - 1)(x^4 + x^2 + 1) \qquad \text{(A)} \end{aligned}$$

From (A) we see that the factor $x^2 - 1 = (x + 1)(x - 1)$ and $x^4 + x^2 + 1 = (x^2 - x + 1)(x^2 + x + 1)$. Thus,

$$\begin{aligned} x^6 - 1 &= (x^2 - 1)(x^4 + x^2 + 1) \\ &= (x + 1)(x - 1)(x^2 - x + 1)(x^2 + x + 1) \end{aligned}$$

It should be noted that $x^6 - 1 = (x^3)^2 - (1)^2$, and that formula (10.2), followed by (10.7) and (10.8), will yield the same result. We now say that all four factors of $x^6 - 1$ are prime factors and that $x^6 - 1$ is factored completely.

Example 10.9

To factor $2ax - bx + 4ay - 2by$, we apply formula (10.9) as follows:

$$\begin{aligned} 2ax - bx + 4ay - 2by &= (2a - b)x + (2a - b)2y \\ &= (2a - b)(x + 2y) \end{aligned}$$

Exercises for Section 10

In Exercises 1 through 50, factor each expression completely.

1. $4x - 10x^2$

2. $3ax^2 + 6ax + 9a$

3. $4x^2y^2 - 12xy$

4. $16x^2 - 9y^2$

5. $x^2 - 9$

6. $25 - x^2$

7. $16 - 25x^2$

8. $x^2 + 7x + 12$

9. $x^2 - 2x - 15$

10. $v^2 + 9v - 36$

11. $4r^2 + 8rs - 5s^2$

12. $x^2 + x + 3$

13. $9x^2 - 6x + 1$

14. $16 - 12x + 9x^2$

15. $x^2 - 12x + 36$

16. $9 - 12x + 4x^2$

17. $x^2 + 2x + 1$

18. $t^3 + 2t^2 + t$

19. $by - ay + bx - ax$

20. $bx^2 + x + bx + 1$

21. $x^2 + bx + xy + by$

22. $ax^2 + axy + bxy + by^2$

23. $3x^3 - 2x^2 - 3x + 2$

24. $(x + 2)y + 3(x + 2)$

25. $(x + y - 1)^2 - 3(x + y - 1)$

26. $x^3 + 3x^2 + 3x + 9$

27. $8x^3 + 1$

28. $x^3y^3 + 27$

29. $8x^3 - 1$

30. $27 - x^3y^3$

31. $v^3 - t^3$

32. $x^4 - x$

33. $16x^4 - 1$

34. $(x + 1)^3 + y^3$

35. $(x - 1)^3 + y^3$

36. $(1 - x)^3 - 8$

37. $x^8 - 1$

38. $x^4 - y^4$

39. $x^6 - 64$

40. $x^4 + 7x^2 + 12$

41. $x^4 - 2x^2 - 15$

42. $x^6 + 12x^3 + 36$

43. $6x^5y - 36xy^5$

44. $x^5y + x^2y^4$

45. $m^4 - 7m^2n^2 + 9n^4$

46. $p^4 + 2p^2 + 9$

47. $x + y + y^2 - x^2$

48. $(x + 1)(2y + 1)^2 - (x + 1)(y + 1)^2$

49. $p^5q^2 - 81pq^2$

50. $(2x - y)^2 - (x - 3y)^2$

51. Technique for factoring trinomials of the form $x^2 + (a + b)x + ab$ or $acx^2 + (ad + bc)x + bd$ [see formulas (10.5) and (10.6)]:

To factor the trinomial $6x^2 + 11x + 3$, we consider the list of all the possible pairs of the integral factors of the product of numerical coefficient 6 of the first term and the numerical coefficient 3 of the third term. From this list we must find a pair whose algebraic sum is equal to the numerical coefficient of the middle term. If we do, we express the middle as this sum and proceed to factor using formula (10.9).

Possible Pairs of Factors	*Sum*
1, 18	19
−1, −18	−19
2, 9	11
−2, −9	−11
3, 6	9
−3, −6	−9

We see that the pair (2, 9) is the proper choice. Now we proceed as follows.

$$\begin{aligned} 6x^2 + 11x + 3 &= 6x^2 + 2x + 9x + 3 \\ &= 2x(3x + 1) + 3(3x + 1) \\ &= (3x + 1)(2x + 3) \end{aligned}$$

Use this method to factor:

(a) $8x^2 + 2x - 1$ (b) $x^2 - 21x - 22$
(c) $10x^2 - x - 2$ (d) $12x^2 + 8x - 15$

SECTION 11

Algebraic Fractions

A rational algebraic fraction is an algebraic expression that can be expressed as the quotient of two polynomials. Since a fractional algebraic expression will become the ratio of two real numbers for allowable real values of the variable, we can establish rules concerning their operations that are similar to rules concerning operations with numerical fractions in arithmetic.

Example 11.1

$$\frac{2x + 1}{x^2 - 4x + 1} \quad \text{and} \quad \frac{x^3 + x^2 + 1}{x^4 - 1}$$

are rational algebraic fractions.

Example 11.2

When $x = 2$, the algebraic fraction

$$\frac{x^3 + x^2 + 1}{x^4 - 1} = \frac{2^3 + 2^2 + 1}{2^4 - 1} = \frac{8 + 4 + 1}{16 - 1} = \frac{13}{15}$$

When $x = 1$,

$$\frac{x^3 + x^2 + 1}{x^4 - 1} = \frac{1^3 + 1^2 + 1}{1^4 - 1} = \frac{3}{0}$$

and it is undefined. Thus, 1 is not an allowable value for x.

When dealing with a numerical fraction of the form a/b, $b \neq 0$, we know that if we multiply or divide both the **numerator** a and the **denominator** b by the same real number (except zero), the resulting fraction is equivalent.

Example 11.3

The numerator and the denominator of $\frac{1}{2}$ can both be multiplied by the number 4 to obtain the equivalent fraction $(4)(1)/(4)(2) = \frac{4}{8}$. The numerator and the denominator of $\frac{18}{24}$ can both be divided by the common factor 6 to obtain the equivalent fraction $\frac{3}{4}$.

In general, we say that $ac/bc = a/b$ when $c \neq 0$. We extend this idea to algebraic fractions with the following examples. We say that an algebraic fraction is **reduced to its lowest terms** when the numerator and denominator have only ± 1 for common factors. We can use factor forms such as those discussed in Section 10 to reduce algebraic fractions to their lowest terms or **simplest form**.

Example 11.4

To reduce $12x/9y$ to lowest terms, we divide both the numerator and the denominator by their common factor 3. We show this as follows:

$$\frac{12x}{9y} = \frac{(\not{3})(4x)}{\not{3}(3y)} = \frac{4x}{3y}$$

The symbol / indicates a process referred to as **cancellation**.

Example 11.5

We reduce $\dfrac{2x + 1}{4x^2 - 1}$ to lowest terms as follows:

$$\frac{2x + 1}{4x^2 - 1} = \frac{\cancel{2x + 1}}{\cancel{(2x + 1)}(2x - 1)} = \frac{1}{2x - 1}$$

It should be noted that we used formula (10.2) to factor the expression $4x^2 - 1$.

Example 11.6

We reduce $\dfrac{3x^2 - x - 2}{x^2 - 2x + 1}$ to its lowest term by factoring both the numerator and the denominator and then dividing each by their common factor $x - 1$ as follows:

$$\frac{3x^2 - x - 2}{x^2 - 2x + 1} = \frac{\cancel{(x - 1)}(3x + 2)}{\cancel{(x - 1)}(x - 1)} = \frac{3x + 2}{x - 1}$$

Example 11.7

The expression $\frac{x^2+1}{x^2-1}$ is already in its lowest term, since the only common factors for the numerator and denominator are ± 1. It should be noted that we *cannot cancel* the x^2 in both the numerator and denominator, because we would be removing a term and thus altering its value. Thus,

$$\frac{x^2+1}{x^2-1} \neq \frac{1}{-1}$$

Example 11.8

We reduce $\frac{x^2-4}{2-x}$ to lowest terms as follows:

$$\frac{x^2-4}{2-x} = \frac{(x+2)\cancel{(x-2)}}{(-1)\cancel{(x-2)}} = \frac{x+2}{-1} = -(x+2)$$

Example 11.9

We reduce $\frac{1-x}{x-1}$ to lowest terms as follows:

$$\frac{1-x}{x-1} = \frac{(-1)\cancel{(x-1)}}{\cancel{x-1}} = -1$$

Examples 11.8 and 11.9 suggest another important observation about fractions. The fraction a/b has three signs associated with it: the sign in front of the numerator, the sign in front of the denominator, and the sign in front of the entire fraction. *Any two of these signs may be changed without altering the value of the fraction.*

We state this rule in general as

$$\frac{a}{b} = +\frac{+a}{+b} = \frac{-a}{-b} = -\frac{-a}{+b} = -\frac{+a}{-b} \tag{11.1}$$

Exercises for Section 11

Reduce (if possible) each fraction to its lowest terms.

1. $\frac{18}{27}$ **2.** $\frac{6a}{12a}$ **3.** $\frac{4x^2y}{-2xy}$

4. $\frac{2x}{x(x+1)}$ **5.** $\frac{3ab}{a(b+c)}$ **6.** $\frac{x-y}{x+y}$

7. $\frac{5x^2-10xy}{x-2y}$ **8.** $\frac{3x^2-12}{x^2+2x}$ **9.** $\frac{4-x^2}{x^2-5x+6}$

10. $\frac{15(m+n)}{18r(m+n)}$ **11.** $\frac{3x^2-3y^2}{4x-4y}$ **12.** $\frac{2p^2-2q^2}{3p+3q}$

13. $\dfrac{p+q}{p^2-q^2}$ **14.** $\dfrac{(x+1)(2-x)}{(x-2)(x+1)}$ **15.** $\dfrac{2x+y+1}{4(2x+y+1)}$

16. $\dfrac{1-3x}{3x-1}$ **17.** $\dfrac{ax+ay}{x^2-y^2}$ **18.** $\dfrac{x^2+5x+4}{x+4}$

19. $\dfrac{x^3+5x^2+6x}{x^2+x}$ **20.** $\dfrac{x^2+5x+6}{x+1}$ **21.** $\dfrac{(1-x)(3-x)}{x^2-4x+3}$

22. $\dfrac{x^2+x-12}{x^2+7x+12}$ **23.** $\dfrac{x^2-x-6}{x^2-5x-6}$ **24.** $\dfrac{2y^2+3y-2}{2y^2-3y+1}$

25. $\dfrac{x^2-9}{x^2-2x-15}$ **26.** $\dfrac{xy-y^2}{x^4y-xy^4}$ **27.** $\dfrac{3(a+b)^3-x(a+b)}{a^2-b^2}$

28. $\dfrac{x^3-1}{x^2-1}$ **29.** $\dfrac{x^2-y^2}{x^3-y^3}$ **30.** $\dfrac{a+b}{a^3+b^3}$

31. $\dfrac{x^4-1}{x^3-1}$ **32.** $\dfrac{4x^2-1}{1-8x^3}$ **33.** $\dfrac{2x^2-8y^2}{40y^3-5x^3}$

SECTION 12

Multiplication and Division of Algebraic Fractions

The product of two or more algebraic fractions yields another algebraic fraction whose numerator is the product of the numerators of the given fractions and whose denominator is the product of the denominators of the given fractions. In actual practice it is generally more efficient to express the respective numerators and denominators in their completely factored forms and then to cancel any common factor before multiplying.

Example 12.1

$$\frac{x^2-6x+5}{x^2-4}\cdot\frac{x-2}{x-1}=\frac{(x-5)\cancel{(x-1)}\cancel{(x-2)}}{(x+2)\cancel{(x-2)}\cancel{(x-1)}}=\frac{x-5}{x+2}$$

Example 12.2

$$\frac{x^3-x}{x+3}\cdot\frac{x^2-9}{x^2-1}=\frac{x\cancel{(x^2-1)}}{\cancel{x+3}}\cdot\frac{\cancel{(x+3)}(x-3)}{\cancel{x^2-1}}=\frac{x(x-3)}{1}$$
$$=x(x-3)$$

Example 12.3

$$\frac{6ay^3}{4b + 4c} \cdot \frac{ab + ac}{2a^3y} = \frac{6ay^3[a(b + c)]}{4(b + c)(2a^3y^2)}$$

$$= \left(\frac{6}{8}\right)\left(\frac{a^2}{a^3}\right)\left(\frac{y^3}{y^2}\right)\left(\frac{b + c}{b + c}\right)$$

$$= \left(\frac{3}{4}\right)\left(\frac{1}{a}\right)(y)(1) = \frac{3y}{4a}$$

The quotient of two algebraic fractions is obtained by inverting the divisor fraction and then multiplying.

Example 12.4

$$\frac{x}{x + 2} \div \frac{x^2}{4 - x^2} = \frac{x}{x + 2} \cdot \frac{4 - x^2}{x^2}$$

$$= \frac{x}{x + 2} \cdot \frac{(2 - x)(2 + x)}{x^2} = \left(\frac{x}{x^2}\right)\left(\frac{2 + x}{2 + x}\right)\left(\frac{2 - x}{1}\right)$$

$$= \frac{2 - x}{x}$$

Example 12.5

$$\frac{2x^2 - x - 3}{3x^2 - 4x + 1} \div \frac{2x - 1}{9x^2 - 1} = \frac{2x^2 - x - 3}{3x^2 - 4x + 1} \cdot \frac{9x^2 - 1}{2x - 1}$$

$$= \frac{(2x - 3)(x + 1)}{(3x - 1)(x - 1)} \cdot \frac{(3x + 1)(3x - 1)}{2x - 1}$$

$$= \frac{(2x - 3)(x + 1)(3x + 1)\cancel{(3x - 1)}}{\cancel{(3x - 1)}(x - 1)(2x - 1)}$$

$$= \frac{(2x - 3)(x + 1)(3x + 1)}{(x - 1)(2x - 1)}$$

Example 12.6

$$\frac{a^2 - 1}{b - 3} \div \frac{4ab + 4b}{9 - b^2} = \frac{a^2 - 1}{b - 3} \cdot \frac{9 - b^2}{4ab + 4b}$$

$$= \frac{(a + 1)(a - 1)}{b - 3} \cdot \frac{(3 + b)(3 - b)}{4b(a + 1)}$$

$$= \frac{\cancel{(a + 1)}(a - 1)(3 + b)(-1)\cancel{(b - 3)}}{4b\cancel{(b - 3)}\cancel{(a + 1)}}$$

$$= \frac{-(a - 1)(b + 3)}{4b} \quad \text{or} \quad -\frac{(a - 1)(b + 3)}{4b}$$

Exercises for Section 12

In Exercises 1 through 25, perform the operation indicated and simplify.

1. $\frac{3}{8} \cdot \frac{64}{9}$

2. $\frac{14}{12} \cdot \frac{6}{7}$

3. $\frac{4}{3} \div \frac{8}{9}$

4. $\frac{3}{4} \div 9$

5. $\frac{3xy}{4} \cdot \frac{12z^2}{9x^2y^2}$

6. $\frac{4y}{3x} \div \frac{8y}{9x}$

7. $\frac{8xyz}{3x^3y^2} \cdot \frac{6x^2y}{z^2}$

8. $\frac{4ab}{3c} \div \frac{8b}{9c^2}$

9. $\frac{3x + 2}{5} \cdot \frac{10}{6x + 4}$

10. $\frac{2x + 4}{3x - 3} \cdot \frac{12x - 12}{x + 2}$

11. $\frac{6}{x + 2} \div \frac{12}{2x + 4}$

12. $\frac{a^2b}{xy + 1} \cdot \frac{x^2y^2 - 1}{ab^2}$

13. $\frac{1 - 9x^2}{x^2 - 4} \cdot \frac{2 - x}{1 - 3x}$

14. $\frac{1 - x^2}{x^2 - 4} \div \frac{1 - x}{x + 2}$

15. $\frac{x^2 - 25}{x^2 - 4} \div \frac{x^2 + 2x - 15}{x^2 + x - 12}$

16. $\frac{x^3 + 8}{x - 2} \cdot \frac{x^3 - 8}{x + 2}$

17. $\frac{x^3 - y^3}{9x^2 - 4y^2} \div \frac{x^2 - y^2}{3x^2 + xy - 2y^2}$

18. $\frac{4x^2 + 8x + 3}{2x^2 - 5x + 3} \div \frac{1 - 4x^2}{6x^2 - 9x}$

19. $\frac{x^2 - 5x + 6}{x^2 + 7x - 8} \cdot \frac{64 - x^2}{9 - x^2}$

20. $(2r^2 - 5r + 2) \div \frac{2r - 1}{4}$

21. $(p^3 - 27) \div \frac{p^2 + 3p + 9}{p - 9}$

22. $\frac{x^3 - y^3}{x^4} \div \frac{x^2 - y^2}{x^4}$

23. $(2x^2 - x - 10) \div \frac{6x^2 - 19x + 10}{4 - x^2}$

24. $\frac{9x^2 - 6x}{6x^2 - 7x + 2} \div \frac{2x^2 + 6x}{2x^2 + 13x - 7}$

25. $\left(\frac{9x^2 - 6x}{6x^2 - 7x + 2} \cdot \frac{4x^2 - 8x + 3}{2x^2 + 3x - 9}\right) \div \frac{2x^2 + 6x}{2x^2 + 13x - 7}$

SECTION 13

Addition of Algebraic Fractions

The addition of two or more algebraic fractions with the same denominator yields another algebraic fraction with the same denominator and a numerator that is the algebraic sum of the numerators of the given fractions.

Example 13.1

(a) $\frac{3}{7} + \frac{4}{7} - \frac{2}{7} = \frac{3 + 4 - 2}{7} = \frac{5}{7}$

(b) $\frac{8}{x} - \frac{4}{x} - \frac{6}{x} = \frac{8 - 4 - 6}{x} = \frac{8 - 10}{x} = \frac{-2}{x} = -\frac{2}{x}$

(c) $\frac{4y}{x - 3} + \frac{2y}{x - 3} + \frac{5}{x - 3} = \frac{4y + 2y + 5}{x - 3} = \frac{6y + 5}{x - 3}$

The problem of adding two or more fractions with different denominators is a little more complicated. To add the fractions $\frac{1}{3}$, $\frac{1}{4}$, and $\frac{1}{5}$ we must first express them as fractions with the same or **common denominator** and then proceed to find their sum. A common denominator is simply an expression that has for its factors all the denominators in question. Thus, a common denominator for the denominators of the fractions $\frac{1}{3}$, $\frac{1}{4}$, and $\frac{1}{5}$ could be the product of 3, 4, and 5, which is $3(4)(5) = 60$. Another common denominator for these fractions is 120. Either choice will enable us to reexpress the given fractions as equal fractions. We demonstrate the use of both choices by finding the sum of these fractions in the following examples.

Example 13.2

If we use 60 as a common denominator for the fraction $\frac{1}{3}$, $\frac{1}{4}$, and $\frac{1}{5}$, we must multiply the numerator and denominator of each by a factor that will make the denominator of each equal to 60. We do this as follows:

$$\frac{1}{3} = \frac{1(20)}{3(20)} = \frac{20}{60} \qquad \frac{1}{4} = \frac{1(15)}{4(15)} = \frac{15}{60} \qquad \frac{1}{5} = \frac{1(12)}{5(12)} = \frac{12}{60}$$

Thus,

$$\frac{1}{3} + \frac{1}{4} + \frac{1}{5} = \frac{20}{60} + \frac{15}{60} + \frac{12}{60} = \frac{47}{60}$$

If we use 120 as a common denominator, we have

$$\frac{1}{3} = \frac{1(40)}{3(40)} = \frac{40}{120} \qquad \frac{1}{4} = \frac{1(30)}{4(30)} = \frac{30}{120} \qquad \frac{1}{5} = \frac{1(24)}{5(24)} = \frac{24}{120}$$

Thus,

$$\frac{1}{3} + \frac{1}{4} + \frac{1}{5} = \frac{40}{120} + \frac{30}{120} + \frac{24}{120} = \frac{94}{120} = \frac{47(2)}{60(2)} = \frac{47}{60}$$

In Example 13.2 we can see that the choice of 60 instead of 120 as a common denominator eliminated the necessity of reducing the final result. It is for this reason that we generally try to find the **least common denominator**. By this we mean the smallest of all the natural numbers that has each of the given denominators as a factor. We extend this concept to polynomials with integer coefficients as follows.

Rules to Find the Least Common Denominator

1. Factor each denominator into the product of its prime factors.
2. Write all the **different** factors that occur in step 1 as a product.
3. Raise each of the factors of the product in step 2 to the highest power to which they occur in step 1. This result is the least common denominator (L.C.D.).

Example 13.3
Find the L.C.D. of $\frac{1}{20}$, $\frac{1}{15}$, and $\frac{1}{30}$.

Solution
Step 1: $20 = 2^2(5)$; $15 = 3(5)$; $30 = 2(3)(5)$
Step 2: $2(3)(5)$
Step 3: $2^2(3)(5) = 4(3)(5) = 60$
Thus, 60 is the L.C.D.

Example 13.4
Find the L.C.D. of $\frac{1}{3x^2}$, $\frac{5}{4xy}$, and $\frac{7}{8x^2y^2}$.

Solution
Step 1: $3x^2 = 3(x)^2$; $4xy = (2)^2(x)(y)$; $8x^2y^2 = (2)^3(x)^2(y)^2$
Step 2: $2(3)(x)(y)$
Step 3: $2^3(3)(x)^2(y)^3 = 24x^2y^2$
Thus, the L.C.D. is $24x^2y^2$.

Example 13.5
Find the L.C.D. for $\frac{1}{3x - 3}$, $\frac{4}{x^2 - 1}$, and $\frac{x}{2x^2 - x - 1}$.

Solution
Step 1: $3x - 3 = 3(x - 1)$
$x^2 - 1 = (x + 1)(x - 1)$
$2x^2 - x - 1 = (x - 1)(2x + 1)$
Step 2: $3(x - 1)(x + 1)(2x + 1)$
Step 3: $3(x - 1)(x + 1)(2x + 1)$ is the L.C.D.

Example 13.6
Convert the fractions $\frac{1}{3x - 3}$, $\frac{4}{x^2 - 1}$, and $\frac{x}{2x^2 - x - 1}$ to equivalent fractions with a least common denominator.

Solution
From Example 13.5 we found the L.C.D. to be $3(x - 1)(x + 1)(2x + 1)$. We now convert each fraction to an equivalent fraction with a denomina-

tor that is equal to the L.C.D., as follows:

$$\frac{1}{3x-3} = \frac{1}{3(x-1)} \cdot \frac{(x+1)(2x+1)}{(x+1)(2x+1)}$$

$$= \frac{(x+1)(2x+1)}{3(x-1)(x+1)(2x+1)}$$

$$\frac{4}{x^2-1} = \frac{4}{(x+1)(x-1)} \cdot \frac{3(2x+1)}{3(2x+1)}$$

$$= \frac{12(2x+1)}{3(x-1)(x+1)(2x+1)}$$

$$\frac{x}{2x^2-x-1} = \frac{x}{(x-1)(2x+1)} \cdot \frac{3(x+1)}{3(x+1)}$$

$$= \frac{3x(x+1)}{3(x-1)(x+1)(2x+1)}$$

Example 13.7

Simplify $\frac{2a}{3xy} + \frac{3a}{3xy} + \frac{c}{3xy}$.

Solution
We see that the fractions all have the same denominator. Thus, we have

$$\frac{2a}{3xy} + \frac{3a}{3xy} + \frac{c}{3xy} = \frac{2a+3a+c}{3xy} = \frac{5a+c}{3xy}$$

Example 13.8

Simplify $\frac{1}{3x-3} + \frac{4}{x^2-1} - \frac{x}{2x^2-x-1}$.

Solution
We can use the results of Example 13.6 to obtain

$$\frac{1}{3x-3} + \frac{4}{x^2-1} - \frac{x}{2x^2-x-1}$$

$$= \frac{1}{3x-3} + \frac{4}{x^2-1} + \frac{-x}{2x^2-x-1}$$

$$= \frac{(x+1)(2x+1)}{3(x-1)(x+1)(2x+1)} + \frac{12(2x+1)}{3(x-1)(x+1)(2x+1)}$$

$$+ \frac{(-3x)(x+1)}{3(x-1)(x+1)(2x+1)}$$

$$= \frac{(x+1)(2x+1) + 12(2x+1) + (-3x)(x+1)}{3(x-1)(x+1)(2x+1)}$$

$$= \frac{2x^2 + 3x + 1 + 24x + 12 - 3x^2 - 3x}{3(x - 1)(x + 1)(2x + 1)}$$

$$= \frac{13 + 24x - x^2}{3(x - 1)(x + 1)(2x + 1)}$$

In Example 13.8 we changed both the signs in front of the third fraction and the sign in front of its numerator so that we could treat the simplification as a sum. This is a convenient method for dealing with the *difference* of two algebraic fractions.

Occasionally, we encounter algebraic fractions that have one or more algebraic fractions in their numerators or denominators or both. We call these fractions *complex*. To simplify such fractions, we first reduce the numerator and denominator to simple fractions and divide.

Example 13.9

Simplify the complex fraction

$$\frac{x - \dfrac{1}{2x + 1}}{\dfrac{x^2 + x}{2x^2 - x - 1}}.$$

Solution

We first simplify the numerator $x - \dfrac{1}{2x + 1}$ to obtain

$$x - \frac{1}{2x + 1} = \frac{x(2x + 1)}{2x + 1} + \frac{-1}{2x + 1} = \frac{2x^2 + x - 1}{2x + 1}$$

$$= \frac{(2x - 1)(x + 1)}{(2x + 1)} \qquad \text{(A)}$$

Now, we simplify the denominator $\dfrac{x^2 + x}{2x^2 - x - 1}$ to obtain

$$\frac{x^2 + x}{2x^2 - x - 1} = \frac{x(x + 1)}{(2x + 1)(x - 1)} \qquad \text{(B)}$$

We substitute the results (A) and (B) into the original fraction and proceed as follows:

$$\frac{x - \dfrac{1}{2x + 1}}{\dfrac{x^2 + x}{2x^2 - x - 1}} = \frac{\dfrac{(2x - 1)(x + 1)}{2x + 1}}{\dfrac{x(x + 1)}{(2x + 1)(x - 1)}}$$

$$= \frac{(2x - 1)(x + 1)}{2x + 1} \cdot \frac{(2x + 1)(x - 1)}{x(x + 1)}$$

$$= \frac{(2x - 1)(x - 1)}{x}$$

Exercises for Section 13

In Exercises 1 through 41, perform the indicated operations and then simplify when necessary.

1. $\frac{3}{5} + \frac{4}{5}$

2. $\frac{5}{7} - \frac{3}{7}$

3. $\frac{4}{9} - \frac{3}{9} - \frac{2}{9}$

4. $\frac{8}{x} + \frac{4}{x}$

5. $\frac{a}{y} + \frac{b}{y}$

6. $\frac{2x}{3} + \frac{x}{2}$

7. $\frac{r}{st} + \frac{s}{t}$

8. $\frac{a}{y^2} + \frac{b}{y}$

9. $\frac{3}{5x} + \frac{1}{x} - \frac{y}{10}$

10. $\frac{3}{x} + \frac{4}{y} - \frac{2}{z}$

11. $\frac{1}{a} + \frac{1}{b} + \frac{1}{c}$

12. $\frac{3}{2x} + \frac{5}{2x^2} - \frac{x}{6x^2}$

13. $\frac{t}{t-3} + \frac{3}{t-2} - \frac{3}{t-3}$

14. $\frac{x}{x-2} - \frac{1}{x+1} + \frac{1}{x-2}$

15. $\frac{1}{p-2} - \frac{1}{p-3} + \frac{1}{p-4}$

16. $\frac{m-2n}{m+2n} - \frac{2m-n}{2m+n}$

17. $\frac{x+y}{x-y} - \frac{x-y}{x+y}$

18. $a + \frac{3}{a-2}$

19. $a - b + \frac{1}{2ab}$

20. $\frac{5}{x-3} + \frac{2}{x^2-9} - \frac{x}{3-x}$

21. $\frac{1}{2x+1} - \frac{2}{4x^2+4x+1}$

22. $\frac{x+3}{x-2} - \frac{x-1}{x^2-2x}$

23. $\frac{4x}{x^2-9} - \frac{3}{x+3} + \frac{4}{x-3}$

24. $\frac{4b^2}{y^2-b^2} - \frac{b-y}{y+b} + \frac{y+b}{b-y}$

25. $\frac{r+1}{r} - \frac{r+2}{s} - \frac{2-r}{rs}$

26. $\frac{x-1}{x^2+x-6} + \frac{x-2}{x^2+4x+3} - \frac{1}{3x-6}$

27. $x - \frac{2x}{x^2-1} + \frac{3}{x+1}$

28. $\frac{6}{y-3} + \frac{4}{3-y} + \frac{4-x}{x+4}$

29. $\frac{p-1}{p^2-9} + \frac{3p+2}{p^2-p-6}$

30. $\frac{m}{m^2-5m+6} - \frac{2m-1}{m^2-4}$

31. $\dfrac{ab}{(c-a)(a-b)}+\dfrac{bc}{(c-b)(a-c)}+\dfrac{ac}{(b-a)(b-c)}$

32. $\dfrac{\frac{1}{4}+\frac{1}{5}}{\frac{1}{3}-\frac{1}{5}}$

33. $\dfrac{\frac{3}{x}+\frac{3}{y}}{\frac{5}{x}+\frac{5}{y}}$

34. $\dfrac{2-\frac{1}{x}}{2+\frac{1}{x}}$

35. $\dfrac{\frac{a}{b}-1}{1-\frac{b}{a}}$

36. $\dfrac{1-\frac{p^2}{q^2}}{\frac{p}{q}-1}$

37. $\dfrac{2y-9+\frac{4}{y}}{2y+3-\frac{2}{y}}$

38. $\dfrac{a+\frac{b^2}{a}+b}{a^2-\frac{b^3}{a}}$

39. $\dfrac{\frac{4}{r-s}-\frac{2}{s^2-r^2}}{\frac{1}{r-s}-\frac{1}{r+s}}$

40. $1+\dfrac{1}{x-\frac{1}{x}}$

41. $1+\dfrac{1}{x+\frac{2}{x}}$

42. Find the value of $\dfrac{x+y}{2}$ when $x=\dfrac{r-s}{s}$ and $y=\dfrac{r+s}{r}$.

REVIEW QUESTIONS FOR CHAPTER 2

1. What is an algebraic expression?
2. What is an algebraic term?
3. Explain the terms "monomial," "binomial," "trinomial," and "multinomial."
4. What are like terms?
5. How are algebraic expressions added?
6. How are algebraic expressions subtracted?
7. How are multinomials multiplied?
8. What is a polynomial?

9. What is the degree of a polynomial?

10. If the remainder polynomial in the division process is equal to zero, what can be said about the divisor polynomial? About the quotient polynomial?

11. What is a prime polynomial?

12. What is a rational algebraic fraction?

13. Why is $\frac{x^2 + 1}{x^2 - 4}$ in its lowest terms?

14. Why is $-\frac{+a}{-b} = \frac{a}{b}$?

15. What is a common denominator?

16. How is the least common denominator found?

17. What is a complex algebraic fraction?

REVIEW EXERCISES FOR CHAPTER 2

Perform the operation indicated and simplify where possible.

1. $2ab + 3ab - 4ab$
2. $2\sqrt{8} + 3\sqrt{18} - \sqrt{32}$
3. $\sqrt{28} - 4\sqrt{7} + 2\sqrt{63}$
4. $x^2 + 2y^2 - 5x + y - x + 3y - x^2 - 5y^2$
5. $3xy - 5yz + 6xz - (xy - yz + 4xz)$
6. $4x - (3x - 2y + 6)$
7. $(a + b) - (a - b)$
8. $-(a + b - 2c) - (-2a + 3b - c)$
9. $(x^2 + 3x + 5) + (2x^2 - 6x + 4) - (x^2 + x - 2)$
10. $a^2b(2ab^3)$
11. $(xy) \cdot (x^2y) \cdot (2x^3y^5)$
12. $(2a - 1)(2a + 1)$
13. $(a - 2)^2$
14. $2ab(a + ab - b)$
15. $(x^2 + y^2)(x - y)$
16. $(x - 1)(x^2 - 2x - 3)$
17. $2x(x - 1)(x^2 - 4x)$
18. $\frac{12x^6}{4x^3}$
19. $\frac{8xy^3}{2x^2y^2}$
20. $\frac{48a^3bc^4}{6ab^2c^3}$
21. $\frac{x^4 - 2x^3 + 6x^2 - 8x}{x}$
22. $\frac{x^2 - 3x + 2}{x - 1}$
23. $\frac{x^2 - x - 12}{x + 3}$
24. $\frac{x^3 + 27}{x + 3}$
25. $\frac{x^4 - 1}{x + 1}$

26. $\dfrac{x^4 - 1}{x^2 + 1}$

27. $\dfrac{4abc}{2a^2b^3} \cdot \dfrac{6a^3b^2}{bc^2}$

28. $\dfrac{8x + 16}{3} \cdot \dfrac{1}{x^2 - 4}$

29. $\dfrac{x^2 + 7x + 12}{3x + 12} \cdot \dfrac{3x - 9}{x^2 - 9}$

30. $\dfrac{3x + 12}{x^2 + 8x + 16} \cdot \dfrac{x^2 - 5x - 36}{3x - 27}$

31. $\dfrac{x + 7}{x^2 + 3x - 28} \cdot \dfrac{x^2 - 16}{x^2 + 5x + 4}$

32. $\dfrac{x^2 - 6x + 5}{x^2 - 4} \div \dfrac{x - 1}{x - 2}$

33. $\dfrac{x^2 - 9}{x^2 - 1} \div \dfrac{x + 3}{x^3 - x}$

34. $\dfrac{2x^2 + x - 3}{9x^2 - 1} \div \dfrac{3x^2 - 4x + 1}{3x - 1}$

35. $\dfrac{5}{x - 1} + \dfrac{2}{x^2 - 1}$

36. $\dfrac{x - 1}{x^2 - 3x - 10} + \dfrac{x + 4}{x - 5}$

37. $\dfrac{x + 7}{x^2 + 3x + 2} + \dfrac{x - 2}{x^2 + 5x + 6}$

38. $\dfrac{x - 2}{2x + 3} - \dfrac{x + 3}{2x^2 - 7x - 15}$

39. $\dfrac{5}{x^2 - 9} - \dfrac{2}{x^2 - 2x - 3}$

40. $\dfrac{x - 1}{x^2 + 5x + 6} - \dfrac{x - 2}{x^2 + 2x - 3}$

41. $\dfrac{x + y}{\dfrac{1}{x} + \dfrac{1}{y}}$

42. $\dfrac{\dfrac{2}{x + 1}}{\dfrac{x^2 + 1}{x^2 + 2x + 1}}$

CHAPTER 3

Equations in One Variable

SECTION 14
Linear Equations

The basic operations for algebraic expressions developed in Chapter 2 can be used to solve **formulas**. A formula is an **equation** that expresses a general fact or rule. An *equation* is a statement of equality between two expressions. A statement of equality such as $a = b$ is an equation. We commonly refer to a as the **left member** and b as the **right member** of the given equation. An equation in one variable involves only one variable. Solutions of such equations are values of the variable that make both members equal.

Example 14.1

The equation $x + 3 = 2x + 5$ has $x + 3$ for its left member and $2x + 5$ for its right member. This is an equation in one variable, x. The equation $x + 3y = 4x - 5y$ is an equation in two variables, x and y.

The solutions to an equation in one variable are all values that make both members equal. We say that the solutions satisfy the equation. In this chapter we are going to deal only with equations involving one variable. Solutions to equations involving one variable are also referred to as the **roots** of the equation. We say that two equations are **equivalent** if they have the same solutions. We shall assume that the possible values for a variable are the real numbers. Since we are dealing with real numbers, the following operations can be used to transform equations into equivalent equations.

Operations Used to Form Equivalent Equations

1. If we add equals or subtract equals from both members of an equation, the resulting equation is equivalent to the original equation. (14.1)

2. If we multiply or divide both members of an equation by equals (with the exception of zero), the resulting equation is equivalent to the original equation. (14.2)

Example 14.2

(a) If $2x + 5 = 7$, we may subtract 5 from both members to obtain the equivalent equation $2x + 5 - 5 = 7 - 5$ or $2x = 2$.

(b) If $1/(x + 2) - 3 = x/(x - 1) + 6$, we may add 3 to both members to obtain the equivalent equation

$$\frac{1}{x+2} - 3 + 3 = \frac{x}{x-1} + 6 + 3 \quad \text{or} \quad \frac{1}{x+2} = \frac{x}{x-1} + 9$$

Example 14.3

(a) If $2x/3 + x/6 = 5$, we can multiply both members by 6 to obtain the equivalent equation $6(2x/3 + x/6) = 6(5)$ or $4x + x = 30$.

(b) If $2x/(x - 1) + 6 = x$, we can multiply both members by $x - 1$ (provided that $x - 1 \neq 0$) to obtain the equivalent equation

$$(x-1)\left(\frac{2x}{x-1} + 6\right) = (x-1)(x) \quad \text{or} \quad 2x + 6(x-1) = x(x-1)$$ ●*

If $3x + a = b$, we can divide both members by 3 to obtain the equivalent equation $x + a/3 = b/3$.

The solution for the equation $x = 2$ is simply 2, since only $2 = 2$. The solution for the equation $3x + 3 = 5x + 6$ could be stated if this equation could be transformed into an equivalent equation of the form $x = a$.

An equation in one variable that contains only the variable to the first power and constants in each of its terms is called a *linear equation* in one variable. Such an equation can be expressed as

$$ax + b = 0 \tag{14.3}$$

In formula (14.3) the variable is x and the constants are a and b. The solution or root of (14.3) can be found by first adding $-b$ to both members and then dividing both members by $a (a \neq 0)$ to obtain $x = -b/a$. If a linear equation is not expressed in the form of (14.3), we can use the operations (14.1) and (14.2) to transform the given equation into the form (14.3).

Example 14.4

Solve $x + 3(x + 2) = 2x - 6$.

Solution

We use the operations (14.1) and (14.2) as follows:

$$x + 3(x + 2) = 2x - 6$$

$$x + 3x + 6 = 2x - 6$$

$$4x + 6 = 2x - 6$$

*Symbol ● designates end of example.

$$4x - 2x = -6 - 6 \quad \text{(subtract } 2x \text{ and 6 from both members)}$$

$$2x = -12$$

$$\frac{2x}{2} = \frac{-12}{2} \quad \text{or} \quad x = -6$$

●

We are frequently required to solve a formula involving several letter symbols for a particular letter or symbol. The particular letter or symbol should then be considered the variable, and the other letters or symbols should be considered as constants. Such formulas can be considered as equations in one variable and be solved accordingly.

Example 14.5

Solve for r when $C = 2\pi r$.

Solution

We consider r to be the variable and the remaining symbols as constants. If we divide both members of the equation by 2π, we obtain

$$\frac{C}{2\pi} = \frac{2\pi r}{2\pi} \quad \text{or} \quad r = \frac{C}{2\pi}$$

Example 14.6

Solve for v_0 when $s = s_0 + v_0 t - 16t^2$.

Solution

We consider v_0 to be the variable and s, s_0, and t to be constants. We wish to isolate the term involving the variable v_0. If we add $16t^2$ to both members and subtract s_0 from both members, we obtain

$$s - s_0 + 16t^2 = s_0 - s_0 + v_0 t - 16t^2 + 16t^2$$

$$= v_0 t$$

Now, we divide by t (assuming that $t \neq 0$) to obtain

$$\frac{s - s_0 + 16t^2}{t} = v_0 \quad \text{or} \quad v_0 = \frac{s - s_0 + 16t^2}{t}$$

Example 14.7

Solve for x when $yx + y = x$.

Solution

We consider x to be the variable and y a constant. We want to collect all the terms involving x on either the left or right side of the equality. Here we arbitrarily choose to collect them on the left side by subtracting both y and x from both the left and right members of the given equations.

$$yx + y = x$$

$$yx + y - y - x = x - x - y \quad \text{(we subtract } x \text{ and } y \text{ from both members)}$$

$$yx - x = -y$$

$$x(y - 1) = -y \quad \text{(we factored the left member)}$$

$$x = \frac{-y}{y - 1} \quad \text{(we divided by } y - 1 \text{ and assumed } y - 1 \neq 0)$$

or

$$x = -\frac{y}{y - 1} \quad \text{or} \quad \frac{y}{1 - y}$$

●

Equations such as those in Examples 14.5, 14.6, and 14.7 are often referred to as *literal equations.*

Exercises for Section 14

In Exercises 1 through 8, solve for the variable given.

1. $2x - 3 = 5$

2. $3t + 4 = 2t + 1$

3. $3y - 4 = 2y$

4. $5(m + 1) = m + (2m + 1)$

5. $2p + 3 - (3 - p) = 2p$

6. $4x - 2(x - 3) = 4 - (3 - x)$

7. $4(x - 5) = 3(x + 1) - 6$

8. $7(x - 1) = 6(3 - x)$

In Exercises 9 through 32, solve for the letter indicated.

9. $bx + b = cx + b$ for x

10. $3x - 4p = 2x + 2p$ for p

11. $4(ax - 2b) = b(x - 3)$ for x

12. $2r - 2s = r - s$ for r

13. $ax = a - x$ for x

14. $l = a + (n - 1)d$ for n

15. $l = a + (n - 1)d$ for d

16. $pv = k$ for v

17. $pv = k$ for p

18. $A = P(1 + rt)$ for r

19. $V = \frac{1}{3}\pi r^2 h$ for h

20. $A = \frac{1}{2}bh$ for b

21. $P = 2l + 2w$ for w

22. $PV = RT$ for T

23. $A + B + C = 180$ for B

24. $S = \frac{1}{2}gt^2$ for g

25. $A = \frac{1}{2}h(b + c)$ for b

26. $V = \frac{4}{3}\pi r^3$ for π

27. $F = \dfrac{Gm_1m_2}{r^2}$ for m_1

28. $C = \frac{5}{9}(F - 32)$ for F

29. $yx + yk = x$ for y

30. $K = \frac{1}{2}r^2\theta$ for θ

31. $E = IR$ for I

32. $E = IR$ for R

33. The relationship between Fahrenheit and Celsius temperature readings can be expressed by the equation $\frac{F - 32}{180} = \frac{C}{100}$,

(a) Solve for C.
(b) Solve for F.
(c) Find F if $C = 0; 10; 42; 100$.
(d) Find the temperature at which $F = C$.

34. The length of a rectangle exceeds the width by 12 ft. If the perimeter is 144 ft, find the dimensions.

35. The perimeter of an isosceles triangle is 74 inches (in.). The base is 8 in. longer than one of the equal sides. Find the sides of the triangle.

36. The length of a rectangle is 3 times the width. The perimeter is 80 ft. Find the dimensions.

37. A given square has a side of length x feet. When each side of the square is increased by 4 ft, the area is increased by 64 ft^2. Determine the dimensions of the original square.

38. The sum of three consecutive integers is 36. Find the integers.

SECTION 15

Quadratic Equations; Solutions by Factoring

If a, b, and c are constants, the equation

$$ax^2 + bx + c = 0 \tag{15.1}$$

is called a *quadratic equation* in the variable x. If $a = 0$, then (15.1) is a linear equation. If $c = 0$, we have

$$ax^2 + bx = 0 \tag{15.2}$$

And if $b = 0$, we have

$$ax^2 + c = 0 \tag{15.3}$$

Both (15.2) and (15.3) are called **incomplete quadratic equations**. Form (15.3) is sometimes called a **pure** quadratic equation.

The solution(s) or root(s) of (15.2) can be found by factoring the left side of the equation and then setting each factor equal to zero. This is allowable since we know that if $a \cdot b = 0$, then $a = 0$ or $b = 0$.

Example 15.1
Solve $4x^2 + 9x = 0$.

Solution
We first observe that $4x^2 + 9x$ can be factored as $x(4x + 9)$. Thus, we have

$$4x^2 + 9x = x(4x + 9) = 0$$

Now, we set each factor equal to zero, to obtain

$$x = 0 \quad \text{or} \quad 4x + 9 = 0$$
$$4x = -9$$
$$x = -\frac{9}{4}$$

Thus, the solutions or roots of the equation $4x^2 + 9x = 0$ are 0 and $-\frac{9}{4}$. ●

The solutions of quadratic equations of the form (15.2) can be found by solving for x^2 and then taking the square root of both sides of the equation as follows:

$$ax^2 + c = 0$$
$$x^2 = -\frac{c}{a}$$
$$x = \pm\sqrt{-\frac{c}{a}} \quad \text{(recall that if } x^2 = a\text{, then } x = \pm\sqrt{a}\text{)}$$

We note that x will be a real number if $-c/a > 0$.

Example 15.2
Solve $4x^2 - 5 = 0$.

Solution
We first solve for x^2 and then extract the square root of both sides of the equation. Thus,

$$4x^2 - 5 = 0$$
$$x^2 = \frac{5}{4}$$
$$x = \pm\sqrt{\frac{5}{4}} = \pm\frac{\sqrt{5}}{2}$$

Example 15.3
Solve $3x^2 + 27 = 0$.

Solution
We first solve for x^2 and then extract the square root of both sides of the equation. Thus,

$$3x^2 + 27 = 0$$

$$x^2 = -\frac{27}{3} = -9$$

$$x = \pm\sqrt{-9}$$

Since $\sqrt{-9}$ is nonreal, $3x^2 + 27 = 0$ has *no real* solutions or roots. ●

The factoring procedure used to solve equations of the form (15.2) can be extended to solve the general quadratic equation (15.1) if we are capable of factoring the left side of that equation. We can now use the appropriate factor forms of Section 10.

Example 15.4
Solve $6x^2 + x - 1 = 0$.

Solution
We see that the left side of the equation can be factored as follows:

$$6x^2 + x - 1 = 0$$

$$(3x - 1)(2x + 1) = 0$$

Then

$$3x - 1 = 0 \quad \text{or} \quad 2x + 1 = 0$$

$$x = \frac{1}{3} \quad \text{or} \quad x = -\frac{1}{2}$$

Thus, the two roots or solutions of the given equation are $\frac{1}{3}$ and $-\frac{1}{2}$.

Example 15.5
Solve $4x^2 = 12x - 9$.

Solution
First we must express the given equation in the form (15.1) and then observe that the new left side can be factored as follows:

$$4x^2 = 12x - 9$$

$$4x^2 - 12x + 9 = 0$$

$$(2x - 3)(2x - 3) = 0 \quad \text{or} \quad (2x - 3)^2 = 0$$

Then $2x - 3 = 0$ or $x = \frac{3}{2}$. Thus, the given equation has a single real root, $x = \frac{3}{2}$.

Quadratic equations of the form (15.1) may have

(a) Two real roots (see Example 15.6).
(b) One real root (see Example 15.5).
(c) No real roots (see Example 15.3).

Example 15.6
Solve $3/x = 2x/4$.

Solution
At first glance this equation does not appear to be quadratic, but if we multiply both sides by the (L.C.D.) $4x$, we obtain

$$(4x)\left(\frac{3}{x}\right) = (4x)\left(\frac{2x}{4}\right)$$

$$12 = 2x^2$$

$$x^2 = \frac{12}{2} = 6$$

$$x = \pm\sqrt{6}$$

Thus, the two roots of the given equations are $\sqrt{6}$ and $-\sqrt{6}$. ●

Formulas or literal equations may be quadratic depending on which letter or symbol is designated as the variable. For example, $s = \frac{1}{2}gt^2$ is quadratic in t if s and g are constants.

Example 15.7
Solve $s = \frac{1}{2}gt^2$ for t.

Solution
We see that $s = \frac{1}{2}gt^2$ fits the form (15.3) when t is the variable. Thus,

$$s = \frac{1}{2}gt^2$$

$$2s = gt^2$$

$$\frac{2s}{g} = t^2 \quad \text{or} \quad t = \pm\sqrt{\frac{2s}{g}}$$

Exercises for Section 15

In Exercises 1 through 8, determine which equations are quadratic in the variable indicated by expressing each into the general form (15.1).

1. $6 = 3x^2 + 2$

2. $x(x + 1) = 4 - x$

3. $\frac{y}{3} = \frac{8}{y}$

4. $\frac{t}{2} = \frac{4}{t - 1}$

5. $x^2 + 3x = x(x + 1)$

6. $p^3 + (p + 1)^2 = p^3 - 8$

7. $V = \pi r^2 h$ (r is variable)

8. $V = \pi r^2 h$ (h is variable)

In Exercises 9 through 30, solve each quadratic equation.

9. $4x^2 - 36 = 0$

10. $5x^2 = 30$

11. $16 - x^2 = 0$

12. $32 - x^2 = 7$

13. $x(3x - 1) = 0$

14. $(2x - 1)(x + 1) = 0$

15. $6x^2 - 4x = 0$

16. $5x - 3x^2 = 0$

17. $2(x + 2) = 6x^2 + 4$

18. $2x^3 + x^2 + 6x = 2x^3$

19. $12x^2 - 7x + 1 = 0$

20. $12x^2 + 7x + 1 = 0$

21. $x^2 + 7x + 6 = 0$

22. $x^2 + 3x - 3 = 10x - 9$

23. $9m^2 + 6m + 1 = 0$

24. $4x^2 + 25 = 20x$

25. $9 + 4p^2 = 12p$

26. $16 - 8t + t^2 = 0$

27. $d^2 - 8d + 15 = 0$

28. $\dfrac{x}{3} = \dfrac{2}{4x}$

29. $\dfrac{1}{w^2} = \dfrac{1}{25}$

30. $\dfrac{1}{r} = \dfrac{r}{9}$

31. Solve for r when $V = \pi r^2 h$.

32. Solve for r when $A = \pi r^2$.

33. Solve for r when $K = \frac{1}{2} r^2 \theta$.

34. Solve for a when $F = ma^2$.

35. The distance an object falls under the influence of gravity (neglecting air resistance) is given by $S = 16t^2$, when S is in feet and t is in seconds. How long would it take an object to fall from a height of:
(a) 144 ft? (b) 256 ft? (c) 288 ft?

36. The width of a rectangle is 4 in. less than the length. If the area is 60 in.^{2}, find the dimensions.

37. A metal worker forms an open gutter by bending up the sides of a sheet of aluminum 10 ft long and 12 in. wide. If the cross-sectional area must be 18 in.^{2}, how much is bent up on each side?

38. Repeat Exercise 37 if the cross-sectional area must be 16 in.^{2}.

39. An open box is made from a square piece of tin by cutting out a 4-in. square from each corner and turning up the sides. If the box contains 144 in.^{3}, find the area of the original square.

40. The formula $S = 2\pi r^2 + 2\pi rh$ gives the total surface area S of a right circular cylinder of radius r and height h. Find r such that a cylinder of height 2 in. will have an area of 48π in.^{2}.

SECTION 16

Solution of Quadratic Equations by Completing the Square; Quadratic Formula

We noted in Section 15 that quadratic equations of the form $x^2 = 9$ were solved by extracting the square root of both sides of the equation. We can extend this method to equations such as $(x - 3)^2 = 5$ as follows.

Example 16.1
Solve $(x - 3)^2 = 5$.

Solution
We can extract the square root of both sides of the equation to obtain

$$(x - 3)^2 = 5$$
$$x - 3 = \pm\sqrt{5}$$
$$x = 3 \pm \sqrt{5}$$

Thus, the roots or solutions of $(x - 3)^2 = 5$ are $3 + \sqrt{5}$ and $3 - \sqrt{5}$. If the equations $(x - 3)^2 = 5$ had been transformed into the standard form, we would have

$$(x - 3)^2 = 5$$
$$x^2 - 6x + 9 = 5$$
$$x^2 - 6x + 4 = 0$$

The left side of this equation is *not factorable* using the techniques of Section 10. ●

Example 16.1 suggests a technique for solving the general quadratic equation when the left side $ax^2 + bx + c$ is *not factorable*. This technique is known as *completing the square*.

Example 16.2
Solve $x^2 + 4x - 2 = 0$.

Solution

First we note from our discussion in Section 10 that $x^2 + 4x - 2$ is not factorable. If we add 2 to both sides of the equation, we obtain

$$x^2 + 4x = 2 \qquad \text{(A)}$$

The left side of (A) resembles the factor form 10 in Section 10 and suggests that if we add 4 to both sides of (A) we can transform the left side of (A) into a perfect square. We do this as follows:

$$x^2 + 4x + 4 = 2 + 4$$

$$(x + 2)^2 = 6 \qquad \text{(B)}$$

We now proceed as in Example 16.1.

$$x + 2 = \pm\sqrt{6}$$

$$x = -2 \pm \sqrt{6}$$

Thus, the solution or roots for the original equation are $-2 + \sqrt{6}$ and $-2 - \sqrt{6}$. ●

The question that arises in Example 16.2 is: How do we know what number to add to both members to transform the left member into a perfect square? The answer lies in the factor forms.

$$(x + a)^2 = x^2 + 2ax + a^2 \qquad (16.1)$$

$$(x - a)^2 = x^2 - 2ax + a^2 \qquad (16.2)$$

of Section 10.

In both forms (16.1) and (16.2) we observe that the numerical coefficient of x^2 is 1 and that the third term a^2 is equal to the *square of one half of the coefficient* of their respective middle terms.

Thus, $x^2 + 4x$ can be transformed into the perfect square $(x + 2)^2$ by adding the square of one half the coefficient of $4x$ to $x^2 + 4x$ as follows.

$$x^2 + 4x + \left(\frac{4}{2}\right)^2 = x^2 + 4x + 2^2 = (x + 2)^2$$

This technique is called ***completing the square***. The discussions above suggest the following general procedure for solving a quadratic equation by completing the square.

Example 16.3

Solve $4x^2 - 2 = 4x$ by completing the square.

Solution

First we express the given equation in its standard form $4x^2 - 4x - 2 = 0$. Now we divide each term by 4 so that the coefficient of x^2 becomes 1.

$$x^2 - x - \tfrac{1}{2} = 0$$

Next, we place the constant term of the right side of the equation by adding $\frac{1}{2}$ to both sides.

$$x^2 - x = \frac{1}{2}$$

Now we divide the coefficient of the x term, -1, by 2 to obtain $-\frac{1}{2}$, and then add the square of $-\frac{1}{2}$ to both sides of the equation.

$$x^2 - x + \left(-\frac{1}{2}\right)^2 = \frac{1}{2} + \left(-\frac{1}{2}\right)^2$$

The left side of the new equation is now the square of $(x - \frac{1}{2})$. Thus,

$$\left(x - \frac{1}{2}\right)^2 = \frac{1}{2} + \frac{1}{4} = \frac{3}{4}$$

$$x = \frac{1}{2} \pm \frac{\sqrt{3}}{2}$$

The roots or solutions are

$$\frac{1}{2} + \frac{\sqrt{3}}{2} = \frac{1 + \sqrt{3}}{2} \quad \text{and} \quad \frac{1}{2} - \frac{\sqrt{3}}{2} = \frac{1 - \sqrt{3}}{2}$$

●

The technique of solving by completing the square can be applied to *any* quadratic equation in its standard form, but it is certainly easier to solve by factoring if the expression $ax^2 + bx + c$ can be factored by inspection.

We can now use the technique of completing the square to develop a formula to solve *all* quadratic equations as follows:

$$ax^2 + bx + c = 0$$

We divide by a, to obtain

$$x^2 + \frac{bx}{a} + \frac{c}{a} = 0$$

We add $-c/a$ to both sides, to obtain

$$x^2 + \frac{b}{a}x = -\frac{c}{a}$$

We now add the square of $\frac{1}{2}$ the coefficient of the term $(b/a)x$ to both sides, to obtain

$$x^2 + \frac{b}{a}x + \left(\frac{b}{2a}\right)^2 = -\frac{c}{a} + \left(\frac{b}{2a}\right)^2$$

$$\left(x + \frac{b}{2a}\right)^2 = \frac{-c}{a} + \frac{b^2}{4a^2}$$

$$= \frac{-4ac + b^2}{4a^2}$$

Now we take the square root of both sides to obtain

$$x + \frac{b}{2a} = \pm\sqrt{\frac{b^2 - 4ac}{4a^2}}$$

$$x = -\frac{b}{2a} \pm \frac{\sqrt{b^2 - 4ac}}{2a} = \frac{-b \pm \sqrt{b^2 - 4ac}}{2a} \tag{16.3}$$

Thus, the roots of the given equation are

$$x = \frac{-b + \sqrt{b^2 - 4ac}}{2a} \quad \text{and} \quad x = \frac{-b - \sqrt{b^2 - 4ac}}{2a}$$

Formula (16.3) is known as the *quadratic formula. If a quadratic equation is not given in its standard form, we must express it in its standard form before determining the values of a, b, and c that are to be used in the quadratic formula.*

Example 16.4
Solve $4x^2 = 3 + 4x$.

Solution
We first express the equation in its standard form,

$$4x^2 - 4x - 3 = 0$$

and then note that $a = 4$, $b = -4$, and $c = -3$. Now we substitute these values into (16.3) to obtain

$$x = \frac{-(-4) \pm \sqrt{(-4)^2 - 4(4)(-3)}}{2(4)}$$

$$x = \frac{4 \pm \sqrt{16 + 48}}{8} = \frac{4 \pm \sqrt{64}}{8}$$

Thus,

$$x = \frac{4 + \sqrt{64}}{8} = \frac{4 + 8}{8} = \frac{12}{8} = \frac{3}{2} \quad \text{and}$$

$$x = \frac{4 - \sqrt{64}}{8} = \frac{4 - 8}{8} = \frac{-4}{8} = -\frac{1}{2}$$

It should be noted that this problem could also have been solved by factoring $4x^2 - 4x - 3$ as $(2x + 1)(2x - 3)$ and then setting each factor equal to zero.

Example 16.5
Solve $x(2x + 3) = -4$.

Solution

First, we write the given equation in its standard form:

$$x(2x + 3) = -4$$

$$2x^2 + 3x + 4 = 0$$

Now we note that $a = 2$, $b = 3$, and $c = 4$. Substitution of these values in (16.3) yields

$$x = \frac{-3 \pm \sqrt{3^2 - 4(2)(4)}}{2(2)} = \frac{-3 \pm \sqrt{9 - 32}}{4} = \frac{-3 \pm \sqrt{-23}}{4}$$

Since $\sqrt{-23}$ is nonreal, we have *no* real roots or solutions.

Exercises for Section 16

In Exercises 1 through 10, solve by completing the square.

1. $x^2 + 6x + 5 = 0$

2. $x^2 - 13x - 30 = 0$

3. $4x^2 - 12x + 7 = 0$

4. $2x^2 + 3 = 7x$

5. $4 - 6t + t^2 = 0$

6. $x^2 = 6x - 2$

7. $s^2 + s + 1 = 0$

8. $y(y - 3) = 4$

9. $12p - 9p^2 = 5$

10. $7k^2 - 56 = 14k$

For Exercises 11 through 22, solve Exercises 9 through 20 of Section 15 by using the quadratic formula.

23. In the formula $i^2R + iE = 4000$, i is the current (amperes), E the voltage (volts), and R the resistance (ohms). If $R = 20$ ohms (Ω) and $E = 50$ volts (V) find i.

SECTION 17
Fractional Equations

We solve fractional equations by multiplying both members of the equation by the least common denominator (L.C.D.) of all the denominators in the given equation. The resultant equation will then be a polynomial equation in one variable. We are prepared to solve the resultant equation if it is linear or quadratic.

Example 17.1

Solve $\dfrac{x}{2} = \dfrac{8}{x - 6}$.

Solution
We note that the L.C.D. for 2 and $x - 6$ is $2(x - 6)$, and we then multiply both members of the equation by $2(x - 6)$, to obtain

$$2(x - 6)\left(\frac{x}{2}\right) = 2(x - 6)\frac{8}{x - 6}$$

$$(x - 6)x = 16$$

$$x^2 - 6x = 16$$

$$x^2 - 6x - 16 = 0$$

Since the resultant equation is quadratic and is factorable, we have

$$(x - 8)(x + 2) = 0$$

Thus, $x = 8$ or $x = -2$.

A check by substituting each root into the original equation shows that: when $x = 8$,

$$\frac{8}{2} = \frac{8}{8 - 6} \quad \text{or} \quad 4 = 4$$

When $x = -2$,

$$\frac{-2}{2} = \frac{8}{-2 - 6} = -1 = -1$$

Thus, both values are roots of the original equation. ●

When we multiply or divide both members of an equation by an expression containing the variable, we might be multiplying or dividing by zero. Thus, the resultant equation may contain roots that do not satisfy the original equation. Such roots are called **extraneous**. When such operations are performed, the solution is not complete until the roots found have been tested in the original equation and the extraneous roots have been rejected.

Example 17.2

Solve $\dfrac{3}{x - 3} = \dfrac{x}{x - 3} - 2$.

Solution
We first multiply both members of the equation by the L.C.D. $x - 3$, to obtain

$$(x - 3)\frac{3}{x - 3} = (x - 3)\frac{x}{x - 3} - 2(x - 3)$$

$$3 = x - 2(x - 3)$$

$$3 = x - 2x + 6$$

$$-3 = -x \quad \text{or} \quad x = 3$$

A check of this value into the original equation shows us that

$$\frac{3}{3-3} = \frac{3}{3-3} - 2 \quad \text{or} \quad \frac{3}{0} = \frac{3}{0} - 2$$

Since $3/0$ is not defined, we have an extraneous root. This occurred since we multiplied by the factor $x - 3$, which equals zero when $x = 3$. Since 3 was the only possible root and it is extraneous, the equation has *no* solution.

Exercises for Section 17

Solve each of the following.

1. $\frac{3}{x} + \frac{4}{3x} = 2$

2. $\frac{2}{x-3} - \frac{3}{x+3} = \frac{6}{x^2-9}$

3. $\frac{3}{x-6} = \frac{4}{2x+7}$

4. $\frac{2}{y^2+y-6} + \frac{1}{y^2+2y-3} = \frac{3}{y^2-y-2}$

5. $\frac{x+1}{x-3} = 5$

6. $\frac{p}{p-2} - 7 = \frac{2}{p-2}$

7. $\frac{4}{5} = \frac{t}{t+3}$

8. $\frac{r}{3} + \frac{r}{4} - \frac{r}{5} = 6$

9. $1 + \frac{x-2}{x-4} = \frac{(x-2)(2x-6)}{x-4}$

10. $\frac{4m-2}{m^2-2m} - \frac{1}{m} = \frac{3}{m-2}$

11. $\frac{6}{x-6} + 5 = \frac{x}{x-6}$

12. $1 + \frac{3x}{x-2} = \frac{2x(2x-1)}{x-2}$

13. $\frac{1}{y-3} - \frac{7}{y^2-y-6} = \frac{1}{y+2}$

14. $\frac{1}{x^2-4} = \frac{2}{x^2+4x+4}$

15. $\frac{1}{m+1} + \frac{1}{m-1} = \frac{4}{m^2-1}$

16. $\frac{x}{x^2-3x+2} = \frac{1}{2x-4} - \frac{1}{2(x-1)}$

17. The combined resistance R of two resistors, R_1 and R_2, connected in parallel is given by

$$R = \frac{1}{1/R_1 + 1/R_2}$$

If $R = 10\,\Omega$ and $R_1 = 12\,\Omega$, find the resistance R_2.

18. The combined capacitance C of two capacitors, C_1 and C_2, connected in series is given by

$$\frac{1}{C} = \frac{1}{C_1} + \frac{1}{C_2}$$

In a particular circuit, $C = \frac{1}{3}$ and $C_1 = 2$. Find C_2.

19. The combined capacitance C of three capacitors, C_1, C_2 and C_3, connected in series is given by

$$\frac{1}{C} = \frac{1}{C_1} + \frac{1}{C_2} + \frac{1}{C_3}$$

Solve for C_2.

20. The focal length f of a lens is given by the equation $1/f = 1/p + 1/q$, where p is the object distance and q is the image distance. Suppose that for a converging lens the focal length is 6.0 cm and the object distance is 18.0 cm. Find the image distance.

SECTION 18

Radical Equations

A **radical equation** is an equation that contains a variable under a radical. The process of eliminating the radical or radicals may lead us to quadratic equations. In this section we shall only consider radical equations involving the square-root symbol.

Our work with exponents and radicals in Sections 4 and 5 will be useful here. We should recall that $\sqrt{x} = x^{1/2}$ and that $(\sqrt{x})^2 = (x^{1/2})^2 = x$ if $x \geq 0$. This fact tells us that the squaring of an expression involving the square-root symbol will eliminate the square-root symbol.

The general procedure for solving such radical equations is to isolate one of the radical terms on one side of the equation and then to square both sides of the equation. This process is continued until all radicals have been eliminated. The equation generated by this procedure is then solved. Its roots or solutions **must be checked in the original equation** since some or all of the roots may be extraneous.

The possibility that squaring both sides of an equation can generate extraneous roots can be seen by considering the simple equation $x = 2$. If we square both sides, we have $x^2 = 4$. But the roots of $x^2 = 4$ are $x = 2$ and $x = -2$, and $x = -2$ does not satisfy the original equation. Thus, $x = -2$ is an extraneous root.

Example 18.1

Solve $\sqrt{x + 3} = 4$.

Solution

We square both sides of the given equation, to obtain

$$(\sqrt{x + 3})^2 = 4^2$$

$$x + 3 = 16$$

$$x = 13$$

If $x = 13$, a check of the original equation shows us that $\sqrt{13 + 3} = \sqrt{16} = 4$. Thus, we accept this root or solution.

Example 18.2

Solve $\sqrt{x - 1} = x - 7$.

Solution

We square both sides of the equation, to obtain

$$(\sqrt{x - 1})^2 = (x - 7)^2$$

$$x - 1 = x^2 - 14x + 49$$

$$0 = x^2 - 15x + 50$$

$$0 = (x - 10)(x - 5)$$

Thus, $x = 10$ or 5.

If $x = 10$, a check of the original equation shows us that $\sqrt{10 - 1} = 10 - 7$ or $\sqrt{9} = 3$. Thus, we accept this root. If $x = 5$, the original equation becomes $\sqrt{5 - 1} = 5 - 7$ or $\sqrt{4} = -2$, which is false. Thus, we reject $x = 5$ as a root or solution and use only $x = 10$.

Example 18.3

Solve $\sqrt{x - 3} - \sqrt{x} = 3$.

Solution

We first isolate the radical $\sqrt{x - 3}$ on the left side by adding $\sqrt{x}$ to both sides of the equation. Then we square both sides, to obtain

$$\sqrt{x - 3} = 3 + \sqrt{x}$$

$$(\sqrt{x - 3})^2 = (3 + \sqrt{x})^2$$

$$x - 3 = 9 + 6\sqrt{x} + x$$

Next we combine like terms and isolate the radical $\sqrt{x}$, to obtain

$$-12 = 6\sqrt{x}$$

$$-2 = \sqrt{x}$$

Finally, we square both sides of this result, to obtain

$$(-2)^2 = (\sqrt{x})^2$$

$$4 = x$$

If $x = 4$, the left side of the original equation becomes $\sqrt{4-3} - \sqrt{4} = \sqrt{1} - \sqrt{4} = 1 - 2 = -1$, which does not equal the right side. Thus, the only possible root is extraneous, and we conclude that there is **no solution.**

Example 18.4

Solve $\sqrt{2y+9} - \sqrt{y+1} = \sqrt{y+4}$.

Solution

We see that the radical $\sqrt{y+4}$ is already isolated on the right side of the equation, so we square both sides as follows:

$$(\sqrt{2y+9} - \sqrt{y+1})^2 = (\sqrt{y+4})^2$$

$$(2y+9) - 2(\sqrt{2y+9})(\sqrt{y+1}) + (y+1) = y+4$$

We now combine like terms and isolate the remaining radical expression, to obtain

$$2y + 9 + y + 1 - 2(\sqrt{2y+9})(\sqrt{y+1}) = y + 4$$

$$-2(\sqrt{2y+9})(\sqrt{y+1}) = -6 - 2y$$

$$(\sqrt{2y+9})(\sqrt{y+1}) = 3 + y$$

Squaring both members of this result yields

$$(2y+9)(y+1) = 9 + 6y + y^2$$

$$2y^2 + 11y + 9 = 9 + 6y + y^2$$

$$y^2 + 5y = 0$$

$$y(y+5) = 0$$

$$y + 5 = 0 \quad \text{or} \quad y = 0$$

Thus, the possible roots are 0 and -5. If $y = 0$, a check of the original equation yields

$$\sqrt{2(0)+9} - \sqrt{0+1} = \sqrt{0+4}$$

or

$$\sqrt{9} - \sqrt{1} = \sqrt{4}$$

or

$$3 - 1 = 2 \quad \text{(which is true)}$$

Therefore, we accept $y = 0$ as a solution to the original equation. If $y = -5$, a check of the original equation yields

$$\sqrt{2(-5)+9} - \sqrt{-5+1} = \sqrt{-5+4}$$

or

$$\sqrt{-1} - \sqrt{-4} = \sqrt{-4}$$

Since this expression contains the nonreal numbers $\sqrt{-1}$ and $\sqrt{-4}$, we reject $y = -5$ as a solution to the original equation. We conclude that there is only one solution to the original equation, 0. ●

In Example 18.4 it was necessary to square the difference between the two radicals expressions. It should be carefully noted that the result was not simply the removal of both radicals; it also gave us the *middle term*, $-2(\sqrt{y+9})(\sqrt{y+1})$.

In summary, when we are solving radical equations, we must *always* check the roots obtained from the newly generated equation in the original equation and reject as extraneous those roots that do not satisfy the original equation.

Exercises for Section 18

Solve each of the following.

1. $\sqrt{x} - 2 = 5$

2. $\sqrt{x} + 2 = 4$

3. $\sqrt{x-2} = 5$

4. $\sqrt{x+2} = 4$

5. $\sqrt{p} + 3 = p$

6. $\sqrt{y} + 2y = 10$

7. $\sqrt{w-3} = \sqrt{4w} - 3$

8. $\sqrt{3y} + 3y = 2$

9. $\sqrt{x+5} - \sqrt{x-4} = 9$

10. $\sqrt{4p-2} + 3 = 0$

11. $\sqrt{s+2} + \sqrt{s+1} + 2 = 0$

12. $\sqrt{9x^2+4} = 3x + 2$

13. $\sqrt{2w} - \sqrt{w^2+6w} = 0$

14. $\sqrt{2x^2-7} = 3 + x$

15. $\sqrt{3x^2 - 6x + 9} = 6$

16. $\sqrt{1-3p} - \sqrt{2p+3} = 1$

17. $\sqrt{r+4} + \sqrt{2r-1} = 3\sqrt{r-1}$

18. $\sqrt{x}(\sqrt{x} - 7) = 12$

19. $\sqrt{17x - \sqrt{x^2-5}} = 7$

20. $\sqrt{8 + 15x - 2x^2} = 6$

21. If $T = 2\pi\sqrt{L/g}$, solve for L.

SECTION 19

Equations in Quadratic Form

Some equations can be made quadratic by making an appropriate change of variable. We say such equations are **quadratic in form**.

Example 19.1

The equation $2x^4 - x^2 - 1 = 0$ is quadratic in x^2. If we let $u = x^2$, the original equation becomes $2(x^2)^2 - x^2 - 1 = 0$ or $2u^2 - u - 1 = 0$.

Example 19.2

The equation $6x^6 + x^3 - 2 = 0$ is quadratic in x^3. If we let $u = x^3$, the original equation becomes $6(x^3)^2 + x^3 - 2 = 0$ or $6u^2 + u - 2 = 0$.

Example 19.3

The equation $x^{1/3} + 6x^{1/6} + 5 = 0$ is quadratic in $x^{1/6}$. If we let $u = x^{1/6}$, the original equation becomes $(x^{1/6})^2 + 6x^{1/6} + 5 = 0$ or $u^2 + 6u + 5 = 0$.

Example 19.4

The equation $(x + 2)^4 + 3(x + 2)^2 + 2 = 0$ is quadratic in $(x + 2)^2$. If we let $u = (x + 2)^2$, the original equation becomes

$$[(x + 2)^2]^2 + 3(x + 2)^2 + 2 = 0 \quad \text{or} \quad u^2 + 3u + 2 = 0$$

●

An equation in one variable is an equation *quadratic in form* if it can be expressed as

$$au^2 + bu + c = 0 \tag{19.1}$$

where u is an algebraic expression in the given variable.

Example 19.5

The equation $x^4 + x^2 - 3 = 0$ can be put into the quadratic form $u^2 + u - 3 = 0$ by letting $u = x^2$. The equation $x^4 + x - 3 = 0$ *cannot* be put into quadratic form, since letting $u = x$ yields $u^2 = x^2$, not x^4. ●

We solve such equations by first solving (19.1) for the values of u. We then solve the equations obtained by substituting the algebraic expression for u.

Example 19.6

Solve $x^4 - 13x^2 + 36 = 0$.

Solution

The equation is quadratic in x^2. If we let $u = x^2$, we have

$$(x^2)^2 - 13x^2 + 36 = 0$$

$$u^2 - 13u + 36 = 0$$

$$(u - 4)(u - 9) = 0$$

$$u - 4 = 0 \qquad u - 9 = 0$$

$$u = 4 \qquad u = 9$$

We now replace u by x^2 and solve the following equations.

$$x^2 = 4 \qquad x^2 = 9$$

$$x = \pm 2 \qquad x = \pm 3$$

The original equation has *four* solutions: 2, −2, 3, and −3.

Example 19.7
Solve $x^{2/3} + x^{1/3} - 6 = 0$.

Solution
The equation is quadratic in $x^{1/3}$. If we let $u = x^{1/3}$, we have

$$(x^{1/3})^2 + x^{1/3} - 6 = 0$$
$$u^2 + u - 6 = 0$$
$$(u - 2)(u + 3) = 0$$
$$u - 2 = 0 \qquad u + 3 = 0$$
$$u = 2 \qquad u = -3$$

We now replace u by $x^{1/3}$ and solve the following equations.

$$x^{1/3} = 2 \qquad x^{1/3} = -3$$
$$(x^{1/3})^3 = 2^3 \qquad (x^{1//3} = (-3)^3$$
$$x = 8 \qquad x = -27$$

Thus, the original equation has two solutions, 8 and -27.

Example 19.8
Solve $4/y^4 + 15/y^2 - 4 = 0$.

Solution
The equation is quadratic in $1/y^2$. If we let $u = 1/y^2$, we have

$$4\left(\frac{1}{y^2}\right)^2 + 15\left(\frac{1}{y^2}\right) - 4 = 0$$
$$4u^2 + 15u - 4 = 0$$
$$(4u - 1)(u + 4) = 0$$
$$4u - 1 = 0 \qquad u + 4 = 0$$
$$u = \frac{1}{4} \qquad u = -4$$

We now replace u by $1/y^2$ and solve the following equations.

$$\frac{1}{y^2} = \frac{1}{4} \qquad \frac{1}{y^2} = -4$$
$$y^2 = 4 \qquad -4y^2 = 1$$
$$y = \pm 2 \qquad y^2 = -\frac{1}{4}$$
$$y = \pm\sqrt{-\frac{1}{4}} \quad \text{(nonreal)}$$

Thus, the only solutions to the original equation are $y = \pm 2$.

Example 19.9

Solve $(x-2)^4 - 9(x-2)^2 + 20 = 0$.

Solution

The equation is quadratic in $(x-2)^2$. If we let $u = (x-2)^2$, we have

$$[(x-2)^2]^2 - 9(x-2)^2 + 20 = 0$$
$$u^2 - 9u + 20 = 0$$
$$(u-4)(u-5) = 0$$
$$u - 4 = 0 \qquad u - 5 = 0$$
$$u = 4 \qquad u = 5$$

We now replace u by $(x-2)^2$ and solve the following equations.

$$(x-2)^2 = 4 \qquad (x-2)^2 = 5$$
$$x - 2 = \pm 2 \qquad x - 2 = \pm\sqrt{5}$$
$$x = 2 \pm 2 \qquad x = 2 \pm \sqrt{5}$$
$$x = 4 \text{ or } 0$$

Thus, the original equation has four solutions or roots: 4, 0, $2+\sqrt{5}$, and $2-\sqrt{5}$.

Exercises for Section 19

Solve the following equations.

1. $x^4 + 2x^2 - 8 = 0$
2. $x^4 - x^2 - 6 = 0$
3. $x^6 + 6x^3 + 5 = 0$
4. $x + 7x^{1/2} - 18 = 0$
5. $x - x^{1/2} - 6 = 0$
6. $x^6 + 28x^3 + 27 = 0$
7. $2x^4 - x^2 - 1 = 0$
8. $x^{10} + 31x^5 - 32 = 0$
9. $8x^{2/3} - 10x^{1/3} - 3 = 0$
10. $8x^{1/2} - 10x^{1/4} - 3 = 0$
11. $x - 5x^{1/2} + 4 = 0$
12. $x^{1/2} - 5x^{1/4} + 4 = 0$
13. $\frac{1}{y^4} + \frac{2}{y^2} - 8 = 0$
14. $\frac{1}{w^4} - \frac{1}{w^2} - 6 = 0$
15. $\frac{2}{p^6} - \frac{1}{p^3} - 1 = 0$
16. $\frac{96}{x^4} - \frac{22}{x^2} + 1 = 0$
17. $x - 19x^{1/2} - 53 = 3x^{1/2} - 5$
18. $x^{1/3} - x^{1/6} = 2$
19. $(x-3)^4 + 2(x-3)^2 - 8 = 0$
20. $(x+1)^4 + 2(x+1)^2 - 8 = 0$
21. $(3x-2)^6 + 6(3x-2)^3 + 5 = 0$
22. $\sqrt{y+2} - \sqrt[4]{y+2} = 6$

23. $(x^2 - 5x)^2 + 10(x^2 - 5x) + 24 = 0$

24. $(x^2 + 3x)^2 - 2(x^2 + 3x) - 8 = 0$

25. $y^{-4/3} - 5y^{-2/3} + 4 = 0$

26. $\left(\frac{1}{x}\right)^2 + \frac{1}{x} - 6 = 0$

27. $x^{16} - 65x^8 + 64 = 0$

28. (a) Solve $y - 5\sqrt{y} + 4 = 0$ as a radical equation.

(b) Solve $y - 5\sqrt{y} + 4 = 0$ as an equation in quadratic form.

SECTION 20

Applications Leading to Linear and Quadratic Equations

Many verbal or statement problems lead to linear or quadratic equations. To solve such problems, we must read the problem carefully and then decide what information is given and what information is wanted. We then use the given information and known formulas or laws from mathematics and the sciences to set up an equation to be solved.

There is no general procedure for solving verbal problems. The comments above should be considered as general suggestions only. These ideas can best be appreciated by considering some examples.

Example 20.1

How much pure copper should be combined with 40 lb of an alloy containing 70% copper to obtain another alloy containing 80% copper?

Solution

Since we wish to find the weight of the pure copper, we let w = the weight (in pounds) of the pure copper. Then the weight of the final alloy can be expressed as

$$w + 40$$

We can apply a formula which states that

(weight of alloy) × (percentage by weight of copper in the alloy)

= (amount of the copper in the alloy)

Since the

(amount of the pure copper) + (amount of copper in 70% alloy)

= (amount of copper in 80% alloy)

we have the following linear equation:

$$w(100\%) + 40(75\%) = (40 + w)(80\%)$$
$$w(1.00) + 40(0.75) = (40 + w)(0.80)$$

We now multiply both sides of the equation by 100, to obtain

$$100w + 40(75) = (40 + w)80$$
$$100w + 3000 = 3200 + 80w$$
$$20w = 200$$
$$w = 10$$

Therefore, the addition of 10 lb of pure copper is necessary.

Example 20.2

Two trains can start from a given point and travel on straight tracks at average speeds of 50 and 70 mph. If the faster of the two trains starts 3 hr after the slower train:

(a) When will they meet?
(b) How far will they have traveled before meeting?

Solution

We first note that

$$\text{distance} = (\text{average speed}) \times (\text{time})$$

Since we wish to find the time, we let t = the number of hours that the 50-mph train traveled before the meeting. Thus, $t - 3$ = the number of hours that the 70-mph train traveled before meeting. The essential point is that *the distance covered by each train must be equal.* Thus, we have the following equation:

$$(\text{distance of 50-mph train}) = (\text{distance of 70-mph train})$$
$$50t = 70(t - 3)$$
$$50t = 70t - 210$$
$$210 = 20t$$
$$\frac{21}{2} = t \quad \text{or} \quad t = 10.5 \text{ hr}$$

We conclude that the two trains meet after 10.5 hr. Since their distances are equal, we can use either $50t$ or $70(t - 3)$ to determine how far they have traveled. This distance is

$$50(10.5) = 525 \text{ miles} \quad \text{or} \quad 70(10.5 - 3) = 525 \text{ miles}$$

●

Examples 20.1 and 20.2 generated linear equations. Problems that generate quadratic equations will frequently generate a pair of solutions. In such cases we must examine the solutions and reject that solution which is inconsistent with physical conditions imposed by the problem.

Example 20.3

The height s, at any time t, of an object thrown up from a roof 176 ft high with an initial velocity of 56 ft/sec is given by the formula $s = 176 + 56t - 16t^2$. When will the object strike the ground?

Solution

We note that the object will strike the ground when height s is equal to zero. Thus, we substitute zero for s in the height equation to obtain

$$0 = 22 + 7t - 2t^2$$

$$0 = (11 - 2t)(2 + t)$$

Thus,

$$11 - 2t = 0 \qquad 2 + t = 0$$

$$t = \frac{11}{2} \qquad t = -2$$

We reject $t = -2$, since t represents time that must be positive in this problem. So we conclude that $t = \frac{11}{2}$, or 5.5 sec, is the time necessary for the object to strike the ground.

Example 20.4

Two input pipes, A and B, can fill a tank together in 4 hr. It takes pipe A 6 hr longer than pipe B to fill the tank alone. How long would it take each pipe alone to fill the tank?

Solution

We shall let h = the number of hours required by pipe A to fill the tank alone and $h - 6$ = the number of hours required by pipe B to fill the tank alone. We also know that pipe A alone can fill $1/h$ of the tank in 1 hr and that pipe B alone can fill $1/(h - 6)$ of the tank in 1 hr. Thus, together they can fill $\frac{1}{h} + \frac{1}{h - 6}$ of the tank in 1 hr. Since together they can completely fill the tank in 4 hr, we can express this as follows:

$$4\left(\frac{1}{h} + \frac{1}{h - 6}\right) = 1 \quad \text{(completely filled tank)}$$

We can solve this fractional equation by multiplying both sides of the equation by the (L.C.D.) $h(h - 6)$, to obtain

$$4h(h - 6)\left(\frac{1}{h} + \frac{1}{h - 6}\right) = h(h - 6)$$

$$4(h - 6) + 4h = h^2 - 6h$$

$$4h - 24 + 4h = h^2 - 6h$$

$$0 = h^2 - 14h + 24$$

We can now factor the resulting quadratic equation, to obtain

$$(h - 12)(h - 2) = 0$$

$$h - 12 = 0 \qquad h - 2 = 0$$

$$h = 12 \qquad h = 2$$

We reject $h = 2$, since $h - 6$ would be negative, and this is supposed to represent a positive time for pipe B. We then conclude that pipe A will take $h = 12$ hr alone and that pipe B will take $h - 6 = 12 - 6 = 6$ hr alone. ●

Examples 20.3 and 20.4 show us that it is *absolutely essential* that we check our results.

Exercises for Section 20

1. A piece of wire 42 cm long is bent into the shape of a rectangle whose width is twice its length. Find the dimensions of the rectangle.

2. A piece of wire 40 cm long is bent into the shape of an isosoceles triangle (a triangle with two equal sides). If the third side is 4 cm longer than one of the two equal sides, find the lengths of the side of the triangle.

3. Find the area of a square whose perimeter equals 20 in.

4. How many cubic centimeters of a 60% solution and a 20% solution must be mixed together to obtain 40 cm^3 of a 25% solution?

5. How many liters of a 70% alcohol solution must be added to 50 liters of a 40% alcohol solution to produce a 50% acid solution?

6. One input pipe can fill a tank in 8 hr alone, another input can fill a tank in 6 hr alone. If the tank is empty, how long will it take the two pipes together to fill the tank?

7. One machine can do a job alone in 8 minutes (min), a second machine can do the job alone in 10 min, and a third machine can do the job alone in 6 min. How long will it take the machines to do the job together?

8. One input pipe can fill a tank alone in 6 hr, and a second input pipe can fill the tank alone in 8 hr. A drain pipe can empty a full tank in 12 hr. If the tank is empty and all three pipes are open, how long will it take to fill the tank?

9. A jet faces a headwind of 50 mph for 3 hr and returns the same distance assisted by a tailwind of 50 mph in $2\frac{1}{3}$ hr. What is the speed of the jet?

10. A boat travels for 3 hr with a current of 3 mph and then returns the same distance against the current in 4 hr. What is the boat's speed in still water? How far did it travel one way?

11. A woman has $8000 invested, part in a savings account at 5.5% simple interest and the remainder in a long-term savings account at 8% simple interest. If her total income from these investments for 1 year was $565, how much did she invest at each rate?

12. A businessman invests a certain amount of money in stocks at 6% simple interest and twice that amount in long term bonds at 9% simple interest. If his total income per year from these investments was $1440, how much does he have invested at each rate?

13. A secretary can type a report in 6 hr less than it takes another secretary alone to type the same report. Both secretaries working together can type the report in 4 hr. How long would it take each to type the report alone?

14. What positive number when added to its reciprocal equals 5.2?

15. If the sum of two numbers is 38 and their product is 360, find the numbers.

16. If the product of two consecutive positive even integers is 224, find the numbers.

17. If the product of two consecutive positive odd integers is 195, find the numbers.

18. The sum of a number and its square is 132. Find the number. Is there more than one answer possible?

19. Two trains can start from a given point and travel on straight tracks at average speeds of 55 and 75 mph. If the faster of two trains starts 2 hr after the slower train, when will the faster train catch up to the slower train? How far must they travel before meeting?

20. A rectangular swimming pool 30 ft long and 20 ft wide has a deck of uniform width built around it. If the area of the deck is 400 ft^2, find its width.

21. A circular swimming pool with a diameter of 28 ft has a deck of uniform width built around it. If the area of the deck is 60π ft^2, find its width.

22. The height h at any time t, of an object thrown down from a roof 160 ft high with an initial velocity of 48 ft/sec is given by the formula $h = 160 - 48t - 16t^2$. When will the object strike the ground?

23. In Exercise 22, when will the object be 80 ft from the ground?

24. The height h, at any time t, of an object thrown upward from a rooftop 200 ft high with an initial velocity of 40 ft/sec is given by the formula

$h = 200 + 40t - 16t^2$. When will the object pass the rooftop on the way down?

25. In Exercise 24, when will the object pass a ledge 144 ft above the ground on its way down?

REVIEW QUESTIONS FOR CHAPTER 3

1. What is a formula?

2. What is an equation?

3. What are the roots of an equation?

4. Describe equivalent equations.

5. State four operations that are used to form equivalent equations.

6. Explain what is meant by the term "linear equations."

7. What is a literal equation?

8. What is a quadratic equation?

9. Describe the technique called completing the square.

10. What is an extraneous root?

11. What is a radical equation?

12. Describe the procedure used for solving radical equations.

13. Why must the roots of a radical equation always be checked in the original equation?

14. Describe an equation "quadratic in form" in one variable.

REVIEW EXERCISES FOR CHAPTER 3

In Exercises 1 through 10, solve each linear equation for the variable indicated.

1. $5x + 4 = 64$

2. $\frac{2x}{3} = \frac{4}{5}$

3. $2 - 3x = 8x + 20$

4. $5x - (2x - 3) = 6$

5. $5x - 3 = 2x + 9$

6. $8x + 3(5 - 2x) = 3$

7. $2(3x - 1) = 5(x + 2) - 7$

8. $\frac{1}{8}R + 7 = R - 3$

9. $ax + bx = 3a + 3b$ (solve for x)

10. $8y + 3(5 - 2y) = 3$

In Exercises 11 through 18, solve each quadratic equation by factoring.

11. $x^2 - 1 = 0$ **12.** $9x^2 - 64 = 0$ **13.** $m^2 - 11m + 24 = 0$

14. $t^2 + 2t = 8$ **15.** $x^2 + 4x + 3 = 0$ **16.** $x^2 = 5x + 6$

17. $x^2 - 4x + 4 = 0$ **18.** $x^2 + 4x + 4 = 0$

In Exercises 19 through 24, solve each quadratic equation by completing the square.

19. $x^2 + 4x + 3 = 0$ **20.** $x^2 - 2x - 24 = 0$

21. $x^2 + 2x - 11 = 0$ **22.** $2x^2 - 6x + 2 = 0$

23. $x^2 + 3x + 1 = 0$ **24.** $3x^2 + 7x - 6 = 0$

In Exercises 25 through 30, solve each quadratic equation by using the quadratic formula.

25. $x^2 - 5x + 4 = 0$ **26.** $8x^2 + 8x + 1 = 0$

27. $x^2 - 2x - 5 = 0$ **28.** $3x^2 - 2x - 2 = 0$

29. $2x^2 - 7x + 3 = 0$ **30.** $4x^2 + 8x - 5 = 0$

In Exercises 31 through 42, solve for x using the methods illustrated in this chapter. Check each result.

31. $\dfrac{3}{x - 7} = \dfrac{2}{x - 5}$ **32.** $\dfrac{4}{x + 3} = \dfrac{2}{x - 3}$

33. $\dfrac{3}{x - 1} - \dfrac{5}{x^2 + 3x - 4} = \dfrac{8}{x + 4}$ **34.** $\dfrac{5}{x + 4} + \dfrac{12}{(x - 2)(x + 4)} = \dfrac{2}{x - 2}$

35. $\sqrt{x - 3} = 2$ **36.** $\sqrt{x - 3} = \sqrt{1 - 3x}$ **37.** $\sqrt{2x - 3} = x - 1$

38. $\sqrt{2x + 4} = 1 - \sqrt{x + 3}$ **39.** $x^4 - 10x^2 + 9 = 0$

40. $x^{-4} - 4x^{-2} + 4 = 0$ **41.** $3y^6 + 7y^3 - 6 = 0$

42. $x^{-4/3} - 5x^{-2/3} + 4 = 0$

43. The total surface area of a right circular cylinder of radius r is given by

$$A = 2\pi r^2 + 2\pi rh$$

If the height of such a cylinder is 20 cm and the total surface area is 192π cm^2, determine the radius.

44. A ball rolls down an inclined plane of length 84 in. in such a manner that its distance, d, from the starting point at any time t is given by

$$d = 12t + \frac{t^2}{3}$$

How long does it take the ball to reach the bottom of the inclined plane? (Assume that it starts at the top.)

45. Solve the following equation for f:

$$2\pi fL = \frac{1}{2\pi fC}$$

46. The reactance of an inductor and capacitor in series is

$$X = wL - \frac{1}{wC}$$

If $L = 4$ and $C = 0.02$, what positive value of w will make $X = 3.5$?

47. When two dc currents flow in a common resistor in the same direction, the power dissipated in the resistor is given by

$$P = R(I_1 + I_2)^2$$

If $P = 300$, $R = 3$, and $I = 6$, find the value of I_2.

48. Separate 40 into two parts such that 3 times the smaller number is 15 less than twice the larger number.

49. The smaller of two numbers is 12 less than the larger. Four times the larger exceeds 3 times the smaller by 90. Find the numbers.

50. Two planes started at the same time from two airports which are 1700 miles apart and flew toward each other. One plane flew 360 mph and the other flew 390 mph. In how many hours were the planes still 200 miles apart?

51. How much pure acid must be added to 20 ounces of an acid solution that is 40% acid to produce a solution that is 50% acid?

52. A large pipe can fill a tank in 2 hr. A smaller pipe can fill the same tank in 3 hr. If both pipes are used together, how many hours are needed to fill the tank?

53. A rectangle is 8 ft long and 6 ft wide. If each dimension is increased by the same number of feet, the area of the new rectangle formed is 32 ft^2 more than the area of the original rectangle. By how many feet was each dimension increased?

CHAPTER 4

Functions and Graphs

SECTION 21
Functions

In science we often ask such questions as:

1. How does the area A of a circle relate to its radius r?
2. How does the current i flowing in a circuit depend upon the resistance R of the circuit?
3. How does the number of bacteria N in a culture depend upon the time t?
4. How does the distance s through which an object falls relate to time t?

The answer to such questions are usually given in terms of a rule (generally a formula) that establishes a pairing (or correspondence) between the two variables in question. These formulas will express some variable quantity as a function of another variable quantity.

If we wish to calculate the area A of a circle for a specific value of its radius r, the following equation applies:

$$A = \pi r^2 \tag{21.1}$$

When we substitute a specific value for r into (21.1) we can calculate the corresponding value for A. We call A the **dependent** variable and r the **independent** variable. In Table 21.1 we have substituted the successive values 1, 2, 3, and 4 for r into equation (21.1) to obtain the corresponding A values π, 4π, 9π, and 16π, respectively. The results shown in Table 21.1 can be described as the set of **ordered pairs** $(1, \pi)$, $(2, 4\pi)$, $(3, 9\pi)$, and $(4, 16\pi)$.

Table 21.1

r	1	2	3	4
A	π	4π	9π	16π

We use the term "ordered" to emphasize that the first element in the pair will always be the independent variable (in this case r) and the second element will always be the dependent variable (in this case A).

A set of ordered pairs of numbers, for example (x, y), will be called a *function* if for each first value x in the ordered pair there corresponds (by some rule) a *unique* second value y in the ordered pair. In such situations we refer to y as a function of x. Unless otherwise stipulated, we shall assume that largest possible set of values for both the independent and the dependent variables will be the set of real numbers.

Example 21.1

The circumference C of a circle can be described as a function of its diameter d by the equation $C = \pi d$. Here C is the dependent variable and d is the independent variable.

Example 21.2

The volume V of a sphere is a function of its radius r and can be described by the equation $V = \frac{4}{3}\pi r^3$. Here V is the dependent variable and r is the independent variable.

Example 21.3

For the equation $V = -32t + 80$, we say that V is a function of t. Here V is the dependent variable and t is the independent variable. In Table 21.2 we show some of the corresponding values for V when we substitute the values 0, 1, 2, and 2.5 for t into the given equation.

Table 21.2

t	0	1	2	2.5
V	80	48	16	0

We often find it convenient to use the notation $y = f(x)$ (read "y equals f of x") for the phrase "y is a function of x." Thus, $y = 2x^2 - x + 1$ may be written as $f(x) = 2x^2 - x + 1$. Then $f(1)$ means the value of $f(x)$ or y when $x = 1$, and $f(1) = 2(1)^2 - 1 + 1 = 2$. Similarly, $f(2) = 2(2)^2 - 2 + 1 = 7$.

Frequently, we will be dealing with more than one function. Then we shall use symbols such as $g(x)$ and $h(x)$ to denote the different functions of x.

Example 21.4

In Example 21.1 we can use the **function notation** to describe the circumference as $f(d) = \pi d$.

In Example 21.2 we can use the function notation to describe the volume as

$$f(r) = \frac{4}{3}\pi r^3$$

Example 21.5

If $f(x) = 2x^2 + 3x + 4$, then

$$f(1) = 2(1)^2 + 3(1) + 4 = 9$$

$$f(-1) = 2(-1)^2 + 3(-1) + 4 = 3$$

$$f(a) = 2(a)^2 + 3a + 4 = 2a^2 + 3a + 4$$

Example 21.6

If $f(x) = \sqrt{2x + 1}$, then

$$f(0) = \sqrt{2(0) + 1} = \sqrt{1} = 1$$

$$f(1) = \sqrt{2(1) + 1} = \sqrt{3}$$

Note that $f(-1) = \sqrt{2(-1) + 1} = \sqrt{-1}$ is not defined, since $\sqrt{-1}$ is not a *real number.*

Example 21.7

If $f(x) = x/(x - 2)$, then

$$f(1) = \frac{1}{1 - 2} = \frac{1}{-1} = -1$$

Note that $f(2) = 2/(2 - 2) = 2/0$ is undefined, since division by zero is not allowed.

Sometimes for a function of the form $y = f(x)$ it is necessary to restrict the choices of the values for the independent variable x to guarantee that the associated value for the dependent variable y will be a real number. We must *reject* any real values of x that would allow:

1. The extraction of the nth root of a negative number when n is even (see Example 21.6).
2. Division by zero (see Example 21.7).

Example 21.8

State any restrictions necessary on the independent variable when

$$f(s) = \frac{s}{(s + 1)(s + 2)}$$

Solution

We note that $f(-1) = \dfrac{-1}{(-1 + 1)(-1 + 2)} = \dfrac{-1}{0}$ is undefined and that

$f(-2) = \dfrac{-2}{(-2+1)(-2+2)} = \dfrac{-2}{0}$ is also undefined. Thus, $f(s)$ will be defined for all real values of s except $s = -1$ and $s = -2$.

Example 21.9

State any restrictions necessary on the independent variable when

$$g(w) = \sqrt{w - 2}$$

Solution

We note that $g(w)$ will only be a real number if $w - 2$ is greater than or equal to zero. Since $w - 2 = 0$ when $w = 2$, we see that $g(w)$ will be defined for all real values of w greater than or equal to 2.

Example 21.10

There are **no** restrictions necessary for the independent variable x when $y = f(x) = \sqrt{x^2 + 1}$, since the expression $x^2 + 1$ will be **positive** for all real value choices of x.

Exercises for Section 21

1. Express the area A of a circle as a function of its diameter d.
2. Express the area A of a circle as a function of its circumference C.
3. Express the radius r of a circle as a function of its area A.
4. Express the radius r of a circle as a function of its circumference C.
5. Express the area A of a square as a function of its side s.
6. Express the perimeter P of a square as a function of its side s.
7. Express the area A of a square as a function of its diagonal d.
8. Express the area A of a rectangle whose width is x units and whose length is $3x$ units as a function of x.
9. Express the area A of an equilateral triangle as a function of its side s.
10. A box is formed from a rectangular piece of tin that is 10 in. long and 20 in. wide by cutting a square out of each corner. Express the volume V of the box as a function of the side x of the cutout squares.
11. Express the area A of a triangle of base b and height 6 in. as a function of its base.
12. Express the perimeter P of an equilateral triangle as a function of its side s.
13. Express the volume V of a rectangular solid as a function of its length l if its width is twice its length and its height or depth is 3 less than its width.

14. A strip of tin 160 in. long and 14 in. wide is to be made into a rain gutter by turning up the edges to form a trough with a rectangular cross section. If the bent-up edge is s, express the volume V of the trough as a function of s.

15. A wire 80 ft long is cut into two pieces; one is bent into a square and the other, into a circle. If p is the perimeter of the square, express the sum of the areas of both figures as a function of p.

16. A window that has the shape of a semicircle sitting atop a rectangle has a perimeter of 80 in. If the side of the rectangle that is not the diameter of the semicircle is denoted as w, express the area of the window as a function of w.

17. A 13-ft ladder leans against a wall. If the base of the ladder is x ft from the base of the wall, express the height h that the top of the ladder is from the ground as a function of x.

18. If $f(x) = 3x - 2$, find $f(1)$, $f(0)$, and $f(\frac{2}{3})$.

19. If $f(t) = 144 + 72t - 16t^2$, find $f(1)$, $f(2)$, and $f(6)$.

20. If $g(w) = 3\sqrt{w + 1} + 2$, find $g(0)$, $g(-1)$, and $g(8)$.

21. If $h(r) = r + 1/r$, find $h(1)$, $h(2)$, and $h(10)$.

22. If $f(x) = x^2 + x$, find $f(h)$ and $f(3h)$.

23. If $f(x) = 2x^2 + x + 1$, find $f(x + h)$ and $f(h)$.

24. If $f(x) = 2 - x - x^2$, find $f(x + h)$, $f(x)$, and $f(x + h) - f(x)$.

25. In Exercise 24, find $\dfrac{f(x + h) - f(x)}{h}$.

In Exercises 26 through 35, state any restrictions necessary on the independent variable.

26. $f(x) = \dfrac{x}{1 - x}$

27. $f(w) = \dfrac{1}{w^2 - 1}$

28. $g(s) = \sqrt{s - 3}$

29. $g(y) = \dfrac{y}{\sqrt{2 - y}}$

30. $h(u) = \dfrac{u}{(u + 1)(u - 2)}$

31. $g(x) = \dfrac{1}{x^2 + 1}$

32. $h(x) = \sqrt{4 + x^2}$

33. $g(x) = x^2 + \dfrac{1}{x}$

34. $h(x) = \dfrac{1}{\sqrt{x^2}} + x$

35. $h(s) = \sqrt{-s^2}$

SECTION 22

Rectangular Coordinate System; Graphs

When we encounter problems dealing with a functional relationship between two variables such as $y = f(x)$, we are often concerned with such questions as:

1. How does y vary with respect to x?
2. What is the largest possible value for y?
3. What is the smallest possible value for y?
4. For what values of x will y equal zero?

Frequently, we can find the answers to such questions if we can obtain a "picture" or visual representation of the functional relationship. We can obtain such a representation by introducing the **rectangular coordinate system**.

Consider two mutually perpendicular lines $X'X$ and $Y'Y$ intersecting each other at the point O (see Figure 22.1). The line $X'X$ is called the *x axis* and the line $Y'Y$ is called the *y axis*. The point O of their intersection is called the *origin*. We now use a convenient unit of length and mark off points on the x axis at successive units to the right and left of the origin. We agree to label the points to the right as 1, 2, 3, ... and the points to the left as $-1, -2, -3, \ldots$. We then do the same thing on the y axis, choosing OY as the positive direction. This is called the **rectangular coordinate system**. We normally associate the values of the independent variable with the horizontal axis and the values of the dependent variable with the vertical axis.

The horizontal and vertical axes divide the plane into four parts, called **quadrants**. We label them I, II, III, and IV, as shown in Figure 22.1.

To each point in this plane we assign a pair of real numbers (x, y). The value of the independent variable x is called the ***x* coordinate** (or abscissa), and the value of the dependent variable y is called the ***y* coordinate** (or ordinate). The x coordinate tells us how to proceed from the origin O along the x axis, and the y coordinate tells us how to proceed from O along the y axis. To locate a point P with coordinates (x, y) we will measure $|x|$ units from the origin along the x axis to the right if x is positive and to the left if x is negative. Then from this point we will measure $|y|$ units upward if y is positive and downward if y is negative. The point P in Figure 22.1 with coordinates $(-2, 3)$ can be located by first measuring 2 units from the origin to the left along the x axis and then from there measuring 3 units upward. The point Q also shown in Figure 22.1 can be located by first measuring 3 units from the origin to the right along the x axis and then from there measuring 1 unit upward. Locating points such as P and Q in Figure 22.1 is called **plotting** the points.

Figure 22.1

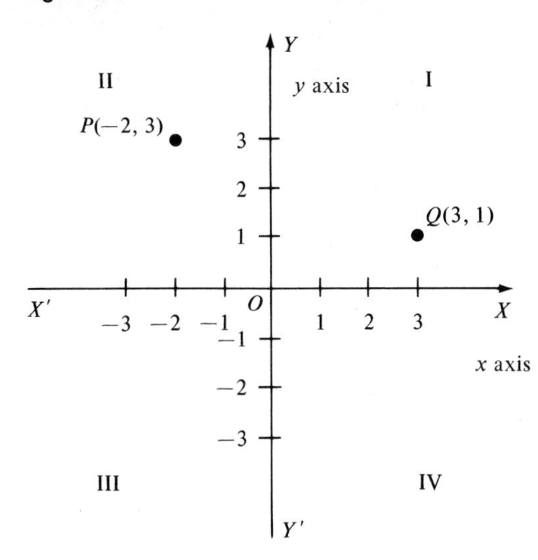

The graph of an equation of the form $y = f(x)$ is the set of all points whose coordinates (x, y) satisfy the equation. Generally, this set of points is uncountable, and thus it would be impossible to plot all the points. In such cases the strategy is to plot a "representative" sample of these points and then sketch in the remaining points by connecting the plotted points by means of a "smooth" curve. We refer to this procedure as **graphing the function**.

Example 22.1

Graph the function $y = 2x + 1$.

Solution

We arbitrarily choose some convenient values for x and we substitute these into the given equation to obtain the values given in Table 22.1. The computations for the entries in the table are obtained as follows:

When $x = -1$, then $y = 2(-1) + 1 = -2 + 1 = -1$.

When $x = -\frac{1}{2}$, then $y = 2(-\frac{1}{2}) + 1 = -1 + 1 = 0$.

When $x = 0$, then $y = 2(0) + 1 = 0 + 1 = 1$.

When $x = 1$, then $y = 2(1) + 1 = 2 + 1 = 3$.

When $x = 2$, then $y = 2(2) + 1 = 4 + 1 = 5$.

Table 22.1

x	-1	$-\frac{1}{2}$	0	1	2
y	-1	0	1	3	5

We then plot the points whose coordinates are shown in Table 22.1 and connect them to form the straight line shown in Figure 22.2.

Figure 22.2

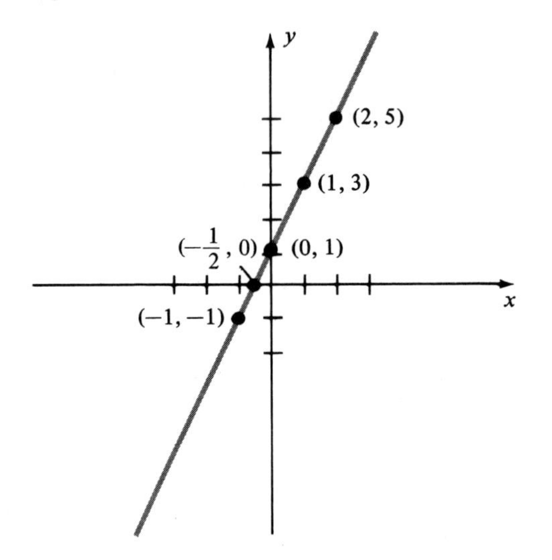

Example 22.2

Graph the function $y = 2 + x - x^2$.

Solution

We again use the given equation and some arbitrary values of x to calculate the values of y shown in Table 22.2.

Table 22.2

x	−2	−1	0	1	2	3
y	−4	0	2	2	0	−4

The computations for the entries shown in the table are obtained as follows:

When $x = -2$, then $y = 2 + (-2) - (-2)^2 = 2 - 2 - 4 = -4$.

When $x = -1$, then $y = 2 + (-1) - (-1)^2 = 2 - 1 - 1 = 0$.

When $x = 0$, then $y = 2 + 0 + 0^2 = 2$.

When $x = 1$, then $y = 2 + 1 - 1^2 = 2 + 1 - 1 = 2$.

When $x = 2$, then $y = 2 + 2 - 2^2 = 2 + 2 - 4 = 0$.

When $x = 3$, then $y = 2 + 3 - 3^2 = 2 + 3 - 9 = -4$.

We then plot these points and connect them to form the curve shown in Figure 22.3. We could use this graph to estimate that the greatest (or maximum) value for y is something a little greater than 2. (Later in this chapter we shall develop a technique for determining the exact maximum value for such functions.

Figure 22.3

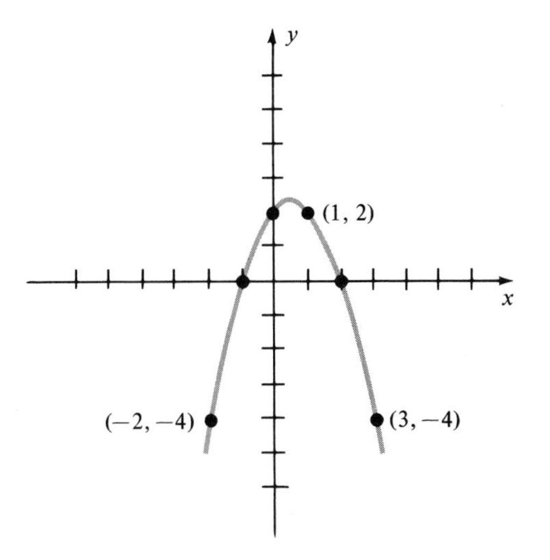

When we consider the rectangular coordinate system, the following observations should be made.

1. Any point that lies on the x axis has a y coordinate equal to zero.
2. Any point that lies on the y axis has a x coordinate equal to zero.
3. Table 22.3 indicates the signs of the coordinates of points in each quadrant.

Table 22.3

Quadrant	*x Coordinate*	*y Coordinate*
I	+	+
II	−	+
III	−	−
IV	+	−

Exercises for Section 22

In Exercises 1 through 4, plot the points.

1. $P(3, 2)$, $Q(-1, 2)$, $R(0, 1)$.

2. $P(-1, -1)$, $Q(2, 2)$, $R(-2, -4)$.

3. $P(\frac{1}{2}, -3)$, $Q(-1, 3)$, $R(3, 0)$.

4. $P(-3, -2)$, $Q(0, -2)$, $R(4, -2)$.

In Exercises 5 through 8, calculate the corresponding y values for the indicated x values.

5. $y = 3x - 2$; $x = -2, -1, 0, \frac{2}{3}, 1$

6. $y = 4 - 2x$; $x = -1, 0, 1, 2, 3$

7. $y = x^2 + 2x - 1$; $x = -3, 2, 1, 4$

8. $y = x^2 - 6x + 5$; $x = -2, -1, 0, 1, 2, 3$

9. Where are all points for which $x > 0$?

10. Where are all points for which $y > 0$?

11. Where are all points for which $x < 0$?

12. Where are all points for which $y < 0$?

13. Where are all points for which $x = -2$?

14. Where are all points for which $y = 3$?

15. If the points A, B, and C form a right triangle, where angle B is the right angle, find the coordinate of C when $A(2, 2)$ and $B(4, 2)$.

16. Three vertices of a rectangle are $(3, 2)$, $(7, 2)$, and $(7, 5)$. What are the coordinates of the fourth vertex?

SECTION 23
Linear Functions

A *linear function* has the form

$$y = ax + b \qquad a \neq 0 \tag{23.1}$$

where x is the independent variable, y is the dependent variable, and a and b are constants. This function is called "linear" because its graph is always a straight line.

We know that two distinct points will determine a straight line. Since there are no restrictions on the independent variable x in (23.1), we can arbitrarily choose any two values for x and then substitute them into the equation to obtain the coordinates of the two needed points.

One of the easiest ways to find two distinct points is to set $x = 0$ and solve for y in the equation and then set $y = 0$ and solve for x in the equation to obtain the respective points $(0, y)$ and $(x, 0)$. These points are known, respectively, as the **y intercept** and the **x intercept**.

Example 23.1

Graph the linear function $y = 4 - 2x$.

Solution

We set $x = 0$ in the equation to obtain $y = 4 - 2(0) = 4$. Thus, the y intercept is located at $(0, 4)$. Now we set $y = 0$ to obtain $0 = 4 - 2x$ or $2x = 4$ or $x = 2$. Thus, the x intercept is located at $(2, 0)$. We show the graph of this function in Figure 23.1 by drawing a line through the two points and extending indefinitely in both directions.

Figure 23.1

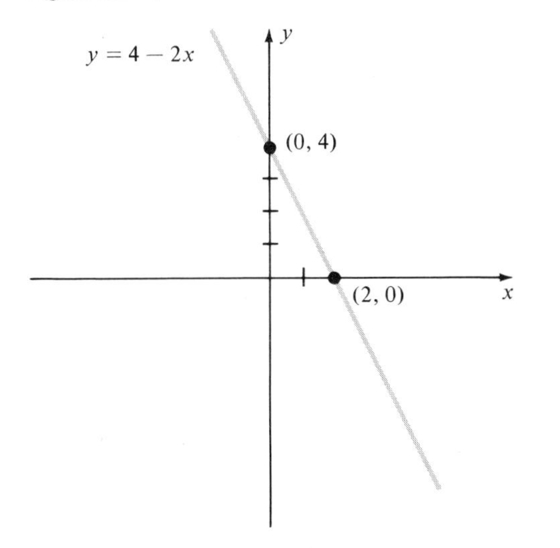

Example 22.1 is another graph of a linear function. Note from Table 22.1 that the respective x and y intercepts are $(-\frac{1}{2}, 0)$ and $(0, 1)$. These two points are sufficient to sketch the graph. However, if we wish to use the resulting graph to obtain additional information about either variable, it is wise to calculate a minimum of three points.

Example 23.2

Graph the linear function $y = 3x$.

Solution

We set $x = 0$ to obtain $y = 3(0) = 0$. Thus, both the x and y intercepts are located at $(0, 0)$. We can easily obtain the additional points $(-1, -3)$ and $(1, 3)$ by setting $x = -1$ and $x = 1$ and solving for y. We show the graph of this function in Figure 23.2.

Figure 23.2

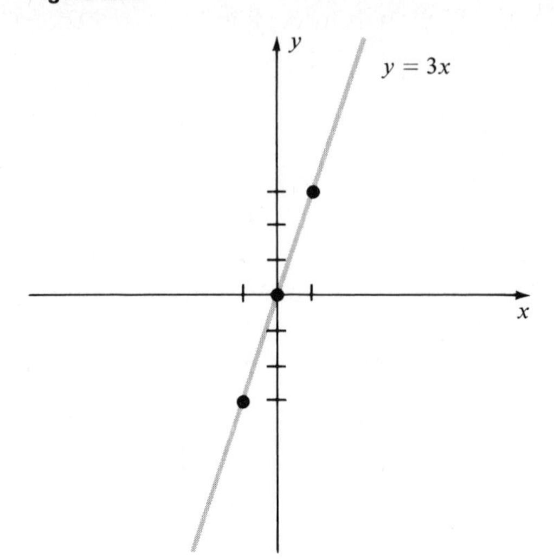

Example 23.3

Graph the linear function $y = 2$ or $f(x) = 2$.

Solution

This particular function tells us that for any choice of the independent variable, we must associate $y = 2$. Thus, when $x = 0$, then $y = 2$; and when $x = 1$, then $y = 2$. We use the points $(0, 2)$ and $(1, 2)$ to determine the graph shown in Figure 23.3. It should be noted that this line is parallel to the x axis and 2 units above it. This type of function is called a **constant function**.

Figure 23.3

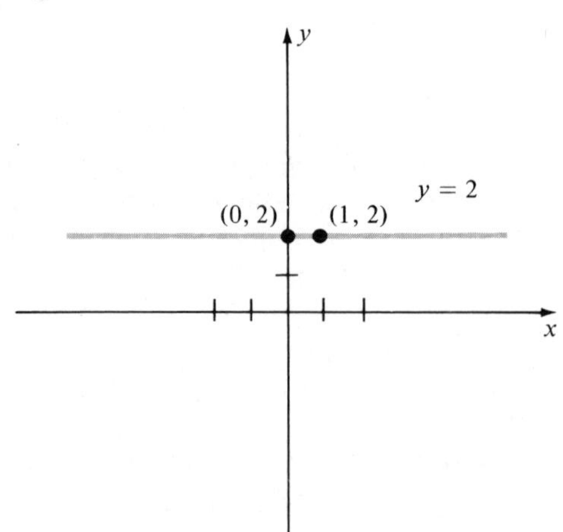

In general, the graph of the constant function $y = k$ or $f(x) = k$, $k \neq 0$, is always a line parallel to the x axis. If $k > 0$, the line is k units above the x axis (as in Example 23.3). If $k < 0$, the line is $|k|$ units below the x axis. If $k = 0$, the line is the x axis. Thus, we say that **the equation of the x axis is $y = 0$.**

The equation $x + 2 = 0$ **is not a linear function**, since it does not fit the form of (23.1). But the graph of this relation can be determined. We interpret $x + 2 = 0$ or $x = -2$ to mean the set of all points whereby the variable x must always equal -2 and the value of y can be any real number. Thus, points such as $(-2 - 1)$, $(-2, 0)$ and $(-2, 1)$ are part of this graph. We plot these points and then connect them to obtain the vertical line shown in Figure 23.4. Notice that this line is parallel to the y axis.

Figure 23.4

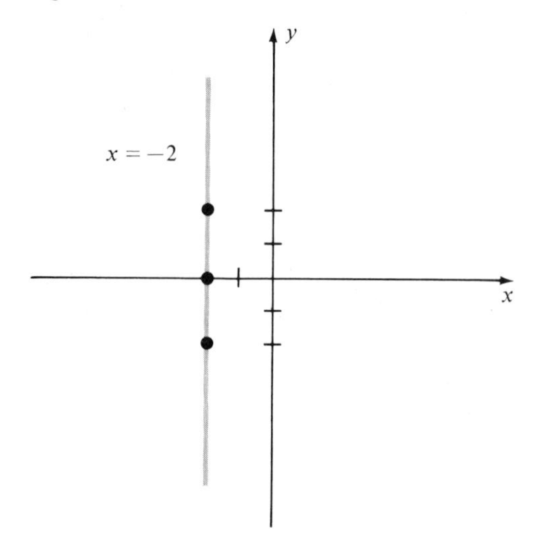

In general, the graph of $x = a$, $a \neq 0$, will always be a line parallel to the y axis. If $a > 0$, the line will be a units to the right of the y axis. If $a < 0$, the line will be $|a|$ units to the left of the y axis. If $a = 0$, the line will be the y axis. Thus, we say that the **equation of the y axis is $x = 0$.**

The x intercept of a function is found by setting $y = 0$ or $f(x) = 0$. The x intercepts are also frequently referred to as the *zeros of the function.* Thus, the instruction "find the x intercepts of a function" is equivalent to "find the zeros of the function." This should not be confused with $f(0)$, which means "evaluate the function when $x = 0$."

We can use this idea to solve linear equations involving one variable graphically.

Example 23.4

Solve the equation $4 - 2x = 0$ graphically.

Solution

This question is equivalent to being asked to find the x intercepts of the associated linear function $y = 4 - 2x$. In Example 23.1 we graphed the function and determined from Figure 23.1 that its x intercept was $x = 2$. Thus, the solution or root of the equation $4 - 2x = 0$ is $x = 2$.

Example 23.5

Under certain conditions, the relationship between the resistance R (in ohms) in a certain wire and the temperature T (in degrees Celsius) is given by the equation $T = 400R - 120$. If $R = 0.3\ \Omega$, then $T = (400°)(0.3) - 120° = 120° - 120° = 0°\text{C}$. If $R = 0.5\ \Omega$, then $T = (400°)(0.5) - 120° = 200° - 120° = 80°\text{C}$. The graph of this function is shown in Figure 23.5. We note that the scale chosen for the units of measure on the horizontal and vertical axes enables us to accommodate the relative differences in the size of the values for R and T.

Figure 23.5

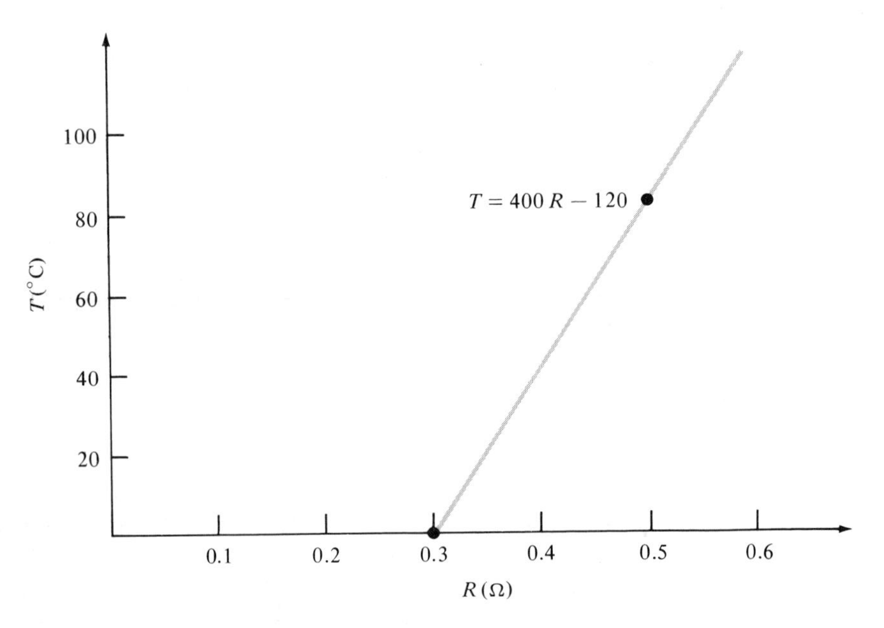

Exercises for Section 23

In Exercises 1 through 6, find algebraically the x and y intercepts for each.

1. $y = 2x + 1$

2. $y = \frac{1}{2}x - 2$

3. $y = 6 - 3x$

4. $y = \frac{3}{2}x + 1$

5. $2x + 3y = 6$

6. $x - 3y = 4$

In Exercises 7 through 16, find the x and y intercepts and then graph each linear function.

7. $y = x + 1$ **8.** $y = 1 - x$

9. $y = 4 - 3x$ **10.** $y = 3x - 4$

11. $y = 4x + 2$ **12.** $y = 2 - (x + 3)$

13. $y = 2x - 3$ **14.** $y = 2(3 - x)$

15. $y = -3$ **16.** $y = 4$

In Exercises 17 through 20, find the zeros of each function algebraically.

17. $y = 4x + 7$ **18.** $y = \frac{1}{2}x + \frac{1}{2}$

19. $y = 3x - 6$ **20.** $y = 10 - 5x$

In Exercises 21 through 26, solve each equation graphically.

21. $3x - 2 = 0$ **22.** $4x + 1 = 0$

23. $5x + 3 = 0$ **24.** $4x = 7$

25. $2x = 4x + 1$ **26.** $5x - 10 = 0$

27. The velocity V (in ft/sec) of an object thrown upward is a linear function of time t (in seconds). The equation is $V = 96 - 32t$.
(a) Find (algebraically) when the velocity will be zero.
(b) Find (graphically) when the velocity will be zero.
(c) Find when the velocity will be 48 ft/sec.

28. The relationship between Fahrenheit temperature F and Celsius temperature C is linear. We can express this relationship as $F = 1.8C + 32$, where C is the independent variable and F is the dependent variable.
(a) Find the value of F when $C = 0$, $C = 100$.
(b) Graph this linear function on a CF axis.
(c) Express the linear relationship as a function of F. That is, let F be the independent variable.
(d) Find the values of C when $F = 32$, $F = 212$.
(e) Use the result of part (c) to graph the linear relationship on an FC axis.

29. A ball is thrown straight up in the air with an initial velocity of 160 ft/sec. The velocity at the end of t seconds is given by $v = 160 - 32t$.
(a) Find v when $t = 1$ sec.
(b) Find v when $t = 3$ sec.
(c) Find v when $t = 5$ sec.
(d) Sketch the function for $0 \leq t \leq 5$.

30. The pressure P of a fixed volume of gas (in centimeters of mercury) and the temperature T (in degrees Celsius) are related by the equation $P = (T/4) + 80$.

(a) Find P when $T = 40$.
(b) Find T when $P = 100$.
(c) Graph the function for $40 \le T \le 80$.

31. (a) In Example 23.5, solve the given equation for R.
(b) Graph the function on a TR axis for $0 \le T \le 120$.

32. As the temperature of a certain gas in a closed container is varied, the pressure is read on a gauge. The relationship is given by the equation $P = 0.08t + 19.2$; P is lb/in.^{2} and t is °C.
(a) Find P when $t = 0$.
(b) Find t when $P = 0$.
(c) Sketch the graph for $-100 \le t \le 100$. Plot t horizontally.

33. The resistance R (in ohms) is defined as a function of temperature t (in °C) by the equation $R = 400(1 - 0.005t)$.
(a) Express t as a function of R.
(b) Graph the function for $-40 \le t \le 400$.
(c) What is the R intercept of the graph?

SECTION 24

Quadratic Functions

The function

$$y = f(x) = ax^2 + bx + c \qquad a \neq 0 \tag{24.1}$$

is called a **quadratic function**, where x is the independent variable; y is the dependent variable; and a, b, and c are constants. The properties of this function can be effectively analyzed by examining its graph.

In Example 22.2 we examined the graph of the quadratic function $y = 2 + x - x^2$ by constructing a table of values (see Table 22.1) and then plotting points to obtain the curve shown in Figure 22.3. We noted from this graph that we could guess at the greatest or maximum value for y.

In this section we shall establish procedures that enable us to sketch the graph of a quadratic function using a limited number of points and then answer the questions concerning the maximum or minimum values of the function. The graph of a quadratic function is called a **parabola**. The lowest (or highest) point on this graph is called the **vertex** of the parabola.

We use the quadratic function $y = 2 + x - x^2$ to develop a method for finding the coordinates of the vertex of a parabola as follows:

$$y = 2 + x - x^2$$

$$y - 2 = -x^2 + x$$

$$y - 2 = -(x^2 - x)$$

Now we use the technique of completing the square on the expression $x^2 - x$ inside the parentheses and then we add $-\frac{1}{4}$ to both sides of the equation, to obtain

$$y - 2 + \left(-\frac{1}{4}\right) = -\left(x^2 - x + \frac{1}{4}\right)$$

$$y - \frac{9}{4} = -\left(x - \frac{1}{2}\right)^2$$

This equation tells us that $y - \frac{9}{4} = 0$ when $x = \frac{1}{2}$ and that $y - \frac{9}{4} < 0$ for all other values of x. So $y - \frac{9}{4} \leq 0$ or $y \leq \frac{9}{4}$. Thus, $(\frac{1}{2}, \frac{9}{4})$ is the highest point and is the vertex of this parabola. The greatest value for y is exactly $\frac{9}{4}$ (the y coordinate of the vertex).

To show that there is a highest (or else a lowest) point on the graph of all quadratic functions of the form (24.1), we complete the square and proceed as follows:

$$y = ax^2 + bx + c \qquad a \neq 0$$

$$y - c = a\left(x^2 + \frac{b}{a}x\right)$$

$$y - c + a\frac{b^2}{4a^2} = a\left(x^2 + \frac{b}{a}x + \frac{b^2}{4a^2}\right)$$

$$y - c + \frac{b^2}{4a} = a\left(x + \frac{b}{2a}\right)^2$$

$$y - \left(c - \frac{b^2}{4a}\right) = a\left(x + \frac{b}{2a}\right)^2$$

$$y - \frac{4ac - b^2}{4a} = a\left(x + \frac{b}{2a}\right)^2 \qquad \text{(A)}$$

In (A), if $a > 0$, the term on the right side is always positive or else equal to zero when $x = -b/2a$. Thus,

$$y - \frac{4ac - b^2}{4a} \geq 0 \quad \text{or} \quad y \geq \frac{4ac - b^2}{4a}$$

This implies that the **lowest** point on this curve will be located at $\left(-\frac{b}{2a}, \frac{4ac - b^2}{4a}\right)$.

In (A), if $a < 0$, the term on the right side is always negative or else equal to zero when $x = -b/2a$. Thus,

$$y - \frac{4ac - b^2}{4a} \leq 0 \quad \text{or} \quad y \leq \frac{4ac - b^2}{4a}$$

This implies that the **highest** point on this curve will be located at $\left(-\frac{b}{2a}, \frac{4ac - b^2}{4a}\right)$.

Both cases yield the same set of coordinates. Thus, $\left(-\frac{b}{2a}, \frac{4ac - b^2}{4a}\right)$ are the coordinates of the vertex of the parabola.

If $a > 0$, the graph of the parabola must open upward.

If $a < 0$, the graph of the parabola must open downward.

The calculation of x and y intercepts (if they exist) for the quadratic function proceeds as for the linear functions.

Example 24.1

Find the x and y intercepts and the coordinates of the vertex, and sketch the graph of the quadratic function $y = x^2 - 2x - 8$.

Solution

We set $x = 0$ in the equation to obtain a y intercept of 8. We set $y = 0$ in the equation, to obtain

$$0 = x^2 - 2x - 8$$

$$0 = (x - 4)(x + 2)$$

$$x - 4 = 0 \qquad x + 2 = 0$$

$$x = 4 \qquad x = -2$$

Thus, 4 and -2 are x intercepts. We read $a = 1$, $b = -2$, and $c = -8$ in the original equation and use the formulas for the coordinates of the vertex as follows:

$$x = -\frac{b}{2a} = -\frac{-2}{2(1)} = 1$$

$$y = \frac{4ac - b^2}{4a} = \frac{4(1)(-8) - (-2)^2}{4(1)} = \frac{-32 - 4}{4} = -9$$

Thus, the coordinates of the vertex are $(1, -9)$. Finally, we note that since $a = 1$ is positive, the parabola will open upward. We show this graph in Figure 24.1.

Remark 24.1. *It is much easier to calculate the y coordinate of the vertex by simply substituting the value for the x coordinate into the original equation. This eliminates memorizing the formula for the y coordinate.*

Example 24.2

Find the x and y intercepts and the coordinates of the vertex, and sketch the graph of $x^2 = 4 - y$.

Solution

First, we must express the equation in its quadratic form, $y = 4 - x^2$. We set $x = 0$ to obtain the y intercept 4. We set $y = 0$ to obtain $0 = 4 - x^2$,

Figure 24.1

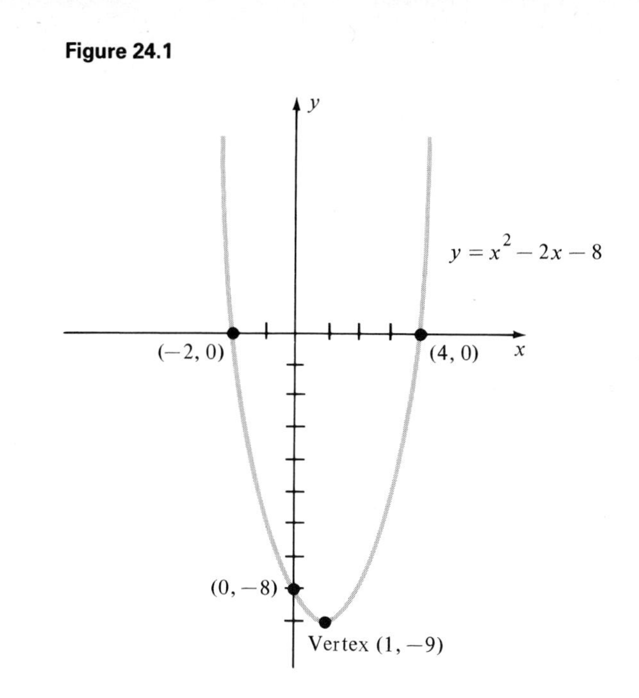

$x^2 = 4$, or $x = \pm 2$. Thus, we have two x intercepts. Next we note that $a = -1$, $b = 0$, and $c = 4$. Now we calculate the coordinates of the vertex as follows:

$$x = -\frac{b}{2a} = -\frac{0}{2} = 0$$

Figure 24.2

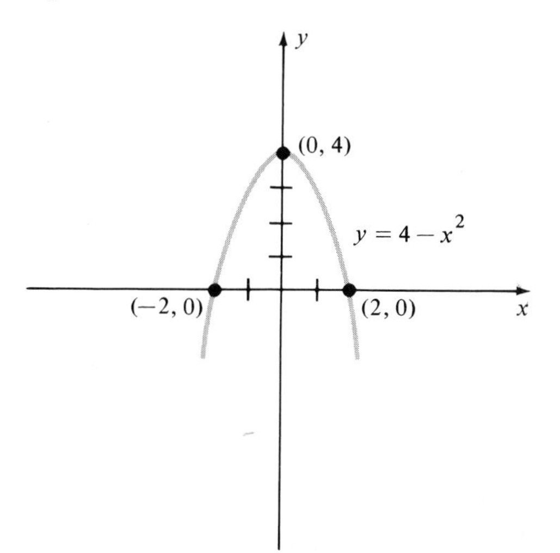

From the equation we have that $y = 4$ when $x = 0$. (It is also the y intercept.) Thus, the coordinates of the vertex are $(0, 4)$. Since a is negative, the parabola breaks downward. The graph is shown in Figure 24.2.

It should be noted that the problem of finding the x intercepts for $y = ax^2 + bx + c$ is identical to the problem of finding the solutions or roots of the quadratic equation $ax^2 + bx + c = 0$ (see Sections 15 and 16). Thus, we could rely on the quadratic formula to solve for the possible x intercepts.

Example 24.3

Find the x and y intercepts for the quadratic function $y = x^2 - 4x + 1$.

Solution

We set $x = 0$ in the equation to obtain the y intercept 1. We set $y = 0$ in the equation and solve for the x intercepts by using the quadratic formula as follows:

$$0 = x^2 - 4x + 1$$

$$x = \frac{-(-4) \pm \sqrt{(-4)^2 - 4(1)(1)}}{2(1)}$$

$$x = \frac{4 \pm \sqrt{16 - 4}}{2} = \frac{4 \pm \sqrt{12}}{2}$$

$$x = \frac{4 + 2\sqrt{3}}{2} \quad \text{or} \quad 2 \pm \sqrt{3}$$

Thus, the x intercepts are $2 + \sqrt{3}$ and $2 - \sqrt{3}$.

Example 24.4

Find the x and y intercepts and the coordinates of the vertex, and sketch the graph of the quadratic function $y = x^2 + x + 1$.

Solution

We set $x = 0$, the equation to find the y intercept 1. We set $y = 0$ in the equation and use the quadratic equation to solve for x as follows:

$$0 = x^2 + x + 1$$

$$x = \frac{-1 \pm \sqrt{1^2 - 4(1)(1)}}{2} = \frac{-1 \pm \sqrt{-3}}{2}$$

Since $\sqrt{-3}$ is nonreal, we have *no x intercepts*. Now we calculate the coordinates of the vertex as follows:

$$x = -\frac{b}{2a} = -\frac{1}{2}$$

and

$$y = \left(-\frac{1}{2}\right)^2 + \left(-\frac{1}{2}\right) + 1 = \frac{1}{4} - \frac{1}{2} + 1 \quad \text{or} \quad y = \frac{3}{4}$$

Thus, the coordinates of the vertex are $(-\frac{1}{2}, \frac{3}{4})$. We note that since $a = 1$ is positive, the parabola opens upward. This is enough information to allow us to sketch the graph. However, if we wish to obtain a more accurate sketch, it is wise to calculate a few more points. We do this in Table 24.1 and then sketch the graph shown in Figure 24.3.

Table 24.1

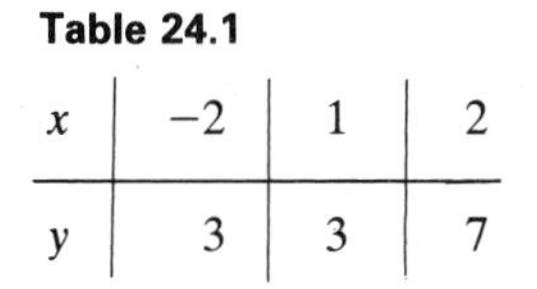

x	-2	1	2
y	3	3	7

Figure 24.3

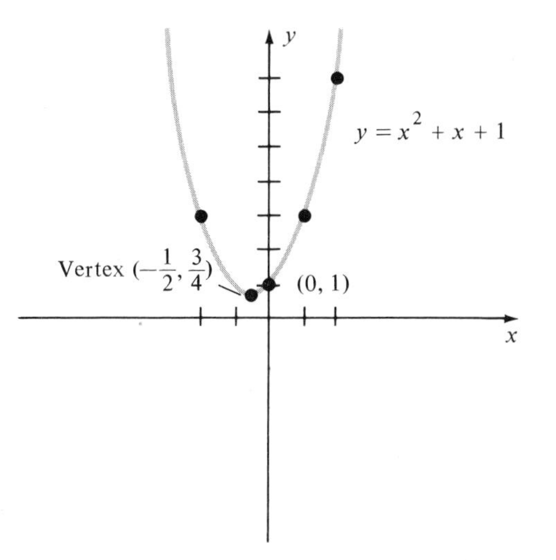

Summary of the Properties of the Graph of the Quadratic Function

1. The y intercept is always $y = c$.
2. The x intercept(s) are the roots of the quadratic equation $ax^2 + bx + c = 0$. We can always use the quadratic formula to find them.
3. The coordinates of the vertex are located at $(-b/2a, (4ac - b^2)/4a)$.
4. If $a > 0$, the parabola breaks upward and the vertex is a minimum point.
5. If $a < 0$, the parabola breaks downward and the vertex is a maximum point.

The x intercepts of the quadratic function are identical to the zeros of the quadratic function or the roots of the associated quadratic equation $ax^2 + bx + c = 0$. We can use this idea to solve quadratic equations graphically.

Example 24.5
Solve $x^2 - 4x + 2 = 0$ graphically.

Solution
We shall graph the associated quadratic function $y = x^2 - 4x + 2$ by using the coordinates of the vertex and some calculated coordinates shown in Table 24.2 to plot the curve. Then we shall approximate the roots of the equation by approximating the x intercepts from the graph of the function. The coordinates of the vertex are

$$x = -\frac{b}{2a} = \frac{-(-4)}{2(1)} = 2$$

$$y = 2^2 - 4(2) + 2 = -2$$

We use this point, the points in Table 24.2, and the fact that $a = 1$ is positive to sketch the graph shown in Figure 24.4. From this figure we see that the x intercepts of the parabola are approximately 0.6 and 3.4. Thus,

Table 24.2

x	−1	0	1	2	3	4
y	7	2	−1	−2	−1	2

Figure 24.4

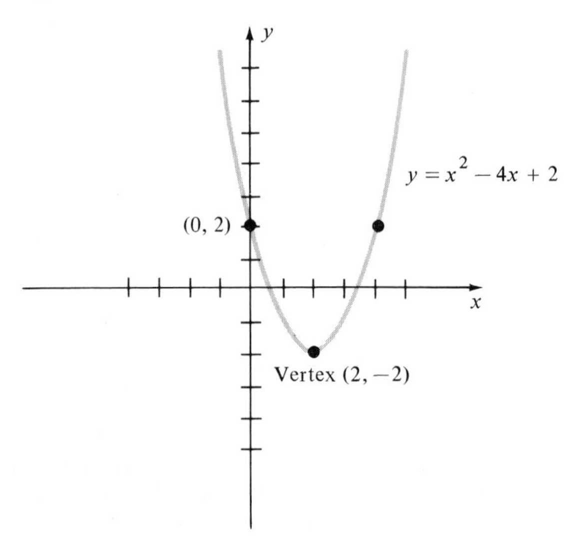

the approximate solutions to the given equation $x^2 - 4x + 2 = 0$ are 0.6 and 3.4.

Example 24.6

An object is thrown vertically upward from the ground with an initial velocity of 80 ft/sec. Its displacement s from its starting point after t seconds is given by $s = f(t) = 80t - 16t^2$. We note that the given function is quadratic. When $s = 0$,

$$80t - 16t^2 = 0$$

$$16t(5 - t) = 0$$

$$t = 0 \quad \text{or} \quad t = 5$$

Thus, we have intercepts at (0, 0) and (5, 0). The coordinates of the vertex are found by noting that

$$t = -\frac{b}{2a} = \frac{-80}{2(-16)} = \frac{5}{2} \quad \text{and}$$

$$s = 80\left(\frac{5}{2}\right) - 16\left(\frac{5}{2}\right)^2 = 200 - 100 = 100$$

The vertex is at $(\frac{5}{2}, 100)$. The graph of the function is shown in Figure 24.5 for $0 \leq t \leq 5$, since the motion takes place only in this time interval. We observe that the maximum value for s is 100 ft and occurs when $t =$ 2.5 sec.

Figure 24.5

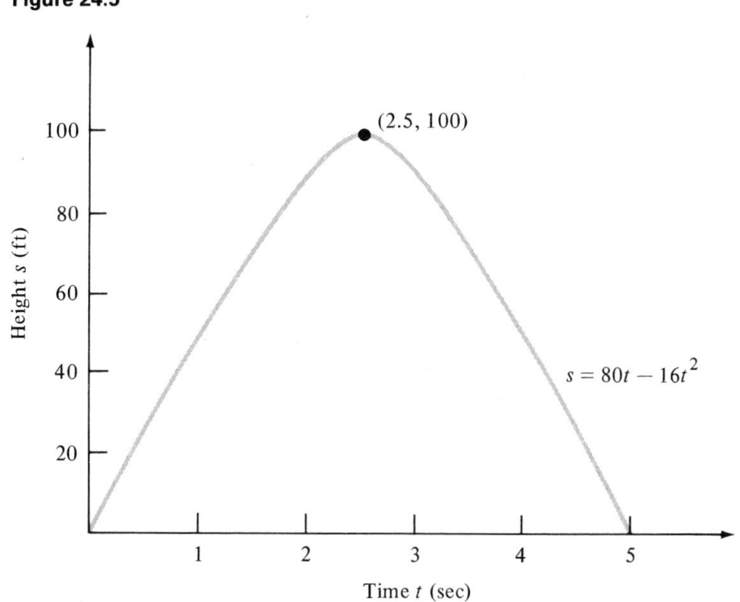

Exercises for Section 24

In Exercises 1 through 14, find the x intercepts (if any), the y intercepts, and the coordinates of the vertex, and graph the quadratic function.

1. $y = 2 - 3x - x^2$

2. $y = x^2 - 2x - 3$

3. $y = x^2 + x - 4$

4. $y = 2x^2 - 3x - 2$

5. $y = 2 + 3x - 2x^2$

6. $y = 2x^2 - 6x + 7$

7. $y = 6x^2 + 13x - 5$

8. $y = x^2 + 6x + 4$

9. $y = 4x^2 - 4x - 2$

10. $y = 2 - 6x + x^2$

11. $y = x^2 - 2x + 4$

12. $y = (3 - x)(x + 4)$

13. $y - 13 = 4x^2 + 12x$

14. $y = -(3 - x)(x + 4)$

In Exercises 15 through 22, solve each quadratic equation graphically.

15. $4x^2 - 4x - 2 = 0$

16. $4x^2 + 5 = 12x$

17. $x^2 + 6x = -4$

18. $0 = 2 - 6x + x^2$

19. $13x = 6x^2 + 6$

20. $4x^2 + 12x - 13 = 0$

21. $(x - 3)^2 = 8$

22. $(3 - 2x)^2 = 6$

In Exercises 23 through 26, graph the physical formula for the interval stated. In each instance use the horizontal axis for the variable that is square.

23. $s = \frac{1}{2}gt^2$ is the formula for the distance s in feet that a body falls in t sec. at a constant acceleration due to gravity ($g = 32$ ft/sec^2). Use $0 \leq t \leq 6$.

24. $s = f(t) = 96t - 16t^2$ is the formula for the displacement s of an object thrown vertically upward from the ground with an initial velocity of 96 ft/sec. Use $0 \leq t \leq 6$. State the maximum value for s.

25. The power P produced by a varying electric current I is given by the equation $P = I^2R$, where the resistance R is constant. Assume that $R = 4\ \Omega$ and use $1 \leq I \leq 5$.

26. A variable voltage is given by the formula $e = t^2 - 6t + 10$, t in seconds. Use the interval $1 \leq t \leq 6$. State the minimum value for e in the given interval.

SECTION 25
Applications of Quadratic Functions

The question of finding the maximum or the minimum value for a function $y = f(x)$ and the value of x for which the value is attained arises frequently in engineering and in business. This general question can be answered in the branch of mathematics called **calculus** and will be examined later.

However, in Section 24 we discovered that the maximum or minimum value of a quadratic function will be the value of the y coordinate of the vertex, depending on whether $a < 0$ or $a > 0$. Thus, we can solve any problem concerning a maximum or minimum value if the function describing the quantity in question is quadratic.

Example 25.1

A rectangular field is to be enclosed by 120 yards (yd) of fencing. What should the dimensions be to assure the largest possible area?

Solution

We let

$$y = \text{length of the field, yd}$$
$$x = \text{width of the field, yd}$$
$$A = \text{area of the field, yd}^2$$
$$P = \text{perimeter of the field, yd}$$

We are told that the perimeter P is equal to 120 yd (the total length of fencing), and we also know that perimeter equals the sum of the sides of the rectangle. Thus, $P = 120 = 2y + 2x$. The area of the rectangle is the product of the length and the width, so $A = xy$. We can express A as a function of either x or y. We solve the equation $120 = 2x + 2y$ for y as follows:

$$120 = 2x + 2y$$
$$60 = x + y$$
$$y = 60 - x$$

Now substitute this result in the equation $A = xy$ to obtain

$$A(x) = x(60 - x) = 60x - x^2$$

We now observe that A is a quadratic function whose independent variable is x. We can calculate the coordinates of the vertex of $A(x)$ by noting that $a = -1$, $b = 60$, and $c = 0$.

$$x = -\frac{b}{2a} = -\frac{60}{2(-1)} = 30$$

$$A(x) = 60(30) - (30)^2 = 1800 - 900 = 900$$

Since $a = -1$ is negative, we know that the graph of the quadratic function $A(x)$ is a parabola that opens downward and has its vertex located at $(30, 900)$. Thus, the area function achieves its maximum value when its width $x = 30$ yd and its length $y = 60 - x = 60 - 30 = 30$ yd.

Example 25.2

A manufacturer offers to deliver 100 lamps to a dealer at a cost of \$20 per lamp. He offers to reduce the price per lamp on the entire order by \$0.05 for each additional lamp over 100. What is the manufacturer's largest possible gross revenue?

Solution

We let x equal the number of lamps ordered. The gross revenue R will be the product of the number of lamps ordered and the cost per lamp. We can represent the cost per lamps as

$$C(x) = \begin{cases} 20 & \text{if } x \leq 100 \\ 20 - 0.05(x - 100) & \text{if } x > 100 \end{cases}$$

Thus, the gross revenue can be expressed as

$$R(x) = \begin{cases} 20x & \text{if } x \leq 100 \\ x[20 - 0.05(x - 100)] = 25x - 0.05x^2 & \text{if } x > 100 \end{cases}$$

Since $R(x)$ is a quadratic function of x (for $x > 100$), where $a = -0.05$, $b = 25$, and $c = 0$, we can determine the coordinates of its vertex as follows:

$$x = -\frac{b}{2a} = -\frac{25}{2(-0.05)} = \frac{-25}{-0.1} = 250$$

and

$$\begin{aligned} R(x) &= 25(250) - 0.05(250)^2 \\ &= 6250 - (0.05)(62{,}500) \\ &= 6250 - 3125 = 3125 \text{ dollars} \end{aligned}$$

Since $a = -0.05$ is negative, we know that the graph of the quadratic function $R(x)$ is a parabola that opens downward and has its vertex located at $(250, 3125)$. Thus, the function has a maximum value of 3125 when $x = 250$. Thus, the maximum gross revenue is \$3125.

Example 25.3

Find the dimensions of a rectangle whose perimeter is 20 ft if the square of its diagonal is to be a minimum.

Solution
We let

$$D = \text{square of the diagonal of the rectangle}$$
$$w = \text{width}$$
$$l = \text{length}$$

The perimeter $P = 2w + 2l$ and we are given that $P = 20$. Thus, we have

$$20 = 2w + 2l$$
$$10 = w + l \quad \text{or} \quad l = 10 - w \tag{1}$$

We recall that the square of the diagonal, D, equals the sum of the squares of the width and the length. Thus, we also have

$$D = w^2 + l^2 \tag{2}$$

Now we substitute $l = 10 - w$ into (2), to obtain

$$\begin{aligned} D &= w^2 + (10 - w)^2 \\ &= w^2 + 100 - 20w + w^2 \\ &= 2w^2 - 20w + 100 \end{aligned}$$

We note that D is a quadratic function of the independent variable w. The graph of this function is the parabola that opens upward; its vertex, located at $(5, 50)$, is a minimum point. Thus, $w = 5$ and $l = 10 - w = 10 - 5 = 5$. Thus, the dimensions of the rectangle should be 5 ft by 5 ft.

Example 25.4
An object is thrown directly upward with an initial velocity of 48 ft/sec. Its height, s, is measured in feet, after t seconds, is given by the formula $s = 48t - 16t^2$. How long does it take the object to reach its maximum height and what is the maximum height?

Solution
Since the height formula is a quadratic function of t, where $a = -16$, $b = 48$, and $c = 0$, we know that its graph will be a parabola breaking downward and the coordinates of its vertex will be the maximum point. We calculate these coordinates as follows:

$$t = -\frac{b}{2a} = -\frac{48}{2(-16)} = \frac{48}{32} = 1.5$$
$$s = 48(1.5) - 16(1.5)^2 = 72 - 36 = 36$$

The coordinates of the vertex are $(1.5, 36)$. We now know that it takes the object 1.5 sec to reach a maximum height of 36 ft.

Example 25.5

The path of a projectile fired from ground level with an initial velocity of 400 ft/sec and direction given by elevation angle 45° is determined by $y = (-x^2/5000) + x$. The distance the projectile travels is called the *range*. To find the range we set $y = 0$.

$$\frac{-x^2}{5000} + x = 0$$

$$x^2 - 5000x = 0$$

$$x(x - 5000) = 0$$

$$x = 0 \quad \text{or} \quad x = 5000$$

Thus, the range is 5000 ft.

The maximum height y can be found by first averaging the zeros of the given function and then solving for y. This is equivalent to finding the coordinates of the vertex point. Thus,

$$x = \frac{0 + 5000}{2} = 2500$$

and

$$y = \frac{-(2500)^2}{5000} + 2500 = -1250 + 2500 = 1250$$

The maximum height is 1250 ft (Figure 25.1).

Figure 25.1

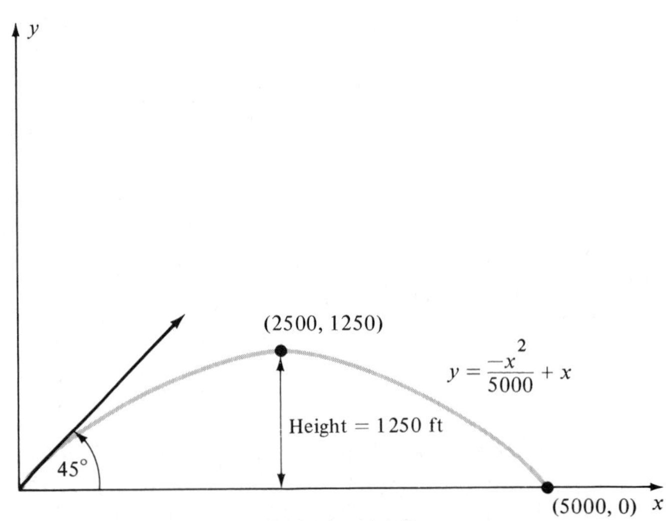

Exercises for Section 25

In Exercises 1 through 8, find the coordinates of the vertex and determine whether the vertex is a maximum point or a minimum point.

1. $y = x^2 - 4x + 6$ **2.** $y = 2x^2 - 4x + 1$

3. $y = 4 - 3x - 3x^2$ **4.** $y - 4 = 6x - 3x^2$

5. $y = 30x - x^2$ **6.** $y = 9 - 6x - x^2$

7. $y = (x - 2)^2 - 8$ **8.** $x^2 + 7x - 3y = 0$

9. Find two numbers whose sum is 28 and whose product is a maximum.

10. Find two numbers whose sum is 30 such that the sum of their squares is a minimum.

11. If a rectangle has a perimeter of 24 ft, find the dimensions of the rectangle with the maximum area. What is the maximum area?

12. Find two numbers whose difference is 14 and whose product is a minimum.

13. A rectangular pen is to be made out of 120 ft of fencing. What should the dimensions be to assure the largest possible area?

14. In Exercise 13, if one side of the pen is the side of a barn, what should the lengths of the other three sides be to assure the largest possible area?

15. A manufacturer offers to deliver 50 sofas to a dealer at a cost of $200 per sofa. He offers to reduce the price per sofa on the entire order by $5 for each additional sofa over 50. How many sofas should be in the order that gives the manufacturer to obtain the greatest revenue?

16. It is known that a local theater with a capacity of 300 will be filled when the admission price is $2, and that the attendance decreases by 30 for each $0.50 added to the admission price. What price of admission will yield the greatest gross receipts?

17. A local telephone company can get 10,000 customers for a monthly service charge of $15 each. It was determined that the company can get 1000 additional customers for each $1 decrease in the charge. What rate will guarantee the greatest monthly gross revenue?

18. A study of the locality and soil type of a certain orchard showed that if 40 apple trees are planted per acre, the yield of apples per tree would be 600. But the yield per tree is reduced by 10 for each additional tree planted per acre. How many trees should be planted per acre to guarantee a maximum yield?

19. A window has the shape of a rectangle surmounted by a semicircle. If the perimeter of the window is 20 ft, find the dimensions that will allow the maximum amount of light to enter.

20. An object is thrown directly upward from the top of a roof 160 ft high with an initial velocity of 48 per second. Its height, s, is measured in feet, after t seconds, and is given by the formula $s = 160 + 48t - 16t^2$. How long does it take the object to reach its maximum height?

21. In Exercise 20, how long will it take the object to fall back to the ground?

22. A wire 40 in. long is cut into two pieces. One piece is bent into the shape of a circle; the other, into the shape of a square. Where should the wire be cut in order that the sum of the areas of the circle and the square be a minimum?

23. A wire 40 in. long is cut into two pieces. One piece is bent into the shape of a circle; the other, into the shape of an equilateral triangle. Where should the wire be cut in order that the sum of the areas of the circle and the triangle be a minimum?

24. In Exercise 22, what cut will yield a maximum?

25. In Exercise 23, what cut will yield a maximum?

26. A wire 24 in. long is cut into two pieces. One piece is bent into a square; the other into an equilateral triangle.
(a) Express the sum of the area of the triangle and the square as a function of x, the part cut off to form the triangle.
(b) Graph the function determined in part (a).
(c) Find x such that the sum of the area is a maximum.
(d) Find x such that the sum of the area is a minimum.
(e) Find x such that the area of the triangle is equal to the area of the square.

27. A rifle bullet is fired directly upward from the ground. Its distance from the ground is given by the formula $s = 1280t - 16t^2$, s given in feet and t in seconds. How long does it take the bullet to reach its maximum height, and what is the maximum height?

28. The equation of the path of a projectile fired at an angle of 45° with the horizontal and with an initial velocity of 80 ft/sec from a hill 150 ft above the ground is given by $y = -\frac{1}{200}x^2 + x + 150$, where x is the horizontal distance on the ground and y is the vertical height.
(a) Find the range; that is, set $y = 0$ and solve for x.
(b) Find the maximum height. ***Hint:*** Average the zeros of the function and then solve for y.

REVIEW QUESTIONS FOR CHAPTER 4

1. What is a function?

2. What is the largest possible set of values for both the independent and dependent variables in a function?
3. How do we "read" $y = f(x)$?
4. State two instances in which it is necessary to restrict the choices of the values for the independent variable for a function.
5. Briefly describe a rectangular coordinate system.
6. What is the abscissa of the point (x, y)? The ordinate?
7. What is the graph of an equation?
8. How do we graph a function?
9. What is a linear function?
10. How do we graph a linear function?
11. What is a constant function?
12. When is a linear equation not a linear function?
13. State the general form of the quadratic function.
14. What is the graph of a quadratic function?
15. What is the highest or lowest point on the graph of a quadratic function called?
16. How do we determine if the graph of a parabola opens upward or downward?
17. How do we find the y intercept of a quadratic function? The x intercept?
18. State the quadratic formula.

REVIEW EXERCISES FOR CHAPTER 4

1. Express the area of a circle whose radius is $r + 2$ units as a function of r.
2. Express the area of a rectangle whose width is $x + 1$ and whose length is $x + 2$ as a function of x.
3. Express the perimeter of a square whose side is s^2 as a function of s.
4. Express the circumference of a circle whose radius is $(2r + 1)$ as a function of r.
5. Express the area of a triangle of base b and height $b + 2$ as a function of b.
6. The width of a rectangular solid is w. The length is 2 more than twice the width and the height is 1 less than the width. Express the volume as a function of the width w.
7. If $f(x) = x + 3$, find $f(0)$, $f(1)$, and $f(-3)$.
8. If $f(x) = 2\sqrt{2x - 3}$, find $f(2)$, $f(6)$, and $f(26)$.
9. If $g(x) = x^2 - 3x + 1$, find $g(0)$, $g(2)$, and $g(a)$.

10. If $h(t) = 64t - 16t^2$, find $h(1)$, $h(2)$, and $h(4)$.

11. If $f(x) = 2x^2 - 4x + 3$, find $f(0)$, $f(1)$, and $f(|-2|)$.

12. If $f(x) = x^2 + x - |x|$, find $f(1)$, $f(-1)$, $f(2)$, and $f(-2)$.

For what real number value(s) of x are the following expressions defined?

13. $f(x) = \dfrac{1}{x - 4}$ **14.** $f(x) = \dfrac{x}{x - 4}$ **15.** $f(x) = \dfrac{1}{x^2 - 4}$

16. $f(x) = \dfrac{1}{x^2 + 4}$ **17.** $g(x) = \sqrt{2x + 3}$ **18.** $g(x) = \dfrac{1}{\sqrt{2x + 3}}$

19. $h(x) = \dfrac{x + 1}{\sqrt{x + 1}}$ **20.** $f(x) = \dfrac{x}{(x + 1)(x - 4)}$

In Exercises 21 through 25, find the x and y intercepts and graph each linear function.

21. $y = x - 2$ **22.** $y = 2x + 5$ **23.** $y = 2 - 3x$

24. $y = 1 - 2(x + 3)$ **25.** $y = 2$

26. Find the zeros of the functions given in Exercises 21 through 25 by algebraic methods.

In Exercises 27 through 32, graph the quadratic function, noting the coordinates of the vertex, the x intercepts, and the y intercepts.

27. $y = x^2 - 2x - 8$ **28.** $y = x^2 + x - 6$

29. $y = -x^2 + 4x - 3$ **30.** $y = x^2 + 2x + 3$

31. $y = 2x^2 - 5x + 3$ **32.** $y = 2 + 3x - 2x^2$

In Exercises 33 through 38, use the results of Exercises 27 through 32 to solve each quadratic equation graphically.

33. $x^2 - 2x = 8$ **34.** $x^2 + x - 6 = 0$

35. $x^2 + 3 = 4x$ **36.** $x^2 + 2x + 3 = 0$

37. $2x^2 - 5x + 3 = 0$ **38.** $2x^2 = 2 + 3x$

39. Determine whether the vertex points of the functions given in Exercises 27 through 32 are maximum points or minimum points.

40. Find two numbers whose sum is 16 and whose product is a maximum.

41. In a certain area, a national manufacturer of razor blades can sell 1000 packages per week at \$1.50 each. A survey determined that the company can sell an additional 50 packages per week for each \$0.10 decrease in price. What selling price will guarantee the greatest monthly income?

42. How long will it take an object to strike the ground if the equation for height as a function of time is given by $h = -16t^2 + 64t + 128$? What is the maximum height achieved by the object?

CHAPTER 5

Inequalities

SECTION 26
Order Properties

In Section 3 we introduced the concept of order by noting that if a real number a is to the left of another real number b on the real number line, a is less than b or, symbolically, $a < b$. We will agree on the following notation:

$a > b$ reads "a is greater than b"

$a < b$ reads "a is less than b"

$a \geq b$ reads "a is greater than or equal to b"

$a \leq b$ reads "a is less than or equal to b"

In this chapter we wish to examine the solutions of inequalities involving one variable. To do this we must first establish some order properties.

The inequalities $a > b$ and $c > d$ have the **same sense**. In the inequality $a < b$ we refer to a as the left side of the inequality and b as the right side of the inequality.

Addition Property

We can add any real number to both sides of an inequality and the sense of the inequality remains the same.

Example 26.1

(a) If $6 < 8$, then $6 + 3 < 8 + 3$.
(b) If $-4 > -7$, then $-4 + 2 > -7 + 2$.
(c) If $x < y$, then $x + 5 < y + 5$.

Subtraction Property

We can subtract any real number from both sides of an inequality and the sense of the inequality remains the same.

Example 26.2

(a) If $4 > 1$, then $4 - 7 > 1 - 7$.
(b) If $-10 < -3$, then $-10 - 6 < -3 - 6$.
(c) If $x < y$, then $x - 4 < y - 4$.

Multiplication Property

If we multiply both sides of an inequality by any **positive** real number, the sense of the inequality remains the same. If we multiply both sides of an inequality by any **negative** real number, the sense of the inequality is **reversed**.

Example 26.3

(a) If $3 < 5$, then $2(3) < 2(5)$.
(b) If $7 > 2$, then $3(7) > 3(2)$.
(c) If $x > y$, then $4x > 4y$.
(d) If $3 < 5$, then $(-2)(3) > (-2)(5)$ or $-6 > -10$.
(e) If $6 > 2$, then $(-4)(6) < (-4)(2)$ or $-24 < -8$.
(f) If $x < y$, then $-2x > -2y$.

Division Property

If we divide both sides of an inequality by any **positive** real number, the sense of the equality remains the same. If we divide both sides of an inequality by any **negative** real number, the sense of the inequality is **reversed.**

Example 26.4

(a) If $8 > 3$, then $\frac{8}{3} > 1$.

(b) If $-4 < -2$, then $-\frac{4}{5} < -\frac{2}{5}$.

(c) If $2x < 6$, then $\frac{2x}{2} < \frac{6}{2}$ or $x < 3$.

(d) If $-3 < 9$, then $\frac{-3}{-3} > \frac{9}{-3}$ or $1 > -3$.

(e) If $-4x > -12$, then $\frac{-4x}{-4} < \frac{-12}{-4}$ or $x < 3$.

In summary, when we perform the basic operations of addition, subtraction, multiplication, and division on both sides of an inequality, we must be aware of the following situations:

1. Multiplication or division by zero is *not permitted.*
2. Multiplication or division by a negative number reverses the sense of the given inequality.

The inequality $a < x < b$ is called a **compound inequality** and is equivalent to the statement $a < x$ and $x < b$. If x is a real number, then $a < x < b$ simply means all the real numbers "between a and b."

Example 26.5

The statements $x < 1$ and $x > -2$ can be expressed as the compound inequality $-2 < x < 1$. The compound inequality $-3 \leq 2x + 1 \leq 5$ can be expressed as $-3 \leq 2x + 1$ and $2x + 1 \leq 5$. The statement $x < 1$ or $x > 3$ *cannot* be expressed as a compound inequality.

Summary of Notation

Symbol	*Meaning*	*Corresponding Graph*
$x > a$	x is greater than a	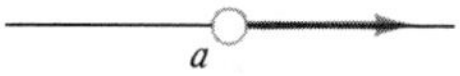 open circle at a, arrow right
$x \geq a$	x is greater than or equal to a	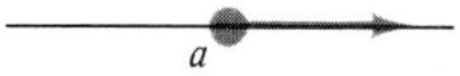 closed circle at a, arrow right
$x < a$	x is less than a	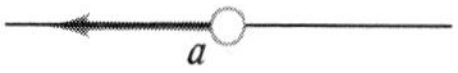 open circle at a, arrow left
$x \leq a$	x is less than or equal to a	 closed circle at a, arrow left
$a < x < b$	x is greater than a and less than b	 open a, open b
$a \leq x < b$	x is greater than or equal to a and less than b	 closed a, open b
$a < x \leq b$	x is greater than a and less than or equal to b	 open a, closed b
$a \leq x \leq b$	x is greater than or equal to a and less than or equal to b	 closed a, closed b

In the summary an open circle indicates that the number is not included and a closed circle indicates that the number is included.

Exercises for Section 26

In Exercises 1 through 15, express the following as inequalities.

1. 3 is less than 9.

2. -5 is greater than -10.

3. -3 is less than or equal to -1.

4. 9 is greater than or equal to 2.

5. $r + 3$ is greater than -4.

6. t is between 1 and 3.

7. t is greater than or equal to 1 and less than or equal to 4.

8. $2s + 3$ is negative.

9. $x - 3$ is nonnegative.

10. $s + r$ is positive.

11. $3x - 4$ is nonpositive.

12. $4x + 1$ is between 2 and -2.

13. $4r$ is greater than or equal to -3 and less than or equal to -1.

14. y is greater than -6 and nonpositive.

15. $x + 1$ is greater than 3 or less than -2.

In Exercises 16 through 22, describe the conditions in terms of inequalities.

16. All points (x, y) that lie in quadrant II.

17. All points (x, y) that lie on or below the x axis.

18. All points (x, y) which lie to the left of the y axis.

19. All points (x, y) which lie on the y axis or to the right of it.

20. All points (x, y) which lie between the vertical lines $x = -1$ and $x = 2$.

21. All points (x, y) which lie between the horizontal lines $y = 3$ and $y = -2$.

22. All points (x, y) lying in quadrants III and IV.

SECTION 27

Linear Inequalities in One Variable

A **linear inequality in one variable** is any inequality whose variable is first degree. Any number for which the inequality is true is a **solution**. When we are instructed to "solve" a given inequality, we are supposed to find the set of *all* solutions to the inequality. In this text we shall assume that the largest possible set of values for the variable will be the real numbers.

An inequality such as $x + 2 > x + 1$ is said to be an **absolute inequality**, since its set of solutions is *all* the real numbers. An inequality such as $x > 1$ is a **conditional inequality,** since not every real value for x is a solution.

Example 27.1

(a) The inequality $3x - 1 > 0$ is **linear**, since the variable x is first degree.

(b) The inequality $x^2 - 3x > 1$ is **not linear**, since the term x^2 is not first degree.

Example 27.2

The inequality $2x - 1 > 0$ has $x = 2$ for a solution, since substitution of 2 for x gives us $2(2) - 1 > 0$ or $3 > 0$, which is true. But $x = 0$ is not a solution for $2x - 1 > 0$, since substitution of 0 for x gives us $2(0) - 1 > 0$ or $-1 > 0$, which is false. Thus, $2x - 1 > 0$ is a conditional inequality.

The procedures for solving inequalities in one variable are similar to those used for solving equations. That is, we use the order properties of Section 26 to obtain equivalent inequalities (inequalities with the same solution) until we obtain one whose solutions are apparent. We illustrate these ideas in the following examples.

Example 27.3

Solve $3x - 1 > 4$, and graph the solution set.

Solution

We add 1 to both sides of the inequality, to obtain

$$3x - 1 + 1 > 4 + 1 \quad \text{or} \quad 3x > 5$$

Now we divide both sides by 3, to obtain

$$\frac{3x}{3} > \frac{5}{3} \quad \text{or} \quad x > \frac{5}{3}$$

Thus, we conclude that the solution set for $3x - 1 > 4$ is $x > \frac{5}{3}$. We show the graph of this solution set in Figure 27.1.

Figure 27.1

$\frac{5}{3}$

Example 27.4

Solve $\dfrac{2x - 3}{4} \leq 2$ and graph the solution set.

Solution

We first multiply both sides of the inequality by 4, to obtain

$$4\left(\frac{2x - 3}{4}\right) \leq 4(2) \quad \text{or} \quad 2x - 3 \leq 8$$

We then add 3 to both sides, to obtain

$$2x - 3 + 3 \leq 8 + 3 \quad \text{or} \quad 2x \leq 11$$

Finally, we divide both sides by 2, to obtain

$$\frac{2x}{2} \leq \frac{11}{2} \quad \text{or} \quad x \leq \frac{11}{2}$$

Thus, we conclude that the solution set for $\frac{2x - 3}{4} \leq 2$ is $x \leq \frac{11}{2}$. We show the graph of this solution set in Figure 27.2.

Figure 27.2

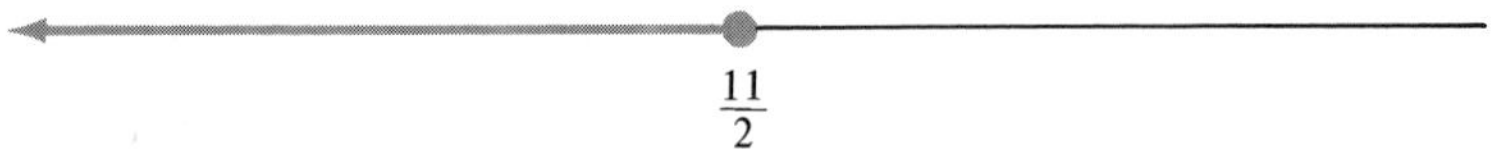

$\frac{11}{2}$

Example 27.5

Solve $2x - 5 \geq 5x$ and graph the solution set.

Solution

We wish to use the appropriate order properties to group the terms involving the variable x on either the right or left side of the inequality and the constants on the opposite side. Here we choose to group the terms involving x on the left by first adding 5 to both sides and then adding $-5x$ to both sides as follows:

$$2x - 5 + 5 \geq 5x + 5$$

$$2x \geq 5x + 5$$

$$2x + (-5x) \geq 5x + (-5x) + 5$$

$$-3x \geq 5$$

Now we divide both sides by -3 (note that the sense of the inequality is reversed), to obtain

$$\frac{-3x}{-3} \leq \frac{5}{-3} \quad \text{or} \quad x \leq -\frac{5}{3}$$

Thus, we conclude that the solution set for $2x - 5 \geq 5x$ is $x \leq -\frac{5}{3}$. We show the graph of this solution set in Figure 27.3.

Figure 27.3

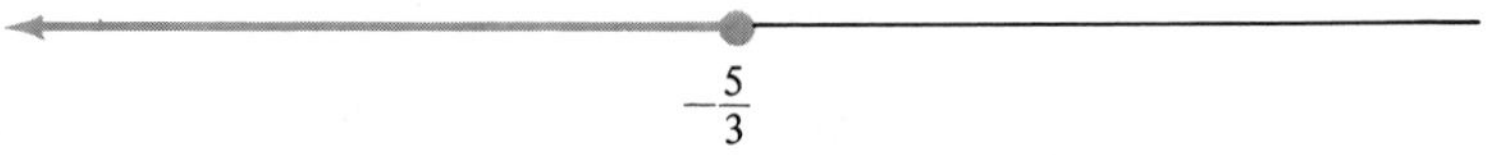

$-\frac{5}{3}$

Example 27.6

The velocity, v (in ft/sec), of an object fired directly upward is given as a function of time, t (in seconds), by the formula $V = 80 - 32t$. For what positive time interval will the velocity be positive?

Solution

We see that the velocity will be positive when $V > 0$, or equivalently when $80 - 32t > 0$. This latter form is a linear inequality in the variable t and we proceed to a solution as follows:

$$80 - 32t > 0$$

$$80 - 32t + (-80) > 0 + (-80) \qquad \text{(added } -80 \text{ to both members)}$$

$$-32t > -80$$

$$\frac{-32t}{-32} < \frac{-80}{-32} \qquad \text{(divided by } -32\text{)}$$

$$t < \frac{5}{2} \text{ sec}$$

Thus, we can conclude that the positive time interval will be when $t < \frac{5}{2}$ and $t > 0$ or equivalently $0 < t < \frac{5}{2}$.

We shall frequently encounter problems involving compound inequalities. The procedures for solving them are identical to those illustrated in Examples 27.3 through 27.6.

Example 27.7

When will the velocity of the object in Example 27.6 be between 32 and 64 ft/sec?

Solution

We can express the statement that the velocity be between 32 and 64 ft/sec as the compound inequality

$$32 < V < 64$$

We can now substitute $80 - 32t$ for V, to obtain

$$32 < 80 - 32t < 64$$

We now solve this compound inequality by using the order properties as follows:

$$32 < 80 - 32t < 64$$

$$32 + (-80) < 80 + (-80) - 32t < 64 + (-80)$$

$$-48 < -32t < -16$$

$$\frac{-48}{-32} > \frac{-32t}{-32} > \frac{-16}{-32} \qquad \text{(note the } \textit{reversed} \text{ senses of the inequalities)}$$

$$\frac{3}{2} > t > \frac{1}{2}$$

Thus, we conclude that 32 ft/sec $< V <$ 64 ft/sec when $\frac{1}{2}$ sec $< t < \frac{3}{2}$ sec.

Exercises for Section 27

In Exercises 1 through 18, solve each inequality.

1. $x + 3 > 1$
2. $2x - 1 \le 0$
3. $3 - 5x < 6 - x$
4. $4x - 7 > 0$
5. $2x - 6 \le 4x + 2$
6. $5x + 7 < 3(x + 1)$
7. $2(x - 1) - 3(x - 2) < 7$
8. $3(x - 2) + 4 \ge 2(2x - 3)$
9. $\dfrac{3 - x}{4} < 1$
10. $\dfrac{x - 3}{2} \ge x$
11. $x \le \dfrac{x - 1}{3}$
12. $\dfrac{5 - 3x}{2} \le x$
13. $\dfrac{x}{2} < \dfrac{2(x - 1)}{3}$
14. $\dfrac{x - 1}{3} \ge \dfrac{2}{3}(x + 1)$
15. $-3 < 2x - 3 < 6$
16. $-6 < 4x + 1 < -4$
17. $4 < 3 - 2x < 12$
18. $-2 < 7 - x < 0$
19. The velocity V (in ft/sec) of an object fired directly upward is given as the function of time t (in seconds) by the formula $V = 72 - 32t$. For what positive time interval will the velocity be positive?
20. In Exercise 19, when will the velocity be between 8 and 32?
21. The relationship between Fahrenheit temperature F and Celsius temperature C is given by the formula $C = \frac{5}{9}(F - 32)$. Find the range of values of F when C is to be between 10 and 30.
22. A student in a particular mathematics course knows that he can earn a grade of B for the course if he has an average less than 90 but not below 80. Suppose that the student's first four examination scores are 76, 84, 88, and 78. What grade on the fifth examination would give him a grade of B for the course?
23. A person wishes to invest part of \$30,000 at 5% simple interest and the remainder at 7%. What is the least amount that can be invested at 7% in order to have a yearly income of at least \$1900 from the two investments?
24. A copper alloy containing at least 46% copper and at most 50% copper is needed. Determine the least and greatest amounts of a 60% copper alloy that should be combined with a 40% copper alloy in order to have 30 lb of the final alloy.
25. An insurance agent receives \$450 monthly plus a 5% commission on sales. How much must his sales for the month be in order for his income to be at least \$600?

26. The daily cost of operating a certain machine is \$3 per hour plus \$15 for maintenance. If the daily cost is not to exceed \$36, how many hours can the machine be operated?

27. Two resistors in parallel have a total resistance R given by $1/R = 1/R_1 + 1/R_2$. If $R_1 = 20\ \Omega$, find the possible values for R_2 so that R is between 10 and 15 Ω.

SECTION 28

Quadratic Inequalities in One Variable

We now wish to introduce a graphic method for solving quadratic inequalities in one variable such as $(x - 3)(x + 1) \geq 0$ and $(2x - 3)(x - 4) < 0$. To do this we must first note that any real number must either be negative or zero or positive.

The expression $x - 2$ will be a real number if x is real. This real number has the following possibilities:

$$x - 2 < 0 \quad \text{or} \quad x - 2 = 0 \quad \text{or} \quad x - 2 > 0$$

A graph of these possibilities is shown on the real number line in Figure 28.1. The negative signs represent the given region of the number line where $x - 2 < 0$. The positive signs represent the region of the number where $x - 2 > 0$. The point $x = 2$ represents the solution to $x - 2 = 0$. We shall refer to the graph in Figure 28.1 as the *sign graph* for $x - 2$. The point $x = 2$ will be called the **critical value**.

Figure 28.1

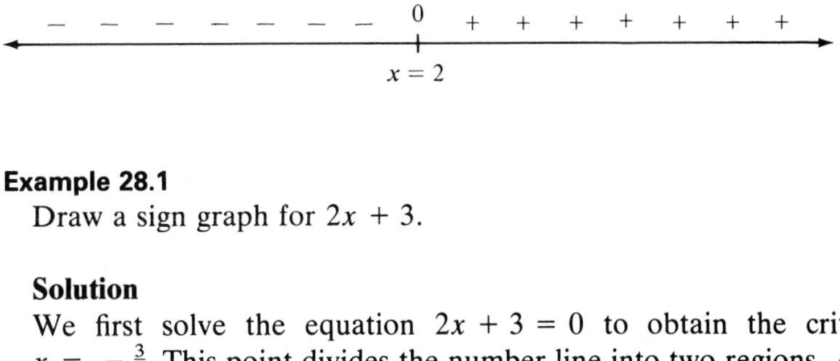

− − − − − − − 0 + + + + + + +

$x = 2$

Example 28.1

Draw a sign graph for $2x + 3$.

Solution

We first solve the equation $2x + 3 = 0$ to obtain the critical value $x = -\frac{3}{2}$. This point divides the number line into two regions. Any choice of values for x in the region to the right of $x = -\frac{3}{2}$ will make $2x + 3$ a positive number. Any choice of x in the region to the left of $x = -\frac{3}{2}$ will make $2x + 3$ a negative number. We show this sign graph in Figure 28.2.

Figure 28.2

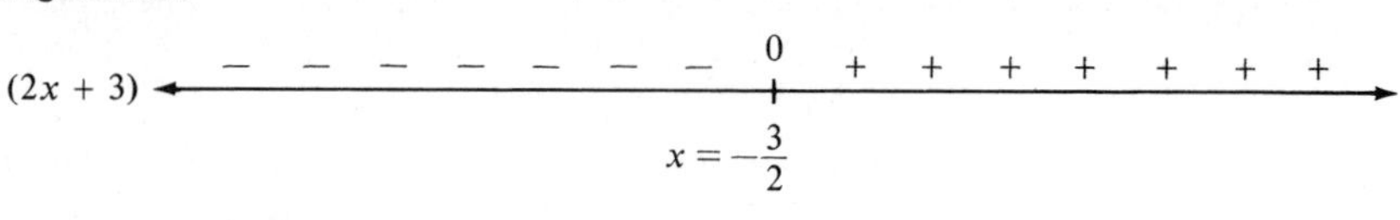

Example 28.2

Draw a sign graph for $3 - x$.

Solution

We first note that the critical value will be $x = 3$. Any choice of a value for x to the right of 3 will make $3 - x$ a negative number, and any choice to the left of 3 will make $3 - x$ a positive number. We show this sign graph in Figure 28.3.

Figure 28.3

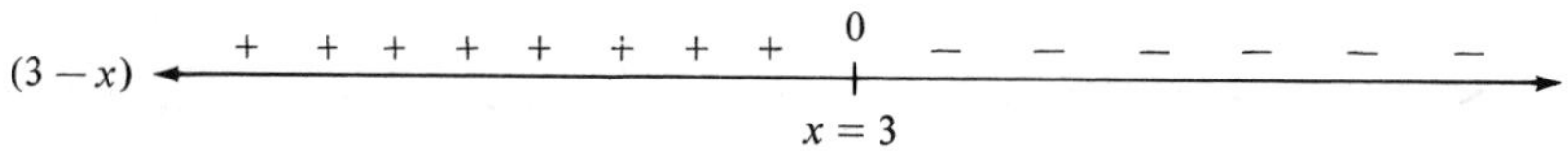

(3 − x) + + + + + + + + 0 − − − − − −

x = 3

We now use the following examples to introduce the **sign graph method** for solving quadratic inequalities.

Example 28.3

Solve the quadratic inequality $(2x - 5)(x + 2) > 0$.

Solution

We first draw a sketch of the sign graphs for each of the two factors $(2x - 5)$ and $(x + 2)$ in Figure 28.4. Next, we mark off both critical values on a number line directly below the two sign graphs. This divides the number line into three regions. We now check each region to see where

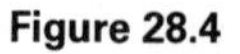

Figure 28.4

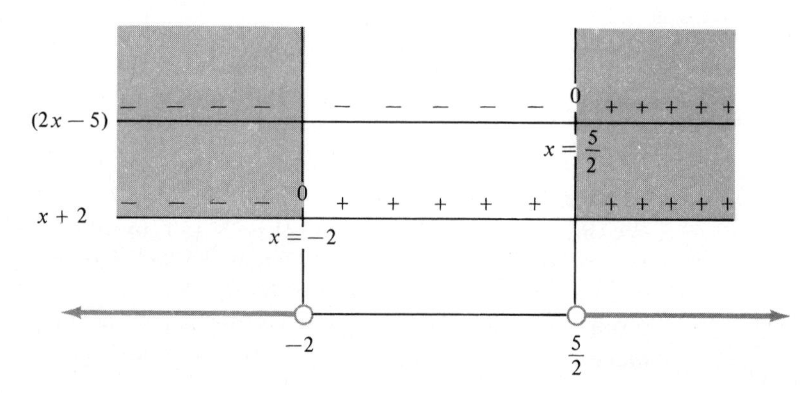

the product $(2x - 5)(x + 2)$ is positive. This will occur if both factors are negative or if both are positive. In Figure 28.4 we have shaded these regions, and the graph of the solution set for $(2x - 5)(x + 2) > 0$ is shown on the number line in Figure 28.4. We interpret this solution set as $x < -2$ or $x > \frac{5}{2}$. Since the critical values are not included, we indicate this by open circles at these points.

Example 28.4

Solve the quadratic inequality $6x^2 + 11x \leq 7$.

Solution

We first subtract 7 from both sides of the inequality in order to make the right side 0. Next we factor the left side, as follows:

$$6x^2 + 11x - 7 \leq 7 - 7$$

$$6x^2 + 11x - 7 \leq 0$$

$$(3x + 7)(2x - 1) \leq 0$$

Now we proceed to draw the sign graphs for each factor shown in Figure 28.5 and then draw a number line directly below them. We now wish to find the regions where the product $(3x + 7)(2x - 1)$ will be negative or zero. In Figure 28.5 we shade the appropriate region and indicate the solution on the number line. Thus, $-\frac{7}{3} \leq x \leq \frac{1}{2}$.

Figure 28.5

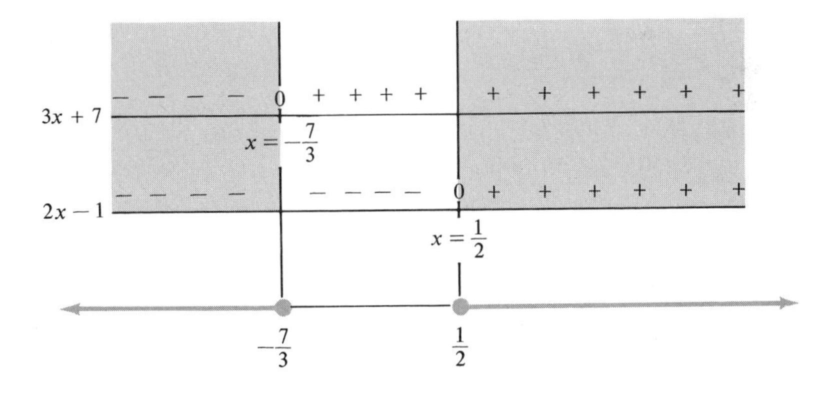

The sign graph method illustrated in Examples 28.3 and 28.4 relies on the ability to recognize the critical values. These critical values are the roots or solutions to the associated quadratic equations and thus can always be found by applying the quadratic formula (see Section 16).

Recall that the roots of the quadratic equation

$$ax^2 + bx + c = 0$$

are

$$\frac{-b + \sqrt{b^2 - 4ac}}{2a} \quad \text{and} \quad \frac{-b - \sqrt{b^2 - 4ac}}{2a}$$

and they will be real if $b^2 - 4ac \geq 0$.

The case where the roots or critical values are nonreal will occur when $b^2 - 4ac < 0$. In this case we state the following theorem.

Theorem 28.1
If $ax^2 + bx + c = 0$, $a > 0$, has nonreal roots, the solution set for:

1. *$ax^2 + bx + c > 0$ is all real values of x.*
2. *$ax^2 + bx + c < 0$ is no real values of x.*

Example 28.5
Solve the quadratic inequality $x^2 + x + 1 > 0$.

Solution
We first find the critical values of $x^2 + x + 1$ by using the quadratic formula to calculate the roots of the associated quadratic equation $x^2 + x + 1 = 0$ as follows. Since $a = 1$, $b = 1$, and $c = 1$, and $b^2 - 4ac = 1^2 - 4(1)(1) = -3$ is negative, the roots are nonreal. We now apply Theorem 28.1 to conclude that the solution set is all real values of x.

Exercises for Section 28

In Exercises 1 through 20, solve each inequality.

1. $(x - 1)(x + 3) > 0$

2. $(x + 2)(x - 6) < 0$

3. $(2x - 7)(x - 2) \leq 0$

4. $(3x - 1)(4 - x) \leq 0$

5. $x^2 + 4x - 5 < 0$

6. $x^2 + 7x + 6 \geq 0$

7. $x^2 - 6x + 5 \geq 0$

8. $2x - 4 \leq x^2$

9. $2x^2 - 5x - 3 > 0$

10. $2x^2 - 5x \leq 3$

11. $x^2 + 2x + 2 > 0$

12. $x^2 - 6x + 9 > 0$

13. $3x^2 + 2x + 6 < 0$

14. $2x^2 + 6 > 7x$

15. $x \leq 2x^2$

16. $\frac{x^2}{4} + \frac{x}{4} \leq 3$

17. $4x - 4 \leq x^2$

18. $9 - 6x - 8x^2 \leq 0$

19. $25 - 4x^2 > 0$

20. $9 + x^2 < 0$

In Exercises 21 through 24, state any necessary restrictions on the independent variable x for each function.

21. $y = \sqrt{(x - 1)(x - 5)}$

22. $y = \sqrt{x^2 - 8x + 7}$

23. $y = \sqrt{x^2 - 9}$

24. $y = \sqrt{x^2 + 9}$

25. Prove Theorem 28.1.

26. The velocity V (in ft/sec), of an object traveling in a straight line is a function of time t (in seconds) and is given by the formula $V = 3t^2 - 24t + 36$. Determine the positive values of t between 0 and 9 where the velocity is (a) positive and (b) negative.

27. The height h (in feet) of an object thrown directly upward is a function of the time, t (in seconds), and is given by the formula $h = 176 + 56t - 16t^2$. Find the time interval when the height is greater than 200 ft.

28. The capacity C of a given carrier is expressed by the equation $C = 6(40S - S^2)$. Find the limits for S that guarantee a minimum value of 1800 for C.

SECTION 29

Inequalities of Higher Degree in One Variable; Fractional Inequalities

The sign graph method of Section 28 can be applied to the solution of higher-degree inequalities in one variable and to the solution of fractional inequalities in one variable. The solution of inequalities such as $(x + 2)(x - 3)(3x + 6) > 0$ and $(2x - 1)(x + 1)(x - 5) < 0$ will rely on our ability to analyze when the product of the factors involved will be positive or negative.

The product of three factors will be positive if two of the factors are negative or if none of the factors are negative. The product of three factors will be negative if all three factors are negative or if one of the factors is negative. The product of four factors will be positive when four factors are negative or two factors are negative or zero factors are negative. The product of four factors will be negative when three factors are negative or one factor is negative. The general case is stated in the following theorem.

Theorem 29.1

The product of n factors will be ***positive*** *if there is an* ***even*** *number of negative factors. Note that zero is an even number.*

The product of n factors will be ***negative*** *if there is an* ***odd*** *number of negative factors.*

The use of Theorem 29.1 and the sign graph method are illustrated in the following examples.

Example 29.1

Find the solution set for $(x + 3)(x - 1)(x - 4) > 0$ and graph this set.

Solution

We first construct sign graphs for each of the three factors $x + 3$, $x - 1$, and $x - 4$ in Figure 29.1. Next, we mark off the critical values -3, 1, and 4 on a number line directly below these sign graphs. These critical values divide the number line into four regions. The product of the three factors must be positive. We use Theorem 29.1 to select the regions where there are an *even* number of negative factors. These regions have been shaded in Figure 29.1. We conclude that the solution set for the given inequality is $-3 < x < 1$ or $x > 4$, and its graph is shown on the number line in Figure 29.1.

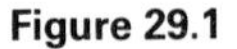

Figure 29.1

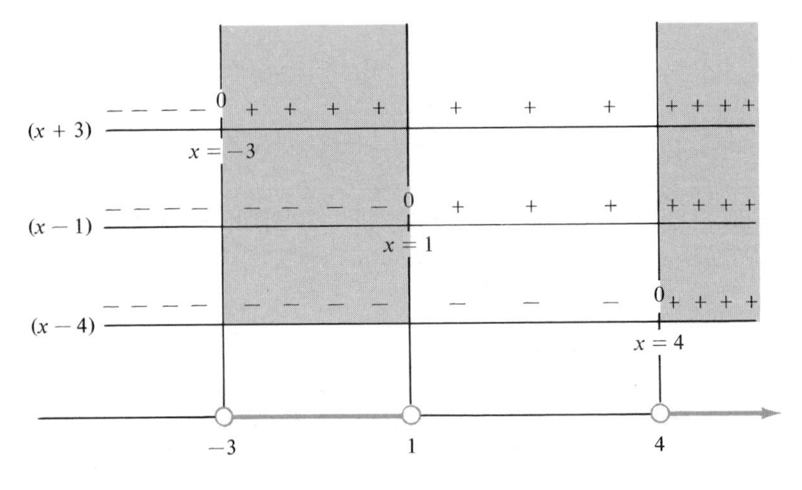

Example 29.2

Find the solution set for $(x + 3)(x - 1)(x - 3)(x - 4) \leq 0$ and graph this set.

Solution

We first construct the sign graph for each of the four factors in Figure 29.2. Next, we mark off the critical values -3, 1, 3, and 4 on a number line directly below these sign graphs. These critical values divide the number line in five regions. The product of the four factors must be negative. We use Theorem 29.1 to select the regions where there are an *odd* number of negative factors. We include the critical values, since the product can also equal zero. These regions have been shaded in Figure 29.2. We conclude that the solution set for the given inequality is $-3 \le x \le 1$ or $3 \le x \le 4$ and its graph is shown on the number line in Figure 29.2.

Figure 29.2

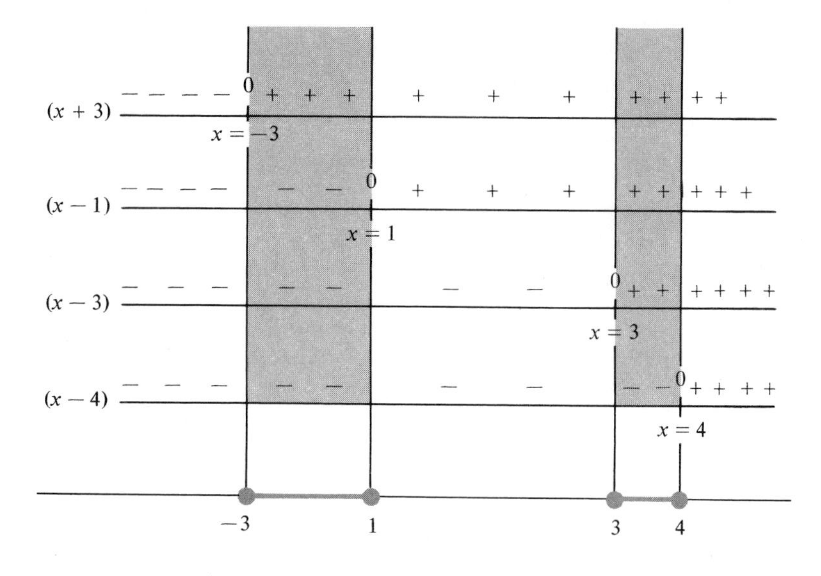

The sign graph method can also be used to solve fractional inequalities in one variable. An algebraic fraction such as $(x - 1)/(x + 2)$ will be positive when both the factor $(x - 1)$ in the numerator and the factor $(x + 2)$ in the denominator are negative or when neither are negative. Thus, we could say that this fraction will be positive if we have even numbers of negative factors in *both* the numerator and the denominator. Similar reasoning shows us that this fraction will be negative if we have an odd number of negative factors in *both* the numerator and denominator. The general case of n factors in both the numerator and denominator is stated as follows.

Theorem 29.2
If the number of negative factors in both the numerator and denominator of a fraction is ***even****, the fraction is positive. If the number of such factors is* ***odd****, the fraction is negative.*

Example 29.3

Find the solution set for $\dfrac{(x + 3)(x - 1)}{x + 2} \ge 0$ and then graph this set.

Solution

We first construct a sign graph for each of the factors that appear in both the numerator and the denominator in Figure 29.3. Next, we mark off the critical values -3, -2, and 1 on a number line directly below these sign graphs. These critical values divide the number line into four regions. The fraction must be positive. We use Theorem 29.2 to select the regions where there are an even number of negative factors. We include the critical values -3 and -1, since the fraction will equal 0. But we do not include the critical value -2, since the denominator of the fractions would equal 0 and the fraction would be undefined. These regions have been shaded in Figure 29.3. We conclude that the solution set for the given inequality is $-3 \leq x < -2$ or $x \geq 1$, and its graph is shown on the number line in Figure 29.3.

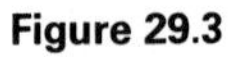

Figure 29.3

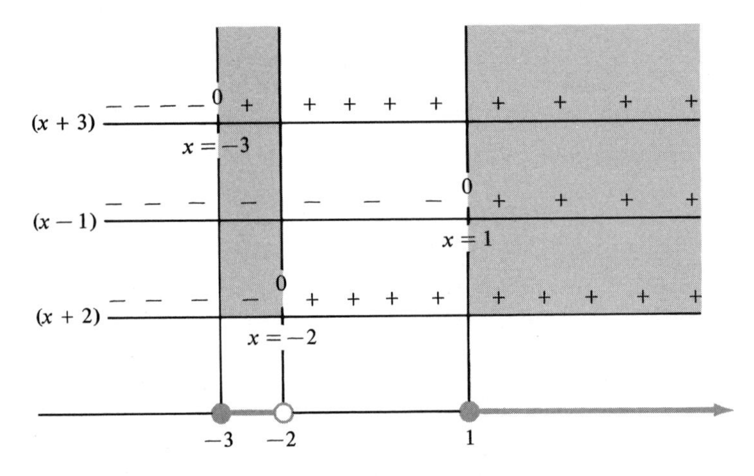

Example 29.4

Find the solution set for $\dfrac{1}{x-2} \geq \dfrac{3}{x}$ and graph this set.

Solution

We must first reexpress $\dfrac{1}{x-2} \geq \dfrac{3}{x}$ as an equivalent inequality with a right side equal to zero. We do this as follows:

$$\frac{1}{x-2} \geq \frac{3}{x}$$

$$\frac{1}{x-2} - \frac{3}{x} \geq 0$$

$$\frac{x - 3(x - 2)}{(x - 2)x} \geq 0$$

$$\frac{x - 3x + 6}{x(x - 2)} \geq 0$$

$$\frac{6 - 2x}{x(x - 2)} \geq 0$$

We now apply the sign graph method to the equivalent inequality $\frac{6 - 2x}{x(x - 2)} \geq 0$. We construct a sign graph for each of the three factors and then mark off the critical values 0, 2, and 3 on the number line in Figure 29.4. The three values divide the number line into four regions. Since the fraction is to be positive or equal to zero, we will select the regions where there are an even number of negative factors. We exclude the critical values 0 and 2, since the fraction would be undefined. These regions have been shaded in Figure 29.4. We conclude that the solution set for the given inequality is $x < 0$ or $2 < x \leq 3$, and its graph is shown on the number line in Figure 29.4.

Figure 29.4

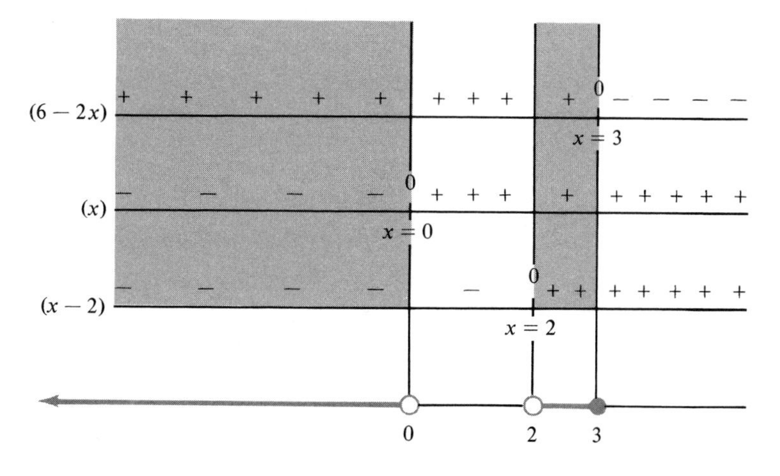

If the factors in the product or quotient are preceded by a negative constant, we must multiply both sides by −1 and then use the sign graph method on the equivalent inequality formed.

The inequality $(-4)(x + 1)(x + 2)(x - 4) > 0$ must be multiplied by −1 in order to apply the sign graph method. Note that the sense of the original inequality will be *reversed*. That is,

$$(-1)(-4)(x + 1)(x + 2)(x - 4) < (-1)(0)$$

or

$$4(x + 1)(x + 2)(x - 4) < 0$$

Exercises for Section 29

Find the solution set and graph this set.

1. $(x - 2)(x + 1)(x - 3) > 0$

2. $(2 - x)(x + 3)(x - 4) \leq 0$

3. $x(2x - 6)(x - 2) \geq 0$

4. $(4x - 3)(3 - x)(x + 4) > 0$

5. $3(x + 1)(x + 2)(x + 3)(x + 4) < 0$

6. $-3(x + 1)(x + 2)(x + 3)(x + 4) > 0$

7. $(2x - 3)(3x + 7)(x - 4) > 0$

8. $x^3 - x \leq 0$

9. $\dfrac{x - 6}{3 - x} > 0$

10. $\dfrac{(x - 1)(x + 2)}{(x + 1)(x - 2)} \leq 0$

11. $\dfrac{-3}{x^2 + 5x + 4} < 0$

12. $\dfrac{(4 - x)(3 + x)}{x^2 - 16} \leq 0$

13. $\dfrac{18 - 5x}{x - 3} > 0$

14. $\dfrac{3}{4x + 1} \geq \dfrac{2}{x - 5}$

15. $\dfrac{1}{x - 1} \leq \dfrac{2}{x}$

16. $\dfrac{4}{x - 3} - \dfrac{16}{x^2 - 9} > 1$

17. $(x - 1)^2(x + 3)(x + 4) > 0$

18. $(x - 1)^2(x + 3)(x + 4) \geq 0$

19. $\dfrac{(x + 3)^2(x - 2)}{x + 6} < 0$

20. $\dfrac{(x + 3)^2(x - 2)}{x + 6} \leq 0$

SECTION 30

Inequalities Involving Absolute Value

In Section 3 we defined the absolute value of x, denoted $|x|$, as the undirected distance between the real number x and 0 (origin) on the real number line. We can now discuss the solution of inequalities involving absolute value. The inequality $|x| < a$, $a > 0$, can be interpreted as the set of all real numbers x whose distance from 0 (the origin) is less than a units. In Figure 30.1 we show a graph of this set of points. We note that this set of points can also be described by the compound inequality $-a < x < a$, and we call this the solution set for $|x| < a$.

Figure 30.1

The inequality $|x| > a$, $a > 0$, can be interpreted as the set of all real numbers x whose distance from 0 is greater than a units. In Figure 30.2 we show a graph of this set of points. We note that this set can also be described by the statement $x < -a$ or $x > a$, and we call this solution set for $|x| > a$.

Figure 30.2

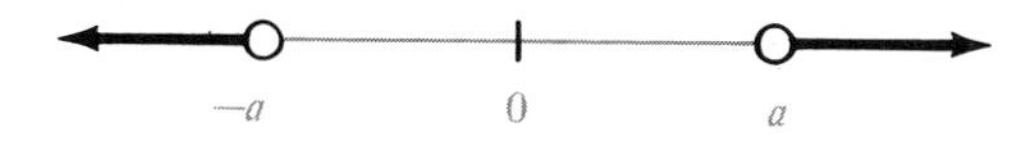

This line of reasoning leads us to the following statements.

$$\text{If } |x| < a, \text{ then } -a < x < a. \tag{30.1}$$

$$\text{If } |x| > a, \text{ then } x < -a \text{ or } x > a. \tag{30.2}$$

Example 30.1

Find the solution set for $|x - 1| < 3$ and graph this set.

Solution

We can interpret $|x - 1|$ to be the distance between x and 1. Thus, $|x - 1| < 3$ is the set of all points x whose distance from 1 is less than 3 units. We show the graph of this set in Figure 30.3 and conclude from the graph that the solution set for $|x - 1| < 3$ is $-2 < x < 4$.

Figure 30.3

Example 30.2

Find the solution set for $|x - 2| \geq 4$ and graph this set.

Solution

We can interpret $|x - 2|$ to be the distance between x and 2. Thus, $|x - 2| \geq 4$ is the set of all points x whose distance from 2 is greater than or equal to 4 units. We show the graph of this set in Figure 30.4 and conclude from the graph that the solution set for $|x - 2| \geq 4$ is $x \leq -2$ or $x \geq 6$.

Figure 30.4

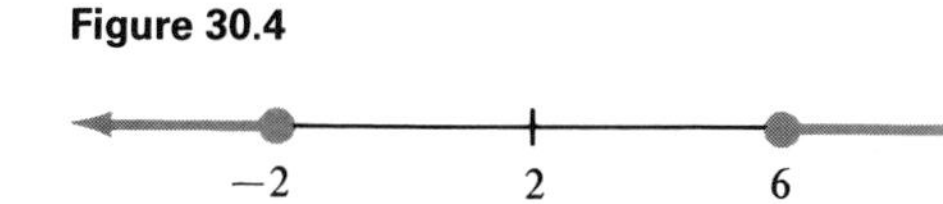

Examples 30.1 and 30.2 can also be solved by using statements (30.1) and (30.2) to reexpress the given inequalities as equivalent inequalities without absolute value symbols.

Example 30.3

Use statement (30.1) to solve $|x - 1| < 3$.

Solution

Statement (30.1) allows us to reexpress $|x - 1| < 3$ as the equivalent inequality.

$$-3 < x - 1 < 3$$

We now add 1 to all members, to obtain

$$-3 + 1 < x - 1 + 1 < 3 + 1$$
$$-2 < x < 4$$

Thus, $-2 < x < 4$ is the solution set for $|x - 1| < 3$.

Example 30.4

Use statement (30.2) to solve $|x - 2| \geq 4$.

Solution

Statement (30.2) allows us to reexpress $|x - 2| \geq 4$ as

$$x - 2 \leq -4 \quad \text{or} \quad x - 2 \geq 4$$

We now solve each inequality, to obtain

$$x \leq -4 + 2 \quad \text{or} \quad x \geq 4 + 2$$
$$x \leq -2 \quad \text{or} \quad x \geq 6$$

Thus, $x \leq -2$ or $x \geq 6$ is the solution set for $|x - 2| \geq 4$.

Example 30.5

Find the solution set for $|x^2 - x - 6| < 6$ and graph this set.

Solution

We use (30.1) to reexpress $|x^2 - x - 6| < 6$ as $-6 < x^2 - x - 6 < 6$. This compound inequality can be written as

$$-6 < x^2 - x - 6 \quad \text{and} \quad x^2 - x - 6 < 6$$

The solution set for the original inequality will be all the values of x that satisfy both of these inequalities at the same time.

The solution for $-6 < x^2 - x - 6$ is obtained as follows:

$$-6 < x^2 - x - 6$$
$$0 < x^2 - x$$
$$0 < x(x - 1)$$

which has $x < 0$ or $x > 1$ for a solution (the reader should use the sign graph method to verify this).

The solution for $x^2 - x - 6 < 6$ is obtained as follows:

$$x^2 - x - 6 < 6$$

$$x^2 - x - 12 < 0$$

$$(x - 4)(x + 3) < 0$$

which has $-3 < x < 4$ for a solution set (the reader should now use the sign method to verify this). The values of x that satisfy both will be those between -3 and 0 or those between 1 and 4. Thus, the solution for $|x^2 - x - 6| < 6$ is $-3 < x < 0$ or $1 < x < 4$.

Exercises for Section 30

In Exercises 1 through 20, find the solution set and graph this set.

1. $|x| \leq 8$

2. $|x| > 8$

3. $|x| \geq 3$

4. $|x| < 3$

5. $|x| < |-3|$

6. $|x| < -3$

7. $|x - 6| < 1$

8. $|3 - x| \leq 1$

9. $|x + 2| \geq 3$

10. $|x - 2| \geq 3$

11. $|3x - 2| < 1$

12. $|3x - 2| \geq 1$

13. $|5x - 4| \geq 6$

14. $|3 - 4x| < 6$

15. $2|x - 3| < 6$

16. $|x| - 4 \geq 3$

17. $|x^2 + 2x - 8| < 7$

18. $|x^2 + 2x - 8| > 7$

19. $|x^2 - 2x - 3| < 5$

20. $|x^2 - 2x - 3| > 5$

21. The speed of an object is the absolute value of its velocity. If the velocity V of an object is given by the formula $V = 96 - 32t$, where t is time in seconds, find the time for which the speed will be less than 64 ft/sec.

REVIEW QUESTIONS FOR CHAPTER 5

1. How do we read $a > b$; $a < b$; $a \geq b$; $a \leq b$?

2. State the addition and subtraction properties of inequalities.

3. State the multiplication and division properties of inequalities if these operations are performed using any positive real numbers.

4. State the multiplication and division properties of inequalities if these operations are performed using any negative real number.

5. Give an example of a compound inequality.

6. What is a linear inequality in one variable?

7. Give an example of an absolute inequality; a conditional inequality.

8. What do we mean by equivalent inequalities?

9. Give an example of a quadratic inequality.

10. Describe the sign graph used to find the solutions for a quadratic inequality.

11. When will the solution set for $ax^2 + bx + c > 0$ be all real values for x?

12. When will the solution set for $ax^2 + bx + c < 0$ be no real values for x?

13. When is the product of three factors positive? Negative?

14. When is the product of n factors positive? Negative?

15. Give an example of Theorem 29.2.

16. Interpret $|x| = a$; $|x| < a$; $|x| > a$.

17. Rewrite $|x| < a$ without absolute value signs.

18. Rewrite $|x| > a$ without absolute value signs.

REVIEW EXERCISES FOR CHAPTER 5

In Exercises 1 through 10, express each statement as an inequality.

1. 4 is greater than 2.

2. x is between -2 and 3.

3. $x + 1$ is less than 4.

4. $2x + 1$ is greater than $x - 2$.

5. $r + 1$ is less than $r - 1$.

6. $x^2 + y^2$ is nonnegative.

7. $x^2 + 1$ is positive.

8. $x + 1$ is between 2 and 5.

9. $x^2 + 1$ is between 0 and 5.

10. $x^2 - 1$ is greater than or equal to 0.

In Exercises 11 through 35, solve each inequality for x.

11. $x + 2 < 3$.

12. $x - 2 \geq 0$

13. $2 + x < 2x$

14. $\dfrac{3 - 2x}{4} \leq 1$

15. $2x + 1 \leq 5$

16. $x \leq \dfrac{x - 1}{2}$

17. $\frac{2 - 3x}{4} \le x$ **18.** $\frac{x - 1}{5} \ge \frac{2}{5}(x - 1)$ **19.** $3 \le 2x - 1 \le 9$

20. $-4 \le 2 - 3x \le 11$ **21.** $(x - 2)(x - 5) > 0$ **22.** $(x - 1)(x + 3) < 0$

23. $x^2 - 4x - 5 \le 0$ **24.** $x^2 + 5x + 4 \ge 0$ **25.** $x^2 + 4 \ge 0$

26. $x^2 - 4 \ge 0$ **27.** $(x - 1)(x - 2)(x + 5) \ge 0$

28. $(2x - 1)(x - 2)(x + 3) < 0$ **29.** $\frac{(x - 3)(x + 2)}{2x - 3} < 0$

30. $(x - 3)(2 - x)(2x + 5) \le 0$ **31.** $|x - 2| < 3$

32. $|x + 1| \le 2$ **33.** $|2x - 3| \ge 5$

34. $|1 - 2x| \le 3$ **35.** $|2 - 3x| > 1$

36. The radius of certain disk is between 2 and 2.01 in. What can be said about the area?

37. The velocity of an object traveling in a straight line is given by $V = 3t^2 + 4t - 4$. Find the values of t between 0 and 6 where the velocity is (a) positive and (b) negative.

38. The height h of an object thrown upward from the ground is a function of the time t and is given by $h = 80t - 16t^2$. Find the time interval when the height is greater than 96 ft.

39. The velocity V of an object thrown upward from the ground is a function of the time t and is given by the formula $V = 80 - 32t$. For what positive time interval will the velocity be positive?

40. Suppose that the length l of a rectangle is between 2 and 5 m and the width of the rectangle is 2 less than 3 times the length. Express the perimeter and the area of this rectangle as inequalities involving l.

CHAPTER 6

Systems of Equations

SECTION 31

Graphical Solution of Two Linear Equations on Two Unknowns

In this chapter we wish to consider the solution of systems of equations involving two or more equations in two unknowns. If all the equations in such a system have a common solution, they are called **simultaneous equations**.

In the first two sections of this chapter we shall consider the solution of a system of two linear equations in two unknowns. In Section 23 we described a linear equation in two unknowns (variables) x and y as

$$Ax + By + C = 0 \tag{31.1}$$

A solution of a system of two simultaneous linear equations that fit form (31.1) is any ordered pair of real values (x, y) that satisfies *both* equations. This pair of values is called a **simultaneous solution** of the given equations.

Example 31.1

The ordered pair (3, 0) is a simultaneous solution of the system

$$\begin{cases} 2x - y = 6 \\ x + y = 3 \end{cases}$$

since substitution of the values 3 for x and 0 for y satisfies both equations.

●

In Section 23 we saw that the graph of (31.1) was always a straight line. Thus the solution of a system of two linear equations in two unknowns is the coordinates (x, y) of the point or points of intersection of the graphs of these equations. We can plot the graph of each equation, and from this graph we can determine the coordinates of the common point or points of intersection. We call this method the **graphical solution** of the system.

Example 31.2

Solve the system of equations

$$\begin{cases} 3x - y = 6 \\ 2x + 3y = 12 \end{cases}$$

Solution

We use the method of Section 23 to determine that the x and y intercepts for $3x - y = 6$ are $(2, 0)$ and $(0, -6)$, respectively, and that the x and y intercepts for $2x + 3y = 12$ are $(6, 0)$ and $(0, 4)$, respectively. We use these intercepts to plot the graph of each equation. The two intersecting lines are shown in Figure 31.1. From this graph we estimate the coor-

Figure 31.1

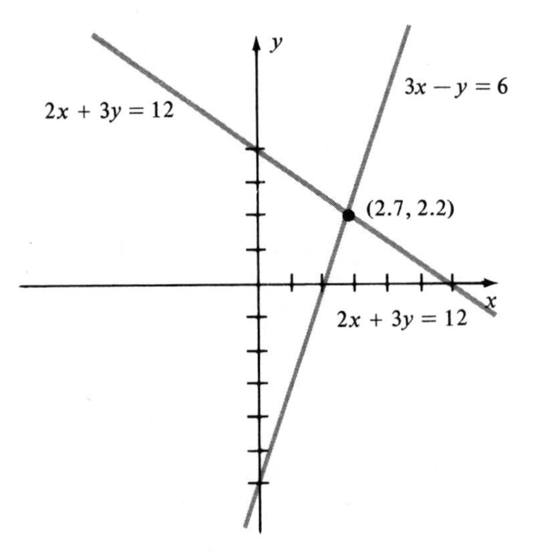

dinates of their point of intersection to be (2.7, 2.2). [We will show in Example 32.1 that the actual coordinates are $(\frac{30}{11}, \frac{24}{11})$.] Thus, the solution for this system is approximately $x = 2.7$ and $y = 2.2$.

Example 31.3

Solve the system of equations

$$\begin{cases} 3x - 5y = 10 \\ 4x - 3y = 6 \end{cases}$$

Solution
We find that the x and y intercepts of $3x - 5y = 10$ are $(\frac{10}{3}, 0)$ and $(0, -2)$, respectively, and that the x and y intercepts of $4x - 3y = 6$ are $(\frac{3}{2}, 0)$ and $(0, -2)$, respectively. We use these intercepts to plot the graph of each equation. The two intersecting lines are shown in Figure 31.2. From this graph we determine that the coordinates of their points of intersection to be exactly $(0, -2)$ (their common y intercept). Thus, the solution for this system is $x = 0$ and $y = -2$.

Figure 31.2

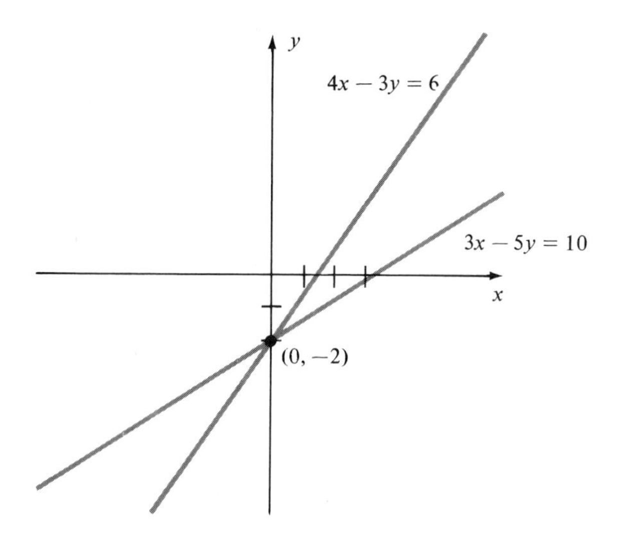

Example 31.4
Solve the system of equations

$$\begin{cases} x + y = 1 \\ x + y = 2 \end{cases}$$

Solution
We find that the intercepts for $x + y = 1$ are $(1, 0)$ and $(0, 1)$ and that the intercepts for $x + y = 2$ are $(2, 0)$ and $(0, 2)$. We use these points to plot the graph of each equation. The two lines are shown in Figure 31.3 (see page 138). We can see from Figure 31.3 that the two lines do not intersect. Thus, the system has no solution. We call this an **inconsistent system**.

Example 31.5
Solve the system of equations

$$\begin{cases} 2x + y = 2 \\ 4x + 2y = 4 \end{cases}$$

Figure 31.3

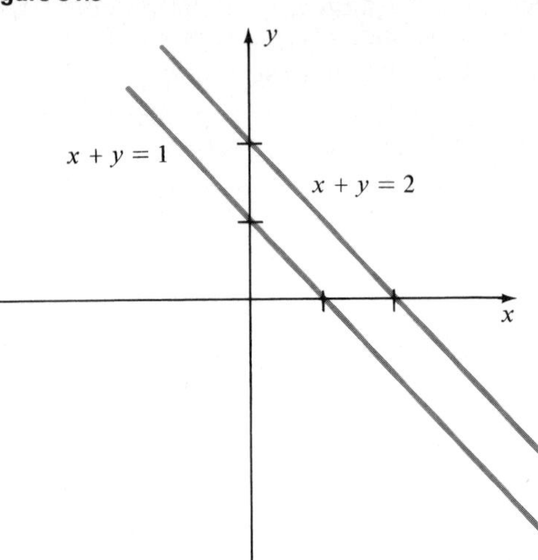

Solution

We see that the intercepts of $2x + y = 2$ are $(1, 0)$ and $(0, 2)$ and that the intercepts of $4x + 2y = 4$ are *also* $(1, 0)$ and $(0, 2)$. As a check we note that $(2, -2)$ satisfies both equations. This means that the two equations describe the *same* line. We show this graph in Figure 31.4. The solution for this system is the coordinates of *all* points on this line. We call this a **dependent system**.

Figure 31.4

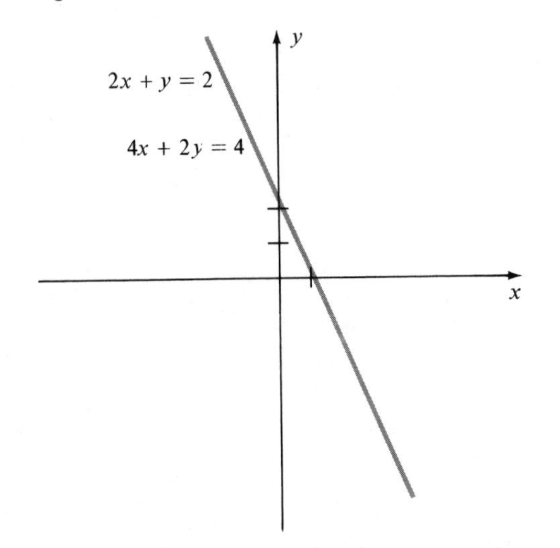

The graphical solutions for a system of equations are usually approximated (see Example 31.2) since they are based on sketches of the

respective graphs. To find accurate solutions in all cases we must develop algebraic methods. Even then it is desirable to plot the graphs as a rough check on the algebraic solution. In addition, a graphical solution for the system is considered sufficiently accurate for many applied problems.

In summary, a system of two linear equations in two unknowns will be:

1. *Consistent and independent* if its solution contains only one point (see Examples 31.2 and 31.3).
2. *Inconsistent* if it has no solution (see Example 31.4).
3. *Dependent* if it has every point on the common line for its solution (see Example 31.5).

Exercises for Section 31

In Exercises 1 through 12, solve each graphically. If necessary, estimate the solution to the nearest tenth.

1. $\begin{cases} 2x - 3y = 9 \\ 4x - y = 8 \end{cases}$

2. $\begin{cases} 3x - y = -6 \\ 2x + 3y = 7 \end{cases}$

3. $\begin{cases} 2x + y = 6 \\ x - y = 2 \end{cases}$

4. $\begin{cases} x + y = 2 \\ x - y + 3 = 0 \end{cases}$

5. $\begin{cases} x - y = 6 \\ x - y = 5 \end{cases}$

6. $\begin{cases} x + 2y = 1 \\ 2x + 4y = 2 \end{cases}$

7. $\begin{cases} 2y - x = 1 \\ 2x + y = 8 \end{cases}$

8. $\begin{cases} x + y = 12 \\ x - y = 6 \end{cases}$

9. $\begin{cases} x - y = 0 \\ x = 0 \end{cases}$

10. $\begin{cases} 2x + 3y = 6 \\ y - 3 = 0 \end{cases}$

11. $\begin{cases} 2x + 3y - 7 = 0 \\ y - 3 = 0 \end{cases}$

12. $\begin{cases} x - 4y = 2 \\ x - 2 = 0 \end{cases}$

13. The length of a rectangle is 3 times its width and the perimeter is 80 ft.
   (a) Express the given statements as a system of equations involving the length and the width w.
   (b) Find the dimensions of the rectangle by solving the system in part (a) graphically.

14. A boat can travel 7 miles downstream in 1 hr and return in 1.5 hr.
   (a) If b is the speed of the boat in still water and c is the speed of the current, express the given statements as a system of equations.
   (b) Solve the system in part (a) graphically for the speed of the boat in still water and the speed of the current.

SECTION 32

Algebraic Solution of Two Linear Equations in Two Unknowns

In this section we wish to develop algebraic methods for solving a system of two linear equations in two unknowns. One such method is called **substitution**. In this method we solve one of the equations for one of the variables in terms of the other, and then we substitute this result into the other equation. This procedure gives us a single equation in one variable. We can then solve this equation for that variable, and then we find the corresponding value for the other variable by substituting back into one of the two original equations.

Example 32.1

Solve the system of equations

$$\begin{cases} 3x - y = 6 \\ 2x + 3y = 12 \end{cases}$$

Solution

We first solve for y in the first equation to obtain $y = 3x - 6$, and then we substitute this result into the second equation, to obtain

$$\begin{aligned} 2x + 3(3x - 6) &= 12 \\ 2x + 9x - 18 &= 12 \\ 11x &= 30 \\ x &= \frac{30}{11} \end{aligned}$$

We can now substitute $x = \frac{30}{11}$ into either of the two original equations to obtain the corresponding value of y. Here we substitute $x = \frac{30}{11}$ into the second equation, to obtain

$$\begin{aligned} 2\left(\frac{30}{11}\right) + 3y &= 12 \\ 3y &= 12 - \frac{60}{11} \\ 3y &= \frac{132}{11} - \frac{60}{11} = \frac{72}{11} \\ y &= \frac{24}{11} \end{aligned}$$

Thus, the solution for this system is $x = \frac{30}{11}$ and $y = \frac{24}{11}$ or the ordered pair $(\frac{30}{11}, \frac{24}{11})$.

Example 32.2

Solve the system of equations

$$\begin{cases} x + y = 1 \\ x + y = 2 \end{cases}$$

Solution

We first solve for x in the first equation to obtain $x = 1 - y$, and then we substitute this result into the second equation, to obtain

$$(1 - y) + y = 2$$
$$1 = 2$$

Since $1 \neq 2$, this latter result is inconsistent. Thus, the system has **no solution**. The graphic interpretation of this situation is that the two lines are parallel. See Example 31.4 and Figure 31.3.

Example 32.3

Solve the system of equations

$$\begin{cases} 2x + y = 2 \\ 4x + 2y = 4 \end{cases}$$

Solution

We can solve for y in the first equation to obtain $y = 2 - 2x$, and then substitute this result into the second equation, to obtain

$$4x + 2(2 - 2x) = 4$$
$$4x + 4 - 4x = 4$$
$$0 = 0$$

Since $0 = 0$ is always true,, the two equations describe the *same* line. Thus, the solution for the system is the coordinates of any point on this line. See Example 31.5 and Figure 31.4. ●

When one of the variables in either equation of a system has a numerical coefficient of 1, that variable can be easily solved for and then used in the method of substitution. However, in a system where the numerical coefficients of the variables in the equations of the system are not equal to 1, the method of solution by **elimination** is generally more efficient than the method of substitution.

In this method we multiply each of the original equations by numbers that will make the coefficients of one of the variables numerically equal. We then subtract one of the new equations from the other to obtain an equation in one variable. We can solve this equation for that variable and then find the corresponding value for the other variable by substituting back into one of the original equations.

Example 32.4

Solve the system of equations

$$\begin{cases} 3x - 5y = 10 \\ 4x - 3y = 6 \end{cases}$$

by use of the elimination method.

Solution

We chose to eliminate the variable y. To do this, we multiply the first equation by 3 and the second equation by 5, to obtain the equivalent equations.

$$9x - 15y = 30$$

$$20x - 15y = 30$$

We now subtract one of the equivalent equations from the other, to obtain the equation

$$-11x = 0 \quad \text{or} \quad x = 0$$

We now solve for y by substituting $x = 0$ into the original first equation, to obtain

$$3(0) - 5y = 10$$

$$-5y = 10 \quad \text{or} \quad y = -2$$

Thus, the solution for this system is $x = 0$ and $y = -2$ or the ordered pair $(0, -2)$. See Example 31.3 and Figure 31.2. ●

In Section 20 we encountered word problems that could be solved by representing the unknown quantity by a single variable. We now apply systems of two equations in two variables to solve such word problems. Some word problems can be solved by either means.

Example 32.5

The resistance R (in ohms) is a linear function of temperature T (in degrees Celsius). If the resistance is 0.45 Ω at 50°C and 0.5 Ω at 80°C, find this linear relationship.

Solution

The linear relationship between R and T can be expressed as

$$R = mT + b$$

We must find the values for m and b. Since $R = 0.45$ when $T = 50$ and $R = 0.5$ when $T = 80$, we substitute these values into $R = mT + b$ to obtain the system of two equations in the two unknowns m and b.

$$\begin{cases} 0.45 = 50m + b \\ 0.5 \ \ = 80m + b \end{cases}$$

We now apply the elimination method to this system by subtracting the second equation from the first, to obtain

$$-0.05 = -30m$$

$$m = \frac{-0.05}{-30} = \frac{1}{600}$$

We now substitute $m = \frac{1}{600}$ into the first equation to find that $b = \frac{220}{600} = \frac{11}{30}$. Thus, the linear relationship is $R = \frac{1}{600}T + \frac{11}{30}$.

Example 32.6

A certain alloy contains 25% copper and 10% tin. How many pounds of copper and tin must be melted with 80 lb of the alloy to produce another alloy that contains 35% copper and 15% tin?

Solution

We let x = the number of pounds of copper and y = the number of pounds of tin. We note that there are 25% of 80, or 20, lb of copper in the given alloy and 10% of 80, or 8, lb of tin in the alloy. The fraction of copper in the new alloy must be the number of pounds of copper $20 + x$ divided by the number of pounds of the new alloy, $80 + x + y$, and this must equal 35%, or 0.35. We can express this as follows:

$$0.35 = \frac{20 + x}{80 + x + y} \tag{1}$$

Similarly, since we need 15% or 0.15 of tin, we have

$$0.15 = \frac{8 + y}{80 + x + y} \tag{2}$$

We can find the solution to our problem by solving equations (1) and (2) simultaneously. We convert equations (1) and (2) to equivalent equations and then solve this system by substitution as follows: The system

$$\begin{cases} 0.35 = \dfrac{20 + x}{80 + x + y} \\ 0.15 = \dfrac{8 + y}{80 + x + y} \end{cases}$$

is equivalent to the system

$$\begin{cases} 80(0.35) + 0.35x + 0.35y = 20 + x \\ 80(0.15) + 0.15x + 0.15y = \ \ 8 + y \end{cases}$$

or to the system

$$\begin{cases} -0.65x + 0.35y = -8 \\ 0.15x - 0.85y = -4 \end{cases}$$

If we multiply both equations in this system by 100, we obtain

$$\begin{cases} -65x + 35y = -800 \\ 15x - 85y = -400 \end{cases} \tag{A}$$

We now solve for y in equation (1) of system (A), to obtain $y = (65x-800)/35$ or $y = (13x-160)/7$, and then substitute this result into equation (2) as follows:

$$15x - 85\left(\frac{13x - 160}{7}\right) = -400$$

$$105x - 1105x + 13{,}600 = -2800$$

$$-1000x = -16{,}400$$

$$x = 16.4$$

We can now substitute $x = 16.4$ into equation (2) of system (A) to obtain the value of y as follows:

$$15(16.4) - 85y = -400$$

$$-85y = -400 - 246$$

$$-85y = -646$$

$$y = \frac{646}{85} \quad \text{or} \quad 7\frac{3}{5}$$

Thus, we need 16.4 lb of copper and 7.6 lb of tin. ●

A common type of stated problem, particularly in mechanics, is the work problem dealing with the lever. The pivot point of a lever at the point at which the lever is balanced is called the *fulcrum*. The distance along the lever from the fulcrum to the point at which the force is applied is called the *lever arm*. In general, a lever is balanced or in equilibrium when the length of one lever arm times the force applied on that side of the fulcrum equals the length of the other lever arm times the force applied on the other side of the fulcrum. We neglect the weight of the lever itself in dealing with this type of problem. We note from Figure 32.1 that $F_1 \times L_1 = F_2 \times L_2$.

Example 32.7

A lever is balanced on a fulcrum with weights of 30 lb at one end and 50 lb at the other end. The lever is still balanced when a 10-lb weight is added to the 30-lb weight and the 50-lb weight is moved 2 ft further from the fulcrum. Find the original lengths of the two arms of the lever.

Figure 32.1

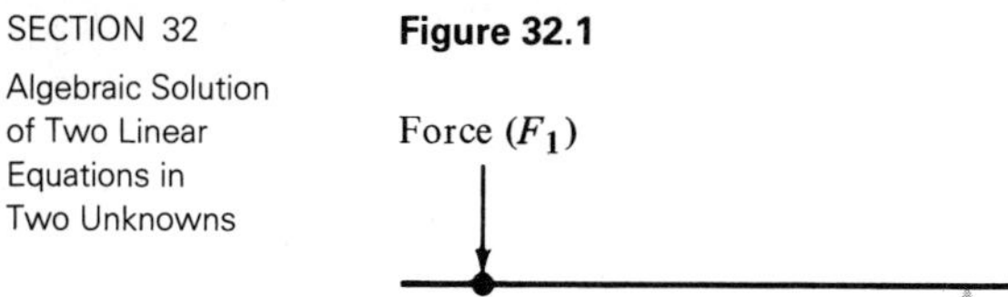

Solution

If we let x equal the distance from the fulcrum to the 30-lb weight and let y equal the distance from the fulcrum to the 50-lb weight, then from Figure 32.2 we have the equations

$$30x = 50y \quad \text{and} \quad (30 + 10)x = 50(y + 2)$$

which is equal to

$$\begin{cases} 30x - 50y = 0 \\ 40x - 50y = 100 \end{cases}$$

Subtracting, we obtain $-10x = -100$ or $x = 10$. Since $30x = 50y$, $300 = 50y$ or $y = 6$. Thus, the 30-lb weight is 10 ft from the fulcrum and the 50-lb weight is 6 ft from the fulcrum. As a check we note that $10 \cdot 30 = 6 \cdot 50$.

Figure 32.2

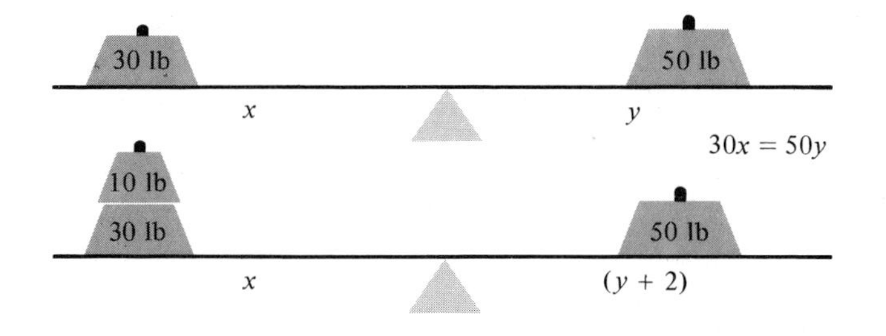

$(30 + 10)x = 50(y + 2)$

Example 32.8

The system

$$\begin{cases} \dfrac{3}{x} - \dfrac{1}{y} = -6 \\[2mm] \dfrac{2}{x} + \dfrac{3}{y} = 7 \end{cases}$$

can be transformed into the linear system

$$\begin{cases} 3u - v = -6 \\ 2u + 3v = 7 \end{cases}$$

if we let $u = 1/x$ and $v = 1/y$. We can then solve this system for the values of u and v and use these results to find the values of x and y.

Exercises for Section 32

For Exercises 1 through 12, solve each of Exercises 1 through 12 of Section 31 by the substitution method.

In Exercises 13 through 20, solve the given system by the substitution or the elimination method.

13. $\begin{cases} x + 2y = 6 \\ 3x - y = 4 \end{cases}$

14. $\begin{cases} 2x + 3y = 2 \\ 3x - 2y = 4 \end{cases}$

15. $\begin{cases} 8y - 4x = 6 \\ x - 3y = 5 \end{cases}$

16. $\begin{cases} \frac{4}{5}x + \frac{2}{5}y = 1 \\ \frac{3}{4}x + \frac{3}{8}y = 1 \end{cases}$

17. $\begin{cases} \frac{1}{4}x - \frac{1}{3}y = -\frac{3}{12} \\ \frac{1}{10}x + \frac{2}{5}y = \frac{2}{5} \end{cases}$

18. $\begin{cases} -6y = 19 \\ 2x - 3y = 1 \end{cases}$

19. $\begin{cases} 7x - 3y = 2 \\ 7x + 3y = 12 \end{cases}$

20. $\begin{cases} y = 4 - 2x \\ x + 5 = 3y \end{cases}$

In Exercises 21 through 24, use the procedure of example 32.8 to transform the system into a linear system and then solve each.

21. $\begin{cases} \frac{2}{x} - \frac{3}{y} = 7 \\ \frac{3}{x} + \frac{1}{y} = 5 \end{cases}$

22. $\begin{cases} \frac{4}{x} + \frac{2}{y} = 5 \\ \frac{5}{x} - \frac{3}{y} = -2 \end{cases}$

23. $\begin{cases} \frac{2}{y} - \frac{1}{x} = 1 \\ \frac{2}{x} + \frac{1}{y} = 8 \end{cases}$

24. $\begin{cases} \frac{2}{x} + \frac{1}{y} + 1 = 0 \\ \frac{3}{x} - \frac{2}{y} + 5 = 0 \end{cases}$

In Exercises 25 through 42, express the problems as systems of two linear equations in two unknowns and then solve algebraically.

25. When the first of two numbers is subtracted from twice the second, the result is 16. When the second number is added to 3 times the first, the sum is 36. Find the two numbers.

26. A freight train traveling at the rate of 25 mph leaves a station 2 hr before a passenger train which travels at the rate of 60 mph. How long will it take the passenger train to catch the freight train? How far from the station will they be?

27. A plane travels 495 miles in $1\frac{1}{2}$ hr with the aid of a tailwind and then makes the return trip in $1\frac{5}{6}$ hr. Find the speed of the wind and the speed of the plane.

28. How much water must be evaporated from 60 lb of a 10% solution of salt in order to obtain a 25% solution?

29. A radiator contains 25 quarts of water and an antifreeze solution which contains 40% antifreeze. How much of this solution should be drained and replaced with antifreeze to produce a solution that will contain 60% antifreeze?

30. A man invests part of $8000 at 6% simple interest and the rest at 4%. If his total annual income from the two investments is $420, how much did he invest at each rate?

31. A collection of nickels and dimes amounts to $2.15. If there are 30 coins in all, how many nickels are there?

32. One tank contains 32 gal of solution which is 25% alcohol. A second tank contains 50 gal of solution which is 40% alcohol. How many gallons should be taken from each tank and combined to make 40 gal of a solution containing 30% alcohol?

33. Find the equation of the linear function that contains the points $(1, 5)$ and $(-2, -4)$.

34. Find the coordinates of the vertices of the triangle formed by the lines $3x + 4y = 5$, $6x + y = 17$, and $x - y = -3$.

35. If the boiling point of water at 1 atmosphere is 212° Fahrenheit and 100° Celsius and the freezing point is 32° Fahrenheit and 0° Celsius, find the linear relationship between Fahrenheit and Celsius.

36. Solve Example 32.5 by expressing temperature T as a linear function of resistance R.

37. The resistance R of certain wire at any temperature T (in Celsius) is given by the formula $R = R_0 + \alpha R_0 T$, where R_0 is the resistance at 0°C and α is the coefficient of resistance. If the resistance is 27.3 Ω at 25°C and 28.4 Ω at 35°C, find R_0 and α.

38. An electrician and his apprentice work for 6 and 5 hr, respectively, and together they receive $77.75. On another occasion, the electrician works 7 hr and the apprentice 8 hr, and together they receive \$97.75. What are the hourly wages of each?

39. Two objects move at different but constant rates along a circle of circumference 330 ft. If they start at the same time and at the same place and move in the same direction, they pass each other every 22 sec. When they move in opposite directions, they pass each other every 6 sec. What are their respective rates?

40. Two input pipes together can fill a pool in 4 hr. If both pipes run for 2 hr and then the second pipe is shut off, it takes 6 more hours for the first pipe to fill the pool. How long will it take each pipe to fill the pool?

41. A lever is balanced on a fulcrum with weights that are placed at the ends. One weight is 6 ft from the fulcrum and the other is 5 ft from the fulcrum. If the lever is still balanced when 4 lb is added to the weight 6 ft from the fulcrum while the weight that was originally 5 ft from the fulcrum is moved 2 ft farther from the fulcrum, find the original weights.

42. When a weight of 50 lb is placed on one end of a fulcrum and a weight of 90 lb is placed on the other end, the lever is balanced. If 25 lb is added to the 50-lb weight, the 90-lb weight must be moved $2\frac{1}{2}$ ft farther away to balance the lever. Find the original lengths of the two arms of the lever.

SECTION 33

Graphical Solution of a System of Nonlinear Equations

In Section 31 we found that we could obtain a graphical solution to a system of two linear equations in two variables by graphing each equation and determining their common points of intersection from the graph. We also noted that frequently our solution would only be an approximate solution.

In this section we wish to discuss the graphical solution of a system of nonlinear equations in two variables. Any system of equations containing a nonlinear equation is called a **nonlinear system.**

Example 33.1

The system

$$\begin{cases} y = 9 - x^2 \\ y = x + 3 \end{cases}$$

is a nonlinear system, since $y = 9 - x^2$ is nonlinear. ●

A system such as the one in Example 33.1 can be solved graphically since the graph of the linear function $y = x + 3$ will be a straight line, and the graph of the quadratic function $y = 9 - x^2$ will be a parabola. The solution of the system will be their points of intersection.

Example 33.2

Solve the system

$$\begin{cases} y = 9 - x^2 \\ y = x + 3 \end{cases}$$

graphically.

Solution

We recognize $y = 9 - x^2$ as a quadratic function, and we use the techniques of Section 24 to determine that this is a parabola whose vertex and y intercept are located at $(0, 9)$ and whose x intercepts are $(\pm 3, 0)$. We use these results to sketch this parabola in Figure 33.1. We also recognize that $y = x + 3$ is a linear function, and we use the techniques of Section 23 to determine that its x and y intercepts are $(-3, 0)$ and $(0, 3)$, respectively. We sketch this straight line in Figure 33.1. From Figure 33.1 we see that the two graphs intersect at the points $(-3, 0)$ and $(2, 5)$. Thus, the solutions for the system are $x = -3$, $y = 0$ and $x = 2$, $y = 5$.

Figure 33.1

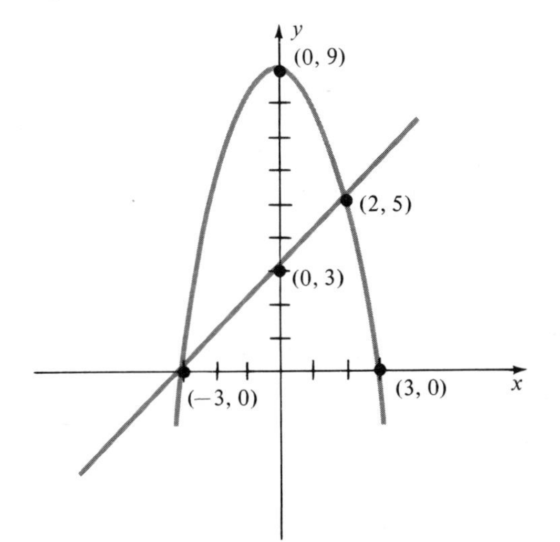

Example 33.3

Solve the system

$$\begin{cases} y = -3x^2 + 18x - 24 \\ y = 4x \end{cases}$$

graphically.

Solution

The graph of $y = 4x$ is a straight line with both its x and y intercepts located at $(0, 0)$. The graph of $y = -3x^2 + 18x - 24$ is a parabola breaking downward with x intercepts at $(2, 0)$ and $(4, 0)$ and vertex at $(3, 3)$. We sketch the graphs of each in Figure 33.2. From the figure we can see that the two graphs do not intersect. Thus, the system has *no solution.*

Figure 33.2

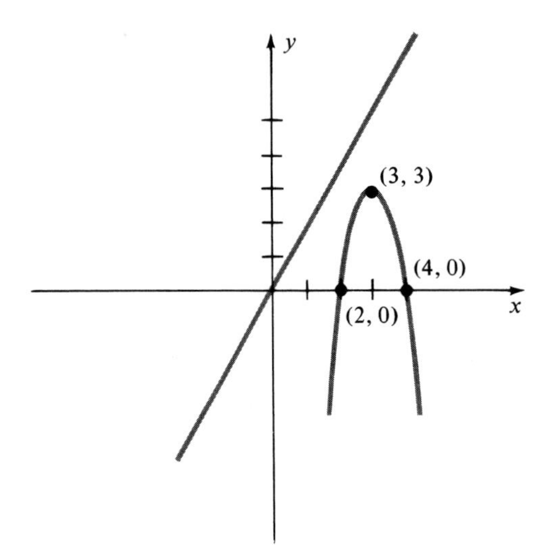

Example 33.4

Solve the system

$$\begin{cases} y = x^2 - 1 \\ y = 2x - 2 \end{cases}$$

graphically.

Solution

The graph of $y = 2x - 2$ is a straight line with intercepts at $(1, 0)$ and $(0, -2)$. The graph of $y = x^2 - 1$ is a parabola breaking upward with x intercepts at $(\pm 1, 0)$ and a vertex at $(0, -1)$. We show a sketch of both these graphs in Figure 33.3. From Figure 33.3 we see that the line is tangent to the parabola at $(1, 0)$. Thus, the solution for this system is $x = 1$ and $y = 0$. ●

If we are given a nonlinear system that contains a linear function and a quadratic function, the possibilities for the solution of the system are:

1. Two solutions (see Example 33.2).

Figure 33.3

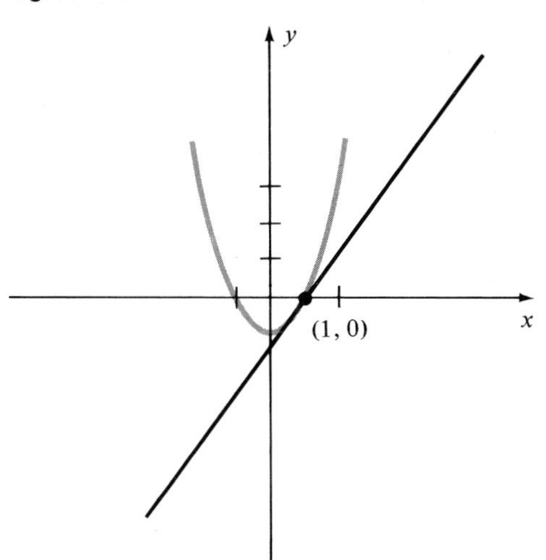

2. One solution (see Example 33.4).
3. No solution (see Example 33.3).

Example 33.5

Solve the system

$$\begin{cases} y = 1 + x^2 \\ y = 8 + 2x - x^2 \end{cases}$$

graphically.

Solution

The graph of $y = 1 + x^2$ is a parabola breaking upward with its vertex at $(0, 1)$. The graph of $y = 8 + 2x - x^2$ is a parabola breaking downward with intercepts at $(-2, 0)$, $(4, 0)$, and $(0, 8)$ and its vertex at $(1, 9)$. We sketch this parabola in Figure 33.4 on page 152. From this figure we estimate the coordinates of the points of intersection to be $(2.5, 7)$ and $(-1.5, 3)$. Thus, the approximate solutions for this system are $x = 2.5$, $y = 7$ and $x = -1.5$, $y = 3$. ●

In the nonlinear system of Example 33.5, both equations were quadratic functions. In general, if a nonlinear system contains two distinct quadratic functions, the possibilities for the solution of the system are:

1. Two solutions (see Figure 33.4).
2. One solution (see Figure 33.5a).
3. No solution (see Figure 33.5b).

Example 33.5 again points out the need to develop algebraic methods for solving such systems if we wish to obtain an exact solution. We shall develop these algebraic methods in the next section.

Figure 33.4

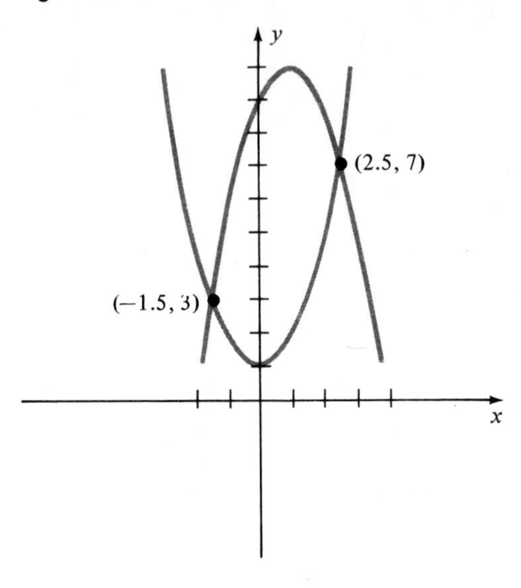

Figure 33.5a

Figure 33.5b

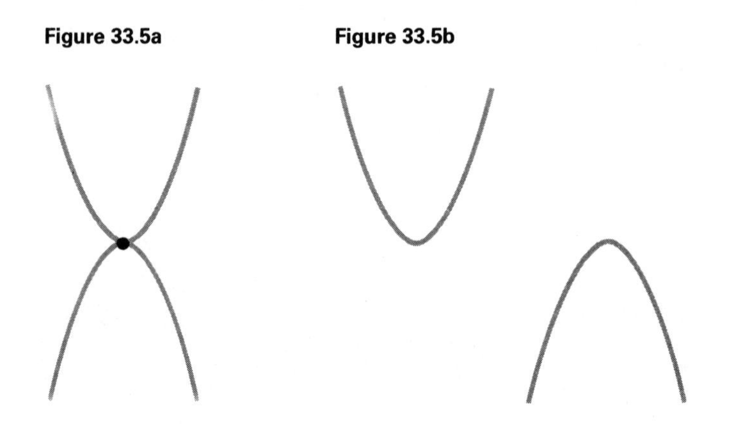

In this book we shall continue to study the graphs of many different types of functions. As our knowledge of graphing increases, so will our ability to solve nonlinear systems graphically. Eventually, we will encounter nonlinear systems that contain equations that are not algebraic and then we must rely on graphic solutions.

Exercises for Section 33

Solve each system graphically.

1. $\begin{cases} y = x^2 - 6x + 5 \\ y = 2x - 7 \end{cases}$

2. $\begin{cases} y = 3x - 3 \\ y = x^2 - 6x + 5 \end{cases}$

3. $\begin{cases} y = 6x - x^2 - 5 \\ y = 3x - 3 \end{cases}$

4. $\begin{cases} y = x^2 + 2x + 1 \\ y = 3 + x \end{cases}$

5. $\begin{cases} y = 3 - x \\ y = x^2 - 2x + 1 \end{cases}$

6. $\begin{cases} y = 4 + x^2 \\ y = 4 - x^2 \end{cases}$

7. $\begin{cases} y = x^2 + 2x + 1 \\ y + x = 0 \end{cases}$

8. $\begin{cases} y = x^2 - 2x + 1 \\ y = x^2 + 2x + 1 \end{cases}$

9. $\begin{cases} 2y + 3x = -6 \\ y = x^2 - 6x + 5 \end{cases}$

10. $\begin{cases} y = x^2 - 4x + 6 \\ y = 2 + 2x - x^2 \end{cases}$

SECTION 34

Algebraic Solution of a System of Nonlinear Equations

We can solve systems of nonlinear equations by the methods of substitution or elimination introduced in Section 32. We illustrate this with the following examples.

Example 34.1

Solve the system

$$\begin{cases} x - y = 3 \\ x = y^2 - 2y + 3 \end{cases}$$

Solution

We use the method of substitution by solving for x in the first equation to obtain $x = y + 3$ and then substituting this result into the second equation, to obtain

$$y + 3 = y^2 - 2y + 3$$
$$0 = y^2 - 3y$$

We can factor this quadratic equation in y and solve as follows:

$$0 = y(y - 3)$$
$$y = 0 \quad \text{or} \quad y - 3 = 0$$
$$y = 3$$

We now substitute each y value obtained into the first equation to find the corresponding x values. When we substitute $y = 0$ into the first equation of the original system, we obtain $x - 0 = 3$ or $x = 3$. When we substitute $y = 3$ into the first equation, we obtain $x - 3 = 3$ or $x = 6$. Thus, the solution for the system is the ordered pairs (3, 0) and (6, 3).

Example 34.2

Solve the system

$$\begin{cases} y = 1 + x^2 \\ y = 8 + 2x - x^2 \end{cases}$$

Solution

We can substitute $y = 1 + x^2$ from the first equation into the second equation, to obtain

$$1 + x^2 = 8 + 2x - x^2$$

or

$$2x^2 - 2x - 7 = 0$$

We now use the quadratic formula to solve this equation as follows:

$$\begin{aligned} x &= \frac{-(-2) \pm \sqrt{(-2)^2 - 4(2)(-7)}}{2(2)} \\ &= \frac{2 \pm \sqrt{60}}{4} \\ &= \frac{2 \pm 2\sqrt{15}}{4} \quad \text{or} \quad \frac{1 \pm \sqrt{15}}{2} \end{aligned}$$

We now substitute each of the x values obtained into the first equation to obtain the following corresponding y values. When

$$\begin{aligned} x &= \frac{1 + \sqrt{15}}{2} \\ y &= 1 + \left(\frac{1 + \sqrt{15}}{2}\right)^2 \\ &= 1 + \frac{1 + 2\sqrt{15} + 15}{4} \\ &= 1 + \frac{16 + 2\sqrt{15}}{4} \\ &= 1 + 4 + \frac{1}{2}\sqrt{15} \\ &= 5 + \frac{1}{2}\sqrt{15} \end{aligned}$$

$$\begin{aligned} x &= \frac{1 - \sqrt{15}}{2} \\ y &= 1 + \left(\frac{1 - \sqrt{15}}{2}\right)^2 \\ &= 1 + \frac{1 - 2\sqrt{15} + 15}{4} \\ &= 1 + \frac{16 - 2\sqrt{15}}{4} \\ &= 1 + 4 - \frac{1}{2}\sqrt{15} \\ &= 5 - \frac{1}{2}\sqrt{15} \end{aligned}$$

Thus, the exact solutions for the system are $x = \frac{1}{2} + \frac{1}{2}\sqrt{15}$, $y = 5 + \frac{1}{2}\sqrt{15}$ and $x = \frac{1}{2} - \frac{1}{2}\sqrt{15}$, $y = 5 - \frac{1}{2}\sqrt{15}$. This solution should be compared to the approximate solution for this system in Example 33.5.

Example 34.3

Solve the system

$$\begin{cases} x^2 + y^2 = 1 \\ y + x = 0 \end{cases}$$

Solution

We use the method of substitution by solving for y in the second equation to obtain $y = -x$ and then substitute this result into the first equation, to obtain

$$x^2 + (-x)^2 = 1$$
$$2x^2 = 1$$
$$x^2 = \frac{1}{2}$$
$$x = \pm\sqrt{\frac{1}{2}} \quad \text{or} \quad \pm\frac{1}{\sqrt{2}}$$

We now substitute each x value into equation $y + x = 0$ to obtain the corresponding y values. When

$$x = \frac{1}{\sqrt{2}} \qquad\qquad x = -\frac{1}{\sqrt{2}}$$

we have | we have

$$y + \frac{1}{\sqrt{2}} = 0 \qquad\qquad y - \frac{1}{\sqrt{2}} = 0$$

or | or

$$y = -\frac{1}{\sqrt{2}} \qquad\qquad y = \frac{1}{\sqrt{2}}$$

Thus, the solutions for the system are $x = 1/\sqrt{2}$, $y = -1/\sqrt{2}$ and $x = -1/\sqrt{2}$, and $y = 1/\sqrt{2}$. ●

The examples above suggest that we use the method of substitution for a nonlinear system containing two equations in two unknowns when one of the equations is linear and the other is quadratic (see Examples 34.2 and 34.3) or when both equations are quadratic in the same unknown (see Example 34.1).

We can also apply the method of elimination to some nonlinear systems.

Example 34.4

Solve the system

$$\begin{cases} 3x^2 - 2y^2 = 10 \\ 2x^2 + 3y^2 = 11 \end{cases}$$

Solution

We shall use the method of elimination. If we multiply the first equation by 2 and the second equation by 3, we obtain the equivalent system

$$\begin{cases} 6x^2 - 4y^2 = 20 \\ 6x^2 + 9y^2 = 33 \end{cases}$$

We now subtract the second equation from the first in this system, to obtain

$$-13y^2 = -13$$

$$y^2 = 1$$

$$y = \pm 1$$

We can now substitute each y value into the original first equation to obtain the following results. When $y = 1$,

$$3x^2 - 2 = 10$$

$$3x^2 = 12$$

$$x^2 = 4 \quad \text{or} \quad x = \pm 2$$

When $y = -1$,

$$3x^2 - 2 = 10$$

$$3x^2 = 12$$

$$x^2 = 4 \quad \text{or} \quad x = \pm 2$$

Thus, the solutions for the system are $x = 2$, $y = 1$; $x = -2$, $y = 1$; $x = 2$, $y = -1$; and $x = -2$, $y = -1$. ●

In general, if the nonlinear system contains two equations that are quadratic in both unknowns, we will solve the system by using the elimination method indicated in Example 34.4.

If we encounter a system of two nonlinear equations in two unknowns that are quadratic in both variables and also contain an xy term, we can use the technique shown in the following example.

Example 34.5

Solve the system

$$\begin{cases} x^2 + 2xy = 16 \\ 3x^2 - 4xy + 2y^2 = 6 \end{cases}$$

Solution

We first eliminate the constant term between both equations by multiplying the first equation by 3 and the second equation by 8 and then subtracting as follows:

$$3x^2 + 6xy = 48 \tag{1}$$

$$24x^2 - 32xy + 16y^2 = 48 \tag{2}$$

When we subtract (2) from (1), we have

$$-21x^2 + 38xy - 16y^2 = 0$$

or

$$21x^2 - 38xy + 16y^2 = 0 \tag{3}$$

We can now factor (3) as

$$(7x - 8y)(3x - 2y) = 0$$

$$7x - 8y = 0 \qquad 3x - 2y = 0$$

$$8y = 7x \qquad 2y = 3x$$

$$y = \frac{7}{8}x \qquad y = \frac{3}{2}x$$

We now substitute $y = \frac{7}{8}x$ into the first equation of the original system, to obtain

$$x^2 + 2x\left(\frac{7}{8}x\right) = 16$$

$$x^2 + \frac{7}{4}x^2 = 16$$

$$\frac{11}{4}x^2 = 16$$

$$x^2 = 16 \cdot \frac{4}{11}$$

$$x = \pm\frac{8}{\sqrt{11}}$$

When $x = \pm 8/\sqrt{11}$, then $y = \frac{7}{8}(\pm 8/\sqrt{11}) = \pm 7/\sqrt{11}$. Thus, two of the solutions are $x = 8/\sqrt{11}$, $y = 7/\sqrt{11}$ and $x = -8/\sqrt{11}$, $y = -7/\sqrt{11}$.

Next, we substitute $y = \frac{3}{2}x$ into the original equation (1), to obtain

$$x^2 + 2x\left(\frac{3}{2}x\right) = 16$$

$$x^2 + 3x^2 = 16$$

$$4x^2 = 16$$

$$x^2 = 4 \quad \text{or} \quad x = \pm 2$$

When $x = \pm 2$, then $y = \frac{3}{2}(\pm 2) = \pm 3$. Thus, we have two more solutions: $x = 2$, $y = 3$ and $x = -2$, $y = -3$. The solutions of the system are $x = 8/\sqrt{11}$, $y = 7/\sqrt{11}$; $x = -8/\sqrt{11}$, $y = -7/\sqrt{11}$; $x = 2$, $y = 3$; and $x = -2$, $y = -3$. ●

Many word problems can be solved by expressing the problem as a nonlinear system of equations and then solving the system.

Example 34.6

At a constant temperature the pressure P (in lb/in.^{2}) and the volume V (in in.^{3}) are related by the formula $PV = K$. The product of the pressure and the volume of a certain gas is 48 in.-lb. If the temperature remains constant as the pressure is decreased by 4 lb/in.^{2}, the volume is increased by 2 in.^{3}. What are the original pressure and volume of this gas?

Solution

We are given that $PV = 48$. We can express the new pressure as $P - 4$ and the new volume as $V + 2$. Thus, their product $(P - 4)(V + 2)$ must equal 48. The two equations we have form the nonlinear system

$$\begin{cases} PV = 48 \\ (P - 4)(V + 2) = 48 \end{cases}$$

or, equivalently,

$$\begin{cases} PV = 48 \\ PV - 4V + 2P - 8 = 48 \end{cases}$$

We can solve this system by substituting $PV = 48$ into the second equation to obtain

$$48 - 4V + 2P - 8 = 48$$
$$-4V + 2P - 8 = 0$$
$$P = 2V + 4$$

We now substitute this result into the first equation, to obtain

$$(2V + 4)V = 48$$
$$2V^2 + 4V - 48 = 0$$
$$(2V + 12)(V - 4) = 0$$
$$2V + 12 = 0 \qquad V - 4 = 0$$
$$V = -6 \qquad V = 4$$

Since volume V is to be positive, we reject the solution $V = -6$ and accept the solution $V = 4$ in.^{3}. Finally, we find the corresponding P value by substituting $V = 4$ into the equation $PV = 48$ to obtain $4P = 48$ or $P = 12$ lb/(in.)^2.

Exercises for Section 34

For Exercises 1 through 10, solve algebraically each of Exercises 1 through 10 of Section 33.

In Exercises 11 through 32, solve each nonlinear system algebraically.

11. $\begin{cases} x - 2y = 10 \\ y = x^2 + 2x - 15 \end{cases}$

12. $\begin{cases} y^2 = 9 - x \\ y = x - 9 \end{cases}$

13. $\begin{cases} y = 9 - x^2 \\ y = x^2 \end{cases}$

14. $\begin{cases} x = 9 - y^2 \\ x = y^2 \end{cases}$

15. $\begin{cases} y = x^2 - 4x + 7 \\ y = 3x - 8 \end{cases}$

16. $\begin{cases} 3x = 8 - 2y - y^2 \\ 5y - 3x = 0 \end{cases}$

17. $\begin{cases} x^2 + y^2 = 1 \\ y - x = 0 \end{cases}$

18. $\begin{cases} x^2 - y^2 = 1 \\ y + x = 0 \end{cases}$

19. $\begin{cases} 2x^2 - y^2 = 14 \\ x - y = 1 \end{cases}$

20. $\begin{cases} x^2 + 2y^2 = 24 \\ x = y^2 \end{cases}$

21. $\begin{cases} x^2 + y^2 = 9 \\ x^2 - y^2 = 9 \end{cases}$

22. $\begin{cases} x^2 - y^2 = 1 \\ x^2 + y^2 = 9 \end{cases}$

23. $\begin{cases} 3x^2 + 2y^2 = 35 \\ 4x^2 - 3y^2 = 24 \end{cases}$

24. $\begin{cases} 2y^2 - 3x^2 = 3 \\ 4y^2 - 5x^2 = 11 \end{cases}$

25. $\begin{cases} x + y = 1 \\ -xy = 12 \end{cases}$

26. $\begin{cases} xy = 1 \\ x + y = 2 \end{cases}$

27. $\begin{cases} y + 10x = 3xy \\ y = x + 2 \end{cases}$

28. $\begin{cases} xy + 2 = 0 \\ 5y = 6 - 4x \end{cases}$

29. $\begin{cases} x^2 - xy + y^2 = 21 \\ x^2 + 2xy - 8y^2 = 0 \end{cases}$

30. $\begin{cases} x^2 + 2xy = 16 \\ 5x^2 + 4y^2 = 48 \end{cases}$

31. $\begin{cases} 2y^2 = x^2 - 2 \\ 4xy = 3x^2 - 4 \end{cases}$

32. $\begin{cases} x^2 + 4 = xy \\ 2x^2 - 2xy + y^2 = 8 \end{cases}$

33. The sum of two numbers is 16 and their product is 63. Find the numbers.

34. The square of one number exceeds 3 times the square of a second number by 4. The sum of the squares of the two numbers is 260. Find the numbers.

35. A rectangle has a perimeter of 40 in. and an area of 96 in.^{2}. Find its dimensions.

36. Find the dimensions of a rectangle if its diagonal is 17 in. and its perimeter is 46 in.

37. The diagonal of a rectangle is 85 in. If the length of the rectangle is decreased by 7 in. and the width is increased by 11 in., the length of the diagonal remains the same. Find the dimensions of the rectangle.

38. Find the values of b so that the straight line $y = 2x + b$ will be tangent to the circle $x^2 + y^2 = 16$.

39. The annual income from an investment is \$600. If \$1000 more is invested and the rate is 1% less, the annual income is \$550. What are the amounts and rate of investment?

40. A rectangular piece of sheet metal has an area of 300 in.^{2}. A 3-in. square is cut from each corner and an open box is formed by turning up the ends and sides. If the volume of the box is 378 in.^{3}, what are the dimensions of the piece of tin?

SECTION 35

Algebraic Solution of Systems of Three Linear Equations in Three Unknowns

A solution to a linear equation in three variables such as $3x + 2y - z = -1$ is an **ordered triple** of numbers (x, y, z) that satisfy the given equation. Thus, the ordered triples $(1, -3, -2)$ and $(0, 0, 1)$ are solutions to the equation above but $(1, 2, 3)$ is not.

The solution of a system of three linear equations in three unknowns is any ordered triple (x, y, z) that satisfies *all* three equations simultaneously. When the solution of such a system consists of a single ordered triple, we say that the system is **consistent** and **independent**. When the system has no solution, we say that it is **inconsistent**, and when the system has an infinite number of solutions, we say that it is **dependent**. In this section we shall consider only the case where the system is consistent (i.e., has a unique solution).

The algebraic methods for finding the solution of such systems are analogous to those developed in Section 32 for finding the solution of a linear system of equations in two unknowns.

Example 35.1

Solve the system

$$\begin{cases} 3x + 2y - z = -1 \\ 2x - y + 2z = 1 \\ x + y - z = 0 \end{cases}$$

Solution

We solve the first equation for z to obtain $z = 3x + 2y + 1$, and then substitute this expression into the second and third equations to obtain the system

$$\begin{cases} 2x - y + 2(3x + 2y + 1) = 1 \\ x + y - (3x + 2y + 1) = 0 \end{cases} \quad \text{or} \quad \begin{cases} 8x + 3y = -1 \\ -2x - y = 1 \end{cases} \qquad \text{(A)}$$

The latter system, involving two equations in the variables x and y, can now be solved by further substitution. That is, we solve for y in the second equation of (A) to obtain $y = -1 - 2x$, and then we substitute this result into the first equation of (A), to obtain

$$\begin{aligned} 8x + 3(-1 - 2x) &= -1 \\ 8x - 3 - 6x &= -1 \\ 2x &= 2 \\ x &= 1 \end{aligned}$$

We can now find the corresponding y value by substituting $x = 1$ into the second equation of (A) to obtain $y = -1 - 2(1) = -3$. Finally, we substitute both $x = 1$ and $y = -3$ into the first equation of the original system, to obtain

$$\begin{aligned} 3(1) + 2(-3) - z &= -1 \\ 3 - 6 - z &= -1 \\ z &= -2 \end{aligned}$$

Thus, the solution for the system is $x = 1$, $y = -3$, and $z = -2$. The reader will find it constructive to check this result. ●

We refer to the method of solution used in Example 35.1 as the substitution method. Other procedures can be used in this method. For example, we could have begun by solving for x in the third equation and then substituted this result into the first two equations to obtain a system of two linear equations in y and z. We can then solve this system for y and z and then substitute these values into any one of the three original equations to find the corresponding x value.

We can also solve systems of three linear equations in three unknowns by a method of elimination that is analogous to the elimination method we used in Section 32. We illustrate this method in the following example.

Example 35.2

Solve the system

$$\begin{cases} 2x + 3y + z = 7 \\ 3x - 2y + 2z = -3 \\ 2x + y + z = 3 \end{cases}$$

Solution

We first eliminate the variable z in the first two equations by multiplying equation one by 2 and then subtracting the second equation from that result, as follows:

$$\begin{array}{r} 4x + 6y + 2z = 14 \\ 3x - 2y + 2z = -3 \\ \hline x + 8y = 17 \end{array} \tag{1}$$

We now eliminate the variable z from the first and third equations by subtracting as follows:

$$\begin{array}{r} 2x + 3y + z = 7 \\ 2x + y + z = 3 \\ \hline 2y = 4 \end{array} \tag{2}$$

Equations (1) and (2) now give us a system of two equations in x and y. We now solve for y in equation (2) to obtain $y = 2$ and then substitute this value into equation (1), to obtain

$$x + 8(2) = 17$$
$$x = 1$$

Finally, we substitute both $x = 1$ and $y = 2$ into the first equation of the original system, to obtain

$$2(1) + 3(2) + z = 7$$
$$z = 7 - 8 = -1$$

Thus, the solution for the system is $x = 1$, $y = 2$, and $z = -1$. ●

The solution of a system of three linear equations by the elimination method can be accomplished by the following steps:

Step 1: Use any pair of the equations to eliminate one of the unknowns.
Step 2: Use any *other* pair of the equations to eliminate the *same* unknown.
Step 3: Solve the system of two equations in two unknowns formed by steps 1 and 2.
Step 4: Substitute the values obtained for the two unknowns in step 3 into *any* one of the three original equations to solve for the third unknown.
Step 5: The solution should be checked by substituting the values into the three equations given.

Example 35.3

If machines A, B, and C are run together, they can complete a job in 2 hr. If only machines A and C are run together, they can complete the same job in $2\frac{2}{5}$ hr. And if only machines A and B are running together, they can complete the job in 4 hr. How long would it take each machine running alone to complete the job?

Solution

We let a = number of hours that machine A alone can complete the job, b = number of hours that machine B alone can complete the job, and c = number of hours that machine C alone can complete the job. Thus, in 1 hr machine A can complete $1/a$ parts of the job, machine B can complete $1/b$ parts of the job, and machine C can complete $1/c$ parts of the job. We can express machines A, B, and C completing the job together in 2 hr by the equation

$$2\left(\frac{1}{a} + \frac{1}{b} + \frac{1}{c}\right) = 1 \qquad \text{(complete job)} \tag{1}$$

Similarly, machines A and C completing the job in $2\frac{2}{5}$ hr can be expressed as

$$\frac{12}{5}\left(\frac{1}{a} + \frac{1}{c}\right) = 1 \qquad \text{(complete job)} \tag{2}$$

Finally, we can describe A and B, completing the job in 4 hr by the equation

$$4\left(\frac{1}{a} + \frac{1}{b}\right) = 1 \qquad \text{(complete job)} \tag{3}$$

We now must solve the system composed of equations (1), (2), and (3). This system can be transformed into a linear system of three equations in three unknowns by letting $U = 1/a$, $V = 1/b$, and $W = 1/c$ as follows:

$$\begin{cases} 2\left(\frac{1}{a} + \frac{1}{b} + \frac{1}{c}\right) = 1 \\ \frac{12}{5}\left(\frac{1}{a} + \frac{1}{c}\right) = 1 \\ 4\left(\frac{1}{a} + \frac{1}{b}\right) = 1 \end{cases} \quad \text{or} \quad \begin{cases} U + V + W = \frac{1}{2} \\ U \qquad + W = \frac{5}{12} \\ U + V \qquad = \frac{1}{4} \end{cases}$$

To solve this system we first eliminate W by subtracting the second equation from the first and we obtain $V = \frac{1}{2} - \frac{5}{12} = \frac{1}{12}$. We now substitute $V = \frac{1}{12}$ into the third equation to obtain $U + \frac{1}{12} = \frac{1}{4}$ or $U = \frac{3}{12} - \frac{1}{12} = \frac{1}{6}$. Finally, we substitute the values $V = \frac{1}{12}$ and $U = \frac{1}{6}$ into equation (1) to obtain $\frac{1}{6} + \frac{1}{12} + W = \frac{1}{2}$ or $W = \frac{1}{2} - \frac{1}{6} - \frac{1}{12} = \frac{3}{12} = \frac{1}{4}$. We now solve for a, b, and c as follows:

$$U = \frac{1}{a} \qquad V = \frac{1}{b} \qquad W = \frac{1}{c}$$

$$\frac{1}{6} = \frac{1}{a} \qquad \frac{1}{12} = \frac{1}{b} \qquad \frac{1}{4} = \frac{1}{c}$$

$$a = 6 \qquad b = 12 \qquad c = 4$$

Thus, machine A can complete the job in $a = 6$ hr; machine B in $b = 12$ hr; and machine C in $c = 4$ hr. We can check this solution by substituting $a = 6$, $b = 12$, and $c = 4$ into equations (1), (2), and (3) to obtain the following true statements:

$$2\left(\frac{1}{6} + \frac{1}{12} + \frac{1}{4}\right) = 2\left(\frac{6}{12}\right) = 1$$

$$\frac{12}{5}\left(\frac{1}{6} + \frac{1}{4}\right) = \frac{12}{5}\left(\frac{10}{24}\right) = 1$$

$$4\left(\frac{1}{6} + \frac{1}{12}\right) = 4\left(\frac{3}{12}\right) = 1$$

Exercises for Section 35

In Exercises 1 through 12, solve each system for the unknowns indicated.

1. $\begin{cases} 2x - y + z = 1 \\ 3x + 2z = -1 \\ 4x + y + 2z = 2 \end{cases}$

2. $\begin{cases} 3x + 4y - 2z = -4 \\ x + 2y + 3z = 9 \\ 2x - y + 2z = 11 \end{cases}$

3. $\begin{cases} U + V + W = 3 \\ 3U + 4V + 2W = 4 \\ 2U + 3V - W = -5 \end{cases}$

4. $\begin{cases} 2a + 3b + c = 2 \\ 6a + 6b + 2c = 5 \\ 3a - 3b - \frac{1}{4}c = 0 \end{cases}$

5. $\begin{cases} x + 3y = 0 \\ z + 3y = 0 \\ x + y + z = -5 \end{cases}$

6. $\begin{cases} 2a - 2b + 3c = 1 \\ a - 3b - 2c = -9 \\ a + b + c = 6 \end{cases}$

7. $\begin{cases} 2s - 2r + 3t = -5 \\ 3s + r + t = 4 \\ 2s - 3r - t = -3 \end{cases}$

8. $\begin{cases} 2U - V + 2W = 6 \\ V + W = 4 \\ W = 1 \end{cases}$

9. $\begin{cases} 2x - y + 2z = -8 \\ x + 2y - 3z = 9 \\ 3x - y - 4z = 3 \end{cases}$

10. $\begin{cases} x + y + z = -2 \\ 2x - y + 2z = -10 \\ x - y - 3z = 3 \end{cases}$

11. $\begin{cases} \frac{2}{a} + \frac{5}{b} + \frac{3}{c} = 7 \\ \frac{3}{a} + \frac{2}{b} - \frac{4}{c} = -2 \\ \frac{5}{a} + \frac{9}{b} - \frac{7}{c} = 5 \end{cases}$

12. $\begin{cases} \frac{2}{x} - \frac{2}{y} + \frac{3}{z} = -5 \\ \frac{3}{x} + \frac{1}{y} + \frac{1}{z} = 4 \\ \frac{2}{x} - \frac{3}{y} - \frac{1}{z} = -3 \end{cases}$

13. Find the quadratic function $y = ax^2 + bx + c$ so that its graph contains the points $(1, -2)$, $(0, -3)$, and $(-1, -6)$.

14. Find the quadratic function $y = ax^2 + bx + c$ so that its graph contains the points (0, 0), (1, −7), and (−2, −40).

15. The sum of three numbers is 34. The second number is equal to the sum of the first and third numbers and the third number exceeds twice the first number by 2. Find the numbers.

16. A collection of 33 coins consisting of nickels, dimes, and quarters has a value of $3.30. If there are three times as many nickels as quarters and one half as many dimes as nickels, how many coins of each kind are there?

17. A portion of $20,000 is invested at 5%, another portion at 6%, and a third at 9%. The annual income from the three investments is $1460 and the income from the 9% investment exceeds the sum of the incomes from the other two investments by $340. Find the amounts invested at each rate.

18. Three pipes, A, B, and C, can fill a pool together in $1\frac{2}{3}$ hr. If only pipes A and C are used, it takes $2\frac{1}{2}$ hr to fill the pool, and if only pipes B and C are used, it takes $3\frac{3}{4}$ hr to fill it. How long will it take each pipe alone to fill the pool?

19. By weight, alloy A is 10% copper, alloy B is 30% copper, and alloy C is 50% copper. If all the alloys are to be mixed, using twice as much of alloy C as alloy B to form 250 lb of an alloy containing 40% copper, how many pounds of each alloy should be used?

20. The height h (feet) of an object above the ground is given by the formula $h = at^2 + bt + c$, where t is time in seconds. The height of the object is 92 when $t = 1$, 140 when $t = 2$, and 156 when $t = 3$. Find the height when $t = 5$.

21. The perimeter of a triangle is 165 in. The side a is 55 in. shorter than the side b, and side b is 20 in. longer than side c. Find the lengths of the sides of the triangle.

22. Three resistors, R_1, R_2, and R_3, in series have a total resistance of 5000 Ω. If R_1 and R_2 in series have a resistance of 3200 Ω, and R_1 and R_3 in series have a resistance of 2400 Ω, find the resistance of each.

23. Three resistors, R_1, R_2, and R_3, in series have a total resistance of 2400 Ω. If the resistance of R_1 is 300 Ω more than 4 times R_2 and the resistance of R_3 is 450 Ω more than R_2, find the resistance of each.

24. Three voltages, E_1, E_2, and E_3, in series with the same polarity have a total voltage of 1000 V. If the polarity of only E_2 is reversed, the total voltage drops to 600 V. If the polarity of only E_3 is reversed, the total voltage is 450 V. Find each voltage.

25. In the analysis of series and parallel electrical circuits, Kirchhoff's two basic laws are:

(a) At any instant, the algebraic sum of the currents flowing toward any point in a circuit is equal to zero.
(b) At any instant, the sum of the voltage (including electromotive forces) around any closed circuit is algebraically equal to zero.

Applying these laws to the electric circuit shown in Figure 35.1 leads to the following system of linear equations. Solve this system of equations. [In Figure 35.1, I signifies current (in amperes), Ω signifies ohms, and V signifies volts.]

$$\begin{cases} I_1 + I_3 = I_2 \\ 2I_1 + 3I_2 = 4 \\ 3I_2 + I_3 = 3 \end{cases}$$

Figure 35.1

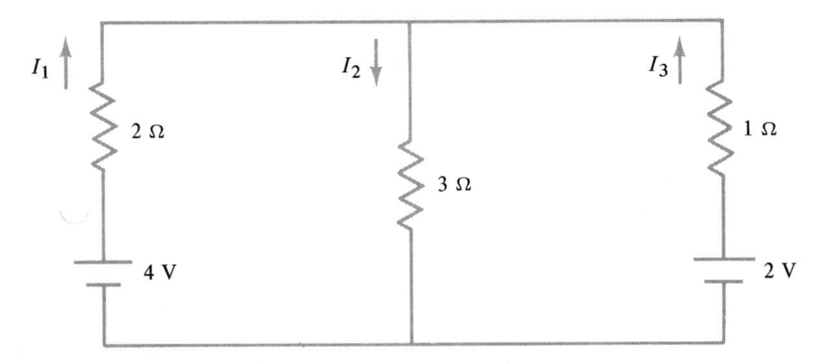

Figure 35.2

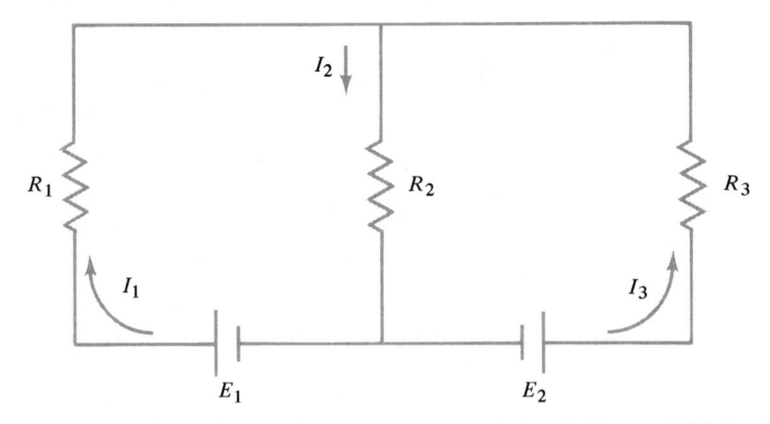

26. Applying Kirchhoff's laws to the circuit of Figure 35.2 leads us to the system of linear equations

$$\begin{cases} I_1 + I_3 = I_2 \\ R_1I_1 + R_2I_2 = E_1 \\ R_2I_2 + R_3I_3 = E_2 \end{cases}$$

If $R_1 = 2$, $R_2 = 4$, and $R_3 = 5\ \Omega$, and $E_1 = 4$ V and $E_2 = 6$ V, find the current in amperes in each branch of the circuit; that is, solve the system for I_1, I_2, and I_3.

27. In the circuit of Problem 26, find I_1, I_2, and I_3 if $R_1 = 4$, $R_2 = 5$, and $R_3 = 10\ \Omega$ and $E_1 = 3$ V and $E_2 = 6$ V.

28. Use Kirchhoff's laws and the equations shown in Exercise 26 to write a system of linear equations that applies to the circuit shown in Figure 35.3.

Figure 35.3

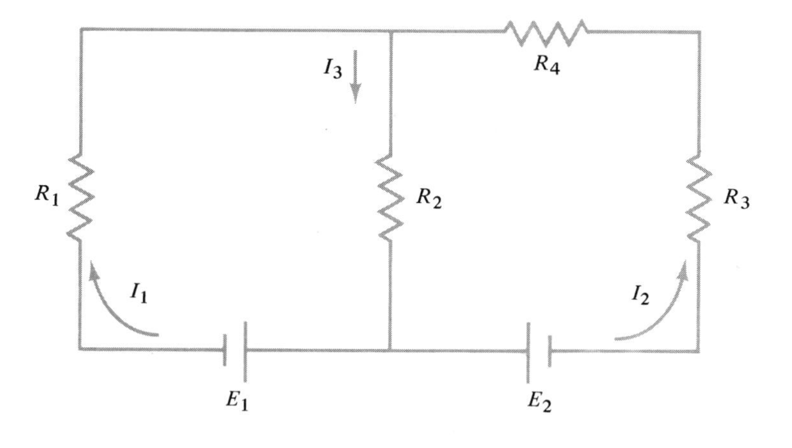

SECTION 36

Solution of a Linear System of Two Equations in Two Unknowns by Using Determinants

In Section 32 we presented two algebraic methods for solving the system of two linear equations in two unknowns. In this section we shall present yet another method which will eventually enable us to solve a system of n linear equations in n unknowns.

Consider the system

$$\begin{cases} a_1x + b_1y = c_1 & (1) \\ a_2x + b_2y = c_2 & (2) \end{cases} \qquad (36.1)$$

We can use the elimination method of Section 32 to solve for x by multiplying equation (1) by b_2 and equation (2) by $-b_1$ and then adding to obtain

$$\begin{array}{r}
a_1b_2x + b_1b_2y = c_1b_2 \\
-a_2b_1x - b_1b_2y = -c_2b_1 \\
\hline
(a_1b_2 - a_2b_1)x = c_1b_2 - c_2b_1
\end{array} \tag{3}$$

Similarly, we can solve for y in the original system by multiplying equation (1) by $-a_2$ and equation (2) by a_1 and then adding, to obtain

$$\begin{array}{r}
-a_1a_2x - a_2b_1y = -a_2c_1 \\
a_1a_2x + a_1b_2y = a_1c_2 \\
\hline
(a_1b_2 - a_2b_1)y = a_1c_2 - a_2c_1
\end{array} \tag{4}$$

If $a_1b_2 - a_2b_1 \neq 0$, we can solve equations (3) and (4) for x and y, respectively, to obtain

$$x = \frac{c_1b_2 - c_2b_1}{a_1b_2 - a_2b_1} \tag{5}$$

$$y = \frac{a_1c_2 - a_2c_1}{a_1b_2 - a_2b_1} \tag{6}$$

Solutions (5) and (6) can be expressed in a convenient way if we introduce a new symbol.

The symbol

$$\begin{vmatrix} a_1 & b_1 \\ a_2 & b_2 \end{vmatrix}$$

consisting of the four numbers a_1, a_2, b_1, b_2 arranged in two horizontal rows and two vertical columns is called a **second-order determinant**. The four numbers are called **elements** of the determinant.

By definition we have

$$\begin{vmatrix} a_1 & b_1 \\ a_2 & b_2 \end{vmatrix} = a_1b_2 - a_2b_1 \tag{36.2}$$

In (36.2) the elements a_1 and b_2 are the elements of the **principal diagonal** and the elements a_2 and b_1 are the elements of the **secondary diagonal**. Thus, formula (36.2) indicates that the value of the second-order determinant is equal to the product of the elements in the principal diagonal minus the product of the elements in the secondary diagonal.

Example 36.1

Evaluate

$$\begin{vmatrix} 3 & 2 \\ 4 & 1 \end{vmatrix}$$

Solution

We use formula (36.2) as follows:

$$\begin{vmatrix} 3 & 2 \\ 4 & 1 \end{vmatrix} = 3(1) - 4(2) = 3 - 8 = -5$$

Example 36.2

Evaluate

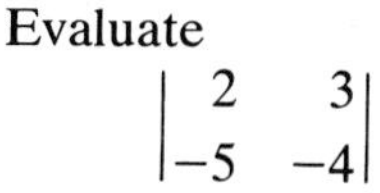

$$\begin{vmatrix} 2 & 3 \\ -5 & -4 \end{vmatrix}$$

Solution

We use formula (36.2) as follows:

$$\begin{vmatrix} 2 & 3 \\ -5 & -4 \end{vmatrix} = 2(-4) - (-5)(3) = -8 + 15 = 7$$

Example 36.3

Evaluate

$$\begin{vmatrix} 4 & 6 \\ 2 & 3 \end{vmatrix}$$

Solution

We use formula (36.2) as follows:

$$\begin{vmatrix} 4 & 6 \\ 2 & 3 \end{vmatrix} = 4(3) - 2(6) = 12 - 12 = 0 \qquad \bullet$$

We can use the determinant symbol to rewrite the solutions (5) and (6) of system (36.1) as follows:

$$x = \frac{\begin{vmatrix} c_1 & b_1 \\ c_2 & b_2 \end{vmatrix}}{\begin{vmatrix} a_1 & b_1 \\ a_2 & b_2 \end{vmatrix}} \qquad y = \frac{\begin{vmatrix} a_1 & c_1 \\ a_2 & c_2 \end{vmatrix}}{\begin{vmatrix} a_1 & b_1 \\ a_2 & b_2 \end{vmatrix}} \tag{36.3}$$

The forms of (36.3) are easy to remember if we note that the determinant of the denominator consists of the coefficients of x and y in (36.1). This determinant is referred to as the **determinant of the coefficients** and is denoted by the symbol Δ. The determinants of the numerator in the solution for either variable are the same as Δ, except that the column of coefficients of the variable to be determined is replaced by the column of constants on the right side of (36.1). The results of (36.3) are a special case of **Cramer's rule**.

We refer to this method as the method of solving systems of equations by determinants.

Example 36.4

Solve the following system of equations using determinants.

$$\begin{cases} 2x - y = 7 \\ 3x + 5y = 4 \end{cases}$$

Solution

We solve for x and y by using the formula of (36.3) as follows:

$$x = \frac{\begin{vmatrix} 7 & -1 \\ 4 & 5 \end{vmatrix}}{\begin{vmatrix} 2 & -1 \\ 3 & 5 \end{vmatrix}} = \frac{7(5) - 4(-1)}{2(5) - 3(-1)} = \frac{35 + 4}{10 + 3} = \frac{39}{13} = 3$$

$$y = \frac{\begin{vmatrix} 2 & 7 \\ 3 & 4 \end{vmatrix}}{\begin{vmatrix} 2 & -1 \\ 3 & 5 \end{vmatrix}} = \frac{2(4) - 3(7)}{2(5) - 3(-1)} = \frac{8 - 21}{10 + 3} = \frac{-13}{13} = -1$$

Thus, the solution for the system is $x = 3$, $y = -1$.

Example 36.5

Solve the following system of equations by using determinants.

$$\begin{cases} 2x = 3y + 1 \\ 5y = 4x - 2 \end{cases}$$

Solution

We must first rewrite the given system in the form (36.1) and apply the formulas of (36.3) as follows. The system is rewritten as

$$\begin{cases} 2x - 3y = 1 \\ -4x + 5y = -2 \end{cases}$$

Now, from (36.3) we have

$$x = \frac{\begin{vmatrix} 1 & -3 \\ -2 & 5 \end{vmatrix}}{\begin{vmatrix} 2 & -3 \\ -4 & 5 \end{vmatrix}} = \frac{1(5) - (-2)(-3)}{2(5) - (-4)(-3)} = \frac{5 - 6}{10 - 12} = \frac{-1}{-2} = \frac{1}{2}$$

$$y = \frac{\begin{vmatrix} 2 & 1 \\ -4 & -2 \end{vmatrix}}{\begin{vmatrix} 2 & -3 \\ -4 & 5 \end{vmatrix}} = \frac{2(-2) - (-4)(1)}{2(5) - (-4)(-3)} = \frac{-4 + 4}{10 - 12} = \frac{0}{-2} = 0$$

Thus, the solution for the system is $x = \frac{1}{2}$, $y = 0$.

Example 36.5 points out the fact that we must *always* express the given system in the standard form of (36.1) before applying the formulas of (36.3).

If both the determinants of the numerator and denominator in (36.3) are zero, the system is **dependent**. To illustrate this, we will solve the dependent system of Example 32.3 by using determinants.

Example 36.6

Solve the following system by using determinants.

$$\begin{cases} 2x + y = 2 \\ 4x + 2y = 4 \end{cases}$$

Solution

We apply formula (36.3) to solve for x and y as follows:

$$x = \frac{\begin{vmatrix} 2 & 1 \\ 4 & 2 \end{vmatrix}}{\begin{vmatrix} 2 & 1 \\ 4 & 2 \end{vmatrix}} = \frac{2(2) - 1(4)}{2(2) - 1(4)} = \frac{4 - 4}{4 - 4} = \frac{0}{0}$$

$$y = \frac{\begin{vmatrix} 2 & 2 \\ 4 & 4 \end{vmatrix}}{\begin{vmatrix} 2 & 1 \\ 4 & 2 \end{vmatrix}} = \frac{2(4) - 4(2)}{2(2) - 1(4)} = \frac{8 - 8}{4 - 4} = \frac{0}{0}$$

●

If the determinant of the numerator is nonzero and the determinant of the denominator is zero, the system is inconsistent. To illustrate this we shall solve the *inconsistent* system of Example 32.3 by using determinants.

Example 36.7

Solve the following system by using determinants.

$$\begin{cases} x + y = 1 \\ x + y = 2 \end{cases}$$

Solution

We apply formulas (36.3) to solve for x and y as follows:

$$x = \frac{\begin{vmatrix} 1 & 1 \\ 2 & 1 \end{vmatrix}}{\begin{vmatrix} 1 & 1 \\ 1 & 1 \end{vmatrix}} = \frac{1(1) - 2(1)}{1(1) - 1(1)} = \frac{1 - 2}{1 - 1} = \frac{-1}{0}$$

$$y = \frac{\begin{vmatrix} 1 & 1 \\ 1 & 2 \end{vmatrix}}{\begin{vmatrix} 1 & 1 \\ 1 & 1 \end{vmatrix}} = \frac{1(2) - 1(1)}{1(1) - 1(1)} = \frac{2 - 1}{1 - 1} = \frac{1}{0}$$

Example 36.8

An investor has an annual income of \$1440 from investments in bonds bearing 6% and 8%. If the amounts invested at 6% and 8% were interchanged, he would earn \$80 less. Find the total amount invested.

Solution

We let a = amount invested at 6% and b = amount invested at 8%. Thus, $0.06a + 0.08b$ represents the annual income and $0.06b + 0.08a$ represents the income when the amounts are reversed. Thus, we have the following system of equations:

$$\begin{cases} 0.06a + 0.08b = 1440 \\ 0.06b + 0.08a = 1440 - 80 \end{cases}$$

We first express this system in standard form as

$$\begin{cases} 0.06a + 0.08b = 1440 \\ 0.08a + 0.06b = 1360 \end{cases}$$

We now use determinants to solve for a and b as follows:

$$a = \frac{\begin{vmatrix} 1440 & 0.08 \\ 1360 & 0.06 \end{vmatrix}}{\begin{vmatrix} 0.06 & 0.08 \\ 0.08 & 0.06 \end{vmatrix}} = \frac{1440(0.06) - 1360(0.08)}{(0.06)(0.06) - (0.08)(0.08)}$$

$$= \frac{86.4 - 108.8}{0.0036 - 0.0064} = \frac{-22.4}{-0.0028} = 8000$$

$$b = \frac{\begin{vmatrix} 0.06 & 1440 \\ 0.08 & 1360 \end{vmatrix}}{\begin{vmatrix} 0.06 & 0.08 \\ 0.08 & 0.06 \end{vmatrix}} = \frac{(0.06)(1360) - (0.08)(1440)}{-0.0028} = \frac{81.6 - 115.2}{-0.0028}$$

$$= \frac{-33.6}{-0.0028} = 12{,}000$$

Thus, the sum of both investments is $a + b = 8000 + 12{,}000 = 20{,}000$ dollars.

Exercises for Section 36

In Exercises 1 through 9, evaluate each determinant.

1. $\begin{vmatrix} 2 & 3 \\ -1 & 4 \end{vmatrix}$

2. $\begin{vmatrix} -4 & -5 \\ 2 & 6 \end{vmatrix}$

3. $\begin{vmatrix} 44 & -4 \\ 11 & -1 \end{vmatrix}$

4. $\begin{vmatrix} 8 & 7 \\ 6 & 5 \end{vmatrix}$

5. $\begin{vmatrix} -4 & -6 \\ -9 & -8 \end{vmatrix}$

6. $\begin{vmatrix} 2 & 3 \\ 3 & 2 \end{vmatrix}$

7. $\begin{vmatrix} 0 & 1 \\ -2 & 0 \end{vmatrix}$ **8.** $\begin{vmatrix} 3 & 5 \\ 6 & 10 \end{vmatrix}$ **9.** $\begin{vmatrix} 4 & 3 \\ 7 & -6 \end{vmatrix}$

For Exercises 10 through 21, solve Exercises 1 through 12 of Section 31 by using determinants.

For Exercises 22 through 30 solve Exercises 13 through 21 of Section 32 by using determinants.

Solve Exercises 31 through 35 by using determinants.

31. Two cars start together and travel in the same direction, one going $1\frac{1}{2}$ times as fast as the other. At the end of 2 hr they are 26 miles apart. How fast is each car traveling?

32. How many pounds of an alloy containing 40% tin must be melted with an alloy containing 60% tin to obtain 80 lb of an alloy containing 52% tin?

33. An investor has twice as much money invested in bonds at 5% as he has invested in stocks at 7%. How much does he have invested in bonds if his annual income from the investments is $510?

34. Find the equation of the line $y = mx + b$ if the line contains the points $(-3, 2)$ and $(4, 7)$.

35. The sum of the reciprocals of two numbers is 8. Three times the reciprocal of the first is 1 less than twice the reciprocal of the second. Find the numbers.

SECTION 37

Solution of a Linear System of Three Equations in Three Unknowns by Using Determinants

In Section 35 we found the solution of a linear system of three equations in three unknowns by using the methods of substitution and elimination. In this section we wish to develop a method for solving these systems by using determinants.

Consider the system

$$\begin{cases} a_1c + b_1y + c_1z = d_1 \\ a_2x + b_2y + c_2z = d_2 \\ a_3x + b_3y + c_3z = d_3 \end{cases} \tag{37.1}$$

If we apply the elimination method to this system, we obtain the following solutions:

$$x = \frac{d_1b_2c_3 + d_2b_3c_1 + d_3b_1c_2 - d_3b_2c_1 - d_2b_1c_3 - d_1b_3c_2}{a_1b_2c_3 + a_2b_3c_1 + a_3b_1c_2 - a_3b_2c_1 - a_2b_1c_3 - a_1b_3c_2} \tag{37.2}$$

$$y = \frac{a_1d_2c_3 + a_2d_3c_1 + a_3d_1c_2 - a_3d_2c_1 - a_2d_1c_3 - a_1d_3c_2}{a_1b_2c_3 + a_2b_3c_1 + a_3b_1c_2 - a_3b_2c_1 - a_2b_1c_3 - a_1b_3c_2} \tag{37.3}$$

$$z = \frac{a_1b_2d_3 + a_2b_3d_1 + a_3b_1d_2 - a_3b_2d_1 - a_2b_1d_3 - a_1b_3d_2}{a_1b_2c_3 + a_2b_3c_1 + a_3b_1c_2 - a_3b_2c_1 - a_2b_1c_3 - a_1b_3c_2} \tag{37.4}$$

The solutions (37.2), (37.3), and (37.4) can be put in a more convenient form if we introduce the following determinant of third order.

The symbol

$$\begin{vmatrix} a_1 & b_1 & c_1 \\ a_2 & b_2 & c_2 \\ a_3 & b_3 & c_3 \end{vmatrix}$$

consisting of nine numbers arranged in three rows and three columns is called a **determinant of third order**. By definition we have

$$\begin{vmatrix} a_1 & b_1 & c_1 \\ a_2 & b_2 & c_2 \\ a_3 & b_3 & c_3 \end{vmatrix} = a_1b_2c_3 + b_1c_2a_3 + c_1a_2b_3 - c_1b_2a_3 - a_1c_2b_3 - b_1a_2c_3 \tag{37.5}$$

We can remember definition (37.5) by using the following scheme. Rewrite the first two columns on the right of the determinant as shown in Figure 37.1. Then:

Step 1. Form the products of the principal diagonal and the two diagonals parallel to it and find their sum.
Step 2. Form the products of the secondary diagonal and the two diagonals parallel to it and find their sum.
Step 3. Subtract the sum of step 2 from the sum of step 1.

Figure 37.1

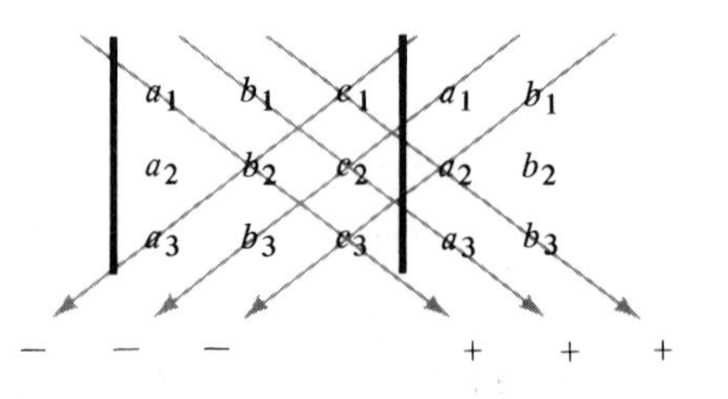

Example 37.1

Evaluate

$$\begin{vmatrix} 1 & -1 & 2 \\ -1 & 3 & 1 \\ 2 & -1 & 1 \end{vmatrix}$$

Solution

We rewrite the first two columns and expand, to obtain

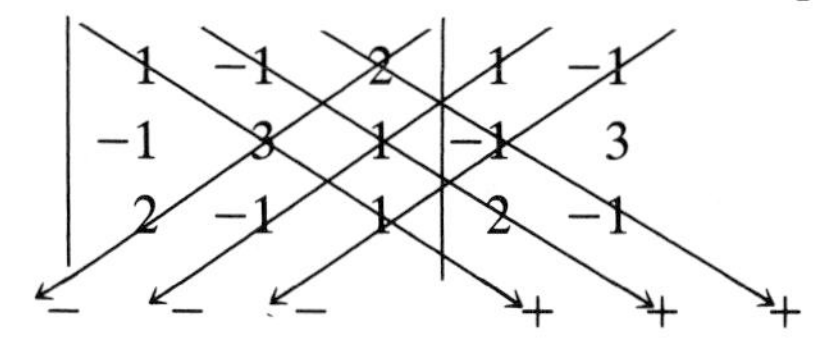

Thus, the value of the given determinant is

$$\begin{aligned}(1)(3)(1) + (-1)(1)(2) + (2)(-1)(-1) - (2)(3)(2) - (1)(1)(-1) \\ - (-1)(-1)(1)\end{aligned}$$

$$= 3 - 2 + 2 - 12 + 1 - 1 = -9$$

●

We can now reexpress the solution [(37.2), (37.3), and (37.4)] of the system (37.1) in terms of third-order determinants.

$$x = \frac{\begin{vmatrix} d_1 & b_1 & c_1 \\ d_2 & b_2 & c_2 \\ d_3 & b_3 & c_3 \end{vmatrix}}{\Delta} \qquad y = \frac{\begin{vmatrix} a_1 & d_1 & c_1 \\ a_2 & d_2 & c_2 \\ a_3 & d_3 & c_3 \end{vmatrix}}{\Delta} \qquad z = \frac{\begin{vmatrix} a_1 & b_1 & d_1 \\ a_2 & b_2 & d_2 \\ a_3 & b_3 & d_3 \end{vmatrix}}{\Delta} \tag{37.6}$$

where

$$\Delta = \begin{vmatrix} a_1 & b_1 & c_1 \\ a_2 & b_2 & c_2 \\ a_3 & b_3 & c_3 \end{vmatrix}$$

is the determinant of the coefficients and $\Delta \neq 0$.

Formula (37.6) is another special case of Cramer's rule. If $\Delta \neq 0$, the solution to the given system is a unique ordered triple of numbers. Formula (37.6) is easy to remember in that the determinant in the numerator for each variable is the same as Δ except that the column of coefficients of the variable to be determined is replaced by the column of constants on the right side of (37.1).

Example 37.2

Solve the following system by using determinants.

$$\begin{cases} 3x - y + 2z = 8 \\ x + 2y - 3z = -7 \\ 2x - y + z = 5 \end{cases}$$

Solution

We first calculate the determinant of the numerical coefficients Δ, to obtain

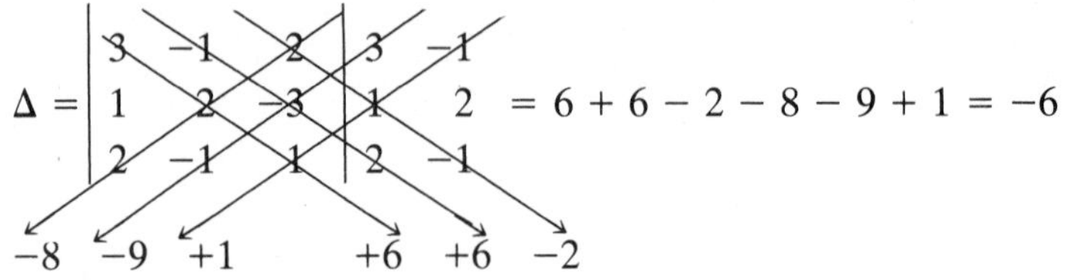

$$\Delta = \begin{vmatrix} 3 & -1 & 2 \\ 1 & 2 & -3 \\ 2 & -1 & 1 \end{vmatrix} \begin{matrix} 3 & -1 \\ 1 & 2 \\ 2 & -1 \end{matrix} = 6 + 6 - 2 - 8 - 9 + 1 = -6$$

$-8 \quad -9 \quad +1 \qquad +6 \quad +6 \quad -2$

Now we use formula (37.5) to calculate x, y, and z, as follows:

$$x = \frac{\begin{vmatrix} 8 & -1 & 2 \\ -7 & 2 & -3 \\ 5 & -1 & 1 \end{vmatrix} \begin{matrix} 8 & -1 \\ -7 & 2 \\ 5 & -1 \end{matrix}}{\Delta}$$

$$= \frac{16 + 15 + 14 - 20 - 24 - 7}{-6} = \frac{-6}{-6} = 1$$

$$y = \frac{\begin{vmatrix} 3 & 8 & 2 \\ 1 & -7 & -3 \\ 2 & 5 & 1 \end{vmatrix} \begin{matrix} 3 & 8 \\ 1 & -7 \\ 2 & 5 \end{matrix}}{\Delta}$$

$$= \frac{-21 - 48 + 10 + 28 + 45 - 8}{-6} = \frac{6}{-6} = -1$$

$$z = \frac{\begin{vmatrix} 3 & -1 & 8 \\ 1 & 2 & -7 \\ 2 & -1 & 5 \end{vmatrix} \begin{matrix} 3 & -1 \\ 1 & 2 \\ 2 & -1 \end{matrix}}{\Delta}$$

$$= \frac{30 + 14 - 8 - 32 - 21 + 5}{-6} = \frac{-12}{-6} = 2$$

Thus, the solution for the system is $x = 1$, $y = -1$, and $z = 2$. ●

It should be noted that the technique described by the schematic in Figure 37.1 for evaluating a third-order determinant is only valid for a third order. Also, the system must be in the standard form of (37.1) before Cramer's rule can be applied.

Example 37.3

Solve the following system by using determinants.

$$\begin{cases} 3u = 2v + 8 \\ 3v = 3 - 2w \\ u + 2v = 0 \end{cases}$$

Solution

We first rewrite the system in the standard form of (37.1), to obtain

$$\begin{cases} 3u - 2v = 8 \\ 3v + 2w = 3 \\ u + 2v = 0 \end{cases}$$

We next calculate Δ as follows:

$$\Delta = \begin{vmatrix} 3 & -2 & 0 \\ 0 & 3 & 2 \\ 1 & 2 & 0 \end{vmatrix} \begin{matrix} 3 & -2 \\ 0 & 3 \\ 1 & 2 \end{matrix} = 0 - 4 + 0 + 0 - 12 + 0 = -16$$

Now we use formula (37.6) to calculate u, v, and w, as follows:

$$u = \frac{\begin{vmatrix} 8 & -2 & 0 \\ 3 & 3 & 2 \\ 0 & 2 & 0 \end{vmatrix} \begin{matrix} 8 & -2 \\ 3 & 3 \\ 0 & 2 \end{matrix}}{\Delta} = \frac{0 + 0 + 0 + 0 - 32 + 0}{-16} = \frac{-32}{-16} = 2$$

$$v = \frac{\begin{vmatrix} 3 & 8 & 0 \\ 0 & 3 & 2 \\ 1 & 0 & 0 \end{vmatrix} \begin{matrix} 3 & 8 \\ 0 & 3 \\ 1 & 0 \end{matrix}}{\Delta} = \frac{0 + 16 + 0 + 0 + 0 + 0}{-16} = \frac{16}{-16} = -1$$

$$w = \frac{\begin{vmatrix} 3 & -2 & 8 \\ 0 & 3 & 3 \\ 1 & 2 & 0 \end{vmatrix} \begin{matrix} 3 & -2 \\ 0 & 3 \\ 1 & 2 \end{matrix}}{\Delta} = \frac{0 - 6 + 0 - 24 - 18 + 0}{-16} = \frac{-48}{-16} = 3$$

Thus, the solution for the system is $u = 2$, $v = -1$, and $w = 3$. ●

From Example 37.3 we should note that if a variable is missing in any of the equations, its coefficient is zero and a zero is placed in the appropriate position in the determinant.

Example 37.4

Kirchhoff's laws for the network in Figure 37.2 (see page 178) yield the following system of equations.

$$\begin{cases} I_1 - I_2 - I_3 = 0 \\ 20I_2 - 10I_3 = 0 \\ 40I_1 + 20I_2 = E_0 \end{cases}$$

Solve for the currents I_1, I_2, and I_3.

Figure 37.2

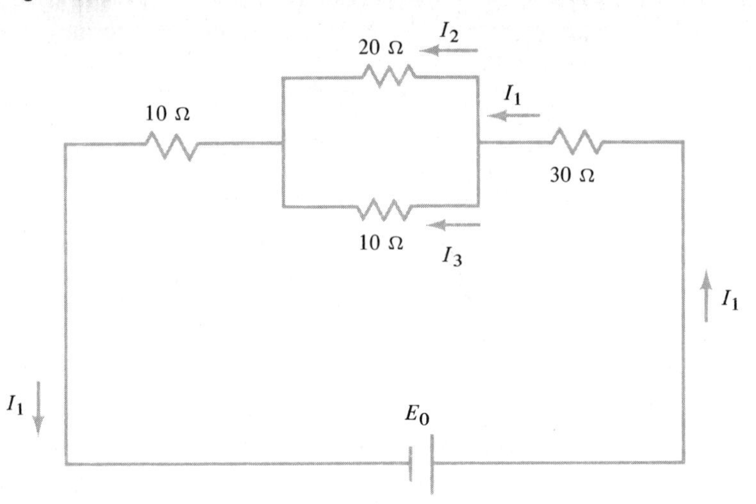

Solution

We first calculate the determinant of the coefficient as follows:

$$\Delta = \left|\begin{matrix} 1 & -1 & -1 \\ 0 & 20 & -10 \\ 40 & 20 & 0 \end{matrix}\right|\begin{matrix} 1 & -1 \\ 0 & 20 \\ 40 & 20 \end{matrix} = 0 + 400 + 0 + 800 + 200 + 0$$
$$= 1400$$

We now use formulas (37.6) to calculate I_1, I_2, and I_3.

$$I_1 = \frac{\left|\begin{matrix} 0 & -1 & -1 \\ 0 & 20 & -10 \\ E_0 & 20 & 0 \end{matrix}\right|\begin{matrix} 0 & -1 \\ 0 & 20 \\ E_0 & 20 \end{matrix}}{\Delta} = \frac{0 + 10E_0 + 0 + 20E_0 + 0 + 0}{1400}$$
$$= \frac{30E_0}{1400} = \frac{3}{140}E_0 \text{ amperes}$$

$$I_2 = \frac{\left|\begin{matrix} 1 & 0 & -1 \\ 0 & 0 & -10 \\ 40 & E_0 & 0 \end{matrix}\right|\begin{matrix} 1 & 0 \\ 0 & 0 \\ 40 & E_0 \end{matrix}}{\Delta} = \frac{0 + 0 + 0 + 0 + 10E_0 + 0}{1400}$$
$$= \frac{10E_0}{1400} = \frac{1}{140}E_0 \text{ amperes}$$

$$I_3 = \frac{\left|\begin{matrix} 1 & -1 & 0 \\ 0 & 20 & 0 \\ 40 & 20 & E_0 \end{matrix}\right|\begin{matrix} 1 & -1 \\ 0 & 20 \\ 40 & 20 \end{matrix}}{\Delta} = \frac{20E_0 + 0 + 0 + 0 + 0 + 0}{1400}$$
$$= \frac{20E_0}{1400} = \frac{1}{70}E_0 \text{ amperes}$$

Thus, the solution for the system is $I_1 = \frac{3}{140}E_0$, $I_2 = \frac{1}{140}E_0$, and $I_3 = \frac{1}{70}E_0$ amperes.

Cramer's rule can be extended to four linear equations in four unknowns and in general to n linear equations in n unknowns. Such systems will have a unique solution if the determinant of the coefficients is not equal to zero.

Exercises for Section 37

For Exercises 1 through 12, solve Exercises 1 through 12 of Section 35 by using determinants.

13. Find the equation of the quadratic function $y = ax^2 + bx + c$ whose graph contains the points (1, 10), (−1, 4), and (0, 1).

14. Three solutions containing 20, 30, and 50% sulfuric acid are to be mixed to form 100 ounces of a solution containing 43% sulfuric acid. Twice as much of the 30% solution as the 20% solution must be used. How many ounces of each solution should be used?

15. A manufacturing company uses three types of machines, A, B, and C. Each machine produces three products, x, y, and z, each day. Total production per machine per day is shown in the following table.

	Machine		
Product	A	B	C
x	2	2	2
y	0	3	4
z	2	1	1

How many machines of each type are in use if the total production per day is 18 units of product x, 25 units of product y, and 11 units of product z?

16. When Kirchhoff's laws are applied to the electrical circuit shown in Figure 37.3 on page 180, the following system of equations is generated.

$$\begin{cases} I_1 - I_2 - I_3 = 0 \\ 5I_1 + 10I_3 = 55 \\ 8I_2 - 10I_3 = 20 \end{cases}$$

Solve for I_1, I_2, and I_3.

17. A delivery company owns three types of trucks. The three trucks together make 130 deliveries per day. The deliveries of the first truck is

Figure 37.3

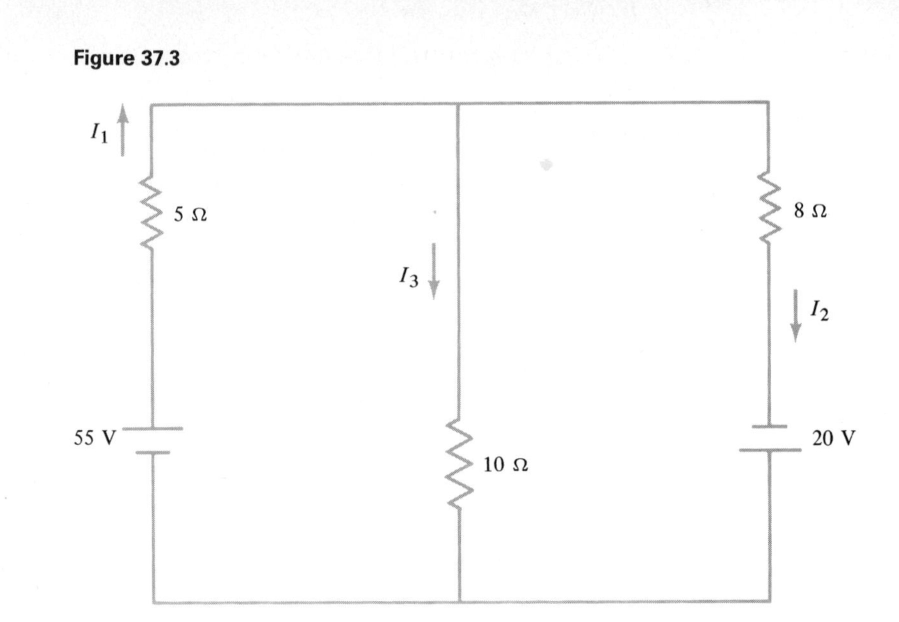

equal to 10 less than the deliveries of the other two together. The first truck makes twice as many deliveries as the third truck. How many deliveries did each truck make?

18. To control ice and protect the environment a certain city determines that the best mixture to be spread on roads consists of 8 units of salt, 4 units of sand, and 2.5 units of a chemical inhibiting agent. Three companies, *A*, *B*, and *C*, sell mixtures of these elements according to the following table.

	Salt	*Sand*	*Inhibiting Agent*
Company *A*	4	2	1
Company *B*	3	2	1
Company *C*	2	0	1

How much should a city purchase from each company in order to spread the best mixture? (Assume a purchase from all three companies.)

REVIEW QUESTIONS FOR CHAPTER 6

1. What is a solution of a system of two simultaneous linear equations in two unknowns?

2. Describe the *graphical* solution of a system of two linear equations in two unknowns.

3. What is a consistent system of two linear equations in two unknowns? An inconsistent system? A dependent system?

4. Explain the method of *substitution* used for solving a system of two linear equations in two unknowns.

5. What can be said about the graph of a system of equations that has no solution?

6. Explain the method of *elimination* used for solving a system of two linear equations in two unknowns.

7. Describe a nonlinear system of equations.

8. State the possibilities for the solution set of a nonlinear system that contains a linear function and a quadratic function or two quadratic functions.

9. What is an ordered triple?

10. Explain the terms "independent," "dependent," and "inconsistent" in regard to a system of three linear equations in three unknowns.

11. Write a second-order determinant.

12. How do we find the value of a second-order determinant?

13. Describe the process of evaluating a third-order determinant.

REVIEW EXERCISES FOR CHAPTER 6

In Exercises 1 through 18, solve each system of equations graphically.

1. $\begin{cases} x - 3y = 9 \\ x + y = 1 \end{cases}$

2. $\begin{cases} 4x - 3y = 12 \\ x + y = -4 \end{cases}$

3. $\begin{cases} 2x - y + 3 = 0 \\ x - y - 1 = 0 \end{cases}$

4. $\begin{cases} x - 3y - 6 = 0 \\ 4x - y + 9 = 0 \end{cases}$

5. $\begin{cases} 2x + y = 5 \\ \dfrac{x}{3} - y = 2 \end{cases}$

6. $\begin{cases} x + 2y = 5 \\ 3x + y = 0 \end{cases}$

7. $\begin{cases} 3x + y = 7 \\ 2x - y = 3 \end{cases}$

8. $\begin{cases} 2x + y - 13 = 0 \\ x - y - 2 = 0 \end{cases}$

9. $\begin{cases} 3x + y + 3 = 0 \\ 2x + y + 1 = 0 \end{cases}$

10. $\begin{cases} 3x + y - 17 = 0 \\ 3x + 5y = 1 \end{cases}$

11. $\begin{cases} 3x - 2y = 6 \\ 2x + 2y - 4 = 0 \end{cases}$

12. $\begin{cases} 4x - 3y - 9 = 0 \\ 6x + y - 19 = 0 \end{cases}$

13. $\begin{cases} 3x + 5y - 7 = 0 \\ x - 4y + 9 = 0 \end{cases}$

14. $\begin{cases} y = -x^2 + x + 2 \\ y = x - 2 \end{cases}$

15. $\begin{cases} y = 2 - x^2 \\ y = x \end{cases}$

16. $\begin{cases} y = x - 2 \\ y = x^2 - 3x + 2 \end{cases}$

17. $\begin{cases} y = 9 - x^2 \\ y = x^2 - 4x - 7 \end{cases}$

18. $\begin{cases} y = 4 - x^2 \\ y = x^2 - 2x \end{cases}$

In Exercises 19 through 36, solve each system of equations algebraically.

19. $\begin{cases} 3x + y - 2 = 0 \\ x + y - 5 = 0 \end{cases}$

20. $\begin{cases} 3x + y = 7 \\ 2x - y = 3 \end{cases}$

21. $\begin{cases} 3x + 2y = 14 \\ 5x - 3y = 17 \end{cases}$

22. $\begin{cases} x - 2y + 4 = 0 \\ 2x - 4y + 5 = 0 \end{cases}$

23. $\begin{cases} 4x - 3y - 1 = 0 \\ 3x + 7y - 10 = 0 \end{cases}$

24. $\begin{cases} x + 3y + 6 = 0 \\ x - 2y - 14 = 0 \end{cases}$

25. $\begin{cases} 2x - y + 3 = 0 \\ 6x - 3y + 9 = 0 \end{cases}$

26. $\begin{cases} x - y - 2 = 0 \\ 2x - y - 9 = 0 \end{cases}$

27. $\begin{cases} y = x^2 - 4x + 5 \\ y = 3x - 1 \end{cases}$

28. $\begin{cases} x^2 + y^2 = 8 \\ y - x = 0 \end{cases}$

29. $\begin{cases} x^2 + y^2 = 4 \\ x^2 - y^2 = 4 \end{cases}$

30. $\begin{cases} 2x^2 - y^2 = 3 \\ x - y = 1 \end{cases}$

31. $\begin{cases} y = x^2 - 4x - 7 \\ y = 9 - x^2 \end{cases}$

32. $\begin{cases} y = x^2 - 4x + 2 \\ y = 3 - 3x - x^2 \end{cases}$

33. $\begin{cases} y = 3x^2 - 4x \\ y = 2x^2 + x - 6 \end{cases}$

34. $\begin{cases} x - 2y + 3z = 3 \\ 3x + y - 2z = 5 \\ 5x + 3y + z = 9 \end{cases}$

35. $\begin{cases} x - 2y + 4z = 1 \\ 5x - 6y + 9z = 9 \\ 2x + 3y + z = 3 \end{cases}$

36. $\begin{cases} x + 2y + z = 0 \\ 3x + 4y + 2z = 1 \\ 2x - 6y - 2z = 3 \end{cases}$

In Exercises 37 through 45, solve each system of equations using determinants.

37. $\begin{cases} x + 2y - 3 = 0 \\ 3x - y - 2 = 0 \end{cases}$

38. $\begin{cases} 2x + y = 9 \\ 3x - 2y = -4 \end{cases}$

39. $\begin{cases} 2x + 2y = 1 \\ 3x + 4y = 3 \end{cases}$

40. $\begin{cases} x - 2y + 3 = 0 \\ 2x - 4y + 5 = 0 \end{cases}$

41. $\begin{cases} x - 2y + 3 = 0 \\ 2x - 4y + 6 = 0 \end{cases}$

42. $\begin{cases} \frac{x}{3} - y - 2 = 0 \\ x - y - 4 = 0 \end{cases}$

43. $\begin{cases} x + 2y + z = 2 \\ 3x - 2y - z = 6 \\ 2x + 4y + 5z = 1 \end{cases}$

44. $\begin{cases} 3x + y - z = 0 \\ 2x + 5y - 2z = 3 \\ 5x + 3y + 2z = 5 \end{cases}$

45. $\begin{cases} 2x + 3y + 4z = 3 \\ 4x + 3y - 8z = 0 \\ 8x - 3y - 4z = 2 \end{cases}$

46. The sum of two numbers is 24. Their difference is 6. Find the numbers.

47. The difference between two numbers is 30. The larger exceeds 3 times the smaller by 6. Find the numbers.

48. A plane flew a distance of 720 miles in 4 hr with the aid of a tailwind. Flying against the wind, the plane covers two thirds of this distance in the same time. Find the speed of the plane in still air and the speed of the wind.

49. Find the equation of the linear function that contains the points (2, 3) and (6, 3).

50. The square of one number exceeds twice the square of a second number by 4. The sum of the squares of the two numbers is 52. Find the numbers.

51. A man won $50,000 in a lottery. He divided $25,000 between his wife and son. The son received $5000 less than the wife. How much did each receive?

52. The surface area of a sphere is given by $S = 4\pi r^2$. The radii of two spheres differ by 4 cm. The surface area of the two spheres differs by 256π cm^2. Find the radius of each sphere.

53. A certain chemistry experiment requires 4 liters of 60% solution of sulfuric acid. Two concentrations are available, an 80% solution and a 50% solution. How many liters of each solution must be mixed to produce the 4 liters of 60% solution?

54. The following equations result when measuring the three individual capacitances of a transistor. Find the values of C_1, C_2, and C_3.

$$C_1 + C_3 = 11$$
$$C_2 + C_3 = 3.06$$
$$C_1 + C_2 = 8.06$$

CHAPTER 7

The Trigonometric Functions

SECTION 38

Angles and Their Measure: Applications of Radian Measure

Trigonometry is concerned with the measurement of parts of a triangle. Many problems can be solved by the use of triangles. In this chapter we shall introduce the trigonometric functions and some of their applications to problems involving triangle solutions.

We began this study by examining the concept of a plane angle. A half-line rotated about its endpoint in a plane generates a plane angle. In Figure 38.1 we generate the angle POQ by rotating the half-line OQ about O from its initial position to the terminal position OP. The point O is called the **vertex** of angle POQ, and OQ and OP are called the **initial** and **terminal sides**, respectively.

An angle such as POQ in Figure 38.1 is called **positive** if the direction of the rotation is **counterclockwise** and **negative** if the direction is **clockwise.** The angle AOB shown in Figure 38.2 is positive while the angle AOB

Figure 38.1 Figure 38.2

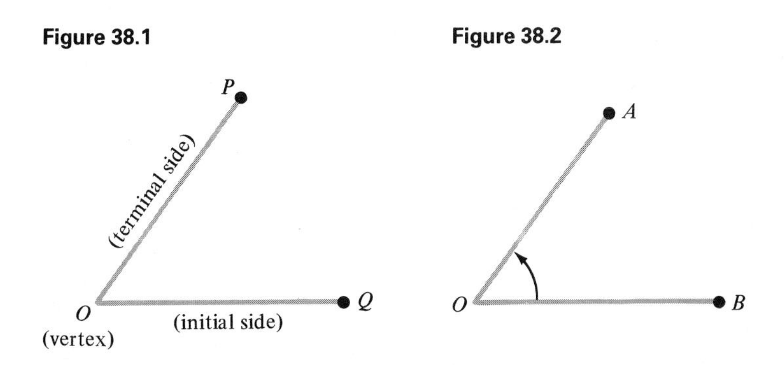

shown in Figure 38.3 is negative. Observe that in both figures we indicate the direction and extent of a rotation by a curved arrow.

Figure 38.3

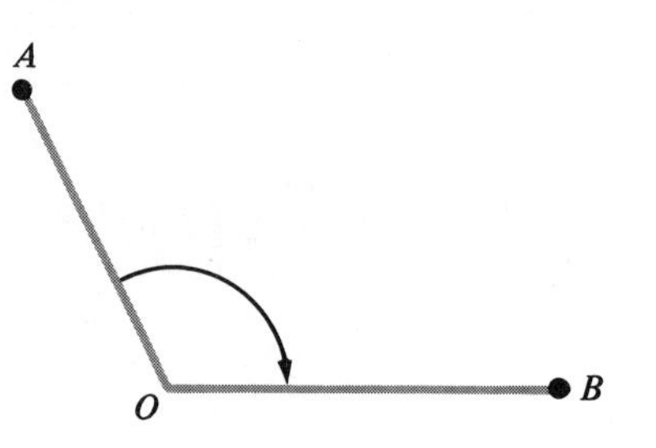

We will generally denote angles by lowercase Greek letters, such as α (alpha), β (beta), and θ (theta). If a terminal side of an angle lies in a given quadrant, we say that the angle lies in that quadrant. If the terminal side of an angle coincides with one of the axes, we call it a **quadrantal angle**. If we place the initial side of an angle on the positive x axis with its vertex at the origin, the angle is said to be in **standard position**. Two angles in standard position which have the same terminal side are called **coterminal**. The angles α and θ in Figure 38.4 are coterminal.

Figure 38.4

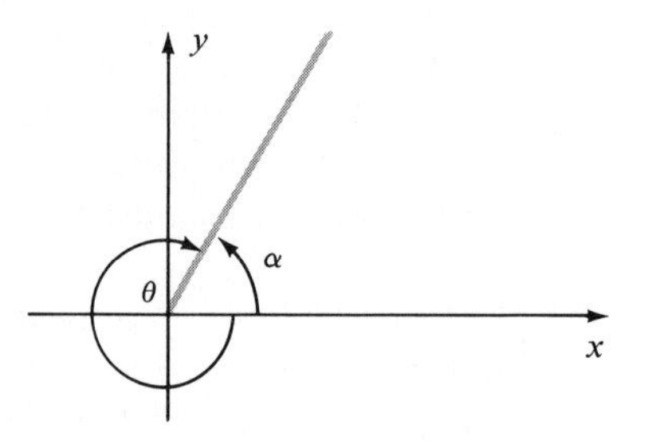

The two most commonly used units of measure of an angle are degree and radian. *A degree (°) is the measure of the central angle subtended by an arc of a circle equal to* $\frac{1}{360}$ *of the circumference of the circle.* Each degree is divided into 60 minutes ($'$), and each minute is divided into 60 seconds ($''$). We say that the measure of an angle POQ is the measure of the arc PQ.

Example 38.1

(a) The measure of a right angle is 90°.
(b) The measure of a straight angle is 180°.
(c) The measure of an angle of two clockwise revolutions is $(-2)(360°) = -720°$.

One radian (rad) is the measure of the central angle subtended by the arc of a circle equal to the radius of the circle (Figure 38.5). Since the circumference of a circle is equal to 2π times the radius and it subtends an angle of 360°, we have

$$2\pi \text{ radians} = 360^\circ$$

Figure 38.5

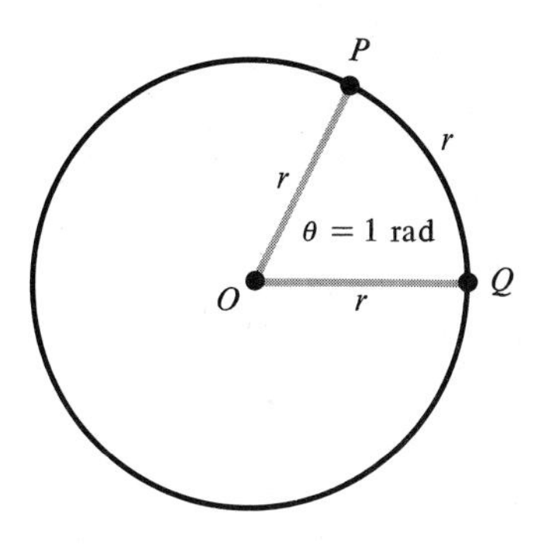

From the above formula we can obtain

$$\left.\begin{aligned} 1 \text{ rad} &= \frac{360^\circ}{2\pi} = \frac{180^\circ}{\pi} = 57.296^\circ \quad \text{or} \quad 57^\circ 17' 45'' \\ 1^\circ &= \frac{2\pi \text{ rad}}{360} = \frac{\pi \text{ rad}}{180} = 0.01745 \text{ rad} \end{aligned}\right\} \tag{38.1}$$

Formula (38.1) gives us a method for converting the measure of an angle from degrees to radians or from radians to degrees.

Example 38.2

Express the angles (a) 60°, (b) 135°, and (c) 30°30′ in radians.

Solution

We use formula (38.1), to obtain

(a) $60^\circ = 60\left(\frac{\pi}{180} \text{ rad}\right) = \frac{\pi}{3} \text{ rad}$

(b) $135^\circ = 135\left(\frac{\pi}{180} \text{ rad}\right) = \frac{3\pi}{4} \text{ rad}$

(c) $30^\circ 30' = (30.5)\left(\frac{\pi}{180} \text{ rad}\right) = 0.532 \text{ rad}$

Example 38.3

Express the angles (a) $\frac{5\pi}{6}$ rad, (b) $\frac{3\pi}{2}$ rad, (c) $\frac{2\pi}{9}$ rad, and (d) 2 rad in degree measure.

Solution

We use formula (38.1), to obtain

(a) $\frac{5\pi}{6} \text{ rad} = \frac{5\pi}{6}\left(\frac{180°}{\pi}\right) = 150°$

(b) $\frac{3\pi}{2} \text{ rad} = \frac{3\pi}{2}\left(\frac{180°}{\pi}\right) = 270°$

(c) $\frac{2\pi}{9} \text{ rad} = \frac{2\pi}{9}\left(\frac{180°}{\pi}\right) = 40°$

(d) $2 \text{ rad} = 2\left(\frac{180°}{\pi}\right) = 114.6°$

Remark 38.1. *From this point on we shall omit the symbol (rad) for radian if an angle is expressed as a number in radians. Thus, we shall write 2 for the angle whose measure is 2 radians and 2° for an angle whose measure is 2 degrees.*

In many applications radian measure is more useful than degree measure. One such application can be obtained if we consider the following fact from plane geometry. The length of the arc on a circle is proportional to the central angle it determines. In Figure 38.6 we denote the arc by s, the central angle by θ, and the radius by r. We know that the length of the arc of

Figure 38.6

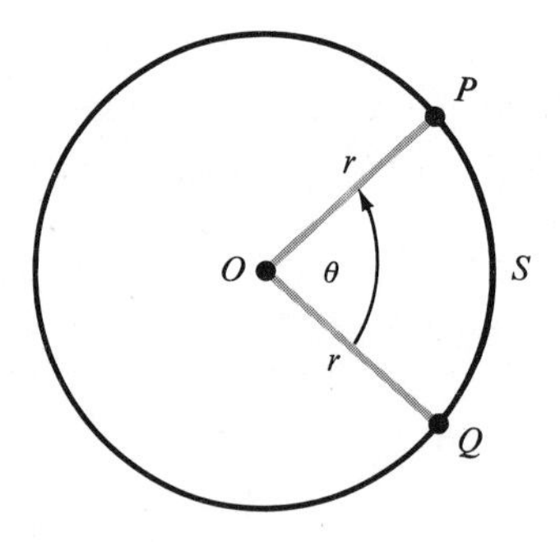

the complete circle (its circumference $= 2\pi r$) determines a central angle θ in radians equal to 2π. Thus, we have

$$\frac{s}{\theta} = \frac{2\pi r}{2\pi} \quad \text{or} \quad s = r\theta \tag{38.2}$$

Formulas (38.2) tells us that

arc length = *radius* × *central angle* (measured in radians)

Example 38.4

Find the length of the arc on a circle whose radius equals 20 cm and whose central angle is 40°.

Solution

Before we can apply formula (38.2) to find s, we must change the measure of the central angle to radian measure. Applying formula (38.1), we have

$$40° = 40\left(\frac{\pi}{180}\right) = 0.698$$

Now, apply formula (38.2) to obtain $s = (20\text{ cm})(0.698) = 13.96\text{ cm}$.

Example 38.5

A train is moving at the rate of 10 mph along a piece of circular track of radius 3000 ft. Through what angle does it turn in 1 min?

Solution

Here we wish to find θ. We are given the radius $r = 3000$ ft and we also know that the train will travel 10 mph $= 10(5280)/60$ ft/min $=$ 880 ft/min. Thus, the train travels over an arc of length $s = 880$ ft in 1 min. We now apply formula (38.2), as follows:

$$\theta = \frac{s}{r}$$

$$\theta = \frac{880}{3000} = 0.29$$

●

Another useful formula concerning the area of a sector of a circle (that portion of a circle bounded by two radii drawn to the arcs extremities; see Figure 38.6) can be developed if we recall from plane geometry that the area of a sector of a circle is proportional to its central angle. If we denote the area of the sector of the circle shown in Figure 38.6 as K, and if we recall that the area of the entire circle is πr^2 and its central angle is 2π, we have

$$\frac{K}{\theta} = \frac{\pi r^2}{2\pi} \quad \text{or} \quad K = \frac{1}{2}r^2\theta \tag{38.3}$$

Formula (38.3) tells us that *the area of a sector of a circle* $=$ $\frac{1}{2} \times$ *(radius)*^{2} $\times$ *(central angle), where the central angle is measured in radians.*

Example 38.6

Find the area of a sector of a circle whose radius is 10 cm and whose central angle is 25°.

Solution
First, we change the central angle to radians, to obtain

$$25° = 25\left(\frac{\pi}{180}\right) = 0.435$$

Now we apply formula (38.3), as follows:

$$\begin{aligned} K &= \frac{1}{2}r^2\theta \\ &= \frac{1}{2}(10 \text{ cm})^2(0.435) \\ &= 21.75 \text{ cm}^2 \end{aligned}$$

●

Another application of radian measure can be found when we deal with the concept of angular velocity. The average linear velocity V is given by the formula

$$V = \frac{s}{t} \tag{38.4}$$

where s is the distance traveled and t is the time elapsed.

If an object is moving around a circular path with a constant speed, the distance traveled is the length of arc traversed; that is, $s = r\theta$. Thus,

$$V = \frac{s}{t} = \frac{r\theta}{t}$$

or

$$V = r\left(\frac{\theta}{t}\right) \tag{38.5}$$

We call θ/t the *angular velocity* and denote it by ω. Thus, we have

$$V = \omega r \tag{38.6}$$

Formula (38.6) gives us the relationship between the linear and the angular velocity of an object moving around a circle of radius r.

Example 38.7
The angular velocity of a grindstone is 2000 revolutions per minute (rpm). The diameter of the grindstone is 18 in. Find the linear velocity of a point on the rim of the grindstone.

Solution
We must first express the angular velocity ω, in radians per minute. Since 1 revolution $= 2\pi$ rad, we have

$$\omega = \left(2000\,\frac{\text{rev}}{\text{min}}\right)\left(\frac{2\pi \text{ rad}}{\text{rev}}\right) = 4000\pi \text{ rad/min}$$

If the diameter is 18 in., the radius is 9 in. We now apply formula (38.6), to obtain

$$\begin{aligned} V &= \omega r \\ &= (4000\pi)\left(\frac{\text{rad}}{\text{min}}\right)(9 \text{ in.}) \\ &= 36{,}000\pi \text{ in./min} = 112{,}040 \text{ in./min} \end{aligned}$$

Example 38.8

A car is traveling at the rate of 50 mph. What is the angular speed in radians per second of the car's wheels if the diameter of a tire is 28 in.?

Solution

We first observe that the radius of the tire is 14 in. We must convert the linear speed of 50 mph to inches per second. Thus,

$$V = \frac{50(5280)(12)}{60(60)} \frac{\text{in.}}{\text{sec}} = 880 \frac{\text{in.}}{\text{sec}}$$

We now apply formula (38.6) to find ω, as follows:

$$\begin{aligned} \omega &= \frac{V}{r} \\ &= \frac{880}{14} \frac{\text{rad}}{\text{sec}} = 62.9 \text{ rad/sec} \end{aligned}$$

Exercises for Section 38

In Exercises 1 through 10, express each angle in radian measure.

1. 45° **2.** 210° **3.** 15°

4. 75° **5.** 120° **6.** 225°

7. 24°30′ **8.** 16°12′ **9.** 390°

10. 44°24′

In Exercises 11 through 20, express each angle in degree measure.

11. $\frac{\pi}{6}$ rad **12.** $\frac{2\pi}{3}$ rad **13.** $\frac{4\pi}{9}$ rad

14. $\frac{3\pi}{4}$ rad **15.** $\frac{7\pi}{6}$ rad **16.** $\frac{7\pi}{12}$ rad

17. π rad **18.** $\frac{1}{3}$ rad **19.** $\frac{7}{3}$ rad

20. 4 rad

21. Find the length of the arc on a circle whose radius equals 30 cm and whose central angle equals 35°.

22. Find the length of the arc on a circle whose radius equals 10 in. and whose central angle equals 40°.

23. Find the area of the circular sector indicated in Exercise 21.

24. Find the area of the circular sector indicated in Exercise 22.

25. The wheel of a car has a diameter of 2.5 ft and it is revolving at 20 rpm. Find the linear velocity of a point on the rim in miles per hour.

26. A flywheel rotates with an angular velocity of 15 rpm. If its radius is 20 in., find the linear velocity of a point on the rim in feet per second.

27. Find the diameter of a pulley that is driven at 360 rpm by a belt moving at the rate of 40 ft/sec.

28. A point on the rim of a turbine wheel of diameter 12 ft moves with a constant linear velocity of 42 ft/sec. Find the angular velocity of the wheel in radians per second.

29. The length of a pendulum is 10 ft. If the pendulum bob makes an arc of 8 in., find the angle of the pendulum's motion in degrees; in radians.

30. A pendulum 5 ft long oscillates through an angle of 8°. Find the distance the end of the pendulum bob travels in going from one end of the arc to the other.

31. If we assume that the earth's orbital path about the sun is circular with a radius of 93,000,000 miles and we let 1 year equal 365 days, find the speed of the earth in miles per second.

32. A train is traveling at the rate of 10 mph on a curve of radius 3000 ft. Through what angle will it turn in 1 min?

33. If the minute hand of a clock is 10 in. long, how far does the tip of the hand move in 30 min? In 40 min?

34. A spaceship orbiting the earth travels along a constant great circle. In the time that it takes the spaceship to make one complete orbit, the earth rotates 22.5° from west to east. Thus, each orbit follows a different path. Determine which orbit matches the first.

SECTION 39

The Trigonometric Functions

We shall now consider the angle θ in the standard position shown in Figure 39.1 (see page 194). Suppose that we choose any two distinct points other than the origin that lie on the terminal side of θ. Let two such points be P and Q, shown in Figure 39.1. We drop perpendicular lines from the points P and Q to the x axis to form the similar triangles shown in Figure 39.1. From plane geometry we know that if two triangles are similar, their corresponding sides are proportional. Thus, from Figure 39.1 we have

$$\left.\begin{aligned}
&\frac{\overline{PR}}{\overline{OR}} = \frac{\overline{QS}}{\overline{OS}} \quad \text{or} \quad \frac{y_1}{x_1} = \frac{y_2}{x_2} \\
&\frac{\overline{PR}}{\overline{OP}} = \frac{\overline{QS}}{\overline{OQ}} \quad \text{or} \quad \frac{y_1}{r_1} = \frac{y_2}{r_2} \\
&\frac{\overline{OR}}{\overline{OP}} = \frac{\overline{OS}}{\overline{OQ}} \quad \text{or} \quad \frac{x_1}{r_1} = \frac{x_2}{r_2} \\
&\text{where} \\
&r_1 = \sqrt{x_1^2 + y_1^2} \quad \text{and} \quad r_2 = \sqrt{x_2^2 + y_2^2}
\end{aligned}\right\} \tag{39.1}$$

The significance of the results in (39.1) is that the ratios y/x, x/r, and y/r do not depend on the choice of the point on the terminal side of θ. These ratios do depend on the angle θ and thus are functions of θ. We call these functions the **trigonometric functions** and define them as follows:

$$\left.\begin{aligned}
&\text{sine } \theta = \sin\theta = \frac{y}{r} \qquad \text{cotangent } \theta = \cot\theta = \frac{x}{y} \\
&\text{cosine } \theta = \cos\theta = \frac{x}{r} \qquad \text{secant } \theta = \sec\theta = \frac{r}{x} \\
&\text{tangent } \theta = \tan\theta = \frac{y}{x} \qquad \text{cosecant } \theta = \csc\theta = \frac{r}{y} \\
&\text{where} \\
&r = \overline{OP} = \sqrt{x^2 + y^2}
\end{aligned}\right\} \tag{39.2}$$

In (39.2) the symbols $\sin\theta$, $\cos\theta$, $\tan\theta$, $\cot\theta$, $\sec\theta$, and $\csc\theta$ are abbreviations for the six trigonometric functions, and r, which is **always positive**, is called the **radius vector**. It should also be noted that if a point $P(x, y)$ on the terminal side of the angle θ lies on either axes (i.e., the angle is quadrantal) then two of the six trigonometric functions will **not be defined**. Specifically, if P lies on the x axis, $x \neq 0$ and $y = 0$ and $\cot\theta$ and $\csc\theta$ are not defined. if P lies on the y axis, $x = 0$ and $y \neq 0$ and $\tan\theta$ and $\sec\theta$ are not defined.

Since r is always positive, the signs of the trigonometric functions of angle θ in the various quadrants will depend upon the signs of x and y. An easy device for determining these signs is shown in Figure 39.2. In this figure only the functions having positive signs are listed.

If we are given a point P other than the origin on the terminal side of an angle θ, we can calculate the values of the trigonometric functions.

Figure 39.1

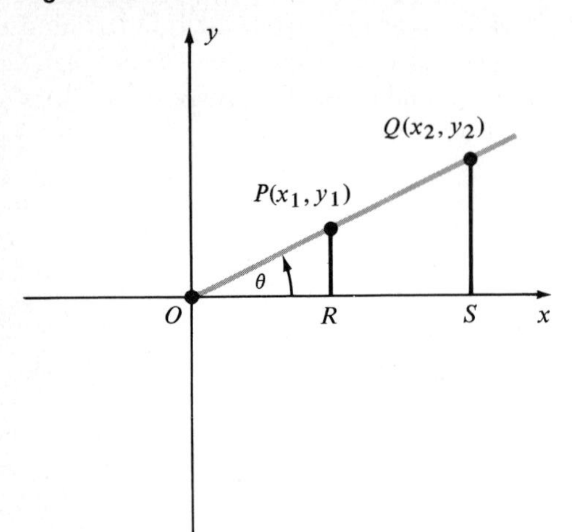

y
$Q(x_2, y_2)$
$P(x_1, y_1)$
θ
O
R
S
x

Figure 39.2

y
Quadrant II
sin θ = +
csc θ = +
Quadrant I
All +
x
Quadrant III
tan θ = +
cot θ = +
Quadrant IV
cos θ = +
sec θ = +

Example 39.1

Find the six trigonometric functions of the angle θ if the point $(-3, 4)$ lies on its terminal side.

Solution

Since $(-3, 4)$ lies on the terminal side, we have $x = -3$ and $y = 4$; see Figure 39.3. We can calculate r as follows:

$$r = \sqrt{x^2 + y^2} = \sqrt{(-3)^2 + 4^2} = \sqrt{25} = 5$$

Figure 39.3

y
P
(−3, 4)
O
x

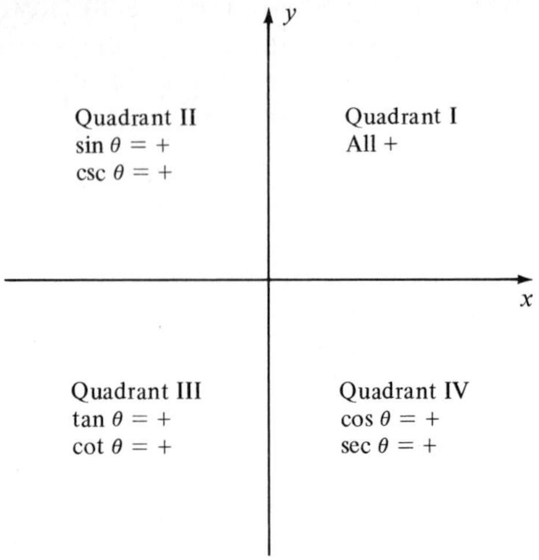

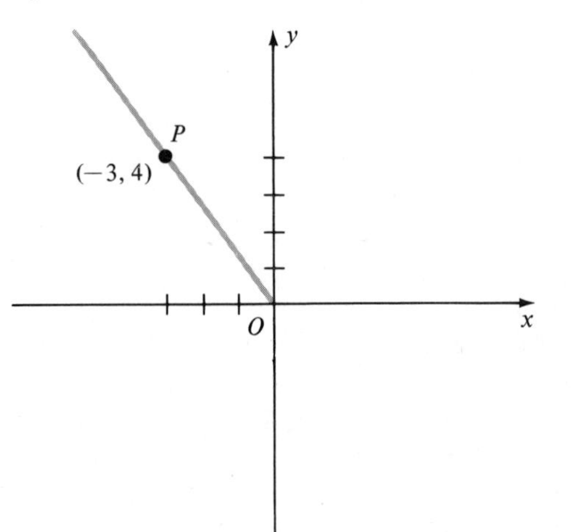

We now use (39.2), to obtain

$$\sin\theta = \frac{4}{5} \qquad \cot\theta = \frac{-3}{4}$$

$$\cos\theta = \frac{-3}{5} \qquad \sec\theta = \frac{5}{-3}$$

$$\tan\theta = \frac{4}{-3} \qquad \csc\theta = \frac{5}{4}$$

Example 39.2

Find the six trigonometric functions of the angle θ if the point $(0, -2)$ lies on its terminal side.

Solution

We note that $x = 0$ and $y = -2$ and that $r = \sqrt{x^2 + y^2} = \sqrt{0^2 + (-2)^2} = 2$. We use (39.2), to obtain

$$\sin\theta = \frac{-2}{2} = -1 \qquad \cot\theta = \frac{0}{-2} = 0$$

$$\cos\theta = \frac{0}{2} = 0 \qquad \sec\theta = \frac{2}{0} \quad \text{(not defined)}$$

$$\tan\theta = \frac{-2}{0} \quad \text{(not defined)} \qquad \csc\theta = \frac{2}{-2} = -1$$

●

If one of the trigonometric functions is known, we can determine the other functions of the angle.

Example 39.3

If $\sin\theta = \frac{3}{4}$ and θ is an angle in the second quadrant, find the other five trigonometric functions of θ.

Solution

Since $\sin\theta = \frac{3}{4} = y/r$, we have $y = 3$ and $r = 4$. We can now calculate x as follows:

$$r^2 = x^2 + y^2$$

$$16 = x^2 + 9$$

$$x^2 = 7 \quad \text{or} \quad x = -\sqrt{7}$$

We choose the negative value for x since θ lies in the second quadrant. We now use (39.2), to obtain

$$\cos\theta = -\frac{\sqrt{7}}{4} \qquad \sec\theta = -\frac{4}{\sqrt{7}} \qquad \tan\theta = -\frac{3}{\sqrt{7}}$$

$$\cot\theta = -\frac{\sqrt{7}}{3} \qquad \csc\theta = \frac{4}{3}$$

Note that Figure 39.2 also tells us that only $\sin\theta$ and $\csc\theta$ will be positive.

Exercises for Section 39

In Exercises 1 through 6, determine the quadrant or quadrants in which the terminal side of the angle θ will lie under the following conditions.

1. $\cos\theta > 0$

2. $\cos\theta > 0$ and $\sin\theta < 0$

3. $\tan\theta < 0$ and $\sin\theta < 0$

4. $\sin\theta > 0$ and $\cos\theta < 0$

5. $\sec\theta > 0$

6. $\tan\theta > 0$

In Exercises 7 through 18, determine the ***algebraic sign*** *for each trigonometric function.*

7. $\sin 115°$

8. $\cos 115°$

9. $\tan 225°$

10. $\cos 315°$

11. $\sin 93°$

12. $\tan 267°$

13. $\sec 110°$

14. $\sin -65°$

15. $\tan -240°$

16. $\sin -1000°$

17. $\tan -1000°$

18. $\cos -1000°$

In Exercises 19 through 28, find the six trigonometric functions of the angle θ, in standard position if the point lies on its terminal side.

19. $P(3, 4)$

20. $P(-3, -4)$

21. $P(-3, 4)$

22. $P(12, -5)$

23. $P(-12, -5)$

24. $P(-1, 3)$

25. $P(2, 0)$

26. $P(0, 2)$

27. $P(4, 0)$

28. $P(-3, 1)$

In Exercises 29 through 40, find the values of the remaining trigonometric functions subject to the conditions.

29. $\sin\theta = \frac{5}{13}$ and θ lies in quadrant I.

30. $\sin\theta = \frac{8}{17}$ and θ lies in quadrant II.

31. $\cos\theta = -\frac{3}{5}$ and θ lies in quadrant III.

32. $\tan\theta = -\frac{3}{4}$ and θ lies in quadrant IV.

33. $\sin\theta = \frac{8}{17}$

34. $\sec\theta = \frac{\sqrt{41}}{4}$

35. $\tan\theta = \frac{3}{4}$

36. $\cos\theta = \frac{1}{4}$

37. $\cos\theta = -\frac{4}{5}$ and $\tan\theta < 0$

38. $\cot\theta = \frac{24}{7}$

39. $\sec\theta = \frac{6}{5}$

40. $\csc\theta = -\frac{3}{2}$

In Exercises 41 through 43, construct two positive angles less than 360° in the standard position that satisfy the conditions.

41. $\cot\theta = \frac{7}{24}$

42. $\cos\theta = -\frac{12}{13}$

43. $\sec\theta = \frac{17}{15}$

SECTION 40

Values of Some Special Trigonometric Functions

In many applications of trigonometry we know the angle in degrees and we are required to find the values of the trigonometric functions of this angle. Generally, the values of the trigonometric functions of an angle are calculated by the use of tables. The values of the trigonometric functions of certain angles can be calculated by geometric methods. In this section we present some of these geometric methods, and in the next section we shall discuss the evaluation of trigonometric functions by use of tables.

We first investigate the values of the trigonometric functions involving the 30° angle. In Figure 40.1 we draw a 30° angle in the standard position

Figure 40.1

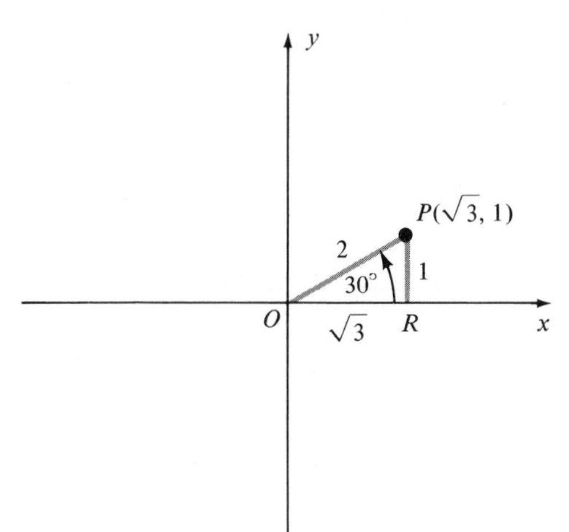

and mark off segment OP equal to 2 units on the terminal side. Next we drop a perpendicular to the x axis at R. The length of PR is 1, since the side opposite the 30° angle in the 30° right triangle OPR is one half the length of the hypotenuse OP. We now use the Pythagorean theorem to calculate the length OR, as follows:

$$(OP)^2 = (OR)^2 + (RP)^2$$
$$4 = (OR)^2 + 1$$
$$(OR)^2 = 3$$
$$OR = \sqrt{3}$$

Thus, the coordinates of point P are $(\sqrt{3}, 1)$. We can now apply formula (39.2) (noting that $r = 2$) to obtain the values of the trigonometric function for the 30° angle, as follows:

$$\sin 30° = \frac{1}{2} \qquad \cot 30° = \frac{\sqrt{3}}{1} = \sqrt{3}$$

$$\cos 30° = \frac{\sqrt{3}}{2} \qquad \sec 30° = \frac{2}{\sqrt{3}} = \frac{2\sqrt{3}}{3}$$

$$\tan 30° = \frac{1}{\sqrt{3}} = \frac{\sqrt{3}}{3} \qquad \csc 30° = \frac{2}{1} = 2$$

To find the values of the trigonometric functions of the 60° angle we draw a 60° angle in the standard position (see Figure 40.2) and mark off segment OP equal to 2 units on the terminal side. Next we drop a perpendicular to the x axis at R. Since OR is the side opposite the 30° angle in the triangle OPR, its length is 1 unit. We can calculate the length of RP

Figure 40.2

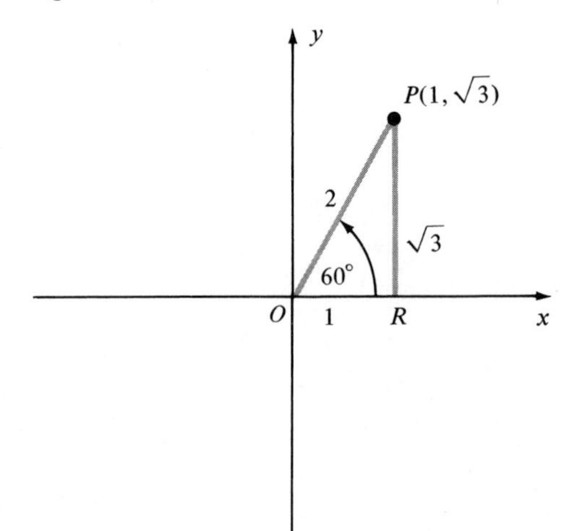

by using the Pythagorean theorem, as follows:

$$(OP)^2 = (OR)^2 + (RP)^2$$

$$4 = 1 + (RP)^2$$

$$(RP)^2 = 3 \quad \text{or} \quad RP = \sqrt{3}$$

Thus, the coordinates of the point P are $(1, \sqrt{3})$. We now apply formula (39.2) (noting that $r = 2$), to obtain

$$\sin 60° = \frac{\sqrt{3}}{2} \qquad \cot 60° = \frac{1}{\sqrt{3}} = \frac{\sqrt{3}}{3}$$

$$\cos 60° = \frac{1}{2} \qquad \sec 60° = \frac{2}{1} = 2$$

$$\tan 60° = \frac{\sqrt{3}}{1} = \sqrt{3} \qquad \csc 60° = \frac{2}{\sqrt{3}} = \frac{2\sqrt{3}}{3}$$

We can also find the function values for the 45° angle. In Figure 40.3 we draw a 45° angle in the standard position and mark off any point P on the terminal side. We then drop a perpendicular from P to the x axis at R.

Figure 40.3

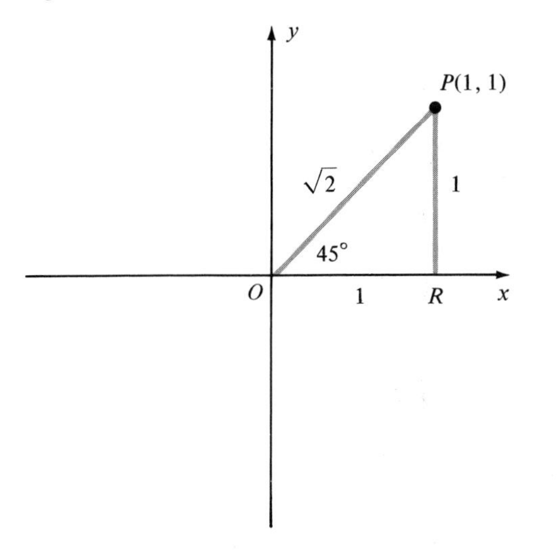

Since angles ROP and OPR are equal, the sides opposite them must be equal. Thus, $OR = RP$. If we choose each of them to be equal to 1 unit, the coordinates of P are $(1, 1)$. We can apply formula (39.2) to calculate the radius vector r as follows:

$$r = \sqrt{x^2 + y^2}$$

$$= \sqrt{1^2 + 1^2} = \sqrt{2}$$

From (39.2) we also have

$$\sin 45° = \frac{1}{\sqrt{2}} = \frac{\sqrt{2}}{2} \qquad \cot 45° = \frac{1}{1} = 1$$

$$\cos 45° = \frac{1}{\sqrt{2}} = \frac{\sqrt{2}}{2} \qquad \sec 45° = \frac{\sqrt{2}}{1} = \sqrt{2}$$

$$\tan 45° = \frac{1}{1} = 1 \qquad \csc 45° = \frac{\sqrt{2}}{1} = \sqrt{2}$$

Now that we have established the values of the trigonometric functions for the 30° and 45° angles, we can find the function values for angles that are multiples of 30° and 45° with the exception of those that are also multiples of 90°.

Example 40.1

Find the values of the trigonometric function of 150°.

Solution

In Figure 40.4 we draw a 150° angle in the standard position, and then we choose a point P on the terminal side, dropping a perpendicular to the x axis at R. If we choose 2 for the length of OP and note that angle ROP is

Figure 40.4

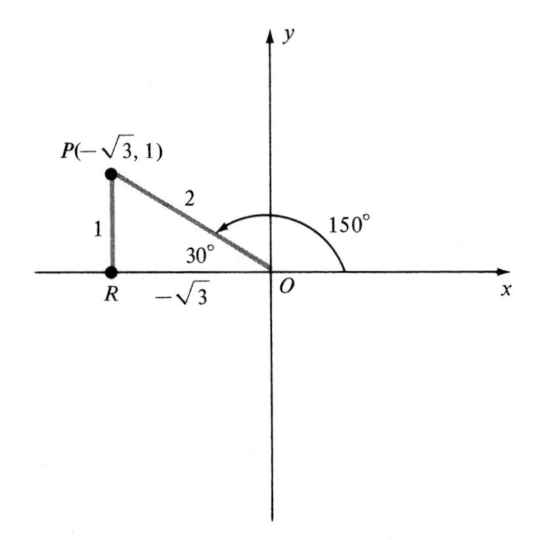

30°, then RP, the side opposite the 30° angle, must have a length of one-half OP or 1, and the length of OR must be $\sqrt{3}$. Since P lies in quadrant II, its coordinates must be $(-\sqrt{3}, 1)$. We now use formula (39.2), to obtain

$$\sin 150° = \frac{1}{2} \qquad \cot 150° = \frac{-\sqrt{3}}{1} = -\sqrt{3}$$

$$\cos 150° = \frac{-\sqrt{3}}{2} = -\frac{\sqrt{3}}{2} \qquad \sec 150° = \frac{2}{-\sqrt{3}} = -\frac{2\sqrt{3}}{3}$$

$$\tan 150° = \frac{1}{-\sqrt{3}} = -\frac{\sqrt{3}}{3} \qquad \csc 150° = \frac{2}{1} = 2$$

Angles that are multiples of 30° and 45° and also multiples of 90° are the quadrantal angles. Their terminal sides will coincide with either the x or y axis. Thus, the choice of any point P on the terminal side will result in either $y = 0$ or $x = 0$ but not both. When we use (39.2) to find the values for the trigonometric functions, we will encounter certain ratios that are not defined (see Example 39.2). For these and all future situations we shall use the symbol ∞ to indicate this.

Example 40.2

Find the values for the trigonometric functions of 270°.

Solution

In Figure 40.5 we draw a 270° angle in the standard position and note that its terminal side coincides with the negative y axis. We now choose a point P on the terminal side so that the length of OP is 1. Since P lies on the negative y axis, its coordinates must be $(0, -1)$, and the radius vector

Figure 40.5

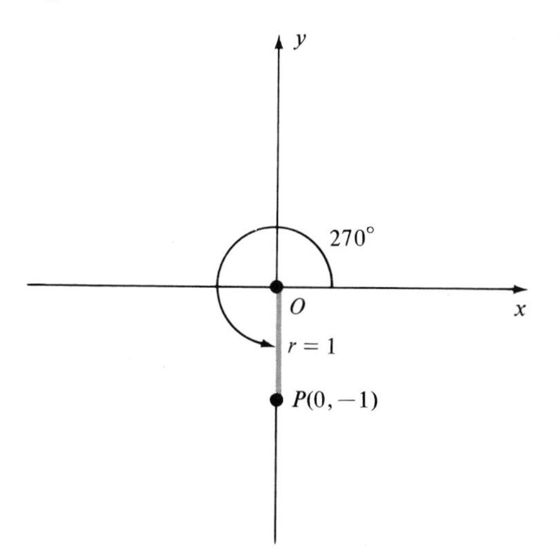

$r = OP$ equals 1. We now use formula (39.2), to obtain

$$\sin 270° = \frac{-1}{1} = -1 \qquad \cot 270° = \frac{0}{-1} = 0$$

$$\cos 270° = \frac{0}{1} = 0 \qquad \sec 270° = \frac{1}{0} = \infty$$

$$\tan 270° = \frac{-1}{0} = \infty \qquad \csc 270° = \frac{1}{-1} = -1$$

Since the values of the trigonometric functions for the angles 30°, 45°, and 60° can be used to calculate the values of most of their multiples, they are worth memorizing. We list these results as well as those for the quadrantal angles in Table 40.1.

Table 40.1

θ	$\sin\theta$	$\cos\theta$	$\tan\theta$	$\cot\theta$	$\sec\theta$	$\csc\theta$
0°	0	1	0	∞	1	∞
30°	$\frac{1}{2}$	$\frac{\sqrt{3}}{2}$	$\frac{\sqrt{3}}{3}$	$\sqrt{3}$	$\frac{2\sqrt{3}}{3}$	2
45°	$\frac{\sqrt{2}}{2}$	$\frac{\sqrt{2}}{2}$	1	1	$\sqrt{2}$	$\sqrt{2}$
60°	$\frac{\sqrt{3}}{2}$	$\frac{1}{2}$	$\sqrt{3}$	$\frac{\sqrt{3}}{3}$	2	$\frac{2\sqrt{3}}{3}$
90°	1	0	∞	0	∞	1
180°	0	−1	0	∞	−1	∞
270°	−1	0	∞	0	∞	−1

Exercises for Section 40

In Exercises 1 through 3, verify the results shown in Table 40.1 by finding the values of the trigonometric functions for the quadrantal angles.

1. 0° **2.** 90° **3.** 180°

In Exercises 4 through 15, find the values of the trigonometric functions of the angle by drawing the angle in standard position and then selecting a point on its terminal side.

4. 120° **5.** 135° **6.** 210°

7. 225° **8.** 240° **9.** 300°

10. $315°$ **11.** $330°$ **12.** $360°$

13. $-30°$ **14.** $-45°$ **15.** $-150°$

In Exercises 16 through 30, determine which statements are true and which are false.

16. $\sin 60° = 2 \sin 30°$

17. $\cos 90° + \cos 270° = 0$

18. $\sin 0° = \cos 180° + \sin 90°$

19. $\cos 30° + \cos 60° = \cos 90°$

20. $4 \tan 45° = \tan 180°$

21. $\csc 90° - \csc 30° = \csc 60°$

22. $1 + \tan^2 30° = \sec^2 30°$

23. $\sin^2 45° + \cos^2 45° = 1$

24. $3 \cos 20° = \cos 60°$

25. $\sin 0° + \cos 0° + \sin 90° + \cos 90° = 2$

26. $\tan 45° - 2 \cot 45° + 2 \sin 270° = 0$

27. $\tan 225° + \tan 90° = \tan 315°$

28. $\sin 120° = 2(\sin 60°)(\cos 60°)$

29. $\cos 90° = \cos^2 45° - \sin^2 45°$

30. $\sin 45° = \sin 135°$

SECTION 41

Values of the Trigonometric Functions for Any Angle; Reduction; Use of Tables

In Section 40 we were able to find the values of the trigonometric functions for some special angles and their multiples. In this section we wish to develop a method using tables* for finding the values of the trigonometric functions for *any* angle.

Table I of Appendix F enables us to find the approximate values to four places for the trigonometric functions of any angle between 0° and 90° inclusive. The angles are listed at intervals of 10′ from 0° to 90° with angles from 0° to 45° in the left-hand side of the table and those from 45° to 90° on the right-hand side of the table. The headings at the top of the table are to be used for the angles in the left column and the headings at the bottom of the table are to be used for the angles in the right column.

Example 41.1

Find the value of $\cos 24°20'$.

* The use of calculators instead of tables is discussed in Appendix E.

Solution

In Table I we look across from 24°20′ under the cos θ heading to find that cos 24°20′ = 0.9903.

Example 41.2

Find the value of tan 76°50′.

Solution

In Table I we look across from 76°50′ under the tan θ heading from the bottom of the table to find that tan 76°50′ = 4.275. ●

Frequently, we have to find the value of a trigonometric function of an angle not listed in Appendix F, Table I. In such cases we use the method known as **linear interpolation**. This method assumes that there is a direct proportion between the difference in two angle values and the difference in the values of their trigonometric functions.

Example 41.3

Find the value of sin 32°44′.

Solution

We must interpolate between the sin 32°40′ and 32°50′. We use the following approach

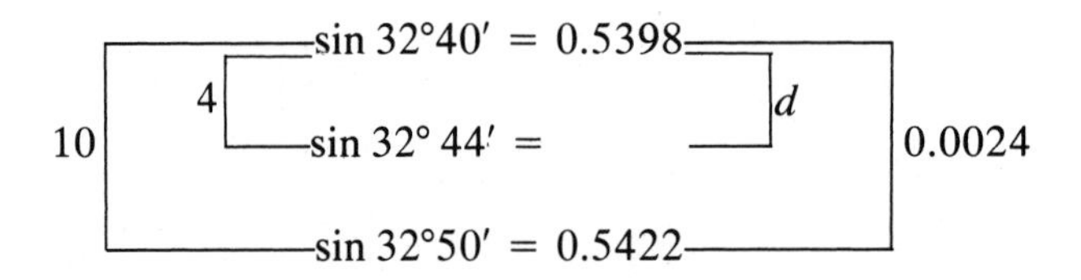

to obtain the ratio

$$\frac{4}{10} = \frac{d}{0.0024}$$

$$d = \frac{4}{10}(0.0024) = 0.0010$$

We see that the value of sin θ is increasing, so we **add** d to the value of sin 32°40′, to obtain

$$\sin 32°44' = 0.5398 + 0.0010 = 0.5408$$

Example 41.4

Find the value of cos 18°26′.

Solution

We must interpolate between the cos 18°20′ and cos 18°30′. We use the following approach

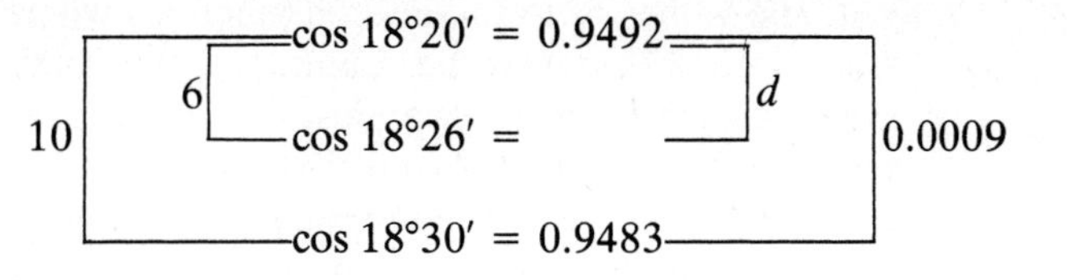

to obtain the ratio

$$\frac{6}{10} = \frac{d}{0.0009}$$

$$d = \frac{6}{10}(0.0009) = 0.0005$$

We see that the value of $\cos\theta$ is decreasing, so we **subtract** d from the value of cos 18°20′, to obtain

$$\cos 18°26' = 0.9492 - 0.0005 = 0.9487$$ ●

When we wish to find the angle when the value of the function is given, we simply reverse the process.

Example 41.5

Find the smallest positive angle θ when $\cot\theta = 5.671$.

Solution

We read directly from Table I that cot 10° = 5.671. Thus, $\theta = 10°$.

Example 41.6

Find the smallest positive angle θ when $\sin\theta = 0.4260$.

Solution

The given value of $\sin\theta$ is not an entry in Table I, so we must interpolate between 25°10′ and sin 25°20′. We use the following diagram

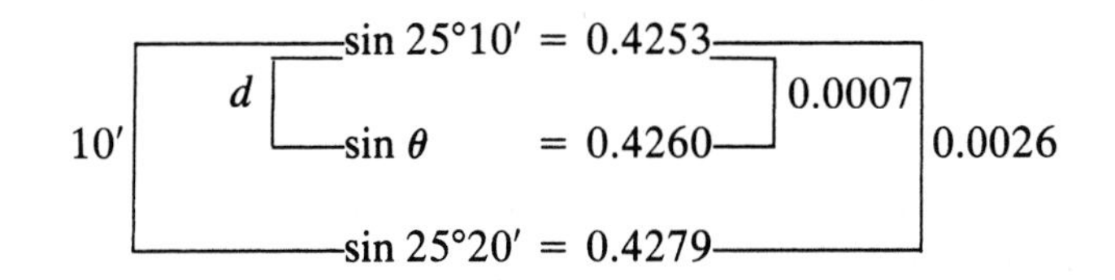

to obtain the ratio

$$\frac{d}{10} = \frac{0.0007}{0.0026}$$

$$d = 3' \text{ (nearest ')}$$

Thus,

$$\theta = 25°10' + 3' = 25°13'$$

The values of trigonometric functions where θ is greater than 90°, such as the sin 123°, cannot be found directly from Table I of Appendix F. To evaluate such functions, we must first express them as functions of a positive acute angle less than 90° and then use Table I of Appendix F. Any angle in the standard position is coterminal with some positive angle less than 360°. Since the function values of two coterminal angles are equal, we only need to express the functions of angles between 0° and 360° in terms of a positive acute angle less than 90°. This can be done by introducing the concept of a **reference angle**.

Example 41.7

The angles 45° and 405° are coterminal. Thus, the trigonometric functions of both angles are equal. For example, tan 45° = tan 405°. ●

For every angle (other than integral multiples of 90°) in the standard position, the positive acute angle formed by the x axis and the terminal side of the given angle is called the **reference angle**. In Figure 41.1 we see that the reference angle for 123° is 57°, and the reference angle for 215° is 35°.

Figure 41.1

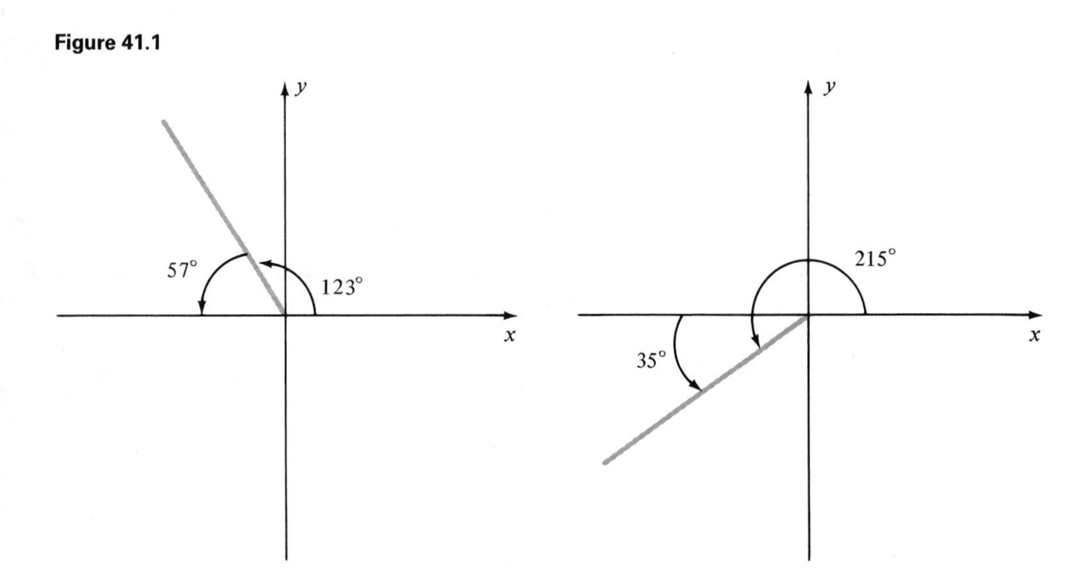

In Figure 41.2 we show the angle θ and its reference angle denoted by $\bar{\theta}$. If the terminal side of an angle θ lies in quadrant I, the terminal side of its reference angle $\bar{\theta}$ is the same, and $\theta = \bar{\theta}$. In Figure 41.3 we select a point $P(x_1, y_1)$ on the terminal of the angle θ, r units from O, and drop a perpendicular to x axis at R. We note that $\sin \theta = \sin \bar{\theta} = y/r$.

If the terminal side of angle θ lies in quadrant II, the terminal side of the reference angle $\bar{\theta}$ is the negative x axis and $\theta = 180 - \bar{\theta}$. In Figure 41.4 we select a point $P_1(x_1, y_1)$ on the terminal side of θ, r units from O, and

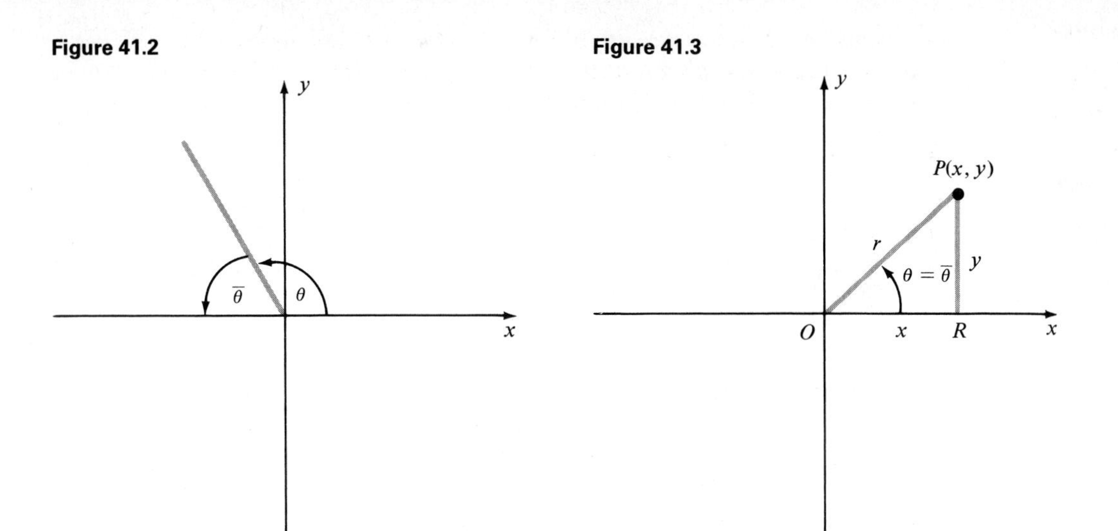

Figure 41.4

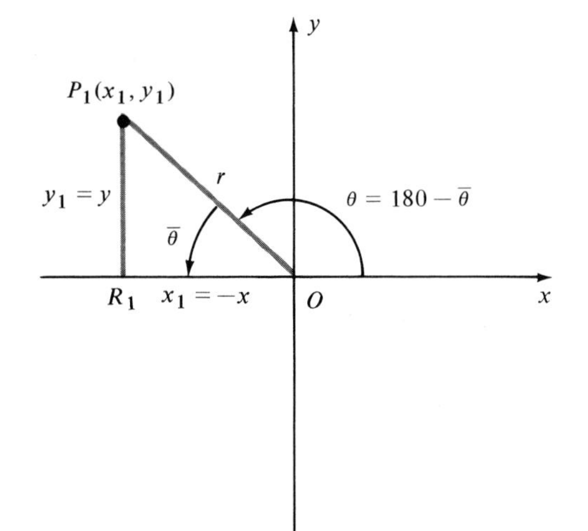

drop a perpendicular to the x axis at R_1. We note that $\sin\theta = \sin(180° - \bar{\theta}) = y_1/r$. We also note from Figures 41.3 and 41.4 that the triangles OPR and OP_1R_1 are congruent. Thus, $y_1 = y$ and $x_1 = -x$ and $\sin(180° - \bar{\theta}) = \sin\theta = \sin\bar{\theta}$. This last observation tells us that the value of sine function for any angle in quadrant II is the *same* as the value of the sine function for the reference angle $\bar{\theta}$.

Similar arguments will show that the value of the sine function for any angle in quadrants III and in IV is equal to the negative of the value of the sine of its respective reference angle.

In general, we say that *the value of the trigonometric function of any given angle is numerically equal to the value of the same trigonometric function of its reference angle. The algebraic sign of each function value is determined by the quadrant in which the given angle lies.* For easy reference we show these results in Table 41.1.

Table 41.1

θ lies in:	*Reference angle*	*Value of $\sin\theta$*	*Value of $\cos\theta$*	*Value of $\tan\theta$*	*Value of $\cot\theta$*	*Value of $\sec\theta$*	*Value of $\csc\theta$*
Quadrant I	$\bar{\theta} = \theta$	$\sin\bar{\theta}$	$\cos\bar{\theta}$	$\tan\bar{\theta}$	$\cot\bar{\theta}$	$\sec\bar{\theta}$	$\csc\bar{\theta}$
Quadrant II	$\bar{\theta} = 180 - \theta$	$\sin\bar{\theta}$	$-\cos\bar{\theta}$	$-\tan\bar{\theta}$	$-\cot\bar{\theta}$	$-\sec\bar{\theta}$	$\csc\bar{\theta}$
Quadrant III	$\bar{\theta} = \theta - 180°$	$-\sin\bar{\theta}$	$-\cos\bar{\theta}$	$\tan\bar{\theta}$	$\cot\bar{\theta}$	$-\sec\bar{\theta}$	$-\csc\bar{\theta}$
Quadrant IV	$\bar{\theta} = 360° - \theta$	$-\sin\bar{\theta}$	$\cos\bar{\theta}$	$-\tan\bar{\theta}$	$-\cot\bar{\theta}$	$\sec\bar{\theta}$	$-\csc\bar{\theta}$

Example 41.8

Find the value of $\cos 132°$.

Solution

Since $\theta = 132°$ is not an entry in Table I, we must first express $\theta = 132°$ in terms of its reference angle $\bar{\theta}$. Since θ lies in quadrant II, we can use Table 41.1 to calculate the reference angle $\bar{\theta} = 180° - 132° = 48°$, and then also observe that since $\cos\theta = -\cos\bar{\theta}$, $\cos 132° = -\cos 48°$. We now can read the value for $\cos 48°$ from Table I, to obtain

$$\cos 132° = -\cos 48° = -0.6691$$

Example 41.9

Find the value of $\tan 213°$.

Solution

Since $\theta = 213°$ is not an entry in Table I, we must first express $\theta = 213°$ in terms of its reference angle $\bar{\theta}$. Since θ lies in quadrant III, we see from Table 41.1 that $\bar{\theta} = 213° - 180° = 33°$. We also note that $\tan\theta = \tan\bar{\theta}$, thus $\tan 213° = \tan 33°$. Finally, we use Table I to determine that

$$\tan 213° = \tan 33° = 0.6494$$

Example 41.10

Find the value of $\sin 320°$.

Solution

Since $\theta = 320°$ lies in quadrant IV, we can use Table 41.1 to determine its reference angle, $\bar{\theta} = 360° - 320° = 40°$. We also see from Table 41.1 that $\sin\theta = -\sin\bar{\theta}$; thus, $\sin 320° = -\sin 40°$. We now use Table I, to obtain

$$\sin 320° = -\sin 40° = -0.6428$$

Example 41.11

Find the value or values of all θ between 0° and 360° when $\cos\theta = 0.8829$.

Solution

From Table I we see that $\cos 28° = 0.8829$; therefore, $\theta = 28°$. Since the value of $\cos\theta$ is positive, we also have another angle θ in the fourth quadrant. Since $\bar{\theta} = 360° - \theta$, we have $\theta = 360° - \bar{\theta}$ or $\theta = 360° - 28° = 332°$. Thus, we have **two** solutions, $\theta = 28°$ and 332°. ●

The question of evaluating a trigonometric function of a negative angle can be handled in a similar fashion. In Figure 41.5 on page 210, we have constructed the angles θ and $-\theta$ in standard position and numerically equal. We have selected the points $P(x, y)$ and $P_1(x_1, y_1)$ on their respective terminal sides. The two triangles in this figure are congruent. Thus, $r = r$, $x = x_1$, and $y = -y_1$. We can now use formula (39.2), to obtain

$$\left.\begin{aligned}
\sin(-\theta) &= \frac{y_1}{r} = \frac{-y}{r} = -\sin\theta \\
\cos(-\theta) &= \frac{x_1}{r} = \frac{x}{r} = \cos\theta \\
\tan(-\theta) &= \frac{y_1}{x_1} = \frac{-y}{x} = -\tan\theta \\
\cot(-\theta) &= \frac{x_1}{y_1} = \frac{x}{-y} = -\cot\theta \\
\sec(-\theta) &= \frac{r}{x_1} = \frac{r}{-y} = \sec\theta \\
\csc(-\theta) &= \frac{r}{y_1} = \frac{r}{-y} = -\csc\theta
\end{aligned}\right\} \tag{41.1}$$

The results of (41.1) will enable us to express the trigonometric function of any negative angle in terms of a trigonometric function of a positive angle. We can then proceed to evaluate this function in the same manner as the previous examples.

Example 41.12

Find the value of $\sin(-85°)$.

Solution

We use (41.1) to reexpress the function in terms of a positive angle and then Table I, to obtain

$$\sin(-85°) = -\sin 85° = -0.9962$$

Figure 41.5

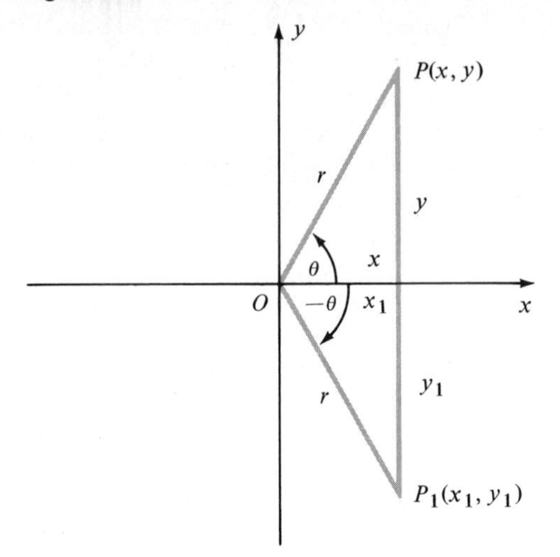

Example 41.13

Find the value of tan (−115°).

Solution

We first use (41.1), to obtain tan (−115°) = −tan 115°. We now proceed to evaluate tan 115° by noting that $\theta = 115°$ lies in quadrant II and its reference angle $\bar{\theta} = 180 - 115° = 65°$. From Table 41.1 and Table I of Appendix F, we have tan 115° = −tan 65° = −2.145. Thus,

$$\tan(-115°) = -\tan 115° = -(-2.145) = 2.145$$

Remark 41.1. *Whenever we wish to evaluate a trigonometric function of a negative angle, we* ***always first apply*** *the appropriate formula from (41.1) to obtain a function of a positive angle, and then we will reduce this angle by using Table 41.1.*

When we are given the value of a trigonometric function and asked to find the angle θ, there exists an infinite number of answers. For example, if $\sin\theta = \frac{1}{2}$, then θ may have any of the following values: $\theta = -330°$, 30°, 150°, 390°, 510°, and so on.

If we wish to find θ, $0° \le \theta < 360°$, we have only to find $\bar{\theta}$ from Table I and make use of Table 41.1 to find the value(s) for θ.

Example 41.14

Find θ, $0° \le \theta < 360°$, if $\sin\theta = 0.9962$; if $\cos\theta = -0.7547$.

Solution

For $\sin\theta = 0.9962$, we note that sine is positive in quadrants I and II, and thus we will have solutions in these quadrants. From Table I of Appendix

F, we find that $\sin\theta = 0.9962$ for $\theta = 85°$. Hence, $\bar{\theta} = 85°$. The solution θ_1 in the first quadrant is given by $\theta_1 = \bar{\theta} = 85°$. The solution θ_2 in the second quadrant is $\theta_2 = 180° - \bar{\theta}$ (see Table 41.1) or $\theta_2 = 180° - 85° = 95°$.

For $\cos\theta = -0.7547$ we note that cosine is negative in quadrants II and III. We now use Table I of Appendix F, to find $\cos\theta = 0.7547$ for $\theta = 41°$, and thus $\bar{\theta} = 41°$. Our two solutions are given by

$$\theta_1 = 180° - \bar{\theta} = 180° - 41° = 139°$$

and

$$\theta_2 = 180° + \bar{\theta} = 180° + 41 = 221°$$

Exercises for Section 41

In Exercises 1 through 8, draw each angle and determine its reference angle.

1. 98° **2.** 114° **3.** 226° **4.** 279°

5. 406° **6.** 850° **7.** −135° **8.** −310°

In Exercises 9 through 20, use Table I of Appendix F to find the values of each trigonometric function. Use interpolation if necessary.

9. sin 26°20′ **10.** cos 48°10′ **11.** tan 72°0′

12. sec 12°40′ **13.** cot 89°50′ **14.** cos 4°10′

15. sin 31°14′ **16.** tan 16°28′ **17.** cos 53°46′

18. sec 71°53′ **19.** csc 21°12′ **20.** cot 43°7′

In Exercises 21 through 32, find each positive acute angle θ. Use interpolation if necessary.

21. $\sin\theta = 0.2812$ **22.** $\cos\theta = 0.9304$ **23.** $\sec\theta = 1.169$

24. $\tan\theta = 0.7766$ **25.** $\csc\theta = 1.504$ **26.** $\sin\theta = 0.3416$

27. $\tan\theta = 4.7165$ **28.** $\cos\theta = 0.5900$ **29.** $\sin\theta = 0.4132$

30. $\tan\theta = 1.210$ **31.** $\cot\theta = 0.7010$ **32.** $\csc\theta = 1.213$

In Exercises 33 through 44, express each problem in terms of an angle θ, $0° < \theta < 90°$, by using Table 41.1, and then evaluate by using Table I of Appendix F.

33. sin 112°10′ **34.** cos 208° **35.** tan 218°

36. sec 303° **37.** cot 323°10′ **38.** cos 92°

39. sin (−102°) **40.** cos (−43°) **41.** sec (−123°)

42. tan (−123°) **43.** cos 12° + cos (−12°)

44. tan 33° + tan (−33°)

In Exercises 45 through 50, find θ, $0° < \theta < 360°$, for each problem.

45. $\sin \theta = 0.0727$ **46.** $\cos \theta = 0.2700$

47. $\tan \theta = 2.0000$ **48.** $\cot \theta = 1.400$

49. $\sin \theta = -0.0727$ **50.** $\cos \theta = -0.7300$.

51. The voltage V in a certain alternating circuit is $V = 120 \cos 15t$, where t is time (in seconds). If $t = 0.2$, find V.

52. The displacement x of a particle moving in simple harmonic motion is $x = 14 \sin 3t$, where t is time (in seconds). If $t = 0.24$, find x.

SECTION 42

Solution of Right Triangles

In many scientific areas we are concerned with determining the unknown sides and angles of triangles when certain angles and sides are given. In this section we shall examine such situations dealing with right triangles.

A right triangle is a triangle one of whose angles is a right angle. The side opposite the right angle is called the **hypotenuse** and the other two sides are called the **legs** of the right triangle. For all discussions involving right triangles we use A and B to designate the acute angles and C to designate the right angle (Figure 42.1). In addition, a will be the side opposite A, b the side opposite B, and c will be the hypotenuse. We shall also refer to a as the opposite side and b the adjacent side with respect to angle A.

Figure 42.1

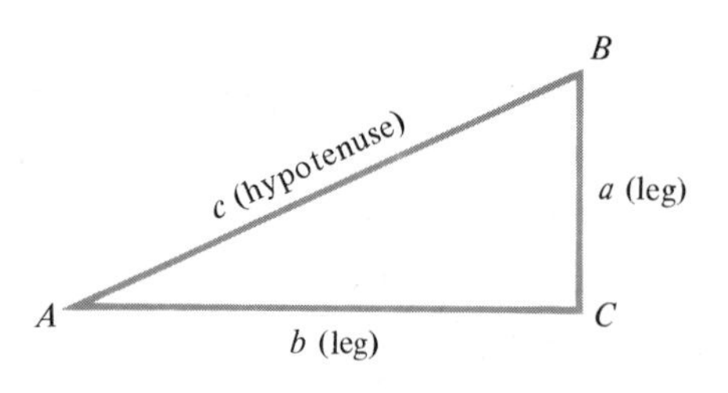

If we place the triangle ABC of Figure 42.1 in the coordinate system so that the acute angle A is in the standard position, we can use (39.2) to describe the trigonometric functions of the angle A in terms of the sides of the right triangle as follows:

$$\left.\begin{aligned}
\sin A &= \frac{\text{opposite side}}{\text{hypotenuse}} = \frac{a}{c} & \cot A &= \frac{\text{adjacent side}}{\text{opposite side}} = \frac{b}{a} \\
\cos A &= \frac{\text{adjacent side}}{\text{hypotenuse}} = \frac{b}{c} & \sec A &= \frac{\text{hypotenuse}}{\text{adjacent side}} = \frac{c}{b} \\
\tan A &= \frac{\text{opposite side}}{\text{adjacent side}} = \frac{a}{b} & \csc A &= \frac{\text{hypotenuse}}{\text{opposite side}} = \frac{c}{a}
\end{aligned}\right\} \quad (42.1)$$

The definitions of (42.1) are only applicable to acute angles of right triangles. Since a triangle can be moved without altering its size or shape, the above ratios will remain the same, regardless of the position of the triangle; thus, the definitions can be used without reference to the coordinate system. The definitions in (42.1) are so useful in working with right triangles that they are worth memorizing.

The **solution of a right triangle** means that we must determine the remaining parts when we are given one side and one acute angle or when we are given two sides.

Example 42.1

If $a = 5$, $c = 13$, and $C = 90°$, find b, A, and B in the right triangle ABC shown in Figure 42.2.

Figure 42.2

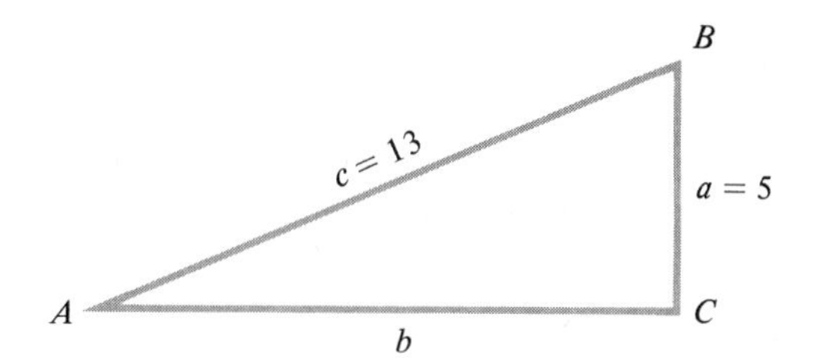

Solution

We can find side b by applying the Pythagorean theorem as follows:

$$\begin{aligned}
c^2 &= a^2 + b^2 \\
169 &= 25 + b^2 \\
b^2 &= 144 \\
b &= \sqrt{144} = 12
\end{aligned}$$

We can find angle A by observing from Figure 42.2 that

$$\sin A = \frac{a}{c} = \frac{5}{13} = 0.3846$$

We find that $A = 22°37'$ by using Table I of Appendix F and interpolation. Finally, we can find B by observing that

$$\tan B = \frac{\text{opposite side}}{\text{adjacent side}} = \frac{b}{a} = \frac{12}{5}$$

or

$$\tan B = 2.4000$$

Again, we find that $B = 67°23'$ by using Table I and interpolation.

Example 42.2

If $b = 6$ and $A = 35°$, find a, c, and B, in the right triangle ABC shown in Figure 42.3.

Figure 42.3

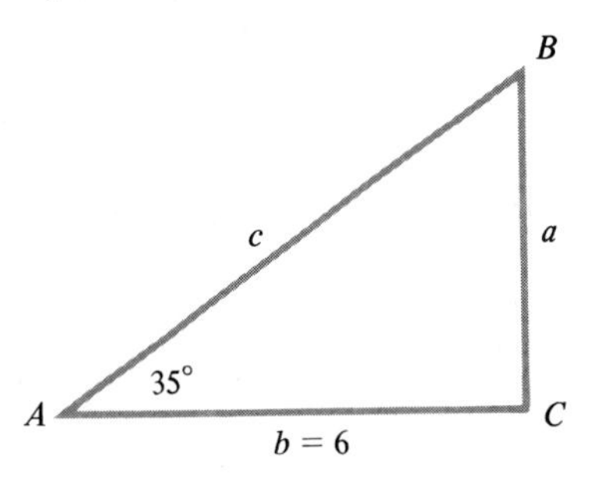

Solution

We can find a by observing in Figure 42.3 that

$$\tan 35° = \frac{a}{6}$$

or

$$a = 6(\tan 35°) = 6(0.7002) = 4.2$$

We also note that

$$\sec 35° = \frac{c}{6}$$

or

$$c = 6(\sec 35°) = 6(1.221) = 7.3$$

Finally, we can find B by observing that since A and B are complementary angles, we have

$$A + B = 90°$$
$$B = 90° - A$$
$$B = 90° - 35° = 55°$$

●

It should be noted that we might have found B by using any one of the trigonometric functions involving B, but the method above is quicker.

Right triangles are useful in solving many types of applied problems. One such type of problem deals with the angles of **elevation** and **depression.** The angle of elevation is the angle formed by the horizontal and the line of sight when the object is above the horizontal. (See Figure 42.4). The angle of depression is the angle formed by the horizontal and the line of sight when the object is below the horizontal (Figure 42.5).

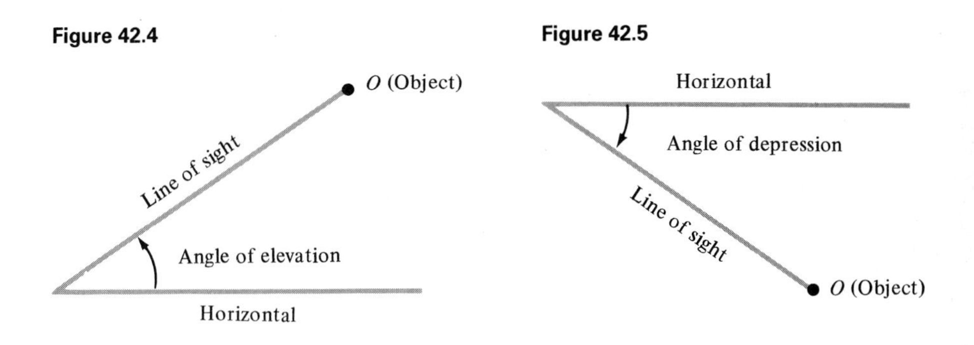

Example 42.3
From a point at ground level 150 ft from the base of a tower, the angle of elevation of the top of the tower is 48°20′. Find the height of the tower.

Solution
In Figure 42.6 we show the right triangle ABC, where A is the angle of elevation and a is the height of the tower. We observe that

$$\tan 48°20' = \frac{a}{150}$$

Figure 42.6

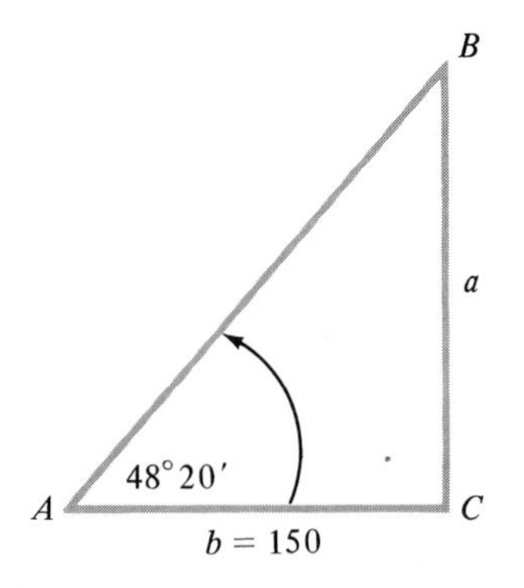

or

$$a = 150(\tan 48°20') = 150(1.124) = 169$$

Thus, the height of the tower is 169 ft.

Example 42.4

From the top of a lighthouse 140 ft above sea level, the angle of depression of a ship sailing directly toward the lighthouse is 18°. How far is the ship from the lighthouse?

Solution

In Figure 42.7 we see that the angle of depression is angle ABD and that angle A = angle ABD. The distance of the ship from the lighthouse is side b and the height of the lighthouse is side a. We observe from Figure 42.7 that

$$\cot A = \frac{b}{a}$$

or

$$\cot 18° = \frac{b}{140}$$

or

$$b = 140(\cot 18°) = 140(3.078) = 431$$

Thus, the ship is 431 ft from the lighthouse.

Figure 42.7

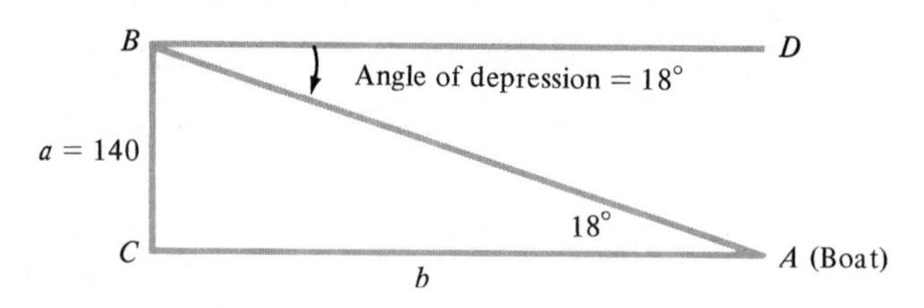

Exercises for Section 42

In Exercises 1 through 12, use Figure 42.1 to solve for the remaining parts of the right triangle ABC. Note that $C = 90°$.

1. $A = 32°, c = 8$

2. $A = 35°, a = 50$

3. $a = 15, b = 24$

4. $A = 52°, b = 5.2$

5. $B = 63°, c = 16$

6. $b = 12, c = 16$

7. $A = 46°, a = 54$

8. $a = 32, b = 40$

9. $B = 63°20'$, $a = 26.7$

10. $A = 32°40'$, $c = 12$

11. $b = 12.2$, $c = 16$

12. $A = 63°10'$, $b = 22.3$

13. Find the height of a telephone pole whose horizontal shadow is 65 ft long when the angle of elevation of the sun is 55°.

14. Find the area of the right triangle when $A = 40°$, $C = 90°$, and $a = 20$.

15. Find the perimeter of an isosceles triangle whose base is 30 in. and whose base angle is 65°.

16. Find the length of the diagonal of a rectangular plot of land that has a length of 120 m and a width of 100 m.

17. An observer on the ground notes that the angle of elevation of a balloon is 48°20′. How high is the balloon if a point on the ground directly below it is 245 ft from the observer?

18. The string on a kite is taut and makes an angle of 25° with the horizontal. Find the height of the kite if 150 ft is let out and the end of the string is 4.5 ft above the ground.

19. A ladder 30 ft long leans against the side of a building, and the angle between the ladder and the building is 13°50′. How far is the bottom of the ladder from the building?

20. Two buildings with flat roofs are 40 ft apart. The angle of elevation from the roof of the shorter building to the edge of the taller building is 36°. If the shorter building is 30 ft high, find the height of the taller building.

21. A road has a rise of 3 miles for every 8 miles. Find the angle of rise.

22. From a point 315 ft above the ground, an observer finds that the angle of depression of an object on the ground is 52°20′. How far is the object on the ground from a point on the ground directly below the observer?

23. From a point 25 ft above ground level it is observed that the angle of depression of the base of a building is 17°20′ and that the angle of elevation of the top of the building is 43°. Find the height of the building.

24. A guy wire is fastened to a pole at a point 18 ft from the ground and to the ground 14 ft from the bottom of the pole. Find the length of the wire and the angle it makes with the pole.

25. If a tower 60 ft high casts a horizontal shadow 100 ft long, find the angle of elevation of the sun.

26. The angle of depression of a boat from the top of a lighthouse is 16° when the boat is 150 m from the lighthouse. Find the height of the lighthouse.

27. The length of a pendulum is 36 in. and the horizontal distance between its extremes is 20 in. Find the angle through which the pendulum swings.

28. A glider is approaching an airport from a distance of 3 miles at a height of 2600 ft. At what angle with the horizontal must the pilot maintain if he is to reach the airport?

SECTION 43

Solution of Oblique Triangles; Laws of Sines and Cosines

In Section 42 we restricted our analysis to the solutions of right triangles. We now wish to develop a method for solving **general** or **oblique triangles**. For all such discussions of the oblique triangle, we shall again use the symbols a, b, and c to represent the lengths of the sides opposite the angles A, B, and C, respectively, as shown in Figure 43.1.

Figure 43.1

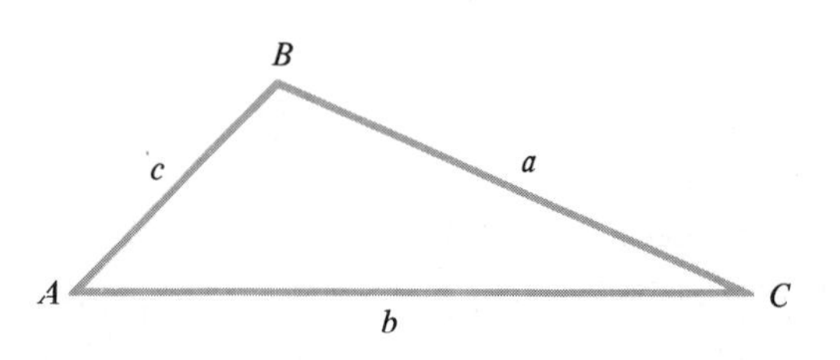

A triangle such as the one shown in Figure 43.1 can be solved if a side and two other parts are given. The following is a list of the possible combinations of given parts that will enable us to solve oblique triangles:

Case 1. Given one side and two angles.
Case 2. Given two sides and the angle opposite one of them.
Case 3. Given two sides and the included angle.
Case 4. Given three sides.

Cases 1 and 2 can be solved by using the **law of sines**. We can develop this law by considering Figure 43.2. In this figure we drop an altitude from vertex B to its opposite side b and label the length of this altitude h. We note that $\sin A = h/c$ or $h = c \sin A$ and that $\sin C = h/a$ or $h = a \sin C$. Equating these two results we obtain $a \sin C = c \sin A$. If we drop altitudes h from each of the other vertices (A and C) to their respective opposite

Figure 43.2

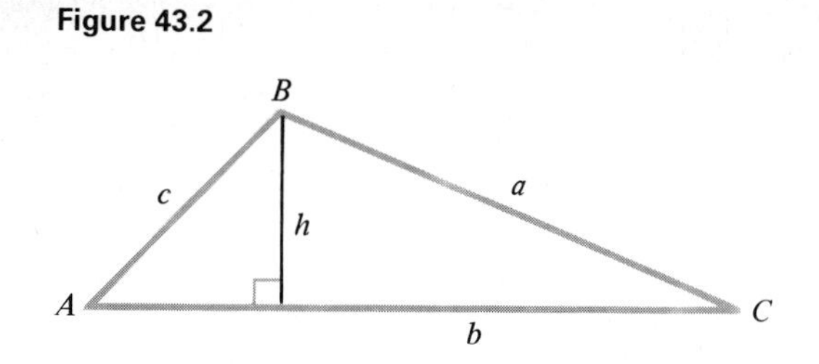

sides (extended if necessary), we would obtain $b \sin C = c \sin B$ and $b \sin A = a \sin B$. We can use these results to obtain

$$c \sin A = a \sin C$$

or

$$\frac{c}{\sin C} = \frac{a}{\sin A} \qquad (1)$$

Also,

$$b \sin C = c \sin B$$

or

$$\frac{b}{\sin B} = \frac{c}{\sin C} \qquad (2)$$

Finally, we can combine (1) and (2) to obtain the **law of sines**, given by

$$\frac{a}{\sin A} = \frac{b}{\sin B} = \frac{c}{\sin C} \qquad (43.1)$$

We now consider the solution for case 1. Suppose that we are given side a and angles A and B. Since the sum of the angles of a triangle is 180°, we have $C = 180° - (A + B)$. We can then use the law of sines to find the sides b and c. From (43.1) we see that

$$b = \frac{a \sin B}{\sin A} \quad \text{and} \quad c = \frac{a \sin C}{\sin A}$$

Example 43.1

Suppose that $a = 120$, $A = 54°$, and $B = 23°$. Find b, c, and C.

Solution

We first find C by noting that $C = 180° - (54° + 23°) = 103°$. From (43.1) we have

$$\frac{120}{\sin 54°} = \frac{b}{\sin 23°} \quad \text{or} \quad b = \frac{120(\sin 23°)}{\sin 54°} = \frac{120(0.3907)}{0.7986} = 59$$

and

$$\frac{120}{\sin 54°} = \frac{c}{\sin 103°} \quad \text{or} \quad c = \frac{120(\sin 103°)}{\sin 54°} = \frac{120(0.9744)}{0.7986} = 146$$ ●

The law of sines can be applied to case 2 also. Suppose that we are given sides a and b and angle A. Then from (42.1) we can use $a/\sin A = b/\sin B$ to obtain $\sin B = b \sin A/a$. Thus, we can find the value of $\sin B$. The value of $\sin B$ depends on the data given, and this value may be less than 1, equal to 1 or greater than 1. These situations will produce no solution, one solution, or two solutions. For this reason we refer to case 2 as the **ambiguous case**. We list the possibilities for this case as follows:

1. If $\sin B > 1$, no angle B can be determined; thus, *no solution.*
2. If $\sin B = 1$, $B = 90°$ and a right triangle is determined; thus, *one solution.*
3. If $\sin B < 1$, two angles are determined: an acute angle B and obtuse angle $B_1 = 180° - B$. Thus, we have the possibility of two solutions. The angle B will always be a solution and the angle B_1 will yield a second solution only if $B_1 + A < 180°$; otherwise, we will only have one solution, B.

Example 43.2

Suppose that $a = 20$, $b = 27$, and $A = 30°$. Find the value of B.

Solution

We use (43.1), to obtain

$$\frac{20}{\sin 30°} = \frac{27}{\sin B}$$

or

$$\sin B = \frac{27(\sin 30°)}{20} = \frac{27(0.5)}{20} = 0.6750$$

Since $\sin B < 1$ we have from Table I two angles: $B = 43°$ and $B_1 = 180° - 43° = 137°$. The angle $B = 43°$ gives us a solution. Since $B_1 + A = 137° + 30° = 167° < 180°$, $B_1 = 137°$ also gives us a solution. Thus, we have *two* solutions.

Example 43.3

Suppose that $a = 8$, $b = 5$, and $A = 40°$. Find the value of B.

Solution

We first use (43.1) to solve for B, as follows:

$$\frac{8}{\sin 40°} = \frac{5}{\sin B}$$

or

$$\sin B = \frac{5(\sin 40°)}{8} = \frac{5(0.6428)}{8} = 0.4018$$

We now use Table I to obtain $B = 23°40'$ and $B_1 = 180° - 23°40' = 156°20'$. Since $A + B_1 = 40° + 156°20' = 196°20' > 180°$, B_1 is not a solution. Thus, we have only *one* solution, $B = 23°40'$.

Example 43.4

Suppose that $a = 5$, $b = 8$, and $A = 40°$. Find the value of B.

Solution

We use (43.1), to obtain

$$\frac{5}{\sin 40°} = \frac{8}{\sin B}$$

or

$$\sin B = \frac{8(\sin 40°)}{5} = \frac{8(0.6428)}{5} = 1.029$$

Since $\sin B > 1$, there is no angle B. Thus, there is no solution. ●

In cases 3 and 4 we do not have enough information to apply the law of sines. Thus, we need a second law for these cases. This law is known as the **law of cosines**.

In Figures 43.3a and 43.3b on page 222, we show the oblique triangle ABC with the altitude h dropped from B to its opposite side b at D. In either figure we see that in the right triangle ABD we have $c^2 = h^2 + (AD)^2$. In the right triangle BDC in Figure 43.3a, we have $\sin C = h/a$ or $h = a \sin C$. We also note from Figure 43.3a that $AD = b - DC$ and that $\cos C = DC/a$ or $DC = a \cos C$. Thus, $AD = b - a \cos C$. We now substitute these results into $c^2 = h^2 + (AD)^2$ as follows, to obtain

$$\begin{aligned} c^2 &= h^2 + (AD)^2 \\ &= (a \sin C)^2 + (b - a \cos C)^2 \\ &= a^2 \sin^2 C + b^2 - ab \cos C + a^2 \cos^2 C \\ &= a^2(\sin^2 C + \cos^2 C) + b^2 - 2ab \cos C \end{aligned} \tag{1}$$

From (39.2) we have $\sin C = y/r$, $\cos C = x/r$, and $r = \sqrt{x^2 + y^2}$ for *any* angle C. Thus,

$$\sin^2 C + \cos^2 C = \frac{y^2}{r^2} + \frac{x^2}{r^2} = \frac{x^2 + y^2}{(\sqrt{x^2 + y^2})^2} = 1 \tag{2}$$

We now substitute the result of (2) into (1), to obtain

$$c^2 = a^2 + b^2 - 2ab \cos C \tag{43.2}$$

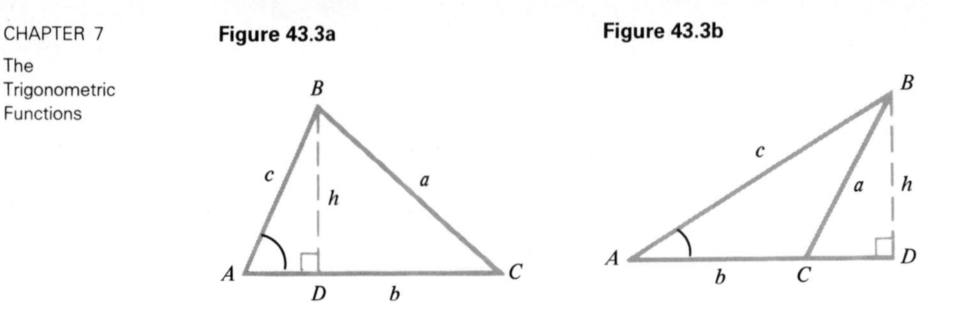

Using similar methods we can also obtain the following two equations:

$$a^2 = b^2 + c^2 - 2bc \cos A \tag{43.3}$$

$$b^2 = a^2 + c^2 - 2ac \cos B \tag{43.4}$$

Formulas (43.2), (43.3), and (43.4) are known as the **law of cosines**. We can express this law in words as follows.

Law of Cosines
The square of any side of a triangle is equal to the sum of the squares of the other two sides minus twice the product of the other two sides and the cosine of the angle between them.

In case 3 when we are given two sides and the included angle of any triangle, we can apply the law of cosines to solve for the side opposite the given angle. We can then proceed to find the remaining angles by using the law of sines.

Example 43.5
Suppose that $a = 344$, $b = 120$, and $C = 43°30'$. Find the remaining parts of triangle ABC.

Solution
We find side c by applying (43.2) as follows:

$$\begin{aligned} c^2 &= a^2 + b^2 - 2ab \cos C \\ &= (344)^2 + (120)^2 - 2(344)(120)(0.7254) \\ &= 118{,}336 + 14{,}400 - 59{,}889 \\ c &= \sqrt{72{,}847} = 270 \end{aligned}$$

Next we can apply the law of sines to find angle B, as follows:

$$\frac{120}{\sin B} = \frac{270}{\sin 43°30'}$$

or

$$\sin B = \frac{120(\sin 43°30')}{270} = \frac{120(0.6884)}{270} = 0.3060$$

From Table I we have $B = 17°50'$. Finally, we note that

$$A = 180° - (B + C) = 180° - (17°50' + 43°30')$$
$$= 180° - 61°20' = 118°40'$$

Example 43.6

Suppose in triangle ABC that $a = 8$, $b = 10$, and $c = 5$. Find the remaining parts of triangle ABC.

Solution

Since we are given the three sides of the triangle, we can apply (43.3) to find angle A as follows:

$$a^2 = b^2 + c^2 - 2bc \cos A$$
$$64 = 100 + 25 - 2(10)(5) \cos A$$
$$-61 = -100 \cos A$$
$$\cos A = \frac{61}{100} = 0.6100$$

From Table I we have $A = 52°$. We can now find angle B by applying (43.4), to obtain

$$b^2 = a^2 + c^2 - 2ac \cos B$$
$$100 = 64 + 25 - 2(8)(5) \cos B$$
$$11 = -80 \cos B$$
$$\cos B = -\frac{11}{80} = -0.1375$$

Since $\cos B = -0.1375$ B lies in quadrant II and the cosine of its reference angle $\bar{B}$ is 0.1375. From Table I we see that $\bar{B} = 82°$ and $B = 180° - \bar{B} = 98°$. Finally, we note that since the sum of the interior angles of triangle ABC must equal 180°, we have

$$C = 180° - (A + B)$$
$$= 180° - (52° + 98°)$$
$$= 180° - 150° = 30°$$

Example 43.7

Two observers on ground level 2150 ft apart measure the angle of elevation of an airplane flying over the line joining them. The angle of elevation at A is found to be 59°40′ and at B it is 52°20′. Find how far the plane is from each observer.

Solution

In Figure 43.4 we let point C represent the position of the airplane. We wish to find AC and CB. We first note that $C = 180° - (A + B) = 180° - (59°40' + 52°20') = 68°$. We can find the distance from observer A to the plane by applying the law of sines as follows:

$$\frac{AC}{\sin 52°20'} = \frac{2150}{\sin 68°}$$

or

$$AC = \frac{2150(0.7916)}{(0.9272)} = 1836 \text{ ft}$$

Figure 43.4

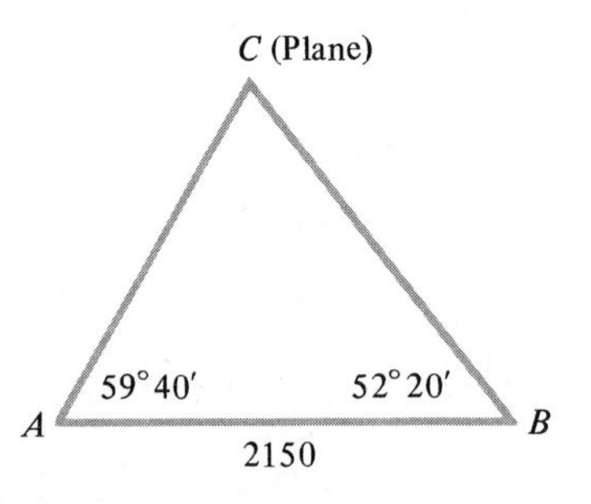

The distance CB from the plane to observer B can also be found by using the law of sines, to obtain

$$\frac{CB}{\sin 59°40'} = \frac{2150}{\sin 68°}$$

or

$$CB = \frac{2150(0.8631)}{0.9272} = 2001 \text{ ft}$$

Example 43.8

A tunnel is to be constructed through a mountain from point A to point B. At a point from which both A and B are visible, the distance to A is 321 ft, and to B, 339 ft. If the angle ACB is $38°30'$, find the length of the tunnel.

Solution

From Figure 43.5 we see that the length of the tunnel is AB. We can find AB by applying the law of cosines as follows:

$$\begin{aligned}(AB)^2 &= (AC)^2 + (CB)^2 - 2(AC)(CB)\cos C \\ &= (321)^2 + (339)^2 - 2(321)(339)(0.7826)\end{aligned}$$

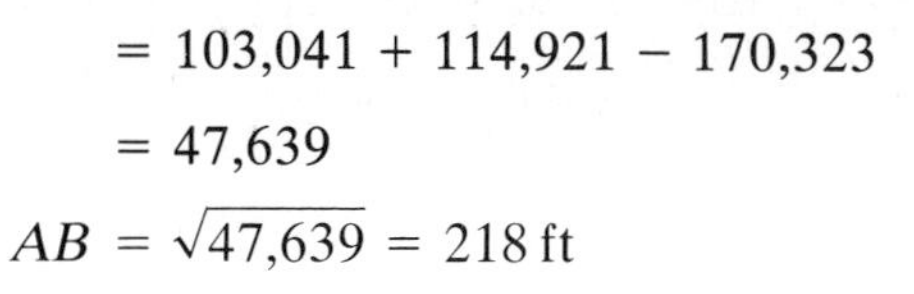

$$= 103{,}041 + 114{,}921 - 170{,}323$$
$$= 47{,}639$$
$$AB = \sqrt{47{,}639} = 218 \text{ ft}$$

Figure 43.5

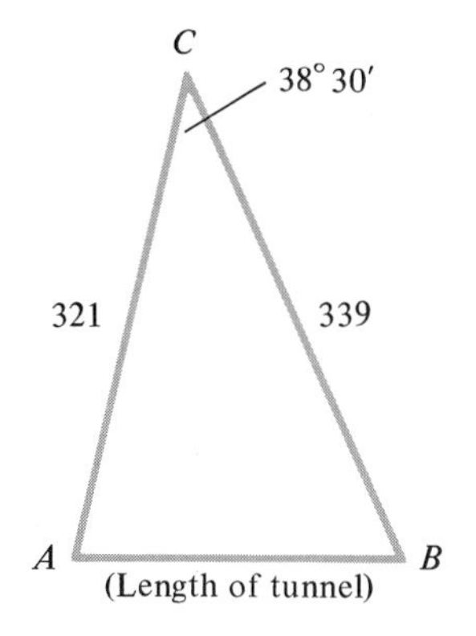

Summary for Solving Oblique Triangles

Case 1. If we are given one side and two angles, we can apply the **law of sines**.

Case 2. If we are given two sides and the angle opposite one of them, we can apply the **law of sines**.

Case 3. If we are given two sides and the included angle, we can apply the **law of cosines** to find the third side. We can then apply the law of sines to find the remaining angles.

Case 4. If we are given three sides, we can apply the **law of cosines** to find one of the angles. The remaining angles can be found by applying the law of sines.

Exercises for Section 43

In Exercises 1 through 20, use either the law of sines or cosines to find the remaining parts of the oblique triangle ABC shown in Figure 43.1.

1. $A = 62°, B = 75°, a = 130$

2. $A = 40°, B = 53°, b = 100$

3. $a = 62.5, b = 143.5, A = 32°$

4. $a = 300, b = 320, A = 35°$

5. $a = 3, b = 4, c = 5$

6. $a = 14.1, b = 16.8, C = 15.0$

7. $a = 20, b = 27, A = 30°$

8. $a = 27, b = 20, A = 30°$

9. $a = 20, b = 25, A = 45°$

10. $a = 24, b = 20, B = 120°$

11. $a = 30, b = 50, c = 65$ **12.** $a = 7.6, c = 9.2, B = 46°$

13. $B = 66°50', A = 75°50', c = 613$

14. $C = 74°10', b = 18.6, a = 24.5$

15. $a = 25.1, b = 50.4, c = 33.3$

16. $A = 39°10', B = 46°40', a = 6.36$

17. $a = 12.2, A = 17°20', c = 16.2$

18. $b = 16.5, c = 21.6, B = 92°20'$

19. $a = 2.52, c = 4.34, B = 103°50'$ **20.** $c = 12.1, b = 23.3, C = 43°$

21. Find the area of the oblique triangle ABC shown in Figure 43.1 in terms of the angle A.

22. Use the results of Exercise 21 to find the area of the oblique triangle ABC if $A = 150°$, $b = 42$, and $c = 32$.

23. Show that the area of *any* triangle is equal to one half the product of any two sides and the sine of their included angle.

24. A surveyor at a point A wishes to locate a point C on a line directly across a swamp. To avoid crossing the swamp he turns an angle 54°20′ through A and then measures 1300 ft to a point B. At B he turns an angle of 63°40′ and then measures directly to the point C. What is the distance between points A and C?

25. A tunnel is to be dug through a mountain from a point A to a point C. From a point B both points A and C are visible. If angle ABC equals 47°20′ and if B is 1124 ft from A and 1324 ft from C, how long is the tunnel?

26. The angle at one corner of a triangular plot of land is 67°20′. If the sides that meet at this corner are 168 yd and 180 yd, find the length of the third side.

27. Find the area of the triangular plot of land in Exercise 26. If 1 acre is approximately 4840 yd^2, how many acres are in this plot?

28. Two buses leave together from point A and travel along straight highways that are at an angle of 83° with one another. If their speeds are 50 and 60 mph, respectively, how far apart will they be after 30 min?

29. A ship leaves a point A and travels 140 miles in the direction 115° and from there travels 200 miles in the direction 135°. How far is the ship from its original point?

30. Find the lengths of the diagonals of a parallelogram whose sides are 18 and 24 and has a vertex of 48°.

SECTION 44
Vectors

In engineering and technology we encounter physical quantities that broadly fall into two categories: quantities that have magnitude or size but are independent of direction, and quantities that have magnitude or size and a specified direction. Thus, it is necessary to establish an efficient notation to describe these quantities. We do so with the following definitions.

A **scalar quantity** is a quantity that has magnitude or size but no direction. Some examples of scalar quantities are length, time, and speed.

A **vector quantity** is a quantity that has magnitude *and* direction. Some examples of vector quantities are velocity, acceleration, and force.

A vector quantity may be represented by a straight-line segment of definite length called a **vector**. A vector can be represented graphically by a directed line segment. In Figure 44.1 we have indicated a directed straight-line segment with an initial point at O and a terminal point at P. In this text such vectors will be indicated by either a letter with the symbol → over it or by using the initial and terminal points of the directed line segment with the symbol → over. Thus, in Figure 44.1 we have the vector $\vec{V}$ or $\overrightarrow{OP}$. The direction of the arrowhead is the direction of the vector, and the length of segment is drawn to arbitrary scale to indicate its magnitude. The **magnitude** of the vector $\vec{V}$ or $\overrightarrow{OP}$ will be denoted by $|\vec{V}|$ or $|\overrightarrow{OP}|$.

Figure 44.1

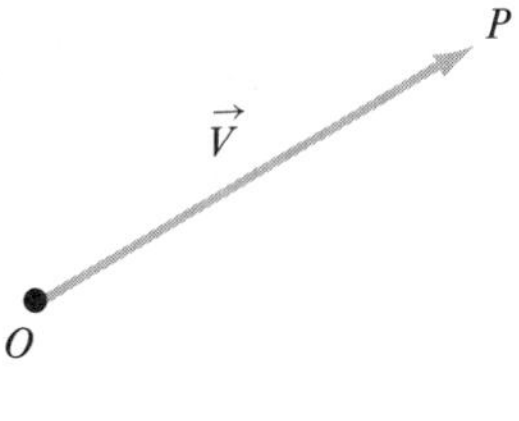

P
$\vec{V}$
O

Example 44.1

In Figure 44.2 on page 228, we show a graphic representation of a velocity of 20 ft/sec that is 40° north of east. The scale is 1 cm = 5 ft/sec. ●

Two vectors are equal if they have the same magnitude and the same direction. In Figure 44.3 $\overrightarrow{V_1} = \overrightarrow{V_2}$, since $|\overrightarrow{V_1}| = |\overrightarrow{V_2}|$ and they have the same direction. In Figure 44.4 $\overrightarrow{V_3} \neq \overrightarrow{V_4}$, since $|\overrightarrow{V_3}| \neq |\overrightarrow{V_4}|$. We also see that $|\overrightarrow{V_3}| = |\overrightarrow{V_5}|$ but $\overrightarrow{V_3} \neq \overrightarrow{V_5}$, since they have different directions.

The **resultant** or **vector sum** of two vectors in the same plane is that vector in the plane which would produce the same effect as that produced by

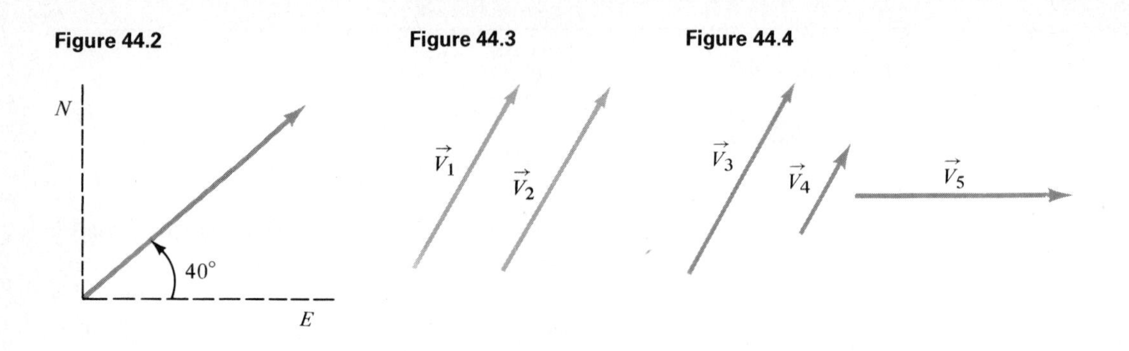

Figure 44.2 **Figure 44.3** **Figure 44.4**

the two vectors acting together. We can find the vector sum of two vectors graphically by observing the following rule.

Vector Addition

To add two vectors $\vec{A}$ and $\vec{B}$ we move (if necessary) one of the vectors parallel to itself until both vectors have the same initial point and then draw the parallelogram determined by the two vectors. The sum or resultant vector is the vector formed by the diagonal of the parallelogram that has the same initial point as A and B. This method is frequently referred to as the **parallelogram method**.

Example 44.2

Find the sum of the vectors $\vec{A}$ and $\vec{B}$ shown in Figure 44.5

Figure 44.5

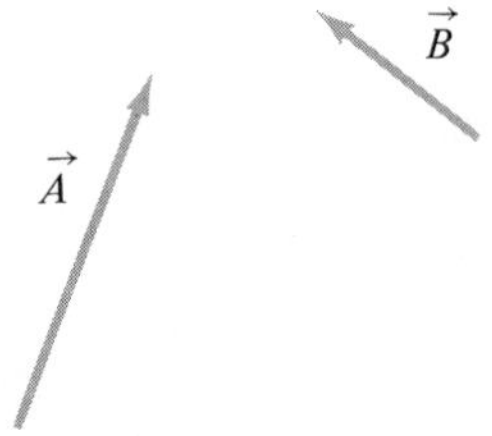

$\vec{B}$

$\vec{A}$

Solution

We shift vector $\vec{B}$ parallel to itself until its initial point coincides with that of vector $\vec{A}$ and then draw the parallelogram formed by the two vectors. This sum is denoted by the vector $\vec{C}$ in Figure 44.6. ●

If two vectors have the same direction, their sum is a vector in the same direction with a magnitude equal to the sum of the magnitudes.

Figure 44.6

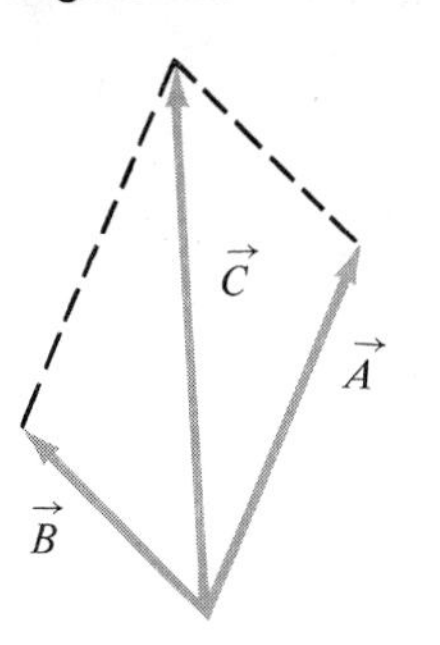

Example 44.3

In Figure 44.7 we see that $\vec{A}$ and $\vec{B}$ have the same direction and that their sum is $\vec{R}$. We also see that $|\vec{R}| = |\vec{A}| + |\vec{B}| = 100 + 50 = 150$.

Figure 44.7

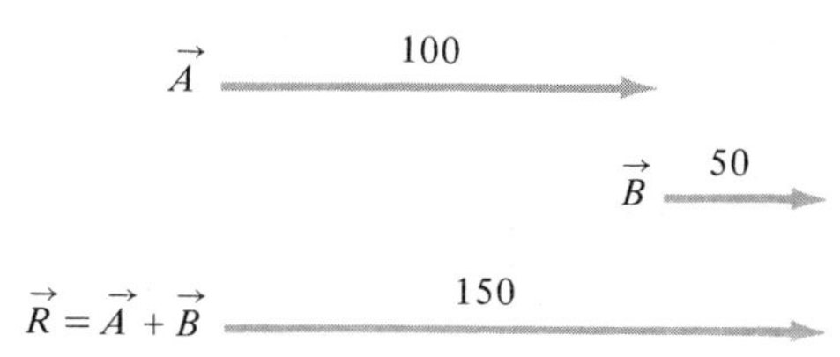

●

If two vectors have opposite directions, their sum is a vector in the direction of the vector with the greater magnitude. The magnitude of this sum is the difference between their magnitudes (larger magnitude minus smaller magnitude).

Example 44.4

In Figure 44.8 we see that $\vec{A}$ and $\vec{B}$ have opposite directions and that their sum $\vec{R}$ has the same direction as $\vec{B}$. We also note that $|\vec{R}| = |\vec{B}| - |\vec{A}| = 150 - 50 = 100$. ●

The graphic addition of any two vectors such as the vectors shown in Figure 44.6 yields only approximate results. The addition of vectors can also

Figure 44.8

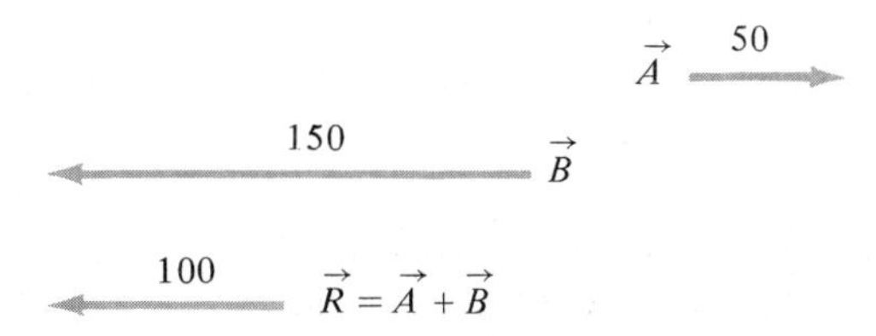

be done by using the Pythagorean theorem and the trigonometric functions. Consider the following examples.

Example 44.5

Find the magnitude and direction of the sum of the vectors $\vec{A}$ and $\vec{B}$ if they are at right angles and $|\vec{A}| = 7.2$ and $|\vec{B}| = 5.4$.

Solution

In Figure 44.9 we show a sketch of the sum $\vec{R} = \vec{A} + \vec{B}$. Note that it is not necessary to make the drawing to scale. The vectors $\vec{A}$, $\vec{B}$, and $\vec{R}$ form a right triangle where $|\vec{R}|$ is the hypotenuse and $|\vec{A}|$ and $|\vec{B}|$ are the legs. We can apply the Pythagorean theorem, to obtain

$$|\vec{R}|^2 = |\vec{A}|^2 + |\vec{B}|^2$$

$$|\vec{R}| = \sqrt{(7.2)^2 + (5.4)^2}$$

$$= \sqrt{51.84 + 29.16} = \sqrt{81} = 9$$

Figure 44.9

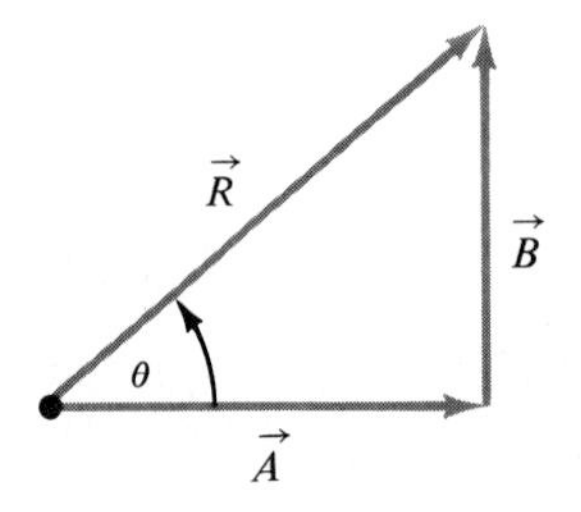

We can find the direction of $\vec{R}$ by finding θ. We see from Figure 44.9 that

$$\tan\theta = \frac{|\vec{B}|}{|\vec{A}|} = \frac{5.4}{7.2} = \frac{3}{4}$$

or $\theta = 36°50'$. Thus, the resultant $\vec{R}$ has a magnitude of 9 and makes a positive angle of $36°50'$ with $\vec{A}$. ●

The addition of two vectors that are not at right angles can be done most efficiently by first reexpressing each vector into its **rectangular components**. In Figure 44.10 we have drawn a vector $\vec{V}$ with its initial point at the origin O in the xy coordinate system and its terminal point at $P(x, y)$. If we drop a perpendicular from P to the x axis at S, we see that $\vec{V} = \overrightarrow{OP} = \overrightarrow{OS} + \overrightarrow{SP}$. From the right triangle OPS we see that

$$\sin\theta = \frac{|\overrightarrow{SP}|}{|\vec{V}|} \quad \text{or, equivalently,} \quad |\overrightarrow{SP}| = |\vec{V}|\sin\theta$$

and

$$\cos\theta = \frac{|\overrightarrow{OS}|}{|\vec{V}|} \quad \text{or, equivalently,} \quad |\overrightarrow{OS}| = |\vec{V}|\cos\theta$$

We call $|\overrightarrow{SP}|$ the **vertical component** of $\vec{V}$ and denote it as V_y, and we call $|\overrightarrow{OS}|$ the **horizontal component** of $\vec{V}$ and denote it as V_x. Thus,

$$V_x = |\vec{V}|\cos\theta \tag{44.1}$$

$$V_y = |\vec{V}|\sin\theta \tag{44.2}$$

The Pythagorean theorem gives us a relationship between $|\vec{V}|$, V_x, and V_y:

$$|\vec{V}| = \sqrt{V_x^2 + V_y^2} \tag{44.3}$$

From Figure 44.10 we also have

$$\tan\theta = \frac{V_y}{V_x} \tag{44.4}$$

Figure 44.10

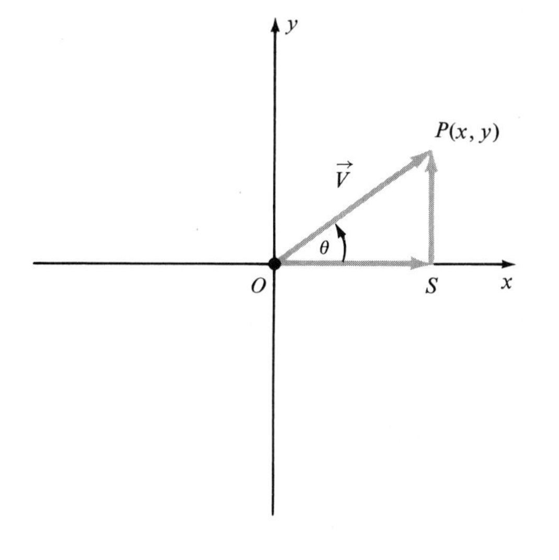

Example 44.6

Find the rectangular components for a vector $\vec{V}$ whose magnitude is 7 and is directed at an angle of 27°.

Solution

We can find V_x and V_y by using formulas (44.1) and (44.2) as follows:

$$V_x = 7\cos 27° = 7(0.8910) = 6.237$$

$$V_y = 7\sin 27° = 7(0.4540) = 3.178$$

●

We can define vector addition in terms of components. If $\vec{R}$ is the sum of the two vectors $\vec{A}$ and $\vec{B}$, then the x and y components of $\vec{R}$ are the respective sums of the x and y components of $\vec{A}$ and $\vec{B}$. That is,

$$R_x = A_x + B_x \tag{44.5}$$

$$R_y = A_y + B_y \tag{44.6}$$

Now from (44.3) and (44.4) we have

$$|\vec{R}| = \sqrt{R_x^2 + R_y^2} \tag{44.7}$$

$$\tan \theta = \frac{R_y}{R_x} \tag{44.8}$$

The formulas above give us a procedure for finding the magnitude and direction for the sum $\vec{R}$ of two vectors $\vec{A}$ and $\vec{B}$. We can find the horizontal component of $\vec{R}$ by summing the horizontal components of $\vec{A}$ and $\vec{B}$. Similarly, we can find the vertical component of $\vec{R}$ by summing the vertical components of $\vec{A}$ and $\vec{B}$. Finally, we use formulas (44.7) and (44.8), respectively, to calculate the magnitude and direction of $\vec{R}$.

Example 44.7

If $\vec{A}$ has a magnitude of 12 and is directed at an angle $\theta_A = 32°$ and $\vec{B}$ has a magnitude of 16 and is directed at an angle $\theta_B = 53°$, find the magnitude and direction of their resultant.

Solution

We use formulas (44.1) and (44.2) to find the x and y components of $\vec{A}$ and $\vec{B}$ as follows:

$$A_x = 12 \cos 32° = 10.18$$

$$A_y = 12 \sin 32° = 6.36$$

$$B_x = 16 \cos 53° = 9.63$$

$$B_y = 16 \sin 53° = 12.78$$

We can now apply formulas (44.5) and (44.6), to obtain

$$R_x = 10.18 + 9.63 = 19.81$$

$$R_y = 6.36 + 12.78 = 19.14$$

Finally, we can use formula (44.7) to find that the magnitude of $\vec{R}$ is

$$|\vec{R}| = \sqrt{(19.81)^2 + (19.14)^2} = 27.5$$

and formula (44.8) to find the directed angle for $\vec{R}$ to be

$$\tan \theta = \frac{19.14}{19.18}$$

or

$$\theta = 45°$$

●

We can find the magnitude and direction of the sum of three or more vectors in an analogous manner.

Example 44.8

Find the magnitude and direction of the sum of the following vectors: $|\vec{A}| = 12$, $\theta_A = 36°$; $|\vec{B}| = 22$, $\theta_B = 47°$; and $|\vec{C}| = 18$, $\theta_C = 108°$.

Solution

We first find the x and y components of the vectors as follows:

$$A_x = 12 \cos 36° = 9.7$$

$$B_x = 22 \cos 47° = 15$$

$$C_x = 18 \cos 108° = -5.6$$

$$A_y = 12 \sin 36° = 7.1$$

$$B_y = 22 \sin 47° = 16$$

$$C_y = 18 \sin 108° = 17$$

Next we find R_x and R_y, as follows:

$$R_x = 9.7 + 15 - 5.6 = 19.1$$

$$R_y = 7.1 + 16 + 17 = 40.1$$

Now we have

$$|\vec{R}| = \sqrt{(19.1)^2 + (40.1)^2} = 44$$

and

$$\tan \theta = \frac{40.1}{19.1} = 2.099$$

or

$$\theta = 64°30'$$

●

We can also define vector subtraction. To do this we denote the **negative of a vector** $\vec{A}$ as $-\vec{A}$ and say that it has the same magnitude as $\vec{A}$ but the *opposite direction.* Thus, we can treat the subtraction of $\vec{B}$ from $\vec{A}$, denoted $\vec{A} - \vec{B}$, as $\vec{A} + (-\vec{B})$. In other words, *to subtract B from A we reverse the direction of B and apply the parallelogram method.*

Example 44.9

Find the difference $\vec{A} - \vec{B}$ for the vectors shown in Figure 44.11.

Solution

Since $\vec{A} - \vec{B} = \vec{A} + (-\vec{B})$, we first **reverse** the direction of $\vec{B}$ and then shift $-\vec{B}$ parallel to itself until its initial point coincides with the point of $\vec{A}$. We then draw the parallelogram formed by $\vec{A}$ and $\vec{B}$ and indicate this difference as $\vec{C}$ in Figure 44.12.

Figure 44.11 Figure 44.12

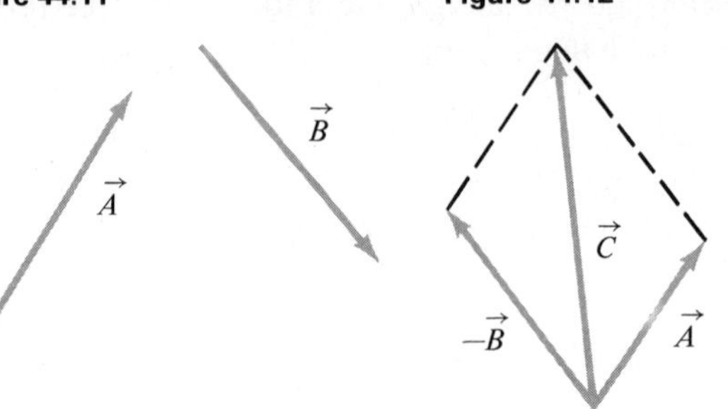

Example 44.10

In Example 44.4 we could describe the resultant vector $\vec{R}$ as $\vec{B} - \vec{A}$.

Exercises for Section 44

In Exercises 1 through 6, determine which quantities are vectors and which are scalars.

1. 15 ft/sec **2.** 15 ft/sec, upward

3. 46° **4.** 20 newtons (N), downward

5. 6.3 m/sec^2 **6.** 6.3 cm/sec^2 at angle of 40° with the horizontal

7. Draw a graphic representation of an upward force of 50 lb that makes an angle of 45° with the positive x axis.

8. Draw a graphic representation of a downward force of 30 N that makes an angle of −45° with the negative y axis.

In Exercises 9 through 16, use the parallelogram method to find the indicated sums and differences. Use the vectors $\vec{A}$, $\vec{B}$, and $\vec{C}$ shown in Figure 44.13.

9. $\vec{A} + \vec{B}$ **10.** $\vec{A} + \vec{B} + \vec{C}$ **11.** $\vec{B} - \vec{A}$ **12.** $\vec{A} - \vec{B}$

13. $\vec{A} + 3\vec{B}$ **14.** $\vec{A} + \vec{B} - \vec{C}$ **15.** $2\vec{B} + \vec{C}$ **16.** $2\vec{B} - \vec{A}$

Figure 44.13

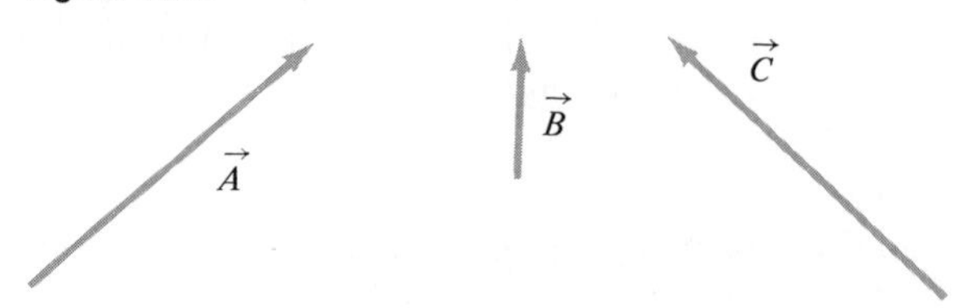

In Exercises 17 through 22, use formulas (44.1) and (44.2) to find the x and y components for each vector.

17. Magnitude 16, $\theta = 43°$ **18.** Magnitude 8.2, $\theta = 118°$

19. Magnitude 120, $\theta = 214°$

20. Magnitude 17.3, $\theta = 308°$

21. Magnitude 1, $\theta = \pi/4$

22. Magnitude 612, $\theta = -40°$

23. Find the rectangular coordinates of a force of 28 N at an angle of 36° with the positive x axis.

24. Find the rectangular coordinates of a force of 100 lb at an angle of −41° with the positive x axis.

25. If a force A has a magnitude of 80 lb at an angle of 45° with the positive x axis and force B has a magnitude of 60 lb at an angle of 225° with the positive x axis. Find the magnitude and direction of their resultant.

In Exercises 26 through 32, find the magnitude and a direction for the ***resultant*** *of each vector.*

26. $|\vec{A}| = 4$, $|\vec{B}| = 3$, $\vec{A}$ and $\vec{B}$ have same direction.

27. $|\vec{A}| = 4$, $|\vec{B}| = 3$, $\vec{A}$ and $\vec{B}$ have opposite directions.

28. $|\vec{A}| = 8$, $|\vec{B}| = 9$, $\vec{A}$ and $\vec{B}$ are at right angles.

29. $|\vec{A}| = 16.2$, $|\vec{B}| = 9.12$, $\vec{A}$ and $\vec{B}$ are at right angles.

30. $|\vec{A}| = 21.6$, $|\vec{B}| = 33.4$, $\vec{A}$ and $\vec{B}$ are at an angle of 48°.

31. $|\vec{A}| = 130$, $|\vec{B}| = 185$, $\vec{A}$ and $\vec{B}$ are at an angle of 21°.

32. $|\vec{A}| = 29.4$, $|\vec{B}| = 8.76$, $\vec{A}$ and $\vec{B}$ are at an angle of 200°

In Exercises 33 through 40, use the methods of Examples 44.7 and 44.8 to find the sum of each vector.

33. $|\vec{A}| = 6$, $\theta_A = 13°$
$|\vec{B}| = 14$, $\theta_B = 63°$

34. $|\vec{A}| = 128$, $\theta_A = 38°$
$|\vec{B}| = 212$, $\theta_B = 144°$

35. $|\vec{A}| = 0.912$, $\theta_A = 200°$
$|\vec{B}| = 0.786$, $\theta_B = 65°$

36. $|\vec{A}| = 5.68$, $\theta_A = 300°$
$|\vec{B}| = 12.3$, $\theta_B = 753°$

37. $|\vec{A}| = 1210$, $\theta_A = -53°$
$|\vec{B}| = 1460$, $\theta_B = 262°$

38. $|\vec{A}| = 16$, $\theta_A = 33°$
$|\vec{B}| = 23$, $\theta_B = 250°$
$|\vec{C}| = 42$, $\theta_C = 192°$

39. $|\vec{A}| = 6.91$, $\theta_A = 20°$
$|\vec{B}| = 8.16$, $\theta_B = 40°$
$|\vec{C}| = 4.12$, $\theta_C = 60°$

40. $|\vec{A}| = 211$, $\theta_A = 45°$
$|\vec{B}| = 479$, $\theta_B = 225°$
$|\vec{C}| = 865$, $\theta_C = 405°$

SECTION 45

Applications of Vectors

We can now apply the techniques developed in Sections 42 and 43 for solving triangles to the solution of some practical problems. The discussion of vectors in Section 44 will also prove helpful in the solution of some of these problems. We demonstrate this with the following examples.

Example 45.1

The velocity of a boat is 20 mph in a direction of 40° north of east. The wind velocity is 4 mph from the west. Determine the magnitude and direction of the resultant velocity of the boat.

Solution

In Figure 45.1 we can represent the resultant velocity of the boat as the side of a triangle in which the two given velocities are the other sides. We label $\overrightarrow{V_W}$ as velocity vector for the wind, $\overrightarrow{V_B}$ as velocity vector for the boat, and $\vec{V}$ as the resultant velocity vector. We can apply the law of cosines to find $|\vec{V}|$ as follows:

$$\begin{aligned} |\vec{V}|^2 &= |\overrightarrow{V_W}|^2 + |\overrightarrow{V_B}|^2 - 2|\overrightarrow{V_W}||\overrightarrow{V_B}| \cos(180° - 40°) \\ &= (20)^2 + (4)^2 - 2(20)(4)\cos 140° \\ &= 400 + 16 + 123 \\ |\vec{V}| &= \sqrt{539} = 23.2 \text{ mph} \end{aligned}$$

Figure 45.1

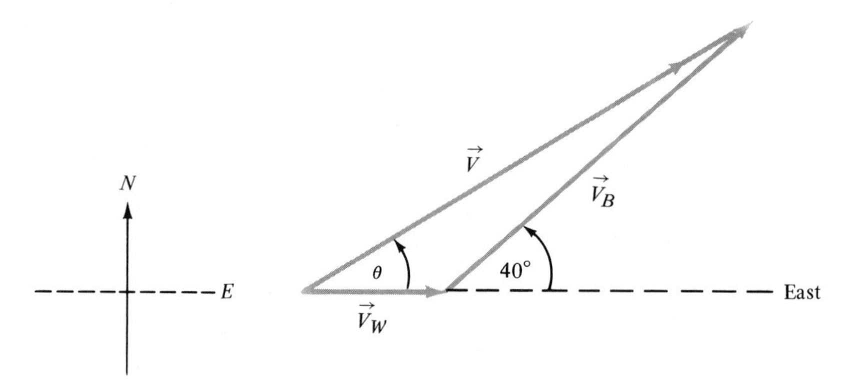

We can find θ by applying the law of sines as follows:

$$\frac{|\overrightarrow{V_B}|}{\sin\theta} = \frac{|\vec{V}|}{\sin 140°}.$$

or

$$\sin\theta = \frac{20(\sin 140°)}{23.2} = 0.5541$$

From Table I of Appendix F, we have

$$\theta = 33°40'$$

Thus, the magnitude of the velocity of the boat is 23.2 mph in the direction of 33°40′ north of east.

Example 45.2

Use the method of x and y components of Section 44 to solve the problem of Example 45.1.

Solution

We first resolve $\overrightarrow{V_B}$ into its x and y components as follows:

$$V_x = |\overrightarrow{V_B}| \cos 40° = 20(0.7660) = 15.3$$

$$V_y = |\overrightarrow{V_B}| \sin 40° = 20(0.6428) = 12.9$$

In Figure 45.1 we note that the x component of $\overrightarrow{V_W}$ is its magnitude, since it is blowing in an easterly direction and the y component of $\overrightarrow{V_W}$ is 0. We can find the x and y components of $\vec{V}$ by summing the respective x and y components of V_B and V_W as follows:

$$V_x = 15.3 + 4 = 19.3$$

$$V_y = 12.9 + 0 = 12.9$$

Thus,

$$|\vec{V}| = \sqrt{V_x^2 + V_y^2}$$

$$= \sqrt{(19.3)^2 + (12.9)^2} = \sqrt{538.9} = 23.2 \text{ mph}$$

We can find θ by observing that

$$\tan \theta = \frac{V_y}{V_x} = \frac{12.9}{19.3} = 0.6684$$

From Table I, we have

$$\theta = 33°50'$$

Note that the computations are simplified considerably by using the method of rectangular components.

Example 45.3

An object on an inclined plane of 25° has a downward force of 50 lb acting on it. Resolve the force into a component parallel to the plane and a component perpendicular to the plane.

Solution

In Figure 45.2 we indicated the downward force as $\vec{F}$, its component parallel to the plane as $\overrightarrow{F_1}$, and its component perpendicular to the plane as $\overrightarrow{F_2}$. The angle between $\vec{F}$ and $\overrightarrow{F_2}$ is 25°, since the two vectors are perpendicular to the two sides of the incline angle. We can now use the definition of the trigonometric functions sine and cosine in the right

Figure 45.2

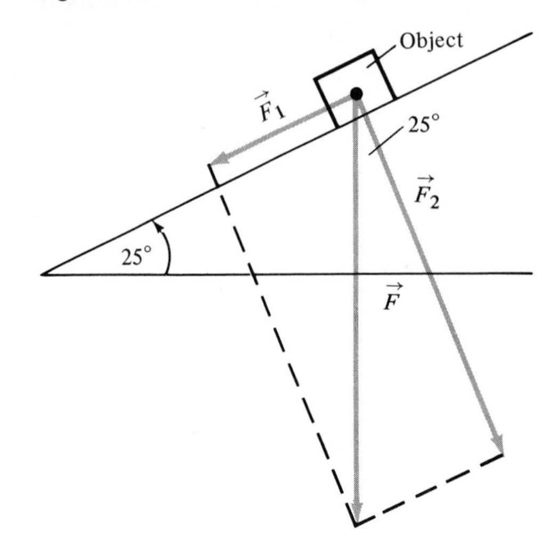

triangle, to obtain

$$\cos 25° = \frac{|\vec{F_2}|}{|\vec{F}|}$$

or

$$|\vec{F_2}| = 50 \cos 25° = 50(0.9063) = 45.3 \text{ lb}$$

and

$$\cos (90 - 25°) = \frac{|\vec{F_1}|}{|\vec{F}|}$$

or

$$|\vec{F_1}| = 50 \cos 75° = 50(0.2588) = 12.9 \text{ lb}$$

Example 45.4

An airplane is flying horizontally at the rate of 400 ft/sec when a bullet is fired with a speed of 2400 ft/sec at right angles to the path of the plane. Find the resultant speed and direction of the bullet.

Solution

In Figure 45.3 we let $\overrightarrow{AB}$ represent the velocity of the airplane, $\overrightarrow{AC}$ the velocity of the bullet, $\overrightarrow{AD}$ the resultant velocity of the bullet, and θ the angle the path of the bullet makes with the path of the airplane. We see that $\overrightarrow{BD} = \overrightarrow{AC}$. From the right triangle ABD, we have

$$|\overrightarrow{AD}|^2 = |\overrightarrow{AB}|^2 + |\overrightarrow{BD}|^2$$

$$|\overrightarrow{AD}| = \sqrt{(400)^2 + (2400)^2} = 2433 \text{ ft/sec}$$

Figure 45.3

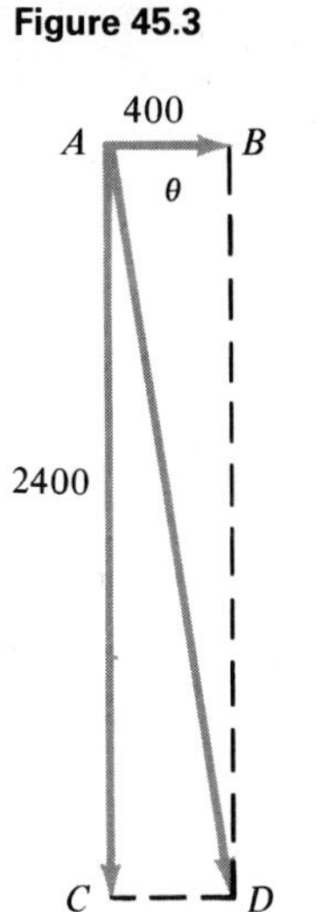

We also have

$$\tan \angle CAD = \frac{|\overrightarrow{BD}|}{|\overrightarrow{AD}|} = \frac{400}{2400} = 0.1667$$

or

$$\angle CAD = 9°30'$$

We see that $\theta = 90° - \angle CAD = 90° - 9°30' = 80°30'$. Thus, the bullet is traveling at a speed of 2433 ft/sec at an angle of 80°30′ with the path of the plane.

Example 45.5

The velocity of an airplane in still air is 300 mph, 38° north of east. If a 50-mph wind blows directly from the south, find the resultant velocity of the plane.

Solution

In Figure 45.4 on page 240, we let $\overrightarrow{AC}$ represent the velocity of the plane, $\overrightarrow{AB}$ the velocity of the wind, and $\overrightarrow{AD}$ the resultant velocity. In the right triangle ACE, angle $ACE = 90° - 38° = 52°$, and angle $ACD = 180° - \text{angle } ACE = 180° - 52° = 128°$, since angles ACE and ACD are supplementary. We also note that $|\overrightarrow{AB}| = |\overrightarrow{CD}| = 50$. We can now apply the law of cosines in triangle ACD to find $|\overrightarrow{AD}|$, as follows:

$$|\overrightarrow{AD}|^2 = |\overrightarrow{AC}|^2 + |\overrightarrow{CD}|^2 - 2|\overrightarrow{AC}||\overrightarrow{CD}| \cos 128°$$

$$= (300)^2 + 50^2 - 2(300)(50)(-0.6157)$$

$$|\overrightarrow{AD}| = \sqrt{110{,}971} = 333 \text{ mph}$$

Figure 45.4

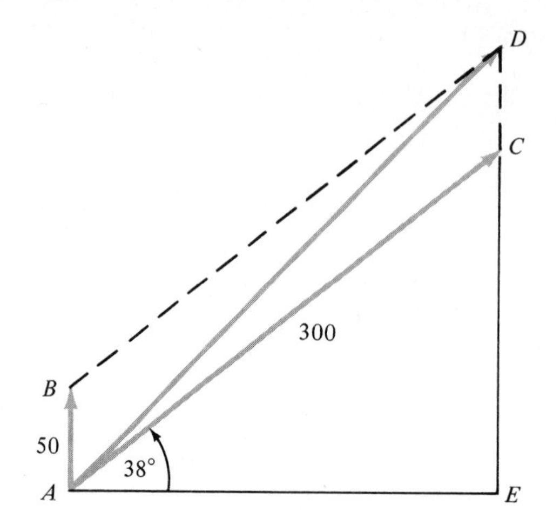

Example 45.6

Determine the magnitude and direction of the resultant displacement (distance and direction an object moves) for the following displacement: 40 miles east; 60 miles, 30° west of north; 20 miles, 40° east of south.

Solution

In Figure 45.5 we represent the given displacements by $\vec{A}$, $\vec{B}$, and $\vec{C}$, respectively. We shall use the method of rectangular components to find the magnitude of their resultant $\vec{R}$. We note that

$$A_x = |\vec{A_x}| = |\vec{A}| = 60$$

$$B_x = |\vec{B_x}| = |\vec{B}| \cos 120° = 40(-0.5) = -20$$

$$C_x = |\vec{C_x}| = |\vec{C}| \cos 310° = 20(0.6428) = 12.9$$

Thus,

$$|\vec{R_x}| = |\vec{A_x}| + |\vec{B_x}| + |\vec{C_x}| = 60 - 20 + 12.9 = 52.9$$

Similarly,

$$A_y = |\vec{A_y}| = 0$$

$$B_y = |\vec{B_y}| = |\vec{B}| \sin 120° = 40(0.8660) = 34.6$$

$$C_y = |\vec{C_y}| = |\vec{C}| \sin 310° = 20(-0.7660) = -15.3$$

Thus,

$$|\vec{R_y}| = |\vec{A_y}| + |\vec{B_y}| + |\vec{C_y}| = 0 + 34.6 - 15.3 = 19.3$$

We can now find $|\vec{R}|$ as follows:

$$|\vec{R}| = \sqrt{|\vec{R_x}|^2 + |\vec{R_y}|^2} = \sqrt{(52.9)^2 + (19.3)^2} = 56.3$$

Figure 45.5

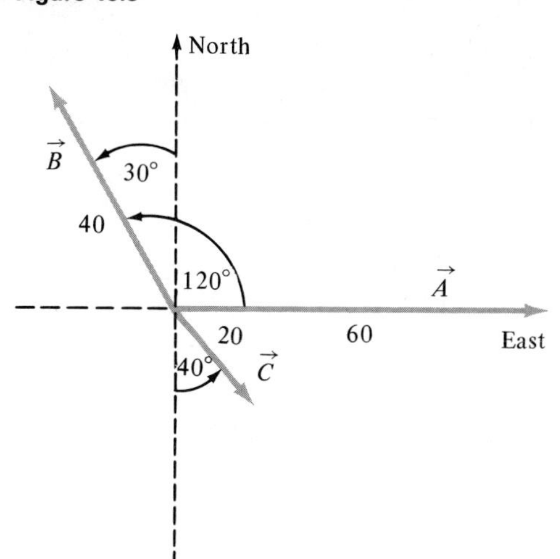

In Figure 45.6 we show $\vec{R}$ and we can find θ as follows:

$$\tan\theta = \frac{|\vec{R_y}|}{|\vec{R_x}|} = \frac{19.3}{52.9} = 0.3648$$

or

$$\theta = 20°$$

Thus, the resultant displacement $\vec{R}$ is 56.3 miles, 20° north of east.

Figure 45.6

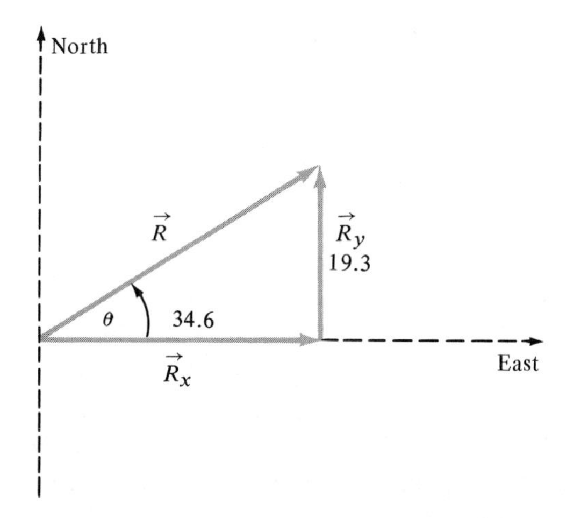

Exercises for Section 45

1. Two forces of 80 lb and 90 lb act at right angles to one another. Find the resultant force and its direction with respect to the 90-lb force.

2. Two forces of 120 lb and 95 lb pull on an object. If the angle between them is 60°, find the resultant of these forces.

3. A force of 50 lb at an angle of 25° to the horizontal is used to move a crate 12 ft on a level floor. If work is force times distance, find the work done.

4. The resultant of two vectors acting at right angles is 400 lb. If one force is 120 lb, find the other force and find the angle it makes with the resultant.

5. If a force has an eastward component of 30 lb and a southward component of 70 lb, find the direction and magnitude of their resultant.

6. Find the rectangular components of the velocity of an object thrown into the air with a velocity of 148 ft/sec at an angle of 43° with the horizontal.

7. If the rectangular components of a velocity vector $\vec{V}$ are $V_x = -7$ ft/sec and $V_y = 9$ ft/sec, find the magnitude and the direction of V.

8. Determine the resultant and direction of a momentum of 12 kg · m/sec upward and a momentum of 16 kg · m/sec along the horizontal to the left.

9. A force $\vec{F_1}$ makes an angle of 35° with the vertical and a force $\vec{F_2}$ makes an angle of 40° with the vertical. Their resultant is force of 120 lb vertically upward. Find the magnitude of $\vec{F_1}$ and $\vec{F_2}$ if they are on the opposite sides of their resultant.

10. The velocity of a boat is 35 mph in a direction 33° north of east. The wind velocity is 7 mph from the west. Find the magnitude and direction of the resultant velocity of the boat.

11. A jet is flying east in still air at a velocity of 600 mph. If the wind is blowing from the south at 75 mph, find the resultant speed of the plane and the direction it is traveling in.

12. A jet traveling 600 mph is climbing at an angle of 32° with the horizontal. How long will it take to climb to an altitude of 15,000 ft?

13. A boat that can travel 15 mph in still water travels directly across a river that is flowing at 6 mph. Find the magnitude and direction of the resultant velocity of the boat.

14. A plane is traveling horizontally at 900 ft/sec when a rocket is fired from it at 2500 ft/sec and at an angle of 35° from the direction of the plane. Find the resultant speed and direction of the rocket.

15. A river flowing northward has a current with a velocity of 6 mph. A boat with a speed of 24 mph leaves the west bank. In what direction should the boat head so as to reach a point on the east bank directly opposite its starting point?

16. Resolve a downward force of 4000 N into a component parallel to an inclined plane of 37° and into a component perpendicular to that plane.

17. Determine the magnitude and direction of the resultant displacement for the following displacements: 75 miles west; 45 miles, 28° west of north; 30 miles, 45° east of south.

18. A support wire 84 ft long runs from the top of a pole 48 ft high to the ground, and pulls on the pole with a force of 300 lb. Find the horizontal pull on the top of the pole.

19. An object that weighs 400 lb is placed on a plane inclined at an angle of 24° with the horizontal and held in place by a rope parallel to the surface that is fastened to a hook in plane. Find the pull on the rope.

20. A force F is applied to a 50 lb crate by a rigid metal rod at an angle of 40° with the horizontal. What is the magnitude of the minimum force necessary to lift the crate from the ground?

REVIEW QUESTIONS FOR CHAPTER 7

1. What is a plane angle?
2. Name the three parts of an angle.
3. When is an angle positive? Negative?
4. What is a quadrantal angle?
5. When is an angle in standard position?
6. What are coterminal angles?
7. What is a degree? A minute? A second?
8. What is a radian?
9. What is the equation that relates degrees and radians?
10. What is the formula that relates the arc length, radius, and central angle of a circle?
11. What is the formula for the area (K) of a sector of a circle if the radius is r and the central angle is θ?
12. What is the relationship between the linear and angular velocity of an object moving around a circle of radius r?
13. State the six trigonometric functions.

14. State the values of the trigonometric function for an angle of 0°, 30°, 45°, 60°, and 90°.

15. What is a reference angle?

16. Express the reference angle $\bar{\theta}$ in terms of θ if θ lies in the first quadrant; the second quadrant; the third quadrant; the fourth quadrant.

17. Express $\sin(-\theta)$ in terms of a positive angle.

18. Express $\cos(-\theta)$ in terms of a positive angle.

19. What is a right triangle?

20. State the relationships between the trigonometric functions and the sides of a right riangle.

21. State the law of sines.

22. State the law of cosines.

23. What is a scalar quantity?

24. What is a vector quantity?

25. What is the resultant of two vectors in the same plane?

26. Explain the parallelogram method for the addition of two vectors.

27. How do we subtract $\vec{B}$ from $\vec{A}$?

REVIEW EXERCISES FOR CHAPTER 7

1. Express each angle in radian measure.
(a) 30° (b) 60° (c) 180°
(d) 240° (e) 165°

2. Express each angle in degree measure.

(a) $\frac{\pi}{3}$ rad (b) $\frac{3\pi}{2}$ rad (c) $\frac{5\pi}{2}$ rad

(d) 2 rad (e) $\frac{2}{3}$ rad

In Exercises 3 through 10, find the values of the six trigonometric functions of the angle θ in standard position if the point lies on the terminal side.

3. $P(4, 3)$ **4.** $P(5, -12)$ **5.** $P(-5, 12)$ **6.** $P(-2, -3)$

7. $P(-3, 0)$ **8.** $P(0, 4)$ **9.** $P(-2, 5)$ **10.** $P(2, 1)$

11. If $\sin\theta = -\frac{3}{5}$ and θ lies in quadrant III, find $\cos\theta$ and $\tan\theta$.

12. If $\cos\theta = \frac{4}{5}$ and θ lies in quadrant IV, find $\sin\theta$ and $\sec\theta$.

13. If $\tan\theta = -\frac{4}{3}$ and θ lies in quadrant II, find $\sin\theta$ and $\csc\theta$.

14. If $\sec\theta = -\frac{17}{8}$ and θ lies in quadrant III, find $\sin\theta$ and $\cot\theta$.

15. Find $\sin\theta$ and $\tan\theta$ if $\cos\theta = \frac{15}{17}$.

16. Find $\sin\theta$, $\cos\theta$, and $\tan\theta$ if $\sec\theta = -\frac{4}{3}$.

17. Find $\cos\theta$ and $\sec\theta$ if $\csc\theta = \frac{5}{4}$.

18. Find $\sec 60°$ and $\tan 60°$.

19. Find $\sin 150°$ and $\cos 150°$.

20. Find $\sin 270°$ and $\cot 270°$.

21. Find $\sin 140°$.

22. Find $\cos 212°$.

23. Find $\sin(-41°)$.

24. Find $\cos(-63°)$.

25. Find $\cos(-131°)$.

26. Find $\sec 14°10'$.

27. Find $\cot 76°12'$.

28. Find $\cos 110°23'$.

29. Find $\sec 214°12'$.

30. Find $\cot 288°36'$.

In Exercises 31 through 40, find each positive acute θ.

31. $\sin\theta = 0.6018$

32. $\cos\theta = 0.8870$

33. $\tan\theta = 1.868$

34. $\csc\theta = 1.111$

35. $\sec\theta = 3.098$

36. $\tan\theta = 0.2432$

37. $\cot\theta = 9.788$

38. $\sin\theta = 0.3907$

39. $\cos\theta = 0.7969$

40. $\sec\theta = 1.445$

41. Find θ such that $0° < \theta < 360°$ if $\sin\theta = 0.3907$.

42. Find θ such that $0° < \theta < 360°$ if $\tan\theta = 0.2432$.

43. Find θ such that $0° < \theta < 360°$ if $\cos\theta = -0.8870$.

44. If ABC is a right triangle with right angle C, solve the triangle if:
(a) $a = 8$, $b = 6$
(b) $A = 24°$, $c = 12$
(c) $c = 10$, $B = 42°40'$
(d) $b = 12.4$, $a = 14.1$
(e) $B = 15°40'$, $b = 52.46$

45. Use the law of sines or cosines to find the remaining parts of the oblique triangle ABC, given that:
(a) $a = 10$, $b = 30$, and $C = 120°$
(b) $A = 30°$, $B = 80°$, and $a = 5$
(c) $a = 4$, $b = 5$, and $A = 30°$

46. Find the sum of each vector.
(a) $|\vec{A}| = 5$, $\theta_A = 30°$
$|\vec{B}| = 8$, $\theta_B = 60°$
(b) $|\vec{A}| = 13$, $\theta_A = 40°$
$|\vec{B}| = 17$, $\theta_B = 70°$
(c) $|\vec{A}| = 27.4$, $\theta_A = 140°$
$|\vec{B}| = 18.6$, $\theta_B = 205°$

47. Find the x and y components of the vector whose magnitude is 8 and makes an angle of $\theta = 70°$ with the positive x axis.

48. A *phasor* is a vector that rotates counterclockwise around the origin at a constant speed. The angular velocity of a phasor is defined as the angle it has traveled through, divided by the time. Thus, $w = \theta/t$ in radians.
(a) A phasor moves through 40π rad in 6 sec. Find its angular velocity.
(b) A phasor makes 10 revolutions in 4 sec. Find its angular velocity.

49. Find the angle of elevation of the top of a 6-ft ladder that rests on the ground if the base of the ladder is 4 ft from a wall.

50. A runner leaves a point A and travels east at the rate of 10 ft/sec. A minute later another runner leaves the point A and runs north at the rate of 8 ft/sec. How far apart are the runners when the first has been running for 2 minutes?

51. Two forces of 100 lb and 60 lb pull on an object. If the angle between the forces is 45°, find the resultant of the forces.

52. A block of weight W rests on a plane whose incline is 40° with the horizontal. If the frictional force is 5 lb and the block remains at rest, find the weight of the block.

CHAPTER 8

Graphs of the Trigonometric Functions

SECTION 46

The Graphs of $y = a \sin x$ and $y = a \cos x$

Today many of the important applications of trigonometry deal with the study of problems involving periodic phenomena, such as alternating electrical current, sound waves, and business cycles. The analysis of such problems requires, among other things, some knowledge of the trigonometric functions and their graphs. In this chapter we shall examine the graphs of the trigonometric functions. We began this study by examining the graphs of the sine and the cosine functions.

When discussing the graphs of the trigonometric functions we shall, hereafter, use x instead of θ to denote the angle. If we agree to express the angle in terms of its radian measure (see Section 38), we can determine the various ordered pairs, (x, y), of real numbers that lie on the graph of the trigonometric function. For example, since $\sin 45° = \sin \pi/4 = \sqrt{2}/2$, the point $(\pi/4, \sqrt{2}/2)$ lies on the graph of $y = \sin x$. Generally, it will be sufficient to generate a table of values by using some of the special values of x such as $\pi/6$, $\pi/4$, $\pi/3$, and so on. If more values for the function are needed, we could use Table I of Appendix F.

We can use the results of Chapter 7 to construct the following table of values of x and y for $y = \sin x$:

x	0	$\frac{\pi}{6}$	$\frac{\pi}{4}$	$\frac{\pi}{3}$	$\frac{\pi}{2}$	$\frac{2\pi}{3}$	$\frac{3\pi}{4}$	$\frac{5\pi}{6}$	π
$y = \sin x$	0	0.5	0.71	0.87	1	0.87	0.71	0.5	0

x	$\frac{7\pi}{6}$	$\frac{5\pi}{4}$	$\frac{3\pi}{2}$	$\frac{5\pi}{3}$	$\frac{7\pi}{4}$	$\frac{11\pi}{6}$	2π
$y = \sin x$	−0.5	−0.71	−1	−0.87	0.71	−0.5	0

We now plot the values from the table above to obtain the graph of the

standard sine curve shown in Figure 46.1. The graph of $y = \sin x$ continues indefinitely in each direction. From this graph we also note that the values for y start to repeat every 2π units for x. This occurs since $\sin(x + 2\pi) = \sin(x + 4\pi)$, or in general $\sin[x + k(2\pi)] = \sin x$, where k is *any* integer. Because of this we say that sine has a *basic period* of 2π units. This observation allows us to obtain a complete sample length 2π. We say that the sine function is **periodic** and note that the graph of the sine function over an interval of one period is called a **cycle** of the curve.

Figure 46.1

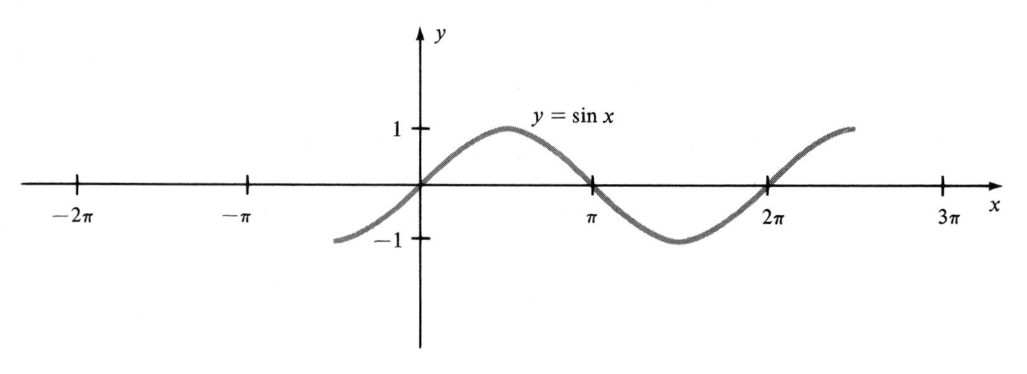

We now construct the following table of values for x and y when $y = \cos x$:

x	0	$\frac{\pi}{6}$	$\frac{\pi}{4}$	$\frac{\pi}{3}$	$\frac{\pi}{2}$	$\frac{2\pi}{3}$	$\frac{3\pi}{4}$	$\frac{5\pi}{6}$	π	$\frac{7\pi}{6}$	$\frac{5\pi}{4}$	$\frac{4\pi}{3}$	$\frac{3\pi}{2}$	$\frac{5\pi}{3}$	$\frac{7\pi}{4}$	$\frac{11\pi}{6}$	2π
y	1	0.87	0.71	0.5	0	−0.5	−0.71	−0.87	−1	−0.87	−0.71	−0.5	0	0.5	0.71	0.87	1

Plotting the values from the table above we obtain the graph of $y = \cos x$ shown in Figure 46.2. From the graph we can see that the basic period of the cosine function is also 2π.

From Figures 46.1 and 46.2 we see that the range of values for both the sine and cosine functions is $-1 \leq y \leq 1$. Thus, we refer to both of these functions as **bounded functions**. Bounded periodic functions such as sine and cosine have **amplitudes**. The amplitude of such functions is one half the absolute value of the difference between the maximum and minimum values. Thus, the amplitude of both $y = \sin x$ and $y = \cos x$ is $\frac{1}{2}|1 - (-1)| = 1$.

In general, the amplitude of $y = a \sin x$ and $y = a \cos x$ is $|a|$.

Example 46.1

Sketch the graph of $y = 3 \sin x$, $-2\pi \leq x \leq 4\pi$.

Figure 46.2

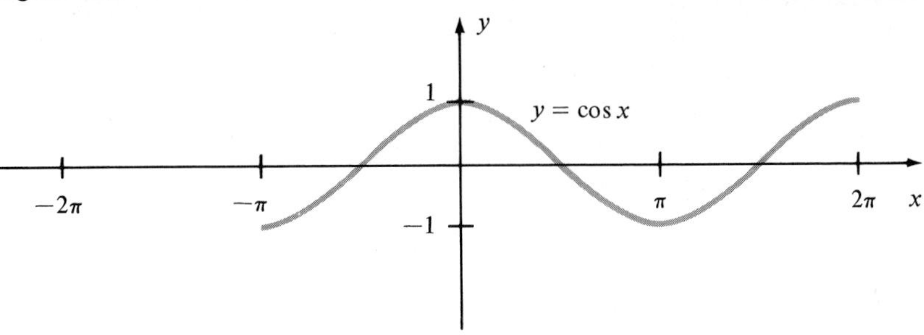

Solution
We first sketch one cycle of the graph of $y = \sin x$ over the interval $0 \le x \le 2\pi$. Then, since the amplitude of $y = 3 \sin x$ is 3, we can sketch the graph of $y = 3 \sin x$ by making each y (or ordinate) value 3 times the corresponding y value of $y = \sin x$. We then extend this cycle to include the given interval $-2\pi \le x \le 4\pi$, as shown in Figure 46.3. Note that three cycles of $y = 3 \sin x$ occur in the interval $-2\pi \le x \le 4\pi$.

Figure 46.3

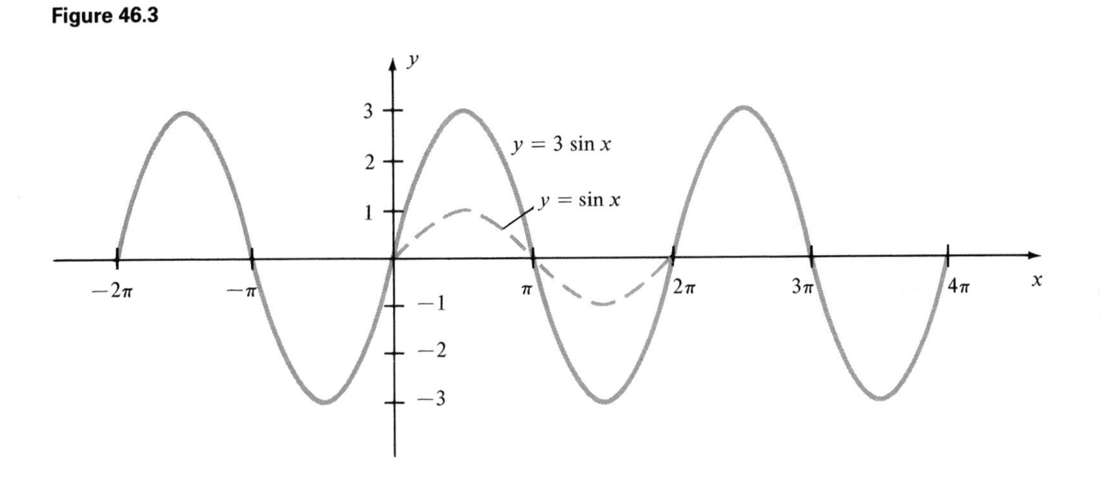

Example 46.2
Sketch the graph of $y = -3 \sin x$, $-\pi \le x \le 3\pi$.

Solution
We first sketch one cycle of the graph of $y = \sin x$ over the interval $0 \le x \le 2\pi$. Then we can sketch the graph of $y = -3 \sin x$ by making each y value -3 times the corresponding y value of $y = \sin x$. We then extend this cycle to include the given interval $-\pi \le x \le 3\pi$ as shown in Figure 46.4. Note that a minus sign in front of 3 **inverts** the graph of the curve. However, the amplitude of $y = -3 \sin x$ **is still equal to 3**.

Figure 46.4

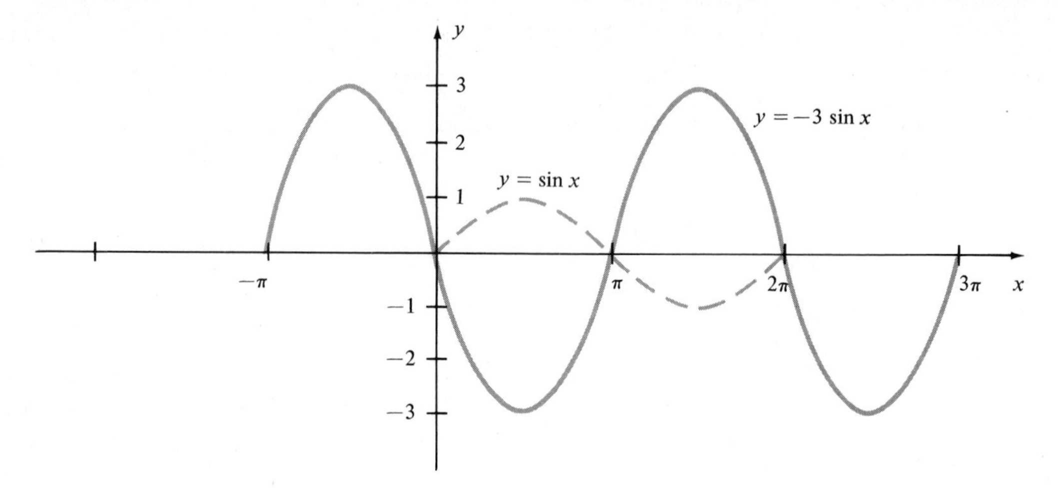

Example 46.3

Sketch the graph of $y = 2 \cos x$, $0 \leq x \leq 4\pi$.

Solution

We first sketch one cycle of the graph of $y = \cos x$ over the interval $0 \leq x \leq 2\pi$. Then we can sketch the graph of $y = 2 \cos x$ by making each y value 2 times the corresponding y value of $y = \cos x$. We then extend this cycle to include the given interval $0 \leq x \leq 4\pi$, as shown in Figure 46.5.

Figure 46.5

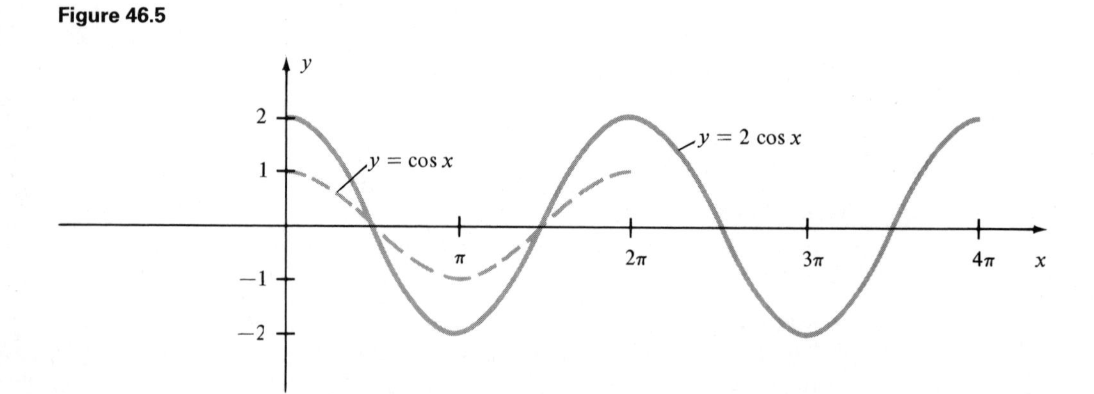

Example 46.4

Sketch the graph of $y = -2 \cos x$, $0 \leq x \leq 4\pi$.

Solution

We note that the minus sign in front of 2 simply has the effect of inverting the graph of $y = 2 \cos x$ of Example 46.3. We show this graph in Figure 46.6.

Figure 46.6

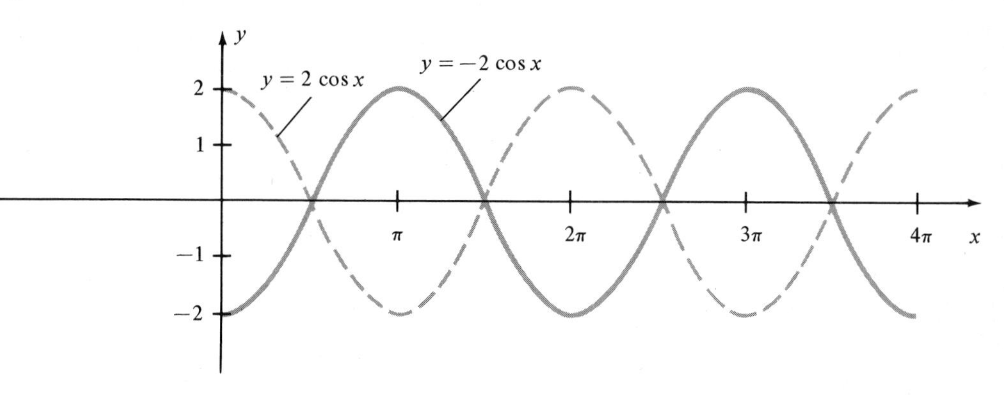

Remark 46.1. *From Figures 46.1 and 46.2 we can readily see that the x intercepts or zeros for the sine function are all x values that are* ***integral multiples of*** π, *and those for the cosine function are all x values that are* ***odd integral multiples of*** $\pi/2$.

Exercises for Section 46

Sketch the graphs for each of the following and state the amplitude.

1. $y = 2 \sin x,\ p \leq x \leq 4\pi$

2. $y = -2 \sin x,\ 0 \leq x \leq 4\pi$

3. $y = 4 \sin x$

4. $y = 5 \sin x$

5. $y = -6 \sin x,\ -4\pi \leq x \leq 2\pi$

6. $y = 3 \cos x$

7. $y = -3 \cos x$

8. $y = \frac{3}{2} \cos x,\ -\pi \leq x \leq \pi$

9. $y = -\frac{7}{2} \cos x,\ -3\pi \leq x \leq -\pi$

10. $y = -\frac{7}{2} \sin x$

11. $y = 6 \cos x$

12. $y = 5 \cos x$

13. $y = 0.9 \sin x$

14. $y = 0.9 \cos x$

15. $y = -0.5 \sin x,\ -2\pi \leq x \leq 6\pi$

SECTION 47

The Graphs of $y = a \sin (bx + c)$ and $y = a \cos (bx + c)$

In this section we first wish to investigate the graphs of $y = a \sin bx$ and $y = a \cos bx$. We can do this by considering the following examples.

Example 47.1

Sketch the graph of $y = \sin 2x$, $0 \le x \le 2\pi$.

Solution

To sketch this curve we can construct a table of values for x and y. In order to calculate the value of y we must first multiply the x value by 2 and then evaluate the sine of this value. We choose some convenient values of x, indicated in the table:

x	0	$\frac{\pi}{12}$	$\frac{\pi}{8}$	$\frac{\pi}{6}$	$\frac{\pi}{4}$	$\frac{\pi}{3}$	$\frac{\pi}{2}$	$\frac{2\pi}{3}$	$\frac{3\pi}{4}$	$\frac{5\pi}{6}$	π
$2x$	0	$\frac{\pi}{6}$	$\frac{\pi}{4}$	$\frac{\pi}{3}$	$\frac{\pi}{2}$	$\frac{2\pi}{3}$	π	$\frac{4\pi}{3}$	$\frac{3\pi}{2}$	$\frac{5\pi}{3}$	2π
$y = \sin 2x$	0	0.5	0.71	0.87	1	0.87	0	−0.87	−1	−0.87	0

We now plot the values from the table to obtain the graph of $y = \sin 2x$, $0 \le x \le \pi$. We can see that the y values of this function start repeating themselves after π units of x. Thus, we say that $y = \sin 2x$ has a period of π and the graph of $y = \sin 2x$ will complete one cycle for $0 \le x \le \pi$. the amplitude of $y = \sin 2x$ is 1. To complete this sketch, we simply repeat this cycle for $\pi \le x \le 2\pi$. This graph is shown in Figure 47.1.

From Example 47.1 we see that there is a definite similarity between the graph of $y = \sin x$ and $y = \sin 2x$. Both have the **same amplitude** but **different periods**. The graph of $y = \sin 2x$ completes two cycles over the interval $0 \le x \le 2\pi$, while $y = \sin x$ completes one cycle over the same interval. This leads us to the following question. How are $y = a \sin x$ and $y = a \sin bx$ related? Example 47.1 suggests that graph of $y = a \sin bx$ will have an amplitude of $|a|$ and achieve one cycle for the interval $0 \le x \le 2\pi/|b|$. This can be shown algebraically as follows. If $0 \le bx \le 2\pi$, then dividing the inequality by $|b|$ gives us $0 \le x \le 2\pi/|b|$. Thus, the period of $y = a \sin bx$ is $2\pi/|b|$. The knowledge that the graphs of all the function of the form $y = a \sin bx$ have a shape similar to the standard sine curve makes the sketching of these curves quite simple.

Figure 47.1

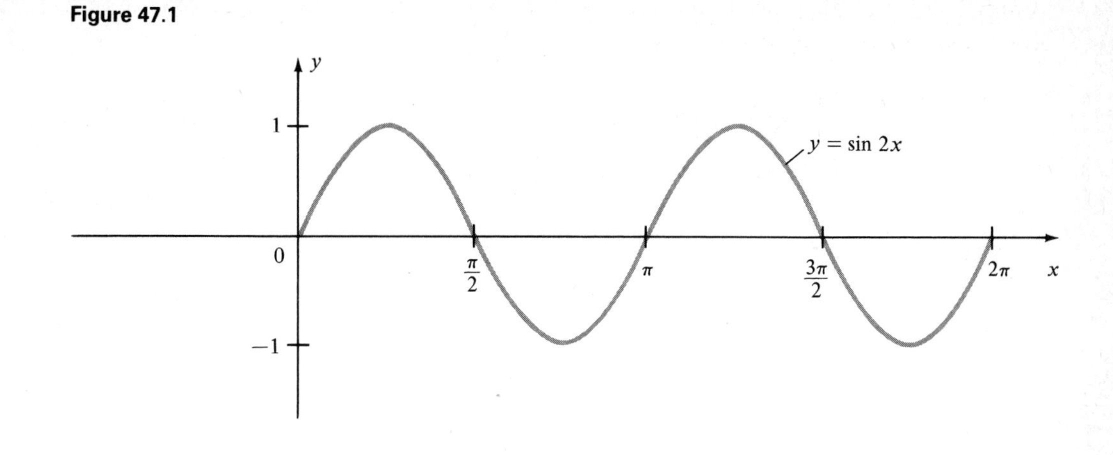

Example 47.2

Sketch the graph of $y = 3 \sin 4x$.

Solution

We note that its amplitude is 3 and its period is $2\pi/4 = \pi/2$. Thus, this graph will achieve one cycle over the interval $0 \leq x \leq \pi/2$. We first sketch $y = 3 \sin x$ as a reference. Next we sketch one complete cycle of $y = 3 \sin 4x$ over the interval $0 \leq x \leq \pi/2$ on the same coordinate system. We then repeat this cycle over and over along both the positive and negative directions of the x axis. We show the sketch of $y = 3 \sin 4x$ in Figure 47.2.

Figure 47.2

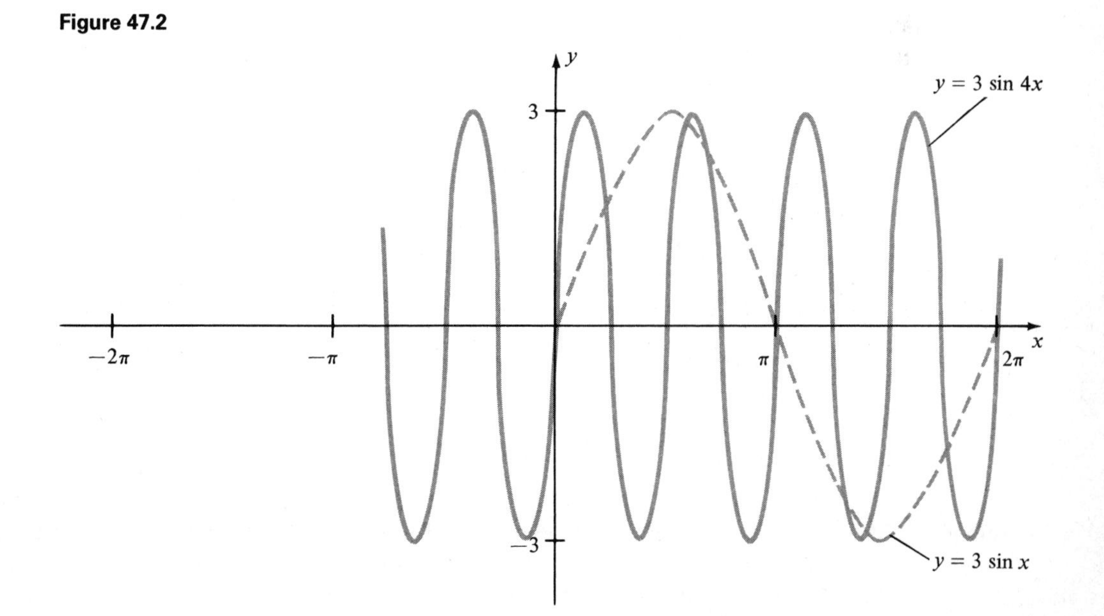

Example 47.3

Sketch the graph of $y = -2 \sin 3x$.

Solution

We note that the amplitude is $|-2| = 2$ and the period is $2\pi/3$. We first sketch the graph $y = 2 \sin x$ as a reference. Next we observe that since $a = -2 < 0$, the graph of $y = -2 \sin 3x$ will be the inversion of $y = 2 \sin x$, and achieve one cycle over the interval $0 \leq x \leq 2\pi/3$. We sketch one complete cycle of $y = -2 \sin 3x$ and then repeat this cycle over and over again in both the positive and negative directions of the x axis. We show the graph of $y = -2 \sin 3x$ in Figure 47.3.

Figure 47.3

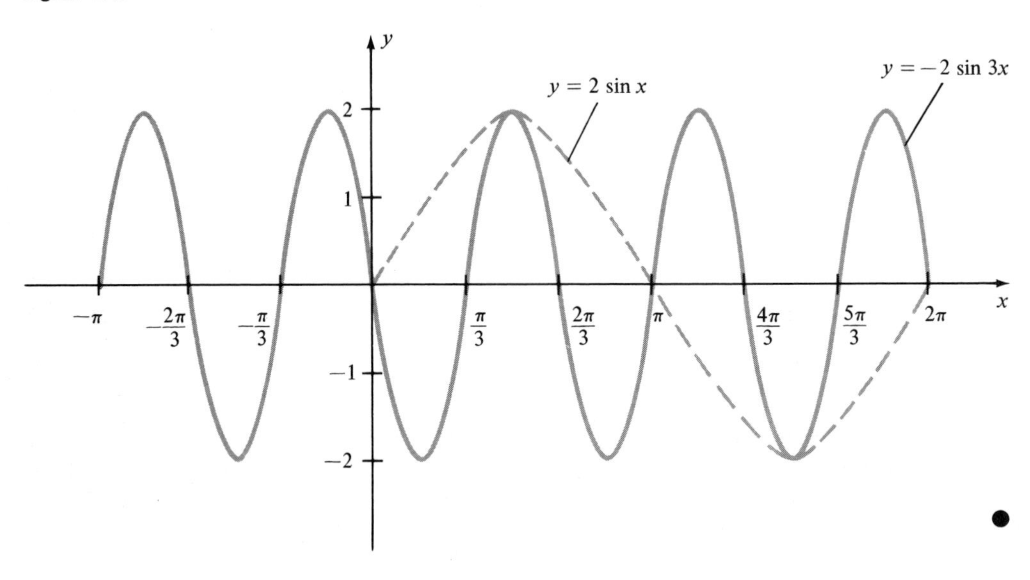

The graph of $y = a \cos bx$ is related to the graph of $y = a \cos x$ is the same manner as the graph of $y = a \sin bx$ is related to the graph of $y = a \sin x$. That is the amplitude of both $y = a \cos bx$ and $y = a \cos x$ is $|a|$ and the period of $y = a \cos bx$ is $2\pi/|b|$.

Example 47.4

Sketch the graph of $y = 3 \cos \frac{1}{2}x$.

Solution

We note that the amplitude of $y = 3 \cos \frac{1}{2}x$ is 3 and that its period is $2\pi/\frac{1}{2} = 4\pi$. We first sketch the graph of $y = 3 \cos x$ as a reference. The graph of $y = 3 \cos \frac{1}{2}x$ will have the same shape as $y = 3 \cos x$ but will achieve one cycle over the interval $0 \leq x \leq 4\pi$ instead of $0 \leq x \leq 2\pi$.

We now sketch one cycle of $y = 3 \cos \frac{1}{2}x$ and then repeat this cycle over and over again in both the positive and negative directions of the x axis. We show this graph of $y = 3 \cos \frac{1}{2}x$ in Figure 47.4.

Figure 47.4

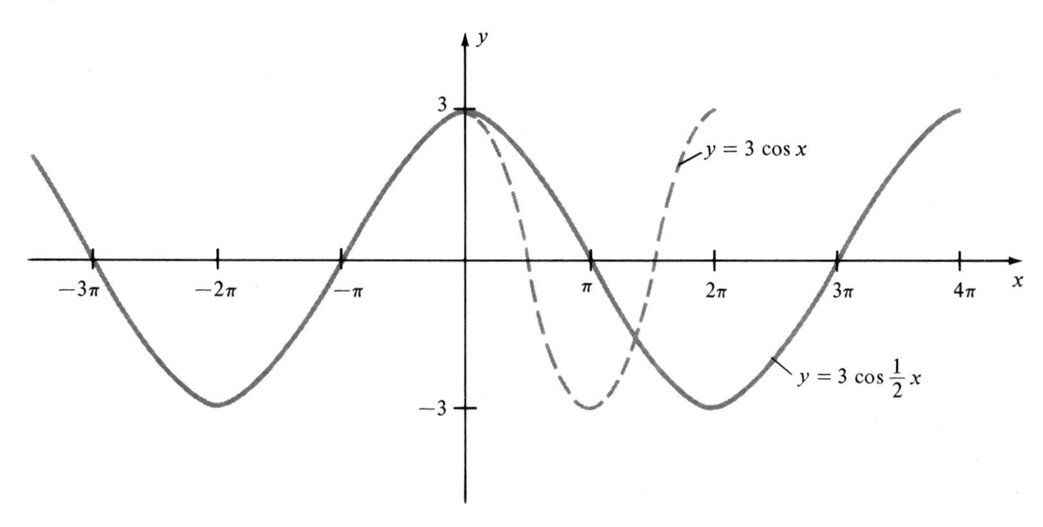

Example 47.5

Sketch the graph of $y = -3 \cos \frac{1}{2}x$, $0 \le x \le 8\pi$.

Solution

We note that the amplitude of this function is $|-3| = 3$ and its period is $2\pi/\frac{1}{2} = 4\pi$. Since $a = -3 < 0$, the graph of this function will simply be the inversion of the graph of $y = 3 \cos \frac{1}{2}x$ from Example 47.4. Since we are only asked to sketch the graph of the given function over the interval $0 \le x \le 8\pi$, we draw two cycles of the graph in Figure 47.5.

Figure 47.5

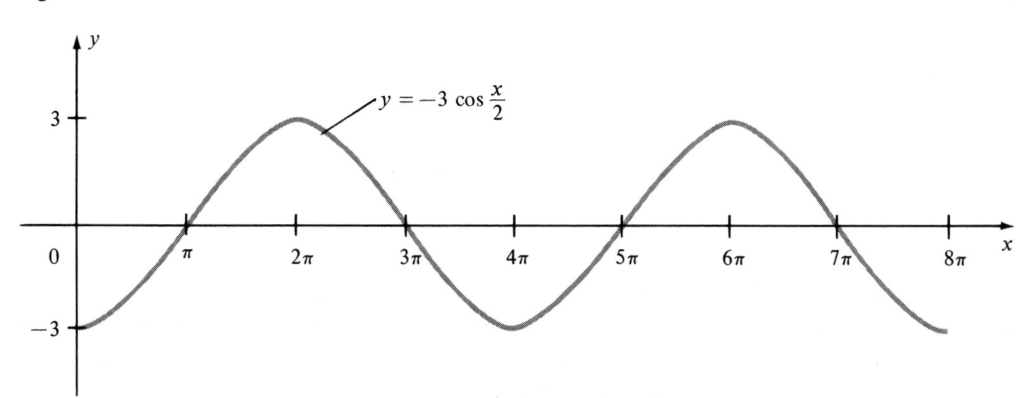

The function $y = a \sin(bx + c)$ is called the **generalized sine function** and its graph is called a **sinusoidal curve**. This function will achieve all its values y, $-a \le y \le a$ as $bx + c$ ranges from 0 to 2π inclusive. Its amplitude is $\frac{1}{2}|a - (-a)| = \frac{1}{2}|2a| = |a|$. We can determine its period as follows. If $bx + c$ ranges from 0 to 2π inclusive, we have

$$0 \le bx + c \le 2\pi$$

$$-c \le bx \le 2\pi - c$$

$$-\frac{c}{|b|} \le x \le \frac{2\pi}{|b|} - \frac{c}{|b|} \qquad \text{(we divided the entire inequality by } |b|\text{)}$$

The last inequality tells us that $y = a \sin(bx + c)$ will complete one cycle over this interval. Since the length of this interval is $2\pi/|b|$, the period of $y = a \sin(bx + c)$ is $2\pi/|b|$. Thus, we see that the function $y = a \sin(bx + c)$ and $y = a \sin bx$ have the *same period*.

These observations lead us to the following technique for sketching the graph of $y = a \sin(bx + c)$.

Procedures for Graphing $y = a \sin(bx + c)$

Step 1: Graph $y = a \sin bx$.

Step 2: The graph of $y = a \sin bx$ is next shifted or slid $-c/b$ units to the right when $-c/b > 0$ or $|-c/b|$ units to the left when $-c/b < 0$. The resulting curve is the graph of $y = a \sin(bx + c)$. The number $-c/b$ is called the **phase shift**.

Example 47.6

Sketch the graph of $y = 4 \sin(2x + 1)$.

Solution

We first analyze the graph of $y = 4 \sin 2x$. This function has an amplitude of 4 and a period of $2\pi/2 = \pi$. We now sketch a portion of the graph of $y = 4 \sin 2x$ in Figure 47.6. Since the phase shift, $-c/b = -\frac{1}{2}$, is negative, the entire graph of $y = 4 \sin 2x$ will be shifted $|-\frac{1}{2}| = \frac{1}{2}$ unit to the left to obtain the graph of $y = 4 \sin(2x + 1)$. We show 3 cycles in Figure 47.6.

Example 47.7

Sketch the graph of $y = -2 \sin(\pi x - 2)$.

Solution

We first analyze the graph of $y = -2 \sin \pi x$. This function has an amplitude of $|-2| = 2$ and a period of $2\pi/\pi = 2$. We now sketch a portion of the graph of $y = -2 \sin \pi x$ in Figure 47.7. Since the phase shift for $y = -2 \sin(\pi x - 2)$ is $2/\pi \approx 0.64$ and is positive, the graph of $y = -2 \sin \pi x$ can be shifted $2/\pi$ units to the right to obtain the graph of $y = -2 \sin(\pi x - 2)$ shown in Figure 47.7.

Figure 47.6

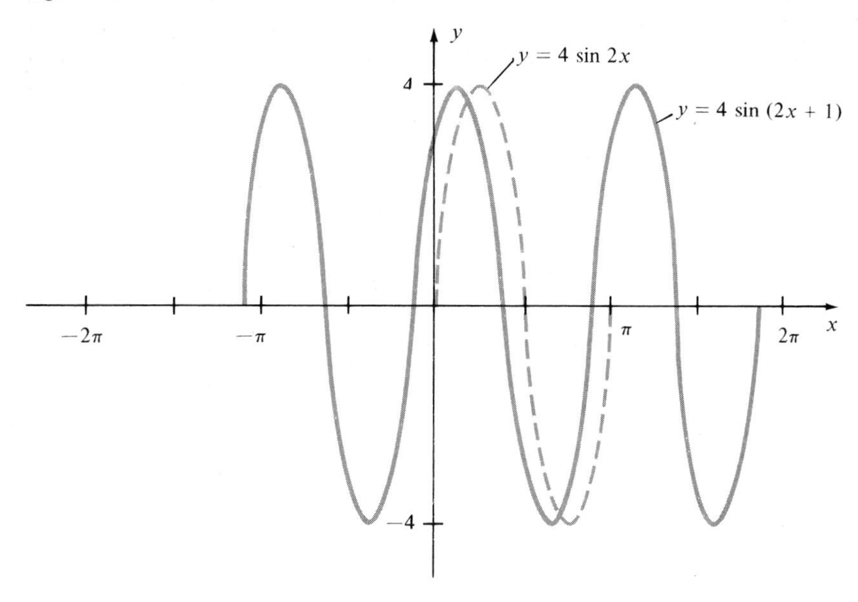

Figure 47.7

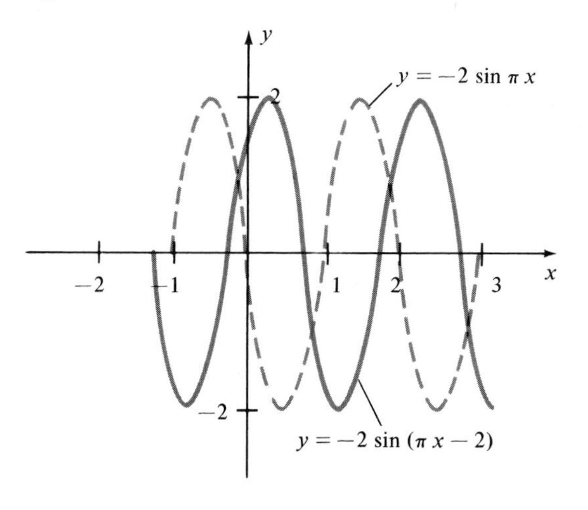

We can use the same information and techniques to sketch the graphs of functions of the form $y = a \cos(bx + c)$.

Example 47.8

Sketch the graph of $y = 3 \cos(\frac{1}{2}x - \pi)$ for $2\pi \le x \le 6\pi$.

Solution

We first analyze the graph of $y = 3 \cos \frac{1}{2}x$. This function has an amplitude of 3 and a period of $2\pi/\frac{1}{2} = 4\pi$. We now sketch a portion of the graph of

$y = 3\cos\frac{1}{2}x$ in Figure 47.8. Since the phase shift $-(-\pi)/\frac{1}{2} = 2\pi$ and is positive, we can sketch the graph of $y = 3\cos(\frac{1}{2}x - \pi)$ by shifting the graph of $y = 3\cos\frac{1}{2}x$ to the right 2π units to obtain one cycle, as shown in Figure 47.8.

Figure 47.8

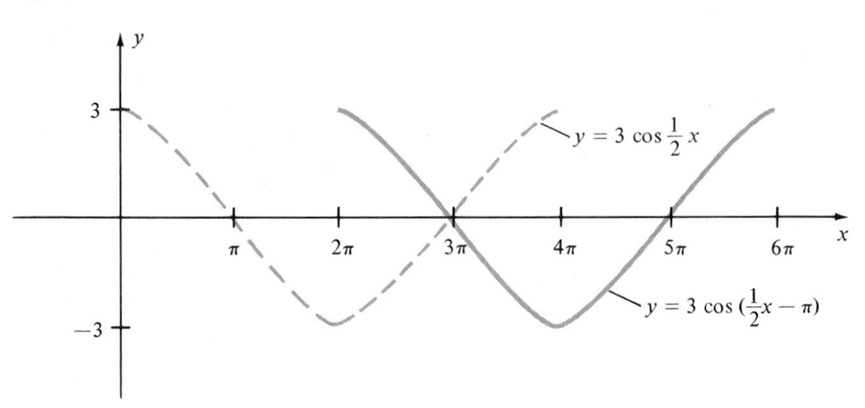

Summary of Properties for Both $y = a\sin(bx + c)$ and $y = a\cos(bx + c)$

1. The amplitude is $|a|$ units.
2. The period is $2\pi/|b|$ units.
3. The phase shift is $-c/b$.

Exercises for Section 47

In Exercises 1 through 8, find the amplitude, period, and phase shift for each function.

1. $y = -2\sin(\pi x + 3)$

2. $y = \dfrac{3}{2}\sin(3x - 2)$

3. $y = 4\sin(x - 2\pi)$

4. $y = -0.7\sin(2 - x)$

5. $y = -3\cos(2\pi - \pi x)$

6. $y = 4\cos(3x - 1)$

7. $y = \cos(x + 1)$

8. $2y = 3\cos(4 - x)$

9. What is the value of b if $y = 3\sin bx$ has a period of $\pi/3$; $\pi/6$; 2; 4?

10. What is the value of b if $y = 2\cos bx$ has a period of $\pi/6$; π; 3; 5?

11. If $y = 3\sin 2x$ is the reference function and the phase shift is π, find the equation of the function.

12. If $y = -2 \cos 3x$ is the reference function and the phase shift is $-\pi/2$, find the equation of the function.

In Exercises 13 through 32, determine the amplitude, period, and phase shift for each function and sketch the graph.

13. $y = 3 \sin 2x$

14. $y = -3 \sin 2x$

15. $y = \sin 2\pi x$

16. $y = \sin \frac{1}{4}x$

17. $y = \frac{1}{2} \sin \frac{1}{2}x$

18. $y = 3 \sin(-2x)$

19. $y = 4 \cos \frac{1}{3}x$

20. $y = -6 \cos(-2x)$

21. $y = \cos\left(\frac{\pi}{2} - x\right)$

22. $y = \sin\left(\frac{\pi}{2} - x\right)$

23. $y = \sin\left(2x - \frac{\pi}{2}\right)$

24. $y = -3 \sin\left(2x - \frac{\pi}{2}\right)$

25. $y = 1.5 \sin\left[\frac{1}{2}(x + 1)\right]$

26. $y = 4 \cos\left(2x + \frac{\pi}{4}\right)$

27. $y = 3 \cos\left(\frac{\pi}{2} - 2x\right)$

28. $y = -1.5 \cos\left(\frac{\pi}{2} - 2x\right)$

29. $y = 4 \cos\left[2\left(x + \frac{\pi}{4}\right)\right]$

30. $y = 2 \sin\left[3\left(x + \frac{\pi}{4}\right)\right]$

31. $y = \cos(4\pi x + 1)$

32. $y = \sqrt{2} \cos(\pi^2 x + \pi)$

33. The equation of a tranverse traveling wave on a string is

$$y = 2 \cos[\pi(0.5x - 200t)]$$

where x and y are in centimeters and t is in seconds. Sketch the shape of the string when $t = 0$, $t = 0.0025$, and $t = 0.005$ sec.

34. The equation of a transverse wave in a stretched string is

$$y = 4 \sin 2\pi\left(\frac{t}{0.02} - \frac{x}{400}\right)$$

where x and y are in centimeters and t is in seconds. Sketch the shape of the string when $t = 0$ and $t = 0.2$ sec.

35. The flow of electricity is dependent upon the *electromotive force*, E, measured in volts. Sketch the graph of E as a function of t when $E = 10 \sin 120\pi t$.

36. Compare the graphs of $y = \sin x$ and $y = \cos(\pi/2 - x)$. What conclusion can be drawn?

37. Compare the graph of $y = \sin(\pi/2 + x)$ and $y = \cos x$. What conclusion can be drawn?

In Exercises 38 through 41, use the technique of Section 33 to solve each system graphically.

38. $\begin{cases} y = x \\ y = \cos x \end{cases}$

39. $\begin{cases} y = x^2 \\ y = \cos x \end{cases}$

40. $\begin{cases} y = x^2 \\ y = \sin 2x \end{cases}$

41. $\begin{cases} y = 2x \\ y = \sin \pi x \end{cases}$

42. Find all the zeros for $y = \sin 2x$ when $= -2\pi \leq x \leq 4\pi$.

43. Find all the zeros for $y = 4 \cos 3x$ when $-\pi \leq x \leq 3\pi$.

44. Find all the zeros for $y = 2 \sin(4x - \pi)$ when $-\pi \leq x \leq 6\pi$.

SECTION 48

Graphs of Functions Involving sin *bx* and cos *bx*: Applications

In many physical applications it is common to encounter functions that are defined in terms of the sine and cosine functions, such as $y = \sin x + \cos x$, $y = \sin x + 2 \sin 2x$, and $y = \sin 2x + 2 \cos x$.

To determine a sketch of the graph of such functions as $y = \sin 2x + 2 \cos x$, we first sketch the graph of the equations $y_1 = \sin 2x$ and $y_2 = 2 \cos x$ on the same coordinate axes. Then $y = y_1 + y_2$. This means that the ordinate of a point on the graph of $y = \sin 2x + 2 \cos x$ with an abscissa x_0 is the algebraic sum of the corresponding ordinates of the points on the graphs of y_1 and y_2. After drawing a vertical line at the point $(x_0, 0)$, the ordinates of y_1 and y_2 can be added geometrically by using a ruler or a compass. This graphing technique is called **addition of ordinates**. We illustrate this method in the following examples.

Example 48.1

Sketch the graph of $y = \sin 2x + 2 \cos x$ by using the method of addition of ordinates.

Solution

We use dashes to show the sketches of the curves $y_1 = \sin 2x$ and $y_2 = 2 \cos x$ on the same coordinate axes in Figure 48.1. We now apply the method of addition of ordinates. We find most of the points for the given curve by summing ordinates at x values where each of the curves has a zero or amplitude value. Thus, in this example we would construct

Figure 48.1

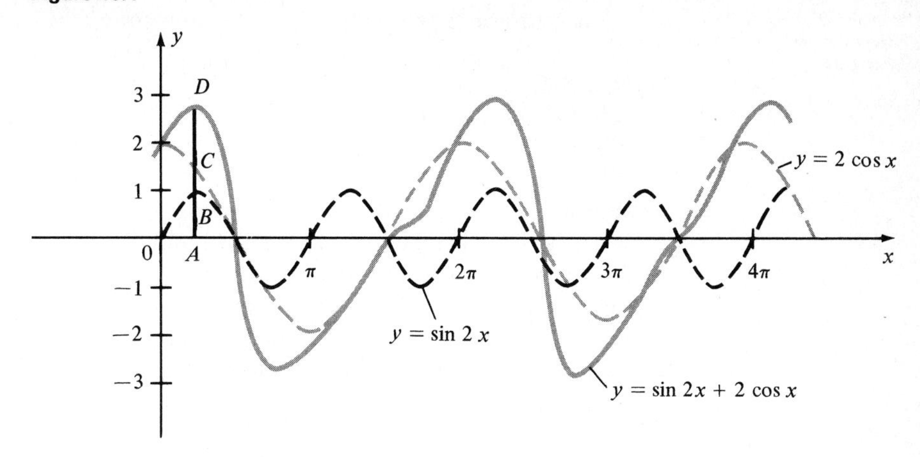

Note: When $x = OA = \frac{\pi}{4}$, then $y_1 = AB$ and $y_2 = AC$, and $y = y_1 + y_2 = AB + AC = AD$.

vertical lines at such points as $(\pi/4, 0)$, $(\pi/2, 0)$, and $(\pi, 0)$ in order to find the geometric sum of the ordinates of y_1 and y_2. We show the graph of the equation in Figure 48.1. It should be noted that as a *check* it is wise to obtain some y values on the curve by substituting a few values for x into the given equation. From the graph we also note that the given function has a period of 2π.

Example 48.2

Sketch the graph of $y = \sin x + \cos x$ by using the method of addition of ordinates.

Solution

We use dashed curves to indicate the sketches of the curves $y_1 = \sin x$ and $y_2 = \cos x$ on the same coordinate axes as shown in Figure 48.2. We now apply the method of addition of ordinates to sketch the graph of $y = \sin x + \cos x$ shown in Figure 48.2. From the graph we also note that the function has a period of 2π.

Figure 48.2

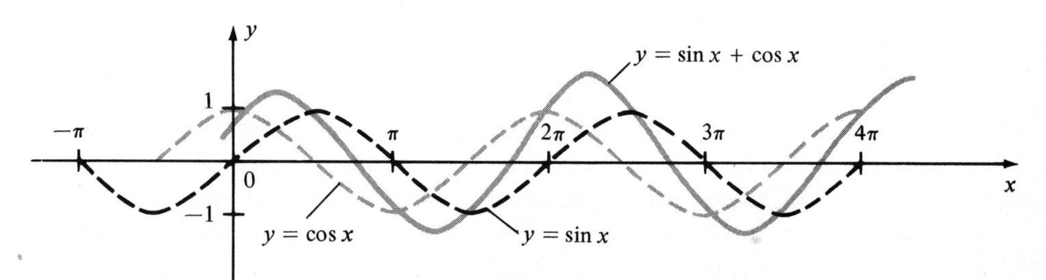

The method of addition of ordinates can be applied to any type of functions that are expressed as sums of other functions.

Example 48.3

Sketch the graph of $y = x - 2 \sin x$ by using the method of addition of ordinates.

Solution

We show the sketches of the straight line $y_1 = x$ and the curve $y_2 = -2 \sin x$ on the same coodinate axes in Figure 48.3. We now apply the method of adding ordinates to obtain points on the given curve, which is also shown in Figure 48.3. Again it should be noted that the easiest ordinates to find are those obtained by using the x values where $y_2 = -2 \sin x$ is zero or achieves its amplitude value.

Figure 48.3

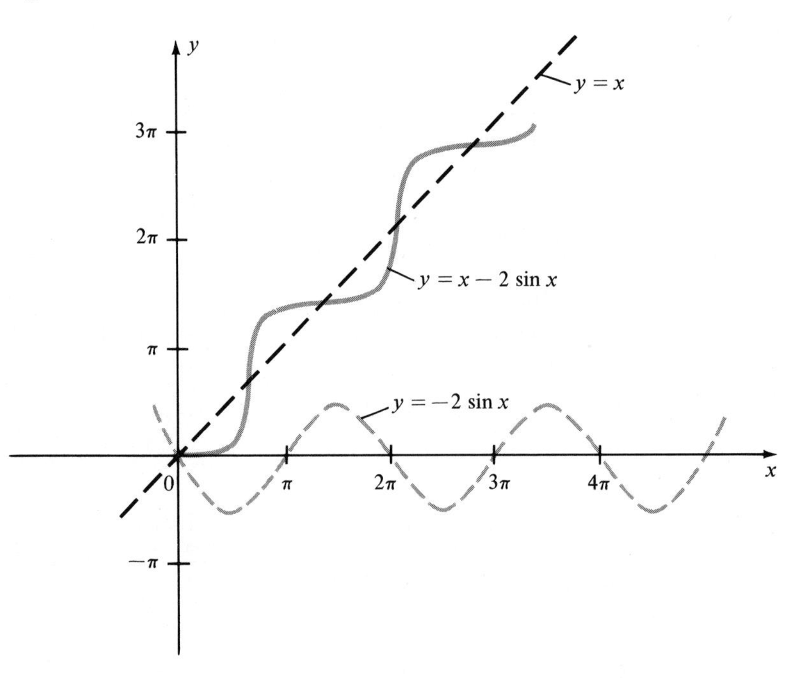

Remark 48.1. *The use of the method of addition of ordinates requires that the sketches of the individual functions involved be done with some care and accuracy. The use of the term "algebraic sum of the ordinates" allows us to add y_1 and y_2 to obtain y when either or both are negative. For example, in Example 48.2, if we draw a vertical line at the point $(5\pi/6, 0)$ we see that the ordinate, y, which is the algebraic sum of $y_1 = \sin x$, which is positive, and $y_2 = \cos x$, which is negative, is negative.*

The graphical analysis of functions that are defined in terms of the trigonometric functions plays an important role in technical applications. One such application is that of **simple harmonic motion**. If a point P moves along the circumference of a circle with radius r at a constant angular velocity, the projection of P on any diameter will produce a type of motion called **simple harmonic motion**. In Figure 48.4 we show the vector $\overrightarrow{OP}$ revolving about O with a constant angular velocity of ω. The projections of P onto the x and y axes, respectively, are designated by P_1 and P_2. These projections will oscillate back and forth along the x and y axes. When $\overrightarrow{OP}$ has rotated through 2π rad, both the projections will have traversed one complete cycle. From the right triangle OP_1P we see that

$$\sin\theta = \frac{P_1P}{r} = \frac{y}{r} \quad \text{and} \quad \cos\theta = \frac{OP_1}{r} = \frac{x}{r}$$

Figure 48.4

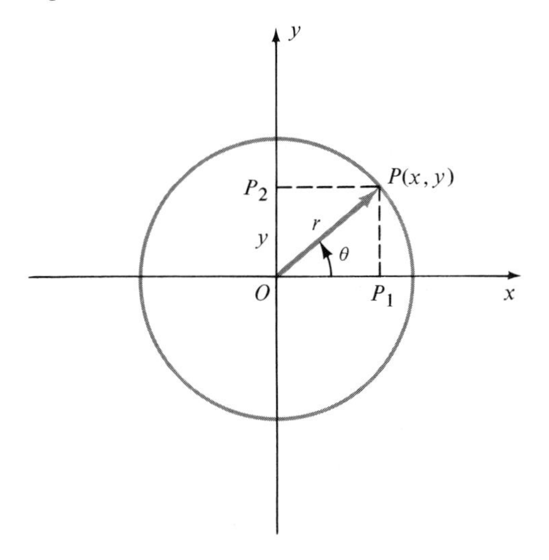

This result and the fact that $\theta/t = \omega$ or $\theta = \omega t$ enables us to describe the coordinates of P as

$$\begin{aligned} OP_1 &= x = r\cos\omega t \\ OP_2 &= y = r\sin\omega t \end{aligned} \qquad (48.1)$$

In (48.1) both x and y are expressed as a function of time t. Equations of the form of (48.1) are called **parametric equations** and t is called the **parameter**.

Example 48.4

Sketch the graph of the motion of the projection P_2 in Figure 48.4, where $r = 2$ and $\omega = \pi/3$ rad/sec.

Solution

From (48.1) we have

$$OP_2 = y = r \sin \omega t$$

$$y = 2 \sin \frac{\pi}{3} t$$

We can sketch this curve in a ty coordinate system by noting that the amplitude is 2 and the period is $2\pi/(\pi/3) = 6$. We show a sketch of a cycle of this curve in Figure 48.5.

Figure 48.5

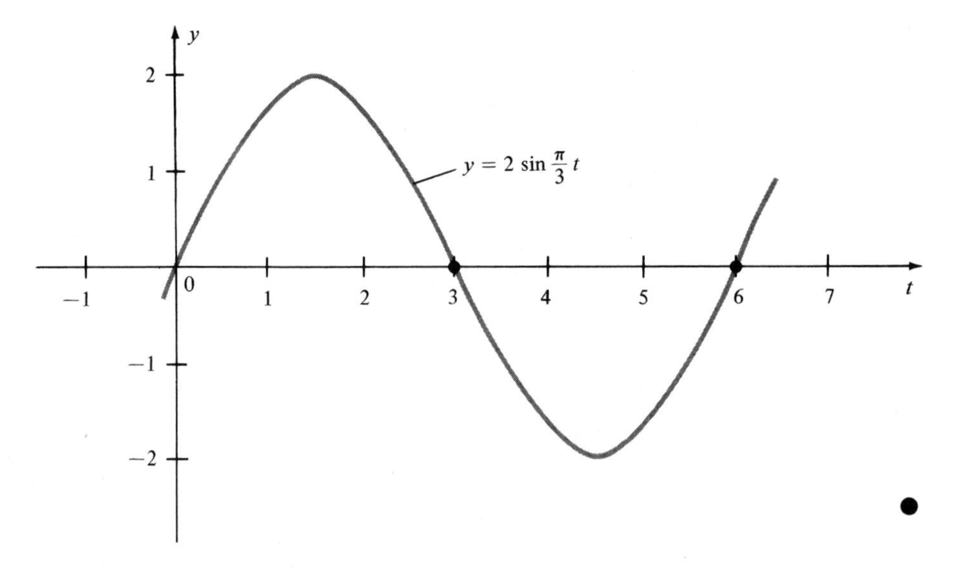

In general, any point whose position l on a straight line is described as a function of time t by the equation $l = a \sin \omega t$ is said to describe **simple harmonic motion**. The analysis of the graph of such motion in an lt coordinate system is equivalent to our analysis of the graph of $y = A \sin bx$ in the xy coordinate system (see Section 47).

Thus, we can see that $l = a \sin \omega t$ has an **amplitude** of a. The time required for one complete cycle is called the **period** and is equal to $2\pi/\omega$. The number of cycles per unit time is called the **frequency** and equals $1/(2\pi/\omega) = \omega/2\pi$. The **phase** at any time t is the fractional part of the period which has elapsed since the point passed through its central position in a positive direction. The position of a point does not have to be given from time $t = 0$, but can be given from any time t. Thus, a more general equation for l is $l = a \sin \omega(t - t_1)$ or $l = a \sin (\omega t - \omega t_1)$. The latter form represents the same curve as $l = a \sin \omega t$ with a phase shift of t_1 along the t axis. Finally, since $l = a \sin \omega t$ and $l = a \cos \omega t$ differ only in phase shift, we can also use $l = a \cos \omega t$ to describe simple harmonic motion.

Example 48.5

For a simple generator the quantity of current I measured in amperes is a function of time t, where $I = a \sin \omega t$. If the amplitude is 8 and the current is 60 hertz (Hz), sketch the graph of this expression.

Solution

If the current is 60 Hz, then $\omega = 120\pi$ rad/sec. Thus, $I = 8 \sin 120\pi t$. We can sketch the graph of I by noting that its amplitude is 8 and its period $2\pi/120\pi = \frac{1}{60}$ sec. We show a graph of I in Figure 48.6.

Figure 48.6

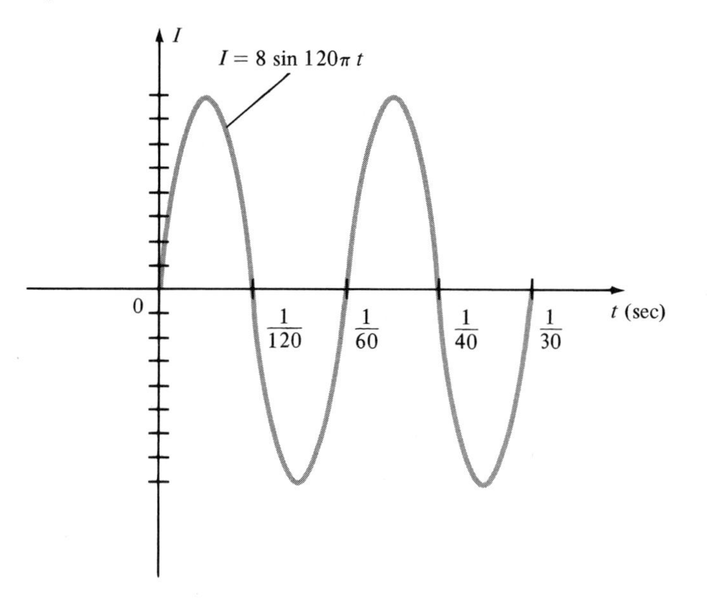

Another important application of the graphs of the trigonometric functions is the graphs of **Lissajou figures**. If a point P with coordinates (x, y) moves in accordance with two simple harmonic motions (that is, $x = a \sin \omega t$ or $a \cos \omega t$ and $y = b \sin \omega t$ or $b \cos \omega t$), at right angles to one another, its path is called a *Lissajous figure.*

Example 48.6

Plot the Lissajous figure described by the parametric equations

$$\begin{cases} x = 2 \cos \pi t \\ y = \sin \pi t \end{cases}$$

Solution

We find the values of the x and y for any point on this figure by assigning values to the parameter t. We select some convenient t values and determine the associated x and y values in Table 48.1. We plot the points obtained from Table 48.1 to determine the graph of Lissajous figure shown in Figure 48.7.

Table 48.1

t	0	$\frac{1}{6}$	$\frac{1}{4}$	$\frac{1}{3}$	$\frac{1}{2}$	$\frac{3}{4}$	$\frac{5}{6}$	1	$\frac{5}{4}$	$\frac{3}{2}$	$\frac{7}{4}$	2
x	2	$\sqrt{3}$	$\sqrt{2}$	1	0	$-\sqrt{2}$	$-\sqrt{3}$	-2	$-\sqrt{2}$	0	$\sqrt{2}$	2
y	0	$\frac{1}{2}$	$\frac{\sqrt{2}}{2}$	$\frac{\sqrt{3}}{2}$	1	$\frac{\sqrt{2}}{2}$	$\frac{1}{2}$	0	$-\frac{\sqrt{2}}{2}$	-1	$-\frac{\sqrt{2}}{2}$	0

Figure 48.7

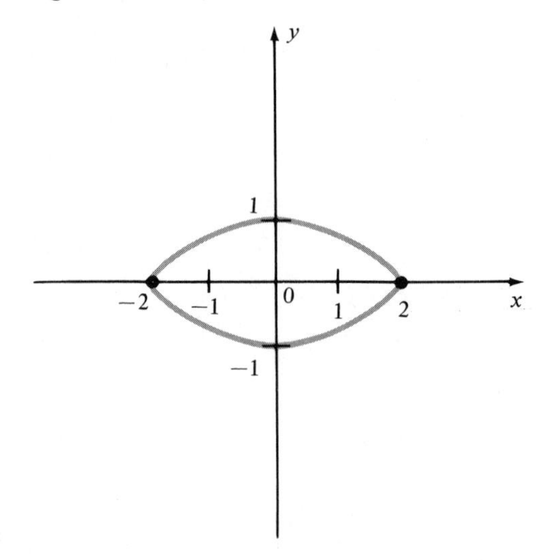

Exercises for Section 48

In Exercises 1 through 16, use the method of addition of ordinates to sketch the graph of each.

1. $y = 2 \sin x + \cos x$

2. $y = 2 \sin x - \cos x$

3. $y = \sin 2x + \cos x$

4. $y = \sin x + \cos 2x$

5. $y = 2 \sin x + 2 \cos x$

6. $y = \cos x + \cos 2x$

7. $y = \sin 2x + \sin x$

8. $y = 2 \sin 2x + 1$

9. $y = 2 \cos x + 1$

10. $y = 2 - \sin x$

11. $y = x - 2\cos x$

12. $y = \frac{1}{2}x - \cos x$

13. $y = x + \cos x$

14. $y = x + \sin x$

15. $y = \frac{1}{2}\cos 2x - x$

16. $y = 3\sin x + 4\cos x$

In Exercises 17 through 24, plot the graph of each Lissajous figure.

17. $y = 2\sin t, x = 2\cos t$

18. $y = \sin 2t, x = 3\sin 2t$

19. $y = \sin 2\pi t,\ x = \cos 2\pi t$

20. $y = 4\sin \pi t,\ x = 3\sin \pi t$

21. $y = 3\sin t,\ x = 2\sin t$

22. $y = \sin 2t,\ x = 3\cos t$

23. $y = \sin 2\pi t,\ x = \sin \pi t$

24. $y = \cos 2\pi t,\ x = \sin \pi t$

25. In a simple generator the electric current I (measured in amperes) can be expressed as a function of time, t as $I = 6\sin 120\omega t$. Find the amplitude and period of this function and sketch its graph in an It coordinate system.

26. In a simple generator the **electromotive force,** E (measured in volts), can be expressed as a function of time, t, as $E = a\sin \omega t$. If the amplitude is 10 and the current is 60 Hz, express E as a function of time. Then sketch the graph of E.

In Exercises 27 through 30, consider a point P with coordinate (x, y) *moving along the circumference of a circle of radius r with a linear velocity v. Express the coordinates of P as simple harmonic functions of time t (in seconds), and then determine the amplitude, period, and frequency for each of the following, subject to the given conditions.* (Recall that $v = \omega r$.)

27. Linear speed of 2 units/sec and a radius of 1 unit.

28. Linear speed of 4 units/sec and a radius of 1 unit.

29. Linear speed of 2 units/sec and a radius of 4 units.

30. Linear speed of 4 units/sec and a radius of 4 units.

SECTION 49

Graphs of the Other Trigonometric Functions

In this section we wish to determine the basic shapes of the graphs of $y = \tan x$, $y = \cot x$, $y = \sec x$, and $y = \csc x$. We can then use the knowledge of the shapes of these basic curves to sketch the graphs of others involving these functions. We will start by examining the graph of $y = \tan x$.

We can find the x intercepts for $y = \tan x$ by setting $y = 0$ and recalling that $\tan x = \sin x/\cos x$ to obtain $0 = \sin x/\cos x$ or $\sin x = 0$. Thus, all x values whereby $\sin x = 0$ will be x intercepts of $y = \tan x$. We can describe these intercepts as all *integral* multiples of π (see Section 46). We also note that $y = \tan x = \sin x/\cos x$ will be undefined when $\cos x = 0$ or equivalently for all x values that are *odd multiples* of $\pi/2$.

Suppose that we now consider the graph of $y = \tan x$ for $-\pi/2 \leq x \leq 3\pi/2$. We immediately observe that this function has x intercepts at $(0, 0)$ and $(\pi, 0)$. We also note that $y = \tan x$ is undefined for $x = -\pi/2, \pi/2$, and $3\pi/2$. We use the values and properties for $\tan x$ established in Chapter 7 to construct Table 49.1. We use these values to determine the graph shown in

Table 49.1

x	$-\frac{\pi}{2}$	$-\frac{\pi}{4}$	$-\frac{\pi}{3}$	0	$\frac{\pi}{4}$	$\frac{\pi}{3}$	$\frac{\pi}{2}$	$\frac{3\pi}{4}$	$\frac{5\pi}{6}$	π	$\frac{7\pi}{6}$	$\frac{5\pi}{4}$	$\frac{4\pi}{3}$	$\frac{3\pi}{2}$
$y = \tan x$	∞	-1	$-\sqrt{3}$	0	1	$\sqrt{3}$	∞	-1	$-\frac{\sqrt{3}}{3}$	0	$\frac{\sqrt{3}}{3}$	1	$\sqrt{3}$	∞

Figure 49.1. The dashed lines shown in this figure are called **vertical asymptotes**, since the value of the $\tan x$ grows very large when x approaches one of these values. It should be carefully noted that there are *no* corresponding y values for any of these x values. From Figure 49.1 we see that the graph of $y = \tan x$ has a period of π. That is, this graph will repeat itself every π units.

Figure 49.1

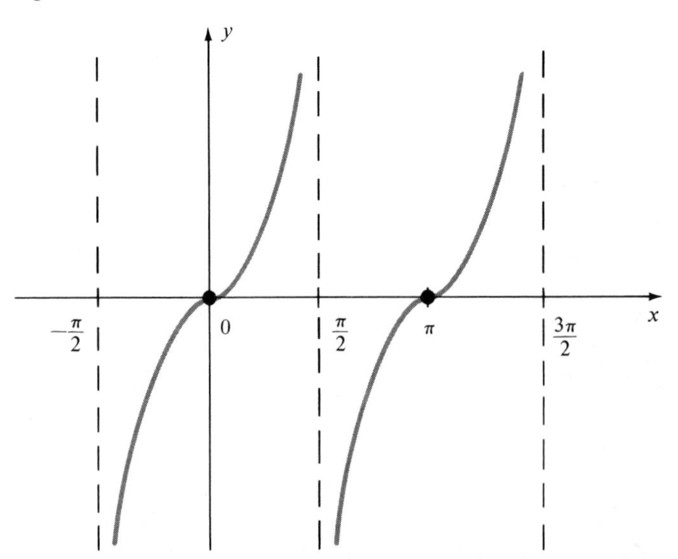

Since $y = \cot x = \cos x / \sin x$ we see that the x intercepts for $y = \cot x$ will occur when $0 = \cos x/\sin x$ or when $\cos x = 0$. Thus, all the x intercepts for $\cot x$ are odd integral multiples of $\pi/2$. We also note that $y = \cot x$ will be undefined when $\sin x = 0$ or when x is an integral multiple of π. Thus, we will have vertical asymptotes at these values. We construct Table 49.2 for values of $0 \le x \le 2\pi$. We use some of the values from this table to sketch the graph of one cycle. We see that the period of $y = \cot x$ is π. We show three periods of this curve in Figure 49.2.

Table 49.2

x	0	$\frac{\pi}{6}$	$\frac{\pi}{4}$	$\frac{\pi}{3}$	$\frac{\pi}{2}$	$\frac{3\pi}{4}$	$\frac{5\pi}{6}$	π	$\frac{5\pi}{4}$	$\frac{4\pi}{3}$	$\frac{3\pi}{2}$	$\frac{7\pi}{4}$	$\frac{5\pi}{3}$	2π
$y = \cot x$	∞	$\sqrt{3}$	1	$\frac{\sqrt{3}}{3}$	0	-1	$-\sqrt{3}$	∞	1	$\frac{\sqrt{3}}{3}$	0	-1	$-\frac{\sqrt{3}}{3}$	∞

Figure 49.2

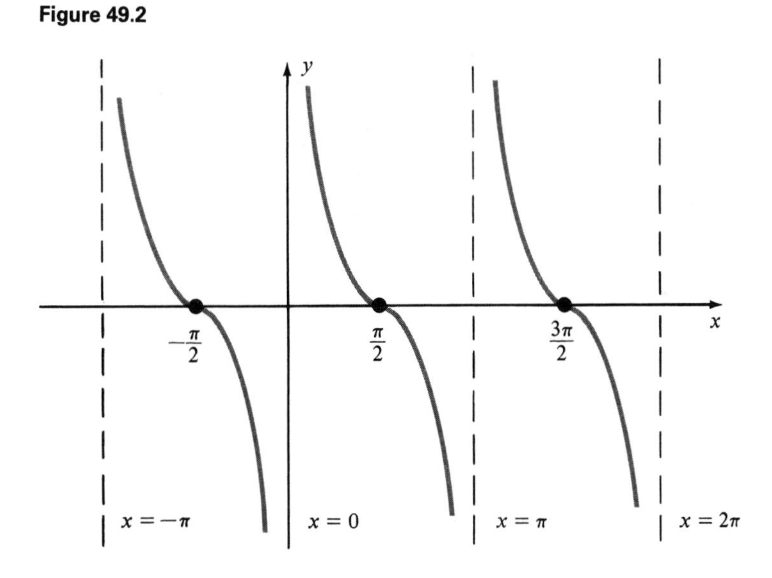

The graphs of $y = \sec x$ and $y = \csc x$ can be conveniently sketched if we recall their definitions in terms of $\sin x$ and $\cos x$. We first consider $y = \sec x = 1/\cos x$ and note that for each point (x, y) on the graph of $y = \cos x$ that there is a corresponding point $(x, 1/y)$, $y \neq 0$, on the graph of $y = \sec x$. This observation allows us to first sketch the graph of the reciprocal function $\cos x$, and then from this graph determine the graph of $y = \sec x$.

Example 49.1

Sketch the graph of $y = \sec x$.

Solution

We first sketch two periods of $y = \cos x$ shown as a dashed curve in Figure 49.3. We see that $\cos x$ will equal 1 when $x = -2\pi$, 0, and 2π. Thus, $\sec x$ will equal $1/1$, or 1, at these same x values. We also see that $\cos x = 0$ when $x = -3\pi/2$, $-\pi/2$, $\pi/2$, and $3\pi/2$. Thus, the lines $x = -3\pi/2$, $x = -\pi/2$, $x = \pi/2$, and $x = 3\pi/2$ are vertical asymptotes. Additional points can be obtained by observing that since the points $(\pi/3, \frac{1}{2})$ and $(2\pi/3, -\frac{1}{2})$ lie on the graph of $\cos x$, then $(\pi/3, 2)$ and $(2\pi/3, -2)$ are points on the graph of $\sec x$. In Figure 49.3 we show two periods of $y = \sec x$. Note that $y = \sec x$ has the same basic period as $y = \cos x$.

Figure 49.3

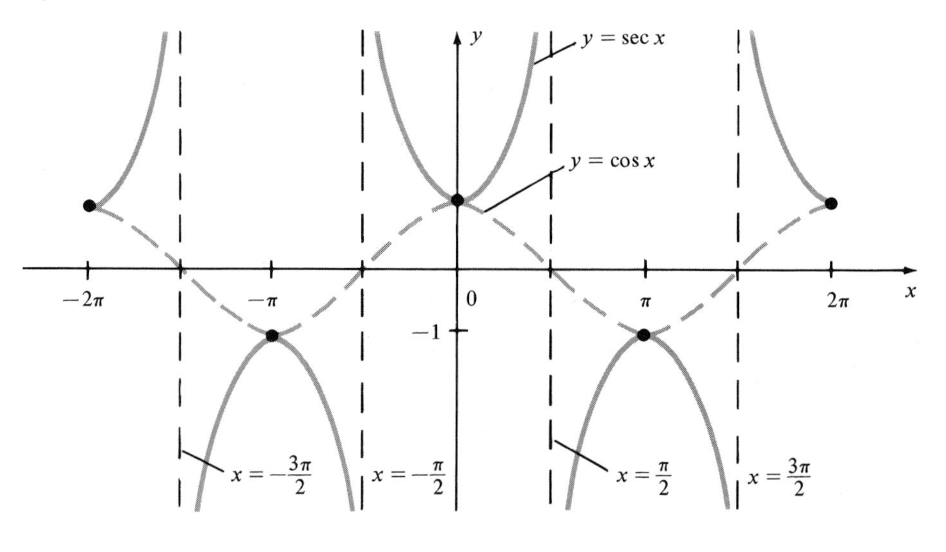

Example 49.2

Sketch the graph of $y = \csc x$.

Solution

We first sketch two periods of $y = \sin x$ shown in Figure 49.4 as the dashed curve. Next we select the necessary points on the graph of $y = \sin x$ to obtain the corresponding points on the graph of $y = \csc x$. We see that this curve has vertical asymptotes at $x = 0$, $x = \pi$, $x = 2\pi$, $x = 3\pi$, and $x = 4\pi$, in the interval shown in Figure 49.4. We show two periods of this curve and note that it has the *same* period as $y = \sin x$. ●

In summary, to sketch the graph of any variation of secant or cosecant functions, we first sketch the graph of the associated sine or cosine function and then use the appropriate reciprocal relationship.

Example 49.3

Sketch the graph of $y = -2\csc(2x - \pi)$.

Figure 49.4

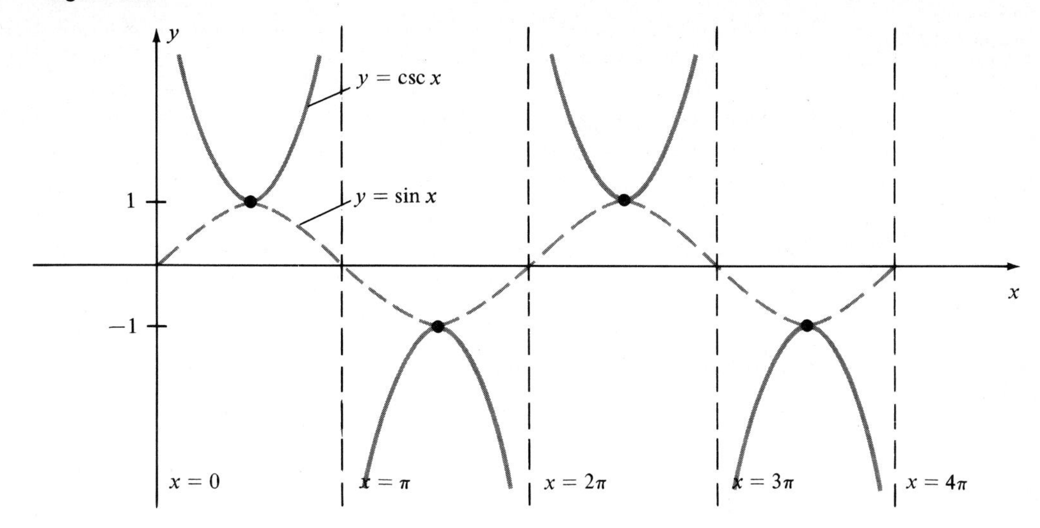

Solution

We first sketch the graph of one period of the associated reciprocal function $y = -\frac{1}{2}\sin(2x - \pi)$. We note that this function has an amplitude of $\frac{1}{2}$, a period of π, and a phase shift equal to $\pi/2$. We show the graph of this function as a dashed curve in Figure 49.5. We then use the

Figure 49.5

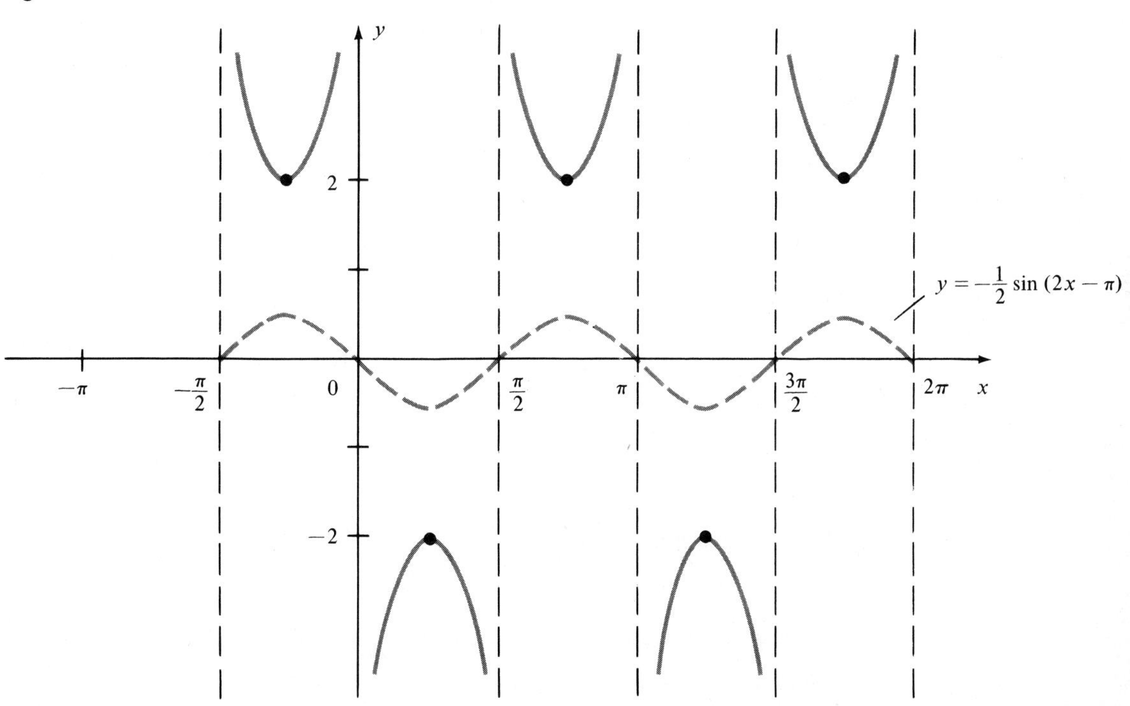

reciprocal relationship

$$y = -2 \csc (2x - \pi) = \frac{1}{-\frac{1}{2} \sin (2x - \pi)}$$

to obtain some points on the graph of $y = -2 \csc (2x - \pi)$. We then sketch the graph of the given function for $-\pi < x < 2\pi$. The graph of this function can then be repeated infinitely in each direction of the x axis.

Remark 49.1. *It should be noted that none of the functions $y = \tan x$, $y = \cot x$, $y = \sec x$, and $y = \csc x$ have an amplitude.*

Exercises for Section 49

In Exercises 1 through 22, sketch the graphs of each function.

1. $y = 2 \tan x$

2. $y = -\tan x$

3. $y = 3 \cot x$

4. $y = -\cot x$

5. $y = \tan 2x$

6. $y = \cot 2x$

7. $y = -3 \cot 3x$

8. $y = -4 \tan \frac{1}{2}x$

9. $y = \tan (2x - \pi)$

10. $y = \cot \left(\frac{\pi}{2} - 2x\right)$

11. $y = -\csc x$

12. $y = -\sec x$

13. $y = \csc (-x)$

14. $y = \sec (-x)$

15. $y = \sec \pi x$

16. $y = \csc \pi x$

17. $y = \sec (x + \pi)$

18. $y = -2 \csc (x + \pi)$

19. $y = 2 \sec \left(2x - \frac{\pi}{4}\right)$

20. $y = 3 \csc \left(\frac{1}{2}x - \frac{\pi}{6}\right)$

21. $y = \tan \left(x - \frac{\pi}{4}\right)$

22. $y = -\cot \left(2x - \frac{\pi}{2}\right)$

23. Does $y = \csc 2x$ have any x intercepts?

24. Does $y = \sec 3x$ have any x intercepts?

REVIEW QUESTIONS FOR CHAPTER 8

1. What is the period of the sine function?

2. What does the term "cycle" mean?

3. What is the amplitude of a function?

4. What is the phase shift of $y = a \sin(bx - c)$?

5. Briefly explain the graphing technique called addition of ordinates.

6. For what values of x does the graph of $y = \tan x$ have vertical asymptotes?

7. What is the period of $y = \tan x$?

8. What is the reciprocal function of $\sin x$? Of $\cos x$? Of $\tan x$?

9. How are the periods of a given trigonometric function and its reciprocal function related?

REVIEW EXERCISES FOR CHAPTER 8

In Exercises 1 through 20, determine the amplitude, period, and phase shift for each function and sketch the graph.

1. $y = 2 \sin x$

2. $y = -2 \sin x$

3. $y = 3 \cos(-x)$

4. $y = -3 \cos(-x)$

5. $y = 2 \sin 3x$

6. $y = 3 \cos 2x$

7. $y = 2 \sin(-3x)$

8. $y = \frac{1}{2} \sin(x + \pi)$

9. $y = 5 \sin(2x - \pi)$

10. $y = -2 \sin(3x + 2\pi)$

11. $y = 4 \cos(2x - \pi)$

12. $y = 2 \sin(\pi - x)$

13. $y = -2 \cos(\pi - 2x)$

14. $y = \frac{3}{2} \sin(1 - x)$

15. $y = 5 \sin\left(\frac{x}{2} - \frac{\pi}{8}\right)$

16. $y = 3 \cos\left[2\left(x + \frac{\pi}{2}\right)\right]$

17. $y = 2 \sin\left(\pi x - \frac{\pi}{2}\right)$

18. $y = 1.5 \sin(\pi - 2x)$

19. $y = 1.5 \cos(\pi - 2x)$

20. $y = 3 \cos\left(\frac{x}{3} + \frac{\pi}{6}\right)$

In Exercises 21 through 30, sketch the graph of each function, noting period and phase shift.

21. $y = -\tan x$

22. $y = \tan(-x)$

23. $y = 2 \tan(2x + \pi)$

24. $y = 2 \cot(2x + \pi)$

25. $y = 2 \sec(2x - \pi)$

26. $y = -3 \csc\left(\frac{x}{2} + \frac{\pi}{4}\right)$

27. $y = -\sec(x - \pi)$ **28.** $y = -\cot\left(\pi x - \frac{\pi}{2}\right)$

29. $y = 2\csc(3x - \pi)$ **30.** $y = -2\csc \pi x$

31. Use the method of addition of ordinates to sketch:
(a) $y = \sin x + 2\csc x$ (b) $y = 2\sin x - 2\cos x$

32. Plot the graphs of the following Lissajous figures.
(a) $y = \sin t,\ x = 2\cos t$ (b) $y = 2\sin \pi t,\ x = \cos \pi t$

33. When an electromotive force, in volts, is applied across a 127-Ω resistance, the relationship between emf in volts and time in milliseconds is given by the equation emf $= 15\sin \pi t$. Graph this equation for $0 \le t \le 40$, noting the amplitude and the period.

34. The equation of a transverse wave traveling on a string is given by

$$y = 2\sin[\pi(2x - 40t)]$$

where x and y are in centimeters and t is in seconds. Sketch the shape of the string when $t = 1$ sec.

CHAPTER 9

Exponential Functions and Logarithms

SECTION 50

Exponential Functions and Equations

Two functions of basic importance in the applications of mathematics are the exponential function and the logarithmic function. We shall first consider the mathematical properties of the exponential function and then discuss some applications. Exponential functions are used in relationships involving growth, radioactive substances, heat, light, and population problems as well as interest and annuity payments.

We begin by defining the exponential function to be any function of the form $y = b^x$, where $b > 0$ and not equal to 1. The function is called **exponential**, since the exponent is variable. The restrictions on b are easily explained. Since 1 raised to any power yields 1, this choice for b would result in the form $y = 1^x = 1$, and we have a constant linear function. For $b < 0$, we generate nonreal values for y if x is any fractional exponent with an even integer in the denominator only. As we have noted, the most convenient way to deal with these situations is to restrict the value for b.

In Examples 50.1 and 50.2, we examine some of the properties of the exponential function for $b > 1$.

Example 50.1

Sketch the graph of $y = 2^x$.

Solution

We shall graph the function by first assuming only integral values for the exponent and then connecting the resulting points with a smooth, continuous curve. We assign arbitrary integral values to x and determine the corresponding values for y as shown in Table 50.1. The graph of $y = 2^x$ is

Table 50.1

x	-3	-2	-1	0	1	2	3
y	$\frac{1}{8}$	$\frac{1}{4}$	$\frac{1}{2}$	1	2	4	8

shown in Figure 50.1. From the graph we note that through negative values for x, 2^x tends toward 0. Hence, the x axis is an asymptote for the graph. In addition, as x increases, 2^x increases without bound. We also note that the graph never lies below the x axis; that is, the value of the function is never negative. The curve has only one intercept point and that is at $(0, 1)$.

Figure 50.1

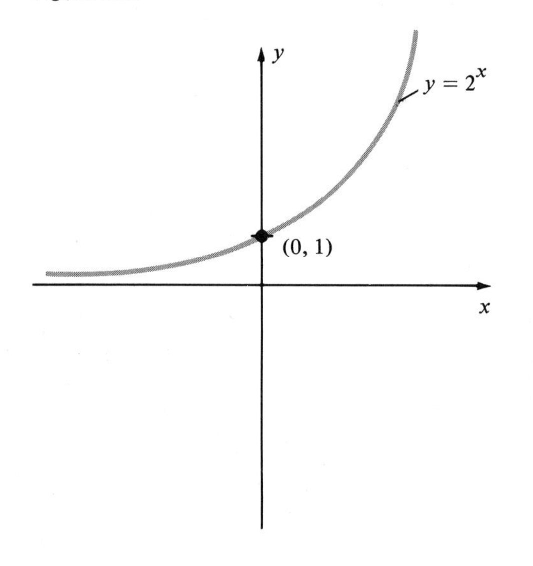

Example 50.2

Graph $y = 3^x$ and $y = 4^x$ on the same set of axes.

Solution

Table 50.2

x	-2	-1	0	1	2
$y = 3^x$	$\frac{1}{9}$	$\frac{1}{3}$	1	3	9

Table 50.3

x	-2	-1	0	1	2
$y = 4^x$	$\frac{1}{16}$	$\frac{1}{4}$	1	4	16

We follow the procedure outlined in Example 50.1 and construct Tables 50.2 and 50.3. We now use the values in these tables to draw the graphs shown in Figure 50.2. We note that the graphs of $y = 3^x$ and $y = 4^x$ are

Figure 50.2

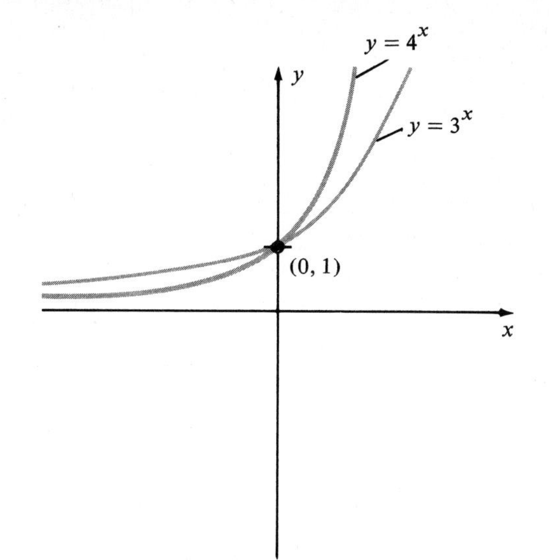

also asymptotic with respect to the x axis, increasing, and lie above the x axis. In addition, their rate of ascent varies with the value of the base b. Both curves intersect the y axis at (0, 1).

The behavior of the exponential function for $0 < b < 1$ is quite different from what we have seen in the previous illustrative examples.

Example 50.3
Graph $y = (\frac{1}{2})^x$.

Solution

Table 50.4

x	-3	-2	-1	0	1	2	3
y	8	4	2	1	$\frac{1}{2}$	$\frac{1}{4}$	$\frac{1}{8}$

We use the values in Table 50.4 to sketch the graph shown in Figure 50.3 on page 278. In this case the graph is asymptotic with respect to the x axis, has an intercept point at (0, 1), is decreasing, and lies above the x axis. We note that since $2^{-x} = 1/2^x = (\frac{1}{2})^x$, the graph of $y = 2^{-x}$ is identical to that of $y = (\frac{1}{2})^x$.

Every exponential function is of the form $y = b^x$. If we wish to identify a particular function, it is only necessary to know one point on the curve, provided that point is not the y intercept, (0, 1).

Figure 50.3

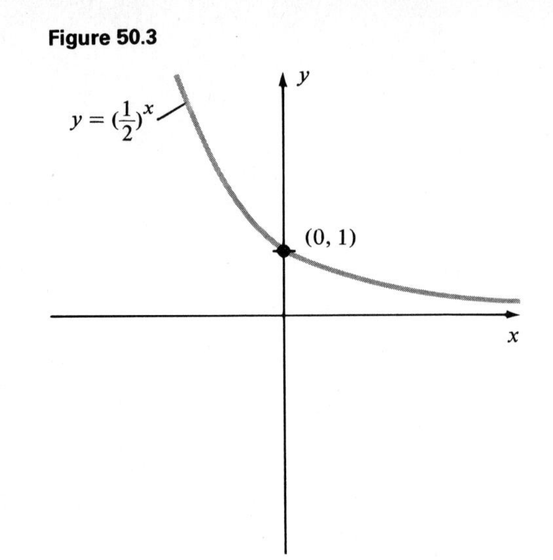

Example 50.4

The graph of an exponential function contains the point (2, 9). What is the base?

Solution

Since the function is exponential, it is of the form $y = b^x$. The point (2, 9) lies on the graph, and thus we have $9 = b^2$ or $b = 3$. We note that although $b = -3$ satisfies the equation $b^2 = 9$, -3 is not the base, because b must be greater than 0. ●

Exponential equations are equations that contain the unknown in the exponent. Some simple exponential equations can be solved by applying the laws of exponents. For others, it is necessary to use the laws of logarithms. Here we shall illustrate some methods used to find solutions that depend only upon laws of exponents. More complex equations will be treated in the following sections.

Example 50.5

Solve for x if $2^x = 32$.

Solution

Since $32 = 2^5$, we have $2^x = 2^5$, so that $x = 5$.

Example 50.6

Solve for x if $3^{2x-1} = 81$.

Solution

Since $81 = 3^4$, it follows that $3^{2x=1} = 3^4$. This equation is satisfied if $2x - 1 = 4$, that is, if $2x = 5$ or $x = \frac{5}{2}$.

Example 50.7

Solve for x if $16^x = \frac{1}{64}$

Solution

Since $16 = 2^4$, $16^x = (2^4)^x = 2^{4x}$. Also, $\frac{1}{64} = 1/2^6 = 2^{-6}$. Thus, $2^{4x} = 2^{-6}$, and this implies that $4x = -6$ or $x = -\frac{3}{2}$ is the desired solution. ●

The ability to solve exponential equations often proves useful in problems involving rates of growth and decay.

Example 50.8

Suppose that the growth of a certain type of bacteria is given by $n = n_0 2^t$, where n is the number of bacteria present at the end of t hours and n_0 is the initial number of bacteria. If there are 20,000 bacteria at the end of 3 hr, find n_0 and also find how long it takes for n_0 to double.

Solution

Since $n = 20{,}000$ when $t = 3$ hr, $20{,}000 = n_0 2^3$ or $8n_0 = 20{,}000$. Thus, $n_0 = 20{,}000/8 = 2500$. To find how long it takes for n_0 to double, we note that $n = 2n_0 = 2(2500) = 5000$. We now have the equation $5000 = 2500 \cdot 2^t$. Thus, $2^t = 5000/2500 = 2$, so $t = 1$ hr.

Exercises for Section 50

In Exercises 1 through 10, graph the function, state if it is increasing or decreasing, and write the intercept point.

1. $y = 5^x$

2. $y = \left(\frac{1}{5}\right)^x$

3. $y = 5^{-x}$

4. $y = 3 \cdot 2^x$

5. $y = -2^x$

6. $y = -3^x$

7. $y = 3^{-x}$

8. $y = -2 \cdot 3^x$

9. $y = \left(\frac{1}{4}\right)^x$

10. $y = \left(\frac{1}{4}\right)^{-x}$

In Exercises 11 through 20, the points lie on the graph of an exponential function. Find the base b

11. $(2, 9)$

12. $(3, 8)$

13. $(2, 100)$

14. $\left(3, \frac{1}{64}\right)$

15. $(-2, 4)$

16. $\left(-2, \frac{1}{9}\right)$

17. $\left(\frac{1}{2}, 4\right)$

18. $\left(\frac{1}{3}, 2\right)$

19. $\left(\frac{2}{3}, 27\right)$

20. $(0, 1)$

In Exercises 21 through 33, solve each exponential equation.

21. $2^x = 16$

22. $3^x = \frac{1}{81}$

23. $5^{-x} = 125$

24. $2^{2x+3} = 8$

25. $3^x = 1$

26. $5^{2x+1} = 1$

27. $4(3^x) = 36$

28. $3^{x+2} = 81^{x-1}$

29. $(2^{x-1})^2 = 16$

30. $3^{1-2x} = \frac{1}{9}$

31. $9(3^{2x}) = 27^{x+1}$

32. $8^{2x-6} = \left(\frac{1}{8}\right)^{4-x}$

33. $\left(\frac{1}{2}\right)^{2x+1} = 8^{-x}$

34. The growth of a certain type of bacteria is given by the equation $n = n_0 2^t$, where n_0 is the initial number of bacteria and t is the time (in hours).
(a) If $n = 200{,}000$ when $t = 2$, find n_0.
(b) Find the number of bacteria present at the end of 4 hr.
(c) How long does it take for n_0 to double?
(d) How long does it take for n_0 to quadruple?

35. The number n of bacteria in a certain culture at the end of t hours is given by equation $n = n_0(3)^{kt}$, where n_0 is the initial population. When $t = 0$, $n = 400$ and when $t = 2\frac{1}{2}$ hr, $n = 1200$.
(a) Find n_0.
(b) Find k.
(c) What is the number of bacteria present at the end of 5 hr?

36. The half-life of radium is approximately 1600 years; that is, half of the original amount remains after 1600 years. The amount A of radium that remains after t years is given by the equation

$$A = A_0\left(\frac{1}{2}\right)^{kt} \qquad \text{where } A_0 \text{ is the initial amount of radium}$$

Suppose that we begin with 100 milligrams (mg) of radium. After 1600 years, 50 mg will remain.
(a) Find the value for k.
(b) How long will it take for the 100 mg to disintegrate to 25 mg?

37. The system of equations $y = x^2$ and $y = 2^x$ has three solutions.
(a) Set up a table of values for x^2 and 2^x using the values $x = -2, -1, 0, 1, 2, 3, 4$. Two of the solutions we seek can be read directly from this table.

(b) On the same set of axes, graph $y = x^2$ and $y = 2^x$ and then use the method of Section 33 (graphic solution to nonlinear systems) to find the third solution.

SECTION 51
Logarithmic Functions

A logarithm is simply an equivalent way of writing an exponential expression. For the exponential function $y = b^x$, b is positive and not equal to 1, x is called the **logarithm** of the number y to the base b and is written

$$x = \log_b y \tag{51.1}$$

Thus, the function $y = b^x$ is equivalent to the function $x = \log_b y$.

In dealing with the variables x and y it is customary to select y as the dependent variable. Thus, in equation (51.1) we would like to express y in terms of x. We accomplish this by interchanging the variables x and y in $y = b^x$ to obtain $x = b^y$. Now $x = b^y$ is equivalent to $y = \log_b x$ and y is in the familiar role of the dependent variable.

The relationship that exists between the exponential function and logarithm function is exhibited as follows:

$$y = b^x \text{ is equivalent to } x = \log_b y \tag{51.2}$$

$$y = \log_b x \text{ is equivalent to } x = b^y \tag{51.3}$$

When we interchange the variable in a given expression, the resulting relations, if they are functions, are called **inverse functions**. The graphs of such functions are reflections or mirror images of each other about the line $y = x$. Thus, $y = b^x$ and $y = \log_b x$ are inverses of one another, since the function $y = b^x$ yields $x = b^y$ when we interchange x and y and equation (51.3) then yields $y = \log_b x$.

The admissible values for x and y in the function $y = \log_b x$ are most easily obtained if we consider the equivalent form, $x = b^y$. For b positive and $b \neq 1$, y can be any real number, while x has the property that it must be positive. Thus, the function $y = \log_b x$ will only be defined for $x > 0$ and will then take on all real values. We are, in effect, stating that the logarithm of a negative number does not exist, while at the same time logarithm values may be negative, zero or positive, that is, any real number. This statement becomes even more apparent when we consider the graphs that follow.

Example 51.1

Graph $y = \log_2 x$.

Solution A
We can find points on the graph by first putting the equation in its exponential form $x = 2^y$ and then assume values for y. The corresponding value for x follow easily from the exponential form. Using the values found in Table 51.1, we construct the graph shown in Figure 51.1.

Table 51.1

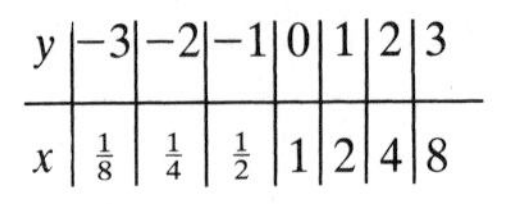

y	-3	-2	-1	0	1	2	3
x	$\frac{1}{8}$	$\frac{1}{4}$	$\frac{1}{2}$	1	2	4	8

Figure 51.1

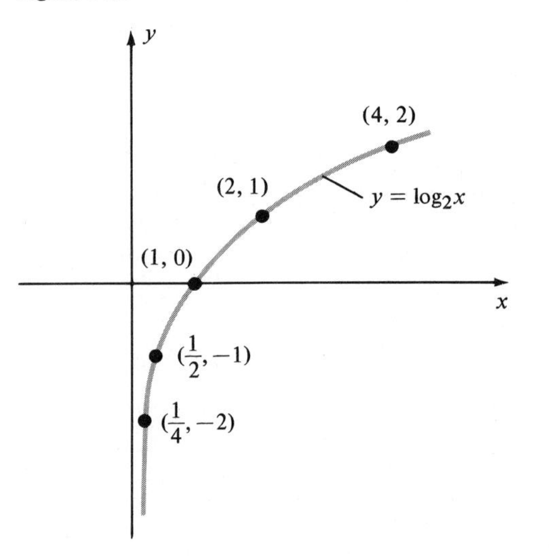

Solution B
Since $y = \log_2 x$ is the inverse of the function $y = 2^x$, we may obtain the graph we seek by reflecting the graph of $y = 2^x$ about the line $y = x$. This reflection is illustrated in Figure 51.2.

The graph of any logarithmic function where $b > 1$ will be similar to the graph of the function in Example 51.1. To observe the behavior of the logarithm curve for $0 < b < 1$, let us consider the following examples.

Example 51.2
Graph $y = \log_{1/2} x$.

Solution
From Example 50.3 we have the graph of $y = (\frac{1}{2})^x$. Since $y = (\frac{1}{2})^x$ is the inverse of $y = \log_{1/2} x$, we obtain the necessary graph by means of a simple reflection in the line $y = x$. The graph is shown in Figure 51.3.

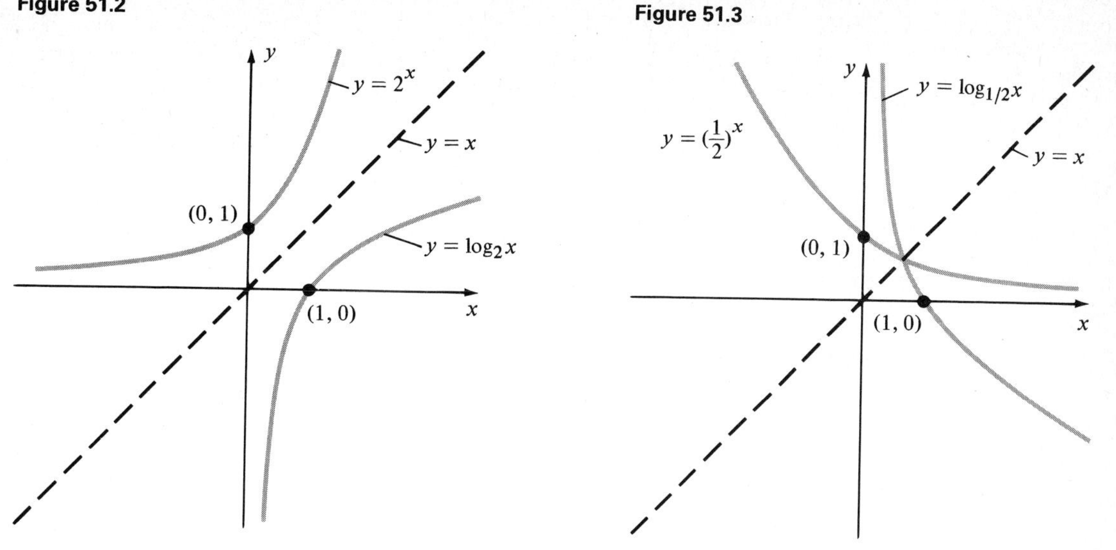

Note that we obtain the same result if we were to use the method of solution A for Example 51.1.

Properties of $y = \log_b x$

For $b > 1$ we observe that the graph of $y = \log_b x$ has the following properties:

1. If $0 < x < 1$, $\log_b x < 0$.
2. If $x = 1$, $\log_b x = 0$.
3. If $x > 1$, $\log_b x > 0$.
4. If $x \leq 0$, $\log_b x$ is **not defined.**
5. For $x > 0$, $y > 0$, $x = y$ if and only if $\log_b x = \log_b y$.

Example 51.3

We use statement (51.2) to express the given exponential statements in logarithmic form.

(a) If $4^2 = 16$, then $\log_4 16 = 2$; that is, the logarithm of 16 to base 4 is 2.
(b) If $10^1 = 10$, then $\log_{10} 10 = 1$.
(c) If $8^{2/3} = 4$, then $\log_8 4 = \frac{2}{3}$.
(d) If $3^{-2} = \frac{1}{9}$, then $\log_3 (\frac{1}{9}) = -2$.
(e) If $5^0 = 1$, then $\log_5 1 = 0$.

Example 51.4

We use statement (51.3) to express the given logarithmic statement in exponential form.

(a) $\log_{10} 1 = 0$ can be written in exponential form as $10^\circ = 1$.

(b) $\log_{25}5 = \frac{1}{2}$ means that $25^{1/2} = 5$.
(c) $\log_{10}(\frac{1}{100}) = -2$ means that $10^{-2} = \frac{1}{100}$.
(d) $\log_{1/2}8 = -3$ means that $(\frac{1}{2})^{-3} = 8$.
(e) $\log_{27}3 = \frac{1}{3}$ means that $27^{1/3} = 3$.
(f) $\log_b 0$ is not defined, since $\log_b 0 = x$ would mean that $b^x = 0$ and there is no value of x for which $b^x = 0$.

Statements (51.2) and (51.3) can also be used to solve simple types of logarithmic equations, as illustrated in the following examples.

Example 51.5
Solve the equation $\log_3 x = 4$ for x.

Solution
$\log_x x = 4$ is equivalent to $x = 3^4$. Thus, $x = 81$.

Example 51.6
Solve $\log_9 81 = x$ for x.

Solution
$\log_9 81 = x$ can be written in the equivalent exponential form as $9^x = 81$. Since $81 = 9^2$, we have $9^x = 9^2$, and by equating the exponents we have $x = 2$.

Example 51.7
Solve $\log_3 3^{2.1} = x$ for x.

Solution
Since $\log_3 3^{2.1} = x$ is equivalent to $3^x = 3^{2.1}$, then $x = 2.1$.

Example 51.8
Solve $\log_x 16 = 2$ for x.

Solution
If $\log_x 16 = 2$, then $x^2 = 16$ is the equivalent exponential form. Thus, $x = 4$. Observe that we have ignored the solution $x = -4$ since $\log_{-4}16$ is undefined.

Example 51.9
Solve $\log_8(\frac{1}{32}) = x$ for x.

Solution
$\log_8(\frac{1}{32}) = x$ can be written as $8^x = \frac{1}{32}$. Now $8 = 2^3$ and $\frac{1}{32} = 1/2^5 = 2^{-5}$. Thus, $(2^3)^x = 2^{-5}$ or $2^{3x} = 2^{-5}$. Equating exponents of this latter form yields $3x = -5$ or $x = -\frac{5}{3}$.

Exercises for Section 51

In Exercises 1 through 5, sketch the graph of each function and state the intercept.

1. $y = \log_3 x$

2. $y = \log_4 x$

3. $y = \log_5 x$

4. $y = 3\log_2 x$

5. $y = \log_{1/3} x$

6. Graph $y = \log_{10} x$.
  (a) For what value of x is $y = 0$?
  (b) For what values of x is $y < 0$?
  (c) For what values of x is $y > 0$?

In Exercises 7 through 20, write the exponential statement in the equivalent logarithmic form.

7. $3^2 = 9$

8. $2^5 = 32$

9. $10^0 = 1$

10. $\left(\frac{1}{2}\right)^3 = 8$

11. $4^{1/2} = 2$

12. $16^{3/4} = 8$

13. $10^{-3} = \frac{1}{1000}$

14. $16^{-1/2} = \frac{1}{4}$

15. $64^{1/3} = \frac{1}{4}$

16. $\left(\frac{9}{25}\right)^{-3/2} = \frac{125}{27}$

17. $10^1 = 10$

18. $10^{-1} = \frac{1}{10}$

19. $e^0 = 1$

20. $e^1 = e$

In Exercises 21 through 35, write the logarithmic statement in the equivalent exponential form.

21. $\log_6 36 = 2$

22. $\log_2 32 = 5$

23. $\log_{10} 100 = 2$

24. $\log_{10} 10 = 1$

25. $\log_{10} 1 = 0$

26. $\log_6 \frac{1}{36} = -2$

27. $\log_{27} 81 = \frac{4}{3}$

28. $\log_{10}(0.001) = -3$

29. $\log_{1/3} 9 = -2$

30. $\log_{0.5} \frac{1}{16} = 4$

31. $\log_8 \frac{1}{64} = -2$

32. $\log_e e = 1$

33. $\log_{10}10^5 = 5$

34. $\log_e e^3 = 3$

35. $\log_e e^{-6} = -6$

In Exercises 36 through 50, find the value of x.

36. $\log_2 x = 5$

37. $\log_8 x = \frac{2}{3}$

38. $\log_{10} x = 3$

39. $\log_3(x - 1) = 2$

40. $\log_5(3x + 1) = 2$

41. $\log_x 0.01 = -2$

42. $\log_{10}10^x = 5$

43. $\log_2 2^5 = x$

44. $\log_x 4 = \frac{2}{5}$

45. $\log_x 8 = \frac{3}{4}$

46. $\log_x 9 = -\frac{2}{3}$

47. $2\log_x 36 = 4$

48. $\log_3 \frac{27}{81} = x$

49. $\log_2(x^2 + 7) = 3$

50. $4\log_x \frac{1}{125} = -12$

SECTION 52

Properties of Logarithms

We have previously noted that $y = \log_b x$ is equivalent to $x = b^y$. Thus, we may write $x = b^y = b^{\log_b x}$ and directly observe that logarithms are exponents. It is not surprising, then, that the properties of logarithms are derived directly from the properties of exponents. Before proceeding to the properties of logarithms, it would be helpful to review the laws of exponents in Section 4. We restate the laws that will be most useful in this section as follows:

$$a^m a^n = a^{m+n} \tag{52.1}$$

$$\frac{a^m}{a^n} = a^{m-n} \tag{52.2}$$

$$(a^m)^n = a^{mn} \tag{52.3}$$

These laws will enable us to develop and use the laws of logarithms.

SECTION 52

Properties of Logarithms

Laws of Logarithms

I. The logarithm of the product of two (or more) positive numbers x and y is equal to the sum of the logarithms of the numbers. That is,

$$\log_b(xy) = \log_b x + \log_b y \tag{52.4}$$

II. The logarithm of the quotient of two positive numbers x and y is equal to the logarithm of the numerator minus the logarithm of the denominator. That is,

$$\log_b \frac{x}{y} = \log_b x - \log_b y \tag{52.5}$$

III. The logarithm of the nth power of a positive number x is equal to n times the logarithm of the number. That is,

$$\log_b x^n = n \log_b x \tag{52.6}$$

The demonstration of equations (52.4) through (52.6) is direct. If we let $u = \log_b x$ and $v = \log_b y$ and write these equations in equivalent exponential form, we have $x = b^u$ and $y = b^v$. Now the product of x and y yields

$$x \cdot y = b^u \cdot b^v = b^{u+v} \qquad \text{[see equation (52.1)]}$$

We now write $xy = b^{u+v}$ in the equivalent logarithmic form $\log_b(xy) = u + v$ or $\log_b(xy) = \log_b x + \log_b y$ and we have equation (52.4).

The quotient of x and y yields the equation

$$\frac{x}{y} = \frac{b^u}{b^v} = b^{u-v}$$

Writing $x/y = b^{u-v}$ in logarithmic form, we have

$$\log_b \frac{x}{y} = u - v$$

or

$$\log_b \frac{x}{y} = \log_b x - \log_b y \qquad \text{[which is equation (52.5)]}$$

For equation (52.6) we note that $x = b^u$, and thus $x^n = (b^u)^n = b^{nu}$. The corresponding logarithmic form yields the equation

$$\log_b x^n = nu$$

or

$$\log_b x^n = n \log_b x \qquad \text{[which is equation (52.6)]}$$

Equation (52.6) has three interesting variations:

1. If we write $\log_b \sqrt[n]{x}$ as $\log_b x^{1/n}$ and use (52.6), we obtain

$$\log_b x^{1/n} = \frac{1}{n} \log_b x \tag{52.7}$$

We note that equation (52.7) states that the logarithm of a root of a number is the logarithm of that number divided by the index of the root.

2. If $n = 0$, $\log_b x^0 = 0\,(\log_b x) = 0$. Since $\log_b x^0 = \log_b 1$, we have

$$\log_b 1 = 0 \tag{52.8}$$

Thus, the logarithm of 1 to *any base* is 0.

3. If $n = -1$, $\log_b x^{-1} = -1 \log_b x$. Since $\log_b x^{-1} = \log_b (1/x)$, we have

$$\log_b \frac{1}{x} = -\log_b x \tag{52.9}$$

Thus, the logarithm of the reciprocal of a number is the negative of the logarithm of that number.

We use the following examples to illustrate the application of the laws of logarithms.

Example 52.1

(a) $\log_2[3(4)] = \log_2 3 + \log_2 4$

(b) $\log_3 \dfrac{15}{47} = \log_3 15 - \log_3 47$

(c) $\log_{10}(24)^3 = 3 \log_{10} 24$

(d) $\log_{10}\sqrt[3]{28} = \log_{10}(28)^{1/3} = \dfrac{1}{3}\log_{10} 28$

(e) $\log_3 \dfrac{21(145)}{312} = \log_3[21(145)] - \log_3 312$

$= \log_3 21 + \log_3 145 - \log_3 312$

(f) $\log_{10} \dfrac{13}{51\sqrt{34}} = \log_{10} 13 - \log_{10}(51\sqrt{34})$

$= \log_{10} 13 - [\log_{10} 51 + \log_{10} 34^{1/2}]$

$= \log_{10} 13 - \log_{10} 51 - \log_{10} 34^{1/2}$

$= \log_{10} 13 - \log_{10} 51 - \dfrac{1}{2}\log_{10} 34$

Example 52.2

Express $\log_{10} 3 + 2 \log_{10} 7$ as a single logarithm.

Solution

$$\begin{aligned}\log_{10} 3 + 2 \log_{10} 7 &= \log_{10} 3 + \log_{10} 7^2 \\ &= \log_{10}(3 \cdot 7^2) \\ &= \log_{10}(3 \cdot 49) = \log_{10} 147\end{aligned}$$

Example 52.3

Express $2 \log_4 3 - 3 \log_4 2$ as a single logarithm.

Solution

$$\begin{aligned} 2 \log_4 3 - 3 \log_4 2 &= \log_4 3^2 - \log_4 2^3 \\ &= \log_4 \frac{3^2}{2^3} \\ &= \log_4 \frac{9}{8} \end{aligned}$$

Example 52.4

Express $\frac{1}{2} \log_5 7 + \log_5 3 - 2 \log_5 4$ as a single logarithm.

Solution

$$\begin{aligned} \tfrac{1}{2} \log_5 7 + \log_5 3 - 2 \log_5 4 &= \log_5 7^{1/2} + \log_5 3 - \log_5 4^2 \\ &= \log_5 (7^{1/2} \cdot 3) - \log_5 4^2 \\ &= \log_5 \frac{7^{1/2} \cdot 3}{4^2} = \log_5 \frac{\sqrt{7} \cdot 3}{16} \end{aligned}$$

Example 52.5

Solve for x if $\log_3(2x + 7) - \log_3 x = 2$.

Solution

$\log_3(2x + 7) - \log_3 x = \log_3[(2x + 7)/x] = 2$. Hence, $(2x + 7)/x = 3^2 = 9$. Now, $2x + 7 = 9x$, and thus $7x = 7$ or $x = 1$. This result can be verified by substituting $x = 1$ into the original equation as follows:

$$\begin{aligned} \log_3(2 + 7) - \log_3 1 &= 2 \\ \log_3 9 - \log_3 1 &= 2 \\ 2 - 0 &= 2 \\ 2 &= 2 \end{aligned}$$

Example 52.6

Solve for x if $\log_5(x - 5) + \log_5(x - 1) = 1$.

Solution

$$\log_5(x - 5) + \log_5(x - 1) = \log_5[(x - 5) \cdot (x - 1)] = 1$$

Thus,

$$\begin{aligned} (x - 5)(x - 1) = 5 \quad \text{or} \quad x^2 - 6x + 5 &= 5 \\ x^2 - 6x &= 0 \\ x(x - 6) &= 0 \quad \text{and} \quad x = 0 \quad \text{and} \quad x = 6 \end{aligned}$$

Hence, our solutions are apparently $x = 0$ and $x = 6$. However, 0 does not satisfy the original equation, so $x = 6$ is the only solution. Note that for $x = 0$, we would have the logarithm of a negative number, and such expressions are not defined.

Example 52.7

Solve for x if $\log_b(3x + 8) - \log_b(2x + 3) = \log_b 2$.

Solution

$\log_b(3x + 8) - \log_b(2x + 3) = \log_b(3x + 8)/(2x + 3)$. Thus, $\log_b[(3x + 8)/(2x + 3)] = \log_b 2$. We now employ property 5 of Section 51, which states that for $x > 0$, $y > 0$ and $x = y$, $\log_b x = \log_b y$. Hence,

$$\frac{3x + 8}{2x + 3} = 2 \quad \text{or} \quad 3x + 8 = 2(2x + 3)$$

$$\text{or} \quad 3x + 8 = 4x + 6$$

$$\text{and} \quad x = 2$$

The result can be verified by direct substitution into the given equation.

Example 52.8

Solve for y in terms of x if $3\log_2 x + \log_2 y - \log_2 5 = 3$.

Solution

For $3\log_2 x$ we can write $\log_2 x^3$. Now $\log_2 x^3 + \log_2 y = \log_2(x^3 y)$ and $\log_2(x^3 y) - \log_2 5 = \log_2(x^3 y/5)$. Thus, $3\log_2 x + \log_2 y - \log_2 5 = \log_2(x^3 y/5) = 3$. In order to express y in terms of x, we write $\log_2(x^3 y/5) = 3$ in the equivalent exponential form $x^3 y/5 = 2^3$. Hence, $x^3 y = 5 \cdot 8 = 40$ and $y = 40/x^3$.

Exercises for Section 52

In Exercises 1 through 10, express each term as a sum, difference, or multiple of logarithm.

1. $\log_{10}(ab)$

2. $\log_{10}(abc)$

3. $\log_{10}\dfrac{x^2}{y}$

4. $\log_3\sqrt[3]{75}$

5. $\log_{10}(xy^2)^3$

6. $\log_5(x^8 y^2)$

7. $\log_{10}\dfrac{x^2 y^3}{z^4}$

8. $\log_2\dfrac{x\sqrt{y}}{2}$

9. $\log_{10}\dfrac{1}{x\sqrt{y}}$

10. $\log_{10}\sqrt{\dfrac{xy^4}{z^2}}$

In Exercises 11 through 20, express each term as a single logarithm.

11. $\log_2 5 + \log_2 7$

12. $\log_5 32 - \log_5 4$

13. $2\log_3 4 - 3\log_3 2$

14. $\log_{10} 36 - \frac{1}{2}\log_{10} 9$

15. $4\log_5 3 - \log_5 27$

16. $\log_{10} 4 + \log_{10} 5 - \log_{10} 6$

17. $\frac{1}{3}\log_2 8 - \frac{1}{4}\log_2 4$

18. $\frac{2}{3}\log_5 64 - \frac{3}{2}\log_5 4$

19. $\frac{1}{2}\log_{10} 16 + \frac{1}{3}\log_{10} 27 - \frac{1}{4}\log_{10} 16$

20. $\frac{1}{3}(\log_{10} 4 + \log_{10} 2) - \frac{1}{2}\log_{10} 16$

In Exercise 21 through 30, solve each equation for x.

21. $\log_3 x = \log_3 2 + \log_3 5$

22. $\log_{10} x = \log_{10} 6 - \log_{10} 4$

23. $\log_5 x = \log_5 56 - 3\log_5 2$

24. $\log_{10}(2x - 3) + \log_{10} 3 = \log_{10} 15$

25. $2\log_{10} x = 4\log_{10} 3$

26. $1\log_{10} x = \log_{10} 4$

27. $\log_{10}(x^2 - 3x) = \log_{10} 8 - \log_{10} 2$

28. $\log_2(x^2 - 4) - \log_2(x + 2) = \log_2 3$

29. $\log_2(x^2 - 4) - \log_2(x + 2) = 3$

30. $\log_{10} x + \log_{10}(x - 1) = \log_{10} 6$

In Exercises 31 through 40, solve each equation for y in terms of x.

31. $2\log_{10} x - \log_{10} y = 1$

32. $\log_2 x + \log_2 y + \log_2 3 = 1$

33. $\log_2 x + \log_2 y - \log_2 3 = 1$

34. $\log_2 x + \log_2 y - \log_2 3 = 0$

35. $\log_2 x - \log_2 y - \log_2 3 = 1$

36. $\log_3 y = \log_3 4 - 2\log_3 x$

37. $\log_{10} y = 3\log_{10} 2 + \log_{10} x - \log_{10}\sqrt[3]{x}$

38. $4\log_2 x - 2\log_2 y = \log_2 81$

39. $2(\log_3 x - \log_3 y) = 4$

40. $\log_{10} y = \log_{10}(2x + 1) + 2\log_{10} x - 3\log_{10} 2$

In Exercises 41 through 46, construct an example that will show that each statement is false.

41. $\log_b \frac{1}{x} = \frac{\log_b 1}{\log_b x}$

42. $\log_b(x \cdot y) = (\log_b x)(\log_b y)$

43. $n\log_b x = (\log_b x)^n$

44. $(\log_b x)(\log_b y) = \log_b x + \log_b y$

45. $\frac{\log_b x}{\log_b y} = \log_b x - \log_b y$

46. $\log_b(x^n) = (\log_b x)^n$

In Exercises 47 through 52, evaluate each by using $\log_{10}3 = 0.4771$ and the appropriate logarithmic properties.

47. $\log_{10}9$ **48.** $\log_{10}300$ **49.** $\log_{10}9000$

50. $\log_{10}\frac{1}{3}$ **51.** $\log_{10}\frac{1}{9}$ **52.** $\log_{10}\frac{1}{3000}$

SECTION 53

Common Logarithms

In theory, there are an infinite number of logarithm systems that could be constructed. In practice, only two are necessary in the study of mathematics: common logarithms and natural logarithms. In this section we shall discuss common logarithms.

Common logarithms are logarithms to the base 10 and are particularly well suited to the simplification of computational problems involving multiplication, divisions, powers, and roots.* When writing common logarithms, the subscript 10 denoting the base of the system is usually omitted. Thus, $\log_{10}N$ is written simply as $\log N$.

Before we investigate the problem of how we will find the logarithm of a nonnegative number N, let us discuss the meaning of scientific notation. In many branches of science we work with very large or very small numbers. For example, the distance between the earth and sun is about 93,000,000 miles, the speed of light is 186,000 miles per second, and the coefficient of expansion of gases is 0.003665. Values such as these are cumbersome to write and operate with, so we employ scientific notation to ease the burden.

In writing a number using scientific notation, we place the decimal point to the right of the first significant digit and then multiply by the appropriate power of 10. This procedure is formulated as follows.

Rules for Scientific Notation

1. For a number written in the usual decimal notation, move the decimal point to the right of the first nonzero digit.
2. Multiply the number in part 1 by a power of 10 whose exponent is:
   (a) n if the decimal point is moved n places to the left.
   (b) $-n$ if the decimal point is moved n places to the left.

Example 53.1

(a) $93{,}000{,}00 = (9.3)(10)^7$

* The role of the calculator for such computations is discussed in Appendix E.

(b) $186{,}000 = (1.86)(10)^5$
(c) $0.003665 = (3.665)(10)^{-3}$
(d) $2.35 = (2.35)(10)^0$
(e) $0.235 = (2.35)(10)^{-1}$ ●

We note, that in general, any positive number N can be written in scientific notation. We shall use the notation $N = x \times 10^n$, where $1 \leq x < 10$ and n an integer. In addition, we observe that

1. If $N < 1$, n is a negative integer.
2. If $1 \leq N < 10$, n is zero.
3. If $N \geq 10$, n is a positive integer.

For common logarithms, $\log 10 = 1$, since $10^1 = 10$. Thus, the common logarithms of all integral powers of 10 are easily found.

Example 53.2

(a) $\log 1 = \log 10^0 = 0 \log 10 = 0 \cdot 1 = 0$
(b) $\log 10 = \log 10^1 = 1 \log 10 = 1 \cdot 1 = 1$
(c) $\log 100 = \log 10^2 = 2 \log 10 = 2 \cdot 1 = 2$
(d) $\log 1000 = \log 10^3 = 3 \log 10 = 3 \cdot 1 = 3$ ●

If the integral powers of 10 are negative, the behavior of the logarithms follows the pattern illustrated in the following example.

Example 53.3

(a) $\log \frac{1}{10} = \log 10^{-1} = -1 \log 10 = -1 \cdot 1 = -1$

(b) $\log \frac{1}{100} = \log 10^{-2} = -2 \log 10 = -2 \cdot 1 = -2$

(c) $\log \frac{1}{1000} = \log 10^{-3} = -3 \log 10 = -3 \cdot 1 = -3$ ●

With the results of Examples 53.2 and 53.3 and the properties of logarithms studied in Section 52, we are able to sketch $y = \log x$. The graph is shown in Figure 53.1 on page 294.

If N is not an integral power of 10, then $N = x \times 10^n$, $1 \leq x < 10$ and n an integer, can be expressed as $\log N = n + \log x$ by taking the logarithm of both sides of this equation, as follows:

$$\begin{aligned}\log N &= \log (x \times 10^n)\\ &= \log x + \log 10^n\\ &= \log x + n \log 10 = \log x + n\end{aligned}$$

or

$$\log N = n + \log x \tag{53.1}$$

Figure 53.1

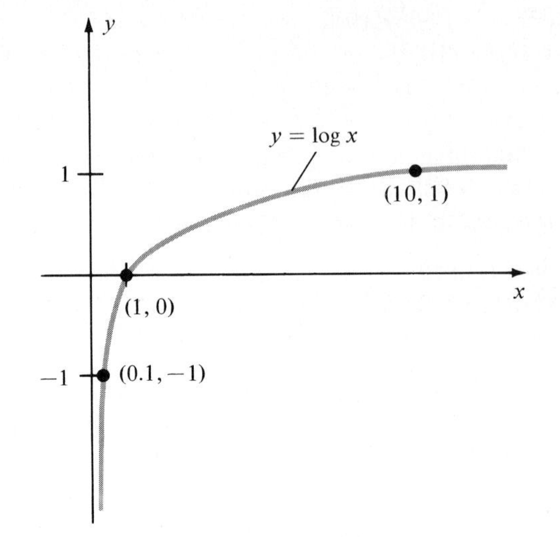

In equation (53.1), n is called the **characteristic** and $\log x$ is called the **mantissa** of $\log N$. Thus, the logarithm of any positive number N is composed of two parts.

The characteristic n can be any integer. However, the value of the mantissa, $\log x$, is restricted by the fact that $1 \leq x < 10$. Since $\log 1 = 0$ and $\log 10 = 1$, the value of the mantissa will always have the property that $0 \leq \log x < 1$.

We observe that for any number N, it is relatively easy to find the characteristic n for $\log N$. The value of the mantissa, $\log x$, must come from a common logarithm table. The use of such a table is shown in the examples that follow.

Example 53.4
Find log 674.

Solution
We first write the number in scientific notation as 6.74×10^2. The characteristic is 2. To find the mantissa, log 6.74, we use Table II of Appendix F. We first look in the column headed by N and find the first two significant digits 67 of our given number. We then read across the table until we are in the column headed by the number 4, the third significant digit, and we find 8287. Actually, the value is 0.8287, but the decimal points have been omitted. Thus, $\log 6.74 = 0.8287$, and therefore

$$\log 674 = 2 + 0.8287 = 2.8287$$

Example 53.5
Find log 42,000.

Solution
Since $42{,}000 = 4.2 \times 10^4$, we have $\log 42{,}000 = 4 + \log 4.2$. We now use Table II to find $\log 4.2$. In the column labeled N we find 42, and then move across to the column headed by 0. The entry 4232 gives us the mantissa, 0.4232. Thus, $\log 42{,}000 = 4 + 0.4232 = 4.4232$. ●

Before proceeding to additional examples, let us formulate the procedure we have outlined for finding common logarithms.

Rules for Finding Common Logarithms of any Nonnegative Number N

Step 1: Express N in scientific notation.

$$N = x \times 10^n \qquad \text{where } 1 \leq x < 10 \text{ and } n \text{ is an integer}$$

Step 2: The characteristic is n, the exponent of 10.
Step 3: Use Table II to find the mantissa, $\log x$.
Step 4: $\log N$ = characteristic + mantissa.

It is important to note that while the characteristic of a common logarithm can be negative, the mantissa, $\log x$, will never be negative.

Example 53.6
Find $\log 0.0431$.

Solution
Since $0.0431 = 4.31 \times 10^{-2}$, $\log 0.0431 = -2 + \log 4.31$. From Table II, $\log 4.31 = 0.6345$, and therefore $\log 0.0431 = -2 + 0.6345$. Here the characteristic is negative and we must be cautious. We cannot write $\log 0.0431 = -2.6345$, since this would imply that the sum of 0.6345 and -2 is -2.6345, a result that is arithmetically unsound. We can cope with this situation in one of two ways. We can rewrite the characteristic -2 by adding and subtracting 10. Thus, $-2 = (10 - 2) - 10$ and

$$\begin{aligned}\log 0.0431 = -2 + 0.6345 &= (10 - 2) + 0.6345 - 10 \\ &= 8 + 0.6345 - 10 \\ &= 8.6345 - 10\end{aligned}$$

What we have done in this case is to simply write a negative characteristic as an appropriate positive number with 10 subtracted. This technique can be used whenever the characteristic is negative. Alternatively, since $\log 0.0431 = -2 + 0.6345$ and $-2 + 0.6345 = -1.3655$, we have

$$\log 0.0431 = -1.3655$$

In this case, we have simply combined -2 and 0.6345 into a single number. However, we must be careful to note that in this form we *cannot* say that the characteristic is -1 and the mantissa is 0.3655. To express

-1.3655 in a form where we can read the characteristic and mantissa directly, we must reverse the process used to combine the original values into one and write

$$\begin{aligned} -1.3655 &= 2 + (-1.3655) - 2 \\ &= 0.6345 - 2 \end{aligned}$$

We have simply rewritten the negative number -1.3655 as a decimal value between 0.0000 and 1.0000, minus an appropriate whole number. Both methods will prove useful in future applications.

Example 53.7
Find $\log 0.000283$.

Solution
We write 0.000283 in scientific notation as 2.83×10^{-4}. Thus, the characteristic is -4, or $6 - 10$. From Table II we have $\log 2.83 = 0.4518$. Hence,

$$\log 0.000283 = -4 + 0.4518 = 6.4518 - 10$$

If we wish to write the characteristic and the mantissa as a single value, we combine -4 and 0.4518 to obtain -3.5482, and therefore

$$\log 0.000283 = -3.5482$$

Example 53.8
The equation $D = 10 \log (I_2/I_2)$ is a formula for sound intensity. Find D if $I_2/I_1 = 0.025$.

Solution

$$D = 10 \log 0.025$$

and

$$\begin{aligned} \log 0.025 &= \log(2.5 \times 10^{-2}) \\ &= -2 + \log 2.5 \\ &= -2 + 0.3118 \end{aligned}$$

Since we must substitute the logarithm into the equation, the entire logarithm is changed into a negative number, and we have $-2 + 0.3118 = -1.6882$. Now

$$\begin{aligned} D &= 10 \log 0.025 \\ &= 10(-1.6882) \\ &= -16.882 \end{aligned}$$

●

In Examples 53.4 through 53.7, we have illustrated how we find the logarithm of a number N. We shall now consider the primary advantage common logarithms have in relation to any other logarithmic system. This advantage is demonstrated in the following example.

Example 53.9

(a) $6.45 = 6.45 \times 10^0$. Thus, $\log 6.45 = 0 + \log 6.45$.
(b) $64.5 = 6.45 \times 10^1$. Thus, $\log 64.5 = 1 + \log 6.45$.
(c) $6450 = 6.45 \times 10^3$. Thus, $\log 6450 = 3 + \log 6.45$.
(d) $0.0645 = 6.45 \times 10^{-2}$. Thus, $\log 0.0645 = -2 + \log 6.45$.
(e) $0.000645 = 6.45 \times 10^{-4}$. Thus, $\log 0.00645 = -4 + \log 6.45$. ●

Example 53.9 shows that the logarithms of number that have the same sequence of digits have identical mantissas; only their characteristics may differ. Thus, the common logarithm of any positive number N reduces to a problem that involves finding the logarithm of a number between 1 and 10, and this number can be found in Table II of Appendix F.

When finding the logarithm of a number that lies between two numbers in the table, we can take the logarithm of the closest number. In following this procedure we reduce the accuracy of our result by one place. Thus, our four-place table is reduced to a three-place table.

If we wish to retain four-place accuracy, we use the method of linear interpolation outlined in Section 41 of Chapter 8. The examples that follow will review the technique.

Example 53.10

Find log 371.6.

Solution

Since $371.6 = 3.716 \times 10^2$ in scientific notation, the characteristic is 2. To find the mantissa, we note that although log 371.6 is not in Table II, log 3.71 and log 3.72 are, and hence we must interpolate. In Table II we find that $\log 3.71 = 0.5694$ and $\log 3.72 = 0.5705$. Thus, as our number increases from 3.71 to 3.72, the logarithm increases from 0.5694 to 0.5705, a change of 0.0011 unit. Since 3.716 is 0.006/0.010, or $\frac{6}{10}$ of the way from 3.71 to 3.72, we assume that log 3.716 is $\frac{6}{10}$ of the way from 0.5694 to 0.5705. Now $\frac{6}{10}$ of $0.0011 = 0.00066 = 0.0007$ to four places. Therefore, $\log 3.716 = 0.5694 + 0.0007 = 0.5701$, and so $\log 371.6 = 2.5701$. The interpolation may be arranged as follows:

$$0.010\left\{\begin{array}{l} 0.006\left\{\begin{array}{l}\log 3.710 = 0.5694 \\ \log 3.716 = \quad ? \end{array}\right\} d \\ \qquad\quad \log 3.720 = 0.5705 \end{array}\right\}0.0011$$

Thus,

$$\frac{0.006}{0.010} = \frac{d}{0.0011}$$

and

$$d = \frac{6}{10}(0.0011) = 0.00066 = 0.0007$$

Hence,

$$\begin{aligned} \log 3.716 &= 0.5694 + d \\ &= 0.5694 + 0.0007 \\ &= 0.5701 \end{aligned}$$

Finally, $\log 371.6 = 2.5701$. ●

The linear interpolation method of Example 53.10 is an approximation method that assumes that changes in $\log N$ are directly proportional to changes in N. Geometrically, we are assuming that the graph of $y = \log x$ is a straight line. We know that this is not the case; that is, our assumption is false. Yet the procedure works because for small changes in N, the amount of error in the system is well within our tolerances.

Example 53.11

Find $\log 0.008193$.

Solution

Since $0.008193 = 8.193 \times 10^{-3}$, we have $\log 0.008193 = -3 + \log 8.193$. We shall use linear interpolation to evaluate $\log 8.193$, as follows:

$$0.010\left\{\begin{matrix} 0.003\left\{\begin{matrix} \log 8.190 = 0.9133 \\ \log 8.193 = \quad ? \end{matrix}\right\}d \\ \log 8.200 = 0.9138 \end{matrix}\right\}0.0005$$

Now

$$\frac{0.003}{0.010} = \frac{d}{0.0005}$$

and

$$d = \frac{3}{10}(0.0005) = 0.00015 = 0.0002$$

Thus,

$$\begin{aligned} \log 8.193 &= 0.9133 + 0.0002 \\ &= 0.9135 \end{aligned}$$

and

$$\log 0.008193 = 0.9135 - 3 \text{ or } 7.9135 - 10$$ ●

A logarithm table will often contain a table of proportional parts, as is the case with Table II of Appendix F. Such tables will greatly simplify the interpolation process because they have predetermined the proportional part d. Now let us illustrate the use of the proportional-parts table with the following examples.

Example 53.12

Find log 21.48 using the table of proportional parts.

Solution

Since $21.48 = 2.148 \times 10^1$, it follows that $\log 21.48 = 1 + \log 2.148$ and $\log 2.148$ lies between $\log 2.140$ and $\log 2.150$, that is, between 0.3304 and 0.3324. Now, $\log 2.148$ lies $\frac{8}{10}$ of the way from $\log 2.14$ to $\log 2.15$. We continue along the line that gave us $\log 2.14$ and $\log 2.15$ to the proportional-parts column headed by 8, and we find the number 16. This number represents $\frac{8}{10}$ of the difference between $\log 2.14$ and $\log 2.15$ without the decimal point. But this is handled easily, for we have only to add the four digits from $\log 2.14$ and the 16 from the proportional-parts table. We have $3304 + 16 = 3320$, and hence $\log 21.48 = 1.3320$.

Example 53.13

Find log 371.6 using the table of proportional parts.

Solution

Since $371.6 = (3.716)10^2$, we have that $\log 371.6 = 2 + \log 3.716$. We note that $\log 3.716$ lies $\frac{6}{10}$ of the way from $\log 3.710$ to $\log 3.720$. Using Table II we note that $\log 3.710 = 0.5694$. Moving along this line to the table of proportional parts, we find the column headed by 6. The value in this column is 7, which represents the proportional part of the distance from $\log 3.71$ to $\log 3.72$. Adding 5694 and 7 yields 5701, and so $\log 3.716 = 0.5701$. Therefore, $\log 371.6 = 2 + 0.5701 = 2.5701$. Note that the value obtained agrees with the result of Example 53.10. ●

Once we know how to find logarithms, they may be used to perform calculations that include multiplication, division, powers, and roots. Such calculations are based upon the laws of logarithms developed in Section 52 and are illustrated in the following example.

Example 53.14

Given $\log 2 = 0.3010$, $\log 3 = 0.4771$, and $\log 5 = 0.6990$, we calculate each of the following without resorting to Table II of Appendix F, that is, by using the laws of logarithms.

(a) $\log 6 = \log (2 \cdot 3) = \log 2 + \log 3 = 0.3010 + 0.4771 = 0.7781$.

(b) $\log 25 = \log (5^2) = 2 \log 5 = 2(0.6990) = 1.3980$.

(c) $\log (0.4) = \log \frac{4}{10} = \log \frac{2}{5} = \log 2 - \log 5 = 0.3010 - 0.6990$ $= -0.3980$. Since this result is negative, we might wish to rewrite the value as $1 - 0.3980 - 1 = 0.6020 - 1$. Note that we can also write $9.6020 - 10$.

(d) $\log \sqrt{3} = \log 3^{1/2} = \frac{1}{2} \log 3 = \frac{1}{2}(0.4471) = 0.2236$.

(e) $\log \sqrt[3]{0.4} = \log (0.4)^{1/3} = \frac{1}{3} \log (0.4)$. From part (c) we have that $\log (0.4) = 9.6020 - 10$. since the index number is 3, the most advantageous form for $\log (0.4)$ is $29.6020 - 30$. Thus

$$\log \sqrt[3]{0.4} = \tfrac{1}{3}(29.6020 - 30) = 9.8673 - 10$$

Note that in choosing to write the characteristic as $29 - 30$, we have guaranteed that the subtracted part will be 10 after the division.

(f)
$$\log 15 = \log \frac{3 \cdot 10}{2} = \log 3 + \log 10 - \log 2$$
$$= 0.4471 + 1 - 0.3010$$
$$= 1.1461$$

We conclude this section by recalling that the logarithm is an exponent. Each time we find the common logarithm of positive number N we find the power of 10 that yields N. Thus, $\log 9.21 = 0.9643$ means that $9.21 = 10^{0.9643}$ and $\log 38.1 = 2.5809$ means that $381 = 10^{2.5809}$.

Exercises for Section 53

In Exercises 1 through 10, express each number in scientific notation.

1. 325 **2.** 4730 **3.** 10,000

4. 0.00234 **5.** 0.001 **6.** 3,142,000

7. 0.124 **8.** 0.0124 **9.** 1,000,000

10. 0.00000002859

For Exercises 11 through 20, write the characteristics of the common logarithms of Exercises 1 through 10.

In Exercises 21 through 40, find the common logarithm of each number. Use linear interpolation or the table of proportional parts wherever necessary.

21. 3.24 **22.** 324 **23.** 0.324

24. 0.0618 **25.** 52.4 **26.** 7590

27. 0.00140 **28.** 0.000347 **29.** 3.47×10^{-4}

30. 3.47×10^{5} **31.** 3.197 **32.** 412.9

33. 6132 **34.** 0.002142 **35.** 0.003094

36. 2619000 **37.** 2.619×10^{6} **38.** 2.619×10^{-5}

39. 8.002×10^{3} **40.** $\sqrt{0.0004129}$

In Exercises 41 through 50, given that $\log 9 = 0.9542$, evaluate each term using the laws of logarithms and the given value.

41. log 90 **42.** log 900 **43.** log 0.9

44. log 0.09 **45.** log 81 **46.** log 3

47. log 27 **48.** $\log \frac{1}{9}$ **49.** $\log \frac{1}{3}$

50. log 0.3

In Exercises 51 through 60, given that log 2 = 0.3010, log 3 = 0.4771, and log 7 = 0.8451, evaluate each term using only the laws of logarithms and the values given.

51. log 21 **52.** log 9 **53.** log 18

54. log 15 **55.** log 98 **56.** log 36

57. $\log \sqrt{7}$ **58.** $\log \sqrt{4}$ **59.** $\log \sqrt{28}$

60. $\log \sqrt{\frac{21}{8}}$

SECTION 54

Antilogarithms

Suppose that $\log N = L$. L is the logarithm of N and N is called the **antilogarithm** or antilog of L. If L is given and we wish to find N, the procedure is called finding the antilogarithm and is the reverse of finding a logarithm.

To illustrate the procedure, recall that for any real number, N can be expressed as

$$N = x \times 10^n \qquad \text{where } 1 \leq x < 10 \text{ and } n \text{ is an integer}$$

Thus,

$$\log N = \log (x \times 10^n) = \log x + \log 10^n$$
$$= n + \log x$$

where n is the characteristic and represents an integral power of 10, $\log x$ is the mantissa, and the sum is the logarithm.

For $\log N = L$, N is always the product of the antilog of the mantissa (always greater than or equal to 1 and less than 10) and 10^n.

Example 54.1

Find N if $\log N = 3.6304$.

Solution

The characteristic is 3 and the mantissa is 0.6304. We look for the value of the mantissa in the body of Table II of Appendix F. We find 6304 in the row headed by 42 and the column headed 7. Therefore, $0.6204 = \log 4.27$ and $N = 4.27 \times 10^3 = 4270$.

Example 54.2

Find the antilog of $9.8028 - 10$.

Solution

The characteristic is $9 - 10$, or -1, and the mantissa is 0.8028. Using Table II, we find that the mantissa $0.8028 = \log 6.35$. Thus, the antilog of $9.8028 - 10 = 6.35 \times 10^{-1} = 0.635$. ●

If we are given a negative number, we can find the antilog by first writing the given number as the difference between a positive decimal value between 0 and 1 and an appropriate integer.

Example 54.3

Find the antilog of -3.4660.

Solution

In this case the entire number is negative, and we cannot conclude that the characteristic is -3 and the mantissa is 0.4660. Since the mantissa is always a positive decimal, we rewrite -3.4660 as follows:

$$\begin{aligned} -3.4660 &= 4 - 3.4660 - 4 \\ &= 0.5340 - 4 \end{aligned}$$

Now, we see that the characteristic is in fact -4 and the mantissa is 0.5340. From Table II we have that $0.5340 = \log 3.42$. Thus, the antilog of $-3.4660 = 3.42 \times 10^{-4} = 0.000342$. ●

In situations where the mantissa lies between the values in the body of Table II of Appendix F, we may proceed in either of two ways: interpolate or use the table of proportional parts. Both methods are illustrated in the following examples.

Example 54.4

Find the antilog of 2.2375.

Solution

The characteristic is 2 and the mantissa is 0.2375. We note that the mantissa is not in Table II of Appendix F, but lies between 0.2355 and 0.2380. Now $0.2355 = \log 1.72$ and $0.2380 = \log 1.73$. Assuming that the antilog of 0.2375 lies between these values, we set up our work as follows.

$$\begin{array}{ccc} & \textbf{Number} & \textbf{Mantissa} \\ 0.01\left\{\begin{array}{l} d\left\{\begin{array}{l}1.72\\ x\end{array}\right. \\ \quad 1.73\end{array}\right. & & \left.\begin{array}{l}\left.\begin{array}{l}0.2355\\0.2375\end{array}\right\}0.0020\\ 0.2380\end{array}\right\}0.0025 \end{array}$$

The appropriate proportion is

$$\frac{d}{0.01} = \frac{0.0020}{0.0025} = \frac{20}{25} = \frac{4}{5}$$

Hence,

$$d = (0.01)\left(\frac{4}{5}\right) = \frac{0.04}{5} = 0.008 \quad \text{and} \quad x = 1.72 + 0.008 = 1.728$$

Since x is the antilog of 0.2375, it follows that the antilog of 2.2375 = $1.728 \times 10^2 = 172.8$. ●

In Example 54.4 the value $d = 0.008$ was obtained by dividing 0.04 by 5, a division that is exact to three decimal places. If the division yields more than three places after the decimal point, we round off to the third place. Thus, 0.0086 would result in our using $d = 0.0009$.

Example 54.5

Find the antilog of 2.2375 using the table of proportional parts.

Solution

The mantissa 0.2375 does not lie in Table II. The next smallest number in the table is 0.2355 and is the log of 1.72. Thus, the first three digits of the antilog are 172, and we seek the fourth. This is found by observing that 0.2375 exceeds 0.2355 by 0.0020, or 20 units. Reading across the row headed 17 until we come to the proportional-parts table, we look for the number closest to 20. In this case the number 20 appears in the table and lies in the column headed 8. Hence, the antilog of 0.2275 = $1.720 + 0.008 = 1.728$, and therefore antilog $2.2375 = 1.728 \times 10^2 = 172.8$. We note that the result is consistent with the result of Example 54.4. ●

Our ability to find logarithms and antilogarithms and to use the laws of logarithms will enable us to simplify what otherwise might be some very tedious computations. Often we will be able to reduce a rather complex expression to a simple equation that involves logarithms. We then have only to solve the logarithmic equation. Example 54.6 is such an equation and illustrates the importance of the antilogarithm.

Example 54.6

Solve $2 + 3 \log N = 7.2175$.

Solution
Subtracting 2 from both sides of the equation yields $3 \log N = 5.2175$. Dividing by 3 yields $\log N = 1.7392$. The characteristic is 1 and the mantissa is 0.7392. Now, 0.7392 does not appear in Table II. We note that the next smallest number in the table is 0.7388, and this number is the log of 5.48. Thus, 548 are the first three digits of N. We find the fourth digit by observing that 7392 is four digits greater than 7388. Reading across the row headed 54, we find the digit in the proportional-parts table in the column headed 5. Hence, the antilog of $0.7392 = 5.485$, and $N = 5.485 \times 10^1 = 54.85$.

Example 54.7
Solve $3 \log x = 3.145 + \log x^2$.

Solution
We first write both terms containing the variable on one side of the equal sign and then use the log properties to write the expression as a single logarithm. Thus,

$$\begin{aligned} 3 \log x - \log x^2 &= 3.145 \\ 3 \log x - 2 \log x &= 3.145 \\ \log x &= 3.145 \end{aligned}$$

We find the antilog by noting that the characteristic is 3 and the mantissa is 0.145. Interpolating, we have antilog $0.145 = 1.397$ and $x = 1.397 \times 10^3 = 1379$.

Exercises for Section 54

In Exercises 1 through 10, find the antilogarithm of each logarithm.

1. 0.9047
2. 0.7427
3. 3.8169
4. $0.3945 - 2$
5. $8.3945 - 10$
6. $7.6684 - 10$
7. $47.6684 - 50$
8. -0.0620
9. -2.1918
10. -1.4989

In Exercises 11 through 20, find the antilogarithm of each logarithm. Use interpolation or the table of proportional parts if necessary.

11. 0.2134
12. 2.3146
13. 3.7129
14. 1.1786
15. $9.1479 - 10$
16. $8.2319 - 10$
17. $18.2319 - 20$
18. 2.2319
19. -1.3149
20. -2.5174

In Exercises 21 through 30, solve for N.

21. $\log N = 2.1818$

22. $\log N = 1.9284$

23. $\log N = 3.8$

24. $\log N = 9.7101 - 10$

25. $\log N = 7.8854 - 10$

26. $1 + \log N = 2.9031$

27. $3 + \log N = 0.8727$

28. $1 + 2 \log N = 2.4824$

29. $\log N = -1.3510$

30. $\log N = 2.4178$

SECTION 55

Computations Using Common Logarithms

Computations involving logarithms make use of the laws of logarithms and antilogarithms set forth in the preceding sections. We use the following examples to illustrate some of the ways logarithms may be used to simplify calculations and aid in the solution of problems.

Example 55.1

Use logarithms to find the value of (17.4)(976)(0.0634).

Solution

We know that the logarithm of a product is equal to the sum of the logarithms of the individual factors. Thus, if $N = (17.4)(976)(0.0634)$, we have $\log N = \log(17.4) + \log 976 + \log(0.0634)$. Now

$$\begin{aligned} \log(17.4) &= 1.2405 \\ \log 976 &= 2.9894 \qquad \text{(add)} \\ \log(0.0634) &= 0.8021 - 2 \\ \hline \log(17.4)(976)(0.0634) &= 5.0320 - 2 \end{aligned}$$

Hence, $\log N = 5.0320 - 2 = 3.0320$. To find N we note that the characteristic is 3 and the mantissa is 0.0320. Interpolating, we find that the antilog of 0.0320 is 1.077, and so

$$N = 1.077 \times 10^3 = 1077$$

Example 55.2

Use logarithms to find the value 0.0195/13.7.

Solution

We know that the logarithm of a quotient is equal to log (numerator) − log (denominator). If $N = 0.0195/13.7$, then

$$\log N = \log(0.0195) - \log(13.7)$$

$$\log(0.0195) = 8.2900 - 10$$

$$\log(13.7) = 1.1367 \qquad \text{(subtract)}$$

$$\log\frac{0.0195}{13.7} = 7.1533 - 10$$

Thus, $\log N = 7.1533 - 10$. The characteristic is -3, the antilog of 0.1533 is 1.423, and

$$N = 1.423 \times 10^{-3} = 0.001423$$

Remark 55.1. *Negative numbers do not have logarithms. If a negative value should appear as part of a problem, we treat the number as though it were positive and determine the sign of the result algebraically.*

Example 55.3

Use logarithms to find N if $N = (-1.64)^7$.

Solution

We begin by noting that our result will be negative. Thus, we find N for $N = (1.64)^7$ and then affix a minus sign. Now

$$\begin{aligned}\log N &= \log(1.64)^7\\ &= 7\log(1.64)\\ &= 7(0.2148)\\ &= 1.5036\end{aligned}$$

The characteristic is 1 and the mantissa is 0.5036. Interpolating, the antilog of 0.5036 is 3.189, and so $N = 3.189 \times 10^1 = 31.89$. Since our result must be negative, $N = -31.89$.

Example 55.4

Find the value of $\sqrt[3]{0.0471}$.

Solution

Let $N = \sqrt[3]{0.0471}$. We can write $\sqrt[3]{0.0471} = (0.0471)^{1/3}$. Thus, $\log N = \log(0.0471)^{1/3} = \frac{1}{3}\log(0.0471)$. Now $\log(0.0471) = 8.6730 - 10$. Since we must multiply by $\frac{1}{3}$ and -10 is not divisible by 3 but -30 is, we rewrite $\log 0.0471 = 8.6730 - 10$ as $\log(0.0471) = 20 + (8.6730 - 10) - 20 = 28.7630 - 30$ by adding and then subtracting 20 from the right side of the expression. We now multiply this result by $\frac{1}{3}$ to obtain $\frac{1}{3}\log(0.0471) = \frac{1}{3}(28.7630 - 30) = 9.5577 - 10$. The antilog of $9.5577 - 10$ is 0.3612, and hence $\sqrt[3]{0.0471} = 0.3612$. ●

Problems involving the use of more than one log property are illustrated in the following examples.

Example 55.5

Find N if $N = \dfrac{475}{(31.4)\sqrt{0.023}}$.

Solution

$\log N = \log 475 - [\log(31.4) + \frac{1}{2}\log(0.23)]$. We start by finding the logarithm of the denominator.

$$\log(31.4) = 1.4969 \qquad (1)$$

$$\log(0.23) = 18.3617 - 20 \qquad (2)$$

$$\frac{1}{2}\log(0.023) = 9.1809 - 10 \qquad (3)$$

Adding (1) and (3) yields $10.6778 - 10 = 0.6778$, which is the logarithm of the denominator. The logarithm of the numerator is given by $\log 475 = 2.6767$. Now,

$$\begin{aligned}\log N &= \log(\text{numerator}) - \log(\text{denominator})\\ &= 2.6767 - 0.6778\\ &= 1.9989\end{aligned}$$

The antilog of 0.9989 is 9.975. Thus, $N = 9.975 \times 10^1 = 99.75$

Example 55.6

Find N if

$$N = \left[\frac{(0.00641)(31.2)}{(875)^3}\right]^{1/5}$$

Solution

Since

$$N = \left[\frac{(0.00641)(31.2)}{(875)^3}\right]^{1/5}$$

we have that

$$\begin{aligned}\log N &= \frac{1}{5}\log\left[\frac{(0.00641)(31.2)}{(875)^3}\right]\\ &= \frac{1}{5}[\log(0.00641) + \log(31.2) - 3\log 875]\end{aligned}$$

We now observe that since $\log(0.00641) = 7.8069 - 10$ and $\log(31.2) = 1.492$, their sum is $9.3011 - 10$ (A). We also note that $3\log 875 = 3(2.9420) = 8.8260$ (B). We now subtract (B) from (A), to obtain

$$\log(0.00641) + \log(31.2) - 3\log 875 = 9.3011 - 10 - 8.8260$$

or

$$\log (0.00641) + \log (31.2) - 3 \log 875 = 0.4751 - 10$$

Thus,

$$\log N = \frac{1}{5}(0.4751 - 10)$$

or

$$\log N = 0.0950 - 2 \qquad \text{(C)}$$

From (C) it follows that $N = 1.245 \times 10^{-2} = 0.01245$. ●

The properties and computational uses of logarithms that have been illustrated in the preceding examples can be applied to exponential equations, problems in physics, chemistry, finance, and, indeed to many other fields. The following examples will enable us to appreciate some of the important applications of logarithms.

Example 55.7

Solve the exponential equation $3^{x-1} = 5$.

Solution

Taking the common logarithm of both sides of the equation, we obtain

$$\log 3^{x-1} = \log 5$$

Using property 3, we have

$$(x - 1) \log 3 = \log 5$$

or

$$x - 1 = \frac{\log 5}{\log 3}$$

The next step in our process involves dividing $\log 5$ by $\log 3$, *not* by subtracting 5 from 3 and then taking the logarithm. In this case we have the quotient of two logarithms, not the logarithm of a quotient. Since

$$\frac{\log 5}{\log 3} = \frac{0.6990}{0.4771} = 1.47$$

we have $x - 1 = 1.47$, or $x = 2.47$.

Example 55.8

The time T for one cycle of a simple pendulum of length l is given by the formula

$$T = 2\pi\sqrt{\frac{l}{g}} \qquad \text{where } g \text{ is the acceleration due to gravity}$$

If $g = 980 \text{ cm/sec}^2$ and $l = 247$ cm, find the period T.

Solution

$$T = 2\pi\sqrt{\frac{l}{g}} = 2\pi\sqrt{\frac{247}{980}}$$

If we take $\pi = 3.14$, then

$$T = 6.28\sqrt{\frac{247}{980}}$$

Taking the common logarithm of both sides of the equation yields

$$\begin{aligned}\log T &= \log(6.28) + \frac{1}{2}(\log 247 - \log 980)\\ &= 0.7980 + \frac{1}{2}(2.3927 - 2.9912)\\ &= 0.7980 - 0.2993\\ &= 0.4987\end{aligned}$$

Interpolating, we obtain $T = 3.153$ sec. ●

In chemistry the pH (potency of hydrogen) of a solution is, by definition, equal to the negative value of the exponent of 10 used to express the molar concentration of H_3O^+ (the molar concentration of H_3O^+ is the number of hydronium ions per liter of solution). Since ion concentrations involve a wide range of numerical values, we define the pH of a solution mathematically to be the negative logarithm of the molar concentration of H_3O^+; that is, pH = −log (concentration H_3O^+).

For pure water the pH is 7 since there is 10^{-7} moles of hydronium ions per liter. Solutions with a pH of 7 are called **neutral**, those with a pH lower than 7 are **acidic**, and those with a pH higher than 7 are called **basic** or **alkaline**.

Example 55.9

Find the pH of a solution that has a H_3O^+ concentration of 4.21×10^{-6}.

Solution

The pH of the solution is given by

$$\begin{aligned}-\log(4.21 \times 10^{-6}) &= -[\log(4.21) + \log 10^{-6}]\\ &= -(0.6243 - 6)\\ &= -(-5.3757)\\ &= 5.375\end{aligned}$$

We note that the solution is acidic, since the pH is less than 7.

Example 55.10

Find the H_3O^+ concentration for a solution whose pH is 7.6.

Solution
Since the pH = −log (concentrated H_3O^+), we have −log (concentrated H_3O^+) = 7.6. Thus, the concentration we seek is the antilog of −7.6. Recalling that the mantissa must be positive, we rewrite −7.6 as $-7.6 + 8 - 8 = 0.4 - 8$, where 0.4 is the mantissa and −8 is the characteristic. The antilog of 0.4 is 2.512 (by interpolation), and thus the concentration is 2.512×10^{-8}. ●

In mathematics of finance, it is shown that if a given principal P is invested at the yearly interest rate of i, the compound amount A of P at the end of t years is given by the formula

$$A = P\left(1 + \frac{i}{n}\right)^{nt}$$

where n is the number of times the money is compounded per year.

Example 55.11
Find the amount present at the end of $3\frac{1}{2}$ years if $5000 is invested at 8% compounded quarterly

Solution
Since the interest is compounded quarterly, the rate per interest period is $i = 0.08/4 = 0.02$ and the number of interest periods $nt = 4(3\frac{1}{2}) = 14$. Thus,

$$\begin{aligned} A &= 5000(1 + 0.02)^{14} \\ &= 5000(1.02)^{14} \end{aligned}$$

Taking the common logarithm of both sides, we have

$$\begin{aligned} \log A &= \log [5000 \cdot (1.02)^{14}] \\ &= \log 5000 + 14 \log (1.02) \\ &= 3.6990 + 0.1204 \\ &= 3.8194 \end{aligned}$$

Thus,

$$\begin{aligned} A &= \text{antilog}\,(3.8194) = \text{antilog}\,(0.8194) \times 10^3 \\ &= 6.592 \times 10^3 \end{aligned}$$

and hence the amount is $6592.

Exercises for Section 55

In Exercises 1 through 20, use logarithms to perform the operations and evaluate the expressions.

1. 260(21.5)

2. (0.4135)(800)

3. $(0.314)(21.3)(2.718)$

4. $\dfrac{840}{60}$

5. $\dfrac{21.4}{0.0137}$

6. $\dfrac{4.52}{21.62}$

7. $\dfrac{(41.3)(26.91)}{0.731}$

8. $(235)^2$

9. $(5.32)^4$

10. $(0.0831)^{1/5}$

11. $\sqrt{18.6}$

12. $\sqrt[3]{234.8}$

13. $(426)^{0.4}$

14. $\sqrt{263}(\sqrt[3]{41.6})$

15. $\sqrt{\dfrac{1.46}{725}}$

16. $\dfrac{(3.14)\sqrt[3]{78.6}}{0.0023}$

17. $\sqrt[4]{\dfrac{4.2}{0.06}}$

18. $\sqrt{235(47.6)^3}$

19. $\left(\dfrac{75.6}{200\sqrt{32.1}}\right)^{0.1}$

20. $\dfrac{(91.4)\sqrt[3]{21.82}}{0.00492}$

In Exercises 21 through 26, use common logarithms and Table II of Appendix F to solve each exponential equation for x.

21. $2^x = 25$ **22.** $5^x = 100$ **23.** $10^x = 2$

24. $(2.72)^{2x} = 3.14$ **25.** $3^{-x} = 0.01$ **26.** $x^3 = 15$

27. The length of a simple pendulum is 735 cm. Use the procedure of Example 55.8 to find the time T of one cycle.

28. The time T for one cycle of a simple pendulum is 2.1 sec. Find the length of the pendulum.

29. In a certain lab in a northern latitude a simple pendulum of length 120.0 cm is found to complete 55 cycles in 2 min. What is the acceleration due to gravity at the lab location?

30. Calculate the pH of solutions that have the following concentration of H_3O^+. Note which are acidic and which are alkaline.
(a) 1.0×10^{-8} (b) 1.0×10^{4} (c) 4.71×10^{-3}
(d) 5.0×10^{-9} (e) 7.0 (f) 0.00471

31. Find the H_3O^+ concentration of solutions with the following pH values.
(a) 6 (b) −4 (c) 3.2436
(d) −5.3472 (e) −0.4430

32. Find the amount present at the end of 6 years if $4000 is invested at 6% compounded quarterly.

33. Find the amount present at the end of 10 years if $10,000 is invested at 7% compounded annually.

34. Find the amount present at the end of 2 years if $5000 is invested at 5% compounded semiannually.

35. A house has a present value of $40,000. If real estate appreciates at the rate of 10% per year, find the value of the house in 5 years.

36. How many years are necessary for an account of $1000 to increase to $2000 if the interest rate is 6% compounded annually?

37. What interest rate compounded annually is necessary if $1000 is to amount to $15000 after 7 years?

38. The velocity of sound in a steel rail of a railroad track is given by the formula $V = \sqrt{y/d}$, where y is Young's modulus for steel and d represents the density of steel. If $y = 20 \times 10^{11}$ dynes/cm^2 and $d =$ 7.9 gm per cm^3, find V.

39. The number N of decibels between two sounds of intensity I_1 and I_2 is given by the equation $N = 10 \log (I_1/I_2)$. Find the intensity difference N when $I_1 = 10^{-16}$ W/cm^2 and $I_2 = 2 \times 10^{-4}$ W/cm^2.

40. Under certain conditions, the velocity of a sound in a cable is given by $V = 400x^2 \log (1/x)$ m/sec. Find V if $x = 0.65$.

41. The pressure and volume of gas expanding adiabatically (no gain or loss of heat in the system) are related by the formula

$$PV^r = c, \; r \text{ and } c \text{ constant}$$

Different conditions for the same weight of gas yields the equations

$$P_1 V_1^r = c \quad \text{and} \quad P_2 V_2^r = c$$

from which we have

$$P_1 V_1^r = P_2 V_2^r$$

(a) For a certain gas, $r = 0.4$. If the gas has an initial volume of 5 liters at a pressure of 1 atm, find the pressure if the gas is compressed adiabatically to a volume of 2 liters.

(b) Suppose thtat we compress air ($r = 1.4$) adiabatically until its final pressure, P_2, is 20 times its original pressure, P_1. Find the ratio of the initial volume, V_1, to the final volume, V_2.

42. Suppose that a man-made satellite orbiting the earth traces out a circular path. The velocity of such an orbit is given by the formula $V = R\sqrt{g/r}$, where R is the radius of the earth, r is the radius of the orbit (measured from the center of the earth), and g is the acceleration due to gravity at the surface of the earth. The radius R of the earth is about 6400 km or 6.4×10^6 m, and the acceleration g due to gravity is 9.8 m/sec. If the satellite is revolving about the earth at a height of 200 km above the surface, find the velocity of the satellite. (*Hint*: r is the radius of earth plus the distance the satellite orbits above the surface of the earth.)

43. Repeat Exercises 42 if the satellite is revolving about the earth at a height of 300 km above the surface.

44. The time T required for one complete revolution of the satellite in Exercise 42 is given by the formula $T = 2\pi r/V$. Using the value for V found in Exercise 42, find T using logarithms.

45. Repeat Exercise 44, using the result of Exercise 43.

46. A radioactive isotope decomposes according to the law $y = (0.6)^{0.02x}$, where yy is the number of grams remaining after x min.
(a) If we begin with 100 g of the isotope, how long will it take to become 75 g?
(b) How many grams will there be after 2 hr?
(c) How much will remain after 45 min?

47. If the area A of a triangle is described by the formula

$$A = \sqrt{s(s-a)(s-b)(s-c)}$$

where a, b, and c are the sides and s is one half the perimeter, use logarithms to find A when $a = 11.6$, $b = 12.4$, and $c = 15.6$.

SECTION 56

Logarithms of Trigonometric Functions

There are many cases in which the use of logarithms simplifies computations that involve the trigonometric functions. We can find the logarithm of a given trigonometric function by using Tables I (four-place values of trigonometric functions) and II (four-place values of common logarithms) of Appendix F and the procedure outlined in the following example.

Example 56.1

Find $\log \sin 41°20'$.

Solution

In Table I we find $\sin 41°20' = 0.6604$. Using Table II of Appendix F and interpolation, we have $\log 0.6604 = 9.8198 - 10$.

When working with the trigonometric functions, our work is made somewhat easier by using Table IV of Appendix F, a four-place table of the logarithms of trigonometric functions. Such a table enables us to find logarithms in a single step. However, we must be careful to note that 10 must be subtracted from each entry in the table. If it is necessary to interpolate, we proceed as before.

Example 56.2

Find log sin 41°20′ and log tan 21°50′, using Table IV of Appendix F.

Solution

Reading directly from Table IV, we have log sin 41°20′ = 9.8198 − 10. This result agrees with the result of Example 56.1 obtained in two steps. Again, using Table IV we find log tan 21°50′ = 9.6028 − 10.

Example 56.3

Find log sin 36°14′.

Solution

In Table IV, log sin 36°20′ = 9.7727 − 10 and log sin 36°10′ = 9.7710 − 10. To find log sin 36°14′ we must interpolate.

$$10\left\{\begin{array}{l} 4\left\{\begin{array}{l}\log\sin 36°10' = 9.7710 - 10 \\ \log\sin 36°14' = \end{array}\right\}d \\ \log\sin 36°20' = 9.7727 - 10\end{array}\right\}0.0017$$

$\frac{4}{10} = d/0.0017$. Thus, $d = 0.4(0.0017) = 0.0007$, and so

$$\begin{aligned}\log 36°14' &= (9.7710 - 10) + 0.0007 \\ &= 9.7717 - 10\end{aligned}$$

●

We can also use Table IV of Appendix F, to find the angle if the value of the trigonometric function is given. The method is especially useful if the given value is expressed as a product, quotient, or power.

Example 56.4

Find θ if $\sin\theta = \sqrt{\dfrac{3.14}{6.82}}$.

Solution

$$\begin{aligned}\log\sin\theta &= \log\left(\frac{3.14}{6.82}\right)^{1/2} \\ &= \frac{1}{2}(\log 3.14 - \log 6.82) \\ &= \frac{1}{2}(0.4969 - 0.8338) \\ &= \frac{1}{2}[(20.4969 - 20) - 0.8338] \\ &= \frac{1}{2}(19.6631 - 20) \\ &= 9.8316 - 10\end{aligned}$$

From Table IV of Appendix F, we find that $\theta = 43°$, to the nearest degree. If more accuracy is needed, we can interpolate as follows:

$$10\left\{\begin{array}{l} d\left\{\begin{array}{l}\log\sin 42°40' = 9.8311 - 10 \\ \log\sin\theta \quad\;\; = 9.8316 - 10\end{array}\right\}0.0005 \\ \log\sin 42°50' = 9.8324 - 10\end{array}\right\}0.0013$$

$$\frac{d}{10} = \frac{0.0005}{0.0013} \quad \text{or} \quad d = 4$$

Thus, $\theta = 42°40' + 4' = 42°44'$. ●

We now turn our attention to computations involved in the solution of triangles, in particular those solutions that use the law of sines (see Section 43).

Example 56.5

Solve the oblique triangle for B given $a = 8$, $b = 5$, and $A = 40°$ by using logarithms.

Solution

We observe that B can be found by employing the law of sines. From the law of sines, we have

$$\sin B = \frac{b \sin A}{a}$$

or

$$\sin B = \frac{5(\sin 40°)}{8}$$

Taking the logarithm of both sides,

$$\log\sin B = \log 5 + \log\sin 40° - \log 8$$

$$\begin{array}{lcl} \log 5 & = & 0.6990 \\ \log\sin 40° & = & 9.8081 - 10 \quad \text{add} \\ \hline & & 10.5071 - 10 \\ \log 8 & = & 0.9031 \quad \text{subtract} \\ \hline \log\sin B & = & 9.6040 - 10 \end{array}$$

Thus, from Table IV of Appendix F, $B = 23°40'$, to the nearest minute. We recall that when two sides and the angle opposite one of them is given, there may be a second solution B_1, where $B_1 = 180° - B = 180° - 23°40' = 156°20'$. Since $A + B_1 = 40° + 156°20' > 180°$, B_1 is not a second solution. Hence, $B = 23°40'$ is the only appropriate value. Compare this result with the result of Example 43.3.

Example 56.6

Solve the oblique triangle given $A = 38°20'$, $B = 65°30'$, and $c = 28.6$.

Solution

We first determine $C = 180° - (A + B) = 180° - 103°50' = 76°10'$ and then note that sides a and b can be found from the law of sines.

$$\frac{a}{\sin A} = \frac{c}{\sin C} \quad \text{and} \quad \frac{b}{\sin B} = \frac{c}{\sin C}$$

Thus,

$$a = \frac{c \sin A}{\sin C} \quad \text{and} \quad b = \frac{c \sin B}{\sin C}$$

$$a = \frac{28.6(\sin 38°20')}{\sin 76°10'} \quad \text{and} \quad b = \frac{28.6(\sin 65°30')}{\sin 76°10'}$$

For a,

$$\log a = \log 28.6 + \log \sin 38°20' - \log \sin 76°10'$$

$$\begin{aligned}
\log 28.6 &= 1.4564 \\
\log \sin 38°20' &= 9.7926 - 10 \quad \text{add} \\
&\ 11.2490 - 10 \\
\log \sin 76°10' &= 9.9872 - 10 \quad \text{subtract} \\
\log a &= 1.2618
\end{aligned}$$

and $a = 18.3$.

For b,

$$\log b = \log 28.6 + \log \sin 65°30' - \log \sin 76°10'$$

$$\begin{aligned}
\log 28.6 &= 1.4564 \\
\log \sin 65°30' &= 9.9590 - 10 \quad \text{add} \\
&\ 11.4044 - 10 \\
\log \sin 76°10' &= 9.9872 - 10 \quad \text{subtract} \\
\log b &= 1.4182
\end{aligned}$$

and $b = 26.2$.

Remark 56.1. *In Example 56.6 the determination of b was simplified somewhat by the use of the values for log 28.6 and log sin 76°10′, values previously used to find a. Both a and b were found to the nearest tenth. Any attempt to obtain greater accuracy would be fruitless, since the given value of c is to the nearest tenth.*

We can also apply the methods and techniques developed in this section to find the magnitude and direction of resultant vectors. This can often be done by making use of the law of sines and the law of consines.

Example 56.7

The velocity of a plane is 360 mph, 40° west of north. The wind is 80 mph from the west. Determine the magnitude and direction of the resultant velocity of the plane.

Solution

In Figure 56.1 we show the relationship of the vectors. We label $\overrightarrow{V_W}$ as the velocity vector for the wind, $\overrightarrow{V_A}$ the velocity vector for the airplane, and $\vec{V}$ as the resultant velocity vector. From the law of cosines the magnitude $|\vec{V}|$ of the resultant vector V is given by

$$|\vec{V}| = \sqrt{|\overrightarrow{V_A}|^2 + |\overrightarrow{V_W}|^2 - 2|\overrightarrow{V_A}|\,|\overrightarrow{V_W}| \cos (90° - 40°)}$$

$$= \sqrt{(360)^2 + (80)^2 - 2(360)(80) \cos 50°}$$

$$= \sqrt{129{,}600 + 6400 - 37{,}000}$$

$$= \sqrt{99{,}000} = 315 \text{ mph}$$

Figure 56.1

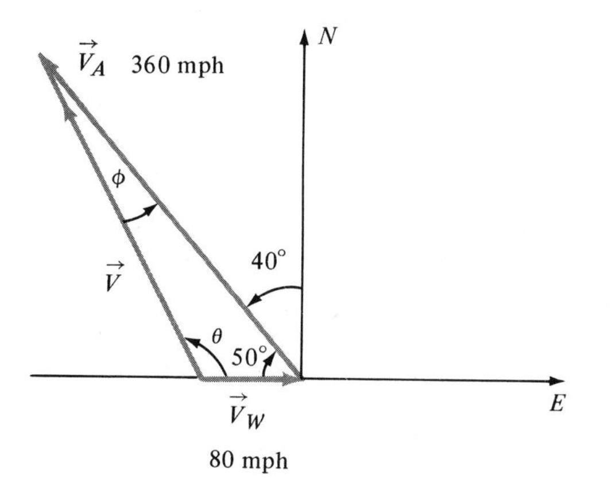

We now find θ by applying the law of sines.

$$\frac{|\overrightarrow{V_A}|}{\sin \theta} = \frac{|\vec{V}|}{\sin 50°}$$

or

$$\sin \theta = \frac{|\overrightarrow{V_A}| \sin 50°}{|\vec{V}|} = \frac{360 \sin 50°}{315}$$

Now,

$$\log \sin \theta = \log 360 + \log \sin 50° - \log 315$$

$$\begin{aligned}
\log 360 \quad &= \quad 2.5563 \\
\log \sin 50° &= \underline{\quad 9.8843 - 10} \quad \text{add} \\
&\quad\ 12.4406 - 10 \\
\log 315 \quad &\quad\ \underline{\quad 2.4923 - 10} \quad \text{subtract} \\
\log \sin \theta \quad &= \quad 9.9523 - 10
\end{aligned}$$

From Table IV of Appendix F, we have $\theta = 63°41'$ or $\theta = 180° - 63°41' = 116°20'$. (Recall that the law of sines always yields two possible results when $\sin \theta < 1$.) Only one of these values for θ can be correct. From Figure 56.1 we suspect that the correct value for θ is $116°20'$. To show that $116°20'$ is, in fact, the correct value, we would find angle ϕ by again using the law of sines. Thus, the magnitude of the velocity of the plane is 315 mph in the direction of $116°20' - 90° = 26°20'$ west of north.

Exercises for Section 56

In Exercises 1 through 20, use Table IV to find the value of each logarithm.

1. $\log \sin 30°$
2. $\log \cos 60°$
3. $\log \tan 45°$
4. $\log \sin 63°20'$
5. $\log \sec 24°10'$
6. $\log \sin 38°24'$
7. $\log \cos 42°13'$
8. $\log \tan 84°26'$
9. $\log \csc 3°48'$
10. $\log \cot 18°56'$
11. $\log \sin 150°$
12. $\log \cos 120°$
13. $\log \tan 135°$
14. $\log \sin 95°$
15. $\log \sin 198°10'$
16. $\log \cos 285°20'$
17. $\log \tan 225°$
18. $\log \sin 315°$
19. $\log \sin 215°26'$
20. $\log \cos 318°18'$

In Exercises 21 through 30, use Table IV to find the smallest positive θ.

21. $\log \sin \theta = 9.5919 - 10$
22. $\log \cos \theta = 9.6093 - 10$
23. $\log \sin \theta = 9.9687 - 10$
24. $\log \cos \theta = 9.8111 - 10$
25. $\log \tan \theta = 9.5650 - 10$
26. $\log \sin \theta = 9.4951 - 10$
27. $\log \sin \theta = 9.7634 - 10$
28. $\log \cos \theta = 9.9636 - 10$
29. $\log \tan \theta = 0.2975$
30. $\log \sec \theta = 0.1865$

In Exercises 31 through 40, solve each triangle using logarithms.

31. $A = 41°20'$, $C = 90°$, $a = 17$

32. $B = 16°10'$, $C = 90°$, $c = 21.5$

33. $C = 90°, a = 24.1, b = 16.3$

34. $A = 38°, B = 76°, c = 17.65$

35. $A = 120°, B = 40°, b = 37.18$

36. $A = 38°20', a = 26.1, b = 19.2$

37. $B = 67°40', C = 55°35', b = 7.46$

38. $A = 32°50', a = 8.46, b = 10.20$

39. $a = 12, b = 14, c = 20$

40. $a = 125, b = 158, c = 192$

In Exercises 41 through 45, solve each problem using logarithms.

41. A force of 125 N makes an angle of 40° with a force of 85 N. Find the magnitude of the resultant and the angle the resultant makes with the 85 N force.

42. The velocity of an airplane in still air is 300 mph, 40° north of east. A 60-mph wind is blowing directly from the west. Find the velocity and direction of the plane.

43. A weight of 50 lb is supported by two chords which make angles of 30° and 45° with the y axis. Find the magnitude of the forces in each chord.

44. A 4350-lb car is parked on a hill that makes the angle of 12′20′ with the horizontal. Find the horizontal and vertical components of the weight of the car.

45. Two forces, of 120 lb and 210 lb, have a resultant of 265 lb. Find the angle the resultant makes with each force.

SECTION 57

Natural Logarithms and Other Bases

In Section 51 we noted that any positive number not equal to 1 can be used as a base for a system of logarithms. We use the base 10 almost exclusively for computation. However, in many situations that involve applications of logarithms, it is advantageous to use a different base, a base denoted by e (an irrational number), where $e = 2.7182818\ldots$.

The symbol e is used in honor of a truly great mathematician, Euler, and the logarithmic system is called the **natural logarithm**. The choice of the word "natural" to describe this system will become apparent when we discuss the uses of e in science and engineering applications.

For common logarithms we write $\log x$ to mean $\log_{10}x$. For natural logarithms, we write $\ln x$ to denoted $\log_e x$. Both expressions have equivalent exponential forms. If $y = \log x$, then $x = 10^y$; and if $y = \ln x$, then $x = e^y$. Using these equivalent forms we can readily determine $\ln 1$ and $\ln e$. If we let $y = \ln 1$, the equivalent exponential form is given by $e^y = 1$, from which $y = 0$. Thus, $\ln 1 = 0$. Similarly, if $y = \ln e$, then $e^y = e$, and so $y = 1$. Hence, $\ln e = 1$. We note that since $y = \ln x$ is equivalent to $x = e^y$, the inverse of the natural logarithm is obtained by interchanging x and y. Thus, the inverse of $y = \ln x$ is the exponential function $y = e^x$.

The properties of exponential functions developed in Section 50 enable us to graph $y = e^x$. The graphs of both $y = e^x$ and $y = \ln x$ are shown in Figure 57.1. Since each is the inverse of the other, the graphs are reflections about the line $y = x$.

In a table of common logarithms it is necessary to have only the values of $\log N$ for $1 \leq N < 10$. For numbers not in this range, we obtain $\log N$ by using the table values and the characteristics, where the characteristics are the integral powers of 10.

Figure 57.1

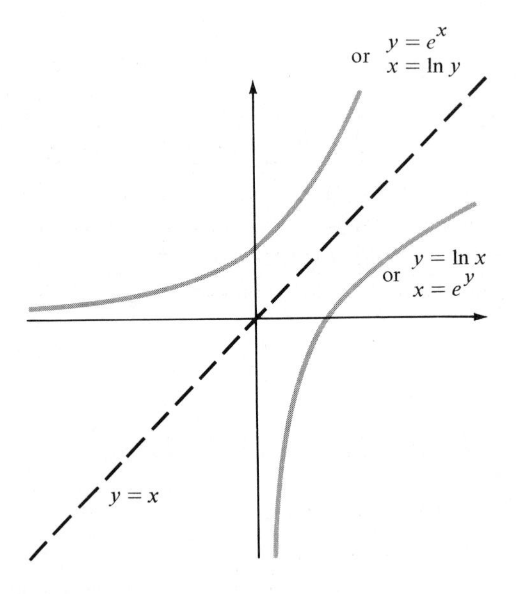

Tables of $\ln N$ are also written for $1 \leq N < 10$. But here we must be more careful in finding $\ln N$ for numbers not in this range, because the natural logarithms of integral powers of 10 are not themselves integers. We note that while $\log 10 = 1$ and thus $\log 10^n = n$ for any integer n, $\ln 10 = 2.3026$, and so $\ln 10^n = n(2.3026)$. However, we will still use scientific notation to aid us in determining ln values. Table III of Appendix F is a four-place table of natural logarithms.

Example 57.1

Here we use scientific notation, Table III of Appendix F, and the fact that $\ln 10 = 2.3026$ to find each of the following values.

(a) $\ln 4.1 = 1.4110$

(b) $\ln 41 = \ln(4.1 \times 10^1) = \ln 4.1 + \ln 10 = 1.4110 + 2.3026$
$= 3.7136$

(c) $\ln 4100 = \ln(4.1 \times 10^3) = \ln 4.1 + \ln 10^3 = 1.4110 + 3 \ln 10$
$= 1.4110 + 6.9078 = 8.3188$

(d) $\ln 0.041 = \ln(4.1 \times 10^{-2}) = \ln 4.1 + \ln 10^{-2} = 1.4110 - 2 \ln 10$
$= 1.4110 - 4.6052 = -3.1942$ ●

Natural logarithms obey the same rules for computations that are applied to common logarithms. These rules are illustrated in the following example.

Example 57.2

We are given $\ln 9 = 2.1972$ and $\ln 10 = 2.3026$. The following values are found by using only these given values and the laws of logarithms.

(a) $\ln 90 = \ln(9 \times 10) = \ln 9 + \ln 10 = 2.1972 + 2.3062 = 4.4998$

(b) $\ln 0.9 = \ln\frac{9}{10} = \ln 9 - \ln 10 = 2.1972 - 2.3026 = -0.2054$

(c) $\ln 3 = \ln 9^{1/2} = \frac{1}{2}\ln 9 = \frac{1}{2}(2.1972) = 1.0986.$ ●

If we are given $\ln N$ and wish to find N, we are attempting to find the antilogarithm or, in abbreviated form, anti-ln. Since $\ln 1 = 0$, $\ln 10 = 2.3026$, and Table III is written for $1 \le N \le 10$, we note that we may find anti-ln directly from the table if $0 \le \ln N \le 2.3026$. However, it may be necessary to interpolate. If $\ln N$ is not in this range, we first rewrite $\ln N$ as some number between 0 and 2.3026 plus or minus the appropriate multiple of 2.3026 and then use Table III of Appendix F. Thus, for $\ln N = \ln x + n(\ln 10)$, where $0 \le \ln x < 2.3026$, $N = x \times 10^n$, where $1 \le x < 10$.

Example 57.3

Find N if $\ln N = 1.4540$.

Solution

The value lies between 0 and 2.3026. Thus, we use Table III of Appendix F, *directly* to find anti-ln $1.4540 = 4.28$ or $N = 4.28$.

Example 57.4

Find N if $\ln N = 8.7291$.

Solution

We observe that the given value of $\ln N$ is outside the range of values that enable us to use Table III directly. Thus, we subtract from 8.7291 as many (2.3026)'s as necessary to bring the given value within the range of the tables. It is necessary to subtract three (2.3026)'s to accomplish our aim. Numerically, $8.7291 - 3(2.3026) = 1.8213$. Thus, we can rewrite the given value, $8.7291 = 1.8213 + 3(2.3026)$. Our evaluation now proceeds as follows:

$$\ln N = 1.8213 + 3(2.3026)$$

Since $\ln 6.18 = 1.8213$ (from Table III of Appendix F) and $\ln 10 = 2.3026$,

$$\ln N = \ln 6.18 + 3 \ln 10$$

Now, using the properties of logarithms,

$$\begin{aligned} \ln N &= \ln 6.18 + \ln 10^3 \\ &= \ln (6.18 \times 10^3) \end{aligned}$$

Therefore, $N = (6.18 \times 10^3) = 6180$.

Example 57.5

Find N if $\ln N = -3.7050$.

Solution

We can express -3.7050 as a positive number between 0 and 2.3026 by adding $2(2.3026)$. Now $-3.7050 + 2(2.3026) = -3.7050 + 4.6052 = 0.9002$. We can now rewite -3.7050 as $0.9002 - 2(2.3026)$ and proceed as we did in Example 57.4, to obtain $\ln N = 0.9002 - 2(2.3026)$. We note from Table III of Appendix F that $\ln 2.46 = 0.9002$. Thus, $\ln N = \ln 2.46 - 2 \ln 10$. Using the properties of logarithms, we have

$$\begin{aligned} \ln N &= \ln 2.46 + \ln 10^{-2} \\ &= \ln (2.46 \times 10^{-2}) \end{aligned}$$

Hence, $N = (2.46 \times 10^{-2}) = 0.0246$. ●

Now let us consider the use of natural logarithms in computations involving the number e.

Example 57.6

Evaluate $27e^{2.47}$ by using natural logarithms.

Solution

Let $N = 27e^{2.47}$. Since $\ln e = 1$, the most efficient manner of solution involves taking the natural logarithm of both sides of the equation. Proceeding in this fashion, we have

$$\ln N = \ln 27e^{2.47} = \ln 27 + \ln e^{2.47}$$

Now, $\ln 27 = \ln (2.7 \times 10^1) = \ln 2.7 + \ln 10 = 0.9933 + 2.3026 = 3.2959$ and $\ln e^{2.47} = 2.47 \ln e = 2.47$. Hence, $\ln N = 3.2959 + 2.47 = 5.7659$. To find N, we rewrite $5.7659 = 1.1607 + 2(2.3026)$. Interpolation and Table III yields $\ln 3.192 = 1.1607$. Now $\ln N =$

$\ln 3.192 + 2 \ln 10$ or $\ln N = \ln 3.192 + \ln 10^2 = \ln (3.192 \times 10^2)$. Therefore, $N = 27e^{2.47} = 3.192 \times 10^2 = 319.2$. Check this result by using Table V.

Remark 57.1. *In Examples 57.4, 57.5, and 57.6 we formed the natural logarithm of 10 (which equals 2.3026) whenever possible so that we could multiply or divide by 10. These operations are easily done in our base 10 number system.*

Example 57.7

Solve the exponential equation $e^{-2x} = 0.71$ for x.

Solution

The equation involves e, and so we begin by taking the ln of both sides of the equation.

$$\ln e^{-2x} = \ln 0.71$$

or

$$-2x = \ln 7.1 + \ln 10^{-1}$$

Now, $\ln 10^{-1} = -1 \ln 10 = -2.3026$ and, from Table III, $\ln 7.1 = 1.9601$. Thus,

$$-2x = 1.9601 - 2.3026$$

or

$$-2x = -0.3425$$

Hence, $x = 0.1713$. ●

An equation of the type given in Example 57.7 might also be solved by using common logarithms. The result would not differ if this approach is used, but the efficiency is certainly diminished, since $\log e \neq 1$; in fact, $\log e = 0.4343$. To illustrate the difference, let us rework Example 57.7.

Example 57.8

Solve the exponential equation $e^{-2x} = 0.71$ for x by using common logarithms.

Solution

We proceed with common logarithms as we did with natural logs.

$$\log e^{-2x} = \log 0.71$$

$$-2x \log e = \log (7.1 \times 10^{-1})$$

Since $\log e = 0.4343$ and $\log (7.1 \times 10^{-1}) = 0.8513 - 1 = -0.1487$, we have

$$-2x(0.4343) = -0.1487$$

or

$$-0.8686x = -0.1487$$

and

$$x = \frac{-0.1487}{-0.8686} = 0.1712$$

If a table of natural logarithms should not be available or if we should have occasion to use a logarithm system with a base other than 10 or e, we need a general method that enables us to convert the logarithm of a number from one base to another.

Suppose that $y = \log_b N$. The equivalent exponential form is given by $N = b^y$. Taking the logarithm of both sides of this expression to the base a yields

$$\log_a N = \log_a b^y$$

or

$$\log_a N = y \log_a b$$

Thus,

$$y = \frac{\log_a N}{\log_a b}$$

and, since $y = \log_b N$, we have

$$\log_b N = \frac{\log_a N}{\log_a b} \qquad (57.1)$$

If we let $b = 10$ and $a = e$, Equation (57.1) produces a formula that enables us to convert common logarithms to natural logarithms. Substituting $b = 10$ and $a = e$ into (57.1) yields

$$\log N = \frac{\ln N}{\ln 10} \qquad (57.2)$$

Since $\ln 10 = 2.3026$, we may write 57.2 as

$$\log N = \frac{\ln N}{2.3026}$$

or

$$\ln N = (2.3026) \log N \qquad (57.3)$$

If we let $b = e$ and $a = 10$, equation (57.1) produces a formula for converting natural logarithms to common logarithms.

$$\ln N = \frac{\log N}{\log e} \qquad (57.4)$$

Noting that $\log e = 0.4343$, we write (57.4) as

$$\ln N = \frac{\log N}{0.4343}$$

or

$$\log N = (0.4343) \ln N \qquad (57.5)$$

We shall use the following examples to illustrate the operations with these equations.

Example 57.9
If $\log 20 = 1.3010$, find $\ln 20$.

Solution
Equation (57.3) states that $\ln N = (2.3026) \log N$. Thus, we have $\ln 20 = (2.3026) \log 20 = (2.3026)(1.3010)$. The product can be found by logarithms:

$$\begin{array}{rl} \log 2.303 & = 0.3623 \\ \log 1.3010 & = 0.1142 \\ \hline \log 2.303 + \log 1.3010 & = 0.4765 \end{array} \quad \left.\right\} \text{add}$$

The antilog of 0.4765 is 2.996. Thus, $\ln 20 = 2.996$. From Table III we see that $\ln 20 = 2.9958$. Thus, our result is accurate to three decimal places. (Note that 2.3026 was rounded off to four digits, since we are using a four-place table.)

Example 57.10
If $\ln 37 = 3.6109$, find $\log 37$.

Solution
From equation (57.5) we have

$$\begin{aligned} \log 37 &= (0.4343) \ln 37 \\ &= (0.4343)(3.6109) \end{aligned}$$

Performing the multiplication by logarithms, we have

$$\begin{array}{ll} \log 0.4343 & = 9.6378 - 10 \\ \log 3.611 & = 0.5576 \\ \hline \log 0.4343 + \log 3.611 & = 10.1954 - 10 \end{array} \quad \left.\right\} \text{add}$$

Thus, $\log (0.4343)(3.611) = 0.1954$. The antilogarithm of $0.1954 = 1.568$. Therefore, $\log 37 = 1.568$. If we check Table II of Appendix F, we find that $\log 37 = 1.5682$. The values agree to three decimal places.

Example 57.11
Find $\log_3 65$.

Solution

We use equation (57.1) with $a = 10$ and $b = 3$. With these substitutions, (57.1) yields

$$\log_3 65 = \frac{\log 65}{\log 3} = \frac{1.8129}{0.4771}$$

Performing the division by logarithms, we have

$$\begin{array}{lr} \log 1.813 = & 10.2584 - 10 \\ \log 0.4771 = & 9.6786 - 10 \\ \hline & 0.5798 \end{array} \Bigg\} \text{subtract}$$

The antilogarithm of 0.5798 is 3.80. Thus, $\log_3 65 = 3.8$.

Exercises for Section 57

In Exercises 1 through 10, find the natural logarithm of each number. Interpolate where necessary.

1. 2.35 **2.** 23.5 **3.** 0.235

4. 25 **5.** 126 **6.** 0.00729

7. 5.614 **8.** 86.32 **9.** 125.6

10. 8765

In Exercises 11 through 20, find the antilogarithms of each number using Table III of Appendix F. Interpolate where necessary.

11. 1.5041 **12.** 2.1041 **13.** 1.6253

14. 0.8329 **15.** 4.8979 **16.** 5.5873

17. −0.8843 **18.** −4.7677 **19.** −2.6460

20. 3.2794

In Exercises 21 through 30, write an equivalent exponential form for N.

21. $\ln N = 3$ **22.** $2 \ln N = 3$

23. $\ln N^2 = 3$ **24.** $2 \ln N + 1 = 3$

25. $\ln N = -2$ **26.** $3 \ln N - 2 = 4$

27. $3 \ln (N - 2) = 4$ **28.** $\ln N - \ln 2 = 3$

29. $3 \ln N - \ln N^2 = \frac{1}{2}$ **30.** $2 \ln N + \ln 2N - 4 \ln N = -2$

31. Given $\ln 2 = 0.6932$ and $\ln 10 = 2.3026$, find each of the following using only the values given and the laws of logarithms.
(a) $\ln 20$ (b) $\ln 200$ (c) $\ln 4$
(d) $\ln 32$ (e) $\ln \sqrt{2}$ (f) $\ln 5$

(g) $\ln 25$ (h) $\ln \frac{1}{2}$ (i) $\ln 0.02$
(j) $\ln 1.25$

32. Given $\ln 2 = 0.6931$, $\ln 3 = 1.0986$, and $\ln 5 = 1.6094$, find each of the following using only the values given and the laws of logarithms.
(a) $\ln 8$ (b) $\ln 10$ (c) $\ln 0.25$
(d) $\ln 0.625$ (e) $\ln 1.2$ (f) $\ln 0.4$
(g) $\ln 81$ (h) $\ln 600$ (i) $\ln 0.03$
(j) $\ln 90$

33. Convert each of the following to natural logarithms.
(a) $\log x = 0.2145$ (b) $\log x = 1.4137$
(c) $\log x = 4.7195$ (d) $\log x = 0.0714$
(e) $\log x = 9.5197 - 10$ (f) $\log x = 7.9127 - 10$

34. Convert each of the following to common logarithms.
(a) $\ln x = 1.3140$ (b) $\ln x = 3.7195$
(c) $\ln x = 0.1349$ (d) $\ln x = 0.00215$
(e) $\ln x = -1.5938$ (f) $\ln x = -4.2176$

35. Use common logarithms to evaluate each of the following expressions.
(a) $\log_3 11$ (b) $\log_5 34$ (c) $\log_2 75$
(d) $\log_7 2$ (e) $\log_6 0.0479$ (f) $\log_{15} 148$

36. Use natural logarithms to evaluate each of the following.
(a) $e^{2.8}$ (b) $e^{-2.1}$ (c) $1.2e^{1.3}$
(d) $24e^{2.5}$ (e) $0.042e^3$ (f) $12e^{-3.2}$

37. Solve each of the following equations for x.
(a) $e^x = 4.6$ (b) $e^{-x} = 0.341$ (c) $3e^{0.2x} = 0.0128$
(d) $e^x = 10$ (e) $3^x = 1.47$ (f) $2^x = 85$

38. Graph each of the following functions.
(a) $y = e^{-x}$ (b) $y = e^{2x}$ (c) $y = e^{-0.4x}$
(d) $y = \dfrac{e^x + e^{-x}}{2}$ (e) $y = \dfrac{e^x - e^{-x}}{2}$

39. (a) Show that $\left(\dfrac{e^x + e^{-2}}{2}\right)^2 - \left(\dfrac{e^x - e^{-x}}{2}\right)^2 = 1$.

(b) Show that $\left(\dfrac{e^x + e^{-x}}{2}\right)^2 + \left(\dfrac{e^x - e^{-x}}{2}\right)^2 = \dfrac{e^{2x} + e^{-2x}}{2}$.

40. The function $y = f(t) = e^{-0.4t} \sin 2\pi t$ represents damped harmonic motion. Let us consider the function for $t \geq 0$ with the intent of sketching the graph. We note that when $t = 0$, $y = e^0 \sin 0 = 0$. Thus the y intercept is 0. The t intercepts are found by setting $y = 0$ and solving the resulting equation for t. Now, $e^{-0.4t} \sin 2\pi t = 0$ only if $\sin 2\pi t = 0$. But, $\sin 2\pi t = 0$ for $2\pi t = 0, \pi, 2\pi, 3\pi, \ldots$. Thus, $t = 0, \frac{1}{2}, 1, \frac{3}{2}, \ldots$. These are the t intercepts. As a further aid to the curve sketching, we note that $\sin 2\pi t = 1$ for $2\pi t = \pi/2, 5\pi/2, 9\pi/2, \ldots$, or

$t = \frac{1}{4}, \frac{5}{4}, \frac{9}{4}, \ldots$ For these values for t, $y = (e^{-0.4t})(1) = e^{-0.4t}$ (use Table V).

Similarly, $\sin 2\pi t = -1$ for $2\pi t = 3\pi/2, 7\pi/2, 11\pi/2, \ldots$, or $t = \frac{3}{4}, \frac{7}{4}, \frac{11}{4}, \ldots$ For these values for t, $y = e^{-0.4t}(-1) = -e^{-0.4t}$ (use Table V of Appendix F). Now use this information to graph the given function for $0 \leq t \leq 4$.

41. Follow the procedure outlined in Exercise 40 to discuss and sketch the function $y = e^{-t/2} \cos 2\pi t$ for $0 \leq t \leq 4$.

42. The modern logarithm scale of stellar intensities has a base of $\sqrt[5]{100} = 2.51$. If the intensity of a is given by I_a and the intensity of b is given by I_b, the formula for the difference between magnitudes is given by

$$b - a = \log_{2.51} \frac{I_a}{I_b}$$

(a) Show that this equation is equivalent to

$$b - a = 2.5 \log \frac{I_a}{I_b}$$

Hint: Use (57.1).

(b) Show that if star a is 100 times as bright as star b, star b is 5 magnitudes greater than star a.

(c) What must I_a/I_b be so that star b is 2 magnitudes greater than star A?

43. Use the **graphic method** of Section 33 to solve each of the following systems of equations.

(a) $\begin{cases} y = x - 2 \\ y = \ln x \end{cases}$ (b) $\begin{cases} y = 1 + 2x - x^2 \\ y = e^x \end{cases}$

SECTION 58

Applications Involving Natural Logarithms

Now, let us consider some of the applications of the natural logarithms and exponential functions. We shall begin by noting that there are many situations in which the rate of change of the amount a given substance is proportional to the amount of the substance present at any given time. Such situations occur in chemistry, physics, biology, and business. If we let A represent the amount of the substance present, at any time t, it can be shown that

$$A = Ce^{Kt} \tag{58.1}$$

where k is constant and C is the amount present when $t = 0$.

Equation (58.1) represents the **law of natural growth** if $K > 0$ and the **law of natural decay** if $K < 0$. We note that in problems involving the law of natural decay, the term *half-life* refers to the time required for one half of the substance to decay. The following examples will illustrate the use of equation (58.1).

Example 58.1

The number of bacteria in a certain culture increases at a rate proportional to the amount present. If there are 1000 bacteria present initially and the amount doubles in 15 min, how many bacteria will there be in 40 min?

Solution

Since the rate of change is proportional to the amount present, we employ equation (58.1). Thus, $A = Ce^{Kt}$. When $t = 0$, $A = 1000$ and so $1000 = Ce^{K(0)} = C$ and we have $A = 1000e^{Kt}$. The amount doubles in 15 min, or $A = 2000$ when $t = 15$. This condition enables us to evaluate K, as follows:

$$2000 = 1000e^{15K}$$

$$2 = e^{15K}$$

Taking the natural logarithm of both sides of the equation, we obtain

$$\ln e^{15K} = \ln 2$$

or

$$15K = 0.6931$$

or

$$K = 0.0462$$

Hence, $A = 1000e^{0.0462t}$, and when $t = 40$,

$$A = 1000e^{1.8480}$$

We might find A by using Table V, a table for e^x, but let us instead illustrate the mechanics involved by employing logarithms.

$$\begin{aligned} \ln A &= \ln 1000e^{1.8480} \\ &= \ln 1000 + \ln e^{1.8480} \\ &= 3 \ln 10 + 1.8480 \ln e \\ &= 3(2.3026) + 1.8480 \end{aligned}$$

To find A,

$$\begin{aligned} \ln A &= 1.8480 + 3(2.3026) \\ &= \ln 6.347 + \ln 10^3 \\ &= \ln (6.347 \times 10^3) \end{aligned}$$

Thus, $A = 6347$ bacteria.

Example 58.2

The amount of a certain radioactive element available at any time t is proportional to the amount present and is given by the equation $A = Ce^{-0.2t}$. If 60 g of the element is present 6 sec after the beginning of its decay, how many grams was present when $t = 0$?

Solution

When $t = 0$, $A = Ce^{(0.2)(0)} = Ce^0 = C$. Thus, we must find C. We are given that $A = 60$ when $t = 6$ sec, so

$$60 = Ce^{(-0.2)(6)} = Ce^{-1.2}$$

Taking the ln of both sides yields

$$\ln 60 = \ln Ce^{-1.2}$$

$$\ln (6)(10^1) = \ln C + \ln e^{-1.2}$$

$$1.7918 + 2.3026 = \ln C + (-1.2)\ln e$$

Combining values on the left and noting that $\ln e = 1$,

$$4.0944 = \ln C - 1.2$$

or

$$\ln C = 4.0944 + 1.2 = 5.2944$$

To find C we rewrite 5.2944 as

$$0.6892 + 2(2.3026)$$

Thus, $\ln C = 0.6892 + 2(2.3026)$. Since $\ln 1.992 = 0.6892$,

$$\ln C = \ln 1.992 + 2 \ln 10$$

$$= \ln 1.992 + \ln 10^2$$

$$= \ln (1.992 \times 10^2)$$

and hence $C = 1.992 \times 10^2 = 199.2$ g.

Example 58.3

The rate of decay of a radioactive substance is proportional to the amount present. If 100 mg of the substance is present now and its half-life is 650 years, how much will be present 100 years from now?

Solution

The substance behaves according to the law of natural decay, and therefore

$$A = Ce^{Kt} \tag{1}$$

Since $A = 100$ when $t = 0$, $100 = Ce^0$ or $C = 100$ and $A = 100e^{Kt}$. Because the half-life is 650 years, $A = 50$ when $t = 650$, and we have

$$50 = 100e^{K(650)} \quad \text{or} \quad 0.5 = e^{650K}$$

Thus,

$$\ln 0.5 = \ln e^{650K}$$

$$\ln 0.5 = 650K$$

Since $\ln 0.5 = -0.6931$, we have

$$K = \frac{\ln 0.5}{650} = \frac{-0.6931}{650} = -0.00106$$

Substituting this result in (1), we have

$$A = 100e^{-0.00106t}$$

When $t = 100$, $A = 100e^{-0.106}$. Using natural logarithms,

$$\begin{aligned} \ln A &= \ln 100e^{-0.106} \\ &= \ln 100 + \ln e^{-0.106} \\ &= \ln 10^2 + -0.106 \ln e \\ &= 2(2.3026) - 0.106 = 4.4992 \end{aligned}$$

In order to find A, we rewrite 4.4992 as $2.1966 + 2.3026$. Thus, $\ln A = 2.1966 + 2.3026$. We use Table IV to find $\ln 8.994 = 2.1966$, and so

$$\begin{aligned} \ln A &= \ln 8.994 + \ln 10 \\ &= \ln (8.994 \times 10^1) \end{aligned}$$

Therefore, $A = 8.994 \times 10^1 = 89.94$, and there will be 89.94 g remaining after 100 years. ●

The law of natural growth can also be used in the area of business. In Section 55 we noted that the formula

$$A = P\left(1 + \frac{i}{n}\right)^{nt} \tag{58.2}$$

gives the amount A of money after t years if P is the amount invested, i the yearly interest rate, and n the number of times per year the interest is compounded. Let us suppose that the number of interest periods per year increases without bound. In this situation, (58.2) becomes, by methods of the calculus,

$$A = Pe^{it} \tag{58.3}$$

Comparing (58.3) with (58.1), we see that they are the same if $P = C$ and $I = K$.

Example 58.4

If \$10,000 is invested at 8% interest compounded continuously, find the amount at the end of 10 years.

Solution

In this case $P = \$10{,}000$, $i = 0.08$, and $t = 10$. Therefore, equation (58.3) becomes

$$A = 10{,}000e^{(0.08)(10)} = 10{,}000e^{0.8}$$

Using natural logarithms,

$$\begin{aligned} \ln A &= \ln 10{,}000 + \ln e^{0.8} \\ &= 4 \ln 10 + 0.8 \ln e \\ &= 4(2.3026) + 0.8 \end{aligned}$$

Since $\ln A = 0.8 + 4(2.3026)$ and $\ln 2.226 = 0.8$, we have

$$\begin{aligned} \ln A &= \ln 2.226 + \ln 10^4 \\ &= \ln (2.226 \times 10^4) \end{aligned}$$

Hence, $A = 2.226 \times 10^4 = \$22{,}260$. ●

The effective rate of interest is the rate that gives the same amount of interest compounded once a year. Thus, we can convert interest compounded continuously to simple interest.

Example 58.5

If \$500 is invested at the rate of 10% per year compounded continuously find the effective rate of interest.

Solution

If we let i be the effective rate of interest, we have

$$500(1 + i) = \mathrm{A}$$

Our calculations can be simplified by noting that since $A = Pe^{it}$, we have $A = 500e^{0.1}$. Thus,

$$\begin{aligned} 500(1 + i) &= 500e^{0.1} \\ 1 + i &= e^{0.1} \end{aligned}$$

We now take the natural logarithm of both sides of the equation $1 + i = e^{0.1}$, to obtain

$$\ln (1 + i) = \ln e^{0.1} = (0.1) \ln e = 0.1$$

Since the antilog of 0.1 is 1.1052, we have

$$1 + i = 1.1052$$

Thus, $i = 1.1052 - 1 = 0.1052 = 10.52\%$. ●

Newton's law of cooling states that the rate of which a body changes temperature is proportional to the difference between its temperature and the surrounding medium. As an illustration of this law, let us consider the following example.

Example 58.6

When a heated object is immersed in a cooling solution, the temperature T of the object at any time t (in seconds) after immersion is given by the formula $T = T_0 + 60e^{-0.4t}$, where T_0 is the temperature of the cooling solution. Find T after 6 sec if $T_0 = 20°C$.

Solution

Since $t = 6$ and $T_0 = 20$, we have

$$T = 20 + 60e^{-2.4}$$

To find T we have only to evaluate $60e^{-2.4}$ and then add 20. Let us find $60e^{-2.4}$ as follows. Let $x = 60e^{-2.4}$; then

$$\begin{aligned} \ln x &= \ln 60e^{-2.4} \\ &= \ln 60 + \ln e^{-2.4} \\ &= 4.0944 - 2.4 = 1.6944 \end{aligned}$$

The antilog of $1.6944 = 5.44$, and thus $x = 5.44$. Hence, $T = 20 + x = 25.44°C$.

Remark 58.1. *In Examples 58.3, 58.5, and 58.6, we might have simplified our work by employing Table V of Appendix F, a table that yields e^x values directly. Our method of solution, although somewhat lengthy, provides repeated illustrations of the mechanics of natural logarithms and antilogarithms.*

Exercises for Section 58

In Exercises 1 through 10, use equation (58.1).

1. The number of bacteria in a certain culture doubles in 3 days. There are 10^5 bacteria present when an experiment begins. How many bacteria are present after 24 hr?
2. A certain population is increasing at a rate proportional to the population present and will double in 30 years. How many years will it take for the population to triple?
3. The half-life of a radioactive substance is 20 min. How much of an original sample of 50 g will remain after 30 min? After 40 min?
4. The half-life of strontium 90 is approximately 25 years. If 40 mg of the isotope is present now, how much remains in 15 years? In how many years will only 10 mg remain?
5. If 10% of a radioactive substance remains after 1 year, find the half-life.

6. The half-life of radium is about 1690 years. If 50 mg of radium is present now, how much radium will be present 100 years from now? 500 years from now?

7. The oil exports of a certain Middle East nation are increasing according to the law of natural growth. If exports increased by 50% from 1972 to 1978, how much will they increase by from 1972 to 1984?

8. A man borrows $1000 at an interest rate of 12% per year compounded continuously. If the loan is to be repaid with one payment at the end of 1 year, how much must the borrower repay? What is the effective rate of interest?

9. How long will it take a sum of money to quadruple at 100% interest compounded continuously?

10. The charge on a spherical surface leaks off at a rate proportional to the amount of the charge. If the charge is 12 coulombs (C) when $t = 0$ and one half leaks off in 30 min, when will there be only 4 C remaining?

11. An object of temperature 150°C is placed in air of temperature 50°C and cools to 75°C in 5 min. Find the temperature T of the body after 15 min if $T = 100e^{Kt} + 50$.

12. A heated object is placed in air of temperature 40°C. The temperature T of the object after t min is given by the formula

$$T = Ce^{Kt} + 40$$

If the body cools from 200°C to 100°C in 20 min, find the temperature of the body after 30 min.

13. The potential in an electric circuit is 120 V. After a switch is thrown, the transient values in volts are given by the equation $v = 120e^{-5t}$, t in seconds. How long after the switch is thrown does the potential drop to 0.0012 V?

14. The atmospheric pressure p lb/ft^3 at a height h ft above sea level is given by the formula

$$p = 2116e^{-0.0000318h}$$

Find the atmospheric pressure outside an airplane at 20,000 ft.

15. There are 100 gal of brine in a tank and the brine contains 75 lb of dissolved salt. Fresh water runs into the tank at the rate of 5 gal/min and the mixture runs out at the same rate. If we assume that the mixture is kept uniform by stirring, the formula $P = Ce^{-0.05t}$ gives the number of pounds P of salt in the tank after t min where t represents the number of minutes that have elapsed since the water began flowing into the tank. How many pounds of salt is there in the tank after 1 hr?

16. Under certain conditions the current I in amperes is given by

$$I = \frac{E(1 - e^{-Rt/L})}{R}$$

If $L = 0.1$ henry (H), $E = 110$ V, and $R = 8\ \Omega$, find the current flowing when $t = 0.05$ sec.

17. The rate constant K for a chemical reaction is given as a function of temperature by the equation

$$\ln K = \frac{-E_a}{RT} + B$$

where E_a is the activation energy and B is a constant. Find K if $T = 40°C$, $B = 1.90$, $R = 5000$ and $E_a = 1.0 \times 10^6$ calories (cal).

18. The intensity I (in lumes) of a light beam, after passing through a thickness x (in centimeters) of a medium having an absorption coefficient of 0.2, is given by the formula

$$I = 2000e^{-0.2x}$$

How may centimeters of the material will reduce the illumination to 600 lumes?

SECTION 59

Logarithmic and Semilogarithmic Graph Paper

The various techniques of graphical analysis that we have discussed up to this point have been concerned with linear and nonlinear functions and relations that could be sketched on rectangular graph paper. The ordinary or rectangular coordinate graph paper has horizontal and vertical rulings that are linear scales. There are a variety of types of graph paper available for special purposes. At this time we shall discuss two additional types of graph paper, types that have nonlinear spacing on the axis.

Type 1: **Semilogarithmic (semilog)** paper. Graph paper in which the horizontal axis has a linear scale and the vertical axis has a logarithmic scale. The plot of a point (x, y) on such paper is equivalent to plotting the point $(x, \log y)$ on ordinary graph paper.

Type 2: **Logarithmic–logarithmic (log–log)** paper. Graph paper in which each axis has a logarithmic scale. Plotting a point (x, y) on this paper is equivalent to plotting the point $(\log x, \log y)$ on ordinary graph paper.

Let us first consider the design and use of semilog paper. The log axis on the semilog graph paper has rulings that are not uniformly spaced. The rulings are marked off at distances that are proportional to the logarithms of numbers from 1 to 10. The linear axis has uniform rulings, rulings that are equally spaced. We note that the log scale will never contain zero, since to do so would imply that the logarithm of zero is defined. In Figure 59.1 we have semilog paper with four blocks on the y axis. In this printed form the scales in each block run from 1 to 10. The relabeling of these scales is generally done in a manner that is determined by the range of values in the given problem.

One use for semilog paper is the graphing of functions that have a wide variance of values in their range. In such instances ordinary rectangular graph paper proves to be inadequate.

Example 59.1

Graph the functions whose set of ordered pairs is given in the following table:

x	1	3	5	6	7	9	10	12
y	50,000	30,000	10,000	5000	1000	600	50	20

Solution

We first note that ordinary graph paper is of little use because of the great differences in magnitude of the values of y. In order to get a picture of the function we use semilog paper with the linear scale on the x axis and the logarithmic scale on the y axis. The logarithmic scale goes by repetitive blocks. The scale representing $100, 200, \ldots, 900$ is repeated for $1000, 2000, \ldots, 9000$, and so on. This is accounted for by the fact that the common logarithms of consecutive powers of 10 increase at a uniform rate. The graph is shown in Figure 59.2 on page 338.

In addition to the ease of handling cumbersome numbers, log and semilog paper afford us the opportunity to simplify, graphically, power functions and exponential functions.

Now we will show that the graph of any exponential function is a straight line if we use semilog paper. Consider the exponential function $Y = B(a^x)$. Taking the logarithm of both sides of the equation yields

$$\begin{aligned} \log Y &= \log B \cdot a^x \\ &= \log B + \log a^x \\ &= \log B + x \log a \end{aligned} \tag{59.1}$$

If we let $y = \log Y$, $b = \log B$ and $m = \log a$, equation (59.1) becomes

$$y = mx + b \tag{59.2}$$

Figure 59.1

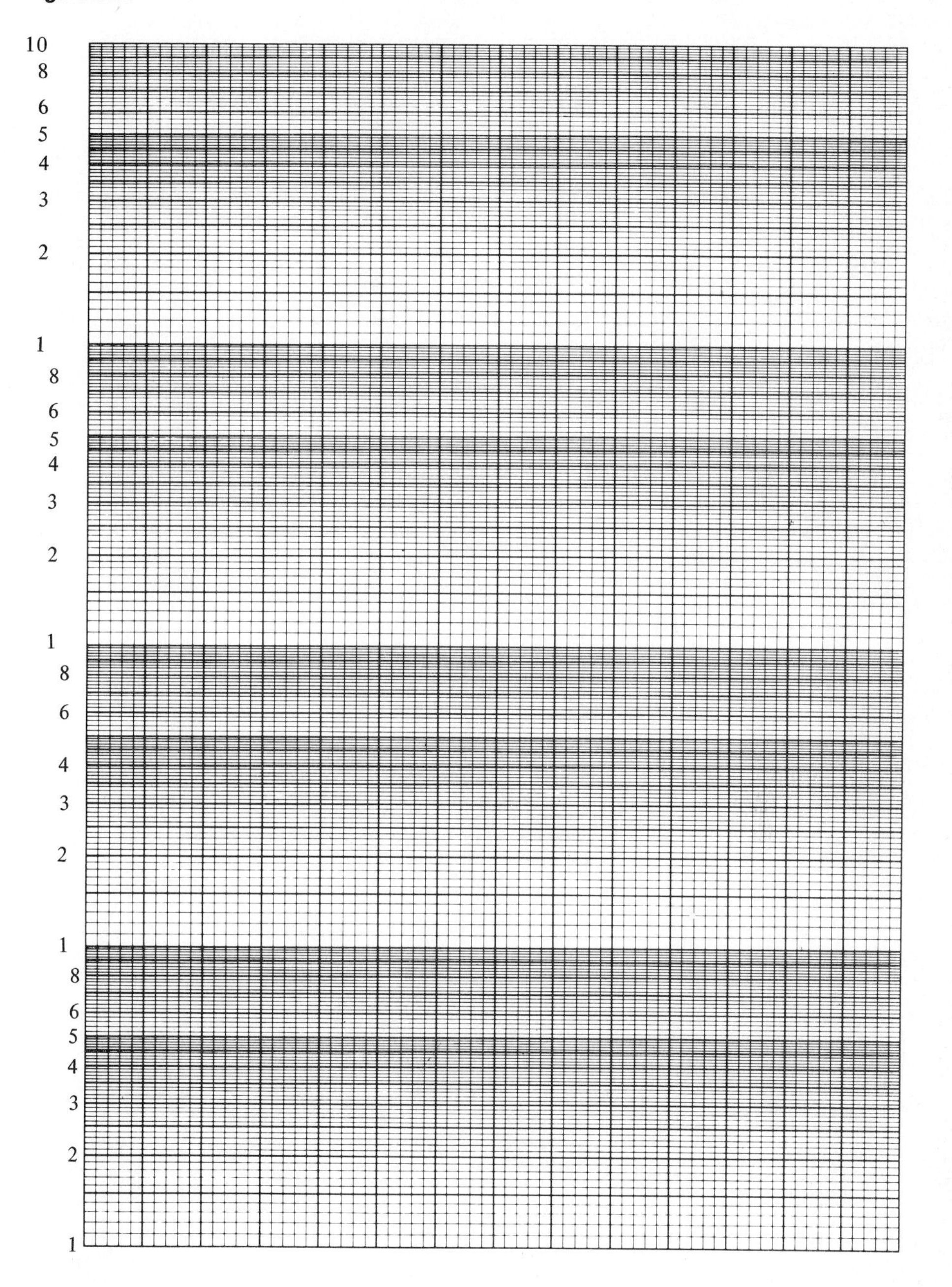

Figure 59.2

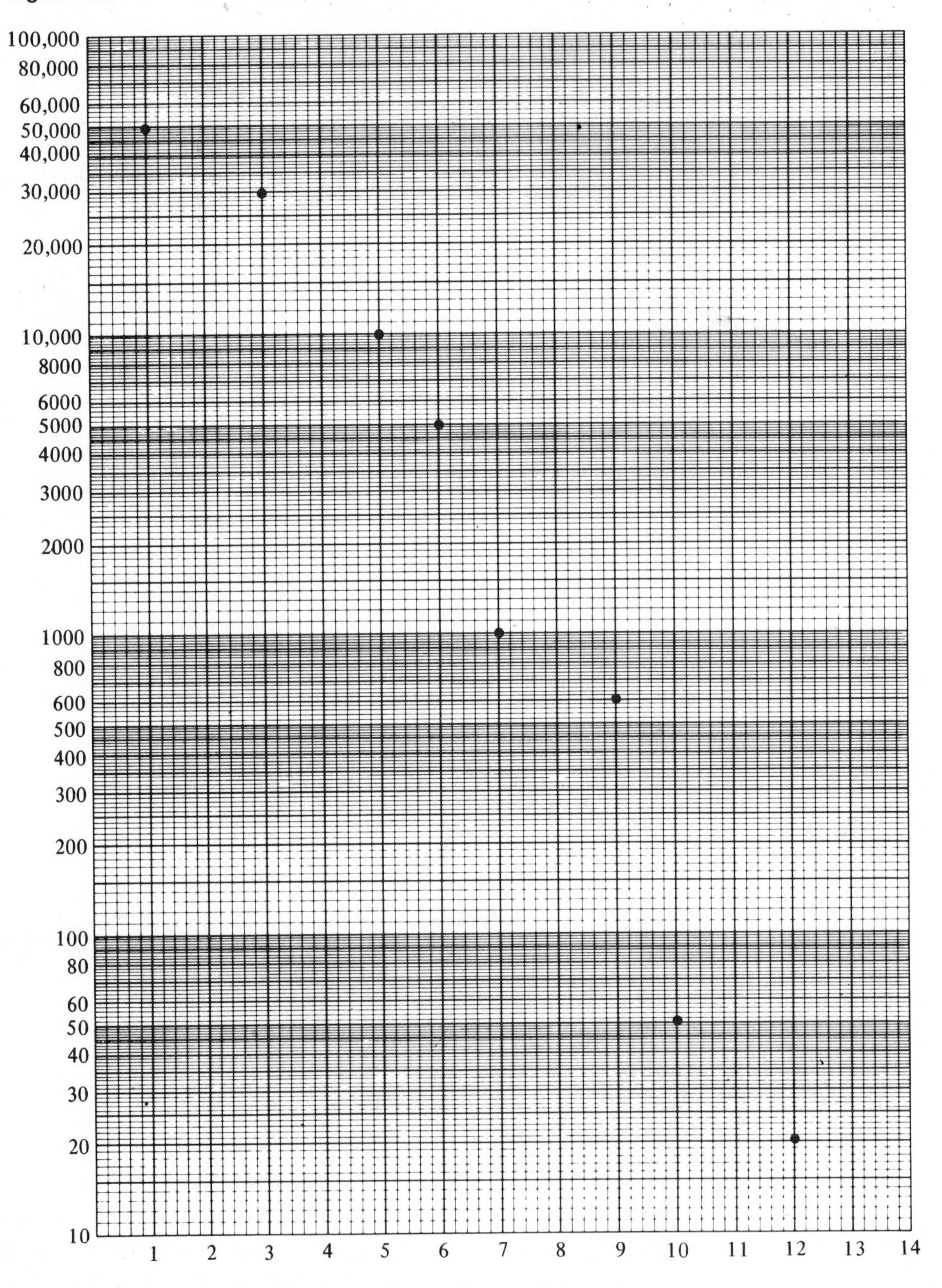

100,000
80,000
60,000
50,000
40,000
30,000
20,000
10,000
8000
6000
5000
4000
3000
2000
1000
800
600
500
400
300
200
100
80
60
50
40
30
20
10
1
2
3
4
5
6
7
8
9
10
11
12
13
14

Equation (59.2) is the equation of a straight line with slope m and y intercept at b (see Section 87). Thus, the graph of an exponential function on semilog paper is a straight line and, conversely, if the graph on semilog paper is a straight line, the equation is exponential in nature.

Example 59.2

Graph $Y = 5(3^x)$ on semilog graph paper.

Solution

We note that the given equation is exponential in form. Thus, on semilog paper the graph will be a straight line. Since we need only two points to graph a straight line, we arbitrarily choose two values for x, find the corresponding value for Y, plot the ordered pairs, and draw the straight line. If $x = 1$, $Y = 15$, and if $x = 4$, $Y = 405$. The graph is shown in Figure 59.3 on page 340. From the graph we read the y intercept to be 5. This is to be expected, since in our given function $Y = 5$ when $x = 0$; that is, $Y = 5(3^0) = 5(1) = 5$.

To find the slope of the straight line, we must exercise care. We do not find the slope by reading the coordinates of points, since our horizontal and vertical scales differ. Rather, we refer to equations (59.1) and (59.2) and observe that the slope m is equal to $\log a$. In this case $m = \log 3$, and using Table II, we have $m = 0.4771$.

Example 59.3

Graph $Y = e^x$ on semilog graph paper.

Solution

The equation is exponential, hence its graph will be a straight line. We need only plot two points. From Table V we note that for $x = 2$, $Y = 7.4$, and for $x = 4$, $Y = 56.4$. The graph is shown in Figure 59.4 on page 341.

From the graph the intercept point on the vertical axis is at $(0, 1)$. This is verified by noting that $e^0 = 1$. The slope of line is given by $m = \log e$ and hence $m = 0.4343$.

Another use for semilog paper involves situations where we may not have an equation, but instead, some bits of data. It may be possible to fit the data to some simple law if we are able to find an appropriate equation. Let us illustrate the technique with the following example.

Example 59.4

Find the equation that relates the following data:

x	1	3	5
Y	60	240	960

Figure 59.3

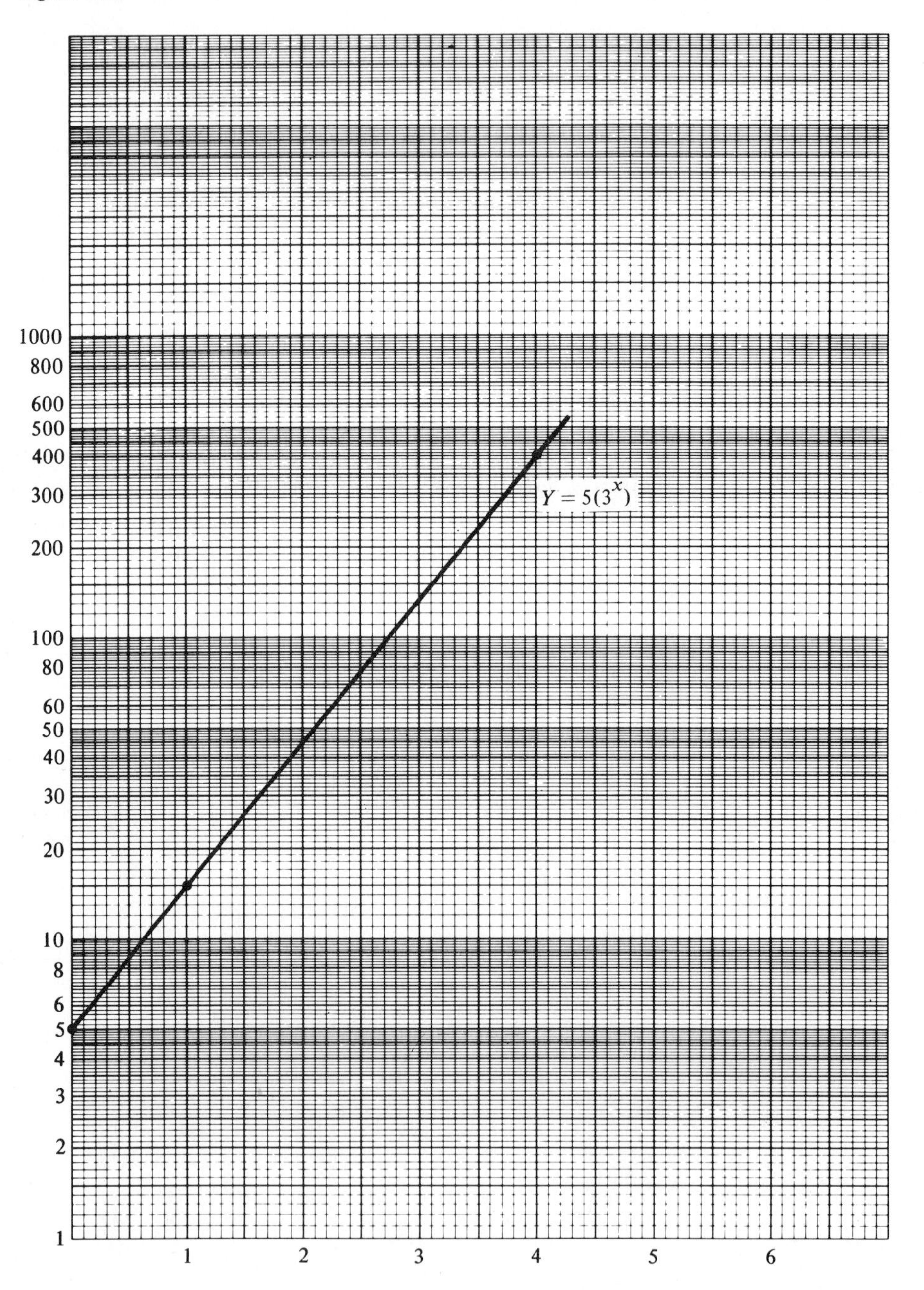

1000
800
600
500
400
300
200
100
80
60
50
40
30
20
10
8
6
5
4
3
2
1
1
2
3
4
5
6
$Y = 5(3^x)$

Figure 59.4

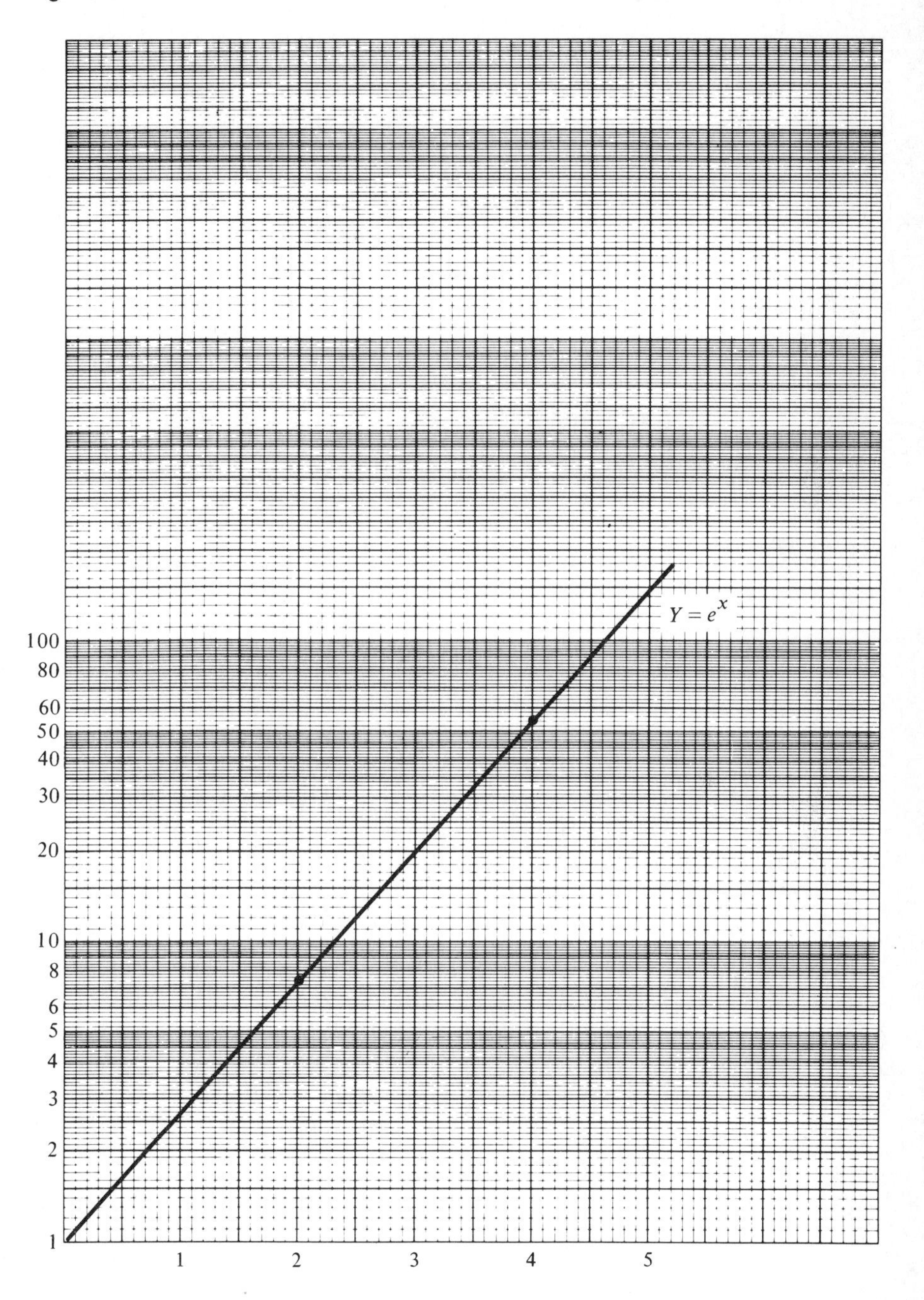

Solution

The values are graphed in Figure 59.5. since they lie on a straight line, the equation we seek is exponential, of the form $Y = B(a^x)$. From the figure we read the vertical intercept, and hence the value for B to be 30. Thus, $Y = 30a^x$. To find a, we substitute $x = 1$ and $Y = 60$ and obtain $60 = 30a^1$. Thus, $a = 2$ and our equation is $Y = 30(2^x)$.

Remark. *In Example 59.4 we plot x against Y on semilog paper and obtain a straight line. Thus, the law is exponential. If the points lies along some other curve, other methods must be used to find the appropriate equation. This process is called* ***curve fitting*** *and is handled by numerous methods.*

As we have noted, log–log paper is graph paper that has one or more logarithmic cycles along each axis. This type of graph paper is very useful when we encounter very large or very small values for either or both variables in a given function. In particular, since log–log paper automatically takes the logarithm of each coordinate, equations of the form $Y = BX^m$ have graphs that are straight lines. If we take the logarithm of both sides of this equation, we have

$$\begin{aligned} \log Y &= \log BX^m \\ &= \log B + \log X^m \\ &= m \log X + \log B \end{aligned} \tag{59.3}$$

If we let $y = \log Y$, $x = \log X$, and $b = \log B$, equation (59.3) becomes

$$y = mx + b \tag{59.4}$$

which is the equation of a straight line with slope m and y intercept at $b = \log B$. Conversely, if the graph is a straight line, the equation relating the two variables is of the form

$$Y = BX^m$$

This type of function is called a **power function**.

Example 59.5

Graph $y = 2x^3$ on log–log paper.

Solution

The equation is of the form $Y = BX^m$, and hence the graph on logarithmic paper will be a straight line. We need only find two points to draw the graph. We note that when $x = 1$, $y = 2$ and when $x = 2$, $y = 16$. In Figure 59.6 we graph the given equation using the points (1, 2) and (2, 16).

Example 59.6

Graph $x^3y^4 = 10$ on logarithmic paper.

Figure 59.5

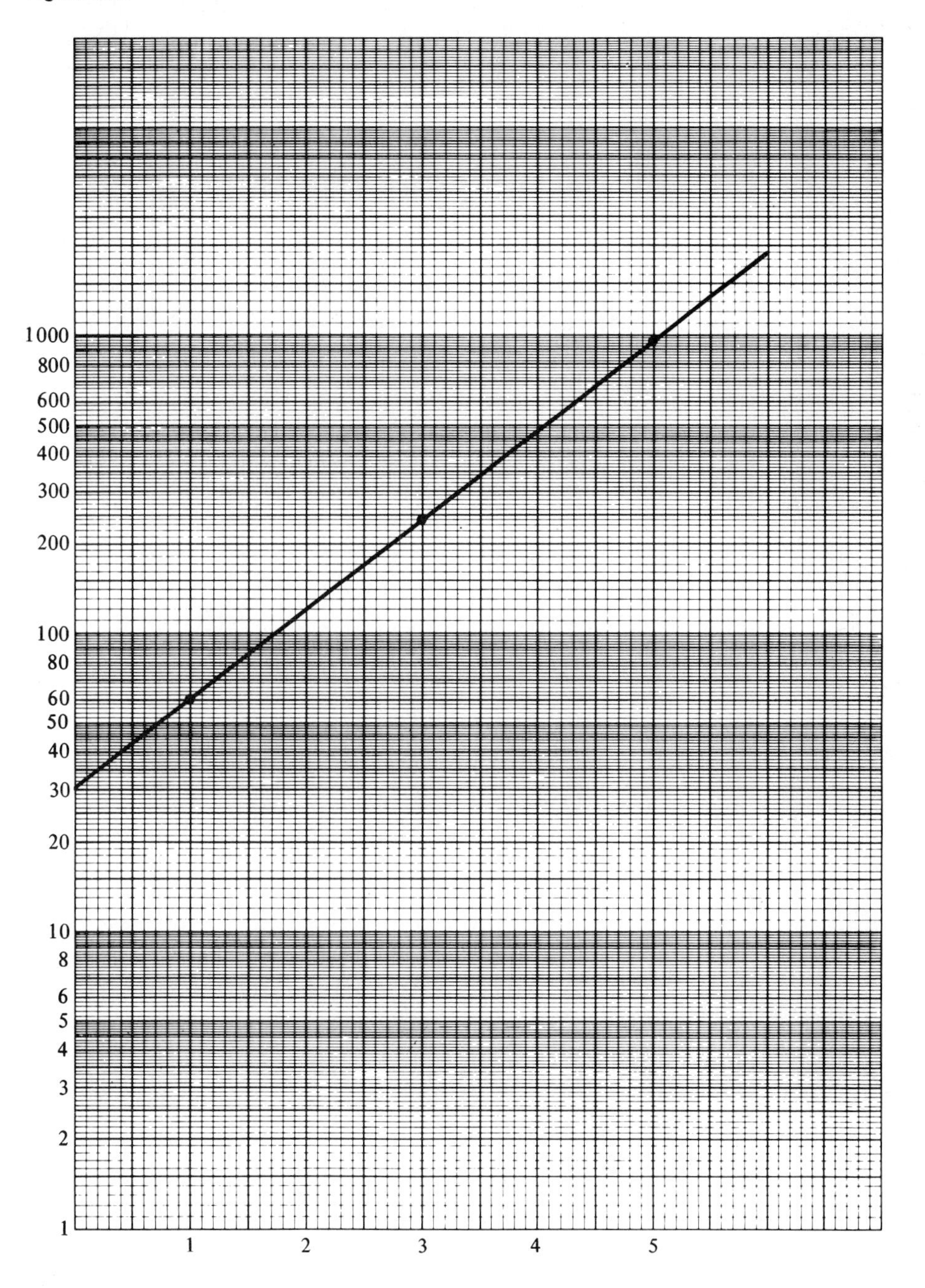
1000
800
600
500
400
300
200
100
80
60
50
40
30
20
10
8
6
5
4
3
2
1
1
2
3
4
5

Figure 59.6

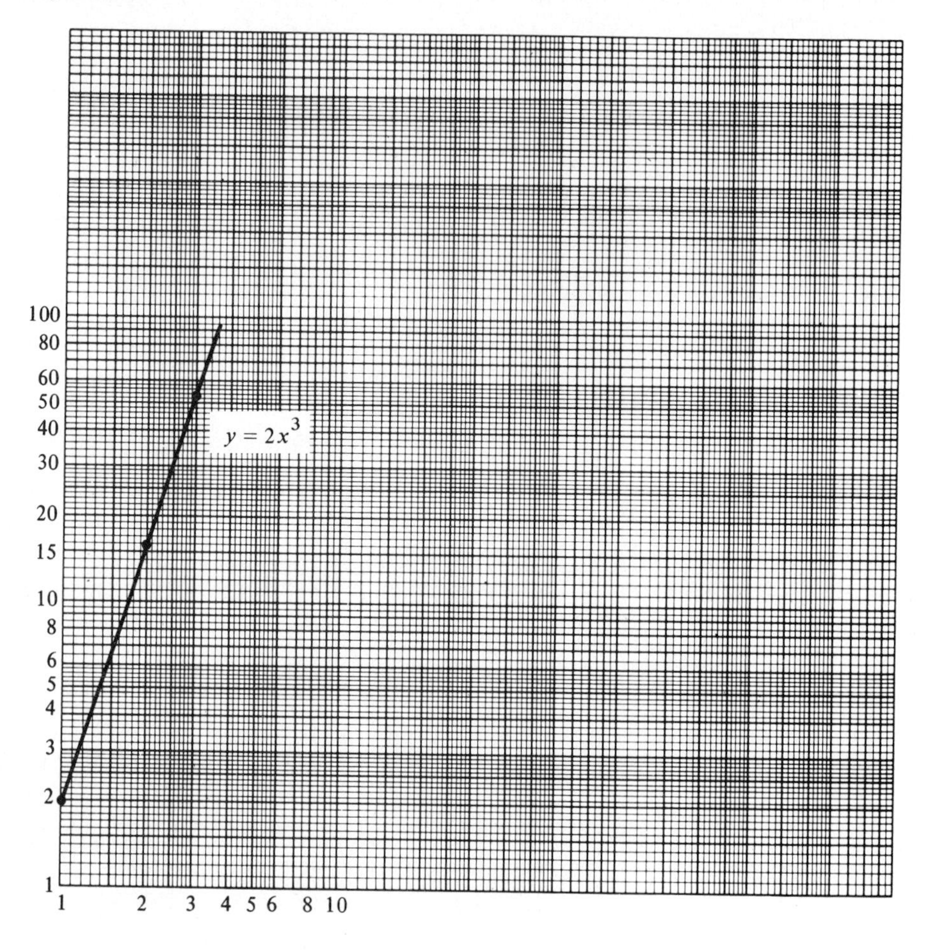

Solution

At first glance it may not be apparent that the equation will graph as a straight line on logarithmic paper. This becomes clear if we take the logarithm of both sides of the equation, to obtain

$$\log (x^3y^4) = \log 10$$

or

$$\log x^3 + \log y^4 = 1$$

and

$$3 \log x + 4 \log y = 1$$

If we let $u = \log x$ and $v = \log y$, the equation takes the form $3u + 4v = 1$, the equation of a straight line. To construct the graph we need only two points. For $x = \frac{1}{2}$, $y = 3$, and for $x = 2$, $y = 1.1$. The graph is shown in Figure 59.7.

Figure 59.7

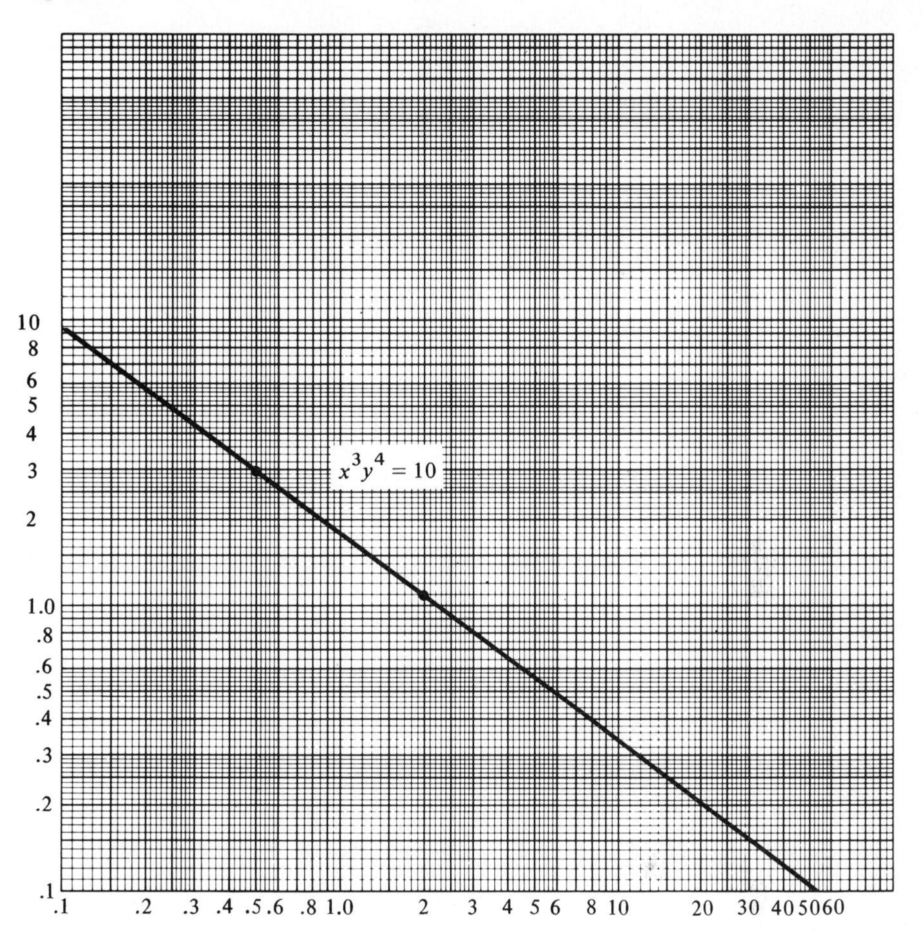

Remark 59.1. *If we plot x against y on semilog paper, we obtain a straight line if the relationship between the variables is exponential. Similarly, if we plot x against y on log–log paper, we obtain a straight line if the relationship between the variables follows a power law.*

Thus, by plotting given data on semilog or log–log paper, it is an easy matter to determine whether the law is exponential or power. If we are given data, logarithmic paper may prove helpful in determining the appropriate law.

Example 59.7

The velocity, v, of an object depends on the time t. The following table gives the velocity (in ft/sec) for corresponding values of t in seconds:

t (sec)	2.5	5	7.5	10	15
v (ft/sec)	3.0	12.5	28.1	50	112.5

Show that the data given obey a power-function law and then determine the equation satisfied by the data.

Solution

We plot the data on logarithmic paper with t along the horizontal axis and v along the vertical axis, as shown in Figure 59.8. Since the graph is a straight line, its equation is of the form $v = Bt^m$, a power function.

Let us determine the values for B and m in the following manner. From the data, $v = 3$ when $t = 2.5$, and $v = 50$ when $t = 10$. Using these values, we generate the following two equations with two unknowns:

$$\left.\begin{aligned} 3 &= B(2.5)^m \\ 50 &= B(10)^m \end{aligned}\right\} \qquad \text{(A)}$$

Figure 59.8

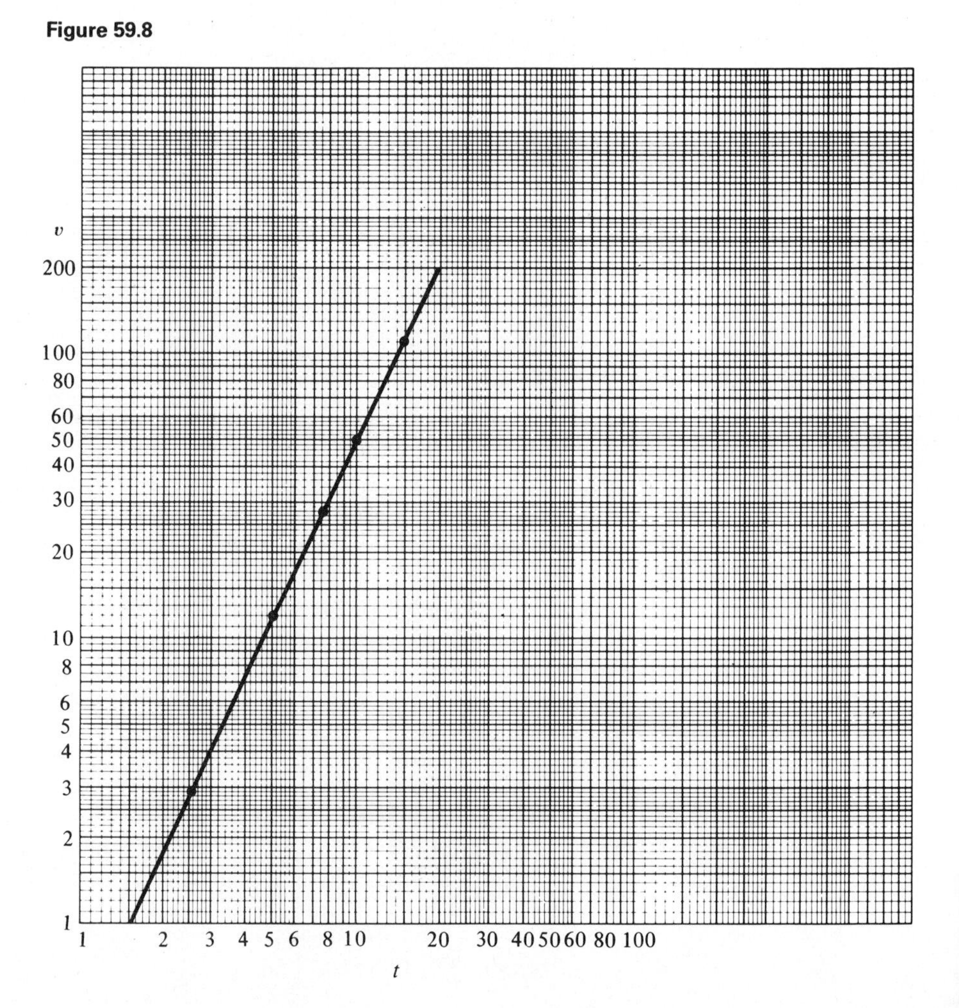

Taking the logarithm of both sides of (A) yields

$$\begin{cases} \log 3 = \log B + \log 2.5^m \\ \log 50 = \log B + \log 10^m \end{cases}$$

or

$$\begin{cases} \log 3 = \log B + m \log 2.5 \\ \log 50 = \log B + m \end{cases}$$

We subtract, to obtain $\log 3 - \log 50 = m \log 2.5 - m$. Factoring, $m(\log 2.5 - 1) = \log 3 - \log 50$, and hence,

$$m = \frac{\log 3 - \log 50}{\log 2.5 - 1}$$

From Table II,

$$m = \frac{0.4771 - 1.6990}{0.3979 - 1} = \frac{-1.2219}{-0.6021} \approx 2$$

We substitute $m = 2$ into either equation in (A) to obtain $B = 0.5$. Thus, the equation that fits the data is $v \approx 0.5t^2$.

Exercises for Section 59

In Exercises 1 through 10, name the kind of paper on which the graph is a straight line.

1. $y = x^2$ **2.** $y = 2e^x$ **3.** $y = 2x^e$

4. $y = e^2$ **5.** $xy^2 = 1$ **6.** $y = e^{-2}$

7. $y = e^{-2x}$ **8.** $x^3y^2 = 1$ **9.** $xy = e$

10. $x + y = e$

In Exercises 11 through 16, graph each function on semilog paper.

11. $y = 2^x$ **12.** $y = 2^{-x}$ **13.** $y = 2e^{3x}$

14. $y = e^{-x}$ **15.** $y = 10^x$ **16.** $y = e^{-2x}$

In Exercises 17 through 22, graph each function on log–log paper.

17. $y = x^4$ **18.** $xy = 1$ **19.** $y = 1/x^2$

20. $x^2y^2 = 10$ **21.** $xy^2 = 10$ **22.** $x^3y = 1$

23. For a certain electric circuit, the voltage is given by $v = e^{-0.2t}$. On semilogarithmic paper, plot v against t for $0 \le t \le 5$.

24. The amount A of an invested sum of money increases with the number of years n, as follows:

n (years)	0	1	5	12	20	30
A (dollars)	500	540	735	1260	2330	5030

Graph the data on semilogarithmic paper and show that they can be expressed as an exponential function.

25. (a) Plot the following data on logarithmic paper:

E	10	50	150	200	500	800
P	600	120	40	30	12	7.5

(b) Why is the relationship between P and E a power function?
(c) Find the equation that relates P and E.

REVIEW QUESTIONS FOR CHAPTER 9

1. What is an exponential function?

2. What is the y intercept for $y = b^x$?

3. For what values of b is $y = b^x$ an increasing function? A decreasing function?

4. What is a logarithm?

5. What is the equivalent logarithm form of $y = b^x$?

6. State the inverse of $y = \log_b x$.

7. For what values of x is the function $y = \log_b x$ defined?

8. What restrictions (if any) are placed on logarithm values?

9. State three laws of logarithms.

10. What is the logarithm of 1 to any base?

11. What are common logarithms?

12. Where is the decimal point placed in writing a number using scientific notation?

13. Can the characteristic of a common logarithm be negative?

14. Can the mantissa, $\log x$, be negative?

15. What is an antilogarithm?

16. What are natural logarithms?

17. What is the exponential form of $y = \ln x$?

18. What is the inverse of $y = \ln x$?

19. Does $\log 1 = \ln 1$?

20. Does $\log 10 = \ln 10$?

21. Distinguish between semilog and log–log paper.

REVIEW EXERCISES FOR CHAPTER 9

1. On the same set of axes, graph $y = 3^x$ and $y = \log_3 x$.

2. Solve each exponential equation for x.

(a) $2^x = 5$ (b) $3^x = \frac{1}{27}$ (c) $4^x = 2$

(d) $e^x = 1$ (e) $16^x = 8$

3. Solve each logarithmic equation for x.

(a) $\log_2 x = 3$ (b) $\log_{16} 8 = x$ (c) $\log_4 2 = x$

(d) $2 \log_x 27 = 6$ (e) $\log_5(3x + 1) = 2$

4. Express $\log_{10}(x^2y^2/\sqrt{z})$ using sums, differences, and multiples.

5. Express $\log_4 32 + \log_4 16 - \log_4 8$ as a single logarithm. What is the value of this logarithm?

6. Solve for x if $\log_3 x = \log_3 4 + \log_3 2$.

7. Solve for x if $2 \log_5 x = 4 \log_5 3$.

8. Solve for x if $\log_{10} x + \log_{10}(x - 1) = \log_{10} 3$.

9. Solve for y in terms of x if $\log_2 x + \log_2 3y + \log_2 5 = 4$.

10. Solve for y in terms of x if $\log_{10} y = \log_{10} 2 - 3 \log_{10} x$.

In Exercises 11 through 16, find the common logarithm of each number. Interpolate where necessary.

11. 234 **12.** 4.15 **13.** 0.156

14. 41750 **15.** 0.0002148 **16.** 0.0135

In Exercises 17 through 22, find the natural logarithm of each number. Interpolate where necessary.

17. 47 **18.** 3.52 **19.** 349

20. 12.71 **21.** 321.5 **22.** 0.0143

In Exercises 23 through 28, find the antilogarithm of each number using Table II. Interpolate where necessary.

23. 2.5866 **24.** 0.7126 **25.** 8.6405 − 10

26. 0.4183 − 1 **27.** −0.4510 **28.** −2.0620

In Exercises 29 through 34, find the antilogarithm of each number using Table III. Interpolate where necessary.

29. 1.0886 **30.** 2.1861 **31.** 4.0037

32. 6.1664 **33.** −1.3985 **34.** −4.3159

In Exercises 35 through 46, solve for x.

35. $\log x = 2.7101$

36. $\log x = -3.3510$

37. $1 + \log x = 0.8727$

38. $\log x = 2.0620$

39. $\ln x = 1.6253$

40. $\ln (2x + 1) = 2.1041$

41. $\ln x = -2.6460$

42. $1 + 2 \ln x = 5.4600$

43. $10^x = 3$

44. $10^{2x} = 6$

45. $e^x = 2.14$

46. $e^{2x} = 4.21$

47. Use common logarithms to evaluate each expression.
(a) $(2.35)^6$ (b) $\dfrac{(4.5)(12.6)^3}{372}$ (c) $\sqrt{\dfrac{42.5}{1.6}}$

48. Use Table IV of Appendix F to find:
(a) log sin 10° (b) log cos 135° (c) log tan 220°
(d) log sin 261°25′ (e) log cos 320°16′

49. Use Table IV of Appendix F to find the *smallest* positive θ if:
(a) $\log \sin \theta = 9.8587 - 10$ (b) $\log \sin \theta = 9.7283 - 10$
(c) $\log \tan \theta = 9.7653 - 10$ (d) $\log \cos \theta = 9.8706 - 10$
(e) $\log \cos \theta = 9.9724 - 10$

50. Solve the following triangle using logarithms: $A = 67°$, $c = 55°$, and $b = 4.3$.

51. In the mathematical analysis of single time-constant circuits, the function $v = 1 - e^{-t/\tau}$ is often encountered. Graph the function by using the following values for t: 0, τ, 2τ, 3τ, 4τ, and 5τ. The graph is called the *unit exponential wave.*

52. The formula for resonant frequency is given by

$$f_r = \frac{1}{2\pi\sqrt{LC}}$$

where f_r is in hertz, L is in henrys, and C in farads. Use logarithms to find f_r if $L = 50\ \mu$H and $C = 1000$ pF. (***Note:*** $\mu = 10^{-6}$ and p $= 10^{-12}$.)

53. A parcel of land appreciates at the rate of 12% per year. If the land has a present value of $25,000, how much will it be worth in 6 years?

54. The sine of the critical angle θ_c for two substances is given by the ratio of their indices of refraction:

$$\sin \theta_c = \frac{n_2}{n_1}$$

Use logarithms to find θ_c if $n_1 = 2.416$ and $n_2 = 1.213$.

CHAPTER 10

Complex Numbers

SECTION 60

Imaginary and Complex Numbers

Thus far, our study of mathematics has dealt only with problems that had solutions in the real number system. There are, however, many instances in which problems in applied mathematics have no solution in this system. In an attempt to arrive at some type of solution, we extend the real number system and in so doing create what is called the **system of complex numbers**. This new number system will enable us to find solutions for problems previously unsolvable. Now, let us investigate what is meant by a complex number and how it comes about.

In Chapter 2, Section 4, we were able to define square roots of positive numbers, since the square of any positive number or negative number is a positive number. Thus, the solution set of the equation $x^2 = 1$ consists of all real numbers x whose square is 1. Since the only real numbers that satisfy this condition are 1 and -1, the solution set consists of these two values.

Now consider the equation $x^2 = -1$. Since the square of any real number is always nonnegative, the equation has no solution in the system of real numbers. We note that another way of stating this is to say that the square root of -1, symbolically $\sqrt{-1}$, is not defined in terms of real numbers. In fact, any even root of a negative number will fail to exist in the real number system.

In order to provide solutions for equations that involve even roots of negative numbers, we will extend the number system. We begin the extension by *imagining* that the $\sqrt{-1}$ does exist. We call $\sqrt{-1}$ the imaginary unit and denote it by the symbol j. The symbol i is also used, but since i represents current in electrical work, we shall use j to avoid confusion.

Note that we need only consider $\sqrt{-1}$, since any negative real number can be written as the product of -1 and a positive real number.

The j operator, denoted simply as j, is the number that when squared is equal to -1; that is,

$$j^2 = -1 \tag{60.1}$$

Equation (60.1) is reasonable in light of what we mean by a solution for $x^2 = -1$. If we take the square root of both sides of the equation, we have $x = \sqrt{-1} = j$. Thus, $x = j$ is a solution and must satisfy the equation. Hence, $j^2 = -1$ and we have (60.1). Since $j^2 = -1$, we express j as $\sqrt{-1}$, and since $j = \sqrt{-1}$ is not a real number, we call it the **imaginary unit**. In fact, any even root of a negative number is called an **imaginary number**.

The word "imaginary" is not a good term for such numbers, because in this context we do not intend the ordinary meaning of the word. "Imaginary" usually refers to something that can only be imagined. However, we shall do much more than simply imagine such numbers. We will, in fact, add, subtract, multiply, divide, and take powers and roots of these numbers, as well as discuss physical applications involving them. In this respect they exist in the same sense that any real number exists.

Now, let us generalize the idea of $\sqrt{-1}$ by showing how we express a square root of any negative number as the product of a real number and j.

If b is any negative real number,

$$\sqrt{b} = j\sqrt{|b|} \tag{60.2}$$

Thus, $\sqrt{-2} = j\sqrt{|-2|} = j\sqrt{2}$. We can think of this result as having been obtained by expressing -2 as the product of -1 and 2. That is, $\sqrt{-2} = \sqrt{(-1)(2)} = \sqrt{-1} \cdot \sqrt{2} = j\sqrt{2}$. Note that numbers of the form bj are called **pure imaginary numbers.**

Example 60.1

The following expressions are examples of pure imaginary numbers.

(a) $\sqrt{-4} = j\sqrt{|-4|} = j\sqrt{4} = 2j$.

(b) $-\sqrt{-25} = -j\sqrt{|-25|} = -j\sqrt{25} = -5j$.

(c) $\sqrt{-\frac{1}{9}} = j\sqrt{\left|-\frac{1}{9}\right|} = j\sqrt{\frac{1}{9}} = \frac{j}{3}$.

(d) $\sqrt{(-4)^3} = j\sqrt{|(-4)^3|} = j\sqrt{64} = 8j$.

(e) $\sqrt{-3^6} = j\sqrt{3^6} = j(3^3) = 27j$. ●

Addition and subtraction with pure imaginary numbers obey the same basic rules that apply to real numbers. Thus, $3j + 5j = 8j$ and $7j - 4j = 3j$. For the operation of multiplication, we must be more careful because the rule which states that $(\sqrt{a})(\sqrt{b}) = \sqrt{ab}$ is *not true* when *both* a and b are negative. If it were true, then $(\sqrt{-1})(\sqrt{-1}) = \sqrt{(-1)(-1)} = \sqrt{1} = 1$. But this result is contrary to equation (60.1). The correct result is obtained by employing the following procedure:

$$(\sqrt{-1})(\sqrt{-1}) = (j)(j) = j^2 = -1$$

In general, we first express square roots of negative numbers in the form bj and then perform the indicated algebraic operations.

Example 60.2

(a) $\sqrt{-4} \cdot \sqrt{9} = j\sqrt{|-4|} \cdot \sqrt{9} = 2j \cdot 3 = 6j$
(b) $\sqrt{-3} \cdot \sqrt{-4} = j\sqrt{|-3|} \cdot j\sqrt{|-4|} = j^2(\sqrt{3})(2) = (-1)(\sqrt{3})2 = -2\sqrt{3}$
(c) $\sqrt{-2} \cdot \sqrt{-3} \cdot \sqrt{-8} = j\sqrt{2} \cdot j\sqrt{3} \cdot j\sqrt{8}$
$= j^3\sqrt{2 \cdot 3 \cdot 8} = j^3\sqrt{48}$
Since $j^3 = j^2 \cdot j = -j$ and $48 = 4\sqrt{3}$, we finally express our result as $-4\sqrt{3}j$. ●

Now let us show that any positive integral power of j can have only one of four possible values: j, -1, $-j$, or 1.

Powers of *j*

(a) $j^1 = j$
(b) $j^2 = -1$
(c) $j^3 = j \cdot j^2 = -j$
(d) $j^4 = j^2 \cdot j^2 = (-1)(-1) = 1$

From this point on we generate only values that have previously occurred. For example, $j^5 = j \cdot j^4 = j \cdot 1 = j$ and $j^6 = j^4 \cdot j^2 = 1 \cdot (-1) = -1$, and so on. By remembering the cycle of powers for j through j^4, it is possible to raise j to any integral power quite easily. We note that the value of j^n, n a positive integer, is found by dividing n by 4 and then raising j to the power equal to the remainder.

Example 60.3

(a) $j^9 = j$, since $9 \div 4$ has a remainder of 1.
(b) $j^{47} = j^3 = -j$, since $47 \div 4$ has a remainder of 3.
(c) $j^{128} = j^0 = 1$, since $128 \div 4$ has a remainder of 0.
(d) $j^{522} = j^2 = -1$, since $522 \div 4$ has a remainder of 2. ●

If we combine the elements of the real number system with pure imaginary numbers, we can defined a new kind of number called a complex number.

A **complex number** is any number that can be written in the form $a + bj$ where a and b are real numbers. The number a is called the **real part** of $a + bj$ and b is called the **imaginary part**. If $a = 0$, we have a number of the form bj, which is a pure imaginary number. If $b = 0$, then $a + bj$ reduces to the form $a + oj = a$, and we have a real number. Thus, the complex numbers include all real numbers and all pure imaginary numbers.

Example 60.4

(a) $2 + 3j$ is a complex number. The real part is 2 and the imaginary part is 3.
(b) $5 - 2j$ is a complex number. The real part is 5 and the imaginary part is -2.
(c) 4 is a complex number, since $4 = 4 + 0j$. The real part is 4 and the imaginary part is 0.
(d) $3j$ is a complex number, since $3j = 0 + 3j$. The real part is 0 and the imaginary part is 3. ●

To ensure as much consistency as possible with the rules of operations with real numbers. we define the equality of complex numbers in the following manner.

We say that $a + bj = c + dj$ if and only if $a = c$ and $b = d$. Equivalently, two complex numbers are **equal** if their real parts are equal and their imaginary parts are equal.

Example 60.5

(a) $3 - 4j = -4j + 3$.
(b) $a + bj = 2 - 7j$ if $a = 2$ and $b = -7$.
(c) $x + yj = -2 + 5j$ if $x = -2$ and $y = 5$.
(d) $2x - 3yj = 4 - 9j$ if $2x = 4$ and $-3y = -9$. Thus, $x = 2$ and $y = 3$.

Example 60.6

What values of x and y satisfy the equation $2 - 3j + x = 2j - jy$?

Solution

We rearrange the equation so that all known terms are on one side of the equation and all terms containing the unknowns x and y are on the other side. This procedure yields

$$2 - 3j - 2j = -x - jy$$

or

$$2 - 5j = -x - yj$$

Using the definition of equality, we have $-x = 2$ and $-y = -5$. Thus, $x = -2$ and $y = 5$.

Example 60.7

What values of x and y satisfy the equation $x + 2(xj - 2y) = -2 + j + 3jy$?

Solution

We rearrange the terms as follows:

$$x + 2xj - 4y - 3jy = -2 + j$$

Factoring j from the two terms on the left and grouping real and imaginary parts yields

$$(x - 4y) + (2x - 3y)j = -2 + j$$

We now use the definition of equality and equate real and imaginary parts. This leads to the following system of equations:

$$x - 4y = -2 \quad \text{and} \quad 2x - 3y = 1$$

Solving the system simultaneously yields $x = 2$ and $y = 1$. ●

The complex numbers $a + bj$ and $a - bj$ are said to be **conjugates** of each other. Thus, the conjugate of $1 + 2j$ is $1 - 2j$, the conjugate of $-2 - 5j$ is $-2 + 5j$, and the conjugate of $3j$ is $-3j$. We note that in general we obtain the conjugate of a complex number by changing the sign of the imaginary part.

Exercises for Section 60

In Exercises 1 through 12, express each number in terms of j.

1. $\sqrt{-4}$ **2.** $\sqrt{-25}$ **3.** $\sqrt{-36}$ **4.** $\sqrt{-225}$

5. $\sqrt{-\frac{1}{4}}$ **6.** $\sqrt{-0.04}$ **7.** $\sqrt{-\frac{1}{100}}$ **8.** $\sqrt{-144}$

9. $\sqrt{-\pi^4}$ **10.** $\sqrt{-8}$ **11.** $\sqrt{-75}$ **12.** $\sqrt{-\frac{9}{16}}$

In Exercises 13 through 24, simplify each expression.

13. j^5 **14.** j^9 **15.** j^{15}

16. j^{16} **17.** j^{51} **18.** $-j^{68}$

19. j^{478} **20.** j^{235} **21.** $j^3 \cdot j^5$

22. $j^2 \cdot j^3 \cdot j^4$ **23.** $-2j^8$ **24.** $j^9 - 2j^{17}$

In Exercises 25 through 36, express each number in the form a + bj.

25. $1 - \sqrt{-4}$ **26.** $-3 + \sqrt{-9}$ **27.** $5 - \sqrt{-49}$

28. $4 - j^3$ **29.** $2j - j^2$ **30.** $\sqrt{16} - \sqrt{-16}$

31. $3j^2 - 2j^3$ **32.** $j^4 - j^2$ **33.** $j^4 + j^2$

34. $j^5 - 4$ **35.** $-j + j^7$ **36.** $j(2 - j^3)$

In Exercises 37 through 48, write the conjugate of each complex number.

37. $-6 + 2j$ **38.** $3 - 2j$ **39.** $3 + j$

40. $2 - \sqrt{-9}$ **41.** 9 **42.** j^2

43. $-j^3$ **44.** $3\sqrt{-27}$ **45.** $-\sqrt{-16}$

46. j^{17} **47.** -3 **48.** 0

In Exercises 49 through 56, solve each equation for x and y.

49. $4x - 3yj = 12 + 6j$ **50.** $2x + yj = -8 - 2j$

51. $2x + 4j = 6 + 2yj$ **52.** $x + (x - 2y)j = 4$

53. $(2x + y) - 3yj = 6j$ **54.** $(x + y) + (x - y)j = 5 + j$

55. $x - 5 + xj - j = yj - y$ **56.** $x + 3xj + jy = 5 - j - 3y$

57. Suppose that two alternating current voltages are given by $(x + 3) - 2j$ and $-2 + (2y + 3)j$. Find x and y so that the voltages are equal.

58. Under what conditions will a complex number equal its conjugate?

SECTION 61

Operations with Complex Numbers; the Quadratic Formula

Algebra operations with complex numbers in the form $a + bj$ (often called the rectangular form) are defined in the same manner as they are for real numbers. For the basic operations of addition, subtraction, multiplication, and division, we have the following rules:

$$\textbf{Addition: } (a + bj) + (c + dj) = (a + c) + (b + d)j \tag{61.1}$$

$$\textbf{Subtraction: } (a + bj) - (c + dj) = (a - c) + (b - d)j \tag{61.2}$$

$$\textbf{Multiplication: } (a + bj)(c + dj) = (ac - bd) + (ad + bc)j \tag{61.3}$$

$$\textbf{Division: } \frac{a + bj}{c + dj} = \frac{a + bj}{c + dj} \cdot \frac{c - dj}{c - dj} = \frac{(ac + bd) + (bc - ad)j}{c^2 + d^2} \tag{61.4}$$

Equations (61.1) and (61.2) state that the sum or difference of two complex numbers is obtained by taking the sum or difference of their real parts and multiplying the sum or difference of their imaginary parts by j.

Example 61.1

Here we perform the indicated operations and express our results in the form $a + bj$.

(a) $(3 + 2j) + (4 + 5j) = (3 + 4) + (2 + 5)j = 7 + 7j$
(b) $(3 - j) + (1 + 3j) = (3 + 1) + (-1 + 3)j = 4 + 2j$
(c) $(6 + 2j) - (2 + 5j) = (6 - 2) + (2 - 5)j = 4 - 3j$
(d) $(2 + 3j) - (-3 + 3j) = [2 - (-3)] + (3 - 3)j = 5 + 0j = 5$ ●

Equation (61.3) is obtained by treating the complex numbers as binomial expressions and then multiplying according to the distributive law. We note that the given complex numbers should be expressed in the form $a + bj$ before performing the indicated operations. We then replace j^2 by -1 whenever it occurs.

Example 61.2

(a) $$\begin{aligned}(2 + 3j)(3 - 5j) &= 2(3 - 5j) + 3j(3 - 5j)\\ &= 6 - 10j + 9j - 15j^2\\ &= 6 - j - 15(-1)\\ &= 6 - j + 15\\ &= 21 - j\end{aligned}$$

(b) $$\begin{aligned}(-1 - 2j)(3 - j) &= -1(3 - j) - 2j(3 - j)\\ &= -3 + j - 6j + 2j^2\\ &= -3 - 5j - 2\\ &= -5 - 5j\end{aligned}$$

(c) $4j(3 - 2j) = 12j - 8j^2 = 8 + 12j$

(d) $$\begin{aligned}(1 + 2j)^2 &= (1 + 2j)(1 + 2j)\\ &= 1(1 + 2j) + 2j(1 + 2j)\\ &= 1 + 2j + 2j + 4j^2\\ &= 1 + 4j - 4\\ &= -3 + 4j\end{aligned}$$

(e) $(2 + 3j)(2 - 3j) = 4 - 9j^2 = 4 + 9 = 13$ ●

Equation (61.4) states that to divide two complex numbers, we multiply both numerator and denominator by the conjugate of the denominator. In following this procedure, the denominator will always be a real number, and thus the quotient can be put in rectangular form.

Example 61.3

For each of the following, we perform the indicated divisions by multiplying the numerator and denominator by the conjugate of the denominator.

(a) $$\frac{3 - 2j}{2 + j} = \frac{3 - 2j}{2 + j} \cdot \frac{2 - j}{2 - j} = \frac{6 - 3j - 4j + 2j^2}{4 - j^2} = \frac{4 - 7j}{5} = \frac{4}{5} - \frac{7}{5}j$$

We note that the real part is $\frac{4}{5}$ and the imaginary part is $-\frac{7}{5}$.

(b) $$\frac{3}{1 - 2j} = \frac{3}{1 - 2j} \cdot \frac{1 + 2j}{1 + 2j} = \frac{3 + 6j}{1 - 4j^2} = \frac{3 + 6j}{5} = \frac{3}{5} + \frac{6}{5}j$$

(c) $$\frac{4}{3j} = \frac{4}{3j} \cdot \frac{-3j}{-3j} = \frac{-12j}{-9j^2} = \frac{-12j}{9} = \frac{-4j}{3}$$

In situations where the denominator is a pure imaginary number, the division can be performed by multiplying numerator and denominator by j.

(d) $$\frac{j^5 + 2j}{1 - j^7} = \frac{j + 2j}{1 - (-j)} = \frac{3j}{1 + j} = \frac{3j}{1 + j} \cdot \frac{1 - j}{1 - j} = \frac{3j - 3j^2}{1 - j^2}$$
$$= \frac{3 + 3j}{2} = \frac{3}{2} + \frac{3}{2}j$$

(e) $$(1 - j)^{-2} = \frac{1}{(1 - j)^2} = \frac{1}{1 - 2j + j^2} = \frac{1}{-2j} = \frac{1}{-2j} \cdot \frac{j}{j} = \frac{-j}{2j^2} = \frac{j}{2}$$

Remark 61.1. *Equations* (61.3) *and* (61.4) *represent the general methods of multiplying and dividing complex numbers. We note that although the equations are always true and will always yield a correct result in actual practice it is easier to follow the methods outlined in Examples 61.2 and 61.3. Thus, there is no need to memorize equations (61.3) and (61.4).*

In Section 15 we noted that the equation $ax^2 + bx + c = 0$ is called a *quadratic equation* in the variable x. The roots of the quadratic equation are always given by the quadratic formula

$$x = \frac{-b \pm \sqrt{b^2 - 4ac}}{2a} \tag{61.5}$$

The quantity $b^2 - 4ac$ under the radical sign is called the **discriminant** because it enables us to discriminate between the different kinds of roots. With our discussion of complex numbers we are now able to consider not only real but also imaginary solutions for quadratic equations.

We note that:

1. If the discriminant $b^2 - 4ac$ is positive and if a, b, and c are real numbers, the quadratic equation has two real and distinct roots.
2. If the discriminant, $b^2 - 4ac$, is zero, the quadratic equation has two real and equal roots.
3. If the disciminant, $b^2 - 4ac$, is negative, the quadratic equation has no real roots or equivalently two imaginary roots. These imaginary roots are conjugate of each other.

The roots of $ax^2 + bx + c = 0$ are the x values of $y = ax^2 + bx + c$ when $y = 0$. With this in mind we can show graphically the three cases for the possible roots of a quadratic equation (see Figure 61.1). Now let us illustrate each of these possibilities by applying (61.5).

Example 61.4

Solve $2x^2 - 5x - 3 = 0$ using the quadratic formula.

Figure 61.1

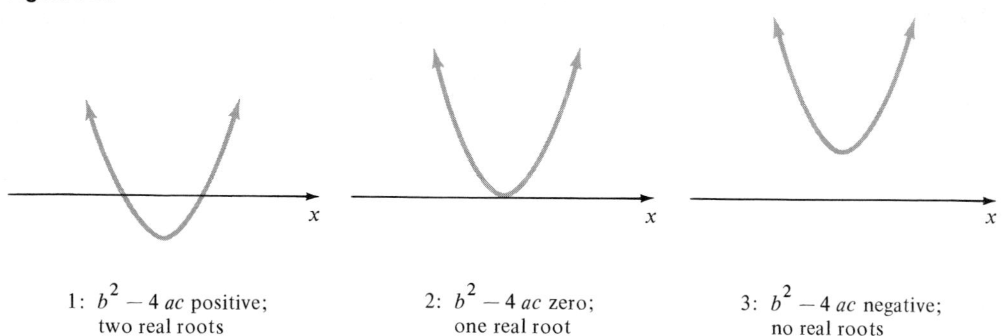

1: $b^2 - 4ac$ positive; two real roots

2: $b^2 - 4ac$ zero; one real root

3: $b^2 - 4ac$ negative; no real roots (two imaginary roots)

Solution

For the given equation $a = 2$, $b = -5$, and $c = -3$. The discriminant $b^2 - 4ac = (-5)^2 - 4(2)(-3) = 25 + 24 = 49$. Thus, the equation has two real and distinct roots. Employing the quadratic formula, we have

$$x = \frac{-b \pm \sqrt{b^2 - 4ac}}{2a} = \frac{-(-5) \pm \sqrt{49}}{2(2)} = \frac{5 \pm 7}{4}$$

Hence,

$$x = \frac{5 + 7}{4} = 3 \quad \text{or} \quad x = \frac{5 - 7}{4} = -\frac{1}{2}$$

Example 61.5

Solve $4x^2 - 4x + 1 = 0$ using the quadratic formula.

Solution

Here $a = 4$, $b = -4$, and $c = 1$. The disciminant $b^2 - 4ac = (-4)^2 - 4(4)(1) = 16 - 16 = 0$. Thus, the equation has two real and equal roots. (We sometimes refer to the equal roots as a root of multiplicity 2.) From the quadratic formula we have

$$x = \frac{-b \pm \sqrt{b^2 - 4ac}}{2a} = \frac{-(-4) \pm \sqrt{0}}{2(4)} = \frac{4}{8} = \frac{1}{2}$$

Example 61.6

Solve $x^2 - 4x + 13 = 0$ using the quadratic formula.

Solution

Here $a = 1$, $b = -4$, and $c = 13$. The discriminant $b^2 - 4ac = (-4)^2 - 4(1)(13) = 16 - 52 = -36$. Thus, the roots of the equation are imaginary numbers. The solutions are given by the quadratic formula, where

$$x = \frac{-b \pm \sqrt{b^2 - 4ac}}{2a} = \frac{-(-4) \pm \sqrt{-36}}{2(1)} = \frac{4 \pm 6j}{2}$$

Thus,

$$x = \frac{4 + 6j}{2} = 2 + 3j \quad \text{or} \quad x = \frac{4 - 6j}{2} = 2 - 3j$$

Note that the roots are complex conjugates of each other.

Exercises for Section 61

In Exercises 1 through 50, perform the operation indicated and express the results in the form a + bj.

1. $2 - 3j + 6j$

2. $(2 + 3j) + (5 - 2j)$

3. $(5 - 3j) + (9 - 6j)$

4. $(1 + j) - (2 - j)$

5. $(-2 + 4j) - (-3 + 3j)$

6. $(2 - 3j) - (3 - 2j)$

7. $(5 - \sqrt{-4}) - (\sqrt{-9} + 3)$

8. $(1 + 2\sqrt{-3}) - (\sqrt{-27} + \sqrt{8})$

9. $2j - (j - 1) - 3$

10. $\sqrt{8} + 2\sqrt{-8} - 3\sqrt{-2} - \sqrt{16}$

11. $2j(1 - j)$

12. $(j)(2j)(3j)$

13. $(3j)^2(-2j)^3$

14. $-2j(3 - j)$

15. $(\sqrt{-4} + 1)(\sqrt{-16})$

16. $(2 - 3j)(1 + 4j)$

17. $(1 - 2j)(2 - j)$

18. $(5 - 3j)(2 + 3j)$

19. $(4 - 2j)(3 + j)$

20. $(2 + 3j)(1 - \sqrt{-1})$

21. $(\sqrt{-4})(\sqrt{-2})(\sqrt{-6})$

22. $(\sqrt{-9})^3$

23. $(2 + 3j)^2$

24. $(2 - j)^2$

25. $(5 - \sqrt{-16})^2$

26. $(\sqrt{9} - \sqrt{-9})^2$

27. $2j^2\sqrt{-8} - \sqrt{-12}$

28. $(1 + j)^2$

29. $(1 + j)^3$

30. $(1 - j)(1 + j)^2$

31. $\dfrac{3}{3 - 2j}$

32. $\dfrac{5}{1 + 2j}$

33. $\dfrac{5}{2 - \sqrt{-1}}$

34. $\dfrac{\sqrt{-9}}{2 + j}$

35. $\dfrac{2 - j}{4 + j}$

36. $\dfrac{3 - 4j}{2 + 3j}$

37. $\dfrac{1 + j}{1 - j}$

38. $\dfrac{2 + j}{2 - j}$

39. $\dfrac{2 + 3j}{3j - 2}$

40. $\dfrac{-3 - 4j}{5j - 1}$

41. $\dfrac{5 + 4j}{2 - 3j}$

42. $\dfrac{j}{j + 1}$

43. $\dfrac{2}{3j}$

44. $\dfrac{5}{2j}$

45. $\dfrac{4}{-5j}$

46. $(1 + j)^{-1}$

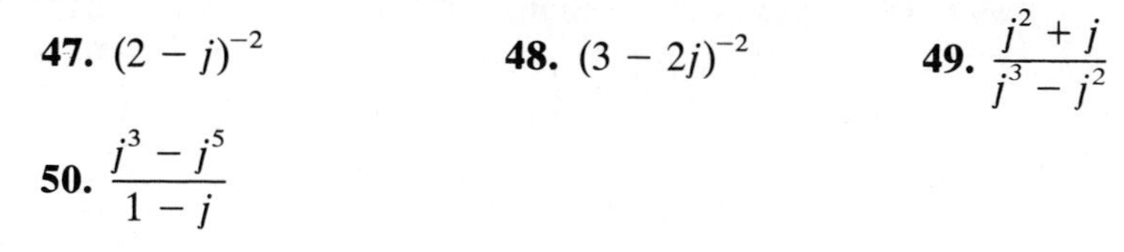

47. $(2 - j)^{-2}$ **48.** $(3 - 2j)^{-2}$ **49.** $\dfrac{j^2 + j}{j^3 - j^2}$

50. $\dfrac{j^3 - j^5}{1 - j}$

In Exercises 51 through 60, find the value of the discriminant of each equation and describe the roots of the equation.

51. $2x^2 + x - 3 = 0$ **52.** $2x^2 + x + 3 = 0$

53. $x^2 - x + \dfrac{1}{4} = 0$ **54.** $4x^2 + 7x - 2 = 0$

55. $9t^2 + 6t - 2 = 0$ **56.** $3x^2 = 6x - 1$

57. $3x^2 = 2$ **58.** $3x^2 - 6x + 4 = 0$

59. $x^2 + 1 = 0$ **60.** $4x^2 + 7x + 3 = 0$

In Exercises 61 through 70, solve for x.

61. $x^2 - 4 = 0$ **62.** $x^2 + 4 = 0$

63. $x^2 - 4x + 4 = 0$ **64.** $x^2 - 3x + 4 = 0$

65. $x^2 + x + 1 = 0$ **66.** $x^2 - 4x + 5 = 0$

67. $x^2 + 2x + 2 = 0$ **68.** $4x^2 + 7x + 3 = 0$

69. $2x^2 + 6x + 17 = 0$ **70.** $3x^2 + 8x + 6 = 0$

71. Find the sum of $a + bj$ and its conjugate.

72. Find the product of $a + bj$ and its conjugate.

73. Under certain conditions the impedance Z (in ohms) in an electrical circuit is given by $Z = E/I$, where E is the voltage and I is the current. If the current is $(6 - 2j)$A and the impedance is $(3 + 2j)\Omega$, find the voltage.

74. For the formula given in Exercise 73, find the current if the voltage is $(4 - 2j)$V and the impedance is $(1 + 2j)\Omega$.

SECTION 62

Graphical Representation of Complex Numbers

It is possible to represent complex numbers in a geometric fashion by the appropriate use of a rectangular coordinate system. In such a coordinate system the x axis is called the **real axis** and the y axis is called the **imaginary**

axis. The reason for the choice of this system and terms is clear if we keep in mind that every complex number of the form $a + bj$ is composed of two parts: the real part a and the imaginary part b. If we now associate with every complex number $a + bj$ a point whose coordinates are (a, b) we can graph the complex number by representing the real parts by x values and the imaginary parts by the y values of the rectangular coordinate system. Since the points in this coordinate system represent complex numbers, the system is called the **complex plane**. Note that any point $(a, 0)$ on the x axis corresponds to the real numbers $a + 0j = a$, while any point $(0, b)$ on the y axis corresponds to the pure imaginary number $0 + bj = bj$ (Figure 62.1).

Figure 62.1

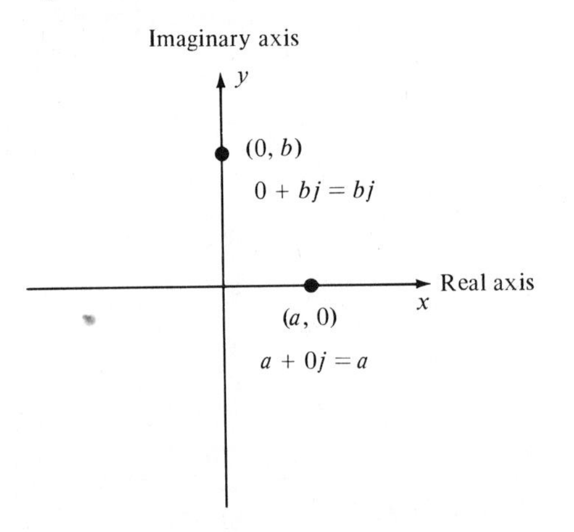

Example 62.1

Plot the points corresponding to the complex numbers $1 + 2j$, $-1 + 3j$, $-2 - j$, and $3 - 2j$.

Solution

We note that the given complex numbers are equivalent to the points $(1, 2)$, $(-1, 3)$, $(-2, -1)$, and $(3, -2)$. These points are shown in Figure 62.2. ●

Complex numbers can be used to represent vectors in the following manner. The x component of the vector is the real part of the complex number and the y component is the imaginary part (the coefficient of j). Thus, the vector $\overrightarrow{OP}$ drawn from the origin to the point $P(a, b)$ has a horizontal component of magnitude a and vertical component of magnitude b (Figure 62.3).

The addition of two complex numbers graphically is equivalent to the addition of vectors by means of the parallelogram method (see Section 44). Thus, to add two complex numbers graphically, we first represent each number as a vector from the origin to a point in the complex plane. We then

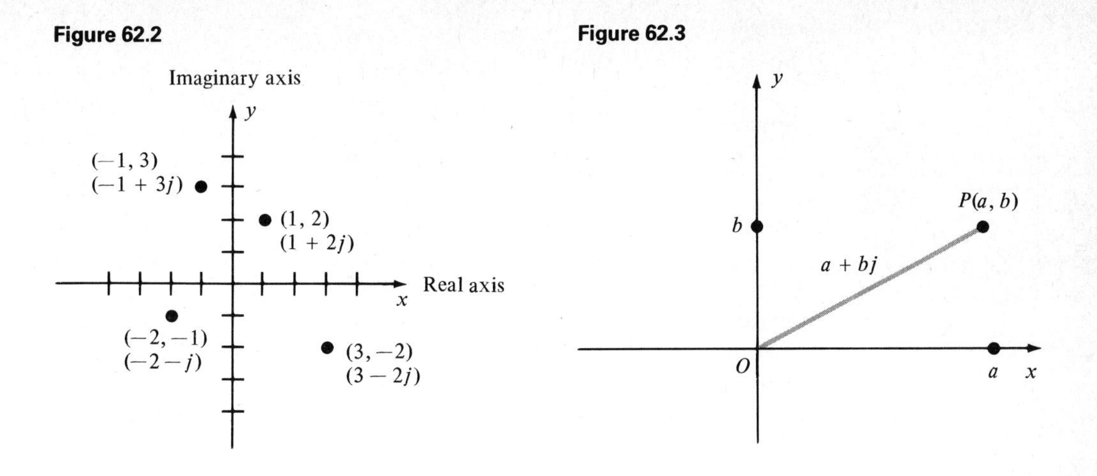

Figure 62.2

Figure 62.3

complete a parallelogram with the two vectors as adjacent sides. The resulting vertex is the point that represents the sum of the two complex numbers.

Example 62.2

Add the complex numbers $2 + j$ and $1 + 3j$ graphically.

Solution

We draw vectors from the origin to the points (2, 1) and (1, 3). Using These vectors as adjacent sides, we construct a parallelogram (Figure 62.4). The fourth vertex of the parallelogram, (3, 4), represents the algebraic sum. Thus, from Figure 62.4 we have $(2 + j) + (1 + 3j) = 3 + 4j$.

Figure 62.4

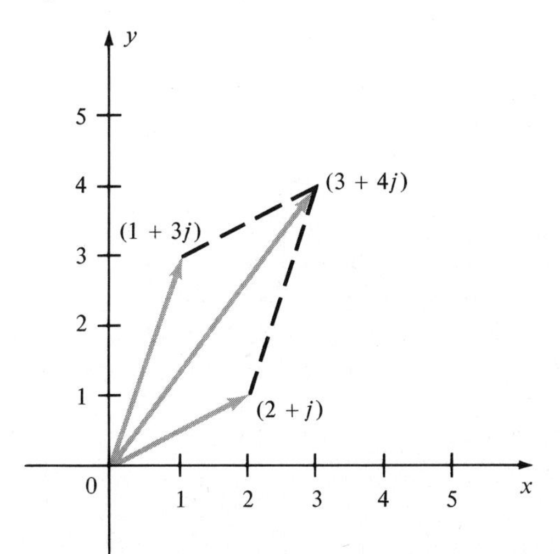

Example 62.3

Subtract $2 + 3j$ from $5 + j$ graphically.

Solution

We note that subtracting $2 + 3j$ from $5 + j$ is equivalent to adding $-2 - 3j$ and $5 + j$. Thus, we graph $-2 - 3j$ and $5 + j$ and then complete the parallelogram, following the procedure of Example 62.2 (See Figure 62.5). Our result is $3 - 2j$.

Example 62.4

Show graphically that the sum of a complex number and its conjugate is a real number.

Solution

Let us choose for our complex number $a + bj$. Then the conjugate is $a - bj$. Graphically, the point (a, b) lies the same distance above the x axis as the point $(a, -b)$ lies below. Thus, the sum of the two vectors representing $a + bj$ and $a - bj$ is given by a point that lies on the x axis (Figure 62.6). Since every point on this axis is a real number, we have shown that the given sum is real.

Remark 62.1 *When the components of a vector are written in the form $a + bj$ of a complex number, the x component becomes the real part of the complex number and the y component becomes the coefficient of j. In such instances the letter j is often placed before its coefficient. Thus, if the x component of a vector is 5 and the y component is -12, we may write $5 - j12$. This form is known as the "j-operator" notation and is often used in work that involves the study of electric circuits.*

Figure 62.5

Figure 62.6

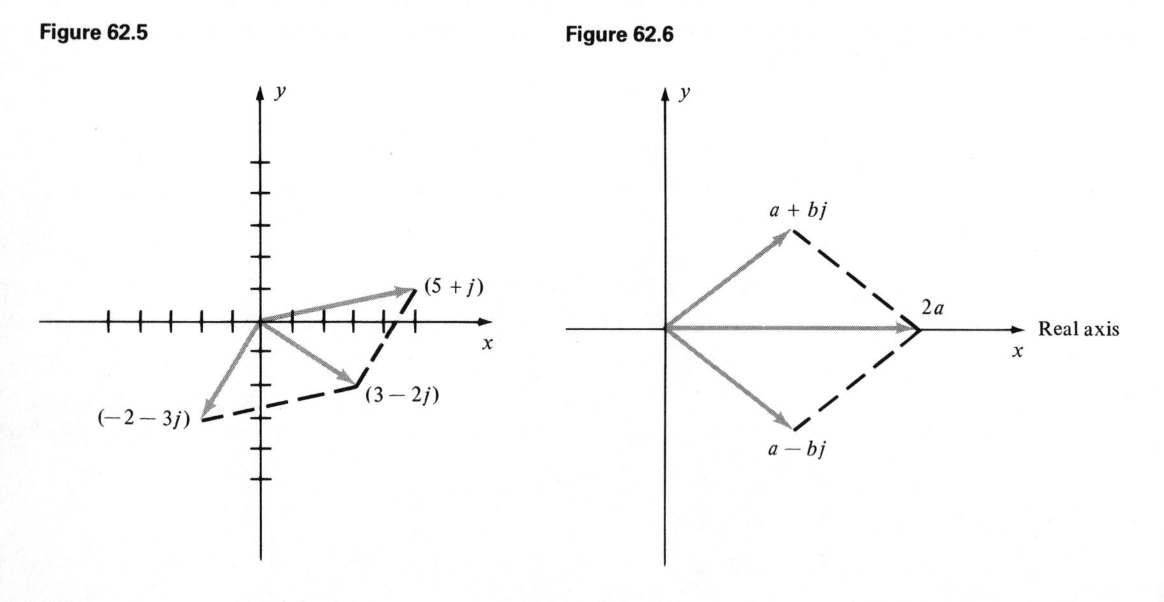

Exercises for Section 62

In Exercises 1 through 10, perform each operation graphically and check the results algebraically.

1. $(1 - j) + (2 + 3j)$

2. $(2 - 3j) + (-1 - 2j)$

3. $(2 - 3j) + (-2 - j)$

4. $(2 + 3j) + 2j$

5. $(4 + j) + (-2 - j)$

6. $(-j - 1) - (-1 + 2j)$

7. $(3 - 2j) - (2 + j)$

8. $(5 + 2j) - (3 + 2j)$

9. $(2 + j) + (j - 1) - 3j$

10. $j - (1 - j) - (1 - 2j)$

In Exercises 11 through 15, plot each number and its conjugate in the complex plane.

11. $2 + 2j$

12. -3

13. $4j$

14. $1 - j$

15. $-2j$

16. Show graphically that the *difference* between a complex number and its conjugate is a pure imaginary number.

SECTION 63

Polar and Exponential Forms of Complex Numbers

We know that the complex number $x + yj$ can be represented as a vector by drawing the vector from the origin to the point $P(x, y)$ in the complex plane (Figure 63.1).

Figure 63.1

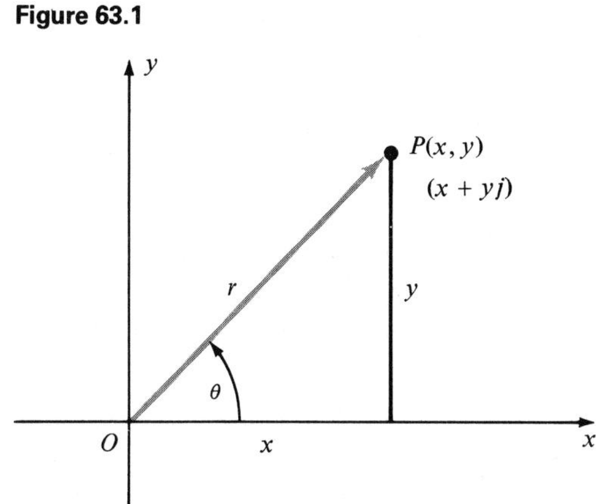

In Figure 63.1 the angle θ that $\overrightarrow{OP}$ makes with the positive real axes is called the **amplitude** or **argument** of $x + yj$. The length r of $\overrightarrow{OP}$ is called the **absolute value** or **modulus** of $x + yj$ and is always a nonnegative number.

In Section 39 we developed the following relationships between x, y, r, and θ.

$$\left.\begin{aligned} \sin\theta &= \frac{y}{r} \\ \cos\theta &= \frac{x}{r} \\ \tan\theta &= \frac{y}{x} \\ r = \overline{OP} &= x^2 + y^2 \end{aligned}\right\} \quad (63.1)$$

We now rewrite these equations in the form that will prove to be most useful in our future work.

$$x = r\cos\theta \qquad y = r\sin\theta \quad (63.2)$$

$$r = x^2 + y^2 \qquad \tan\theta = \frac{y}{x} \quad (63.3)$$

If we substitute equations (63.2) into the rectangular form $x + yj$ of a complex number, we have

$$x + yj = r\cos\theta + j(r\sin\theta)$$

or

$$x + yj = r(\cos\theta + j\sin\theta) \quad (63.4)$$

The expression $r(\cos\theta + j\sin\theta)$ is called the **polar** or **trigonometric form** of a complex number. Let us illustrate how we find the polar form, with the following examples.

Example 63.1

Find the polar form of the complex number $4 + 4j$.

Solution

From the rectangular form of the number we note that $x = 4$ and $y = 4$. We now use equations (63.3) to find $r = \sqrt{4^2 + 4^2} = \sqrt{32} = 4\sqrt{2}$ and $\tan\theta = \frac{4}{4} = 1$. For θ we seek an angle whose tangent is 1. But there are many choices that satisfy this condition. For example, $\theta = 45, 45° + 360°, 45° - 360°$, and so on. In addition, $\theta = 225°, 225° + 260°, 225° - 360°$, and so on. Since $4 + 4j$ lies in the first quadrant, we restrict our choices to $45°, 45° + 360°$, and so on. Customarily, we choose θ such that $0° \leq \theta < 360°$. Thus, we must choose $\theta = 45°$. Hence,

$$4 + 4j = 4\sqrt{2}(\cos 45° + j\sin 45°)$$

We note that the amplitude or argument of $4 + 4j$ is $45°$ and the absolute value or modulus is $4\sqrt{2}$. The graphical representation is shown in Figure 63.2.

Example 63.2

Express $-3\sqrt{3} + 3j$ in trigonometric form.

Solution

We note that $x = -3\sqrt{3}$ and $y = 3$. Thus,

$$r = \sqrt{(-3\sqrt{3})^2 + 3^2} = \sqrt{27 + 9} = 6 \quad \text{and} \quad \tan\theta = \frac{3}{-3\sqrt{3}} = -\frac{1}{\sqrt{3}} = -\frac{\sqrt{3}}{3}$$

Since the point $(-3\sqrt{3}, 3)$ lies in the second quadrant, the appropriate choice for θ is $150°$. Therefore,

$$-3\sqrt{3} + 3j = 6(\cos 150° + j \sin 150°)$$

The graphical representation is shown in Figure 63.3.

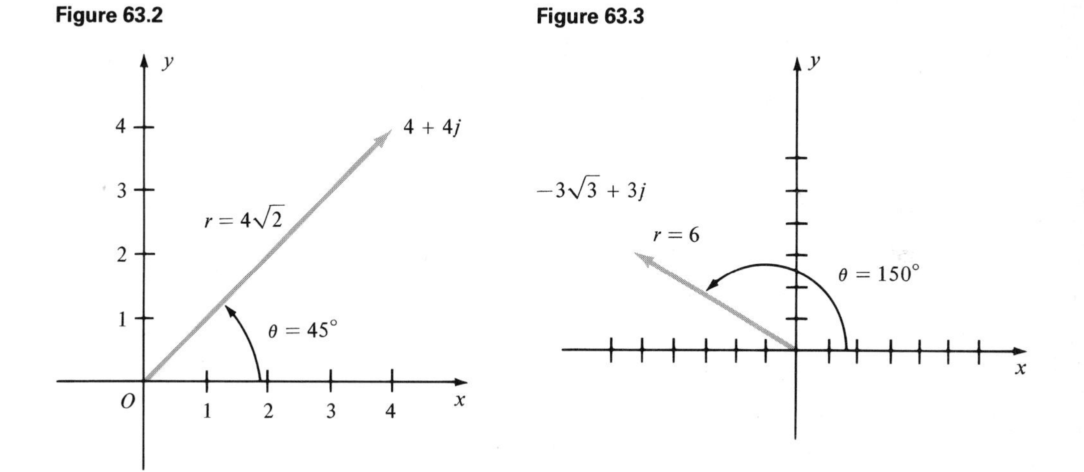

Figure 63.2

Figure 63.3

Example 63.3

Express $-3 - 4j$ in trigonometric form.

Solution

The point $(-3, -4)$ lies in the third quadrant. Since $x = -3$ and $y = -4$,

$$r = \sqrt{(-3)^2 + (-4)^2} = \sqrt{25} = 5 \quad \text{and} \quad \tan\theta = \frac{-4}{-3} = 1.3333$$

We use Table I of Appendix F and the fact that θ lies in the third quadrant to find $\theta = 233°08'$. The graphical representation is shown in Figure 63.4.

Figure 63.4

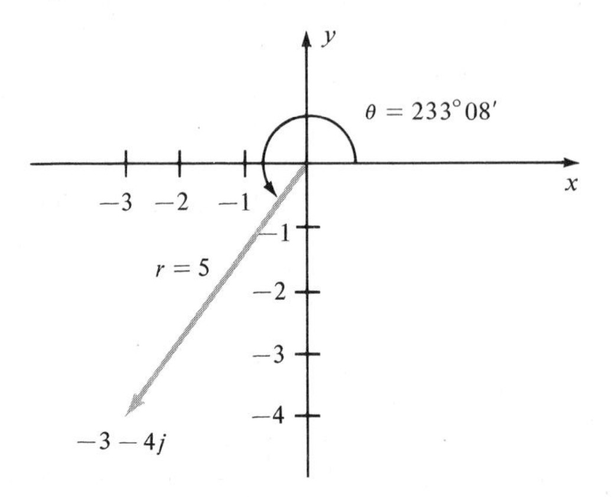

The determination of the polar or trigonometric form of a real number or a pure imaginary number is direct, since these numbers are always represented by points that lie on the coordinate axes. Thus, there are just four possible values for θ: 0°, 90°, 180°, and 270°. Since r is always greater than or equal to zero, we can determine the form we seek by inspection.

Let us show the procedure in the following example.

Example 63.4

Express the complex numbers 5, $3j$, -4, and $-2j$ in polar form.

(a) The point $5 = 5 + 0j$ lies on the positive x axis and is 5 units from the origin. Thus, $\theta = 0°$, $r = 5$, and so $5 = 5(\cos 0° + j \sin 0°)$. See Figure 63.5a.

(b) The point $3j = 0 + 3j$ lies on the positive y axis and is 3 units from the origin. Thus, $\theta = 90°$, $r = 3$, and thus $3j = 3(\cos 90° + j \sin 90°)$. See Figure 63.5b.

(c) The point $-4 = -4 + 0j$ lies on the negative x axis and is 4 units from the origin. Thus, $\theta = 180°$, $r = 4$, and thus $-4 = 4(\cos 180° + j \sin 180°)$. See Figure 63.5c.

(d) The point $-2j = 0 - 2j$ lies on the negative y axis and is 2 units from the origin. Thus, $\theta = 270°$, $r = 2$, and thus $-2j = 2(\cos 270° + j \sin 270°)$. See Figure 63.5d. ●

If we have the polar form of a complex number, we can convert to the rectangular form by using equations (63.2). The procedure is illustrated in the following example.

Figure 63.5

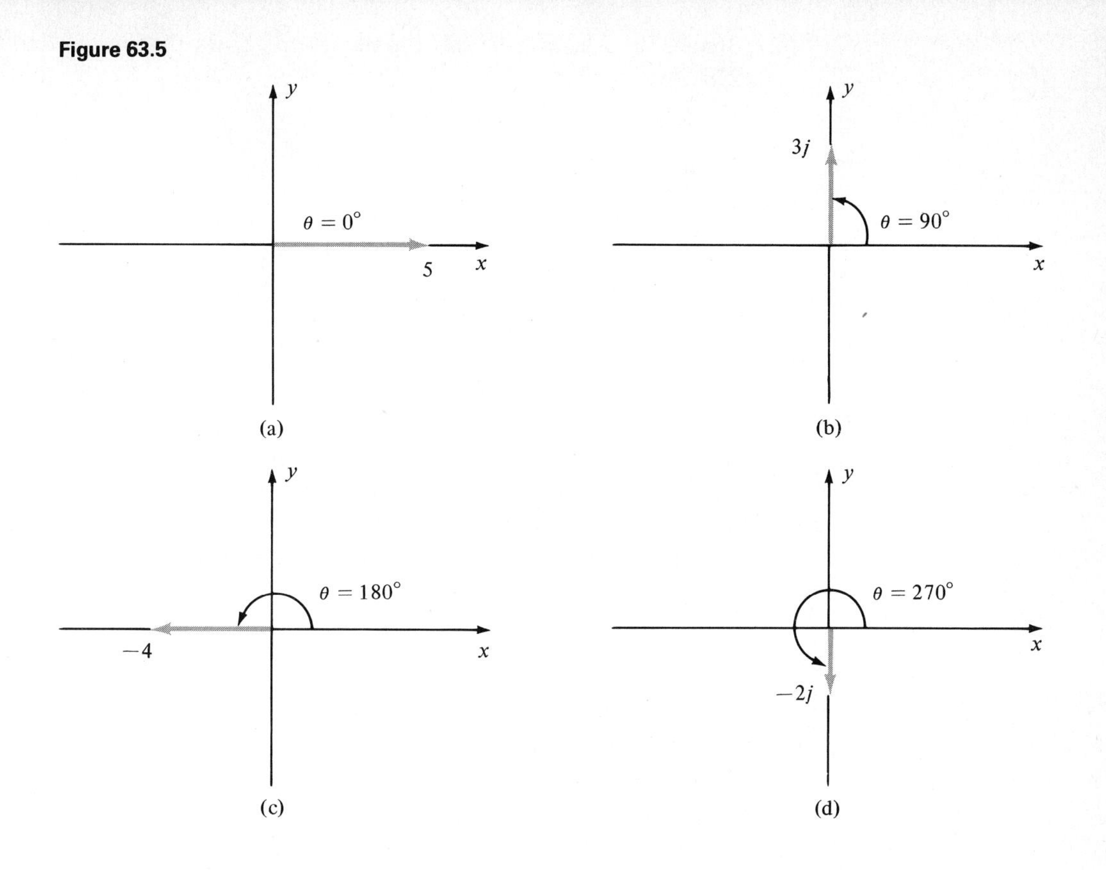

Example 63.5

Express the following complex numbers in rectangular form.

(a) $2(\cos 30° + j \sin 30°)$. From the given polar form, $r = 2$ and $\theta = 30°$. Using equations (63.2), we have

$$x = 2 \cos 30° = 2\left(\frac{\sqrt{3}}{2}\right) = \sqrt{3}$$

$$y = 2 \sin 30° = 2\left(\frac{1}{2}\right) = 1$$

Thus, $2(\cos 30° + j \sin 30°) = \sqrt{3} + j$.

(b) $4(\cos 135° + j \sin 135°)$. From the given polar form, $r = 4$ and $\theta = 135°$. Thus,

$$x = 4 \cos 135° = 4(-\cos 45°) = 4\left(-\frac{\sqrt{2}}{2}\right) = -2\sqrt{2}$$

$$y = 4 \sin 135° = 4 \sin 45° = 4\left(\frac{\sqrt{2}}{2}\right) = 2\sqrt{2}$$

Hence, $4(\cos 135° + j \sin 135°) = -2\sqrt{2} + 2\sqrt{2}j$.

(c) $3(\cos 270° + j \sin 270°)$. The polar form shows that $r = 3$ and $\theta = 270°$. Thus,

$$x = 3 \cos 270° = 3(0) = 0$$

$$y = 3 \sin 270° = 3(-1) = -3$$

Hence, $3(\cos 270° + j \sin 270°) = 0 - 3j = -3j$.

(d) $(\cos 310° + j \sin 310°)$. We note that $r = 1$ and $\theta = 310°$. Thus,

$$x = \cos 310° = \cos (360° - 310°) = \cos 50° = 0.6428$$

$$y = \sin 310° = \sin (360° - 310°) = -\sin 50° = -0.7660$$

Hence, $(\cos 310° + j \sin 310°) = 0.6428 - 0.7660j$. ●

Note that it is not necessary to convert a complex number in polar form to rectangular form in order to locate the number in the complex plane. We need only keep in mind that the absolute value of the number represents the distance the point lies from the origin and that the argument is the angle the vector makes with the positive x axis.

Example 63.6

Graph the number $4(\cos 150° + j \sin 150°)$ in the complex plane.

Solution

In the given form, the absolute value of the number is 4 and the argument is 150°. Hence, the point in the plane is the end point of a vector 4 units long, making an angle of 150° with the positive x axis (Figure 63.6). ●

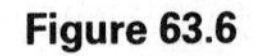

Figure 63.6

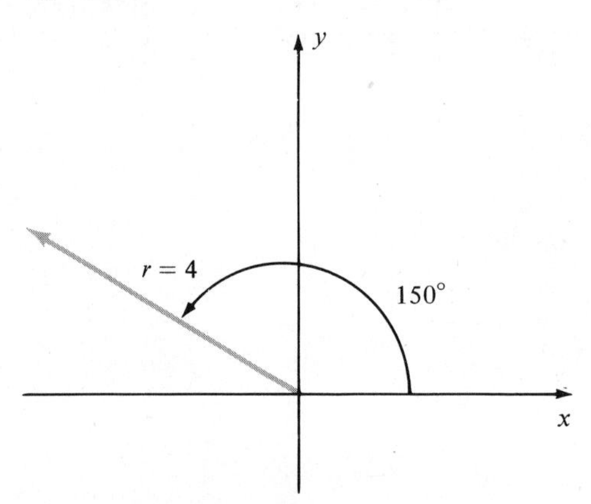

In addition to the polar or trigonometric form, a complex number also has an exponential form. By methods that involve the tools of calculus, it can be shown that

$$re^{j\theta} = r(\cos\theta + j\sin\theta) \tag{63.5}$$

where r is still the modulus of the complex number and θ is the argument in radian measure. We note that a complex number can be described in any one of three forms: rectangular, polar, or exponential. Hence,

$$x + yj = r(\cos\theta + j\sin\theta) = re^{j\theta}$$

Example 63.7

Express $2 + 2j$ in exponential form.

Solution

Since $x = 2$ and $y = 2$, $r = \sqrt{2^2 + 2^2} = \sqrt{8} = 2\sqrt{2}$. For θ we have $\tan\theta = \frac{2}{2} = 1$ or $\theta = 45°$. Thus, the polar form of the given number is $2\sqrt{2}(\cos 45° + j\sin 45°)$. To write the exponential form, we convert 45° to radian measure as follows:

$$45° = 45°\left(\frac{\pi}{180°}\right) = \frac{\pi}{4} = 0.7854$$

So $2 + 2j = 2\sqrt{2}(\cos 45° + j\sin 45°) = 2\sqrt{2}e^{0.7854j}$.

Example 63.8

Express $4e^{0.6109j}$ in polar and rectangular form.

Solution

We first convert 0.6109 rad to 35°. From the given exponential form, we note that $r = 4$. Thus, the polar form of complex numbers is $4(\cos 35° + j\sin 35°)$. Using Table I, we find that $\cos 35° = 0.8192$ and $\sin 35° = 0.5736$. Hence, the rectangular form is $4(0.8192 + 0.5736j) = 3.277 + 2.294j$. We can now write the three equivalent forms as follows:

$$4e^{0.6109j} = 4(\cos 35° + j\sin 35°) = 3.3272 + 2.294j$$

Exercises for Section 63

In Exercises 1 through 16, represent each complex number graphically and find the polar or trigonometric form.

1. $2 + 2j$

2. $2 - 2j$

3. $-\frac{\sqrt{2}}{2} + \frac{\sqrt{2}}{2}j$

4. $-\frac{\sqrt{2}}{2} - \frac{\sqrt{2}}{2}j$

5. $\frac{3\sqrt{3}}{2} + \frac{3}{2}j$

6. $8 - 8\sqrt{3}j$

7. $1 + \sqrt{3}j$

8. $-4 + 4\sqrt{3}j$

9. $8j$

10. -1

11. 27

12. $-16j$

13. $3 - 5j$

14. $4 + 6j$

15. $-2 + 4j$

16. $2 - 4j$

In Exercises 17 through 34, locate each number graphically and express it in rectangular form.

17. $2(\cos 30° + j \sin 30°)$

18. $2(\cos 60° + j \sin 60°)$

19. $3(\cos 0° + j \sin 0°)$

20. $6(\cos 45° + j \sin 45°)$

21. $2(\cos 90° + j \sin 90°)$

22. $2(\cos 120° + j \sin 120°)$

23. $4(\cos 150° + j \sin 150°)$

24. $6(\cos 180° + j \sin 180°)$

25. $2(\cos 130° + j \sin 130°)$

26. $2(\cos 210° + j \sin 210°)$

27. $4(\cos 240° + j \sin 240°)$

28. $2(\cos 225° + j \sin 225°)$

29. $(\cos 270° + j \sin 270°)$

30. $4(\cos 300° + j \sin 300°)$

31. $2(\cos 315° + j \sin 315°)$

32. $2(\cos 330° + j \sin 330°)$

33. $2(\cos 320° + j \sin 320°)$

34. $4(\cos 340° + j \sin 340°)$

35. If the current in a circuit is $(4 - 3j)$A and the impedance is $(5 + 2j)\Omega$, find the absolute value of the voltage. See Exercise 73 in Section 61.

36. If the voltage in a circuit is $(2 - 3j)$V and the impedance is $(3 - j)\Omega$, find the absolute value of the current.

In Exercises 37 through 42, express the given complex number in exponential form.

37. $2(\sin 30° + j \cos 30°)$

38. $4(\cos 90° + j \sin 90°)$

39. $1 + \sqrt{3}j$

40. $2 - 3j$

41. $-3 - 4j$

42. $-4 + 2j$

In Exercises 43 through 50, express each complex number in polar and rectangular form.

43. $3e^{j}$

44. $3e^{-j}$

45. $4e^{1.8j}$

46. $0.2e^{4.25j}$

47. $2e^{7\pi J/6}$

48. $2e^{2j}$

49. $6e^{5j}$

50. $0.2e^{2.50j}$

SECTION 64

Multiplication and Division of Complex Numbers in Polar Form

We have shown that the addition and subtraction of complex numbers in rectangular form is a rather simple operation. We have also seen that complex numbers can be added and subtracted graphically by drawing vectors that represent the numbers and then completing a vector parallelogram.

Multiplication and division, however, can be very tedious when the complex numbers are given in rectangular form. For example, the evaluation of $(3 + 2j)^{10}$ or $(3 - 5j)^3 \div (4 - 2j)^3$ is possible, but extremely lengthy. As we will now show, the operations are comparatively simple when the numbers are expressed in polar form.

Let us begin by investigating the product of two complex numbers in polar form. Suppose that

$$R_1 = r_1(\cos\theta_1 + j\sin\theta_1) \quad \text{and} \quad R_2 = r_2(\cos\theta_2 + j\sin\theta_2)$$

Then

$$\begin{aligned} R_1R_2 &= r_1(\cos\theta_1 + j\sin\theta_1)\cdot r_2(\cos\theta_2 + j\sin\theta_2) \\ &= r_1r_2(\cos\theta_1\cos\theta_2 + j\cos\theta_1\sin\theta_2 + j\sin\theta_1\cos\theta_2 \\ &\qquad + j^2\sin\theta_2\sin\theta_2) \\ &= r_1r_2[\cos\theta_1\cos\theta_2 - \sin\theta_1\sin\theta_2 \\ &\qquad + j(\sin\theta_1\cos\theta_2 + \cos\theta_1\sin\theta_2)] \\ &= r_1r_2[\cos(\theta_1 + \theta_2) + j\sin(\theta_1 + \theta_2)] \end{aligned}$$

Thus,

$$\begin{aligned} &r_1(\cos\theta_1 + j\sin\theta_1)\cdot r_2(\cos\theta_2 + j\sin\theta_2) \\ &\qquad = r_1r_2[\cos(\theta_1 + \theta_2) + j\sin(\theta_1 + \theta_2)] \end{aligned} \tag{64.1}$$

Equation (64.1) is obtained by applying the addition rules for sine and cosine that will be derived in Chapter 14.

We note that equation (64.1) states that the **product** of two complex numbers in polar form is a complex number whose absolute value is the product of the given absolute values and whose amplitude is the sum of the given amplitudes.

Example 64.1

Multiply $2(\cos 50° + j\sin 50°)$ and $3(\cos 85° + j\sin 85°)$ and express the result in rectangular form.

Solution

The product of the absolute values is $2 \cdot 3 = 6$. The sum of the amplitudes is $50° + 85° = 135°$. Thus,

$$2(\cos 50° + j \sin 50°) \cdot 3(\cos 85° + j \sin 85°)$$
$$= 6(\cos 135° + j \sin 135°)$$

Since $\cos 135° = -\cos 45° = -\sqrt{2}/2$ and $135° = \sin 45° = \sqrt{2}/2$, the rectangular form is

$$6\left(-\frac{\sqrt{2}}{2} + j\frac{\sqrt{2}}{2}\right) = -3\sqrt{2} + 3\sqrt{2}j$$

Example 64.2

Find the product of $3(\cos 25° + j \sin 25°)$, $4(\cos 70° + j \sin 70°)$, and $\frac{5}{2}(\cos 85° + j \sin 85°)$.

Solution

We first multiply $3(\cos 25° + j \sin 25°)$ and $4(\cos 70° + j \sin 70°)$ to obtain $12(\cos 95° + j \sin 95°)$. Now

$$12(\cos 95° + j \sin 95°) \cdot \frac{5}{2}(\cos 85° + j \sin 85°)$$
$$= 30(\cos 180° + j \sin 180°) = 30(-1 + 0j) = -30$$

Note that the multiplication of more than two numbers can be accomplished immediately by multiplying the absolute values together and summing the amplitudes involved. Thus,

$$3(\cos 25° + j \sin 25°) \cdot 4(\cos 70° + j \sin 70°) \cdot \frac{5}{2}(\cos 85° + j \sin 85°)$$
$$= 3(4)\left(\frac{5}{2}\right)[\cos (25° + 70° + 85°) + j \sin (25° + 70° + 85°)]$$
$$= 30(\cos 180° + j \sin 180°) = -30$$

Example 64.3

We know that the rectangular form $x + yj$ of any complex number may be written in the polar form $r(\cos \theta + j \sin \theta)$. We also know that $j^0 = 1$ has the polar form $\cos 0° + j \sin 0°$. Similarly,

$$j = \cos 90° + j \sin 90°$$
$$j^2 = -1 = \cos 180° + j \sin 180°$$
$$j^3 = -j = \cos 270° + j \sin 270°$$

Now using our rule for multiplication, we have

$$(x + yj)j^0 = r[\cos (\theta + 0°) + j \sin (\theta + 0°)]$$
$$(x + yj)j = r[\cos (\theta + 90°) + j \sin (\theta + 90°)]$$

$$(x + yj)j^2 = r[\cos(\theta + 180°) + j \sin(\theta + 180°)]$$

$$(x + yj)j^3 = r[\cos(\theta + 270°) + j \sin(\theta + 270°)]$$

Thus, the effect of multiplying a complex number by j to some integral power is to rotate the vector representation of the number through an angle that is a multiple of 90° (Figure 64.1). This is the basis for speaking of j as an operator. ●

Figure 64.1

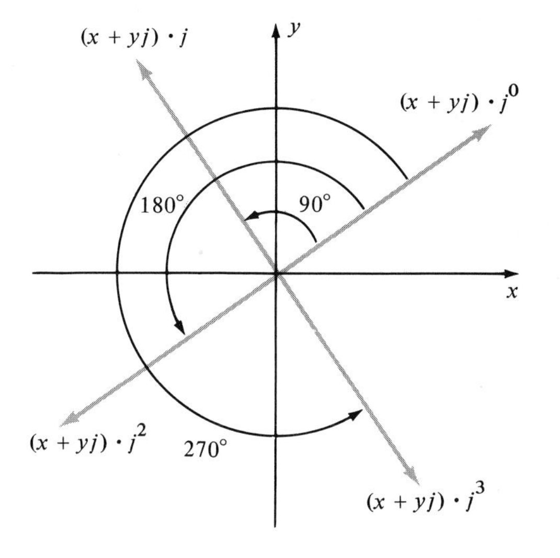

If the complex numbers are in exponential form, the rule for multiplication is still valid. For example, $2e^{1.5j} \cdot 3e^{2.3j} = 2 \cdot 3e^{(1.5j+2.3j)} = 6e^{3.8j}$. In general, $r_1e^{j\theta_1} \cdot r_2e^{j\theta_2} = r_1r_2e^{j(\theta_1+\theta_2)}$.

Now let us investigate the operation of division. As before, suppose that $R_1 = r_1(\cos\theta_1 + j \sin\theta_1)$ and $R_2 = r_2(\cos\theta_2 + j \sin\theta_2)$ and suppose that the quotient R_1/R_2 is the complex number R_3. Then

$$\frac{R_1}{R_2} = R_3$$

or

$$\frac{r_1(\cos\theta_1 + j \sin\theta_1)}{r_2(\cos\theta_2 + j \sin\theta_2)} = r_3(\cos\theta_3 + j \sin\theta_3)$$

Thus,

$$r_1(\cos\theta_1 + j \sin\theta_1) = r_2(\cos\theta_2 + j \sin\theta_2) \cdot r_3(\cos\theta_3 + j \sin\theta_3)$$

and using equation (64.1), we have

$$r_1(\cos\theta_1 + j \sin\theta_1) = r_2 \cdot r_3[\cos(\theta_2 + \theta_3) + j \sin\theta_2 + \theta_3)]$$

Equating absolute values yields $r_1 = r_2 \cdot r_3$, or $r_3 = r_1/r_2$. Similarly, equating amplitudes yields $\theta_1 = \theta_2 + \theta_3$ or $\theta_3 = \theta_1 - \theta_2$. Therefore,

$$\frac{r_1(\cos\theta_1 + j\sin\theta_1)}{r_2(\cos\theta_2 + j\sin\theta_2)} = \frac{r_1}{r_2}[\cos(\theta_1 - \theta_2) + j\sin(\theta_1 - \theta_2)] \qquad (64.2)$$

Equation (64.2) states that the quotient of two complex numbers in polar form is a complex number whose absolute value is the quotient of the absolute values of the given numbers and whose amplitude is the difference of the amplitudes of the given numbers, noting the appropriate order of the division and the subtraction.

Example 64.4

Divide $4(\cos 70° + j\sin 70°)$ by $2(\cos 25° + j\sin 25°)$ and express the result in rectangular form.

Solution

$$\frac{4(\cos 70° + j\sin 70°)}{2(\cos 25° + j\sin 25°} = \frac{4}{2}[\cos(70° - 25°) + j\sin(70° - 25°)]$$

$$= 2(\cos 45° + j\sin 45°)$$

$$= 2\left(\frac{\sqrt{2}}{2} + \frac{\sqrt{2}}{2}j\right)$$

$$= \sqrt{2} + \sqrt{2}j$$

Example 64.5

Divide $3(\cos 20° + j\sin 20°)$ by $12(\cos 80° + j\sin 80°)$ and express the result in rectangular form.

Solution

$$\frac{3(\cos 20° + j\sin 20°)}{12(\cos 80° + j\sin 80°)} = \frac{3}{12}[\cos(20° - 80°) + j\sin(20° - 80°)]$$

$$= \frac{1}{4}[\cos(-60°) + j\sin(-60°)]$$

Now

$$\cos(-60°) = \cos 60° = \frac{1}{2} \quad \text{and} \quad \sin(-60°) = -\sin 60° = -\frac{\sqrt{3}}{2}$$

Thus

$$\frac{1}{4}[\cos(-60°) + j\sin(-60°)] = \frac{1}{4}(\cos 60° - j\sin 60°)$$

$$= \frac{1}{4}\left(\frac{1}{2} - \frac{\sqrt{3}}{2}j\right)$$

$$= \frac{1 - \sqrt{3}j}{8}$$

●

In the following example we encounter a combination of the operations of multiplication and division. As with real numbers, the order in which the operations are performed is not significant.

Example 64.6

$$\frac{2(\cos 40° + j \sin 40°) \cdot 4(\cos 175° + j \sin 175°)}{16(\cos 65° + j \sin 65°)}$$

$$= \frac{2 \cdot 4[\cos (40° + 175°) + j \sin (40° + 175°)]}{16(\cos 65° + j \sin 65°)}$$

$$= \frac{8(\cos 215° + j \sin 215°)}{16(\cos 65° + j \sin 65°)}$$

$$= \frac{8}{16}[\cos (215° - 65°) + j \sin (215° - 65°)]$$

$$= \frac{1}{2}(\cos 150° + j \sin 150°)$$

$$= \frac{1}{2}[-\cos (180° - 150°) + j \sin (180° - 150°)]$$

$$= \frac{1}{2}(-\cos 30° + j \sin 30°) = \frac{1}{2}\left(-\frac{\sqrt{3}}{2} + \frac{1}{2}j\right) = -\frac{\sqrt{3} + j}{4}$$ ●

The division of one complex number by another in exponential form is given by

$$\frac{r_1 e^{j\theta_1}}{r_2 e^{j\theta_2}} = \frac{r_1}{r_2} e^{j(\theta_1 - \theta_2)}$$

For example,

$$4e^{2.5j} \div 2e^{1.7j} = \frac{4}{2}e^{(2.5-1.7)j} = 2e^{0.8j}$$

Exercises for Section 64

In Exercises 1 through 25, perform the operations indicated and express the results in rectangular form.

1. $4(\cos 23° + j \sin 23°) \cdot 2(\cos 37° + j \sin 37°)$

2. $3(\cos 52° + j \sin 52°) \cdot 4(\cos 38° + j \sin 38°)$

3. $5(\cos 72° + j \sin 72°) \cdot 2(\cos 48° + j \sin 48°)$

4. $2(\cos 115° + j \sin 115°) \cdot 3(\cos 65° + j \sin 65°)$

5. $6(\cos 126° + j \sin 126°) \cdot \frac{1}{3}(\cos 99° + j \sin 99°)$

6. $2(\cos 70° + j \sin 70°) \cdot (\cos 40° + j \sin 40°)$

7. $3(\cos 110° + j \sin 110°) \cdot 4(\cos 160° + j \sin 160°)$

8. $(\cos 30° + j \sin 30°) \cdot 2(\cos 45° + j \sin 45°) \cdot (\cos 60° + j \sin 60°)$

9. $2(\cos 75° + j \sin 75°) \cdot 3(\cos 105° + j \sin 105°) \cdot \frac{1}{4}(\cos 220° + j \sin 220°)$

10. $4(\cos 63° + j \sin 63°) \cdot (\cos 84° + j \sin 84°) \cdot 2(\cos 38° + j \sin 38°)$

11. $20(\cos 160° + j \sin 160°) \div 5(\cos 40° + j \sin 40°)$

12. $8(\cos 90° + j \sin 90°) \div 4(\cos 55° + j \sin 55°)$

13. $9(\cos 220° + j \sin 220°) \div 3(\cos 70° + j \sin 70°)$

14. $2(\cos 315° + j \sin 315°) \div (\cos 300° + j \sin 300°)$

15. $16(\cos 215° + j \sin 215°) \div 4(\cos 35° + j \sin 35°)$

16. $4(\cos 340° + j \sin 340°) \div 2(\cos 100° + j \sin 100°)$

17. $2(\cos 220° + j \sin 220°) \div 4(\cos 250° + j \sin 250°)$

18. $\dfrac{(\cos 130° + j \sin 130°) \cdot 4(\cos 250° + j \sin 250°)}{2(\cos 50° + j \sin 50°)}$

19. $\dfrac{3(\cos 75° + j \sin 75°) \cdot 6(\cos 230° + j \sin 230°)}{9(\cos 125° + j \sin 125°)}$

20. $\dfrac{4(\cos 15° + j \sin 15°) \cdot 8(\cos 135° + j \sin 135°)}{6(\cos 90° + j \sin 90°)}$

21. $\dfrac{3(\cos 20° + j \sin 20°) \cdot 4(\cos 140° + j \sin 140°)}{2(\cos 80° + j \sin 80°) \cdot 6(\cos 35° + j \sin 35°)}$

22. $\dfrac{5(\cos 100° + j \sin 100°) \cdot 6(\cos 305° + j \sin 305°)}{3(\cos 25° + j \sin 25°) \cdot 2(\cos 20° + j \sin 20°)}$

23. $\dfrac{5(\cos 100° + j \sin 100°) \cdot 6(\cos 305° + j \sin 305°)}{3(\cos 25° + j \sin 25°) \cdot 2(\cos 200° + j \sin 200°)}$

24. $[2(\cos 20° + j \sin 20°)]^3$

25. $[2(\cos 30° + j \sin 30°)]^4$

In Exercises 26 through 30, convert each number to polar form and perform the operations indicated. Express the result in polar form and check your answer by performing the same operation in rectangular form.

26. $(2 + 2j) \cdot (1 - j)$

27. $(-8\sqrt{3} + 8j) \cdot (1 - \sqrt{3}j)$

28. $\dfrac{1 + 2j}{2 - 3j}$

29. $\dfrac{-3 + 3\sqrt{3}j}{\sqrt{3} + j}$

30. $(-2\sqrt{3} + 2j)^3$

31. If $R_1 = r_1 e^{j\theta_1}$ and $R_2 = r_2 e^{j\theta_2}$, use the properties of exponents to show that $R_1 R_2 = r_1 r_2[\cos(\theta_1 + \theta_2) + j \sin \theta_1 + \theta_2)]$.

SECTION 65

Powers and Roots of Complex Numbers: Demoivre's Theorem

If we wish to raise a complex number to the nth-power, where n is a positive integer, we can obtain the desired result by repeated multiplication. For complex numbers in rectangular form, expansion of powers by repeated multiplication is rather lengthy. Thus, the evaluation of an expression such as $(1 + \sqrt{2}j)^{12}$ would be a tedious and laborious process. Fortunately, this kind of operation is quite direct if the number is expressed in polar form. To show that this is, in fact, the case, let us consider the following situation.

Suppose that we are given a complex number in the form $r(\cos\theta + j\sin\theta)$ and suppose that n is any positive integer. Then the expression $[r(\cos\theta + j\sin\theta)]^n$ can be evaluated by repeated multiplication. That is,

$$(r[\cos\theta + j\sin\theta)]^n = \underbrace{r(\cos\theta + j\sin\theta)\cdot r(\cos\theta + j\sin\theta)\cdot\cdots\cdot r(\cos\theta + j\sin\theta)}_{n \text{ times}}$$

Employing equation (64.1) yields

$$[r(\cos\theta + j\sin\theta)]^n = r^n[\cos \underbrace{\theta + \theta + \cdots + \theta)}_{n \text{ summands}} + j\sin\underbrace{(\theta + \theta + \cdots + \theta)}_{n \text{ summands}}]$$

Thus,

$$[r(\cos\theta + j\sin\theta)]^n = r^n(\cos n\theta + j\sin n\theta) \tag{65.1}$$

Equation (65.1) is known as **DeMoivre's theorem**. We note that although we assumed n to be a positive integer, the theorem is true for *any* integral value of n.

Example 65.1

(a) $$[2(\cos 15° + j \sin 15°)]^6 = 2^6(\cos 6 \cdot 15° + j \sin 6 \cdot 15°) = 64(\cos 90° + j \sin 90°) = 64(0 + j) = 64j$$

(b) $$[3(\cos 20° + j \sin 20°)]^3 = 3^3(\cos 3 \cdot 20° + j \sin 3 \cdot 20°) = 27(\cos 60° + j \sin 60°) = 27\left(\frac{1}{2} + \frac{\sqrt{3}}{2}j\right) = \frac{27 + 27\sqrt{3}j}{2}$$

(c) $$[4(\cos 120° + j \sin 120°)]^{-2} = 4^{-2}(\cos (-2 \cdot 120°) + j \sin (-2 \cdot 120°)] = \frac{1}{4^2}[\cos (-240°) + j \sin (-240°)] = \frac{1}{16}(\cos 240° - j\ 240°)$$

Now

$$\cos 240° = -\cos (180° + 60°) = -\cos 60° = -\frac{1}{2}$$

and

$$\sin 240° = -\sin (180° + 60°) = -\sin 60° = \frac{-\sqrt{3}}{2}$$

Thus,

$$[4(\cos 120° + j \sin 120°)]^{-2} = \frac{1}{16}\left(-\frac{1}{2} + \frac{\sqrt{3}}{2}j\right) = \frac{-1 + \sqrt{3}j}{32}$$

●

If the complex number is in rectangular form and is to be raised to an integral power, we first convert to polar form and then employ DeMoivre's theorem.

Example 65.2

Evaluate $(1 + j)^6$.

Solution

We first convert $1 + j$ to polar form by noting that $r = \sqrt{1^2 + 1^2} = \sqrt{2}$ and $\theta = 45°$. Thus,

$$(1 + j) = \sqrt{2}(\cos 45° + j \sin 45°)$$

and

$$\begin{aligned}(1+j)^6 &= [\sqrt{2}(\cos 45° + j\sin 45°)]^6\\ &= (\sqrt{2})^6[\cos(6\cdot 45°) + j\sin(6\cdot 45°)]\\ &= 8(\cos 270° + j\sin 270°)\\ &= 8(0-j) = -8j\end{aligned}$$

Example 65.3

Use DeMoivre's theorem to evaluate $\dfrac{(2\sqrt{3}+2j)^3}{(1+\sqrt{3}j)^2}$.

Solution

We convert each of the complex numbers to polar form as follows. For $2\sqrt{3}+2j$, $r = \sqrt{(2\sqrt{3})^2+2^2} = 4$, $\tan\theta = 2/2\sqrt{3} = 1/\sqrt{3}$, and $\theta = 30°$. Thus, $2\sqrt{3}+2j = 4(\cos 30° + j\sin 30°)$. For $1+\sqrt{3}j$, $r = \sqrt{1^2+(\sqrt{3})^2} = 2$, $\tan\theta = \sqrt{3}/1$, and $\theta = 60°$. Thus, $1+\sqrt{3}j = 2(\cos 60° + j\sin 60°)$. Now

$$\begin{aligned}\frac{(2\sqrt{3}+2j)^3}{(1+\sqrt{3}j)^2} &= \frac{[4(\cos 30° + j\sin 30°)]^3}{[2(\cos 60° + j\sin 60°)]^2}\\ &= \frac{4^3[\cos(3\cdot 30°) + j\sin(3\cdot 30°)]}{2^2[\cos(2\cdot 60°) + j\sin(2\cdot 60°)]}\\ &= \frac{64(\cos 90° + j\sin 90°)}{4(\cos 120° + j\sin 120°)}\\ &= 16[\cos(90° - 120°) + j\sin(90° - 120°)]\\ &= 16[\cos(-30°) + j\sin(-30°)]\end{aligned}$$

Since $\cos(-30°) = \cos 30° = \sqrt{3}/2$ and $\sin(-30°) = -\sin 30° = -\frac{1}{2}$, $16[\cos(-30°) + j\sin(-30°)] = 16(\sqrt{3}/2 - \frac{1}{2}j)$. Thus,

$$\frac{(2\sqrt{3}+2j)^3}{(1+\sqrt{3}j)^2} = 8\sqrt{3} - 8j$$

●

With DeMoivre's theorem we are able to discuss the problem of determining the roots of complex numbers. We recall that a number r is said to be an nth root of a, a positive integer, if $r^n = a$. Thus, 3 is a fourth root of 81, since $3^4 = 81$ and $2(\cos 45° + j\sin 45°)$ is a square root of $4(\cos 90° + j\sin 90°)$, since $[2(\cos 45° + j\sin 45°)]^2 = 4(\cos 90° + j\sin 90°)$.

In a general sense, suppose that $R(\cos\alpha + j\sin\alpha)$ is an nth root of $r(\cos\theta + j\sin\theta)$. Then $[R(\cos\alpha + j\sin\alpha)]^n = r(\cos\theta + j\sin\theta)$. Now, by DeMoivre's theorem,

$$R^n(\cos n\alpha + j\sin n\alpha) = r(\cos\theta + j\sin\theta) \tag{65.1a}$$

Equation (65.1a) is true if $R^n = r$ and $n\alpha = \theta$. For $R^n = r$, $R = \sqrt[n]{r}$, where $\sqrt[n]{r}$ is the positive real nth root of r.

With regard to the relationship between the angles, we must recall that the argument of a complex number is not unique. Since coterminal angles are equal, we note that we may add or subtract any multiple of 360° to θ without changing the value of the amplitude.

Thus, $r(\cos\theta + j\sin\theta) = r[\cos(\theta + k\cdot 360°) + j\sin(\theta + k\cdot 360°)]$ for any integer k. Therefore,

$$n\alpha = \theta + k\cdot 360° \quad \text{or} \quad \alpha = \frac{\theta}{n} + k\cdot\frac{360°}{n}$$

For $k = 0$, $\alpha = \theta/n$ and one nth root of $r(\cos\theta + j\sin\theta)$ is $r^{1/n}[\cos(\theta/n) + j\sin(\theta/n)]$. The remaining roots are obtained by letting k take on the values $1, 2, 3, \ldots, n-1$. It is easy to show that for $k > n - 1$, the roots repeat. In general, the nth roots of a complex number in polar form are given by

$$[r(\cos\theta + j\sin\theta)]^{1/n} = r^{1/n}\left[\cos\left(\frac{\theta}{n} + k\cdot\frac{360°}{n}\right) + j\sin\left(\frac{\theta}{n} + k\cdot\frac{360°}{n}\right)\right] \tag{65.2}$$

Example 65.4

Find the four fourth roots of -16.

Solution

We first express -16 in polar form by noting that $r = 16$ and $\theta = 180°$. Thus, we write $-16 = 16(\cos 180° + j\sin 180°)$. Using equation (65.2),

$$[16(\cos 180° + j\sin 180°)]^{1/4} = 16^{1/4}\left[\cos\left(\frac{180°}{4} + k\cdot\frac{360°}{4}\right) + j\sin\left(\frac{180°}{4} + k\cdot\frac{360°}{4}\right)\right]$$

where $k = 0, 1, 2, 3$

Let us denote the four roots by r_1, r_2, r_3, and r_4.

$$r_1 = 2(\cos 45° + j\sin 45°) = \sqrt{2} + j\sqrt{2}$$
$$r_2 = 2(\cos 135° + j\sin 135°) = -\sqrt{2} + j\sqrt{2}$$
$$r_3 = 2(\cos 225° + j\sin 225°) = -\sqrt{2} - j\sqrt{2}$$
$$r_4 = 2(\cos 315° + j\sin 315°) = \sqrt{2} - j\sqrt{2}$$

If we were to continue in this manner, we would obtain no new roots, only a repetition of those already found. In Figure 65.1 we note that the roots lie on a circle with center at the origin and radius equal to 2. The roots form a square inscribed in a circle. We further note that this problem is equivalent to solving the equation $x^4 + 16 = 0$.

Figure 65.1

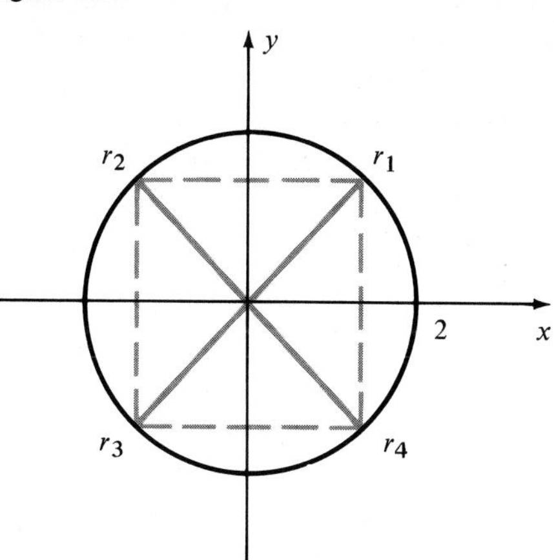

Example 65.5

Find the three cube roots of $-4 + 4\sqrt{3}j$.

Solution

We note that $r = \sqrt{(-4)^2 + (4\sqrt{3})^2} = \sqrt{64} = 8$ and $\tan\theta = 4\sqrt{3}/-4 = -\sqrt{3}$. Since the point $(-4, 4\sqrt{3})$ lies in the second quadrant, $\theta = 120°$. Thus, the polar form of $-4 + 4\sqrt{3}j$ is $8(\cos 120° + j\sin 120°)$. Using equation (65.2),

$$[8(\cos 120° + j\sin 120°)]^{1/3} = 8^{1/3}\left[\cos\left(\frac{120°}{3} + k\cdot\frac{360°}{3}\right) + j\sin\left(\frac{120°}{3} + k\cdot\frac{360°}{3}\right)\right] \quad k = 0, 1, 2$$

If we denote the three roots as r_1, r_2, and r_3, then

$$r_1 = 2(\cos 40° + j\sin 40°)$$

$$r_2 = 2(\cos 160° + j\sin 160°)$$

$$r_3 = 2(\cos 280° + j\sin 280°)$$

Note that the three roots represent the solutions of the equations $x^3 = -4 + 4\sqrt{3}j$. The graph of the roots is shown in Figure 65.2 on page 384.

Remark 65.1. *In general, if $r(\cos\theta + j\sin\theta)$ is any complex number not equal to zero and n is any positive integer, there exists exactly n different complex numbers each of which is an nth root of $r(\cos\theta + j\sin\theta)$. These nth roots lie on a circle centered at the origin of the complex plane and are uniformly spaced about the circumference of the circle and are separated by a central angle of $360°/n$. The circle has a radius equal to $\sqrt[n]{r}$.*

Figure 65.2

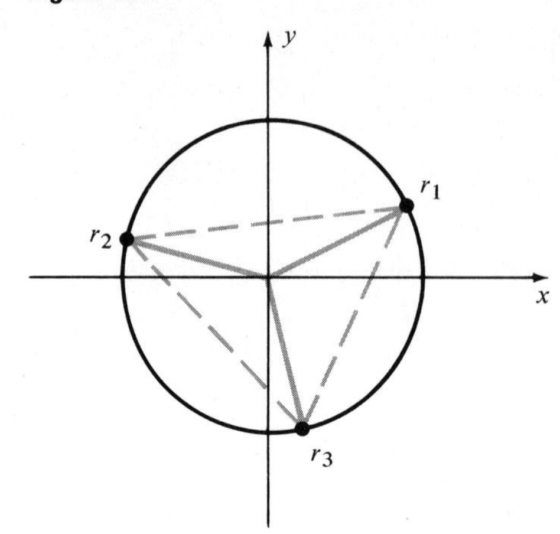

Exercises for Section 65

In Exercises 1 through 20, find the powers indicated and express each result in rectangular form.

1. $[2(\cos 10° + j \sin 10°)]^6$

2. $\left[\frac{1}{2}(\cos 15° + j \sin 15°)\right]^{10}$

3. $[2(\cos 60° + j \sin 60°)]^5$

4. $(\cos 30° + j \sin 30°)^{-9}$

5. $[\cos 25° + j \sin 25°]^6$

6. $[2(\cos 72° + j \sin 72°)]^5$

7. $[2(\cos 216° + j \sin 216°]^5$

8. $[2(\cos 210° + j \sin 210°)]^4$

9. $[\sqrt[3]{2}(\cos 160° + j \sin 160°)]^3$

10. $[2(\cos 45° + j \sin 45°)]^{-2}$

11. $(1 + j)^5$

12. $(\sqrt{3} - j)^4$

13. $\left(-\frac{1}{2} - j\frac{\sqrt{3}}{2}\right)^3$

14. $\left(-\frac{1}{2} + j\frac{\sqrt{3}}{2}\right)^3$

15. $\left(\frac{1}{2} + \frac{\sqrt{3}}{2}j\right)^3$

16. $(-2\sqrt{3} + 2j)^3$

17. $(\sqrt{3} + j)^{10}$

18. $\dfrac{(\sqrt{3} + j)^{10}}{(1 + \sqrt{3}j)^2}$

19. $\dfrac{(1 + j)^6}{(-1 + j\sqrt{3})^2}$

20. $\dfrac{(-\sqrt{3} - j)^3}{(1 + j)^6}$

In Exercises 21 through 30, find the roots indicated, in polar form. If the angle is such that a table is not needed, express the result in rectangular form.

21. The cube roots of 1.

22. The cube roots of $-27j$.

23. The fourth roots of -16.

24. The square roots of j.

25. The square roots of $-j$.

26. The cube roots of -8.

27. The square roots of j^5.

28. The cube roots of $-4 + 4\sqrt{3}j$.

29. The fourth roots of $-\frac{1}{2} + \frac{\sqrt{3}}{2}j$.

30. The square roots of $-\frac{1}{2} + \frac{\sqrt{3}}{2}j$.

In Exercises 31 through 38, solve each equation.

31. $x^3 - 8 = 0$

32. $x^4 - 16 = 0$

33. $x^6 - 1 = 0$

34. $x^2 - 1 - \sqrt{3}j = 0$

35. $x^5 + \frac{\sqrt{2}}{2} + \frac{\sqrt{2}}{2}j = 0$

36. $jx^3 - 1 = 0$

37. $x^3 + j^2 = 0$

38. $x^4 + 5x^2 + 4 = 0$

REVIEW QUESTIONS FOR CHAPTER 10

1. What is the imaginary unit?

2. What are pure imaginary numbers?

3. What are the four possible values for any integral power of j?

4. What is a complex number?

5. When are two complex numbers equal?

6. What is the conjugate of $a + bj$?

7. How do we add or subtract complex numbers?

8. How do we divide complex numbers?

9. State the quadratic formula.

10. What is the discriminant of the quadratic formula?

11. If the discriminant of the quadratic equation is positive, what can be said about the roots?

12. If the discriminant of the quadratic equation is negative, what can be said about the roots?

13. If the discriminant of the quadratic equation is zero, what can be said about the roots?

14. What is the amplitude or argument of a complex number?

15. What is the absolute value or modulus of a complex number?

16. State the polar or trigonometric form of the complex number $x + yj$.

17. How do we multiply two complex numbers in polar form?

18. How do we divide two complex numbers in polar form?

19. State DeMoivre's theorem.

20. If $R(\cos\alpha + j\sin\alpha)$ is an nth root of $r(\cos\theta + j\sin\theta)$, what can we say about $[R(\cos\alpha + j\sin\alpha)]^n$?

REVIEW EXERCISES FOR CHAPTER 10

Exercises 1 through 14, perform the operations indicated and express the result in the form $a + bj$.

1. $(3 + 5j) + (2 - 3j)$

2. $4 - 3j) - (8 - 2j)$

3. $\left(\frac{1}{2} - \frac{2j}{3}\right) - (1 - j)$

4. $(3 + 5j)(7 - 2j)$

5. $(-2 + 3j)(3 + 5j)$

6. $j^5 \cdot j^8$

7. $j^7 - (-j)^6$

8. $(2 + 3j)^3$

9. $\frac{2 + 3j}{1 - j} \cdot \frac{2 + j}{6 + 5j}$

10. $(2 - \sqrt{-6})(1 + \sqrt{-8})$

11. $(2 - 3j)^{-2}$

12. $j(j^2 - j^5)$

13. $j(j^2 - j^5)$

14. $\frac{j}{j - 1}$

15. Perform the indicated operations graphically then check the results algebraically.
(a) $(1 - j) + (1 + 5j)$
(b) $(2 - 3j) - (2 - j)$
(c) $(2 + 3j) + (1 + 2j) - j$

In Exercises 16 through 20, solve for x and y.

16. $2x - 3yj = 8 + 9j$

17. $x + (x + 2y)j = 3 + 5j$

18. $(x - y) + (2x - 3y)j = 1$

19. $3x + (y - 2)j = 6 - 5j$

20. $x + 3j + jy = 5 - 2yj$

In Exercises 21 through 25, solve for x.

21. $x^2 + 36 = 0$

22. $2x^2 - 2x + 1 = 0$

23. $x^2 - 4x + 8 = 0$

24. $4x^2 + 8x + 5 = 0$

25. $3x^2 + 4x + 8 = 0$

In Exercises 26 through 31, represent each complex number in polar form.

26. $-10j$

27. $-6 - 6j$

28. $-1 + \sqrt{3}j$

29. $-4 - 4\sqrt{3}j$

30. $\frac{\sqrt{2}}{2} - \frac{\sqrt{2}}{2}j$

31. $-2 + \sqrt{5}j$

In Exercises 32 through 37, perform the operations indicated and express the results in rectangular form.

32. $2(\cos 17° + j \sin 17°) \cdot (\cos 43° + j \sin 43°)$

33. $2(\cos 135° + j \sin 135°) \cdot 4(\cos 90° + j \sin 90°)$

34. $(\cos 310° + j \sin 310°) \div (\cos 70° + j \sin 70°)$

35. $\dfrac{2(\cos 40° + j \sin 40°) \cdot (\cos 165° + j \sin 165°)}{6(\cos 55° + j \sin 55°)}$

36. $2(\cos 25° + j \sin 25°) \div 4(\cos 55° + j \sin 55°)$

37. $(\cos 13° + j \sin 13°) \cdot (\cos 128° + j \sin 128°) \cdot 2(\cos 39° + j \sin 39°)$

In Exercises 38 through 48, use De Moivre's theorem to find each expression.

38. $(1 - \sqrt{3}j)^5$

39. $\left(\frac{1}{2} - \frac{\sqrt{3}}{2}j\right)^4$

40. $(-2 + 2j)^3$

41. $\dfrac{2}{(1 + j)^4}$

42. $(\sqrt{3} - j)^5$

43. $(\cos 48° + j \sin 48°)^5$

44. $[3(\cos 75° + j \sin 75°)]^3$

45. $2(\cos 30° + j \sin 30°)^4$

46. $(\cos 25° + j \sin 25°)^6$

47. $2(\cos 110° + j \sin 110°)^{-2}$

48. $(\cos 65° + j \sin 65°)^{-3}$

49. Find the cube roots of 64.

50. Find the fourth roots of $16(\cos 120° + j \sin 120°)$.

51. Find x if $x^6 + 64 = 0$.

52. Find x if $x^3 = -\sqrt{2} + j\sqrt{2}$.

53. The impedance Z in an ac circuit is given

$$Z = R + j\left(\omega L - \frac{1}{\omega c}\right)$$

where R is the resistance, L is the inductance, C is the capacitance, and $\omega = 2\pi f$. The quantity f represents frequency.

(a) Find the impedance if $R = 90\ \Omega, L = 0.5$ H, $f = 60$ Hz, and $C = 30$ F.

(b) Find the impedance if $R = 100\ \Omega$, $L = 0.5$ H, $f = 60$ Hz and $C = 40$ F.

(c) Find the impedance if $R = 1000\ \Omega$, $L = 0.3$ H, $f = 60$ Hz, and $C = 2 \times 10^{-6}$ F.

54. (a) Show that the imaginary part of the impedance Z given in Exercise 53 is zero when

$$f = \frac{1}{2\pi\sqrt{LC}}$$

(b) Find f is $L = 0.5$ H and $C = 2.5 \times 10^{-6}$ F. This value for f is called the **resonance frequency**.

CHAPTER 11

Ratio, Proportion, and Variation

SECTION 66
Ratio and Proportion

In mathematics we are often concerned with comparing various quantities. The comparison of similar quantities can be done in different ways. For example, when we say that 10 in. is 4 in. more than 6 in., we are comparing the two quantities by subtraction. In fact, whenever we state that one quantity is greater than or less than another, we are comparing by subtraction. We note that the difference between the quantities has the same units as the original quantities.

A more useful kind of comparison is obtained by finding the quotient of the two quantities provided that they have the same units. This comparison by division is called the *ratio* between the two quantities.

Example 66.1

(a) Since \$12 − \$8 = \$4, we say that \$12 is \$4 more than \$8.
(b) Since 2 ft − 8 in. = 16 in., we say that 2 ft is 16 in. more than 8 in.
(c) The length of a sheet of graph paper is 12 in. and the width is 8 in. The ratio of the length to the width is $\frac{12}{8}$ or $\frac{3}{2}$.
(d) The ratio of 18 V to 24 V is $\frac{18}{24}$ or $\frac{3}{4}$. ●

We note that when comparing two quantities by division the result is a pure number. That is, the answer has no units and is said to be **dimensionless.**

A statement of equality between two ratios is called a **proportion**. Thus, 3 ft/2 ft = \$24/\$16 is a proportion and is read "3 ft is to 2 ft as \$24 is to \$16." In general, we denote a proportion as

$$a:b = c:d \qquad \text{which is read "}a\text{ is to }b\text{ as }c\text{ is to }d\text{"}$$

The four quantities *a*, *b*, *c*, and *d* are called the *terms* of the proportion. The first and fourth terms are called the **extremes** and the second and third terms are called the **means** of the proportion.

If we write the proportion $a:b = c:d$ in fractional form, we have $a/b = c/d$. If we multiply both sides of the equation by bd, we obtain $ad = bc$. We observe that the term ad is the product of the extremes, and the term bc is the product of the means. Thus, the product of the means equals the product of the extremes.

If we divide both sides of the equation $ad = bc$ by ac, we have $d/c = b/a$. If we divide by cd, then $a/c = b/d$. We formulate these principles with the following list.

For the Proportion $a/b = c/d$

1. The product of the means is equal to the product of the extremes. Thus, $ad = bc$.
2. The terms are proportional if we invert both sides. Thus, $b/a = d/c$.
3. The means and the extremes of the proportion may be interchanged. Thus,

$$\frac{a}{c} = \frac{b}{d} \quad \text{and} \quad \frac{d}{b} = \frac{c}{a}$$

Example 66.2

Find the missing term in the following proportions.
(a) $6:x = 4:12$ (b) $2:7 = 3:x$

Solution

In (a) the product of the means is $4x$ and the product of the extremes is 72. Thus, we have the equation $4x = 72$ or $x = 18$. For (b) the product of the means is 21 and the product of the extremes is $2x$. Thus, $2x = 21$ and $x = \frac{21}{2}$. ●

If the means of a proportion are equal, as in the proportion $2:4 = 4:8$, the expression is called a **mean proportion**. The number 4 is said to be the **mean proportional** between 2 and 8. In general, the proportion $a:b = b:c$ is a mean proportion and b is the mean proportional between a and c. We can solve for b by noting that the product of the means is b^2 and the product of the extremes is ac. Thus, $b^2 = ac$ or $b = \pm\sqrt{ac}$.

Example 66.3

Find the mean proportional between 3 and 27.

Solution

We first set up the proportion $3:x = x:27$. Then $x^2 = 81$ and $x = \pm\sqrt{81} = \pm 9$. Hence, the mean proportional is either 9 or -9. We can check these results by observing that $3:9 = 9:27$ is a true proportion, since $\frac{3}{9} = \frac{9}{27} = \frac{1}{3}$. For -9, we note that $3:-9 = -9:27$ is also true, since $-\frac{3}{9} = -\frac{9}{27} = -\frac{1}{3}$. ●

In the following examples we illustrate the use of proportions in solving word problems.

Example 66.4

In an electric circuit, the current I is proportional to the voltage E. If $I = 1.6\ A$ when $E = 24$ V, what is the voltage when the current is 3.6 A?

Solution

If we let x represent the value of the voltage we seek, we can set up the proportion

$$\frac{1.6\text{ A}}{24\text{ V}} = \frac{3.6\text{ A}}{x\text{ V}} \quad \text{or} \quad 1.6x = 3.6(24)$$

Solving the latter equation, we find that $x = 54$ V.

Example 66.5

Two men realize a profit of \$3675.60 in a certain business venture. They agree to divide the money in the ratio 2:3. How much does each receive?

Solution

We can represent the amounts by $2x$ and $3x$, since $2x:3x = 2:3$. Thus,

$$2x + 3x = \$3675.60$$
$$5x = 3675.60$$
$$x = 735.12$$

Hence, $2x = \$1470.24$ and $3x = \$2205.36$.

Exercises for Section 66

In Exercises 1 through 10, express each ratio in simplest form.

1. \$10 to \$45

2. 16 lb to 48 lb

3. 8 ft to 4 yd

4. 2 ft to 8 in.

5. 80 sec to 2 min

6. 7 pints to 2 gal

7. 1 in. to 1 cm

8. 8 oz to 3 lb

9. 6 V to 4 V

10. 2640 ft to 2 miles

11. The side of one square is 6 in. and the side of a second square is 8 in. Find the ratio of their perimeters, diagonals, and areas.

12. The radius of one circle is 4 in. and the radius of a second circle is 8 in. Find the ratio of their circumferences and areas.

In Exercises 13 through 20, find the value of each unknown in the proportions given.

13. $2:x = 3:9$

14. $2:5 = 4:x$

15. $x:6 = 9:15$

16. $25:x = x:4$

17. $x:25 = 4:x$

18. $3:4 = 5:2x$

19. $(x + 1):2 = 6:3$

20. $(2x + 1):2 = (x + 2):5$

21. Find the mean proportional between 2 and 128.

22. Find the mean proportional between 3 and 12.

23. Find the mean proportional between -3 and -12.

24. Find the mean proportional between -3 and 12.

25. Find the mean proportional between $\frac{3}{2}$ and $\frac{3}{8}$.

In Exercises 26 through 36, use proportions to find the solutions.

26. If 12 in. = 30.48 cm, how many centimeters are there in 30 in.?

27. A metal bar of length 10 ft weighs 128 lb. What is the weight of a bar 2 ft 4 in.?

28. The tax on a property assessed at $7000 is $1100. What is the assessed value of a property if the tax is $1400?

29. A plane travels 680 km in 1 hr 40 min. If the plane maintains a constant speed, what distance does it cover in 2 hr 30 min?

30. A picture is 3.5 in. by 5 in. If the width of an enlargement is 8 in., what is the length?

31. Water is flowing over a dam at the rate of 2500 ft^3/min. How much water flows over the dam in 2 min 16 sec?

32. A car can travel 46 miles on 4 gal of gas. How far can it travel on 14 gal of gas?

33. How many feet per second are equivalent to 60 mph?

34. A piece of wire 21 m long is to be cut into two pieces. Find the length of each piece if the lengths are in the ratio 2:5.

35. A pole 20 ft high casts a shadow 16 ft long at the same time a building casts a shadow 84 ft long. find the height of the building.

36. If 1 kW is equal to 1.34 hp, how many kilowatts are in 1 hp?

SECTION 67
Direct Variation

Suppose that the two variables x and y are related by the equation $y = kx$, where k is a constant. In such an equation, we say that y **varies directly as** x **or** y **is proportional to** x. The constant k is called the **constant of proportionality** or the **constant of variation**. For example, in the equation $y = 2x$ we say that y varies directly as x and observe that the proportionality constant k is 2.

As another example of direct variation, let us consider the following example.

Example 67.1

A car is traveling at the rate of 60 km/hr. If we represent the distance by d and the time in hours by t, we can express this relationship by the equation $d = 60t$. Thus, we note that d varies directly as t and that the proportionality constant k is 60. ●

For the equations $y = 2x$ and $d = 60t$, we note that the ratio of the variables is a constant. That is, $y/x = 2$ and $d/t = 60$. It is this constant ratio that characterizes direct variation. If we know that y is directly proportional to x, we can always determine the proportionality constant k if we are given any pair of corresponding values for x and y (except when both x and y equal zero).

Example 67.2

If y varies directly as x and $y = 4$ when $x = 6$, find y when $x = 15$.

Solution

Since y varies directly as x, we have $y = kx$. Substituting the given values for x and y yields

$$4 = 6k \quad \text{or} \quad k = \frac{2}{3}$$

Thus, $y = \frac{2}{3}x$ and when $x = 15$, $y = \frac{2}{3}(15) = 10$. ●

Direct variation has many uses in science. One such use involves Hooke's law.

Example 67.3

Hooke's law states that if the elastic limit is not exceeded, the force F causing the extension of a spring is proportional to x, the displacement from equilibrium. If a force of 20 lb causes a displacement of 5 in. what displacement will be caused by an 8-lb force?

Solution

Since F is proportional to x or F varies directly as x, we have

$$F = kx$$

Substituting $F = 20$ lb and $x = 5$ in. into this equation yields

$$20 \text{ lb} = k(5 \text{ in.}) \quad \text{or} \quad k = \frac{20 \text{ lb}}{5 \text{ in.}} = 4 \text{ lb/in.}$$

Thus, the constant of proportionality is 4 lb/in. and

$$F = 4x(\text{lb/in.})$$

For $F = 8$ lb,

$$8 \text{ lb} = 4x(\text{lb/in.}) \quad \text{or} \quad x = \frac{8 \text{ lb}}{4 \text{ lb/in.}} = 2 \text{ in.}$$

Hence, the 8-lb force causes a displacement of 2 in. We note that the units in this example were treated in an algebraic fashion to arrive at the appropriate units. ●

In many situations, direct variation may occur between variables which have different powers. For example, in the formula for the area of a circle, $A = \pi r^2$, we say that the area varies directly as the square of the radius. The constant of proportionality is π.

In general, we say that the variable y varies directly or is directly proportional to the nth power of the variable x if $y = kx^n$. The constant k is the constant of proportionality or the constant of variation.

When we work with equations that are derived from physical problems, the description of the physical quantities requires a number and a unit if they are to be adequately described. The combination of the number and unit is called the *dimension* of the quantity. For example, the area A of a surface may be described in square feet, square inches, square centimeters, square miles, or square meters. In each case A has the dimension of area even though the units used to express the area are all different.

We note that when performing operations with dimensional symbols, we follow the usual laws of algebra. See appendix C.

Example 67.4

(a) The area of a square varies directly as the square of the side and is given by the equation $A = s^2$. The proportionality constant k is equal to 1.

(b) The volume of a sphere varies directly as the cube of the radius and is given by the equation $V = \frac{4}{3}\pi r^3$. The proportionality constant k is equal to $\frac{4}{3}\pi$.

Example 67.5

The surface area A of a sphere varies directly as the square of the radius r. If the area is 36π when the radius is 3, find the area when the radius is 5.

Solution

Since A is directly proportional to r, we have $A = kr^2$. Using the given values for A and r yields

$$36\pi = 9k \quad \text{or} \quad k = 4\pi$$

Thus, the constant of the variation is 4π and

$$A = 4\pi r^2$$

For $r = 5$, $A = 4\pi(25) = 100\pi$.

Exercises for Section 67

In Exercises 1 through 10, express the relations indicated as an equation of direct variation. In each case use a constant of variation and appropriate literal symbols.

1. The circumference of a circle is proportional to the radius.

2. The area of a square varies directly as the length of the diagonal.

3. The speed of a falling object varies directly as the time in flight.

4. The amount of light in a room varies directly as the number of candlepower.

5. The weight of a sphere is directly proportional to the cube of the radius.

6. The surface area of a cube varies directly as the square of the length of an edge.

7. The solubility of a certain chemical salt varies directly as the temperature.

8. The length of a pendulum varies directly as the square of the period.

9. The rate of heat generated in an electrical conductor varies directly as the square of the current.

10. The velocity of a transverse wave in a stretched wire varies directly as the square root of the tension in the wire.

In Exercises 11 through 20, state the equation relating the variables, evaluate the constant of proportionality, and find the indicated unknown.

11. If y varies directly as x, and $y = 10$ when $x = 4$, find y when $x = 6$.

12. If y varies directly as x, and $y = 2$ when $x = 3$, find y when $x = 5$.

13. If y varies directly as $3x$, and $y = 8$ when $x = 2$, find y when $x = 5$.

14. If y varies directly as $x^{1/2}$, and $y = 2$ when $x = 9$, find y when $x = 36$.

15. If y varies directly as $x^{1/2}$, and $y = 4$ when $x = 4$, find y when $x = 9$.

16. If y varies directly as x^2, and $y = 8$ when $x = 2$, find y when $x = 1$.

17. If y varies directly as the square of x, and $y = 9$ when $x = 3$, find y when $x = \frac{5}{2}$.

18. If v is directly proportional to the cube of y, and $v = 12$ when $y = 2$, find v when $y = 3$.

19. If y varies directly as the cube of x, and $y = 4$ when $x = 2$, find y when $x = 4$.

20. If y varies directly as $x^{3/2}$, and $y = 9$ when $x = 9$, find y when $x = 4$.

21. The distance an object falls under the influence of gravity varies directly as the square of the time. If the distance it falls is 256 ft at the end of 4 sec, find how far it will fall in 6 sec.

22. The distance traversed by a ball rolling down an inclined plane varies directly as the square of the time. After 4 sec, the distance traveled is 36 in. How far has the ball rolled after 7 sec?

23. Under certain conditions, the flow of water through a pipe varies directly as the square of the diameter. If water flows through a 1-in. pipe at the rate of 50 gal/min, what will be the rate of flow through a 5-in. pipe?

24. According to Hooke's law, the force needed to stretch a spring is proportional to the amount the spring is stretched. If 50 lb stretches a spring 5 in., how much will the spring be stretched by a force of 120 lb?

25. The weight of a cube varies directly as the length of one edge. If a cube with a 4-in. edge weighs 12 lb, what is the weight of a cube if the edge is 1 ft?

26. The volume of a sphere varies directly as the cube of the radius. If the volume is 36π cubic units when the radius is 3 units, find the volume when the radius is 4 units.

27. In Exercise 26, how does the volume of the sphere change if the radius is doubled? Tripled?

28. The volume of a cylindrical can of height 8 in. and radius r varies directly as the radius. If the volume is 32π in.^{3} when $r = 2$ in., find the volume when the radius is 5 in.

29. In a certain conical reservoir the radius of the section of the cone at the water line is one half of the depth of the water. Under these conditions, the volume of water is proportional to the cube of the depth of the water. If the volume is 18π ft^3 when the depth is 6 ft, find the volume when the depth is 12 ft.

30. The velocity of an object falling under the influence of gravity is proportional to the square root of the distance s the object has fallen. If the velocity is 160 ft/sec when s is 100 ft, find the velocity when s is 256 ft.

31. *Kepler's third law of planetary motion* states that the square of the time required for a planet to make one revolution about the sun varies directly as the cube of the average distance of the planet from the sun. If we assume that Mars is $1\frac{1}{2}$ times as far from the sun as the earth, find the approximate length of a Martian year.

32. The pressure p exerted by a liquid at a given point varies directly as the depth d of the point beneath the surface of the liquid. If a certain liquid exerts a pressure of 45 lb/ft^2 at a depth of 9 ft, find the pressure at 50 ft.

SECTION 68

Joint and Inverse Variation

In many instances we encounter situations where one variable depends on two or more other variables. For example, the area of a triangle depends on the base, b, and the altitude h and is given by the equation $A = \frac{1}{2}bh$. We say that A varies jointly or is directly proportional to the product of the base b and the height h.

In general, we say that y **varies jointly** or is **directly proportional** to the product of x^m and z^n if

$$y = kx^m z^n \tag{68.1}$$

We note that joint variation **always** implies a product of variables.

Example 68.1

If y varies jointly as $\sqrt{x}$ and z^2, find y when $x = 4$ and $z = 3$ if $y = 8$ when $x = 9$ and $z = 2$.

Solution

Since y varies jointly as $\sqrt{x}$ and z^2, we have $y = k\sqrt{x} \cdot z^2$. Substituting the given alues $y = 8$, $x = 9$, and $z = 2$ into the equation yields

$$8 = k\sqrt{9}(2)^2$$

or

$$8 = 12k$$

or

$$k = \frac{2}{3}$$

Thus,

$$y = \frac{2}{3}\sqrt{x} \cdot z^2$$

When $x = 4$ and $z = 3$,

$$y = \frac{2}{3} \cdot \sqrt{4} \cdot 3^2 = \frac{2}{3} \cdot 2 \cdot 9 = 12$$

Example 68.2

The weight of a cylinder of a certain material varies jointly as the height of the cylinder and the square of the radius. A cylinder with a radius of 10 cm and a height of 50 cm weighs 75 kg. What is the weight of a cylinder of the same material, if the radius is 8 cm and the height is 30 cm?

Solution

Since the weight w is directly proportional to the height h and the square of the radius r, we have $w = khr^2$. Using the given values for w, h, and r,

$$\begin{aligned} 75\text{ kg} &= k(50\text{ cm})(10\text{ cm})^2 \\ &= k(5000\text{ cm}^3) \end{aligned}$$

or

$$k = \left(\frac{75}{5000}\right)(\text{kg/cm}^3) = \left(\frac{3}{200}\right)(\text{kg/cm}^3)$$

Thus,

$$w = \left(\frac{3}{200}\text{ kg/cm}^3\right)\text{hr}^2$$

For $h = 30$ cm and $r = 8$ cm, we have

$$w = \left(\frac{3}{200}\text{ kg/cm}^3\right)(30\text{ cm})(8\text{ cm})^2$$

or

$$w = \left(\frac{3}{200}\text{ kg/cm}^3\right)(30\text{ cm})(64\text{ cm}^2)$$

or

$$w = \frac{144}{5}\text{ kg}$$

Example 68.3

The kinetic energy, K.E., of a moving body varies jointly as the mass m and the square of the velocity v. In computing the kinetic energy, we must use consistent units for m and v. In the **mks system** m is in kilograms and

v is in m/sec. In the **cgs system**, m is grams and v is in cm/sec. For a body with mass 1 kg moving at a speed of 10 m/sec, the kinetic energy is 50 $(\text{kg} \cdot \text{m}^2)/\text{sec}^2$. Find the kinetic energy of the same body when its speed is 20 m/sec.

Solution

We note from the description of kinetic energy that K.E. $= kmv^2$. Using the given values to solve for k, we have

$$50\ (\text{kg} \cdot \text{m}^2)/\text{sec}^2 = k \cdot (1\ \text{kg})(10\ \text{m/sec})^2$$

or

$$k = \frac{50\ (\text{kg} \cdot \text{m}^2)/\text{sec}^2}{100\ (\text{kg} \cdot \text{m}^2)/\text{sec}^2} = \frac{1}{2}$$

Thus, K.E. $= \frac{1}{2}mv^2$. We find the desired value for K.E. by substituting the values $m = 1$ kg and $v = 20$ m/sec. Hence,

$$\text{K.E.} = \frac{1}{2}(1\ \text{kg})(20\ \text{m/sec})^2$$

or

$$\text{K.E.} = 200\,\frac{\text{kg} \cdot \text{m}^2}{\text{sec}^2}$$

Kinetic energy is usually expressed in units of work. Noting that $1\ \text{kg} \cdot \text{m}^2/\text{sec}^2 = 1\ \text{J}$, we have that K.E. $= 200$ J.

Remark 68.1. *In Example 68.3 the final result was expressed in terms of joules. We note that the unit of kinetic energy in any system is equal to the unit of work in that system. The units of work that are generally used to express kinetic energy are joules, ergs, or foot-pounds. The units occur as follows.*

If m is in kilograms and v in meters/sec, our work unit is given by

$$1\ \text{kg} \cdot \text{m}^2/\text{sec}^2 = 1\ \text{J}$$

If m is in grams and v in cm/sec, we have $1\ \text{g} \cdot \text{cm}^2/\text{sec}^2 = 1\ \text{erg}$ as our work unit.

If m is slugs and v in ft/sec, our work unit is given by

$$1\ \text{slug} \cdot \text{ft}^2/\text{sec}^2 = 1\ \text{ft-lb}$$

Now let us consider what is meant by the type of relationship called **inverse variation**. We say that y varies inversely as or is inversely proportional to the nth power of x if

$$y = \frac{k}{x^n} \tag{68.2}$$

Example 68.4

If y varies inversely as x and $y = 2$ when $x = 8$, find y when $x = 6$.

Solution

We can write $y = k/x$ since y varies inversely as x. Substituting $y = 2$ and $x = 8$ into the equation yields $2 = k/8$ or $k = 16$. Thus, the constant of proportionality is 16 and we have $y = 16/x$. Now, for $x = 6$, $y = \frac{16}{6} = \frac{8}{3}$.

Example 68.5

The length of time required to fill a swimming pool varies inversely as the square of the diameter of the pipe used to fill the pool. If it takes 18 hr for a 2-in. pipe to fill the pool, how long will it take a 3-in. pipe?

Solution

Since the time t is inversely proportional to the square of the diameter d of the pipe, we have

$$t = \frac{k}{d^2}$$

When $t = 18$ hr, $d = 2$ in. Thus,

$$18 \text{ hr} = \frac{k}{4 \text{ in.}^2}$$

or

$$k = 72 \text{ hr} \cdot \text{in.}^2$$

and

$$t = \frac{72 \text{ hr} \cdot \text{in.}^2}{d^2}$$

For

$$d = 3 \text{ in.,} \qquad t = \frac{72 \text{ hr} \cdot \text{in.}^2}{9 \text{ in.}^2}$$

and therefore, $t = 8$ hr.

Example 68.6

Boyle's law states that the volume v of a mass of gas at a given temperature varies inversely as the pressure p. If the volume is 40 ft^3 when the pressure is 16 lb, find the volume when the pressure is 24 lb.

Solution

Since the volume varies inversely as the pressure, Boyle's law states that $v = k/p$. We are given that $v = 40 \text{ ft}^3$ when $p = 16$ lb. Thus, $40 \text{ ft}^3 = k/16 \text{ lb}$ or $k = 640 \text{ ft}^3 \cdot \text{lb}$. The solution is obtained by substituting $p = 24$ lb into the equation

$$v = \frac{640 \text{ ft}^3 \cdot \text{lb}}{p}$$

Hence,

$$v = \frac{640\ \text{ft}^3 \cdot \text{lb}}{24\ \text{lb}} = \frac{80}{3}\text{ft}^3$$

Exercises for Section 68

In Exercises 1 through 10, express each statement as an equation.

1. y varies jointly as x and z.
2. y varies inversely as the square of x.
3. y varies jointly as the square root of x and the cube root of z.
4. p varies jointly as r and s^3.
5. s varies inversely as the square root of $2t$.
6. w varies jointly as the square of x and the cube root of y.
7. v is inversely proportional to the cube root of s.
8. y is directly proportional to x and the square of z.
9. y varies inversely as the fourth power of x.
10. t varies jointly as the square of x, the cube of y, and the square of z.

In Exercises 11 through 20, find the value indicated.

11. Find y when $x = 2$ and $z = \sqrt{3}$ if y varies directly as x and z^2 and $y = 4$ when $x = 1$ and $z = 2$.
12. *If* y varies directly as x and z and $y = 5$ when $x = 3$ and $z = 4$, find y when $x = 2$ and $z = 3$.
13. If y varies inversely as the square root of x and $y = 4$ when $x = 64$, find y when $x = 8$.
14. If y is inversely proportional to the cube root of x, and $y = 1$ when $x = 8$, find y when $x = 27$.
15. If y varies jointly as $\sqrt{x}$ and $\sqrt{z}$ and $y = 8$ when $x = 4$ and $z = 9$, find y when $x = 16$ and $z = 12$.
16. Find p when $q = 4$ if p varies inversely as the square root of q and $p = 0.01$ when $q = 25$.
17. Find s when $t = 4$ if s is inversely proportional to t and $s = 0.4$ when $t = 2.5$.
18. Find y when $x = 2$ and $z = 8$ if y varies jointly as x and the cube root of z and $y = 2$ when $x = 1$ and $z = 27$.
19. If y varies jointly as v and w^2 and $y = 2$ when $v = 2$ and $w = 1$, find y when $v = 4$ and $w = 3$.

20. Find r where $s = 2$ and $t = 3$ if r varies jointly as s and the cube of t and $r = 2$ when $s = 3$ and $t = 1$.

21. The time t needed to travel a given distance varies inversely with the speed s. If $t = 3$ hr and $s = 40$ mph, find t if $s = 60$ mph.

22. The force of repulsion F between two charges of like sign is inversely proportional to the square of the distance d between them. If F is 10 units when $r = 5$ cm, find F when $r = 2$ cm.

23. The power P in electric circuit varies directly as the resistance R and the square of the current i. If the power is 10 W when the current is 0.4 A and the resistance is 80 Ω, find the power when the current is 2 A and the resistance is 60 Ω.

24. The intensity of illumination is inversely proportional to the square of the distance from the light source. If the illumination is 50 units when the distance is 80 cm, find the illumination when the distance is 30 cm.

25. The length of time needed to empty a reservoir varies inversely as the square of the diameter of the drain pipe. If it takes 6 hr for the reservoir to empty using a 5-in. drain pipe, how long will it take using a 3-in. pipe?

26. The kinetic energy of a body varies jointly as the mass m and the square of the speed v (see Example 68.3). Find the kinetic energy of a body if its mass is 200 g and its speed is 500 cm/sec.

27. Use the method of Exercise 26 to find the kinetic energy of a 4000-lb car traveling at the rate of 90 ft/sec.

28. The force of attraction F between two charges of opposite sign varies inversely as the square of the distance between them. If the force is 50 units when the distance is 8 cm, find the force when the distance is 5 cm.

29. At a constant temperature, the volume V of an ideal gas varies inversely as the pressure p. If the volume is 200 units when the pressure is 2 atm, find the volume when the pressure is 5 atm.

30. Compare the kinetic energy of a car moving at 30 mph with that of the same car moving at 60 mph.

31. Boyle's law states that for an enclosed gas at a constant temperature, the pressure p varies inversely as the volume v. If $v = 100 \text{ in.}^3$ when $p = 20 \text{ lb/in.}^2$, find the volume when the pressure is 80 lb/in.^2.

32. The force of the wind on a sail varies jointly as the area of the sail and the square of the velocity of the wind. If the force on 1 ft^2 of sail is 5 lb when the wind velocity is 20 mph, find the force of a 40-mph wind on a sail of 9 yd^2.

33. The weight of a body varies inversely as the square of its distance from the center of the earth. If the radius of the earth is 4000 miles, how much would a 200-lb man weight 100 miles above the surface of the earth? 1000 miles? 4000 miles?

34. Under certain conditions, the thrust T of a propeller varies jointly as the fourth power of its diameter d and the square of the number n of revolutions per second. Show that if n is doubled and d is halved, the thrust T is decreased by 75%.

SECTION 69

Combined Variation

In situations where a variable varies directly as one quantity and inversely as another, we have *combined* variation. The equation we use to express combined variation is

$$y = \frac{kx^m}{z^n} \tag{69.1}$$

Equation (69.1) states that y varies directly as x^m and inversely as z^n. The following examples illustrate the uses and applications of combined variation.

Example 69.1

If y varies directly as the square of x and inversely as z, and if $y = 30$ when $x = 10$ and $z = 5$, find y when $x = 4$ and $z = 8$.

Solution

The combined variation can be expressed by the equation $y = kx^2/z$. Substituting the given values in this equation yields

$$30 = \frac{k \cdot 10^2}{5}$$

or

$$k = 1.5$$

Thus, $y = 1.5x^2/z$ and when $x = 4$ and $z = 8$, we have $y = (1.5)4^2/8 = 3$.

Example 69.2

Suppose that T varies jointly as x and the square of y and inversely as the cube root of z. If $T = 12$ when $x = 3$, $y = 2$, and $z = 27$, find T when $x = 5$, $y = 4$, and $z = 8$.

Solution
The combined variation equation is

$$T = \frac{kxy^2}{\sqrt[3]{z}}$$

Substituting the given values and solving for k, we have

$$12 = \frac{k \cdot 3 \cdot 2^2}{\sqrt[3]{27}} \quad \text{or} \quad k = 3$$

Thus, $T = 3xy^2/\sqrt[3]{z}$. When $x = 5$, $y = 4$, and $z = 8$,

$$T = \frac{3 \cdot 5 \cdot 4^2}{\sqrt[3]{8}} = 120$$

Example 69.3
Newton's law of gravitation states that the force F exerted by two bodies varies jointly as their masses m_1 and m_2 and inversely as the square of the distance between them. Write this formula symbolically.

Solution
The conditions indicate that the equation involves combined variation. Since F varies jointly as m_1 and m_2, the numerator is of the form m_1m_2. Similarly, the denominator is of the form d^2, since F varies inversely as the square of d. Using k as our proportionality constant, we have

$$F = \frac{km_1m_2}{d^2}$$

Example 69.4
The resistance of a wire to the flow of electricity varies directly as its length and inversely as the square of its diameter. If a wire of length 1000 cm and diameter 0.5 cm has a resistance of 0.4 Ω, what will the resistance be for the same wire if the length is 400 cm and the diameter is 0.2 cm?

Solution
Since the resistance R varies directly as the length l and inversely as the square of the diameter d, we have

$$R = \frac{kl}{d^2}$$

Now $R = 0.4$ when $l = 1000$ and $d = 0.5$. Thus,

$$0.4 = \frac{1000 \cdot k}{(0.5)^2}$$

or

$$k = \frac{(0.4)(0.25)}{1000} = 0.0001 = 10^{-4}$$

Hence, our variation equation is

$$R = \frac{10^{-4}l}{d^2}$$

When $l = 400$ and $d = 0.2$,

$$R = \frac{10^{-4}(400)}{(0.2)^2} = 1\ \Omega$$

Exercises for Section 69

1. If y varies directly as x^2 and inversely as z and $y = 4$ when $x = 2$ and $z = 3$, find y when $x = 5$ and $z = 2$.

2. If z varies directly as x and inversely as the square of y and $z = 10$ when $x = 2$ and $y = 3$, find z when $x = 5$ and $y = 3$.

3. If T varies jointly as x and y and inversely as the square of z and $T = 5$ when $x = 2$, $y = 3$, and $z = 4$, find T when $x = 5$, $y = 2$, and $z = 3$.

4. If W varies directly as the cube of x and inversely as the square of y and $W = 200$ when $x = 4$ and $y = 6$, find W when $x = 2$ and $y = 5$.

5. If y varies directly as x and inversely as z^3 and $y = 24$ when $x = 12$ and z = 1, find z when $x = 4$ and $y = 36$.

6. If T varies directly as x^2 and y and inversely as the square root of z and $T = 40$, when $x = 3$, $y = 5$, and $z = 9$, find T when $x = 2$, $y = 3$, and $z = 16$.

7. If y varies directly as x^2 and inversely as z^3, what change in y will result if we double x and halve z?

8. Use the results of Example 69.3 to find the effect on the force F between two bodies if the masses of each are doubled and the distance between them is halved.

9. The strength of a horizontal beam supported at both ends varies jointly as the width and square of the depth and inversely as the length.
(a) What is the effect on the strength of the beam if the width is doubled and the depth is halved?
(b) What change in the depth will increase the strength by 40% if the width and length remain unchanged?

10. If a wire of length 200 cm and diameter 0.1 cm has a resistance of 0.02 Ω, what will the resistance be if the length of the wire is 600 cm and the diameter is 0.25 cm? See Example 69.4.

11. The number of hours h that it takes m men to assemble x machines varies directly as the number of machines and inversely as the number of men. If 4 men can assemble 12 machines in 4 hr, how many men are needed to assemble 36 machines in 8 h?

12. The altitude h of a right circular cone varies directly as the volume v and inversely as the square of the radius r. If $h = 6$ when $r = 2$ and $v = 8\pi$, find h when $r = 3$ and $v = 12\pi$.

13. Under certain conditions the illumination E of a surface varies directly as the intensity I of the light source and inversely as the square of the distance d of the surface from the light source. What is the effect on the illumination if the intensity is doubled and the distance is halved?

14. The current I in a wire varies directly as the electromotive force E and inversely as the resistance R. If $I = 30$ A when $E = 120$ V and $R = 5\ \Omega$, find I when $E = 210$ V and $R = 15\ \Omega$.

REVIEW QUESTIONS FOR CHAPTER 11

1. What is the comparison of two quantities by division called?
2. What does the term "dimensionless" mean?
3. What is a proportion?
4. What are the *extremes* of a proportion? The means?
5. State three properties for the proportion $a/b = c/d$.
6. What is a mean proportion?
7. Write an equation such that y varies directly as x.
8. Write an equation such that y varies directly as the nth power of x.
9. Describe what we mean by joint variation.
10. What do we mean by inverse variation?
11. What is combined variation?
12. Write an equation used to express combined variation.

REVIEW EXERCISES FOR CHAPTER 11

1. Find x if $2:x = 4:12$.

2. Find x if $x:3 = 6:9$.

3. Find x if $3:8 = 5:4x$.

4, Find x if $x:25 = 3:x$.

5. Find the mean proportional between 4 and 9.

6. Find the mean proportional between 6 and 24.

7. Find the mean proportional between -12 and -6.

8. If y varies directly as x and $y = 2$ when $x = 3$, find y when $x = 12$.

9. If y varies directly as $2x$ and $y = 8$ and $x = 4$, find y when $x = 2.5$.

10. If y varies directly as x^2 and $y = 6$ when $x = 3$, find y when $x = 9$.

11. If y is proportional to the cube of x and $y = 2$ when $x = 3$, find y when $x = 4$.

12. If y varies jointly as x and z^2, find y when $x = 5$ and $z = 3$ if $y = 10$ when $x = 1$ and $z = 2$.

13. If y varies directly as x and z and $y = 5$ when $x = 2$ and $z = 1$, find y when $x = 1$ and $z = 3$.

14. If y varies inversely as the cube of x and $y = \frac{1}{2}$ when $x = 2$, find y when $x = 5$.

15. If y is inversely proportional to the cube root of x and $y = 3$ when $x = 8$, find y when $x = 81$.

16. If y varies jointly as x^2 and v and $y = 2$ when $x = 1$ and $v = 2$, find y when $x = 2$ and $v = 3$.

17. If y varies directly as x and inversely as the square of z, and if $y = 10$ when $x = 2$ and $z = 3$, find y when $x = 1$ and $z = 2$.

18. If s varies jointly as x and the square of y and inversely as z, and $s = 8$ when $x = 1$, $y = 2$, and $z = 3$, find s when $x = 5$, $y = 1$, and $z = 2$.

19. If y varies directly as x and inversely as z^2, what change in y will result if we double x and halve z?

20. Y varies directly as x and inversely as the product $v \cdot w$. If $y = 4$ when $x = 2$, $v = 3$, and $w = 5$, find y when $x = 5$, $v = 1$, and $w = 2$.

21. A spring is stretched 24 cm by a force of 50 kg. How far will it stretch if a force of 104 kg is applied? (Use Hooke's law.)

22. A force of 3 lb stretches a spring 6 in. What force is required to stretch the spring 7.5 in.? (Use Hooke's law.)

23. The area of a circle varies directly as the radius. If the area is 25π square units when the radius is 5, find the area when the radius is 2.

24. The flow of liquid through a pipe is proportional to square of the radius. If water flows through a 3-in. pipe at the rate of 250 gal/min, what will be the rate of flow through a 5-in. pipe?

25. The surface area of a cube varies directly as the square of the length of an edge. If the area is 24 square units when the length of the edge is 2, find the area when the length of the edge is 5.

26. The weight of a sphere is directly proportional to the cube of the radius. A sphere with a radius of 12 cm weighs 2 kg. How much does a sphere of radius 20 cm weigh?

27. The volume of a cone varies jointly as the height h and the square of the radius. If the volume is 25π cubic units when $r = 5$ and $h = 3$, find the volume when $r = 4$ and $h = 10$.

28. The force of the wind on a window varies jointly as the area of the window and the square of the velocity of the wind. If the force on a window 2 ft by 4 ft is 12 lb when the wind velocity is 10 mph, find the force of a 30-mph wind on a window whose area is 40 ft^2.

29. The altitude h of a right circular cone varies directly as the volume and inversely as the square of the radius. If $h = 6$ when $r = 2$ and $v = 24\pi$, find h when $v = 20$ and $r = 4$.

30. In Exercise 29, how is the altitude affected if the radius is doubled? Tripled?

31. *Coulomb's law* states that the force between two charges Q_1 and Q_2 is directly proportional to the product of the charges and inversely proportional to the distance r between the charges. Express the force F in terms of Q_1, Q_2, and r.

32. The temperature T of a certain hot metal plate at a point (x, y) is directly proportional to the fourth power of the point's distance from the origin. If the temperature at the point $(3, 4)$ is 200°, find a formula for the temperature in terms of x and y. Find T at the point $(5, 12)$.

33. The temperature T of a certain hot metal plate at a point (x, y) is inversely proportional to the point's distance from the origin. If the temperature at the point $(8, 6)$ is 150°, find a formula for the temperature in terms of x and y. Find T at the point $(4, 3)$.

CHAPTER 12

Higher-Degree Equations

SECTION 70

Synthetic Division

In Sections 14 through 16 we found the real roots of linear and quadratic equations in one variable. We now wish to discuss the problem of finding the roots of a polynomial equation of any degree.

The function defined by

$$P(x) = a_0x^n + a_1x^{n-1} + a_2x^{n-2} + \cdots + a_{n-1}x + a_n \tag{70.1}$$

where n is a nonnegative integer and $a_0, a_1, \ldots, a_n$ are constants is called a **polynomial function** in x. If $a_0 \neq 0$, the polynomial is said to be of degree n (see Section 9). Throughout this chapter we shall consider only polynomial functions. Our primary goal is that of finding the zeros of $P(x)$. We note that the zeros of the polynomial function $P(x)$ are the solutions or roots of the polynomial equation $P(x) = 0$.

The roots of any polynomial equation $P(x) = 0$ can be readily determined if we know or can find the linear or quadratic factors of $P(x)$. This is true because any value for x that makes one of the factors of $P(x)$ equal to zero will be a root of the equation. For example, if $P(x) = (x - 1)(x - 2)(x + 3)$, the roots of $P(x) = 0$ are $x = 1$, $x = 2$, and $x = -3$, since each of these values makes one of the factors equal to zero.

If $P(x)$ has linear factors of the form $x - r$, they can be found by using **synthetic division**. Synthetic division is simply an abbreviated form of the long-division process. The mechanics of this type of division are illustrated in the following examples.

A careful examination of long division will provide the basis for the steps used in synthetic division.

Example 70.1

Divide $2x^4 - 3x^3 - 5x^2 + 3x + 8$ by $x - 2$.

Solution

$$
\begin{array}{rrrrrrl}
 & 2x^3 & +\ x^2 & -\ 3x & -\ 3 & & \rightarrow \text{quotient} \\
\text{divisor} \rightarrow x - 2 \overline{\big)} & 2x^4 & -\ 3x^3 & -\ 5x^2 & +\ 3x & +\ 8 & \rightarrow \text{dividend} \\
 & 2x^4 & -\ 4x^3 & & & & \\ \hline
 & & x^3 & -\ 5x^2 & & & \\
 & & x^3 & -\ 2x^2 & & & \\ \hline
 & & & -\ 3x^2 & +\ 3x & & \\
 & & & -\ 3x^2 & +\ 6x & & \\ \hline
 & & & & -\ 3x & +\ 8 & \\
 & & & & -\ 3x & +\ 6 & \\ \hline
 & & & & & 2 & \rightarrow \text{remainder}
\end{array}
$$

Thus,

$$\frac{2x^4 - 3x^3 - 5x^2 + 3x + 8}{x - 2} = 2x^3 + x^2 - 3x - 3 + \frac{2}{x - 2} \qquad x \neq 2$$

●

In general, if $P(x)$ denotes the dividend, $D(x)$ the divisor, $Q(x)$ the quotient, and $R(x)$ the remainder, our result may be written in the form

$$\frac{P(x)}{D(x)} = Q(x) + \frac{R(x)}{D(x)} \tag{70.2}$$

If we examine the division process of Example 70.1 carefully, we notice that the initial coefficient of 2 in the quotient is the same as the initial coefficient in the dividend. Each succeeding coefficient of the quotient results from multiplying the coefficient that precedes it by -2 and then subtracting that product from the corresponding coefficient of the dividend. Thus the coefficient of the x^2 term in the quotient is obtained by subtracting $2 \cdot (-2)$ from -3. For the coefficient of x in the quotient we subtract $1 \cdot (-2)$ from -5. Successive coefficients in the quotient and the remainder are determined in a similar fashion.

To further analyze the division, we will omit writing the variables and write only the coefficients of the terms. Then the division appears as follows:

$$
\begin{array}{rrrrrr}
 & 2 & 1 & -\ 3 & -\ 3 & \\
1 - 2 \overline{\big)} & 2 & -\ 3 & -\ 5 & +\ 3 & +\ 8 \\
 & \mathbf{2} & -\ 4 & & & \\ \hline
 & & 1 & -\ 5 & & \\
 & & \mathbf{1} & -\ 2 & & \\ \hline
 & & & -\ 3 & +\ 3 & \\
 & & & \mathbf{-\ 3} & +\ 6 & \\ \hline
 & & & & -\ 3 & +\ 8 \\
 & & & & \mathbf{-\ 3} & +\ 6 \\ \hline
 & & & & & 2
\end{array}
$$

The numbers in boldface type are repetitions of the numbers immediately above and are also repetitions of the coefficients of the variables in the quotient. If we omit these repetitions and also omit the coefficient 1 of x in the divisor, we can write

$$\begin{array}{r|l}
-2 & 2 - 3 - 5 + 3 + 8 \\
 & \quad\; - 4 \\ \cline{2-2}
 & \quad\;\; 1 - 5 \\
 & \qquad\; - 2 \\ \cline{2-2}
 & \qquad\; - 3 + 3 \\
 & \qquad\qquad + 6 \\ \cline{2-2}
 & \qquad\qquad - 3 + 8 \\
 & \qquad\qquad\quad\; + 6 \\ \cline{2-2}
 & \qquad\qquad\qquad\;\; 2
\end{array}$$

If we compress the process by moving the terms up near the first line, we have

$$\begin{array}{r|rrrrrl}
-2 & 2 & -3 & -5 & +3 & +8 & \text{(line 1)} \\
 & & -4 & -2 & +6 & +6 & \text{(line 2)} \\ \hline
 & 2 & 1 & -3 & -3 & +2 & \text{(line 3)}
\end{array}$$

The last number in line 3 is the remainder. The other numbers in line 3, from left to right, are the coefficients of the quotient $2x^3 + x^2 - 3x - 3$. We note that the numbers in line 3 are obtained by subtracting the coefficients in line 2 from the coefficients of the terms having the same degree in line 1.

We can obtain the same result if we replace -2 with 2 in the divisor, and then add at each step instead of subtracting. In schematic form, we have the following:

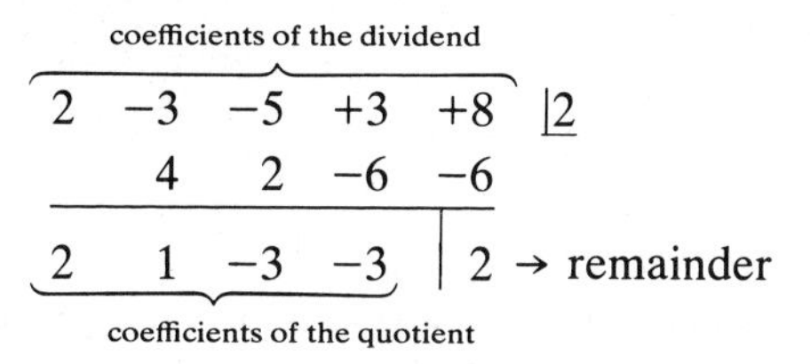

We write the number 2 at the upper right because we are dividing by $x - 2$. If we were dividing by $x + 2$, we would write -2 since $x + 2 = x - (-2)$. Division done using this arrangement is called **synthetic division.**

Example 70.2

Use synthetic division to divide $2x^3 + 3x^2 - 7x + 3$ by $x + 3$.

Solution

Since $x + 3 = x - (-3)$, we have

$$\begin{array}{rrrr|l}
2 & 3 & -7 & 3 & -3 \\
 & -6 & 9 & -6 & \\ \hline
2 & -3 & 2 & -3 &
\end{array}$$

The quotient is $2x^2 - 3x + 2$ and the remainder is -3. Hence,

$$\frac{2x^3 + 3x^2 - 7x + 3}{x + 3} = 2x^2 - 3x + 2 + \frac{-3}{x + 3} \qquad x \neq -3$$

Example 70.3

Use synthetic division to find the quotient and remainder if $P(x) = 2x^4 - 3x^3 + x - 4$ and $D(x) = x - 2$.

Solution

We note that the coefficient of the x^2 term in $P(x)$ is 0. Thus, $P(x) = 2x^4 - 3x^3 + 0x^2 + x - 4$. Since we are dividing by $x - 2$, we use the number 2 in the upper right and we have

$$\begin{array}{rrrrr|l} 2 & -3 & 0 & 1 & -4 & \underline{2} \\ & 4 & 2 & 4 & 10 & \\ \hline 2 & 1 & 2 & 5 & 6 & \end{array}$$

Thus, the quotient is $2x^3 + x^2 + 2x + 5$ and the remainder is 6.

Now let us summarize the process of synthetic division. To divide a polynomial $P(x)$ by $x - r$ synthetically:

1. *Write the coefficients of $P(x)$ in order of descending powers, inserting zero for the coefficient of any missing power and write r on the right. This is the first line.*
2. *Bring down the first coefficient of $P(x)$ to the first position on the third line.*
3. *Write the product of r and this coefficient in the second line beneath the second coefficient in the first line and add, putting the sum in the third line, and so on until a product has been added to the final coefficient of $P(x)$.*
4. *The last sum in the third line represents the remainder. The preceding numbers represent the coefficients of the powers of x in the quotient, written in descending order.*

Remark. *When we divide a polynomial $P(x)$ by an expression of the form $x - r$, the quotient is always a polynomial of degree one less than the degree of $P(x)$.*

Example 70.4

Use synthetic division to divide $3x^3 + 2x^2 - 4x + 2$ by $x - \frac{1}{3}$.

Solution

In this example we have a fractional value for r. However, the division proceeds as before, and we have

$$\begin{array}{rrrr|l} 3 & 2 & -4 & 2 & \underline{\frac{1}{3}} \\ & 1 & 1 & -1 & \\ \hline 3 & 3 & -3 & 1 & \end{array}$$

Thus the quotient is $3x^2 + 3x - 3$ and the remainder is 1.

Synthetic division can only be used when the divisor is a binomial of the form $x - r$, where r is a constant and the coefficient of x is 1. If the coefficient of x is some number other than 1, we may still employ synthetic division, but it is necessary to make certain adjustments in the divisor. If, for example, our divisor is $2x - 3$, we would first divide by the coefficient of x, in this case 2, making it $x - \frac{3}{2}$. If we do noting else, the quotient will be twice as large as it should be. The correct result is obtained by dividing the quotient by 2.

Example 70.5

Use synthetic division to divide $2x^3 + 3x^2 - 5x - 1$ by $2x - 3$.

Solution

We change the divisor to $x - \frac{3}{2}$ by dividing by 2 and perform the synthetic division as follows:

$$
\begin{array}{rrrr|l}
2 & 3 & -5 & -1 & \underline{\frac{3}{2}} \\
 & 3 & 9 & 6 & \\
\hline
2 & 6 & 4 & \multicolumn{1}{|r}{5} &
\end{array}
$$

Since we have divided the divisor by 2, the coefficients 2, 6, and 4 are twice as large as they should be. We divide these numbers by 2 and get 1, 3, and 2, the coefficients of the quotient. Thus, $Q(x) = x^2 + 3x + 2$ and the remainder is not affected by adjusting the coefficients of the divisor.

Exercises for Section 70

Find the quotient and remainder by using synthetic division.

1. $(3x^3 - 2x^2 + 2x - 4) \div (x - 3)$

2. $(3x^2 + 2x - 4) \div (x - 3)$

3. $(x^4 + 2x^3 - 3x^2 - 4x - 8) \div (x - 2)$

4. $(x^4 - 2x^3 - 3x^2 - 4x - 8) \div (x - 1)$

5. $(x^3 + 4x - 7) \div (x - 3)$

6. $(x^4 - 4x^2 + x^2 - 2x - 6) \div (x - 1)$

7. $(2x^3 + x^2 - x + 4) \div (x + 1)$

8. $(x^3 - 2x^2 + 3x - 5) \div (x - 3)$

9. $(x^4 - 5x^3 + 3x^2 - 6x + 4) \div (x - 1)$

10. $(2x^4 - 3x^3 - 5x^2 + 7x + 4) \div \left(x - \frac{1}{2}\right)$

11. $(x^4 - 1) \div (x - 1)$

12. $(x^4 - 1) \div (x + 1)$

13. $(2x^4 - x^2 + 3x - 5) \div (x - 2)$

14. $(x^6 - x^4 - 4) \div (x - 1)$

15. $(3x^3 + 2x^2 - 7x + 2) \div \left(x - \frac{1}{3}\right)$

16. $(x^4 - 5x^3 + 3x^2 - 2x + 6) \div (x - 4)$

17. $(x^4 + 4x^3 - x^2 - 16x - 12) \div (x - 2)$

18. $(2x^5 - x^2 + 8x - 75) \div (x - 2)$

19. $(3x^3 + 4x^2 + 6x + 8) \div \left(x + \frac{4}{3}\right)$

20. $(x^4 - x^2 - 12) \div (x - 2)$

21. $(2x^3 - 3x^2 + 5x + 6) \div (2x - 1)$

22. $(3x^3 + 5x^2 + x + 2) \div (3x - 1)$

23. $(12x^4 + x^3 - 12x^2 + x + 2) \div (3x - 2)$

24. $(6x^3 + x^2 - 9) \div (3x - 4)$

25. $(4x^4 + 2x^3 + 8x^2 - x - 6) \div (2x + 3)$

SECTION 71

The Remainder Theorem and the Factor Theorem

If we divide the polynomial $P(x) = x^3 - 3x^2 + 6x - 4$ by $x - 2$, we can obtain the quotient and the remainder by using synthetic division as follows:

$$\begin{array}{rrr|r|l} 1 & -3 & 6 & -4 & \underline{2} \\ & 2 & -2 & 8 & \\ \hline 1 & -1 & 4 & 4 & \end{array}$$

Thus, the quotient is $Q(x) = x^2 - x + 4$ and the remainder $R = 4$. We also note that $P(2) = 2^3 - 3(2)^2 + 6(2) - 4 = 4$, the same value as the remainder in the division. This illustration suggests that the remainder R and the value of $P(x)$ for $x = r$ are equal.

In order to show that this is, in fact, the case, we recall from equation (70.2) that

$$\frac{P(x)}{D(x)} = Q(x) + \frac{R(x)}{D(x)}$$

If we multiply the equation by $D(x)$, we have

$$P(x) = D(x) \cdot Q(x) + R(x) \tag{71.1}$$

where the degree of $R(x)$ is less than the degree of $D(x)$.

If $D(x)$ is of the form $(x - r)$, the divisor is of degree 1 and the remainder must be of degree zero, that is, a constant. Hence, we may write

$$P(x) = (x - r) \cdot Q(x) + R \tag{71.2}$$

where R is a constant. Since (71.2) is an identity that is true for all values of x, it is true when $x = r$. Substituting $x = r$ into (71.2) we have

$$\begin{aligned} P(r) &= (r - r) \cdot Q(r) + R \\ &= 0 \cdot Q(r) + R \\ &= R \end{aligned}$$

The discussion above leads us to the remainder theorem.

Theorem 71.1. The Remainder Theorem
If a polynomial $P(x)$ is divided by $(x - r)$ until a constant remainder R is obtained, then $P(r) = R$.

Example 71.1
Find the remainder when $x^3 + 2x^2 - x + 5$ is divided by $x - 2$.

Solution
According to the remainder theorem, $R = P(2)$. Since $P(2) = 2^3 - 2 \cdot 2^2 - 2 + 5 = 19$, $R = 19$. Observe that the same result is obtained by means of synthetic division.

$$\begin{array}{rrrr|l} 1 & 2 & -1 & 5 & \underline{2} \\ & 2 & 8 & 14 & \\ \hline 1 & 4 & 7 & |19 & = R \end{array}$$

Example 71.2
Determine the remainder when $-2x^5 + 6x^4 + 10x^3 - 6x^2 - 9x + 4$ is divided by $x - 4$.

Solution
In this example we observe that it is simpler to divide synthetically than to substitute. Since either procedure yields the remainder, we choose the most direct method. The division is as follows:

$$\begin{array}{rrrrrr|l} -2 & 6 & 10 & -6 & -9 & 4 & \underline{4} \\ & -8 & -8 & 8 & 8 & -4 & \\ \hline -2 & -2 & 2 & 2 & -1 & |\ 0 & \end{array}$$

The remainder is zero and thus $P(4) = 0$.

In Example 71.2 the remainder is zero, and so the division is exact. In such cases we say that the divisor is a factor of the dividend. Hence, $x - 4$ is a factor of $-2x^5 + 6x^4 + 10x^3 - 6x^2 - 9x + 4$. Example 71.2 illustrates a second important theorem, called the factor theorem.

Theorem 71.2. The Factor Theorem
If $P(r) = R$ is zero, then $(x - r)$ is a factor of $P(x)$.

The theorem is easily verified by substituting $R = 0$ into equation (71.2) to obtain $P(x) = (x - r) \cdot Q(x)$. Thus, $(x - r)$ is a factor of $P(x)$. In addition, we note that if $(x - r)$ is a factor of $P(x)$, then r is a zero of $P(x)$ and a root of the equation $P(x) = 0$.

Example 71.3
Show that $(x - 3)$ is a factor of $f(x) = x^3 - x^2 - 7x + 3$.

Solution
By the factor theorem, $x - 3$ is a factor of $f(x) = x^3 - x^2 - 7x + 3$ if $f(3) = 0$. To find $f(3)$ we use synthetic division (substitution would work equally well) and then apply the remainder theorem.

$$\begin{array}{rrrr|l} 1 & -1 & -7 & 3 & \underline{3} \\ & 3 & 6 & -3 & \\ \hline 1 & 2 & -1 & 0 & \end{array}$$

Since the remainder is 0, $f(3) = 0$. Hence, $(x - 3)$ is a factor of $f(x)$ and 3 is a root of $x^3 - x^2 - 7x + 3 = 0$.

Example 71.4
Show that $(x + 2)$ is not a factor of $f(x) = x^4 - 2x^3 + 3x - 5$.

Solution
If $f(-2) \neq 0$, then $x + 2$ is not a factor of $f(x)$. Since $f(-2) = (-2)^4 - 2(-2)^3 + 3(-2) - 5 = 21 \neq 0$, $(x + 2)$ is not a factor of $f(x)$.

Example 71.5
Show that $x - 3$ is a factor of $x^4 - 5x^3 + 8x^2 - 5x - 3$ and find one other factor.

Solution
We use synthetic division to show that the remainder is zero and hence $x - 3$ is a factor of the given polynomial as follows:

$$\begin{array}{rrrrr|l} 1 & -5 & 8 & -5 & -3 & \underline{3} \\ & 3 & -6 & 6 & 3 & \\ \hline 1 & -2 & 2 & 1 & 0 & \end{array}$$

Another factor of $f(x)$ is the quotient $x^3 - 2x^2 + 2x + 1$ obtained by the division of $f(x)$ by $x - 3$. Thus, $x^4 - 5x^3 + 8x^2 - 5x - 3 = (x - 3)(x^3 - 2x^2 + 2x + 1)$.

Exercises for Section 71

In Exercises 1 through 10, find the remainder R, using the remainder theorem. Check the results using synthetic division.

1. $(x^3 + 2x^2 - 6x + 1) \div (x - 1)$ **2.** $(x^4 - 2x^2 + 3x - 5) \div (x - 1)$

3. $(x^3 - 3x^2 + 6x - 4) \div (x - 2)$ **4.** $(x^3 + 3x^2 + 3x + 1) \div (x + 1)$

5. $(x^3 - 2x^2 + 4x + 2) \div (x - 2)$ **6.** $(x^3 + 2x + 3) \div (x - 1)$

7. $(x^3 + 2x - 3) \div (x - 1)$

8. $(2x^4 + 3x^3 - 8x^2 - 5x + 3) \div (x + 2)$

9. $(8x^4 + 4x^3 + 2x + 1) \div \left(x + \frac{1}{2}\right)$

10. $(9x^3 + 6x^2 + 4x + 2) \div (3x + 1)$

In Exercises 11 through 20, use the factor theorem to determine whether or not the second polynomial is a factor of the first. If it is a factor, write the first polynomial in factored form.

11. $x^3 + 2x^2 - 25x - 50, x - 5$ **12.** $4x^3 - 4x^2 - 11x + 6, x - 2$

13. $4x^4 - 12x^3 + x^2 + 12x + 4,\ x + 1$ **14.** $x^3 - 4x^2 + x + 6,\ x + 1$

15. $x^5 - 2x^4 - 3x^3 + 4x^2 - 6x - 12,\ x—2$

16. $3x^3 + 14x^2 + 8x - 1, x + 4$

17. $4x^3 - 4x^2 - 11x + 6,\ x + \frac{3}{2}$

18. $5x^3 - 12x^2 - 36x - 16, x + 2$

19. $x^4 + 4x^3 - x^2 - 16x - 12,\ x - 2$ **20.** $9x^3 + 6x^2 + 4x + 1,\ 3x + 1$

In Exercises 21 through 30, determine whether each statement is true or false.

21. 2 is a root of $2x^3 - 3x^2 - 4 = 0$.

22. -1 is a root of $3x^3 - 4x^2 + 3x - 2 = 0$.

23. 1 is a root of $3x^2 - 4x^2 + 3x - 2 = 0$.

24. -2 is a root of $x^3 + 6x^2 + 11x + 6 = 0$.

25. 2 is a root of $x^4 - 2x^3 - 7x^2 + 8x + 12 = 0$.

26. -1 is a root of $x^4 - 2x^3 - 7x^2 + 8x + 12 = 0$.

27. $\frac{1}{2}$ is a root of $2x^4 + 5x^3 - x^2 + 5x - 3 = 0$.

28. $\frac{3}{4}$ is a root of $8x^3 - 10x^2 - x + 3 = 0$.

29. $\frac{1}{3}$ is a root of $3x^4 - 40x^3 + 130x^2 - 120x + 27 = 0$.

30. $\frac{1}{2}$ is a root of $2x^4 - 5x^2 - 8x - 2 = 0$.

31. Find the value of k so that when $x^3 + 2x^2 - 3kx + 5$ is divided by $x - 1$, the remainder is 2.

32. For what value of p is $x^2 + 5x + p$ divisible by $x - p$?

33. Find a value for p such that $x + 1$ is a factor of $x^3 - 9x^2 + 14x + p$.

34. Show that $x - 1$ is a factor of $x^{100} - 1$.

35. Show that $x = -1$ is a root of $17x^{99} + 36x^{66} - 19 = 0$.

36. Show that $x + y$ is a factor of $x^5 + y^5$.

37. Show that $x + y$ is not a factor of $x^6 + y^6$.

38. Show that $x + y$ is a factor of $x^8 - y^8$.

39. Show that if n is a positive even integer, $x + y$ is a factor of $x^n - y^n$.

40. Show that if n is a positive odd integer, $x + y$ is a divisor of $x^n + y^n$.

SECTION 72

Factors and Roots of Polynomial Equations

In finding solutions for polynomial equations, it is advantageous to know the number and character of the roots before hand. There are two theorems that are particularly helpful in this situation. The first of these is so important that it is called the fundamental theorem of algebra.

Theorem 72.1. The Fundamental Theorem of Algebra
Every polynomial equation has at least one root. The root may be real or imaginary.

The proof of this theorem is dependent upon a knowledge of higher mathematics, and so we will accept the theorem without proof. However, we take advantage of the theorem in the following discussion.

Suppose that $f(x) = 0$ is a polynomial of degree $n \geq 1$. The fundamental theorem of algebra states that $f(x)$ has at least one root, r_1. By the factor theorem it follows that $(x - r_1)$ is a factor of $f(x)$. Thus, we can write

$$f(x) = (x - r_1) \cdot g(x) = 0$$

where $g(x)$ is a polynomial of degree $(n - 1) \geq 0$. If $n - 1 = 0$, $f(x) = a(x - r_1) = 0$. If $(n - 1) > 0$, then $g(x) = 0$ has at least one root, say r_2, and hence $g(x)$ has the factor $(x - r_2)$. Thus, $f(x) = (x - r_1)(x - r_2)h(x) = 0$, where $h(x)$ is a polynomial of degree $n - 2 \geq 0$. We continue this use of the fundamental theorem of algebra until we have obtained n factors of $f(x)$. Thus, we can write $f(x)$ in the factored form

$$f(x) = a(x - r_1)(x - r_2)(x - r_3) \cdots (x - r_n)$$

It follows then that $f(x) = 0$ has n roots. We have shown the following theorem.

Theorem 72.2. The Number-of-Roots Theorem
Every polynomial equation of degree n has exactly n roots.

We note that the theorem does not state that the roots are distinct. This means that if any factor $(x - r)$ of $f(x)$ occurs m times in the factored form of $f(x)$ then r is called a **root of multiplicity *m***. For example, $(x - 4)^2(x + 5)^3 = 0$ is a polynomial equation of degree 5 where 4 is called a **double root** or a **root of multiplicity 2** and -5 is called a **triple root** or **root of multiplicity 3.**

Example 72.1
Find a polynomial equation that has 2 as a double root and 3 as a triple root.

Solution
One such polynomial equation which has the required roots is

$$(x - 2)^2 \cdot (x - 3)^3 = 0$$

Note that if we multiply the equation by any number $a \neq 0$, the stated conditions are still satisfied. Thus, another solution is $a(x - 2)^2 \cdot (x - 3)^3 = 0$.

Example 72.2
Find the roots of the equation $f(x) = (x^2 - 1)(x^2 - 5x + 6) = 0$.

Solution
Since $(x^2 - 1) = (x - 1)(x + 1)$ and $(x^2 - 5x + 6) = (x - 2)(x - 3)$, $f(x) = (x - 1)(x + 1)(x - 2)(x - 3)$. Now $f(x)$ will be zero for any value of x that makes a factor equal to zero. Hence, by inspection the roots are -1, 1, 2, and 3. ●

Polynomial equations with real coefficients may have roots that are not real numbers. For example, we know that the equation $x^2 + x + 1 = 0$ has two imaginary roots since the discriminant $b^2 - 4ac = -3 < 0$ (see Section 61). Recalling that the roots of an equation of the form $ax^2 + bx + c = 0$ are given by the quadratic formula

$$x = \frac{-b \pm \sqrt{b^2 - 4ac}}{2a}$$

we find that the roots of $x^2 + x + 1 = 0$ are

$$\frac{-1 + \sqrt{3}j}{2} \quad \text{and} \quad \frac{-1 - \sqrt{3}j}{2}$$

We further observe that these roots are conjugates of one another, because they only differ by the sign before the radical. This is one example of the following theorem.

Theorem 72.3
If $a + bj$, where a and b are real numbers, is a root of $f(x) = 0$, its conjugate $a - bj$ is also a root of $f(x) = 0$.

Example 72.3
Find the roots of the equation $(x - 1)^2(x^2 + 2x + 2) = 0$.

Solution
We observe that $x = 1$ is a double root or a root of multiplicity 2 of the equation. To find the remaining roots, we factor the expression $x^2 + 2x + 2$ by using the quadratic formula. In so doing we are finding those values for which $x^2 + 2x + 2 = 0$. Substitution into the quadratic formula yields

$$x = \frac{-2 \pm \sqrt{4 - 8}}{2} = \frac{-2 \pm \sqrt{-4}}{2} = \frac{-2 \pm 2j}{2}$$

Thus, $x = -1 + j$ and $x = -1 - j$.
Therefore, the roots of the equation are 1, 1, $-1 + j$, and $-1 - j$. ●

We have seen that if $f(x) = 0$ is an nth-degree polynomial equation and has a root $x = r_1$, then $x - r_1$ is a factor of $f(x)$ and we may write the equation as

$$f(x) = (x - r_1) \cdot g(x) = 0 \qquad \text{where } g(x) \text{ is of degree } n - 1$$

Since the product of two factors can be zero only if one or both of the factors is equal to zero, the roots of $f(x) = 0$ must satisfy either $x - r_1 = 0$ or $g(x) = 0$. The only root of $x - r_1 = 0$ is r_1, which we knew at the start. Hence, any additional roots of $f(x) = 0$ must be roots of $g(x) = 0$. The equation $g(x) = 0$ is called the **depressed** or **reduced equation** since its degree is one lower than that of the original equation. As we shall see, the reduced equation is often easier to solve than the original equation. If r_1 is known, we can find the depressed equation by synthetic division. If the depressed equation is quadratic, we can find its roots by factoring or applying the quadratic formula.

Example 72.4

If 3 is a root of $f(x) = x^3 - 4x^2 + x + 6 = 0$, solve the equation.

Solution

Since 3 is a root of $f(x) = 0$, $x - 3$ is a factor of $f(x)$. Dividing synthetically,

$$\begin{array}{rrrr|l} 1 & -4 & 1 & 6 & \underline{3} \\ & 3 & -3 & -6 & \\ \hline 1 & -1 & -2 & 0 & \end{array}$$

we find that the quotient is $g(x) = x^2 - x - 2$. Thus, $f(x) = (x - 3) \times (x^2 - x - 2)$ and the depressed equation is $x^2 - x - 2 = 0$. Factoring $x^2 - x - 2$ yields $(x - 2)(x + 1) = 0$. Hence, the roots of the equation are 3, 2, and -1.

Example 72.5

If 1 is a root of the equation $f(x) = x^3 - x^2 + x - 1 = 0$, solve the equation.

Solution

Since $x = 1$ is a root of $f(x) = 0$, $x - 1$ is a factor of $f(x)$. Using synthetic division, we have

$$\begin{array}{rrrr|l} 1 & -1 & 1 & -1 & \underline{1} \\ & 1 & 0 & 1 & \\ \hline 1 & 0 & 1 & 0 & \end{array}$$

Hence, the quotient is $x^2 + 1$, and so $f(x) = (x - 1)(x^2 + 1)$. The depressed equation is $x^2 + 1 = 0$. Solving for x yields $x = \pm\sqrt{-1}$ or $x = j$ and $x = -j$. Therefore, the roots of the equation are 1, j, and $-j$.

Remark 72.1. *Whenever the depressed equation is of degree 2, we can always complete the solution by either factoring or using the quadratic formula.*

If more than one root is known, the depressed equation may be further depressed. The process is illustrated in the following example.

Example 72.6

If -2 and $\frac{1}{2}$ are roots of $f(x) = 2x^4 - 9x^3 + 6x^2 + 51x - 26 = 0$, solve the equation.

Solution

Since -2 is a root of $f(x) = 0$, $x + 2$ is a factor. We divide synthetically:

$$\begin{array}{rrrrr|l} 2 & -9 & 6 & 51 & -26 & \underline{-2} \\ & -4 & 26 & -64 & 26 & \\ \hline 2 & -13 & 32 & -13 & 0 & \end{array}$$

and find that $f(x) = (x + 2)\cdot(2x^3 - 13x^2 + 32x - 13)$. But $\frac{1}{2}$ is also a root of $f(x) = 0$ and is therefore a root of the depressed equation $2x^3 - 13x^2 + 32x - 13 = 0$. We again divide synthetically, in this case by the factor $x - \frac{1}{2}$, to obtain

$$\begin{array}{rrrr|l} 2 & -13 & 32 & -13 & \underline{\frac{1}{2}} \\ & 1 & -6 & 13 & \\ \hline 2 & -12 & 26 & 0 & \end{array}$$

Hence, our second depressed equation is $2x^2 - 12x + 26 = 0$. Dividing the equation by 2 yields $x^2 - 6x + 13 = 0$. Solving by means of the quadratic formula, we have

$$x = \frac{6 \pm \sqrt{-16}}{2} = \frac{6 \pm 4j}{2} = 3 \pm 2j$$

Therefore, the roots of the equation are -2, $\frac{1}{2}$, $3 + 2j$, and $3 - 2j$.

Example 72.7

Solve the equation $x^4 + 2x^3 + 5x^2 + 2x + 4 = 0$ given that j is a root.

Solution

Since j is a root, we know that the conjugate $-j$ is also a root. We now use synthetic division twice to reduce the equation to a quadratic equation. Thus,

$$\begin{array}{rrrrr|l} 1 & 2 & 5 & 2 & 4 & \underline{j} \\ & j & -1 + 2j & -2 + 4j & -4 & \\ \hline 1 & j + 2 & 2j + 4 & 4j & 0 & \end{array}$$

and

$$\begin{array}{rrrr|l} 1 & j + 2 & 2j + 4 & 4j & \underline{-j} \\ & -j & -2j & -4j & \\ \hline 1 & 2 & 4 & 0 & \end{array}$$

The depressed quadratic equation $x^2 + 2x + 4 = 0$ is solved by employing the quadratic formula as follows:

$$x = \frac{-2 \pm \sqrt{-12}}{2} = \frac{-2 \pm 2\sqrt{3}j}{2} = -1 \pm \sqrt{3}j$$

Therefore, the roots of the equation are j, $-j$, $-1 + \sqrt{3}j$, and $-1 - \sqrt{3}j$.

Exercises for Section 72

In Exercises 1 through 5, find the roots of each equation by inspection and state the multiplicity.

1. $(x - 1)(x - 2)^2(x - 3)^3 = 0$

2. $(x + 5)(2x + 3)^4 = 0$

3. $x^3(x + 1)^4(x - 2)^5 = 0$

4. $(3x + 2)^3(x^2 - 4x + 4)^2 = 0$

5. $(x^2 - 6x + 9)^2(x^2 - 4x + 3) = 0$

In Exercises 6 through 10, find a polynomial equation with integral coefficients having the numbers as roots.

6. $1, 2, -3$

7. $1, 2, 2, 4$

8. $0, \frac{1}{2}, 1, 2$

9. $2, 1 + j$

10. $1, 2, 1 - j$

In Exercises 11 through 25, solve each equation given the roots indicated.

11. $x^3 - 4x^2 + x + 6 = 0, r_1 = 3$

12. $3x^3 - 4x^2 + 2x + 4 = 0, r_1 = -\frac{2}{3}$

13. $x^3 + 6x^2 + 11x + 6 = 0, r_1 = -2$

14. $x^3 - 4x^2 + 5x - 2 = 0, r_1 = 1$

15. $2x^3 - 3x^2 - x + 1 = 0, r_1 = \frac{1}{2}$

16. $4x^3 - 4x^2 - 11x + 6 = 0, r_1 = -\frac{3}{2}$

17. $x^3 - x^2 + 4x - 4 = 0, r_1 = 2j$

18. $x^3 + 5x^2 + 9x + 5 = 0, r_1 = -2 + j$

19. $2x^4 - 3x^3 - 20x^2 + 27x + 18 = 0, r_1 = -3, r_2 = -\frac{1}{2}$

20. $3x^4 + 5x^3 + x^2 + x - 10 = 0, r_1 = 1, r_2 = -2$

21. $x^4 - 4x^3 + 3x^2 + 4x - 4 = 0, r_1 = 1, r_2 = -1$

22. $4x^4 + x^2 - 3x + 1 = 0, \frac{1}{2}$ is a double root

23. $x^4 - 7x^3 + 12x^2 + 4x - 16 = 0$, 2 is a double root

24. $x^4 - 2x^3 - 7x^2 + 8x + 12 = 0, r_1 = -1, r_2 = 2$

25. $4x^5 - 16x^4 + 17x^3 - 19x^2 + 13x - 3 = 0, r_1 = 3, r_2 = j$

SECTION 73

Rational Roots

In this section we shall consider a method for finding the rational roots of equations with **integral coefficients**. In preparation, let us recall that a real number is either rational or irrational and that a rational number is one that can be expressed as a quotient or ratio of two integers. We can determine whether any particular rational number r is a root of the equation $f(x) = 0$ by substituting r for x in the equation. It is most convenient to do this by synthetic division if we recall that the last number in the third line of the synthetic division is $f(r)$. The number r is a root if $f(r) = 0$.

If we attempt to find the rational roots of a given equation by trial and error, the process might be somewhat lengthy, since there are infinitely many rational numbers. In fact, an equation such as $x^2 - 5 = 0$ has no rational roots, and thus we could never reach a conclusion concerning the roots by this method. The following theorem shows us how to determine the possible rational roots.

Theorem 73.1. The Rational-Root Theorem
If the coefficients of the equation

$$f(x) = a_0x^n + a_1x^{n-1} + a_2x^{n-2} + \cdots + a_{n-1}x + a_n = 0 \quad (a_0 \neq 0)$$

are all integers, and if p/q is a rational root in lowest terms, p is a factor of the constant term a_n and q is a factor of the leading coefficient a_0.

The proof of the theorem is direct, and so we shall give a brief outline of how to proceed. Since p/q is a root of $f(x) = 0$, we have

$$a_0\left(\frac{p}{q}\right)^n + a_1\left(\frac{p}{q}\right)^{n-1} + \cdots + a_{n-1}\left(\frac{p}{q}\right) + a_n = 0$$

Multiplying by q^n, we obtain

$$a_0p^n + a_1p^{n-1}q + \cdots + a_{n-1}pq^{n-1} + a_nq^n = 0$$

Transposing the last term and taking the common factor p out of what remains on the left yields

$$p[a_0p^{n-1} + a_1p^{n-2}q + a_2p^{n-3}q^2 + \cdots + a_{n-1}q^{n-1}] = -a_nq^n$$

Both sides of the equation are integers. Now p is a factor of the left-hand side, it must also be a factor of the right-hand side, a_nq^n. Since p/q is in lowest terms, p cannot be a factor of q^n and must therefore be a factor of a_n. Similarly, if

$$a_1p^{n-1}q + \cdots + a_{n-1}pq^{n-1} + a_nq^n = -a_0p^n$$

then q is a factor of the left-hand side and also the right-hand side, $-a_0p^n$. Since q cannot be a factor of p^n, it must therefore be a factor of a_0.

We must be careful to note that the rational-root theorem says nothing about the existence of rational roots. The theorem merely states how we determine the *possible* rational roots if $f(x) = 0$ is a polynomial equation with integral coefficients.

The following theorem is a direct consequence of the rational-root theorem and characterizes the roots of an equation if the leading coefficient is equal to 1.

Theorem 73.2
Any rational root of the equation

$$f(x) = x^n + a_1x^{n-1} + a_2x^{n-2} + \cdots + a_{n-1}x + a_n = 0$$

with integral coefficients, must be an ***integer*** *that is a factor of the constant term* a_n.

Example 73.1
List the possible rational roots of:
(a) $3x^3 - 4x^2 + x - 2 = 0$
(b) $x^4 - 2x^2 + 6x - 12 = 0$

Solution
For $3x^2 - 4x^2 + x - 2 = 0$, the possible numerators p are ± 1 and ± 2 (the factors of 2); the possible denominators q are ± 1 and ± 3 (the factors of 3). Thus, the possible rational roots p/q are ± 1, ± 2, $\pm\frac{1}{3}$, and $\pm\frac{2}{3}$.

For $x^4 - 2x^2 + 6x - 12 = 0$, the leading coefficient is 1, and hence the possible rational roots are the factors of the constant term -12 (see Theorem 73.2). These factors are ± 1, ± 2, ± 3, ± 4, ± 6, and ± 12.

Example 73.2
Find the rational roots of $2x^3 - 3x^2 - 11x + 6 = 0$.

Solution
If the equation has rational roots of the form p/q, then p must be a factor of 6 and q must be a factor of 2. The possible values for p are ± 1, ± 2, ± 3, ± 6, and the possible values for q are ± 1 and ± 2. Hence, the *possible* values of p/q are $\pm\frac{1}{1}$, $\pm\frac{2}{1}$, $\pm\frac{3}{1}$, $\pm\frac{6}{1}$, $\pm\frac{1}{2}$, $\pm\frac{2}{2}$, $\pm\frac{3}{2}$, and $\pm\frac{6}{2}$. When we eliminate duplications, the set of possible rational roots is ± 1, ± 2, ± 3, ± 6, $\pm\frac{1}{2}$, and $\pm\frac{3}{2}$. Testing the numbers in the preceding line in order, we find that -2 is a root, since

$$\begin{array}{rrrr|r}
2 & -3 & -11 & +6 & \underline{-2} \\
 & -4 & 14 & -6 & \\
\hline
2 & -7 & 3 & 0 &
\end{array}$$

Any further roots of the given equation must be roots of the depressed equation $2x^2 - 7x + 3 = 0$, which is quadratic. By factoring, we have $2x^2 - 7x + 3 = (2x - 1)(x - 3) = 0$. Thus, $x = \frac{1}{2}$ and $x = 3$. Our complete solution is $x = -2, \frac{1}{2}$, and 3. We note that since we know the roots of the equation, we also know the factors of the polynomial. Hence, $2x^3 - 3x^2 - 11x + 6 = (x + 2)(2x - 1)(x - 3)$ expresses the given polynomial as a product of factors with integral coefficients. ●

In Section 29 we saw how the sign graph method could be applied to the solution of higher-degree inequalities in one variable. The solution of such inequalities depends upon knowing or being able to find the factors of the given algebraic expressions. If the expression is given in factored form, we construct a sign graph for each of the factors, mark off the critical values, and then determine the solution set (see Examples 29.1 and 29.2).

If the algebraic expression $f(x)$ is not given in factored form, we use the rational-root theorem to find the factors. This procedure assumes that $f(x)$ is a polynomial and that $f(x) = 0$ has rational roots. Let us consider the procedure in the following example.

Example 73.3

Find the solution set for $2x^3 - 3x^2 - 11x + 6 > 0$.

Solution

We must first factor $2x^3 - 3x^2 - 11x + 6$. Since the equation $2x^3 - 3x^2 - 11x + 6 = 0$ has three rational roots, $-2, \frac{1}{2}$, and 3 (see Example 73.2), the factors of the given polynomial are $(x + 2)$, $(2x - 1)$, and $(x - 3)$. Thus, the solution set for $2x^3 - 3x^2 - 11x + 6 > 0$ is equivalent to the solution set for $(x + 2)(2x - 1)(x - 3) > 0$, since $2x^3 - 3x^2 - 11x + 6 = (x + 2)(2x - 1)(x - 3)$. If we construct a sign graph for each of the three factors $(x + 2)$, $(2x - 1)$, and $(x - 3)$ and mark off the critical values $-2, \frac{1}{2}$, and 3 on a number line directly below these sign graphs and use Theorem 29.1, we find that the solution set is $-2 < x < \frac{1}{2}$ or $x > 3$.

Example 73.4

Find the rational roots of $x^4 - 4x^3 - 2x^2 + 21x - 18 = 0$ and then solve the equation completely.

Solution

Since the leading coefficient of the equation is 1, the possible rational roots are the factors of the constant term -18. These factors are $\pm 1, \pm 2, \pm 3, \pm 6, \pm 9$, and ± 18. Trying these numbers in order of increasing magnitude, we find that 2 is a root, as follows:

$$\begin{array}{rrrrr|l} 1 & -4 & -2 & 21 & -18 & \underline{2} \\ & 2 & -4 & -12 & 18 & \\ \hline 1 & -2 & -6 & 9 & 0 & \end{array}$$

Any further roots of the given equation must be roots of the depressed equation $x^3 - 2x^2 - 6x + 9 = 0$. The possible rational roots of this equation are the factors of the constant term 9. These factors are ±1, ±3, and ±9. We find that 3 is a root, as follows:

$$\begin{array}{rrrr|l} 1 & -2 & -6 & 9 & \underline{3} \\ & 3 & 3 & -9 & \\ \hline 1 & 1 & -3 & 0 & \end{array}$$

To complete the solution, we solve the depressed equation $x^2 + x - 3 = 0$ by the quadratic formula and find that $x = (-1 \pm \sqrt{13})/2$. Thus, the complete solution is $x = 2, 3, (-1 \pm \sqrt{13})/2$. We note that the equation has four real roots, two rational and two irrational.

Example 73.5

Find the rational roots of $x^3 + 6x - 3 = 0$.

Solution

The possible rational roots of the equation are ±1 and ±3. We find that none of these is a solution. Therefore, the three roots are not rational. Let us carefully note that this does not imply that all three roots are irrational. We must keep in mind the possibility of imaginary roots. ●

The use of the rational-root theorem is restricted to polynomial equations with integral coefficients. If the equation has fractional coefficients, we must first multiply the equation by the lowest common denominator and then apply the theorem. Let us illustrate the procedure in the following example.

Example 73.6

Find the rational roots of $x^3 - 3x^2 - \frac{7}{2}x - 2 = 0$ and solve completely.

Solution

It would be incorrect to state that the possible rational roots are the factors of the constant term −2 because not every coefficient of the equation is an integer. We note that every coefficient will become integral is we multiply the equation by the lowest common denominator, which is 2. Thus, the equation is equivalent to $2x^3 - 6x^2 - 7x - 4 = 0$. Hence, the possible rational roots are ±1, ±2, ±4, and $\pm\frac{1}{2}$. We find that 4 is a root as follows:

$$\begin{array}{rrrr|l} 2 & -6 & -7 & -4 & \underline{4} \\ & 8 & 8 & 4 & \\ \hline 2 & 2 & 1 & 0 & \end{array}$$

The remaining roots are the roots of the depressed equation $2x^2 + 2x + 1 = 0$. We use the quadratic formula to find that the roots of $2x^2 + 2x + 1 = 0$ are $(-1 \pm j)/2$. Thus, the roots of the equation are 4, $(-1 + j)/2$, and $(-1 - j)/2$.

The method for finding rational roots by means of the rational-root theorem and the use of synthetic division may involve a considerable number of trials if the coefficients of the given equation are large. For example, in the equation $x^3 + x^2 - 60x + 48 = 0$, the possible rational roots are the factors of 48. These factors are ±1, ±2, ±3, ±4, ±6, ±8, ±12, ±16, ±24, and ±48, twenty factors in all. In an attempt to simplify the process of finding the roots, we shall present some rules that cut down the work.

Theorem 73.2. Upper and Lower Bounds for the Roots
Suppose that $f(x) = a_0x^n + a_1x^{n-1} + \cdots + a_n = 0$ *has real coefficients and* $a_0 > 0$. *Then:*

1. *If we divide* $f(x)$ *synthetically by* $x - a$, *where* $a \geq 0$, *and find that all the numbers obtained in the third row are nonnegative, no real root of* $f(x) = 0$ *can be greater than* a. *That is,* a *is an upper bound for the real roots of* $f(x) = 0$.
2. *If we divide* $f(x)$ *synthetically by* $x - b$, *where* $b < 0$, *and find that all the numbers obtained in the third row are alternately positive and negative (or zero), no real root of* $f(x) = 0$ *can be less than* b. *That is,* b *is a lower bound for the real roots of* $f(x) = 0$.

Example 73.7
Find the rational roots of $x^4 - 4x^3 + 20x^2 - 64x + 64 = 0$ using the upper and lower bounds for the roots to cut down the number of trials and solve completely.

Solution
The possible rational roots of the given equation are the factors of 64. These are ±1, ±2, ±4, ±8, ±16, ±32, and ±64. We find that -1 is a lower bound for the roots, since the numbers in the third line of the synthetic division are alternately positive and negative, as follows:

$$\begin{array}{rrrrr|l} 1 & -4 & 20 & -64 & 64 & \underline{-1} \\ & -1 & 5 & -25 & 89 & \\ \hline 1 & -5 & 25 & -89 & 153 & \end{array}$$

Similarly, we find that 4 is an upper bound for the roots, since no number in the third line of the synthetic division is negative, as follows:

$$\begin{array}{rrrrr|l} 1 & -4 & 20 & -64 & 64 & \underline{4} \\ & 4 & 0 & 80 & 64 & \\ \hline 1 & 0 & 20 & 16 & 128 & \end{array}$$

Thus, if the equation has rational roots, they are *not* less than negative 1 or greater than 4. This limits our choice to the only numbers not included in these restrictions, the values 1 and 2. Synthetic division shows that 2 is a *multiple root*, since

$$\begin{array}{rrrrr|l} 1 & -4 & 20 & -64 & 64 & \underline{|2} \\ & 2 & -4 & 32 & -64 & \\ \hline 1 & -2 & 16 & -32 & 0 & \end{array}$$

and

$$\begin{array}{rrrr|l} 1 & -2 & 16 & -32 & \underline{|2} \\ & 2 & 0 & 32 & \\ \hline 1 & 0 & 16 & 0 & \end{array}$$

Hence, the only rational root of the equation is 2, and this is a root of multiplicity 2. The remaining roots of the equation are the roots of the depressed equation $x^2 + 16 = 0$ and are $\pm 4j$. The complete solution set for the equation is $x = 2$, 2, and $\pm 4j$. ●

In Example 73.7 we saw that -1 was a lower bound and 4 was an upper bound for the roots. We observe that any number less than -1 is a lower bound, Thus, -100 is also a lower bound. Similarly, any number greater than 4 is an upper bound. Thus, 1000 is also an upper bound for the roots. It is, of course, advantageous to locate as low an upper bound and as high a lower bound as is conveniently possible in order to reduce the number of trials needed in the search for real roots.

Another rule that is often helpful in finding the real roots of an equation is **Descartes' rule of signs**. This rule deals with the variation of signs in an equation. Before stating the rule, we note that a **variation** of sign occurs when two consecutive terms in a polynomial, with the terms in order of decreasing degree, differ by sign. Thus, $x^3 - x^2 + 5x - 6$ has three variations of sign, while $x^4 + 6x^2 - 7x + 8$ has two variations of sign. Now let us state the rule.

Theorem 73.3. Descartes' Rule of Sign
If $f(x) = 0$ is a polynomial with real coefficients, the number of positive roots of $f(x) = 0$ is either equal to the number of variations of sign of $f(x)$ or is this number less an even positive integer. The number of negative roots of $f(x) = 0$ is equal to the number of variations in sign $f(-x)$ or is this number less an even positive integer.

Example 73.8
Use Descartes' rule of signs to discuss the possible number of positive, negative, and imaginary roots of $2x^3 - 4x^2 + x - 2 = 0$. Solve the equation completely.

Solution
The number of positive roots of $f(x) = 2x^3 - 4x^2 + x - 2 = 0$ is either 3 or 1, since $f(x)$ has three variations in sign. We see that there are no negative roots, since $f(-x) = -2x^3 - 4x^2 - x - 2$ has no variations in

sign. Thus, $f(x) = 0$ must have three positive real roots or one positive real root and two conjugate imaginary roots. We find that 2 is a root, as follows:

$$\begin{array}{rrrr|l} 2 & -4 & 1 & -2 & \underline{|2} \\ & 4 & 0 & 2 & \\ \hline 2 & 0 & 1 & 0 & \end{array}$$

The remaining roots of the given equation are the roots of the depressed equation $2x^2 + 1 = 0$. These roots are $\pm j\sqrt{2}/2$. Thus, the complete solution is $x = 2, \pm j\sqrt{2}/2$.

Example 73.9

Use the information presented in this section to find the roots of $f(x) = x^4 - x^3 - 19x^2 + 49x - 30 = 0$.

Solution

Since the degree of the equation is 4, there will be four roots. We note that $f(x)$ has three variations of sign and $f(-x) = x^4 + x^3 - 19x^2 - 49x - 30$ has one variation of sign. Thus, by Descartes' rule of signs, $f(x) = 0$ has three positive real roots and one negative root, or one positive real root, one negative real root, and two conjugate imaginary roots. In any case we note that the equation *must* have one negative root and at least one positive root. Using the rational-root theorem, we note that the possible rational roots are the factors of -30. These factors are ± 1, ± 2, ± 3, ± 5, ± 6, ± 10, ± 15, and ± 30. We first check the negative values and find that -5 is a root, as follows:

$$\begin{array}{rrrrr|l} 1 & -1 & -19 & 49 & -30 & \underline{|-5} \\ & -5 & 30 & -55 & 30 & \\ \hline 1 & -6 & 11 & -6 & 0 & \end{array}$$

The remaining roots of the given equation are the roots of the depressed equation $x^3 - 6x^2 + 11x - 6 = 0$. The depressed equation cannot have negative roots, since $f(x) = 0$ has only one negative root and that root has been found. Hence, we need check only positive values. We quickly find that 1 is also a root, as follows:

$$\begin{array}{rrrr|l} 1 & -6 & 11 & -6 & \underline{|1} \\ & 1 & -5 & 6 & \\ \hline 1 & -5 & 6 & 0 & \end{array}$$

The two remaining roots are roots of the second depressed equation $x^2 - 5x + 6 = 0$. Since $x^2 - 5x + 6 = (x - 2)(x - 3) = 0$, $x = 2$ and $x = 3$. Thus, the complete solution is $x = -5, 1, 2$, and 3.

Exercises for Section 73

In Exercises 1 through 20, find all rational roots. If a quadratic depressed equation is obtained, solve the equation completely.

1. $2x^3 - x^2 - 4x + 2 = 0$

2. $x^3 - 2x^2 - x + 2 = 0$

3. $2x^3 - 7x^2 + 10x - 6 = 0$

4. $24x^3 - 2x^2 - 5x + 1 = 0$

5. $x^3 - 2x^2 - 5x + 6 = 0$

6. $2x^3 - 13x^2 + 27x - 18 = 0$

7. $8x^3 - 10x^2 - x + 3 = 0$

8. $9x^4 - 3x^3 + 7x^2 - 3x - 2 = 0$

9. $x^3 + 3x^2 - 10x - 24 = 0$

10. $x^4 - 2x^3 + x^2 - 2x = 0$

11. $2x^4 + 5x^3 - 11x^2 - 20x + 12 = 0$

12. $x^4 - 2x^3 + x^2 + 2x - 2 = 0$

13. $2x^4 + 5x^3 - x^2 + 5x - 3 = 0$

14. $3x^4 - 40x^3 + 130x^2 - 120x + 27 = 0$

15. $x^4 - 6x^2 - 8x - 3 = 0$

16. $x^3 - \frac{17}{2}x^2 + \frac{9}{2}x - 4 = 0$

17. $x^3 - \frac{9}{2}x^2 + \frac{7}{4}x + 1 = 0$

18. $x^3 + 3x^2 - 5x - 39 = 0$

19. $\frac{x^3}{3} - x - \frac{2}{3} = 0$

20. $5x^3 + 12x^2 - 36x - 16 = 0$

In Exercises 21 through 25, use the results of the exercises indicated and the sign graph method to find the solution set for the given inequality.

21. $x^2 - 2x^2 - x + 2 \le 0$, Exercise 2

22. $x^3 - 2x^2 - 5x + 6 > 0$, Exercise 5

23. $8x^3 - 10x^2 - x + 3 \ge 0$, Exercise 7

24. $2x^4 + 5x^3 - 11x^2 - 20x + 12 < 0$, Exercise 11

25. $x^4 - 6x^2 - 8x - 3 \le 0$, Exercise 15

In Exercises 26 through 30, without solving the equation, state whatever you can concerning positive, negative, and imaginary roots. Use any of the results of this section.

26. $x^3 + 4x - 4 = 0$

27. $3x^3 + 5x + 1 = 0$

28. $x^3 - x^2 + 5x - 2 = 0$

29. $x^4 - 3x^2 + 1 = 0$

30. $x^4 + 3x^2 + 1 = 0$

31. A variable electric current is given by the formula $i = 2t^3 - 13t^2 + 27t - 15$. If t is in seconds, at what time is the current equal to 3 A?

32. A variable voltage is given by the formula $e = 2t^3 - 13t^2 + 22t - 3$. At what time t (in seconds) is the voltage equal to 5?

33. The lengths of the sides of a rectangular box, measured in centimeters, are three consecutive integers. If the volume is 720 cm^3, find the length of the sides.

34. An open box is to be made from a square sheet of metal 18 cm on each side by cutting out equal squares from the corners and folding up the sides. How long should the edge of the cutout square be if the volume of the box is to be 392 cm^3?

35. A slice 1 cm thick is cut off from one side of a cube. The volume that remains is 100 cm^3. Find the side of the original cube.

SECTION 74

Real Roots by Linear Approximation

There are many procedures that can be used to approximate the real roots of a polynomial equation that are irrational. In this section we shall discuss the method of *linear interpolation.* This method assumes that if two points are close to one another, the straight line joining the two points is a good approximation to the actual curve between the points.

To illustrate the procedure we will follow, let us consider the equation $x^3 + 3x - 1 = 0$. The possible rational roots are ± 1. Synthetic division or the remainder theorem will show that neither of these values is a root. Thus, the equation has no rational roots. If we use Descartes' rule of signs, we find that the equation has one real positive root, no negative roots, and therefore two conjugate imaginary roots. The bounds for the real root of the equation can be found by noting that for $f(x) = x^3 + 3x - 1$, $f(0) = -1$, and $f(1) = 3$. We notice that the points $(0, -1)$ and $(1, 3)$ lie on opposite sides of the x axis, and we therefore assume that the graph of $y = f(x)$ crosses the axis somewhere between $x = 0$ and $x = 1$. If this is the case, then $f(x)$ has a zero between 0 and 1, and hence $x^3 + 3x - 1 = 0$ has a root between 0 and 1. This is the basic idea of the following theorem.

Theorem 74.1
If $f(x)$ is a real polynomial and a and b are distinct numbers such that $f(a)$ and $f(b)$ have different signs, there exists at least one number r between a and b such that $f(r) = 0$.

Theorem 74.1 has the effect of determining the interval in which we have at least one real root. Repeated use of the theorem shortens the interval and hence allows us to compute an approximate value for the root to any degree of accuracy required.

Example 74.1

Find the real root of $f(x) = x^3 + 3x - 1$ to the nearest tenth.

Solution

Since $f(0) = -1$ and $f(1) = 3$, Theorem 74.1 tells us that the equation has a real root between 0 and 1. We now construct a scale drawing of the straight line between the point $(0, -1)$ and $(1, 3)$. By estimating where the line crosses the x axis, we obtain a linear approximation to the root in the tenths place (see Figure 74.1). The line crosses the x axis between 0.2 and 0.3. On this basis we use synthetic division to test 0.3 as follows:

$$\begin{array}{cccc|l} 1 & 0 & 3 & -1 & \underline{0.3} \\ & 0.3 & 0.09 & 0.927 & \\ \hline 1 & 0.3 & 3.09 & -0.073 & \end{array}$$

Figure 74.1

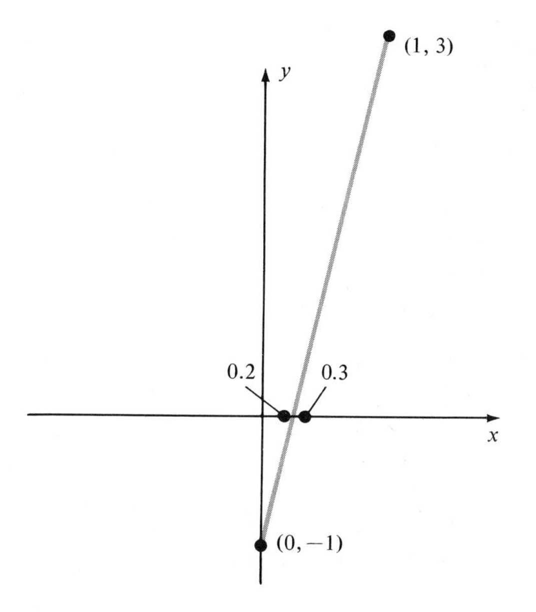

The root must be between 0.3 and 1 since $f(0.3) = =0.073$ and $f(1) = 3$ have opposite signs. To narrow the interval, we test 0.4 as follows:

$$\begin{array}{cccc|l} 1 & 0 & 3 & -1 & \underline{0.4} \\ & 0.4 & 0.16 & 1.264 & \\ \hline 1 & 0.4 & 3.16 & 0.264 & \end{array}$$

We now know the root lies between 0.3 and 0.4. Since the remainder for $x = 0.3$ is numerically smaller, we conclude that this is the value of the root to the nearest tenth. We note that this process may be continued to obtain a greater degree of accuracy.

The graph of $y = f(x) = x^3 + 3x - 1$ can be roughly sketched from the information we have obtained in Example 74.1 and some additional points. We know that the curve crosses the x axis in only one point, and we have determined the value of this point to the nearest tenth. In addition to this, we know that the points $(0, -1)$ and $(1, 3)$ lie on the graph. With this information and the points shown in Table 74.1, we sketch the curve shown in Figure 74.2. Note that scales on the x and y axes are conveniently choosen to fit our table values.

Table 74.1

x	-2	-1	0	1	2
y	-15	-5	-1	3	13

Figure 74.2

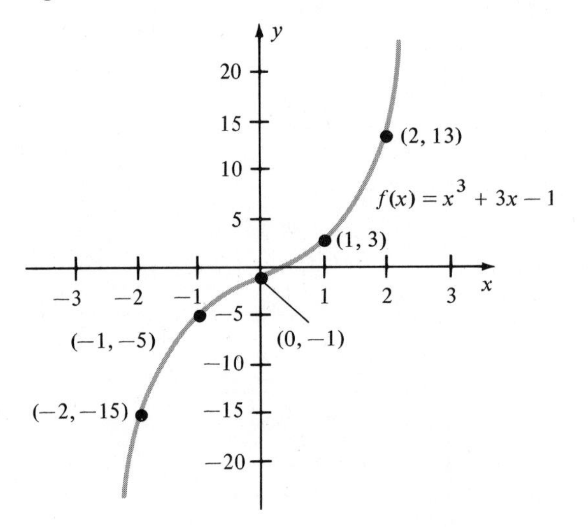

Example 74.2

Discuss the character of the roots of $x^3 - x^2 - 2x + 1 = 0$ and then approximate to the nearest hundredth the least positive root.

Solution

We first note that $f(x) = x^3 - x^2 - 2x + 1 = 0$ has two variations of sign, and hence by Descartes' rule of signs must have either two positive real roots or no positive real roots. Since $f(-x) = -x^3 - x^2 + 2x + 1$ has one variation of sign, the equation has exactly one negative root. Since we are trying to approximate the least positive root, we check the function for a sign change in the interval from 0 to 1. Now $f(0) = 1$ and $f(1) = -1$. Hence, by Theorem 74.1, the equation has at least one real root in this interval. If we combine this fact with Descartes' rule of signs, we know that the equation has one negative root and two positive roots.

To obtain our first approximation, we draw a straight line between the points (0, 1) and (1, −1) (Figure 74.3). In the figure the line crosses the x axis near 0.5. Hence, we try this value by using synthetic division, as follows:

$$\begin{array}{rrrr|l} 1 & -1 & -2 & 1 & \underline{0.5} \\ & 0.5 & -0.25 & -1.125 & \\ \hline 1 & -0.5 & -2.25 & \multicolumn{1}{|r}{-0.125} & \end{array}$$

Figure 74.3

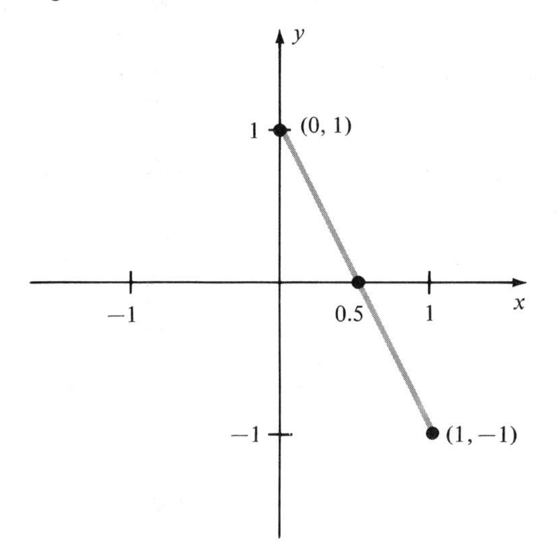

Since $f(0) = 1$ and $f(0.5) = -0.125$, the root lies between $x = 0$ and $x = 0.5$. We now try 0.4 as follows:

$$\begin{array}{rrrr|l} 1 & -1 & -2 & 1 & \underline{0.4} \\ & 0.4 & -0.24 & -0.896 & \\ \hline 1 & -0.6 & -2.24 & \multicolumn{1}{|r}{0.104} & \end{array}$$

We now know that the root lies between 0.4 and 0.5 and we use the scale drawing in Figure 74.4 on page 436 to approximate the hundredths digit. The root is apparently between 0.44 and 0.45. We try these values by synthetic division and round off to three significant digits, as follows:

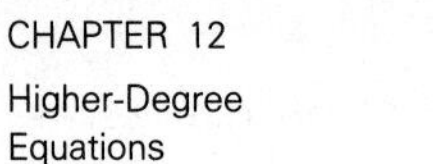

Figure 74.4

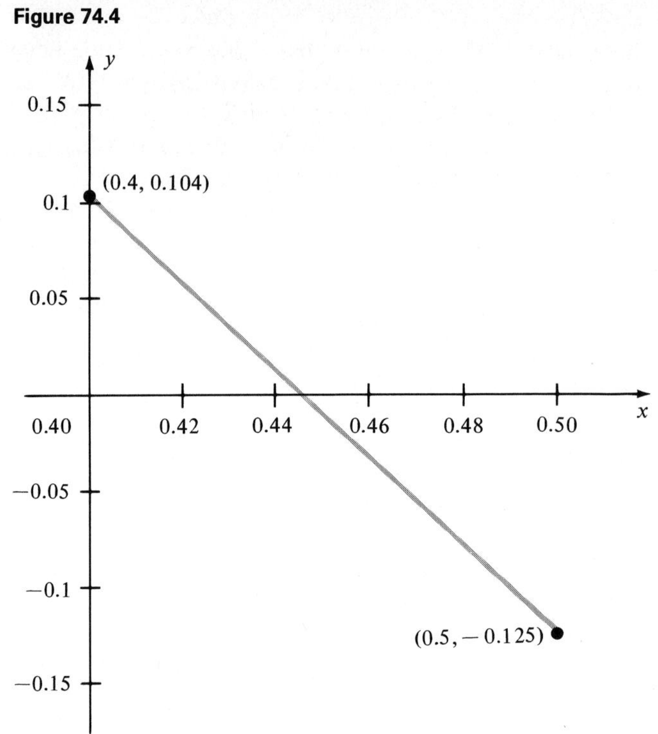

$$
\begin{array}{rrrr|l}
1 & -1 & -2 & 1 & \underline{0.44} \\
 & 0.44 & -0.25 & -0.99 & \\
\hline
1 & -0.56 & -2.25 & \multicolumn{1}{|r}{0.01} &
\end{array}
$$

$$
\begin{array}{rrrr|l}
1 & -1 & -2 & 1 & \underline{0.45} \\
 & 0.45 & -0.25 & -1.01 & \\
\hline
1 & -0.55 & -2.25 & \multicolumn{1}{|r}{-0.01} &
\end{array}
$$

Since the signs of the remainders are not the same, the root must be between 0.44 and 0.45. Since the remainders are numerically equal, we cannot base a conclusion upon these values. Hence, we check the midpoint of the interval $x = 0.445$, rounding off to four significant digits.

$$
\begin{array}{rrrr|l}
1 & -1 & -2 & 1 & \underline{0.445} \\
 & 0.445 & -0.247 & -0.999 & \\
\hline
1 & -0.555 & -2.247 & \multicolumn{1}{|r}{0.001} &
\end{array}
$$

Thus, the root lies between 0.445 and 0.45, and hence $x = 0.45$ is the value of the root of the nearest hundreth.

Example 74.3

Graph $y = f(x) = x^3 - x^2 - 2x + 1$.

Solution
In Example 74.2 we note that the given function has three real zeros. Thus, the curve crosses the x axis three times. We use this observation and the values listed in Table 74.2 to draw the graph shown in Figure 74.5. We note from the table of values that the zeros of the function are in the intervals -2 to -1, 0 to 1, and 1 to 2.

Table 74.2

x	-3	-2	-1	0	1	2	3
y	-29	-7	1	1	-1	1	13

Figure 74.5

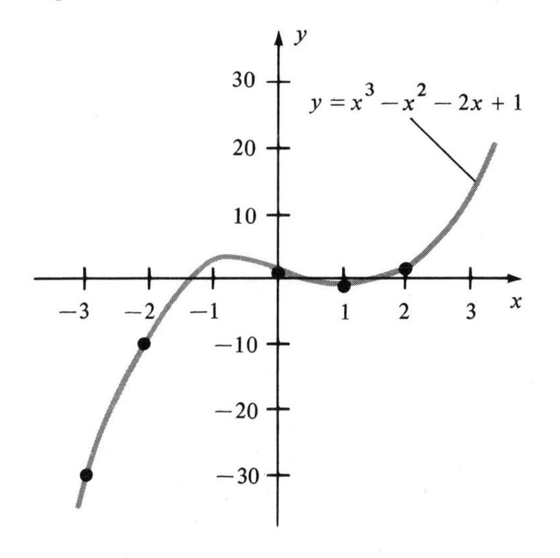

Exercises for Section 74

In Exercises 1 through 10, find the real root of each polynomial equation to the nearest tenth for the indicated interval.

1. $x^3 + 18x - 30 = 0, 1 < x < 2$

2. $x^3 + x - 3 = 0, 1 < x < 2$

3. $2x^3 + 3x^2 - 9x - 7 = 0, 1 < x < 2$

4. $x^3 - 3x^2 - 3x + 18 = 0, -3 < x < -2$

5. $x^4 - 2x^2 - 1 = 0, 1 < x < 2$

6. $x^4 + x^2 - 3x - 1 = 0, 1 < x < 2$

7. $x^3 - 5x^2 + 7x - 2 = 0, 0 < x < 1$

8. $x^4 - x^3 - 2x^2 - x - 3 = 0, -2 < x < -1$

9. $x^3 - x^2 - 3x + 1 = 0, 2 < x < 3$

10. $x^3 - 7x + 7 = 0, 1 < x < 1.5$

In Exercises 11 through 20, find the value of each real root to the nearest hundredth.

11. $x^3 - x^2 - 3x + 1 = 0, 2 < x < 3$

12. $x^3 - 7x + 7 = 0, 1.5 < x < 2$

13. $x^4 - 2x^3 + x^2 - 1 = 0, -1 < x < 0$

14. $x^3 + 3x - 5 = 0, 1 < x < 2$

15. $x^4 - 2x^3 + x^2 - 1 = 0, 1 < x < 2$

16. $x^4 - 5x^2 + 2x^2 + x + 7 = 0, 1 < x < 2$

17. $x^4 - 2x^3 - 3x^2 - 2x - 4 = 0, -2 < x < -1$

18. $x^4 + x^3 - x^2 - 3x - 6 = 0, 1 < x < 2$

19. $x^4 + 4x^3 + 2x^2 - 12x - 15 = 0, 1 < x < 2$

20. $2x^4 + 4x^3 + 1 = 0, -1 < x < 0$

21. Find $\sqrt[3]{6}$ to the nearest hundredth. ***Hint:*** $\sqrt[3]{6}$ is the positive root of $x^3 - 6 = 0$.

22. Find $\sqrt[4]{2}$ to the nearest hundredth (see Exercise 21).

23. The dimensions of a box are 1 m by 3 m by 5 m. The volume is to be doubled by increasing each side by the same amount. Find, to the nearest tenth, how much each side should be increased.

REVIEW QUESTIONS FOR CHAPTER 12

1. Write an nth-degree polynomial function in x.

2. What are the zeros of the polynomial function $P(x)$?

3. What is synthetic division?

4. If we divide $P(x)$ by $x - r$, what can be said about the degree of the quotient?

5. State the remainder theorem.

6. State the factor theorem.

7. State the fundamental theorem of algebra.

8. State the number-of-roots theorem.

9. What is a multiple root?

10. State the rational-root theorem.

11. State Descartes' rule of signs.

REVIEW EXERCISES FOR CHAPTER 12

In Exercises 1 through 6, find the quotient and remainder using synthetic division.

1. $(2x^3 - 3x^2 - 4x - 6) \div (x - 2)$
2. $(4x^3 - 7x^2 + 4x - 12) \div (x - 2)$
3. $(3x^4 + 5x^3 - 9x - 13) \div (x + 2)$
4. $(2x^3 - 5x^2 + 4x - 1) \div (x - \frac{1}{2})$
5. $(x^5 - 3x^3 + 6x^2 + 7x + 5) \div (x + 1)$
6. $(2x^3 - 5x^2 + 4x - 5) \div (x - \frac{3}{2})$
7. Show that $(x - 2)$ is a factor of $4x^3 - 7x^2 + 4x - 12$.
8. Show that $(x + 1)$ is a factor of $x^5 - 3x^3 + 6x^2 + 7x - 1$.
9. Show that $(2x - 1)$ is a factor of $2x^3 - 5x^2 + 4x - 1$.
10. Show that $(3x + 1)$ is a factor of $3x^3 - 8x^2 + 3x + 2$.
11. Show that $(x + 1)$ and $(x - 2)$ are factors of $x^4 - 2x^2 - 3x - 2$.
12. Use the results of Exercise 11 to write $x^4 - 2x^2 - 3x - 2$ in factored form.
13. Show that $x = -4$ is a root of $x^3 + 2x^2 - 23x - 60 = 0$.
14. Show that $x = -\frac{3}{2}$ is a root of $2x^3 + x^2 - 7x - 6 = 0$.
15. Show that $x = 2$ is a root of $x^3 - 4x^2 + 14x - 20 = 0$.
16. Show that $x = \frac{5}{2}$ is a root of $2x^3 - 11x^2 + 17x - 5 = 0$.
17. Show that $x = -\frac{1}{3}$ is a root of $3x^3 - 8x^2 + 3x + 2 = 0$.
18. Use the results of Exercise 13 to find all roots of $P(x) = x^3 + 2x^2 - 23x - 60 = 0$. Write $P(x)$ in factored form.
19. Use the results of Exercises 14 to find all roots of $P(x) = 2x^3 + x^2 - 7x - 6 = 0$. Write $P(x)$ in factored form.
20. Use the results of Exercise 15 to find all roots of $P(x) = x^3 - 4x^2 + 14x - 20 = 0$.
21. Use the results of Exercise 16 to find all roots of $P(x) = 2x^3 - 11x^2 + 17x - 5 = 0$. Write $P(x)$ in factored form.
22. Use the results of Exercise 17 to find all roots of $P(x) = 3x^3 - 8x^2 + 3x + 2 = 0$.
23. Find the value of k such that $(x - 1)$ is a factor of $x^3 + 2x^2 - 4kx + 5$.

24. Write a polynomial equation with integral coefficients whose roots are 0, 1, and 4.

25. Write a polynomial equation with integral coefficients whose roots are 1, $\frac{1}{2}$, and $-\frac{2}{3}$.

26. Write a polynomial equation with integral coefficients whose roots are 1, -1, and j.

In Exercises 27 through 32, find all rational roots. If a quadratic depressed equation is obtained, solve the equation completely.

27. $x^3 - 3x^2 + 4x - 2 = 0$

28. $x^3 - 2x^2 - 5x + 6 = 0$

29 $2x^3 - x^2 - 2x + 1 = 0$

30. $x^4 - 3x^2 + 3x^2 - 3x + 2 = 0$

31. $x^4 + 2x^3 - 7x^2 - 8x + 12 = 0$

32. $x^4 + 6x^3 + 13x^2 + 12x + 4 = 0$

33. Use the results of Exercise 27 and the sign graph method to find the solution set for $x^3 - 3x^2 + 4x - 2 \leq 0$.

34. Use the results of Exercise 31 and the sign graph method to find the solution set for $x^4 + 2x^3 - 7x^2 - 8x + 12 \geq 0$.

35. Find the real root of $x^3 + 5x + 2 = 0$ to the nearest tenth.

36. Find the positive real root of $x^3 - 2x^2 - 2x - 7 = 0$ to the nearest tenth.

37. Show that the equation $x^4 + 2x^2 + 5 = 0$ has no real roots.

38. A variable electric current is given by the formula $i = t^3 - 2t^2 - 5t - 9$, t in seconds. At what time is the current equal to 3 A?

39. The velocity of a point P moving on a coordinate line is given by $v(t) = t^3 - 9t^2 + 20t - 12$, where t is measured in seconds and $v(t)$ in ft/sec.
(a) Find the values of t for which $v(t) = 0$.
(b) Use the sign graph method to find the intervals where $v(t) > 0$ for $t \geq 0$.

40. The position function s of a point moving rectilinearly is given by $s(t) = t^3 - 8t^2 + 20t + 84$, where t is in seconds and $s(t)$ in meters. For what values of t will $s(t)$ be 100 m?

41. A company determines that the cost $C(x)$ of manufacturing x units of a certain commodity is given by $C(x) = 5x^3 - 60x^2 + 5x + 440$. If the company allocates $500 daily to the production of this commodity, how many units can be produced?

42. If a rectangular metal box has a length of l in., and a width that is 1 in. more than the length, and a height that is 4 in. less than the length, find a formula for the volume V in terms of l. Find the dimensions of the box when its volume is 288 in.^{3}.

CHAPTER 13

Sequences and Series

SECTION 75

Sequences

A sequence of numbers is generally a set of numbers arranged in a definite order. The phrase **definite order** means that there is a first term, a second term, a third term, and so on. Let us call the first term in the sequence a_1, the second term a_2, the third term a_3, and so on, until all terms have been enumerated. We note that in such a listing of elements, to each positive integer n there corresponds the element a_n. This correspondence can be shown as follows:

$$\begin{array}{ccccccc} 1 & 2 & 3 & \cdots & n & \cdots \\ \downarrow & \downarrow & \downarrow & & \downarrow & \\ a_1 & a_2 & a_3 & \cdots & a_n & \cdots \end{array}$$

Formally, an infinite sequence is a function f whose domain is the set of natural numbers 1, 2, 3, ... A **finite sequence** is a function whose domain is a subset of the natural numbers 1, 2, 3, . . . , n. The infinite sequence can be described by its ordered pairs as follows:

$$f = (1, f(1)), (2, f(2)), (3, f(3)), \ldots, (f, (f(n)), \ldots \qquad \text{(A)}$$

Since the domain of the sequence is always the ordered set of natural numbers, we can describe the sequence (A) by listing only its functional values,

$$f(1), f(2), f(3), \ldots, f(n), \ldots \qquad \text{(B)}$$

in the order of the natural numbers with which they are associated.

If we let $f(1) = a_1$, $f(2) = a_2$, $f(3) = a_3, \ldots, f(n) = a_n, \ldots$, we obtain from form (B)

$$a_1, a_2, a_3, \ldots, a_n, \ldots \qquad \text{(C)}$$

which is the usual form for an infinite sequence. We note that in (C), a_1 is called the **first term** of the sequence, a_2 the **second term**, a_3 the **third term**, and a_n the nth **term** or the **general term**. The nth or general term usually denotes the rule of function. For convenience we use the symbol $\{a_n\}$ to denote the sequence.

Example 75.1

Determine the first five terms of the infinite sequence having the general term $a_n = 2n + 1$.

Solution

We find the first five terms of the infinite sequence by successively replacing n in the expression $2n + 1$ by the integers 1, 2, 3, 4, and 5. For

$$n = 1: \quad a_1 = 2(1) + 1 = 3$$

$$n = 2: \quad a_2 = 2(2) + 1 = 5$$

$$n = 3: \quad a_3 = 2(3) + 1 = 7$$

$$n = 4: \quad a_4 = 2(4) + 1 = 9$$

$$n = 5: \quad a_5 = 2(5) + 1 = 11$$

Thus, the first five terms of the sequence are 3, 5, 7, 9, and 11. We can describe the entire sequence as $\{2n + 1\} = 3, 5, 7, 9, 11, \ldots, 2n + 1, \ldots$.

Example 75.2

Find the first four terms of the infinite sequence whose nth term is $a_n = n(n + 2)$.

Solution

Replacing n in the expression $n(n + 2)$ by the integers 1, 2, 3, and 4, we find that for

$$n = 1: \quad a_1 = 1(1 + 2) = 3$$

$$n = 2: \quad a_2 = 2(2 + 2) = 8$$

$$n = 3: \quad a_3 = 3(3 + 2) = 15$$

$$n = 4: \quad a_4 = 4(4 + 2) = 24$$

Thus, the first four terms of the sequence are 3, 8, 15, and 24. We can describe the entire sequence as $\{n(n + 2)\} = 3, 8, 15, 24, \ldots, n(n + 2), \ldots$.

Example 75.3

Find the first four terms of the infinite sequence whose nth term is $a_n = (-1)^{n-1}(n^2 - 1)$.

Solution

Substituting the integers 1, 2, 3, and 4 for n, we find that for

$$n = 1: \quad a_1 = (-1)^0(1^2 - 1) = 0$$

$$n = 2: \quad a_2 = (-1)^1(2^2 - 1) = -3$$

$$n = 3: \quad a_3 = (-1)^2(3^2 - 1) = 8$$

$$n = 4: \quad a_4 = (-1)^3(4^2 - 1) = -15$$

Thus, the first four terms of the sequence are 0, -3, 8, and -15. We can describe the entire sequence as $0, -3, 8, -15, \ldots, (-1)^{n-1}(n^2 - 1), \cdots$.

Example 75.4

Find by inspection a general term for an infinite sequence whose first five terms are $\frac{1}{2}$, $-\frac{1}{3}$, $\frac{1}{4}$, $-\frac{1}{5}$, and $\frac{1}{6}$.

Solution

We observe that the terms in the sequence are alternately positive and negative. The numerators are constant 1 and the denominators begin at 2 and increase by 1. Thus, a general term is given by

$$a_n = (-1)^{n+1}\frac{1}{n + 1}$$

Remark 75.1. *If we are given several successive terms of a sequence such as in Example 75.4, it is sometimes possible to find an expression for a general term by inspection or by trial and error. If a general term is found, it is **not unique**. The reader should verify that*

$$a_n = (-1)^{n+1}\frac{1}{n + 1} + (n - 1)(n - 2)(n - 3)(n - 4)(n - 5)$$

would also be a general term for the given sequence in Example 75.4.

Example 75.5

Let the sequence a_n be defined by the conditions

$$a_1 = 3$$

$$a_n = 2a_{n-1} - 1$$

Find the first four terms of the sequence.

Solution

We are given the first term $a_1 = 3$. We can find the next three terms by replacing n by the integers 2, 3, and 4 in the formula $a_n = 2a_{n-1} - 1$ as follows:

$$a_2 = 2a_1 - 1 = 2 \cdot 3 - 1 = 5$$

$$a_3 = 2a_2 - 1 = 2 \cdot 5 - 1 = 9$$

$$a_4 = 2a_3 - 1 = 2 \cdot 9 - 1 = 17$$

Thus, the first four terms of this sequence are 3, 5, 9, and 17. We note that an equation which enables us to obtain any term (other than the first) from the preceding term is called a **recursion formula**. Thus, $a_n = 2a_{n-1} - 1$ is an example of a recursion formula.

Example 75.6
The first 20 terms of the infinite sequence whose nth term is $a_n = 3n + 2$ constitutes a finite sequence that can be described as $5, 8, 11, \ldots, 3n + 2, \ldots, 62$. We note that a finite sequence has both a first term *and* a last term.

Exercises for Section 75

In Exercises 1 through 16, write the first four terms and the tenth term of the sequence for each general term.

1. $a_n = 3n$

2. $a_n = 2n + 1$

3. $a_n = \dfrac{1}{n+1}$

4. $a_n = \dfrac{n}{n+1}$

5. $a_n = \dfrac{n}{n(n+1)}$

6. $a_n = \dfrac{n-1}{n+1}$

7. $a_n = \dfrac{n}{2^n}$

8. $a_n = \dfrac{n^2-4}{n^2-3}$

9. $a_n = \dfrac{3^n}{n^2+1}$

10. $a_n = \dfrac{(-1)^n}{n^2}$

11. $a_n = \dfrac{(-1)^n}{n+1}$

12. $a_n = (-1)^n 3^n$

13. $a_n = \dfrac{e^n}{n}$

14. $a_n = (-1)^{n+1}$

15. $a_n = n^n$

16. $a_n = (-1)^n (2)^{n-2}$

In Exercises 17 through 30, find, by inspection, an nth or general term for each infinite sequence.

17. $4, 7, 10, 13, \ldots$

18. $1, 8, 27, 64, \ldots$

19. $-1, \frac{1}{2}, -\frac{1}{3}, \frac{1}{4}, \ldots$

20. $4, 8, 12, 16, \ldots$

21. $5, 8, 11, 14, \ldots$

22. $2, 5, 10, 17, \ldots$

23. $1, -1, 1, -1, \ldots$

24. $0, -1, -2, -3, \ldots$

25. $\frac{1}{2}, \frac{2}{3}, \frac{3}{4}, \frac{4}{5}, \ldots$

26. $1, \frac{1}{2}, \frac{1}{4}, \frac{1}{8}, \ldots$

27. $\frac{1}{3}, \frac{1}{9}, \frac{1}{27}, \frac{1}{81}, \ldots$

28. $1, \frac{3}{2}, 2, \frac{5}{2}, \ldots$

29. $1, \sqrt{2}, \sqrt{3}, 2, \ldots$

30. $1, 2, 4, 8, \ldots$

In Exercises 31 through 36, find the first four terms of each sequence for the conditions given.

31. $a_1 = 2, a_n = a_{n-1} + 1$

32. $a_1 = 2, a_n = (a_{n-1})^2$

33. $a_1 = 1, a_n = na_{n-1}$

34. $a_1 = 5, a_n = \frac{a_{n-1}}{2}$

35. $a_1 = 5, a_n = 4a_{n-1} - 1$

36. $a_1 = 2, a_n = \frac{1}{-a_{n-1}}$

37. A ball is dropped from a height of 10 ft. If it rebounds one half of the distance it has fallen after each fall, how far will it rebound the fourth time? The tenth time?

38. A force is applied to an object moving in a straight line in such a way that each second it moves only one third of the distance it moved the preceding second. If the object moves 18 cm the first second, how far will it move during the third second? The fifth second?

39. The number of bacteria in a culture triples every 4 hr. If there are 10,000 bacteria present at 8:00 A.M., how many will be present exactly 1 day later?

40. A certain machine depreciates in value over the years. If we assume that its value when new is 1 and if at the end of any year, it has only 75% of the value it had at the beginning of the year, find its value after 2 years; after 3 years; after n years.

SECTION 76
Series

Associated with every sequence is a series that is formed by taking the algebraic sum of the corresponding terms of the given sequence.

Example 76.1

(a) With the finite sequence $1, 3, 5, \ldots, 2n - 1$, we can associate the finite series

$$S_n = 1 + 3 + 5 + \cdots + (2n - 1)$$

(b) With the finite sequence, $1, \frac{1}{2}, \frac{1}{3}, \ldots, 1/n$, we can associate the finite series

$$S_n = 1 + \frac{1}{2} + \frac{1}{3} + \cdots + \frac{1}{n}$$ ●

In general, the **finite series**

$$S_n = a_1 + a_2 + a_3 + \cdots + a_n$$

is obtained by summing the terms of the finite sequence $a_1, a_2, a_3, \ldots, a_n$.

The **infinite series**

$$a_1 + a_2 + a_3 + \cdots + a_n + \cdots$$

is obtained by summing the terms of the infinite sequence $a_1, a_2, a_3, \ldots, a_n, \ldots$.

The nth or general term of a series is the nth or general term of the corresponding sequence. We note that a finite series will always have for its sum a finite number. The meaning, if any, of the "sum" of an infinite series will be discussed in Section 79.

A series for which the general term is known can be written in a compact form by means of **sigma** or **summation notation**. The symbol $\sum_{k=1}^{n} a_k$ represents the sum of the first n terms of the finite series $a_1 + a_2 + a_3 + \cdots + a_n$. That is,

$$\sum_{k=1}^{n} a_k = a_1 + a_2 + a_3 + \cdots + a_n \tag{76.1}$$

The Greek capital letter $\sum$ (sigma) is used to denote the sum of the series, the symbol a_k represents the kth or general term of the series, the letter k is called the **summation variable**, and the numbers 1 and n indicate the extremes of the set of **natural numbers** that we sum over.

Example 76.2

Evaluate $\sum_{k=1}^{4} (2k - 1)$.

Solution

We have $a_k = 2k - 1$ and the summation variable k ranges over the natural numbers 1, 2, 3, and 4. To obtain the sum indicated, we substitute in succession the values of k and add the resulting terms, as follows:

$$\sum_{k=1}^{4} (2k - 1) = (2 \cdot 1 - 1) + (2 \cdot 2 - 1) + (2 \cdot 3 - 1) + (2 \cdot 4 - 1)$$
$$= 1 + 3 + 5 + 7 = 16$$

Example 76.3

Evaluate $\sum_{k=2}^{5} k(k + 1)$.

Solution

We have $a_k = k(k + 1)$ and the summation variable k ranges over the natural numbers 2, 3, 4, and 5. To obtain the indicated sum, we substitute in succession the values of k and add the resulting terms, as follows:

$$\begin{aligned}\sum_{k=2}^{5} &= 2(2 + 1) + 3(3 + 1) + 4(4 + 1) + 5(5 + 1) \\ &= 6 + 12 + 20 + 30 = 68\end{aligned}$$

We note that the summation range may begin with a value other than 1.

Example 76.4

Evaluate $\sum_{j=3}^{6} (j^2 + 1)$.

Solution

The general term $a_j = j^2 + 1$ and the summation variable j ranges over the natural numbers 3, 4, 5, and 6. To obtain the indicated sum, we substitute in succession the values of j and then add the resulting terms, as follows:

$$\begin{aligned}\sum_{k=3}^{6} (j^2 + 1) &= (3^2 + 1) + (4^2 + 1) + (5^2 + 1) + (6^2 + 1) \\ &= 10 + 17 + 26 + 37 = 90\end{aligned}$$

We note that the letter used for the summation variable (in this case j) is arbitrary.

Example 76.5

Use sigma notation to describe a finite series associated with each of the following finite sequences.

(a) $1, \frac{1}{2}, \frac{1}{3}, \ldots, \frac{1}{25}$

(b) $\frac{1}{2}, \frac{2}{3}, \frac{3}{4}, \ldots, \frac{n}{n + 1}$

(c) $3^2, 4^2, 5^2, \ldots, 100^2$

(d) $-1, \frac{1}{2}, -\frac{1}{4}, \ldots, \frac{(-1)^n}{2^{n-1}}$

Solution

(a) We have $a_k = 1/k$. Thus, an associated series

$$1 + \frac{1}{2} + \frac{1}{3} + \cdots + \frac{1}{25} = \sum_{k=1}^{25} \frac{1}{k}$$

(b) We have $a_k = k/(k + 1)$. Thus, an associated series

$$\frac{1}{2} + \frac{2}{3} + \frac{3}{4} + \cdots + \frac{n}{n + 1} = \sum_{k=1}^{n} \frac{k}{k + 1}$$

(c) We have $a_k = k^2$. Thus, an associated series

$$3^2 + 4^2 + 5^2 + \cdots + 100^2 = \sum_{k=3}^{100} k^2$$

(d) We have $a_k = (-1)^k/2^{k-1}$. Thus, an associated series

$$-1, +\frac{1}{2} - \frac{1}{4} + \cdots + \frac{(-1)^n}{2^{n-1}} = \sum_{k=1}^{n} \frac{(-1)^k}{2^{k-1}}$$

●

If the general term of the series is constant, say $a_k = c$, we have

$$\sum_{k=1}^{n} a_k = \sum_{k=1}^{n} c = \underbrace{c + c + c = \cdots + c}_{n \text{ terms}} = cn \tag{76.2}$$

Example 76.6

(a) If $a_k = c = 1$, then, by equation (76.2), we have

$$\sum_{k=1}^{n} c = \sum_{k=1}^{n} 1 = 1 \cdot n = n$$

(b) $\sum_{k=1}^{4} 3 = 3 \cdot 4 = 12$, since $c = 3$ and $n = 4$.

(c) $\sum_{k=1}^{12} 5 = 5 \cdot 12 = 60$, since $c = 5$ and $n = 12$.

Remark 76.1. *When we express the finite series* $\sum_{k=1}^{n} a_k$ *as*

$$\sum_{k=1}^{n} a_k = a_1 + a_2 + a_3 + \cdots + a_k$$

we refer to the right side, $a_1 + a_2 + a_3 + \cdots + a_n$, *as the* ***expanded form*** *of* $\sum_{k=1}^{n} a_k$.

Exercises for Section 76

In Exercises 1 through 16, express each sum using summation notation.

1. $1 + 2 + 3 + \cdots + 10$

2. $2 + 4 + 6 + 8$

3. $1 + 2 + 3 + \cdots + (n - 1)$

4. $2 + 2 + 2 + 2 + 2 + 2 + 2$

5. $1 + 4 + 9 + \cdots + 900$

6. $1^2 + 2^2 + 3^2 + \cdots + 50^2$

7. $3 + 6 + 9 + \cdots + 300$

8. $\frac{1}{3} + \frac{1}{9} + \frac{1}{27} + \cdots + \frac{1}{3^k}$

9. $2 + 4 + 8 + \cdots + 128$

10. $2 + 4 + 8 + \cdots + 2^n$

11. $1 + 4 + 9 + 16 + 25 + 36$

12. $1 + \sqrt{2} + \sqrt{3} + 2 + \cdots + 10$

13. $1 - 1 + 1 - 1 + 1 - 1 + 1 - 1 + 1$

14. $1 + 4 + 9 + 16 + \cdots + n^2$

15. $\frac{2}{2} + \frac{4}{4} + \frac{6}{8} + \frac{8}{16} + \frac{10}{32} + \frac{12}{64}$

16. $4 + 7 + 10 + 13 + \cdots + (3n + 1)$

In Exercises 17 through 24, find each sum.

17. $\sum_{k=1}^{6} k$

18. $\sum_{k=2}^{4} (2k - 1)$

19. $\sum_{k=1}^{4} \frac{1}{k + 1}$

20. $\sum_{j=0}^{2} 3^j$

21. $\sum_{k=1}^{3} \frac{2^k}{k}$

22. $\sum_{k=1}^{100} 1^{2k}$

23. $\sum_{j=1}^{5} \frac{1}{2^j}$

24. $\sum_{k=1}^{1000} 3$

In Exercises 25 through 36, write each sum in expanded form.

25. $\sum_{k=2}^{6} k$

26. $\sum_{t=1}^{6} \frac{1}{t}$

27. $\sum_{i=2}^{100} i(i - 1)$

28. $\sum_{k=1}^{n} (2k)^k$

29. $\sum_{k=1}^{n} 2k^2$

30. $\sum_{k=5}^{8} 3$

31. $\sum_{k=1}^{4} \frac{(-1)^k}{2^k}$

32. $\sum_{i=1}^{5} \frac{1}{2^i}$

33. $\sum_{k=1}^{5} \frac{1}{k(k + 1)}$

34. $\sum_{k=1}^{5} \frac{1}{2^{k-1}}$

35. $\sum_{k=1}^{10} \frac{3^k}{k}$

36. $\sum_{k=3}^{4} \frac{2^k}{k^2}$

SECTION 77

Arithmetic Sequences and Series

An **arithmetic progression** (abbreviated A.P.) is a sequence of numbers in which each term after the first is obtained by adding a fixed number d to the preceding term. The fixed number d is called the **common difference**.

If a is the first term of an arithmetic progression, with a common difference d, then the first n terms of this sequence are

$$a, a + d, a + 2d, a + 3d, \ldots, a + (n - 1)d$$

where the nth or last term, denoted l, is given by the formula

$$l = a + (n - 1)d \tag{77.1}$$

Example 77.1

The first four terms of the arithmetic progression, with a first term equal to 2 and a common difference equal to 3 are 2, 5, 8, and 11. The nth or last term l is given by equation (77.1). We find l by replacing a by 2 and d by 3 in (77.1) to obtain $l = 2 + (n - 1)3 = 2 + 3n - 3$ or $l = 3n - 1$.

Example 77.2

If the first four terms of an arithmetic progression are $-11, -8, -5$, and -2, find the twentieth term.

Solution

We can compute the common difference d by subtracting any two successive terms. If we subtract the first term from the second, we find that $d = -8 - (-11) = 3$. We now use (77.1), where $n = 20$, to find the twentieth term. Hence, $l = -11 + (20 - 1)3 = -11 + 57 = 46$. ●

Associated with the finite arithmetic sequence $a, a + d, a + 2d, \ldots, a + (n - 1)d$ is the finite arithmetic series.

$$S_n = a + (a + d) + (a + 2d) + \cdots + a + (n - 1)d$$
$$= \sum_{k=1}^{n} a + (k - 1)d$$

We can express the sum of this arithmetic progression or series denoted S_n in terms of n, the number of terms; a, the first term; and l, the last term, as follows:

$$S_n = a + (a + d) + (a + 2d) + \cdots + l \tag{A}$$

We can also express S_n in terms of a, l, and d by **reversing** the order of the terms in (A), as follows:

$$S_n = l + (l - d) + (l - 2d) + \cdots + a \tag{B}$$

We now add (A) and (B), to obtain

$$2S_n = (a + l) + [(a + d) + (l - d)] + \cdots + (l + a)$$
$$= (a + l) + (a + l) + \cdots + (a + l)$$
$$= n(a + l)$$

$$S_n = \frac{n(a + l)}{2} \tag{77.2}$$

If we wish to express S_n in terms of a, n, and d, we substitute (77.1) into (77.2), to obtain

$$S_n = \frac{n}{2}[2a + (n - 1)d] \tag{77.3}$$

Example 77.3

Find the sum of the first 15 terms of the arithmetic progression 4, 7, 10, ...

Solution

We first find d by subtracting the first term, 4, from the second term, 7. Hence, $d = 7 - 4 = 3$. We now use (77.1) to find the fifteenth term, $l = 4 + (15 - 1)3 = 46$. Finally, we use the fact that $l = 46$, and (77.2), to obtain

$$S_{15} = \frac{n(a + l)}{2} = \frac{15(4 + 46)}{2} = 375$$

Example 77.4

Find the number of terms in the progression for which $a = 3$, $l = 213$, and $d = 5$.

Solution

We know from (77.1) that $l = a + (n - 1)d$. Substitution of the given values into (77.1) yields

$$\begin{aligned} 213 &= 3 + (n - 1)5 \\ &= 5n - 2 \\ 5n &= 215 \\ n &= 43 \end{aligned}$$

Thus, there are 43 terms in the progression.

Example 77.5

Find the sum of the first 100 odd positive integers.

Solution

The positive odd integers form the sequence $1, 3, 5, \ldots, 2n - 1, \ldots$. The first odd integer is 1 and the 100th is given by $l = 2(100) - 1 = 199$. Hence, $a = 1$, $l = 199$, and $n = 100$. Substituting these values in (77.2), we have

$$S_n = \frac{100(1 + 199)}{2} = 50(200) = 10{,}000$$

Example 77.6

A body falls approximately 16 ft during the first second, 48 ft during the second second, 80 ft during the third second, and so on. How far will it fall during the tenth second?

Solution

In this example each value represents a term in the arithmetic progression 16, 48, 80, ... To find the distance the object falls in the tenth second, we use equation (77.1) with $a = 16$, $n = 10$, and $d = 48 - 16 = 32$. Thus, $l = a + (n - 1)d$ or $l = 16 + (10 - 1)32 = 304$ ft, the distance the object falls in the tenth second.

Example 77.7

A car costs $8000 when new. It depreciates 18% the first year, 16% the second year, 14% the third year, and so on, for 6 years. Assuming that all depreciations apply to the original cost, find the value of the car at the end of 6 years.

Solution

The depreciations form an arithmetic series with $a = 18\%$, $n = 6$, and $d = -2\%$. Employing equation (77.3) we see that the sum of the depreciations is given by $S = \frac{6}{2}[2(18\%) + 5(-2\%)]$. Hence, $S = 3(36\% - 10\%) = 78\%$. Since the car has depreciated 78%, its value at the end of 6 years is 22% of $8000, or $1760.

Exercises for Section 77

In Exercises 1 through 10, find the indicated last term and the sum for each arithmetic progression.

1. 2, 5, 8, . . . , tenth term

2. 11, 5, −1, . . . , ninth term

3. 3, 0, −3, . . . , seventh term

4. 7, 5, 3, . . . , eleventh term

5. −3, −1, 1, . . . , sixth term

6. 1, 5, 9, . . . , twelfth term

7. 5, 9, 13, . . . , twentieth term

8. 1, −1, −3, . . . , sixteenth term

9. 4, 10, 16, . . . , eighth term

10. 7, 2, −3, . . . , tenth term

11. If $l = 43$, $n = 10$, and $S_n = 250$, find a and d.

12. If $S_n = 48$, $a = 3$, and $l = 5$, find n.

13. If $l = 27$, $n = 11$, and $d = 6$, find a and S_n.

14. If $l = 239$, $n = 27$, and $d = 9$, find a and S_n.

15. If $S_n = 210$, $a = -2$, and $l = 37$, find n.

16. If $S_n = 150$, $n = 10$, and $l = 24$, find a and d.

17. If $S_n = 150$, $n = 15$, and $l = 24$, find a and d.

18. The first term of an arithmetic progression is 4. The tenth term is 31. Find the common difference d.

19. The fourth term of an arithmetic progression is 26 and $d = 5$. Find the tenth term.

20. If $a = 5$, $l = 45$, and $d = 8$, find n and S_n.

21. If $a = -2$, $n = 40$, and $l = 28$, find d and S_n.

22. If $a = \frac{5}{3}$, $n = 20$, and $S_n = \frac{40}{3}$, find l and d.

23. If $n = 17$, $S_n = 187$, and $l = 27$, find a and d.

24. The third term of an arithmetic progression is 6 and the fortieth term is -68. What is the tenth term?

25. Find the sum of the first 100 integers.

26. An auditorium has 30 rows with 23 seats in the first row, 25 seats in the second row, 27 seats in the third row, and so on. How many seats are in the auditorium?

27. A man borrows $1980 and pays off the debt by paying $20 the first month, $22 the second month, $24 the third month, and so on. How long will it take to pay off the debt?

28. A construction company purchases a new piece of equipment for $30,000. The equipment depreciates 15% the first year, 13.5% the second year, 12% the third year, and so on for 6 years. Assuming that all depreciations apply to the original cost, what is the value of the equipment at the end of 6 years?

29. The first swing of a pendulum is 10 ft and each successive swing is 3 in. shorter.
(a) What is the length of the ninth pendulum swing?
(b) Find the total distance traveled by the pendulum in the first nine swings.
(c) How many swings of the pendulum are completed before it comes to rest?

30. A body in motion has an initial speed of 10 m/sec and travels in a straight line with an acceleration of 3 m/sec.^{2}. The values of the speed of the body in meters per second at positive integral values of time, t, in seconds, form the arithmetic progression $10, 13, 16, \ldots, 10 + 3t, \ldots$. Find the speed of the body at $t = 30$ sec.

31. A man saves $100 the first year he works, $150 the second year, $200 the third year, and so on.
(a) How much will he save the tenth year he works?
(b) How much will his entire savings be after 10 years?

SECTION 78

Geometric Sequences and Series

A *geometric progression* (denoted G.P.) is a sequence of numbers in which each term after the first term is obtained from the preceding one by multiplying the preceding term by a fixed number r called the **common ratio**.

If a is the first term of a geometric progression with a common ratio r, the first n terms of the sequence are

$$a, ar, ar^2, ar^3, \ldots, ar^{n-1}$$

where the nth or last term, denoted l, is given by

$$l = ar^{n-1} \tag{78.1}$$

Example 78.1

Find the first four terms of a geometric progression whose first term is 3 and whose common ratio is $\frac{1}{3}$. Then find the nth term.

Solution

We are given that $a = 3$ and $r = \frac{1}{3}$. Thus, the first four terms of the geometric progression are 3, 1, $\frac{1}{3}$, and $\frac{1}{9}$. The nth term is given by $l = ar^{n-1}$. Thus, when $a = 3$ and $r = \frac{1}{3}$, $l = 3(\frac{1}{3})^{n-1}$.

Example 78.2

If the first three terms of a geometric progression are 18, 6, and 2, find the common ratio r and then find the fifteenth term.

Solution

We can compute the common ratio r, for any G.P. by dividing any term in the sequence by the preceding one. To determine r for this sequence we divide the second term 6 by the first term 18, to obtain $r = \frac{6}{18} = \frac{1}{3}$. We now compute the fifteenth term by substituting $r = \frac{1}{3}$, $a = 18$, and $n = 15$ into (78.1) to obtain $l = 18(\frac{1}{3})^{15-1} = 18(\frac{1}{3})^{14}$

Associated with the finite geometric sequence $a, ar, ar^2, \ldots, ar^{n-1}$ is the finite **geometric series**

$$S_n = a + ar + ar^2 + \cdots + ar^{n-1} = \sum_{k=1}^{n} ar^{k-1} \tag{A}$$

We can express the sum of this geometric series, denoted S_n or $\sum_{k=1}^{n} ar^{k-1}$, in terms of a, the first term; r, the common ratio; and n, the number of terms, as follows. We must first multiply S_n in (A) by r, to obtain

$$rS_n = ar + ar^2 + ar^3 + \cdots + ar^n \tag{B}$$

We next subtract (B) from (A) and find that

$$S_n - rS_n = a - ar^n$$

$$S_n(1 - r) = a(1 - r^n)$$

$$S_n = \frac{a(1 - r^n)}{1 - r} \qquad r \neq 1 \tag{78.2}$$

If we wish to express S_n in terms of a, r, and l, we first write (78.2) as

$$S_n = \frac{a - r(ar^{n-1})}{1 - r}$$

Since $l = ar^{n-1}$, we have

$$S_n = \frac{a - rl}{1 - r} \tag{78.3}$$

Example 78.3

Find the sum of the first 10 terms of the geometric progression $\frac{1}{2}, 1, 2, \ldots$.

Solution

Here we have $a = \frac{1}{2}$, $r = 1/\frac{1}{2} = 2$, and $n = 10$. Using formula (78.2), we obtain

$$S_{10} = \frac{\frac{1}{2}[1 - (2)^{10}]}{1 - 2}$$

$$= \frac{\frac{1}{2}(1 - 1024)}{1 - 2} = \frac{1023}{2}$$

Example 78.4

The first term of a geometric sequence is $\frac{1}{2}$ and the tenth term is 256. Find the twelfth term of the sequence and the sum of the first 12 terms.

Solution

We are given that $a = \frac{1}{2}$, $l = 256$, and $n = 10$. We can find the common ratio r by using (78.1). Hence, $256 = \frac{1}{2}r^9$. Since $256 = 2^8$, we have

$$2^8 = \frac{1}{2}r^9$$

or

$$2^9 = r^9$$

Thus,

$$r = 2$$

To find the twelfth term of the sequence, we again use 78.1, with $a = \frac{1}{2}$, $r = 2$, and $n = 12$. Therefore, the twelfth term is $\frac{1}{2}(2)^{11} = 2^{10} = 1024$. We now use the results above and (78.2) to find the sum of the first 12 terms, as follows:

$$S_{12} = \frac{\frac{1}{2}(1 - 2^{12})}{1 - 2} = \frac{4095}{2}$$

We note that if any three of the elements a, n, r, l, and S_n are given, the remaining two can be found by use of formulas (78.1) through (78.3).

Example 78.5
The sum of the first seven terms of a G.P. is $\frac{463}{81}$. If the common ratio is $-\frac{2}{3}$, find the first term and the seventh term.

Solution
Here $r = -\frac{2}{3}$, $n = 7$, and $S_7 = \frac{463}{81}$. Using equation (78.2), we have

$$\frac{463}{81} = \frac{a\left[1 - \left(-\frac{2}{3}\right)^7\right]}{1 - \left(-\frac{2}{3}\right)}$$

$$\frac{463}{81} = \frac{463a}{729} \text{ or } 81a = 729$$

Thus, $a = 9$ and the seventh term

$$l = ar^{n-1} = 9\left(-\frac{2}{3}\right)^6 = \frac{64}{81}$$

Example 78.6
A man invests \$1000 per year in real estate. How much is the investment worth after 6 full years if it appreciates at the rate of 10% per year?

Solution
At the end of the first year the investment is worth \$1000 + (0.10)\$1000 = \$1000(1 + 0.10). At the end of the second year, the investment is worth \$1000(1 + 0.10) + \$1000(1 + 0.10)^2, and so on. Thus, we wish to sum the G.P.:

$$\$1000(1 + 0.10) + \$1000(1 + 0.10)^2 + (1 + 0.10)^3 + \cdots + \$1000(1 + 0.10)^6$$

We note that for this G.P., $a = 1000$, $r = 1.1$, and $n = 6$. Therefore, using equation (78.2), we find that

$$S_6 = \frac{1000[1 - (1.1)^6]}{1 - 1.1} = \frac{1000(-0.772)}{-0.1} = \$8492$$

Hence, the total value of the investment at the end of 6 years is \$8492. Of this amount, \$2492 is due to the rate of appreciation. Logarithms were used to find $(1.1).^6$

Exercises for Section 78

In Exercises 1 through 10, find the indicated nth term and the sum of the first n terms for each geometric progression.

1. $2, 1, \frac{1}{2}, \ldots$, seventh term

2. $2, 4, 8, \ldots$, sixth term

3. $\frac{1}{2}, -\frac{5}{2}, \frac{25}{2}, \ldots$, fifth term

4. $1, 2, 4, \ldots$, eighth term

5. $27, 9, 3, \ldots$, ninth term

6. $16, 4, 1, \ldots$, tenth term

7. $\frac{1}{125}, -\frac{1}{25}, \frac{1}{5}, \ldots$, seventh term

8. $\frac{1}{16}, \frac{1}{4}, 1, \ldots$, eighth term

9. $\frac{1}{36}, \frac{1}{30}, \frac{1}{25}, \ldots$, eighth term

10. $\frac{1}{3}, \frac{1}{4}, \frac{3}{16}, \ldots$, sixth term

In Exercises 11 through 20, three of the elements a, l, r, n, and S_n of the geometric progression are given. In each case, find the missing elements.

11. $a = 3, r = \frac{2}{5}, S_n = \frac{609}{125}$

12. $a = 1, n = 3, S_n = 13$

13. $a = \frac{1}{9}, r = 3, n = 8$

14. $a = \frac{1}{9}, r = 3, l = 243$

15. $a = 2, n = 4, l = 16$

16. $a = -1, r = -2, l = 32$

17. $a = 3, r = -\frac{\sqrt{3}}{3}, n = 8$

18. $S_n = 1020, a = 4, n = 8$

19. $a = \frac{1}{2}, r = \frac{2}{3}, n = 6$

20. $S_n = \frac{463}{81}, r = -\frac{2}{3}, l = \frac{64}{81}$

21. If the second term of a geometric progression is 3 and the fifth term is $\frac{81}{8}$, find the tenth term.

22. Find the first term of a G.P. if the sixth term is 729 and the common ratio is 3.

23. The first term of a G.P. is 3 and the fourth term is 24. Find r.

24. If the first term of a G.P. is $\frac{1}{243}$ and the eighth term is 9, find the sum of the first six terms.

25. If the sum of the first five terms of a G.P. is 121 and $r = \frac{1}{3}$, find a and l.

26. If the first term of a G.P. is 1 and the sixth term is 32, find the common ratio r and the value of S_6.

27. Find the tenth term of a G.P. if second term is 3 and the fourth term is 9.

28. A ball is dropped from a height of 20 ft. If it rebounds one half the distance it has fallen after each fall, how far will it rebound the fifth time? Through what distance has it traveled when it strikes the ground the sixth time?

29. An automobile purchased for $6000 depreciates in value 15% each year. Find the value of the car after 4 years.

30. A man accepts a position with an initial salary of $9000 per year. If he receives a 5% increase every year, what will his salary be after 10 years of service?

31. The population of a certain town is 10,000. If it increases 4% every year, what will the approximate population be at the end of 10 years?

32. At the beginning of an experiment, there are four bacteria present in a culture. If each bacteria divides into two every hour, how many bacteria will be present at the end of 8 hr?

33. A tank full of alcohol is emptied of one fourth of its contents and filled up with water and mixed. If this procedure is followed five times, what fraction of the original volume of alcohol remains?

34. A sum of $1000 is invested at an annual rate of 8% per year. What amount will be accumulated at the end of 5 years?

35. The tip of a pendulum moves 20 cm the first second, 10 cm the next, 5 cm the next, and so on. Find the total distance the tip moves in 10 sec.

36. Suppose that you had the following choice. An employer will pay you $1000 per day for 30 days *or* 1 cent for the first day's work, 2 cents for the second, 4 cents for the third; and each day thereafter for the remaining 27 days your salary would be doubled. Which choice results in the larger salary?

SECTION 79

Infinite Geometric Series

We shall now investigate the **infinite geometric series** associated with the infinite geometric sequence

$$a, ar, ar^2, \ldots, ar^{n-1}, \ldots$$

This infinite series can be described by the equation

$$S = a + ar + ar^2 + ar^3 + \cdots + ar^{n-1} + \cdots$$

where the symbol S stands for the "sum" of this series, if it exists. We carefully note that the word "sum" is not being used in the ordinary sense, but rather to imply an entirely new concept.

In Section 78 we found that we could describe the sum of a finite geometric series as

$$S_n = \frac{a - ar^n}{1 - r} \qquad r \neq 1$$

We rewrite this as

$$S_n = \frac{a}{1-r} - \frac{ar^n}{1-r}$$

or

$$S_n = \frac{a}{1-r} - \left(\frac{a}{1-r}\right)r^n \qquad r \neq 1 \tag{79.1}$$

If we consider the values in the interval $-1 < r < 1$, or equivalently all r such that $|r| < 1$, we find that when r is raised to increasing positive integral powers n, r^n grows smaller and smaller. For example, if $r = \frac{1}{2}$, then

$$r^2 = \left(\frac{1}{2}\right)^2 = \frac{1}{4}, \quad r^3 = \left(\frac{1}{2}\right)^3 = \frac{1}{8}, \quad r^4 = \left(\frac{1}{2}\right)^4 = \frac{1}{16}$$

and so on. If $r = -\frac{1}{2}$, then $r^2 = (-\frac{1}{2})^2 = \frac{1}{4}$, $r^3 = (-\frac{1}{2})^3 = -\frac{1}{8}$, $r^4 = (-\frac{1}{2})^4 = \frac{1}{16}$, and so on.

It would seem that as n gets larger and larger, the term r^n becomes smaller and smaller. In fact, if n is "sufficiently large," the term r^n is **effectively zero**. The term "effectively zero" means that although we cannot find any value for n large enough to make $r^n = 0$, r^n gets closer and closer to 0 as n becomes very, very large (approaches infinity).

The notation for this is as follows:

$$\lim_{n\to\infty} r^n = 0 \qquad |r| < 1$$

read as "the limit as n approaches infinity of r to the nth power is zero." The symbol ∞ is read "infinity" and the number 0 is called the **limit**.

Since r^n approaches zero, the expression $(a/1 - r)r^n$ in (79.1) also approaches zero as the values of n become sufficiently large. Thus, using these observations and equation (79.1) we can *define* the sum of an **infinite geometric series** as follows:

$$S_\infty = \frac{a}{1-r} \qquad \text{if } |r| < 1 \tag{79.2}$$

(We do not define the sum S_∞ for $|r| \geq 1$.)

Example 79.1

Find the sum (if it exists) for the infinite geometric series $1 + \frac{1}{3} + \frac{1}{9} + \cdots$.

Solution

Here $a = 1$ and $r = \frac{1}{3} \div 1$ or $\frac{1}{9} \div \frac{1}{3} = \frac{1}{3}$. Since $|r| < 1$, we can use (79.2) to obtain

$$S_\infty = \frac{1}{1 - \left(\frac{1}{3}\right)} = \frac{1}{\frac{2}{3}} = \frac{3}{2}$$

If we use summation notation, we have, equivalently,

$$\sum_{n=1}^{\infty} \left(\frac{1}{3}\right)^{n-1} = \frac{3}{2}$$

Example 79.2

Find the sum of each of the following infinite geometric series (if it exists).

(a) $\sum_{n=1}^{\infty} 5\left(\frac{1}{2}\right)^n$ (b) $\sum_{n=1}^{\infty} 8\left(-\frac{3}{5}\right)^n$ (c) $\sum_{n=1}^{\infty} 2\left(\frac{5}{3}\right)^n$

Solution

(a) $\sum_{n=1}^{\infty} 5\left(\frac{1}{2}\right)^n = \frac{5}{2} + \frac{5}{4} + \frac{5}{8} + \cdots$

Here $a = \frac{5}{2}$ and $r = \frac{1}{2}$. Since $|r| < 1$, we apply (79.2), to obtain

$$S = \frac{\frac{5}{2}}{1 - \frac{1}{2}} = \frac{\frac{5}{2}}{\frac{1}{2}} = 5$$

(b) $\sum_{n=1}^{\infty} 8\left(-\frac{3}{5}\right)^n = \frac{-24}{5} + \frac{72}{25} - \frac{216}{125} + \cdots$

Here $a = -\frac{24}{5}$ and $r = -\frac{3}{5}$. Since $|r| < 1$, we use (79.2), to obtain

$$S = \frac{-\frac{24}{5}}{1 - \left(-\frac{3}{5}\right)} = \frac{-\frac{24}{5}}{\frac{8}{5}} = -3$$

(c) $\sum_{n=1}^{\infty} 2\left(\frac{5}{3}\right)^n = \frac{10}{3} + \frac{50}{9} + \frac{250}{27} + \cdots$

Here $a = \frac{10}{3}$ and $r = \frac{5}{3}$. Since $|r| > 1$, there is *no sum.*

Example 79.3

Express the decimal 0.2222 . . . in fractional form.

Solution

We can write the given decimal form as $0.2 + 0.02 + 0.002 + 0.0002 + \cdots$. But this is an infinite geometric series with $a = 0.2$ and $r = 0.1$. Again, applying (79.2), we have

$$S = \frac{0.2}{1 - 0.1} = \frac{2}{9}$$

Thus, the fraction $\frac{2}{9}$ and the decimal 0.222 . . . are equivalent.

Example 79.4

Express the decimal 0.232323 . . . in fractional form.

Solution

The given decimal form is equivalent to the infinite geometric series $0.23 + 0.0023 + 0.000023 + \cdots$. Since $a = 0.23$ and $r = 0.01$, we use (79.2) to find $S = 0.23/(1 - 0.01) = \frac{23}{99}$. Thus, the fraction $\frac{23}{99}$ is equivalent to the repeating decimal 0.232323 . . . .

Example 79.5

If a ball rebounds $\frac{3}{5}$ as far as it falls, how far will it travel before coming to rest if it is dropped from a height of 40 m?

Solution

We consider the successive drops and rebounds of the ball as two infinite geometric sequences as follows:

Drops	*Rebounds*
40 m	24 m
24 m	$\frac{72}{5}$ m
$\frac{72}{5}$ m	$\frac{216}{25}$ m
⋮	⋮

The sum of the drop distances form an infinite geometric series with $a = 40$ and $r = \frac{3}{5}$. Hence, the sum is $40/(1 - \frac{3}{5}) = 100$ m. The sum of the distances the ball rebounds also forms an infinite geometric series. In this case $a = 24$, $r = \frac{3}{5}$, and the sum is $24/(1 - \frac{3}{5}) = 60$ m. Therefore, the total distance is 100 m + 60 m = 160 m.

Exercises for Section 79

In Exercises 1 through 20, find the sum of each infinite geometric series, if it exists. If the sum does not exist, state why.

1. $1 + \frac{1}{2} + \frac{1}{4} + \cdots$

2. $1 + \frac{1}{5} + \frac{1}{25} + \cdots$

3. $4 + 2 + 1 + \cdots$

4. $1 - \frac{1}{2} + \frac{1}{4} + \cdots$

5. $6 + 2 + \frac{2}{3} + \cdots$

6. $\frac{1}{3} - \frac{2}{9} + \frac{4}{27} - \cdots$

7. $2 + 0.2 + 0.02 + \cdots$

8. $3 + \frac{3}{2} + \frac{3}{4} + \cdots$

9. $\frac{5}{3} + \frac{1}{6} + \frac{1}{60} + \cdots$

10. $\frac{3}{4} + \frac{3}{4^2} + \frac{3}{4^3} + \cdots$

11. $-4 + 2 - 1 + \cdots$

12. $100 - 10 + 1 - \cdots$

13. $\frac{1}{2}, -1 + 2 - \cdots$

14. $\sum_{n=1}^{\infty} \left(\frac{4}{5}\right)^n$

15. $\sum_{n=1}^{\infty} 3^{n-1}$

16. $\sum_{n=1}^{\infty} 2^n$

17. $\sum_{n=1}^{\infty} \left(\frac{2}{3}\right)^n$

18. $\sum_{n=1}^{\infty} 2\left(\frac{1}{3}\right)^n$

19. $\sum_{n=1}^{\infty} 5\left(\frac{3}{4}\right)^{n-1}$

20. $\sum_{n=1}^{\infty} (-1)^n\left(\frac{4}{9}\right)^{n-1}$

In Exercises 21 through 30, find the fraction equivalent to each decimal.

21. 0.4444 . . .

22. 0.3333 . . .

23. 0.7777 . . .

24. 0.9999 . . .

25. 0.141414 . . .

26. 0.595959 . . .

27. 0.123123 . . .

28. 0.375375 . . .

29. 2.1313 . . .

30. 3.1424242 . . .

31. If a ball rebounds two thirds as far as it falls, how far will it travel before coming to rest if it is dropped from a height of 30 ft?

32. Each swing of a pendulum bob is three fourths as long as the preceding swing. If the first swing is 64 cm, how far does the bob travel before it comes to rest?

33. The initial oscillation of a body suspended on a vertical spring is 24 cm long. Each succeeding oscillation decreases by 20%. How far does the body travel before coming to rest?

34. The sum of an infinite geometric series is $\frac{16}{3}$ and the first term is 4. Find the common ratio.

35. The sum of an infinite geometric series is $\frac{2}{3}$ and the first term is 1. Find the common ratio.

36. An object is projected up an inclined plane. It is observed to move 25 cm the first second, and in succeeding seconds it moves 80% as far as it did in the preceding second. How far does the object travel before coming to rest?

SECTION 80

The Binomial Theorem

If we expand some of the positive integral powers of $a + b$, we obtain the following results:

$$\begin{aligned}
(a+b)^1 &= a+b \\
(a+b)^2 &= a^2+2ab+b^2 \\
(a+b)^3 &= a^3+3a^2b+3ab^2+b^3 \qquad (80.1)\\
(a+b)^4 &= a^4+4a^3b+6a^2b^2+4ab^3+b^4 \\
(a+b)^5 &= a^5+5a^4b+10a^3b^2+10a^2b^3+5ab^4+b^5
\end{aligned}$$

From these identities we observe that the expansion of $(a + b)^n$ for $n = 1$, 2, 3, 4, and 5 has $(n + 1)$ terms and the following properties:

1. The first term of the expansion is $a^n b^0 = a^n$. The exponents of a then decrease by 1 for each successive term.
2. The second term is $na^{n-1}b^1$. The exponents of b then increase by 1 for each successive term.
3. The last term is $a^0 b^n = b^n$.
4. The sum of the exponents of a and b in each term is n.
5. The exponent of b is always one less than the number of the term in which it appears.

In order to find a way in which to determine the coefficient of the terms in the expansion, we shall establish a convenient notation known as **factorial notation**. The symbol $n!$ (read n factorial) is the product of all positive integers between 1 and n inclusive. That is,

$$n! = 1 \times 2 \times 3 \times \cdots \times n \qquad (80.2)$$

Example 80.1

$$5! = 1 \times 2 \times 3 \times 4 \times 5 = 120$$

$$4! = 1 \times 2 \times 3 \times 4 = 24$$

$$\frac{5!}{4!} = \frac{1 \times 2 \times 3 \times 4 \times 5}{1 \times 2 \times 3 \times 4} = 5$$

●

We note that $n!$ is given by the equation $n! = n(n-1)!$. If we substitute 1 for n in this equation, we obtain $1! = 1 \cdot 0!$. Since $1! = 1$, the only possible real value for $0!$ that makes the equation true is 1. Thus, we define $0!$ to be equal to 1.

The symbol $\binom{n}{r}$, where n and r are positive integers and $n \geq r$, is called the **binomial coefficient symbol**. The value of the symbol is given as follows:

$$\binom{n}{r} = \frac{n!}{(n-r)!\,r!} \qquad (80.3)$$

Example 80.2

Evaluate $\binom{5}{2}$, $\binom{3}{3}$, and $\binom{4}{0}$.

Solution

$$\binom{5}{2} = \frac{5!}{(5-2)!2!} = \frac{5!}{3!2!} = 10$$

$$\binom{3}{3} = \frac{3!}{(3-3)!3!} = \frac{3!}{0!3!} = 1$$

$$\binom{4}{0} = \frac{4!}{(4-0)!0!} = \frac{4!}{4!0!} = 1$$

●

We observe that $\binom{3}{3} = 1$ and $\binom{4}{0} = 1$. This suggests that $\binom{n}{n} = 1$ and $\binom{n}{0} = 1$. We can show this by noting that

$$\binom{n}{n} = \frac{n!}{(n-n)!n!} = \frac{n!}{0!n!} = 1$$

and

$$\binom{n}{0} = \frac{n!}{(n-0)!0!} = \frac{n!}{n!0!} = 1$$

We can now state the following theorem.

Theorem 80.1. The Binomial Theorem
If n is a positive integer, then

$$(a+b)^n = \binom{n}{0}a^n + \binom{n}{1}a^{n-1}b + \binom{n}{2}a^{n-2}b^2 + \cdots + \binom{n}{r}a^{n-r}b^r + \cdots + \binom{n}{n}b^n \tag{80.4}$$

If we use the method of evaluation of $\binom{n}{r}$ shown in (80.3) to expand the coefficients in (80.4), we obtain

$$(a+b)^n = a^n + na^{n-1}b + \frac{n(n-1)}{2}a^{n-2}b^2 + \frac{n(n-1)(n-2)}{2\cdot 3}a^{n-3}b^3 + \cdots + b^n \tag{80.5}$$

In this form the coefficients of the first and last terms are 1 and the coefficient of any other term can be found by multiplying the coefficient of the preceding term by the exponent of a in the preceding term and then dividing this by the number of terms that precede it. The rth term in the binomial expansion is

$$\binom{n}{r-1} a^{n-r+1} b^{r-1}$$

Example 80.3

Expand and simplify $(x + 2y)^4$.

Solution

By (80.5) we have

$$(x + 2y)^4 = x^4 + 4x^3(2y) + \frac{4(3)}{2}x^2(2y)^2 + \frac{4(3)(2)}{2(3)}x(2y)^3 + (2y)^4$$

$$= x^4 + 8x^3y + 24x^2y^3 + 32xy^3 + 16y^4$$

Example 80.4

Expand and simplify $(2x - y^2)^5$.

Solution

We note that $(2x - y^2)^5 = [2x + (-y^2)]^5$. Employing (80.5), we have

$$[2x + (-y)^2]^5 = (2x)^5 + 5(2x)^4(-y^2) + \frac{5(4)}{2}(2x)^3(-y^2)^2$$

$$+ \frac{5(4)(3)}{2 \cdot 3}(2x)^2(-y^2)^2$$

$$+ \frac{5(4)(3)(2)}{2 \cdot 3 \cdot 4}(2x)(-y)^4 + (-y^2)^5$$

or

$$(2x - y^2)^5 = 32x^5 - 80x^4y + 80x^3y^2 - 40x^2y^6 + 10xy^8 - y^{10}$$

Example 80.5

Find and simplify the sixth term in the expansion of $(x^2 - 2y)^8$.

Solution

In the expansion, the sixth term will involve b^5, where $b = -2y$. Thus, the sixth term will be

$$\binom{8}{5}(x^2)^3(-2y)^5 = 56(x^6)(-32y^5) = -1792x^6y^5$$ ●

If we set $a = 1$, $b = x$, and let n be any real number in the binomial formula, we generate the infinite binomial series

$$(1 + x)^n = 1 + \frac{nx}{1!} + \frac{n(n-1)}{2!}x^2 + \cdots$$

$$+ \frac{n(n-1)\cdots(n-r+1)x^r}{r!} + \cdots \tag{80.6}$$

It can be shown that equation (80.6) is valid for any real number n if $|x| < 1$. When n is not a positive integer, the series is unending, but useful. Let us illustrate one of the uses with the following example.

Example 80.6
Approximate $\sqrt[4]{1.02}$ to three decimal places.

Solution
We can write $\sqrt[4]{1.02}$ as $(1 + 0.02)^{1/4}$. We now use (80.6) with $x = 0.02$, to obtain

$$\sqrt[4]{1.02} = (1 + 0.02)^{1/4} = 1 + \frac{1}{4}(0.02) - \frac{3}{32}(0.02)^2 + \cdots$$

Thus, $(1 + 0.02)^{1/4} = 1 + 0.005 - 0.00004 + \cdots$. We can readily see that no term after the third has any effect on the first three decimal places. Thus, $\sqrt[4]{1.02} = 1.005$.

Example 80.7
Write the first four terms of $(1 + x)^{-1}$ and $(1 + x)^{-2}$, $|x| < 1$.

Solution
In (80.6) we let $n = -1$ and obtain

$$(1 + x)^{-1} = \frac{1 + (-1)x}{1!} + \frac{(-1)(-2)x^2}{2!} + \frac{(-1)(-2)(-3)x^3}{3!} + \cdots$$

Thus, $(1 + x)^{-1} = 1 - x + x^2 - x^3 + \cdots$. If we now let $n = -2$ in (80.6), we have

$$(1 + x)^{-2} = 1 + \frac{(-2)x}{1!} + \frac{(-2)(-3)x^2}{2!} + \frac{(-2)(-3)(-4)x^3}{3!} + \cdots$$

Thus, $(1 + x)^{-2} = 1 - 2x + 3x^2 - 4x^4 + \cdots$.

Exercises for Section 80

1. Evaluate the following.

(a) $\binom{6}{2}$ (b) $\binom{6}{4}$ (c) $\binom{3}{3}$ (d) $\binom{20}{0}$ (e) $\binom{100}{99}$

In Exercises 2 through 15, expand and simplify using the binomial formula.

2. $(x + y)^6$ **3.** $(x - y)^6$ **4.** $(x - 3)^5$

5. $(x - 2y)^6$ **6.** $(1 - 2x)^5$ **7.** $(2x - 3)^4$

8. $(1 - 2x)^4$ **9.** $(2x - 3y)^4$ **10.** $(2x^2 - 1)^5$

11. $(xy - 2x)^6$ **12.** $(1 + 0.1)^4$ **13.** $(x/y + y/x)^4$

14. $(x/y - y/x)^4$ **15.** $(x^3 - y^2)^5$

In Exercises 16 through 30, find the term indicated and simplify without expanding completely.

16. The fifth term of $(x^2 + y^3)^8$

17. The seventh term of $(a - 2b)^9$

18. The fourth term of $(\sqrt{x} + y)^6$

19. The eighth term of $(x^3 - 2y)^{12}$

20. The fifth term of $(2x - 3y)^{12}$

21. The sixth term of $(x + y)^{15}$

22. The term involving b^8 in $(2a^2 - 3b^2)^6$

23. The term involving y^6 in $\left(3x^2 + \frac{y}{3}\right)^8$

24. The term involving x^2 in $\left(\frac{x}{2} - \frac{4}{x}\right)^8$

25. The middle term of $(x^2 - y^2)^{10}$

26. The middle term of $(2x + y^3)^8$

27. The middle term of $(2x - xy)^{12}$

28. The middle term of $(x - y)^{20}$

29. The term involving x^6 in $(3x^2 - 2y)^{15}$

30. The term involving x^7 in $(2x - y)^{12}$

In Exercises 31 through 36, use the binomial series to find the value of each number to an accuracy of three decimal places.

31. $(1.06)^6 = (1 + 0.01)^6$

32. $\sqrt[3]{1.02} = (1 + 0.02)^{1/3}$

33. $(0.98)^8 = (1 - 0.02)^8$

34. $\sqrt[4]{0.98} = (1 - 0.02)^{1/4}$

35. $\sqrt{50} = (49 + 1)^{1/2}$

36. $\sqrt[4]{80} = (81 - 1)^{1/4}$

REVIEW QUESTIONS FOR CHAPTER 13

1. What is a sequence of numbers?

2. What is an infinite sequence?

3. What is a finite sequence?

4. What is the relation between a sequence and a series?

5. What is an arithmetic progression?

6. What is a geometric progression?

7. What is an infinite geometric sequence?

8. What is an infinite geometric series?

9. How do we define the sum (S_∞) of an infinite geometric series?

10. Explain what the symbol $n!$ means.

11. What is the symbol $\binom{n}{r}$ called?

12. State the binomial theorem.

REVIEW EXERCISES FOR CHAPTER 13

In Exercises 1 through 6, write the first three terms and the eighth term of the sequence whose general term is as given.

1. $a_n = 3n - 2$

2. $a_n = \dfrac{2n}{n + 3}$

3. $a_n = \dfrac{2^n}{2n}$

4. $a_n = (-1)^n n^2$

5. $a_n = (-1)^{n+1}3^{n-1}$

6. $a_n = (-1)^n 3^{n-1}$

7. Write the first three terms of the sequence if $a_1 = 2$ and $a_n = 2a_{n-1} + 1$.

8. Write the first four terms of the sequence if $a_1 = 3$ and $a_n = \dfrac{a_{n-1} + 3}{2}$.

9. Find the indicated sums.

(a) $\sum_{i=1}^{100} 2$ (b) $\sum_{k=1}^{5} (3k - 2)$ (c) $\sum_{j=2}^{4} j^j$

In Exercises 10 through 18, determine whether the progression is arithmetic or geometric, find the indicated nth term, and the sum of the first n terms.

10. 15, 11, 7, . . . , eighth term

11. 3, $\frac{1}{2}$, $\frac{1}{12}$, . . . , seventh term

12. 1, $\frac{3}{2}$, $\frac{9}{4}$, . . . , sixth term

13. 6, −3, $\frac{3}{2}$, . . . , eighth term

14. −3, −1, 1, . . . , tenth term

15. 4, −1, $\frac{1}{4}$, . . . , twelfth term

16. −4, 2, −1, . . . , tenth term

17. 1, 4, 7, . . . , ninth term

18. $\frac{1}{3}$, $-\frac{5}{3}$, $\frac{25}{3}$, . . . , sixth term

19. The fourth term of an arithmetic progression with $d = 14$ is 16. Find the eighth term.

20. The first term of an arithmetic progression is 1. The tenth term is 55. Find d.

21. The sixth term of a geometric progression is 16. The seventh term is 12. Find the first term.

22. The first term of a geometric progression is 8 and the eighth term is $-\frac{1}{2048}$. Find r.

23. Find the sum of each of the following series.

(a) $12 + 4 + \frac{4}{3} + \cdots$ (b) $2 + 1 + \frac{1}{2} + \cdots$

(c) $-6 + 3 - \frac{3}{2} + \cdots$ (d) $\sum_{n=1}^{\infty} 2\left(\frac{3}{5}\right)^n$

(e) $\sum_{n=1}^{\infty} \left(\frac{4}{9}\right)^n$

24. The sum of an infinite geometric series is $\frac{36}{5}$ and the first term is 6. Find the common ratio.

25. The sum of an infinite geometric series is 8 and the first term is 4. Find the sixth term.

26. Expand the following expressions by using the binomial formula.
(a) $(2x - 1)^4$ (b) $(x - 2y)^5$ (c) $(3x + 2y)^3$

27. Use the binomial series to find the value of the following numbers to an accuracy of three decimal places.

(a) $\sqrt{26}$ (b) $\sqrt[3]{29}$

28. A woman saves \$10 the first week she works, \$12 the second week, \$14 the third week, and so on. How long will it take her to save \$580?

29. A ball is dropped from a height of 320 ft. The ball rebounds one half of the distance it has fallen after each fall.
(a) How far will it rebound the fourth time?
(b) How far will it rebound the 12th time?
(c) How far will the ball travel before coming to rest?

30. The cost of a new car is \$8000. The car depreciates 15% the first year, 12% the second year, 9% the third year, and so on. Assuming that all depreciations apply to the original cost, what is the value of the car at the end of 5 years?

31. In a certain experiment, the number of bacteria present at the start is 50,000. If the bacteria increases at the rate of 5% per hour, find the approximate population at the end of 6 hours.

32. A certain radioactive substance decays, losing half of its mass in 1 hour. How much is left after 3 hours? 5 hours? n hours?

33. A certain machine depreciates in value over the years. If at the end of any year it has only 70% of the value it had at the beginning of the years, find its value after 2 years; 4 years; n years. (Assume that its value was 1 when the machine was new.)

CHAPTER 14

Trigonometric Formulas, Equations, and Inverse Functions

SECTION 81

Fundamental Trigonometric Identities

In Section 39 we defined the trigonometric functions of the angle θ in standard position. We restate these relationships now for easy reference.

$$\left.\begin{aligned} \sin\theta &= \frac{y}{r} & \csc\theta &= \frac{r}{y} \\ \cos\theta &= \frac{x}{r} & \sec\theta &= \frac{r}{x} \\ \tan\theta &= \frac{y}{x} & \cot\theta &= \frac{x}{y} \end{aligned}\right\} \tag{81.1}$$

where

$$r = \sqrt{x^2 + y^2}$$

From these definitions we can develop the following basic identities:

$$\sin\theta = \frac{1}{\csc\theta} \qquad \csc\theta = \frac{1}{\sin\theta} \qquad \sin\theta\csc\theta = 1 \tag{81.2}$$

$$\cos\theta = \frac{1}{\sec\theta} \qquad \sec\theta = \frac{1}{\cos\theta} \qquad \cos\theta\sec\theta = 1 \tag{81.3}$$

$$\tan\theta = \frac{1}{\cot\theta} \qquad \cot\theta = \frac{1}{\tan\theta} \qquad \tan\theta\cot\theta = 1 \tag{81.4}$$

$$\frac{\sin\theta}{\cos\theta} = \tan\theta \qquad \frac{\cos\theta}{\sin\theta} = \cot\theta \tag{81.5}$$

The relationships in (81.2) through (81.5) are called **identities** since they are true for all **permissible values of θ.** We must be careful to note that specific values that would indicate a division by zero are to be excluded from our identities. For example, $\theta = 90°$ is not a permissible value in the identity $\sec\theta = 1/\cos\theta$, since $\sec 90° = 1/\cos 90°$ would indicate a division by $\cos 90°$, which equals zero.

We can use the relationships in (81.1) through (81.5) to develop other relationships or identities. From (81.1) we see that

$$r = \sqrt{x^2 + y^2} \quad \text{or} \quad x^2 + y^2 = r^2 \tag{A}$$

Now if we divide (A) by r^2, we obtain

$$\frac{x^2}{r^2} + \frac{y^2}{r^2} = 1 \quad \text{or} \quad \left(\frac{x}{r}\right)^2 + \left(\frac{y}{r}\right)^2 = 1$$

From (81.1) we can substitute $\cos\theta$ for x/r and $\sin\theta$ for y/r, to obtain

$$\sin^2\theta + \cos^2\theta = 1 \tag{81.6}$$

Next, we can divide (A) by x^2, to obtain

$$1 + \frac{y^2}{x^2} = \frac{r^2}{x^2} \quad \text{or} \quad 1 + \left(\frac{y}{x}\right)^2 = \left(\frac{r}{x}\right)^2$$

From (81.1) we can substitute $\tan\theta$ for y/x and $\sec\theta$ for r/x, to obtain

$$1 + \tan^2\theta = \sec^2\theta \tag{81.7}$$

Finally, we can divide (A) by y^2, to obtain

$$\frac{x^2}{y^2} + 1 = \frac{r^2}{y^2} \quad \text{or} \quad \left(\frac{x}{y}\right)^2 + 1 = \left(\frac{r}{y}\right)^2$$

Again, from (81.1) we can substitute $\cot\theta$ for x/y and $\csc\theta$ for r/y, to obtain

$$\cot^2\theta + 1 = \csc^2\theta \tag{81.8}$$

It should be noted that the foregoing identities are often seen in slightly different forms. For instance, instead of $\sin^2\theta + \cos^2\theta = 1$, we may have

$$\sin^2\theta = 1 - \cos^2\theta \quad \text{or} \quad \cos^2\theta = 1 - \sin^2\theta$$

Similarly, for $1 + \tan^2\theta = \sec^2\theta$, we may have

$$\tan^2\theta = \sec^2\theta - 1 \quad \text{or} \quad \sec^2\theta - \tan^2\theta = 1$$

and for $\cot^2\theta + 1 = \csc^2\theta$, we may have

$$\cot^2\theta = \csc^2\theta - 1 \quad \text{or} \quad \csc^2\theta - \cot^2\theta = 1$$

In addition, we observe that the difference between two squares may be written in factored form. Thus, for $1 - \cos^2\theta$ we can write $(1 + \cos\theta)(1 -$

$\cos\theta$) and so $\sin^2\theta = 1 - \cos^2\theta = (1 + \cos\theta)(1 - \cos\theta)$. Similar results may be obtained for any identities that involve the difference of squares.

In equations (81.6) through (81.8) we note the standard use of the notation $\sin^2\theta = (\sin\theta)^2$. This is true for all trigonometric functions. The expression $\sin^2\theta$ is read "sine square theta" and means that we square the sine of the angle theta.

In using any of the identities we have stated, θ may represent any angle or algebraic expression.

Example 81.1

(a) $\sin(x^2 + 1) = \dfrac{1}{\csc(x^2 + 1)}$

(b) $\tan 27° = \dfrac{\sin 27°}{\cos 27°}$

(c) $\sin^2\dfrac{\pi}{3} + \cos^2\dfrac{\pi}{3} = 1$

(d) $\sec 18° = \dfrac{1}{\cos 18°} = \dfrac{1}{0.9511} = 1.051$, a value that checks with the entry for sec 18° in Table I of Appendix F. ●

The identities (81.2) through (81.8) are called the **fundamental identities** for the trigonometric functions. These identities are so named since they can be used to establish or prove other identities. Unfortunately, there is no "set" or "standard" procedure that guarantees a correct proof for every problem. However, the proofs of most identities can be accomplished by using any of the following methods:

1. We can transform the right member of the equality into the exact form of the left member.
2. We can transform the left member of the equality into the exact form of the right member.
3. We can transform each side separately into the same form.

If the initial attempts to find a solution fail, it may help to express the trigonometric functions in terms of the sine and the cosine and then apply one of these three methods.

Example 81.2

Prove that $\tan\theta + \cot\theta = \sec\theta \cdot \csc\theta$.

Solution

Let us prove the identity by changing all terms to some form involving sine or cosine. Then we have

$$\frac{\sin\theta}{\cos\theta} + \frac{\cos\theta}{\sin\theta} = \frac{1}{\cos\theta} \cdot \frac{1}{\sin\theta}$$

Multiplying both sides of the equation by the quantity $\sin\theta\cos\theta$, we obtain

$$\sin\theta\cos\theta\left(\frac{\sin\theta}{\cos\theta}+\frac{\cos\theta}{\sin\theta}\right)=\sin\theta\cos\theta\frac{1}{\sin\theta\cos\theta}$$

or

$$\sin^2\theta+\cos^2\theta=1$$

Since this last form is a known identity [see (81.6)], the proof is complete.

Example 81.3

Prove that $\dfrac{1+\tan^2\theta}{\tan^2\theta}=\csc^2\theta$.

Solution

We shall prove this identity by transforming the left member into the exact form of the right member, as follows. We first divide $\tan^2\theta$ into each term in the numerator, to obtain

$$\frac{1+\tan^2\theta}{\tan^2\theta}=\frac{1}{\tan^2\theta}+1 \tag{A}$$

We now use the identity $\tan\theta=1/\cot\theta$ and substitution in (A), to obtain

$$\frac{1+\tan^2\theta}{\tan^2\theta}=\frac{1}{\tan^2\theta}+1=\cot^2\theta+1 \tag{B}$$

Using (81.8) and substitution into (B), we obtain

$$\frac{1+\tan^2\theta}{\tan^2\theta}=\frac{1}{\tan^2\theta}+1=\cot^2\theta+1=\csc^2\theta \tag{C}$$

In (C) we have expressed the left member of the expression into the exact form of the right member. Thus, the proof is complete. We note that the identity is true for all values of θ where $\tan\theta\neq 0$.

Example 81.4

Prove that $\sec\theta-\cos\theta=\sin\theta\tan\theta$.

Solution

We shall prove this identity by transforming each side separately into the same form. We first express the left side in terms of $\cos\theta$ and then simplify, to obtain

$$\sec\theta-\cos\theta=\frac{1}{\cos\theta}-\cos\theta=\frac{1-\cos^2\theta}{\cos\theta}=\frac{\sin^2\theta}{\cos\theta} \tag{A}$$

We now express the right sides in terms of $\sin\theta$ and $\cos\theta$ and then simplify, to obtain

$$\sin\theta\tan\theta=\sin\theta\frac{\sin\theta}{\cos\theta}=\frac{\sin^2\theta}{\cos\theta} \tag{B}$$

Since (A) and (B) are identical, the proof is complete. We note that $\sin^2\theta/\cos\theta$ will not be defined if $\cos\theta = 0$. Thus, the identity is true for all values of θ where $\cos\theta \neq 0$. ●

We can also use the fundamental identities to express all the trigonometric functions in terms of any one of them. Let us illustrate the procedure with the following example.

Example 81.5

Express $\csc\theta$ in terms of $\cos\theta$ by initially employing formula 81.8.

Solution

From formula (81.8), we have

$$\csc^2\theta = 1 + \cot^2\theta$$

Thus,

$$\csc^2\theta = 1 + \frac{\cos^2\theta}{\sin^2\theta}$$

We find a common denominator, by expressing the right side as

$$\frac{\sin^2\theta + \cos^2\theta}{\sin^2\theta}$$

and since $\sin^2\theta + \cos^2\theta = 1$, we have

$$\csc^2\theta = 1 + \frac{\cos^2\theta}{\sin^2\theta} = \frac{\sin^2\theta + \cos^2\theta}{\sin^2\theta} = \frac{1}{\sin^2\theta}$$

Since $\sin^2\theta = 1 - \cos^2\theta$,

$$\csc^2\theta = \frac{1}{1 - \cos^2\theta}$$

or

$$\csc\theta = \pm\sqrt{\frac{1}{1 - \cos^2\theta}} = \frac{\pm 1}{\sqrt{1 - \cos^2\theta}}$$

where the sign depends on the quadrant in which θ is located.

Example 81.6

Given that $\sin\theta = \frac{3}{5}$ (θ in the second quadrant), find the value of the other five trigonometric functions.

Solution

We begin by finding the corresponding value for $\cos\theta$. Since $\cos^2\theta = 1 - \sin^2\theta$ and $\sin\theta = \frac{3}{5}$, we have

$$\cos^2\theta = 1 - \left(\frac{3}{5}\right)^2$$

$$\cos^2\theta = 1 - \frac{9}{25} = \frac{16}{25}$$

Thus

$$\cos\theta = \pm\sqrt{\frac{16}{25}} = \pm\frac{4}{5}$$

But θ was given to be in the second quadrant. In this quadrant cosine is negative, so we have $\cos\theta = -\frac{4}{5}$. From equation (81.5),

$$\tan\theta = \frac{\sin\theta}{\cos\theta} = \frac{\frac{3}{5}}{-\frac{4}{5}} = -\frac{3}{4}$$

To find the values for the remaining three trigonometric functions, we employ the reciprocal identities given in equations (81.2) through (81.4). Hence,

$$\csc\theta = \frac{1}{\sin\theta} = \frac{1}{\frac{3}{5}} = \frac{5}{3}$$

$$\sec\theta = \frac{1}{\cos\theta} = \frac{1}{-\frac{4}{5}} = -\frac{5}{4}$$

$$\cot\theta = \frac{1}{\tan\theta} = \frac{1}{-\frac{3}{4}} = -\frac{4}{3}$$

Note that all algebraic signs are consistent with the behavior of the trigonometric functions in the second quadrant.

Exercises for Section 81

In Exercises 1 through 30, prove each identity.

1. $\tan\theta\cot\theta\sec\theta\cos\theta = 1$

2. $\tan\theta\sin\theta + \cos\theta = \sec\theta$

3. $\sin\theta\sec\theta = \tan\theta$

4. $\dfrac{\sin\theta\sec\theta}{\tan\theta} = 1$

5. $\cos\theta\tan\theta = \sin\theta$

6. $\dfrac{\sin^2\theta}{1 - \sin^2\theta = \tan^2\theta}$

7. $\dfrac{\cos^2\theta}{1 - \cos^2\theta} = \cot^2\theta$

8. $1 - \tan\theta = \dfrac{\cos\theta - \sin\theta}{\cos\theta}$

9. $(1 - \cos^2\theta)(1 + \cot^2\theta) = 1$

10. $\dfrac{1}{1 + \tan^2\theta} = \cos^2\theta$

11. $\dfrac{1 + \tan^2 \theta}{\csc \theta} = \dfrac{\sin \theta}{1 - \sin^2 \theta}$

12. $\sin \theta(1 + \cot^2 \theta) = \csc x$

13. $\dfrac{1 + \tan^2 \theta}{\csc^2 \theta} = \tan^2 \theta$

14. $\sin^2 \theta(\cot^2 \theta + 1) = 1$

15. $\tan^2 \theta - \sin^2 \theta = \tan^2 \theta \sin^2 \theta$

16. $\dfrac{\sin \theta}{\csc \theta} + \dfrac{\cos \theta}{\sec \theta} = 1$

17. $1 - 2 \sin^2 \theta = \cos^2 \theta - \sin^2 \theta$

18. $\dfrac{2}{1 - \cos x} = \dfrac{2 \sec x}{\sec x - 1}$

19. $\sec x + \cot x + \tan x = \dfrac{\sin x + 1}{\cos x \sin x}$

20. $\dfrac{\sin x}{1 - \cos x} = \csc x(1 + \cos x)$

21. $\dfrac{1 + \sin x}{1 - \sin x} = \dfrac{\csc x + 1}{\csc x - 1}$

22. $\dfrac{1 - \sin x}{\cos x} = \sec x - \tan x$

23. $\dfrac{1 + \cos x}{\sin x} = \dfrac{\sin x}{1 - \cos x}$

24. $\dfrac{1}{1 + \sin x} + \dfrac{1}{1 - \sin x} = 2 \sec^2 x$

25. $\dfrac{\cos x}{2(1 + \sin x)} + \dfrac{\cos x}{2(1 - \sin x)} = \sec x$

26. $\dfrac{\sin x}{2(1 + \cos x)} + \dfrac{\sin x}{2(1 - \cos x)} = \csc x$

27. $4 \sin x + \tan x = \dfrac{4 + \sec x}{\csc x}$

28. $\csc^4 x - \cot^4 x = \csc^2 x + \cot^2 x$

29. $\dfrac{\cos x}{\tan x + \sec x} - \dfrac{\cos x}{\tan x - \sec x} = 2$

30. $\dfrac{1}{\sec x - \tan x} - \dfrac{1}{\sec x + \tan x} = 2 \tan x$

31. If $\sin x = \frac{2}{3}$, find the values of the other five trigonometric functions in the first quadrant.

32. If $\sin x = \frac{2}{3}$, find the values of the other five trigonometric functions in the second quadrant.

33. If $\tan x = -\frac{1}{2}$, find the values of the other five trigonometric functions in the fourth quadrant.

34. If $\cos x = -\frac{1}{5}$, find the values of the other five trigonometric functions in the third quadrant.

35. Express the other five circular functions in terms of $\sin x$.

36. Express the other five circular functions in terms of $\tan x$.

37. Express each of the following in terms of $\sin x$ only.

(a) $\cos^2 x$ (b) $\sec^2 x$ (c) $\dfrac{\sec x + 1}{\sin x + \tan x}$

38. Prove the following identities by noting that the left hand side of the given equation is an infinite geometric series.
(a) $1 + \cos^2 x + \cos^4 x + \cdots = \csc^2 x$
(b) $1 + \sin^2 x + \sin^4 x + \cdots = \sec^2 x$

SECTION 82

Sum and Difference Formulas of Two Angles

We shall often wish to consider trigonometric functions of two angles. We will see that such functions can be expressed in terms of the functions of each angle separately.

Before beginning the development of these formulas, we note that the trigonometric function of the *sum* of two angles is *not* equal to the sum of the trigonometric functions of each angle. For example, the sin of 75° cannot be found by simply adding sin 30° to sin 45°. If this were so, then

$$\sin 75° = \sin 45° + \sin 30° = \frac{\sqrt{2}}{2} + \frac{1}{2} = 1.2071$$

But this is not possible, since the values of the sine function never exceeds 1.

We shall now develop the expressions for the sine and cosine of the sum and difference of two angles and then use them in the following section to develop double- and half-angle formulas. We begin with the sum formula for sine function.

In Figure 82.1 angle α is in standard position. Angle β is placed so that its initial side lies along the terminal side of α. Thus, the angle $(\alpha + \beta)$ is in standard position. The sine of $(\alpha + \beta)$ is the same as the sine of angle COD. To reexpress the sine of angle COD, we take a point P on the terminal side of $\overline{OD}$ and draw the perpendicular $\overline{PM}$ to the x axis. We then have

$$\sin(\alpha + \beta) = \frac{\overline{MP}}{\overline{OP}}$$

To generate the sum formula for sine, we must express the lengths $\overline{MP}$ and $\overline{OP}$ as functions of the angles α and β. To this end, we perform three constructions.

1. Draw $\overline{PQ}$ perpendicular to $\overline{OE}$.
2. Draw $\overline{QN}$ perpendicular to the x axis.

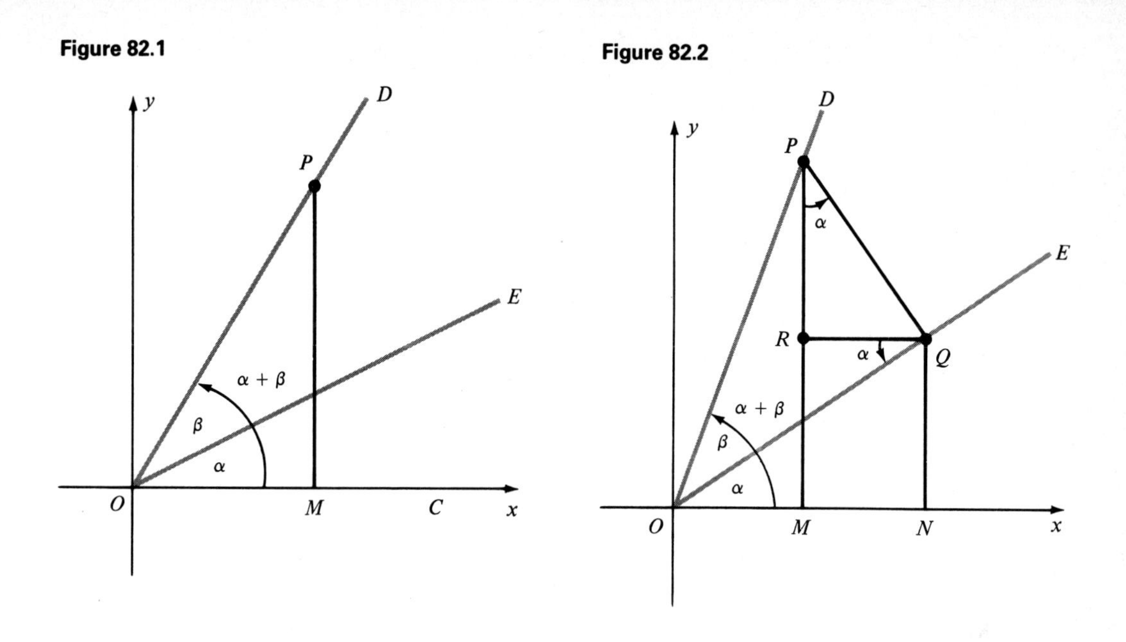

3. Draw $\overline{QR}$ perpendicular to $\overline{MP}$ (Figure 82.2).

By this construction, $\overline{MR} = \overline{NQ}$. Furthermore,

$$\angle RQO + \angle PQR = 90° \text{ and } \angle RPQ + \angle PQR = 90°$$

Thus, $\angle RQO = \angle RPQ$. However, $\angle RQO = \angle\alpha$, since they are alternate interior angles formed by parallel lines cut by a transversal. Hence, $\angle RPQ = \angle\alpha$. Since $\overline{MP} = \overline{MR} + \overline{RP}$, we can write

$$\sin(\alpha + \beta) = \frac{\overline{MP}}{\overline{OP}} = \frac{\overline{MR} + \overline{RP}}{\overline{OP}} = \frac{\overline{MR}}{\overline{OP}} + \frac{\overline{RP}}{\overline{OP}}$$

Substituting $\overline{NQ}$ for $\overline{MR}$ yields

$$\sin(\alpha + \beta) = \frac{\overline{NQ}}{\overline{OP}} + \frac{\overline{RP}}{\overline{OP}} \qquad \text{(A)}$$

We note that

$$\frac{\overline{NQ}}{\overline{OQ}} = \sin\alpha \quad \text{or} \quad \overline{NQ} = \overline{OQ}\sin\alpha$$

and

$$\frac{\overline{RP}}{\overline{PQ}} = \cos\alpha \quad \text{or} \quad \overline{RP} = \overline{PQ}\cos\alpha$$

If we substitute for $\overline{NQ}$ and $\overline{RP}$ in (A), we have

$$\sin(\alpha + \beta) = \frac{\overline{OQ}\sin\alpha}{\overline{OP}} + \frac{\overline{PQ}\cos\alpha}{\overline{OP}} \qquad \text{(B)}$$

From Figure 82.2 we observe that

$$\frac{\overline{OQ}}{\overline{OP}} = \cos\beta \quad \text{and} \quad \frac{\overline{PQ}}{\overline{OP}} = \sin\beta$$

Substituting the values above in (B), we obtain the desired formula

$$\sin(\alpha + \beta) = \sin\alpha\cos\beta + \cos\alpha\sin\beta \tag{82.1}$$

Similarly,

$$\cos(\alpha + \beta) = \frac{\overline{OM}}{\overline{OP}} = \frac{\overline{ON} - \overline{MN}}{\overline{OP}}$$

Since $\overline{MN} = \overline{RQ}$, we have

$$\cos(\alpha + \beta) = \frac{\overline{ON} - \overline{RQ}}{\overline{OP}} = \frac{\overline{ON}}{\overline{OP}} - \frac{\overline{RQ}}{\overline{OP}}$$

A procedure similar to that used to develop formula (82.1) will show that

$$\cos(\alpha + \beta) = \cos\alpha\cos\beta - \sin\alpha\sin\beta \tag{82.2}$$

Formulas (82.1) and (82.2) were developed using the case where the two angles had a sum less than 90°. That is, the terminal side of the sum $(\alpha + \beta)$ lies in the first quadrant. These formulas can also be established for the cases where the sum of the two angles place the terminal side of the angle $(\alpha + \beta)$ in quadrants II, III, and IV by using similar reasoning.

Example 82.1

Find the value of sin 75° by noting that $\sin 75° = \sin(45° + 30°)$.

Solution

Using formula (82.1) we find that

$$\begin{aligned}\sin(45° + 30°) &= \sin 45°\cos 30° + \cos 45°\sin 30° \\ &= \frac{\sqrt{2}}{2}\cdot\frac{\sqrt{3}}{2} + \frac{\sqrt{2}}{2}\cdot\frac{1}{2} \\ &= \frac{\sqrt{6}}{4} + \frac{\sqrt{2}}{4} = \frac{\sqrt{6} + \sqrt{2}}{4} = 0.9659\end{aligned}$$

Example 82.2

Show that $\cos(180° + \theta) = -\cos\theta$.

Solution

Substituting in formula (82.2), we obtain

$$\cos(180° + \theta) = \cos 180°\cos\theta - \sin 180°\sin\theta$$

Since $\cos 180° = -1$ and $\sin 180° = 0$, we have

$$\cos (180° + \theta) = (-1)\cos \theta - (0)\sin \theta = -\cos \theta$$

●

The sum formula for the tangent function can be developed as follows:

$$\tan (\alpha + \beta) = \frac{\sin (\alpha + \beta)}{\cos (\alpha + \beta)}$$

$$= \frac{\sin \alpha \cos \beta + \cos \alpha \sin \beta}{\cos \alpha \cos \beta - \sin \alpha \sin \beta}$$

Dividing numerator and denominator by $\cos \alpha \cos \beta$ yields

$$\tan (\alpha + \beta) = \frac{\dfrac{\sin \alpha \cos \beta}{\cos \alpha \cos \beta} + \dfrac{\cos \alpha \sin \beta}{\cos \alpha \cos \beta}}{\dfrac{\cos \alpha \cos \beta}{\cos \alpha \cos \beta} - \dfrac{\sin \alpha \sin \beta}{\cos \alpha \cos \beta}}$$

Therefore,

$$\tan (\alpha + \beta) = \frac{\tan \alpha + \tan \beta}{1 - \tan \alpha \tan \beta} \qquad (82.3)$$

Equations (82.1) through (82.3) are called the **addition formulas.** We note that although we could derive corresponding formulas for $\cot (\alpha + \beta)$, $\sec (\alpha + \beta)$, and $\csc (\alpha + \beta)$, they are seldom used.

Example 82.3

Given $\sin \alpha = \frac{2}{3}$, $\cos \beta = \frac{1}{2}$, α is a second quadrant angle and β is a fourth quadrant angle. Find $\sin (\alpha + \beta)$.

Solution

Since $\sin (\alpha + \beta) = \sin \alpha \cos \beta + \cos \alpha \sin \beta$, we must find $\cos \alpha$ and $\sin \beta$. Using the identity $\sin^2 \theta + \cos^2 \theta = 1$, we find that for $\sin \alpha = \frac{2}{3}$,

$$\frac{4}{9} + \cos^2 \alpha = 1$$

$$\cos^2 \alpha = 1 - \frac{4}{9} = \frac{5}{9}$$

and

$$\cos \alpha = \pm\sqrt{\frac{5}{9}} = \pm\frac{\sqrt{5}}{3}$$

We are given that α is a second quadrant angle, and since the cosine function is negative in the second quadrant, we have $\cos \alpha = -\sqrt{5}/3$.

We find $\sin \beta$ in a similar manner, as follows:

$$\sin^2 \beta + \cos^2 \beta = 1$$

$$\sin^2 \beta + \frac{1}{4} = 1$$

$$\sin^2 \beta = \frac{3}{4}$$

or

$$\sin \beta = \pm \frac{\sqrt{3}}{2}$$

Since β is a fourth quadrant angle, we must choose the negative value. Hence, $\sin \beta = -\sqrt{3}/2$. Now

$$\begin{aligned}\sin(\alpha + \beta) &= \left(\frac{2}{3}\right)\cdot\left(\frac{1}{2}\right) + \left(-\frac{\sqrt{5}}{3}\right)\left(\frac{-\sqrt{3}}{2}\right)\\ &= \frac{2 + \sqrt{15}}{6}\end{aligned}$$

●

In Section 41 we found that we could express the trigonometric function of any negative angle in terms of a trigonometric function of a positive angle. We restate these relationships here for easy reference.

$$\left.\begin{aligned}\sin(-\theta) &= -\sin\theta \qquad & \csc(-\theta) &= -\csc\theta\\ \cos(-\theta) &= \cos\theta & \sec(-\theta) &= \sec\theta\\ \tan(-\theta) &= -\tan\theta & \cot(-\theta) &= -\cot\theta\end{aligned}\right\} \tag{82.4}$$

We can now easily derive the formulas for the functions of the difference of two angles. To find $\sin(\alpha - \beta)$ we substitute $-\beta$ for β in (82.1), and we obtain

$$\sin[\alpha + (-\beta)] = \sin\alpha\cos(-\beta) + \cos\alpha\sin(-\beta)$$

By (82.4) we conclude that

$$\sin(\alpha - \beta) = \sin\alpha\cos\beta - \cos\alpha\sin\beta \tag{82.5}$$

Following the same procedure for equations (82.2) and (82.3) yields, respectively,

$$\cos(\alpha - \beta) = \cos\alpha\cos\beta + \sin\alpha\sin\beta \tag{82.6}$$

and

$$\tan(\alpha - \beta) = \frac{\tan\alpha - \tan\beta}{1 + \tan\alpha\tan\beta} \tag{82.7}$$

Equations (82.5) through (82.7) are called the **difference formulas**.

Example 82.4

Find $\cos 15°$ by using a difference formula.

Solution

We begin by noting that $15° = 60° - 45°$. Thus, $\cos 15° = \cos(60° - 45°)$. By formula (82.6),

$$\cos 15° = \cos(60° - 45°) = \cos 60° \cos 45° + \sin 60° \sin 45°$$
$$= \frac{1}{2} \cdot \frac{\sqrt{2}}{2} + \frac{\sqrt{3}}{2} \cdot \frac{\sqrt{2}}{2}$$
$$= \frac{\sqrt{2} + \sqrt{6}}{4}$$

Thus, $\cos 15° = (\sqrt{2} + \sqrt{6})/4$. We can also find the sec 15° without having a difference formula for secant by noting that $\sec 15° = 1/\cos 15°$ and then use the above result for cos 15° as follows:

$$\sec 15° = \frac{1}{\frac{\sqrt{2} + \sqrt{6}}{4}} = \frac{4}{\sqrt{2} + \sqrt{6}}$$
$$= \frac{4}{\sqrt{2} + \sqrt{6}} \cdot \frac{\sqrt{2} - \sqrt{6}}{\sqrt{2} - \sqrt{6}}$$
$$= \frac{4(\sqrt{2} - \sqrt{6})}{-4} = \sqrt{6} - \sqrt{2}$$

●

The formulas derived in this section can be used to prove certain trigonometric identities. The following examples will illustrate these uses.

Example 82.5

Show that $[\sin(x + y)][\sin(x - y)] = \sin^2 x - \sin^2 y$.

Solution

We can transform the left side by using (82.1) and (82.5), to obtain

$$\sin(x + y) = \sin x \cos y + \cos x \sin y \qquad \text{(A)}$$

and

$$\sin(x - y) = \sin x \cos y - \cos x \sin y \qquad \text{(B)}$$

The right sides of (A) and (B) are conjugates. Thus, the product

$$\sin(x + y)\sin(x - y) = (\sin x \cos y)^2 - (\cos x \sin y)^2$$

Hence,

$$\sin(x + y)\sin(x - y) = \sin^2 x \cos^2 y - \cos^2 x \sin^2 y \qquad \text{(C)}$$

Since $\cos^2 y = 1 - \sin^2 y$ and $\cos^2 x = 1 - \sin^2 x$, we can substitute these results into the right side of (C), to obtain

$$\sin^2 x(1 - \sin^2 y) - (1 - \sin^2 x)\sin^2 y$$

or

$$\sin^2 x - \sin^2 x \sin^2 y - \sin^2 y + \sin^2 x \sin^2 y = \sin^2 x - \sin^2 y$$

Example 82.6

Express $\sin x \cos 2x - \sin 2x \cos x$ as a single term.

Solution

If we let $x = \alpha$ and $2x = \beta$, the given expression becomes $\sin\alpha\cos\beta - \sin\beta\cos\alpha$, which is the form for $\sin(\alpha - \beta)$. Replacing α and β by x and $2x$, respectively, we have

$$\sin x \cos 2x - \sin 2x \cos x = \sin(x - 2x) = \sin(-x) = -\sin x$$

Exercises for Section 82

In Exercises 1 through 12, use the appropriate addition or subtraction formula to evaluate each function without the use of trigonometric tables.

1. $\cos 75° = \cos(30° + 45°)$

2. $\sin 15° = \sin(60° - 45°)$

3. $\cos 15° = \cos(45° - 30°)$

4. $\sin 105° = \sin(60° + 45°)$

5. $\cos 105° = \cos(60° + 45°)$

6. $\tan 105° = \tan(60° + 45°)$

7. $\tan 15° = \tan(45° - 30°)$

8. $\cos 165° = \cos(120° + 45°)$

9. $\cos 195° = \cos(225° - 30°)$

10. $\sin 285° = \sin(330° - 45°)$

11. $\csc 285° = 1/\sin 285°$ and the result of Exercise 10

12. $\sec 195° = 1/\cos 195°$ and the result of Exercise 9

13. Evaluate $\sin(\alpha + \beta)$ if $\sin\alpha = \frac{3}{5}$ and $\sin\beta = \frac{5}{13}$, where α and β are in the first quadrant.

14. Evaluate $\cos(\alpha - \beta)$ given the conditions in Exercise 13.

15. Evaluate $\sin(\alpha + \beta)$ if $\sin\alpha = \frac{8}{17}$ (α in the first quadrant) and $\sin\beta = \frac{4}{5}$ (β in the second quadrant).

16. Evaluate $\cos(\alpha + \beta)$ given the conditions in Exercise 15.

17. Evaluate $\sin(\alpha - \beta)$ if $\sin\alpha = \frac{15}{17}$ (α in the second quadrant) and $\cos\beta = -\frac{3}{5}$ (β in the third quadrant).

18. Evaluate $\cos(\alpha - \beta)$ given the conditions in Exercise 17.

In Exercises 19 through 25, prove each identity.

19. $\sin(90° + x) = \cos x$

20. $\cos(90° + x) = -\sin x$

21. $\tan(180° + x) = \tan x$

22. $\sin(90° - x) = \cos x$

23. $\sin(270° + x) = -\cos x$

24. $\tan(x + 90°) = -\cot x$

25. $\cos(x - 45°) = \frac{\sqrt{2}}{2}(\cos x + \sin x)$

In Exercises 26 through 30, use sum and difference formulas to simplify each function and then graph one cycle of the curve defined by the equation.

26. $y = 3\sin(x + \pi)$

27. $y = 2\cos(x + \pi)$

28. $y = \cos\left(\frac{\pi}{2} - x\right)$

29. $y = 2\sin\left(x + \frac{\pi}{2}\right)$

30. $y = -\sin(x - \pi)$

31. Show that $\cos(x + y)\cos(x - y) = \cos^2 x - \sin^2 y$.

32. Show that $\cos(x + y) + \cos(x - y) = 2\cos x \cos y$.

33. Show that $\sin x \cos y = \frac{1}{2}\sin(x + y) + \frac{1}{2}\sin(x - y)$.

34. Show that $\cos x \sin y = \frac{1}{2}\sin(x + y) - \frac{1}{2}\sin(x - y)$.

35. Show that $\cos x \cos y = \frac{1}{2}\cos(x + y) + \frac{1}{2}\cos(x - y)$.

36. Show that $\sin x \sin y = \frac{1}{2}\cos(x - y) - \frac{1}{2}\cos(x + y)$.

37. The index n of refraction of glass is given by

$$n = \frac{\sin\left[\frac{1}{2}(\alpha + \beta)\right]}{\sin\frac{\beta}{2}}$$

where α is the angle of the prism and β the minimum angle of the refraction. If $\beta = 30°$, find an equivalent expression for n.

SECTION 83

Double- and Half-Angle Formulas

If we let $\beta = \alpha$ in equations (82.1) and (82.2), we can derive **double-angle formulas** for sine and cosine. These formulas will enable us to express the trigonometric functions of twice an angle in terms of functions of the angle itself.

From equation (82.1), the substitution $\beta = \alpha$ yields

$$\sin(\alpha + \alpha) = \sin\alpha\cos\alpha + \cos\alpha\sin\alpha$$

or

$$\sin 2\alpha = 2\sin\alpha\cos\alpha \tag{83.1}$$

From equation (82.2), the same substitution yields

$$\begin{aligned}\cos(\alpha + \alpha) &= \cos\alpha\cos\alpha - \sin\alpha\sin\alpha \\ &= \cos^2\alpha - \sin^2\alpha\end{aligned}$$

Since $\sin^2\alpha = 1 - \cos^2\alpha$, we have

$$\cos^2\alpha - \sin^2\alpha = \cos^2\alpha - (1 - \cos^2\alpha) = 2\cos^2\alpha - 1$$

Similarly, since $\cos^2\alpha = 1 - \sin^2\alpha$, we can write

$$\cos^2\alpha - \sin^2\alpha = (1 - \sin^2\alpha) - \sin^2\alpha = 1 - 2\sin^2\alpha$$

Thus,

$$\begin{aligned}\cos 2\alpha &= \cos^2\alpha - \sin^2\alpha \\ \cos 2\alpha &= 2\cos^2\alpha - 1 \\ \cos 2\alpha &= 1 - 2\sin^2\alpha\end{aligned} \tag{83.2}$$

The double-angle formula for the tangent function can be derived by either using equation (82.3) or the fact that $\tan 2\alpha = \sin 2\alpha/\cos 2\alpha$. In either case we have

$$\tan 2\alpha = \frac{2\tan\alpha}{1 - \tan^2\alpha} \tag{83.3}$$

Example 83.1

Use formula (83.1) to evaluate $\sin 90°$.

Solution

If we let $2\alpha = 90°$, then $\alpha = 45°$, Thus, using (83.1), we have

$$\begin{aligned}\sin 90° = \sin(2 \cdot 45°) &= 2\sin 45°\cos 45° \\ &= 2 \cdot \left(\frac{\sqrt{2}}{2}\right)\left(\frac{\sqrt{2}}{2}\right) = 1\end{aligned}$$

Example 83.2

Use formula (83.2) to evaluate $\cos 60°$.

Solution

If we let $2\alpha = 60°$, then $\alpha = 30°$, and since

$$\cos 60° = \cos(2 \cdot 30°) = \cos^2 30° - \sin^2 30°$$
$$= \left(\frac{\sqrt{3}}{2}\right)^2 - \left(\frac{1}{2}\right)^2 = \frac{1}{2}$$

Equivalently, $\cos 2\alpha = 2\cos^2 \alpha - 1$. Thus,

$$\cos 60° = 2\cos^2 30° - 1$$
$$= 2\left(\frac{\sqrt{3}}{2}\right)^2 - 1 = \frac{1}{2}$$

or

$$\cos 2\alpha = 1 - 2\sin^2 \alpha$$
$$\cos 60° = 1 - 2\sin^2 30°$$
$$= 1 - 2\left(\frac{1}{2}\right)^2 = \frac{1}{2}$$

Example 83.3

If θ is a second quadrant angle and $\sin\theta = \frac{4}{5}$, find $\sin 2\theta$, $\cos 2\theta$, and $\tan 2\theta$.

Solution

To evaluate $\sin 2\theta$, we must know the values for $\sin\theta$ and $\cos\theta$. Since $\sin\theta = \frac{4}{5}$, we find $\cos\theta$ by using equation (81.6). Hence,

$$\cos^2\theta = 1 - \sin^2\theta$$
$$= 1 - \left(\frac{4}{5}\right)^2 = \frac{9}{25}$$

Now,

$$\cos\theta = \pm\sqrt{\frac{9}{25}} = \pm\frac{3}{5}$$

Since θ is a second quadrant angle and cosine is negative in the second quadrant, $\cos\theta = -\frac{3}{5}$. Therefore,

$$\sin 2\theta = 2\sin\theta\cos\theta$$
$$= 2\left(\frac{4}{5}\right)\left(-\frac{3}{5}\right) = -\frac{24}{25}$$

We now use the fact that $\cos\theta = -\frac{3}{5}$ and formula (83.2) to evaluate $\cos 2\theta$ and $\tan 2\theta$ as follows:

$$\cos 2\theta = \cos^2\theta - \sin^2\theta$$
$$\cos 2\theta = \left(-\frac{3}{5}\right)^2 - \left(\frac{4}{5}\right)^2 = -\frac{7}{25}$$

and

$$\tan 2\theta = \frac{\sin 2\theta}{\cos 2\theta} = \frac{-\frac{24}{25}}{-\frac{7}{25}} = \frac{24}{7}$$

●

The formulas developed in this section can also be used to prove additional identities.

Example 83.4

Prove that $\sin 2x = \dfrac{2 \tan x}{1 + \tan^2 x}$.

Solution

We begin by converting the terms on the right side to expressions involving sine and cosine. Hence,

$$\frac{2 \tan x}{1 + \tan^2 x} = \frac{\dfrac{2 \sin x}{\cos x}}{1 + \dfrac{\sin^2 x}{\cos^2 x}} \tag{A}$$

If we multiply numerator and denominator by $\cos^2 x$, we obtain

$$\frac{\left[2\dfrac{\sin x}{\cos x}\right] \cdot \cos^2 x}{\left[1 + \dfrac{\sin^2 x}{\cos^2 x}\right] \cdot \cos^2 x} = \frac{2 \sin x \cos x}{\cos^2 x + \sin^2 x} \tag{B}$$

Since $\cos^2 x + \sin^2 x = 1$, the right side of (B) reduces to $2 \sin x \cos x = \sin 2x$. Hence, we have proved the given identity. ●

The half-angle formulas are used to express a function of half an angle in terms of the angle itself. If we let $\alpha = \theta/2$ in the double-angle formula $\cos 2\alpha = 1 - 2 \sin^2 \alpha$, we have

$$\cos \left[2\left(\frac{\theta}{2}\right)\right] = 1 - 2 \sin^2 \frac{\theta}{2}$$

or

$$\cos \theta = 1 - 2 \sin^2 \frac{\theta}{2}$$

If we solve for $\sin (\theta/2)$, we obtain

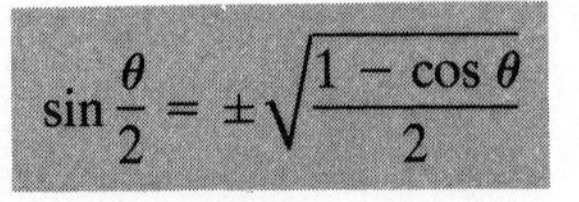

$$\sin \frac{\theta}{2} = \pm\sqrt{\frac{1 - \cos \theta}{2}} \tag{83.4}$$

The same substitution in the formula $\cos 2\alpha = 2\cos^2\alpha - 1$ yields

$$\cos\frac{\theta}{2} = \pm\sqrt{\frac{1+\cos\theta}{2}} \tag{83.5}$$

In equations (83.4) and (83.5), the sign chosen depends on the quadrant in which $\theta/2$ lies. For $\sin\theta/2$, we choose the positive sign if $\theta/2$ is a first or second quadrant angle, the negative sign if $\theta/2$ is a third or fourth quadrant angle. For $\cos\theta/2$ we choose the positive sign if $\theta/2$ is a first or fourth quadrant angle; the negative sign if $\theta/2$ is a second or third quadrant angle.

The half-angle formula for tangent is obtained by using the fact that

$$\tan\frac{\theta}{2} = \frac{\sin\frac{\theta}{2}}{\cos\frac{\theta}{2}}$$

and formulas (83.4) and (83.5), to obtain

$$\tan\frac{\theta}{2} = \pm\sqrt{\frac{1-\cos\theta}{1+\cos\theta}} \tag{83.6}$$

The following examples illustrate the uses of these identities in evaluations and other identities.

Example 83.5

Use a half-angle formula to determine $\sin 15°$.

Solution

We use formula (83.4) with $\theta = 30°$. Since $\sin 15°$ is positive, we use the positive sign before the radical. Thus,

$$\sin 15° = \sin\frac{30°}{2} = \sqrt{\frac{1-\cos 30°}{2}}$$

$$= \sqrt{\frac{1-\frac{\sqrt{3}}{2}}{2}}$$

$$= \sqrt{\frac{2+\sqrt{3}}{2}}$$

Example 83.6

Use a half-angle formula to determine $\cos 150°$.

Solution

We use formula (83.5) with $\theta = 300°$. Since cos 150° is negative, we use the minus sign before the radical. Thus,

$$\cos 150° = \cos\frac{300°}{2} = -\sqrt{\frac{1 + \cos 300°}{2}}$$

$$= -\sqrt{\frac{1 + \frac{1}{2}}{2}}$$

$$= -\frac{\sqrt{3}}{2}$$

Example 83.7

Use a half-value formula to determine tan 165°.

Solution

We use formula (83.6) with $\theta = 330°$. Since tan 165° is negative, we use the minus sign before the radical. Thus,

$$\tan 165° = \tan\frac{330°}{2} = -\sqrt{\frac{1 - \cos 330°}{1 + \cos 330°}}$$

$$= -\sqrt{\frac{1 - \frac{\sqrt{3}}{2}}{1 + \frac{\sqrt{3}}{2}}}$$

$$= -\sqrt{\frac{2 - \sqrt{3}}{2 + \sqrt{3}}}$$

Example 83.8

Find the value of $\cos(\theta/2)$ if $\sin\theta = -\frac{3}{5}$ and θ is a third quadrant angle.

Solution

Since θ is a third quadrant angle, we have $180° < \theta < 270°$ or $90° < \theta/2 < 135°$. Thus, $\theta/2$ is a second quadrant angle and we will use formula (83.5) with a negative sign in front of the radical. Since

$$\cos\frac{\theta}{2} = -\sqrt{\frac{1 + \cos\theta}{2}}$$

we need to find the value of $\cos\theta$ in order to evaluate $\cos(\theta/2)$. We use Figure 83.1 to determine that $\cos\theta = -\frac{4}{5}$. Thus,

$$\cos\left(\frac{\theta}{2}\right) = -\sqrt{\frac{1 + \left(-\frac{4}{5}\right)}{2}}$$

$$= -\sqrt{\frac{1}{10}} = -\frac{\sqrt{10}}{10}$$

It should be noted that we could have calculated $\cos\theta$ by using the identity $\sin^2\theta + \cos^2\theta = 1$, but the figure is more convenient. ●

Figure 83.1

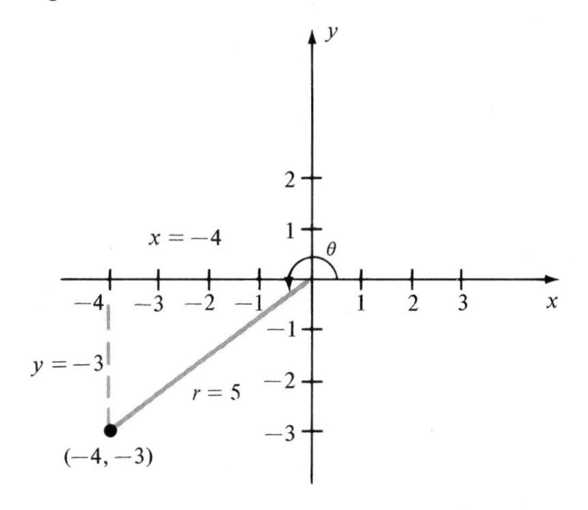

Remark 83.1. *In Example 83.8 we assumed that since θ was an angle in the third quadrant, $180° < \theta < 270°$. If $540° < \theta < 630°$, then θ is still a third quadrant angle but $270° < \theta/2 < 315°$. Thus, $\theta/2$ would be a fourth quadrant angle and $\cos(\theta/2)$ would also be positive. Hence, the answer for Example 83.8 would be $\cos(\theta/2) = \pm\sqrt{10}/10$. In all further discussions, we will assume (unless otherwise stated) that $0° < \theta < 90°$ is an angle in quadrant I; $90° < \theta < 180°$ is an angle in quadrant II; $180° < \theta < 270°$ is an angle in quadrant III; $270° < \theta < 360°$ is an angle in quadrant IV.*

Example 83.9

Express $\cos^4 x$ as an expression involving only functions of x raised to the first power.

Solution

From formula (83.5) we note that if we square both sides, we obtain

$$\cos^2\frac{x}{2} = \frac{1 + \cos x}{2} \tag{A}$$

Thus, we see that by using form (A), the cosine function can be reduced from a power of 2 to a power of 1 by doubling x. Hence, form (A) is equivalent to

$$\cos^2 x = \frac{1 + \cos 2x}{2} \tag{B}$$

Therefore,

$$\cos^4 x = (\cos^2 x)^2 = \left(\frac{1 + \cos 2x}{2}\right)^2$$

or

$$\cos^4 x = \frac{1 + 2\cos 2x + \cos^2 2x}{4} \tag{C}$$

We now reduce the $\cos^2 2x$ term in (C) by means of (B). Thus,

$$\cos^2 2x = \frac{1 + \cos 4x}{2}$$

We combine this result with (C), to obtain

$$\begin{aligned} \cos^4 x &= \frac{1}{4} + \frac{2\cos 2x}{4} + \frac{1}{4}\left(\frac{1 + \cos 4x}{2}\right) \\ &= \frac{1}{4} + \frac{\cos 2x}{2} + \frac{1}{8} + \frac{\cos 4x}{8} \\ &= \frac{3}{8} + \frac{\cos 2x}{2} + \frac{\cos 4x}{8} \end{aligned}$$

This type of transform is of particular importance in more advanced mathematics.

Exercises for Section 83

In Exercises 1 through 6, use a double-angle formula to evaluate each expression. Do not use trigonometric tables.

1. $\sin 120°$

2. $\sin 90°$

3. $\cos 120°$

4. $\cos 240°$

5. $\tan 240°$

6. $\tan 120°$

In Exercises 7 through 12, use a half-angle formula to evaluate each expression. Do not use trigonometric tables.

7. $\cos 15°$

8. $\sin 22.5°$

9. $\sin 75°$

10. $\tan 112.5°$

11. $\cos 67.5°$

12. $\sin 165°$

13. Find $\sin 2x$ if $\cos x = \frac{3}{5}$ (x is in the first quadrant).

14. Find $\cos 2x$ if $\cos x = \frac{3}{5}$ (x is in the first quadrant).

15. Find $\cos 2x$ if $\sin x = -\frac{5}{13}$ (x is in the third quadrant).

16. Find $\sin 2x$ if $\sin x = -\frac{5}{13}$ (x is in the fourth quadrant).

17. Find $\sin 2x$ and $\cos 2x$ if $\tan x = \frac{3}{4}$ (x is in the third quadrant).

18. Use the results and conditions of Exercise 17 to find $\sin 4x$.

19. Find $\cos(x/2)$ if $\cos x = -\frac{3}{5}$ (x is in the third quadrant).

20. Find $\sin(x/2)$ if $\sin x = -\frac{4}{5}$ (x is in the third quadrant).

21. Find $\tan(x/2)$ if $\sin x = -\frac{4}{5}$ (x is in the third quadrant).

22. Find $\sin(x/2)$ if $\tan x = \frac{7}{24}$ (x is in the third quadrant).

23. Express $\sin^2 x$ as an expression involving only functions of x raised to the first power.

24. Repeat Exercise 23 for $\sin^4 x$.

In Exercises 25 through 33, prove each identity.

25. $\dfrac{2}{1 - \cos 2x} = \csc^2 x$

26. $\cos 2x = \dfrac{1 - \tan^2 x}{1 + \tan^2 x}$

27. $\sin 2x = \dfrac{1}{\tan x + \cot 2x}$

28. $\sin 3x = 3 \sin x - 4 \sin^3 x$

***Hint*:** Let $3x = 2x + x$.

29. $\sin \dfrac{x}{4} \cos \dfrac{x}{4} = \dfrac{1}{2} \sin \dfrac{x}{2}$

30. $\cos 2x = \cos^4 x - \sin^4 x$

31. $\tan \dfrac{x}{2} = \dfrac{\sin x}{1 + \cos x}$

32. $2 \sin \dfrac{x}{2} = \dfrac{1 - \cos x}{\sin(x/2)}$

33. $(\sin x + \cos x)^2 = 1 + \sin 2x$

34. A projectile fired from ground level at an angle α with the horizontal with an initial speed V_0 has a horizontal range given by

$$R = V_0 t \cos \alpha$$

where t = the time in flight. If $t = V_0 \sin \alpha / g$, g the acceleration due to gravity, show that

$$R = \frac{V_0^2 \sin 2\alpha}{g}$$

35. The approximate time t of the swing of a pendulum of length L is

$$t = 2\pi \sqrt{\frac{L}{g}} \left(1 + \frac{1}{4} \sin^2 \frac{\theta}{2}\right)$$

Find t if $L = 50$ cm, $g = 980\ \text{cm/sec}^2$, and $\theta = 30°$, without the use of trigonometric tables.

SECTION 84

Trigonometric Equations

An important application of trigonometric identities occurs in the solution of equations involving trigonometric functions. A trigonometric equation is an equation that involves trigonometric functions of unknown angles. A solution of a trigonometric equation is a value of the angle, in degrees or radians, which satisfies the equation. For our purposes we shall agree to consider as solutions only those values for x in the interval $0 \leq x < 360°$ if x is in degrees or $0 \leq x < 2\pi$ for x given in radians unless otherwise stated.

Unfortunately, there is no general method that will enable us to solve all equations. The following examples illustrate methods that prove to be effective.

Example 84.1

Solve $2 \cos x - 1 = 0$.

Solution

$$2 \cos x - 1 = 0$$

$$2 \cos x = 1$$

$$\cos x = \frac{1}{2}$$

Since $\cos x$ is positive in the first and fourth quadrants, $x = \pi/3$ and $x = 2\pi - \pi/3 = 5\pi/3$ are the only two solutions to this equation for $0 \leq x < 2\pi$.

Example 84.2

Solve $\sin^2 x = \sin x$.

Solution

We solve the equation $\sin^2 x = \sin x$ as follows:

$$\sin^2 x = \sin x$$

$$\sin^2 x - \sin x = 0$$

$$\sin x (\sin x - 1) = 0$$

Setting each factor equal to zero, we obtain

$$\sin x = 0 \quad \text{or} \quad \sin x = 1$$

For $\sin x = 0$, $x = 0$ and π. For $\sin x = 1$, $x = \pi/2$. Thus, the complete solution is $x = 0$, $x = \pi/2$, and $x = \pi$. Direct substitution in the given equation will verify the solutions.

Example 84.3

Solve $\sin 2x + \cos x = 0$.

Solution

We use the double-angle formula for $\sin 2x$ to write the equation in the form

$$2 \sin x \cos x + \cos x = 0$$

Factoring, we have

$$\cos x(2 \sin x + 1) = 0$$

Setting each factor equal to zero, we have

$$\cos x = 0 \quad \text{or} \quad \sin x = -\tfrac{1}{2}$$

For $\cos x = 0$, $x = \pi/2$ or $3\pi/2$. For $\sin x = -\frac{1}{2}$, $x = 7\pi/6$ or $11\pi/6$. Therefore, $x = \pi/2, 7\pi/6, 3\pi/2$, and $11\pi/6$ are the solutions.

Example 84.4

Solve $\cos 2x - \sin x = 0$.

Solution

We use the formula $\cos 2x = 1 - 2 \sin^2 x$ to rewrite the given equation as

$$1 - 2 \sin^2 x - \sin x = 0$$

or

$$2 \sin^2 x + \sin x - 1 = 0 \tag{A}$$

We factor (A) to obtain

$$(2 \sin x - 1)(\sin x + 1) = 0 \tag{B}$$

We set each factor of (B) equal to zero, to obtain

$$\sin x = \frac{1}{2} \quad \text{or} \quad \sin x = -1$$

Since $\sin x = \frac{1}{2}$ when $x = \pi/6$ or $5\pi/6$ and $\sin = -1$ when $x = 3\pi/2$, the complete solution is $x = \pi/6, 5\pi/6$, and $3\pi/2$.

Example 84.5

Solve $\sin 3x = \sqrt{3}/2$.

Solution

We note that $\sin 3x = \sqrt{3}/2$ when $3x = \pi/3$ and when $3x = 2\pi/3$. Thus, $x = \pi/9$ and $x = 2\pi/9$. But there are other solutions in the interval $0 \le x < 2\pi$. Let us recall that the function $y = \sin 3x$ has a period of $2\pi/3$ units. Thus, the graph of $y = \sin 3x$ goes through three complete cycles in the interval $0 \le x < 2\pi$ (Figure 84.1). Hence, our initial values $\pi/9$ and $2\pi/9$ have corresponding values in the second and third cycles of

Figure 84.1

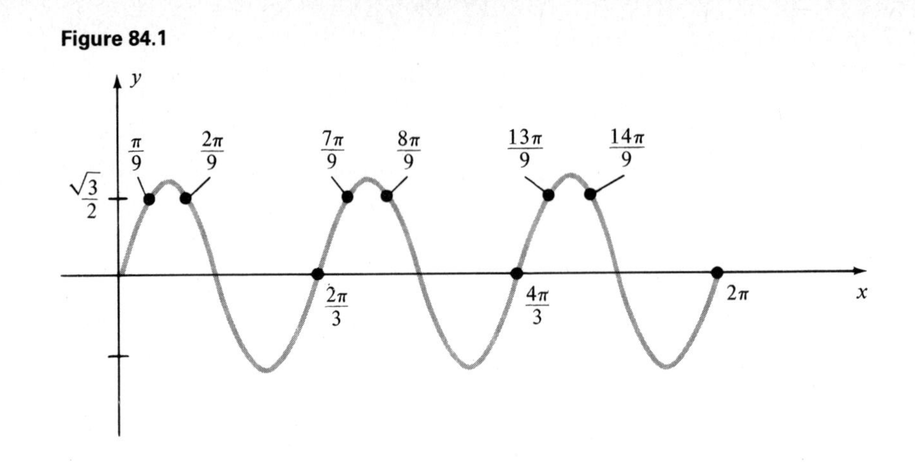

the curve. These values are obtained algebraically by adding the period $2\pi/3$ to $\pi/9$ and $2\pi/9$, respectively, until we have accounted for *all* solutions in the interval $0 \le x < 2\pi$. Therefore,

$$x = \frac{\pi}{9} \qquad\qquad x = \frac{2\pi}{9}$$

$$x = \frac{\pi}{9} + \frac{2\pi}{3} = \frac{7\pi}{9} \quad \text{and} \quad x = \frac{2\pi}{9} + \frac{2\pi}{3} = \frac{8\pi}{9}$$

$$x = \frac{7\pi}{9} + \frac{2\pi}{3} = \frac{13\pi}{9} \qquad x = \frac{8\pi}{9} + \frac{2\pi}{3} = \frac{14\pi}{9}$$

These six solutions are shown in Figure 84.1.

Example 84.6

Solve $\sin x + \cos x = 1$.

Solution

We rewrite $\sin x + \cos x = 1$ as

$$\sin x = 1 - \cos x \tag{A}$$

We now square both sides of (A), to obtain

$$\sin^2 x = 1 - 2\cos x + \cos^2 x \tag{B}$$

Substituting $1 - \cos^2 x$ for $\sin^2 x$ in (B) yields

$$1 - \cos^2 x = 1 - 2\cos x + \cos^2 x$$

$$2\cos^2 x - 2\cos x = 0$$

$$2\cos x(\cos x - 1) = 0 \tag{C}$$

We set each factor in (C) equal to zero and solve. Hence,

$$2\cos x = 0 \qquad \cos x - 1 = 0$$

$$\cos x = 0 \qquad \cos x = 1$$

$$x = \frac{\pi}{2}, \frac{3\pi}{2} \qquad x = 0$$

It appears that the complete solution set is $x = 0$, $\pi/2$, and $3\pi/2$. However, a check of the original equation shows that while $x = 0$ and $x = \pi/2$ are solutions, $x = 3\pi/2$ is not a solution, since

$$\text{for } x = 0: \quad \sin 0 + \cos 0 = 0 + 1 = 1$$

$$\text{for } x = \frac{\pi}{2}: \quad \sin\frac{\pi}{2} + \cos\frac{\pi}{2} = 1 + 0 = 1$$

$$\text{for } x = \frac{3\pi}{2}: \quad \sin\frac{3\pi}{2} + \cos\frac{3\pi}{2} = -1 + 0 \neq 1$$

Thus, the solution is $x = 0$ and $\pi/2$. We note that since we squared both sides of the given equation, we must check all solutions. This is necessary, since the squaring of both sides of an equation may introduce extraneous solutions.

Exercises for Section 84

In Exercises 1 through 26, solve for x where $0 \leq x < 2\pi$.

1. $2\sin x + 1 = 0$

2. $2\sin x - \sqrt{3} = 0$

3. $\cos x - 1 = 0$

4. $\cos x - 2 = 0$

5. $(\sin x + 1)(2\sin x + 1) = 0$

6. $(\cos x - 1)(\sin x - 1) = 0$

7. $2\sin^2 x + 3\sin x + 1 = 0$

8. $\cos^2 x - \cos x = 2$

9. $1 + \tan x = \sec x$

10. $\sin x = 1 - 2\cos x$

11. $\sin 2x + \sin x = 0$

12. $2\sin x\cos x = \frac{\sqrt{3}}{2}$

13. $2\sin^2 x + \sin x = 1$

14. $\sin 2x = -\frac{\sqrt{3}}{2}$

15. $\cos 2x = 0$

16. $\tan 2x = 1$

17. $\tan 2x = -\sqrt{3}$

18. $\sin 2x + \cos 2x = 0$

19. $2\sin\frac{x}{2} = 1$

20. $\sin 2x \sin x + \cos x = 0$

21. $2\sin x - \tan x = 0$

22. $\sin\frac{x}{2} = \sin x$

23. $\cos\frac{x}{2} = 1 + \cos x$

24. $\sin^2 x\cos x - \sin x\cos x = 0$

25. $\sin 2x - 2\cos x = 0$

26. $\sin x = \cos x - 1$

27. Solve the following system of equations graphically.

$$\begin{cases} y = x \\ y = \sin \pi x \end{cases}$$

In Exercises 28 through 33, solve for x in the interval indicated.

28. $2 \sin x + 1 = 0, -2\pi \le x \le 4\pi$

29. $2 \sin x - \sqrt{3} = 0, \quad 0° \le x \le 540°$

30. $\cos^2 x - \cos x = 2, 0 \le x \le 6\pi$

31. $\cos 2x = 0, -4\pi \le x \le \pi$

32. $2 \sin 3x = 1, 0° \le x \le 540°$

33. $2 \sin 3x = 1, 0° \le x \le 1080°$

SECTION 85
Inverse Trigonometric Functions

In Section 51 we noted that a function and its inverse (if it exists) are related geometrically by the fact that each is the reflection of the other in the line $y = x$ (see Figures 51.1 and 51.2). When we say "if it exists," we imply that not all functions have inverses. In fact, only those functions that are "one-to-one" have inverses. Geometrically, this means that any vertical *or* horizontal line will intersect the graph in at most one point. If the function is not one-to-one, it does not have an inverse (Figure 85.1).

Figure 85.1

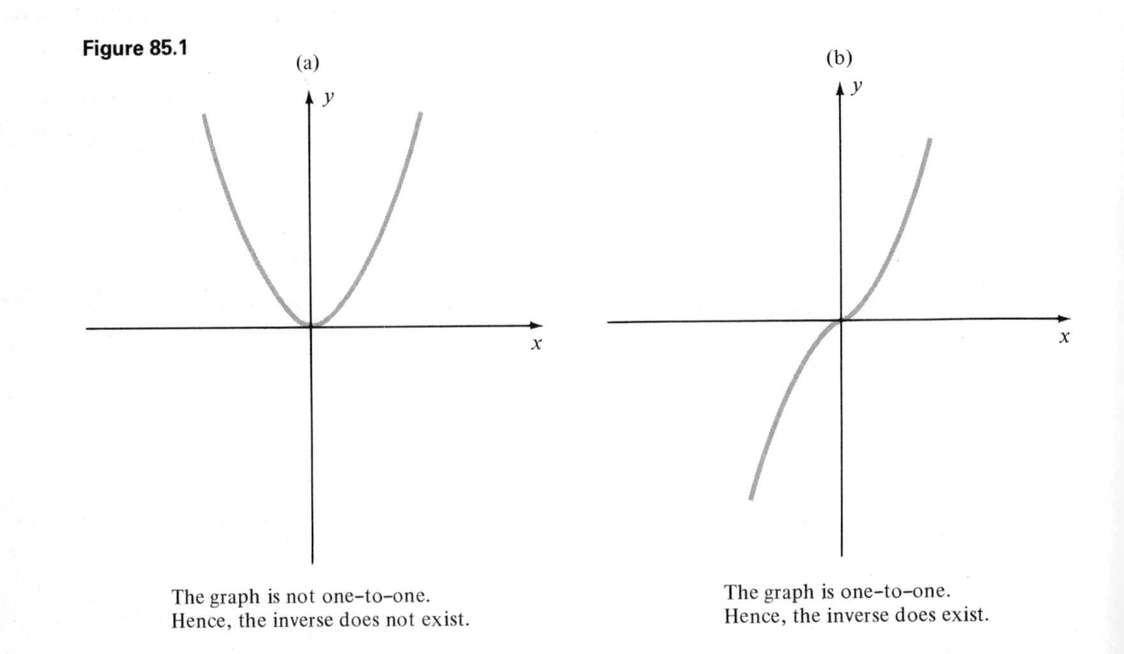

(a) The graph is not one-to-one. Hence, the inverse does not exist.

(b) The graph is one-to-one. Hence, the inverse does exist.

The graph of the inverse of the curve shown in Figure 85.1(b) is obtained by reflecting the graph about the line $y = x$. Both curves are shown in Figure 85.2.

Figure 85.2

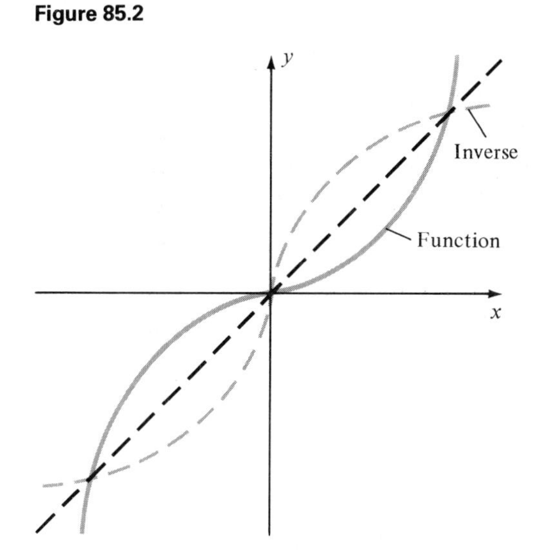

Procedure for Finding the Equation of the Inverse Function

If $y = f(x)$ is the equation of any one-to-one function, the inverse may be found as follows:

Step 1: Interchange x and y.
Step 2: Solve for y in terms of x *if possible*. The resulting equation is the inverse of the given function and is denoted $y = f^{-1}(x)$. [The term $y = f^{-1}(x)$ is read "y is equal to f inverse of x.]

Example 85.1
Find the inverse of $y = f(x) = 2x + 1$ and graph both $f(x)$ and $f^{-1}(x)$.

Solution
The graph of $y = f(x) = 2x + 1$ is a straight-line function. Hence, $f(x)$ is one-to-one and does have an inverse. Following the procedure outlined, we interchange x and y, to obtain

$$x = f(y) = 2y + 1$$

We now solve for y in terms of x and find that

$$y = f^{-1}(x) = \frac{x - 1}{2}$$

The graph of $f(x) = 2x + 1$ and $f^{-1}(x) = (x - 1)/2$ are shown in Figure 85.3. The graph of $f^{-1}(x)$ is obtained by reflecting the graph of $f(x)$ about the line $y = x$.

Figure 85.3

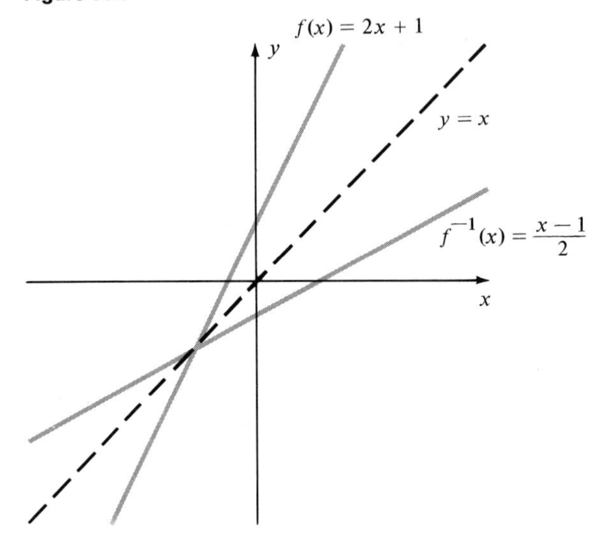

If a function is not one-to-one, it does not have an inverse, as we have noted. If the function is of particular interest, we may restrict the interval of values for the independent variable and in so doing "create" a one-to-one function. For example, the graph of $y = x^2 + 1$ is not one-to-one. If, however, we restrict the values of x to the interval $x \geq 0$, the function is one-to-one and has an inverse that can be described both algebraically and geometrically. The mechanics are shown in the following example.

Example 85.2

Find the inverse of $y = f(x) = x^2 + 1$, $x \geq 0$. Graph $f(x)$ and $f^{-1}(x)$.

Solution

The graph of $f(x)$, $x \geq 0$, is shown in Figure 85.4a. Since $f(x)$ is one-to-one, the inverse does exist and is found as follows:

1. Interchanging x and y, we have

$$x = f(y) = y^2 + 1$$

2. We now solve for y, to obtain

$$y = f^{-1}(x) = \sqrt{x - 1}$$

The graphs of $f(x)$ and $f^{-1}(x)$ are shown in Figure 85.4b. The graph of $f^{-1}(x)$ is obtained by reflecting the graph of $f(x)$ about the line $y = x$. We note that $f^{-1}(x)$ is defined only in the interval $x \geq 1$. ●

Figure 85.4

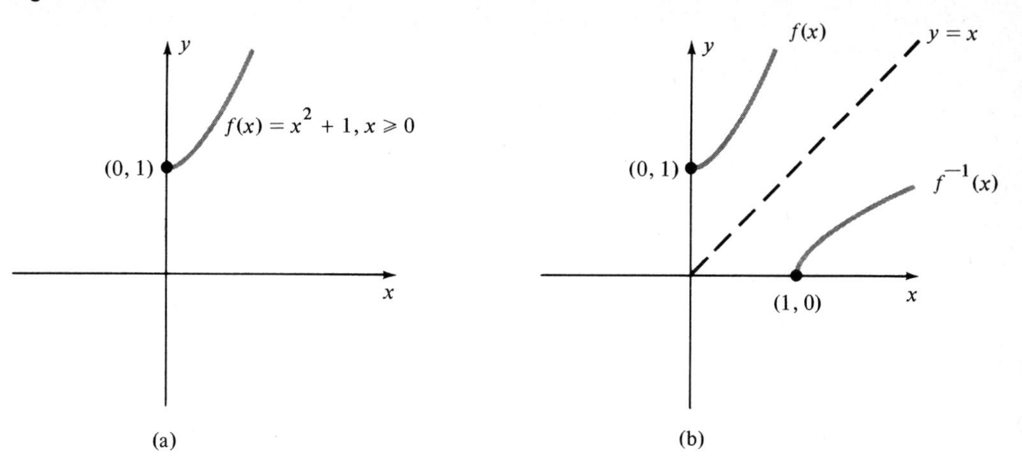

Now let us consider the inverse trigonometric functions. We begin by examining the possibility of an inverse for $y = \sin x$. From Figure 46.1 we see that $y = \sin x$ is not a one-to-one function unless we restrict the values of the independent variable. For the sine curve it is customary to choose x in the interval $-\pi/2 \leq x \leq \pi/2$. The curve thus obtained is called the **principal part** of the sine curve, is one-to-one, and hence has an inverse. We will call our inverse function the **inverse sine** or the **arcsine function** and denote it by $y = \sin^{-1} x$ or $y = \arcsin x$. We read $y = \sin^{-1} x$ as "y equals the inverse sine of x." For $y = \arcsin x$, we read "y equals the arcsin of x." This is equivalent to the statement "y equals an angle whose sine is x."

The graphs of $y = \sin x$, $-\pi/2 \leq x \leq \pi/2$, and $y = \arcsin x$ are shown in Figure 85.5 on page 502. We note from the graph that for $y = \sin x$,

$$-\frac{\pi}{2} \leq x \leq \frac{\pi}{2} \quad \text{and} \quad -1 \leq y \leq 1$$

while for $y = \arcsin x$,

$$-1 \leq x \leq 1 \quad \text{and} \quad -\frac{\pi}{2} \leq y \leq \frac{\pi}{2}$$

An algebraic discussion of the arcsin function proceeds as follows. We begin with $y = \sin x$, $-\pi/2 \leq x \leq \pi/2$. Interchanging x and y, we obtain

$$x = \sin y \qquad \text{(A)}$$

Now we would like to solve for y in terms of x in (A), but this is impossible. Hence, we "invent" a way of writing y as a function of x. We write

$$y = \arcsin x \quad \text{or} \quad y = \sin^{-1} x \qquad \text{(B)}$$

It is important to note that (A) and (B) are equivalent expressions. Thus, $y = \arcsin x$ may be written $x = \sin y$.

Figure 85.5

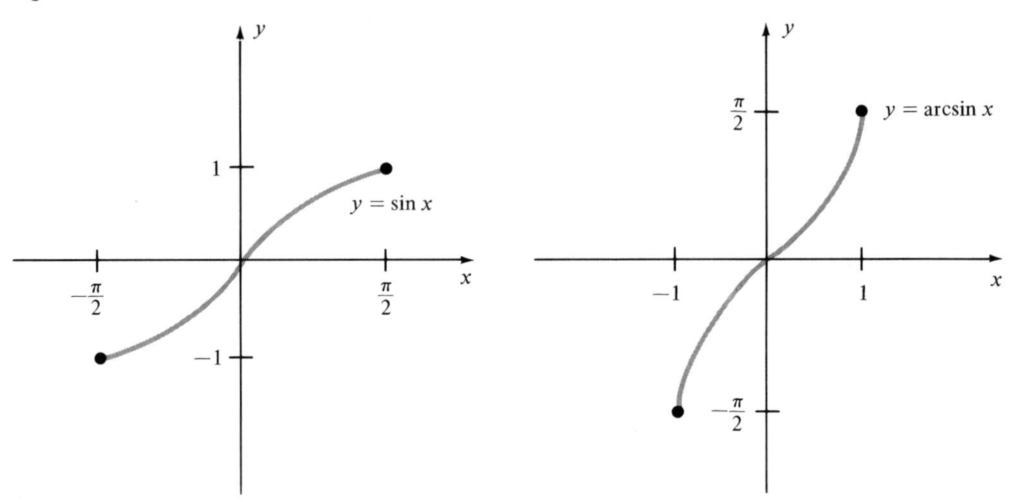

Example 85.3

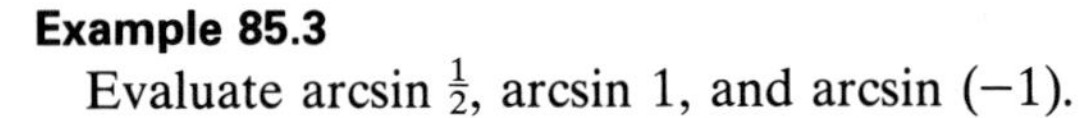

Evaluate arcsin $\frac{1}{2}$, arcsin 1, and arcsin (-1).

Solution

We read arcsin $\frac{1}{2}$ as the angle y between $-\pi/2$ and $\pi/2$ whose sine is $\frac{1}{2}$ or, equivalently, $\sin y = \frac{1}{2}$. Since $\sin \pi/6 = \frac{1}{2}$, we have $y = \pi/6$ or arcsin $\frac{1}{2} = \pi/6$. To evaluate arcsin 1, we look for an angle, y, again in the interval between $-\pi/2$ and $\pi/2$, whose sine is equal to 1. Since $\sin \pi/2 = 1$, arcsin $1 = \pi/2$. To evaluate arcsin (-1) we look for an angle y between $-\pi/2$ and $\pi/2$ whose sine is -1. Since $\sin(-\pi/2) = -1$, $y = -\pi/2$ or arcsin $(-1) = -\pi/2$. ●

Now, let us consider the inverse cosine and tangent functions. For our purposes we will, for now, omit discussion of the inverses of the other three trigonometric functions. Our approach to the cosine and tangent inverses will parallel the discussion of the inverse sine function.

For $y = \cos x$, we choose x to lie in the interval $0 \le x \le \pi$. The curve obtained is called the **principal part** of the cosine curve. Since it is one-to-one, the function has an inverse called the **arc cosine function,** denoted $y = \cos^{-1} x$ or $y = \arccos x$. The statement $y = \arccos x$ is equivalent to "y equals an angle whose cosine is x." The graphs of $y = \cos x$, $0 \le x \le \pi$, and $y = \arccos x$ are shown in Figure 85.6. We note from the graph that for $y = \cos x$, we have

$$0 \le x \le \pi \quad \text{and} \quad -1 \le y \le 1$$

and for $y = \arccos x$, we have

$$-1 \le x \le 1 \quad \text{and} \quad 0 \le y \le \pi$$

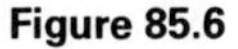
Figure 85.6

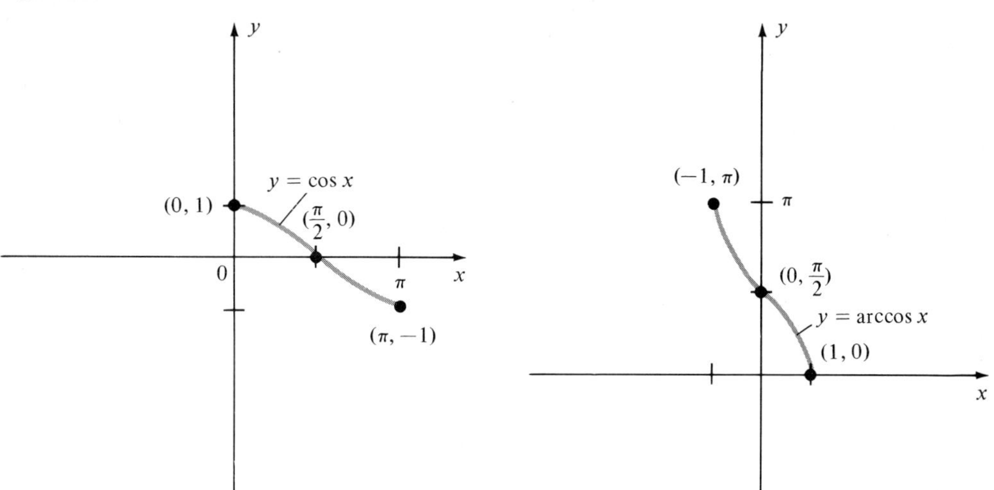

The algebraic development of the inverse cosine proceeds in the manner employed for the inverse sine.

For $y = \cos x$, $0 \le x \le \pi$, we interchange x and y, to obtain

$$x = \cos y \tag{A}$$

We now write

$$y = \cos^{-1} x \quad \text{or} \quad y = \arccos x \tag{B}$$

Again, keep in mind that forms (A) and (B) are equivalent.

Example 85.4

Evaluate $\arccos \frac{1}{2}$ and $\arccos -\frac{1}{2}$.

Solution

$\text{Arccos } \frac{1}{2}$ is the angle y between 0 and π whose cosine is $\frac{1}{2}$ or, equivalently, $\cos y = \frac{1}{2}$. Since $\cos \pi/3 = \frac{1}{2}$, $y = \pi/3$ or $\arccos(\frac{1}{2}) = \pi/3$. To evaluate $\arccos(-\frac{1}{2})$, we look for an angle y in the interval between 0 and π whose cosine is equal to $-\frac{1}{2}$. Since $\cos(2\pi/3) = -\frac{1}{2}$, $y = 2\pi/3$ or $\arccos(-\frac{1}{2}) = 2\pi/3$. Note that $-2\pi/3$ is not a correct result, even though $\cos(-2\pi/3) = -\frac{1}{2}$ since $-2\pi/3$ is not in the appropriate interval. ●

We note that $y = \tan x$ is a one-to-one function in the interval $-\pi/2 < x < \pi/2$. Thus, the inverse tangent function $y = \arctan x$ or $y = \tan^{-1} x$ is graphed by reflecting the graph of $y = \tan x$, $-\pi/2 < x < \pi/2$, about the line $y = x$. These graphs are shown in Figure 85.7 on page 504. Note that the graph of $y = \arctan x$ has the lines $y = \pi/2$ and $y = -\pi/2$ as horizontal asymptotes.

Figure 85.7

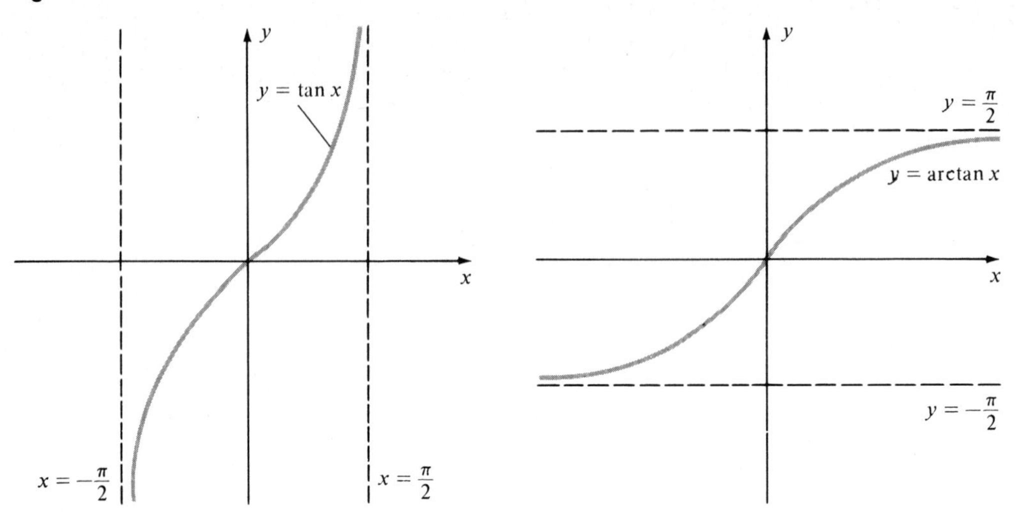

We note from that graph that for $y = \tan x$, we have

$$-\frac{\pi}{2} < x < \frac{\pi}{2} \quad \text{and} \quad -\infty < y < \infty$$

while for $y = \arctan x$, we have

$$-\infty < x < \infty \quad \text{and} \quad -\frac{\pi}{2} < y < \frac{\pi}{2}$$

The statement $y = \arctan x$ is equivalent to the statement "y equals an angle whose tangent is x." Notationally, $y = \arctan x$ is equivalent to $x = \tan y$.

Example 85.5

Evaluate arctan (-1).

Solution

Arctan (-1) is the angle y between $-\pi/2$ and $\pi/2$ whose tangent is -1. Since $\tan(-\pi/4) = -1$, we have $y = -\pi/4$ or $\arctan(-1) = -\pi/4$. ●

A close inspection of Figures 85.5 through 85.7 reveals an interesting property of the inverse function. If x is a positive value, then arcsin x, arccos x, and arctan x are positive. If however, x is negative, then arcsin x and arctan x are negative, while arccos x remains positive.

Example 85.6

Evaluate $\cos(\sin^{-1} \frac{3}{5})$.

Solution

We will evaluate the expression using two different methods.

Solution A

If we let $\arcsin \frac{3}{5} = y$, then $\cos(\arcsin \frac{3}{5}) = \cos y$ and our problem reduces to that of finding $\cos y$. This is done by noting that since $\text{arcsine} \frac{3}{5} = y$, we can write $\sin y = \frac{3}{5}$. We now use the identity $\sin^2 y + \cos^2 y = 1$ and the fact that $-\pi/2 \leq y \leq \pi/2$ to find $\cos y$ as follows:

$$\cos^2 y = 1 - \sin^2 y$$

$$\cos y = \pm\sqrt{1 - \left(\frac{3}{5}\right)^2}$$

$$= \pm\sqrt{1 - \frac{9}{25}} = \pm\sqrt{\frac{16}{25}} = \pm\frac{4}{5}$$

Since $-\pi/2 \leq y \leq \pi/2$ and the cosine function is positive in this interval, we must choose the positive value. Hence, $\cos(\arcsin \frac{3}{5}) = \cos y = \frac{4}{5}$.

Solution B

When we write $\arcsin \frac{3}{5}$ we are stating the angle whose sine is $\frac{3}{5}$. If we draw a right triangle and label one of the acute angles θ, the side opposite θ is 3 units and the hypotenuse is 5 units (Figure 85.8). Using the Pythagorean theorem, we find that the adjacent side is 4. Since $\sin\theta = \frac{3}{5}$ is equivalent to $\theta = \arcsin \frac{3}{5}$, the problem of evaluating $\cos(\arcsin \frac{3}{5})$ reduces to that of finding $\cos\theta$. From Figure 85.8 we see that $\cos\theta = \cos(\arcsin \frac{3}{5}) = \frac{4}{5}$.

Figure 85.8

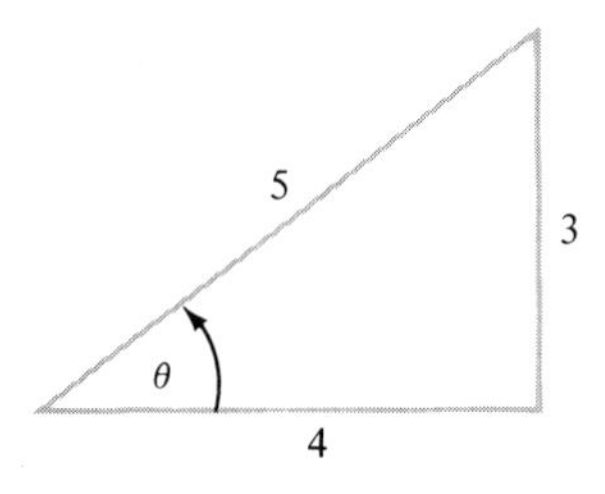

Example 85.7

Express $\sin(2 \arccos x)$ in terms of x.

Solution

We let $y = \arccos x$. Then $\sin(2 \arccos x) = \sin 2y$. From the double-angle formula for sine, we have $\sin 2y = 2 \sin y \cos y$. Now $y = \arccos x$ is equivalent to $\cos y = x$, $0 \leq y \leq \pi$. We can find $\sin y$ by using the identity $\sin^2 y + \cos^2 y = 1$. Thus,

$$\sin^2 y = 1 - \cos^2 y$$

$$= 1 - x^2$$

$$\sin y = \pm\sqrt{1 - x^2}$$

We choose the positive value since $\sin y$ is positive for $0 \leq y \leq \pi$. Therefore,

$$\begin{aligned}\sin(\arccos x) &= \sin 2y \\ &= 2 \sin y \cos y \\ &= 2\sqrt{1 - x^2} \cdot x = 2x\sqrt{1 - x^2}\end{aligned}$$

We observe that $\sin y$ could also be found in the following manner. We draw a right triangle and label one of the acute angles y (Figure 85.9). Since $\cos y = x$, the adjacent side of angle y has length x, the hypotenuse has a length of 1, and by the Pythagorean theorem, the opposite side is $\sqrt{1 - x^2}$. Hence, $\sin y = \sqrt{1 - x^2}/1 = \sqrt{1 - x^2}$, as before.

Figure 85.9

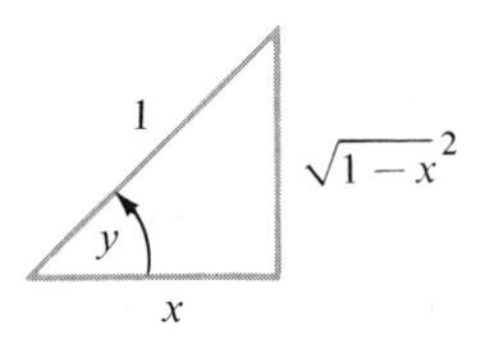

Exercises for Section 85

In Exercises 1 through 30, evaluate each expression.

1. $\arcsin \frac{\sqrt{3}}{2}$

2. $\arcsin \left(-\frac{\sqrt{3}}{2}\right)$

3. $\arccos \frac{\sqrt{3}}{2}$

4. $\arccos \left(-\frac{\sqrt{3}}{2}\right)$

5. $\arctan 0$

6. $\arcsin \frac{\sqrt{2}}{2}$

7. $\arccos \left(-\frac{\sqrt{2}}{2}\right)$

8. $\arccos \frac{\sqrt{2}}{2}$

9. $\arctan \sqrt{3}$

10. $\arctan (-\sqrt{3})$

11. $\arctan 2$

12. $\arctan 1$

13. $\arcsin 0.3$

14. $\arcsin (-0.3)$

15. $\arccos 0.3$

16. $\arccos (-0.3)$

17. $\arctan \frac{\sqrt{3}}{3}$

18. $\arctan \left(-\frac{\sqrt{3}}{3}\right)$

19. $\sin (\arccos \frac{4}{5})$

20. $\cos (\arcsin \frac{1}{2})$

21. $\cos\,[\arcsin\,(-\frac{1}{2})]$

22. $\sin\,(\arcsin\,1)$

23. $\sin\,[\arccos\,(-1)]$

24. $\cos\,(\arctan\,\frac{2}{3})$

25. $\cos\,[\arccos\,(-0.2)]$

26. $\arccos\left(\tan\dfrac{\pi}{4}\right)$

27. $\arcsin\left(\sin\dfrac{\pi}{3}\right)$

28. $\arccos\left(\cos\dfrac{2\pi}{3}\right)$

29. $\sin\,[\arccos\,(-1)]$

30. $\cos\left[\arccos\left(-\dfrac{\sqrt{3}}{2}\right)\right]$

In Exercises 31 through 40, rewrite each expression in terms of x.

31. $\cos\,(\arcsin x)$

32. $\sin\,(\arccos x)$

33. $\sin\,(\arcsin x)$

34. $\cos\,(2\arcsin x)$

35. $\sin\,(2\arcsin x)$

36. $\cos\,(2\arccos x)$

37. $\tan\,(\arcsin x)$

38. $\sin\,(\arctan x)$

39. $\tan\,(2\arctan x)$

40. $\sin\,(2\arctan x)$

41. Evaluate $\sin\,(\arccos\frac{3}{5} + \arcsin\frac{2}{3})$. ***Hint***: Let $\alpha = \arccos\frac{3}{5}$, $\beta = \arcsin\frac{2}{3}$, and then use the formula, for $\sin\,(\alpha + \beta)$.

42. Evaluate $\cos\,(\arcsin\frac{3}{5} + \arccos\frac{12}{13})$. See the hint in Exercise 41.

REVIEW QUESTIONS FOR CHAPTER 14

1. What is an identity?

2. State five trigonometric identities.

3. Explain why $\theta = 0°$ is not a permissible value in the identity $\csc\theta = 1/\sin\theta$.

4. Write the identity $\sin^2\theta + \cos^1\theta = 1$ in three different forms.

5. State three methods that can be used to establish the validity of given identities.

6. What is the formula for $\sin\,(\alpha + \beta)$? $\cos\,(\alpha + \beta)$? $\tan\,(\alpha + \beta)$?

7. For what functions are the values of any negative angle equal to the values of the corresponding positive angles?

8. State the difference formulas for sin, cos, and tan.

9. Express $\sin 2\alpha$ in terms of α.

10. Express $\cos 2\alpha$ in terms of α in three different ways.

11. Express $\tan 2\alpha$ in terms of α.

12. Express $\sin\,(\theta/2)$ in terms of θ.

13. Express $\cos(\theta/2)$ in terms of θ.

14. What is a trigonometric equation?

15. If $y = f(x)$ is the equation of any function, how do we find the inverse?

16. What is the principal part of the sine curve?

17. How are the graphs of $y = \sin x$ and $y = \arcsin x$ related?

18. For what values of x is $\arcsin x$ negative?

19. For what values of x is $\arccos x$ negative?

20. For what values of x is $\arctan x$ positive? Negative?

REVIEW EXERCISES FOR CHAPTER 14

1. If $\cos x = \frac{2}{3}$, find the values of the other five trigonometric functions in the first quadrant.

2. If $\sin x = -\frac{3}{5}$, find the values of the other five trigonometric functions in the third quadrant.

3. If $\sec x = 5$, find the values of the other five trigonometric functions in the fourth quadrant.

4. If $\tan x = -\frac{3}{2}$, find the values of the other five trigonometric functions in the second quadrant.

5. Use the appropriate sum formula to find (without tables) $\cos 255° = \cos(30° + 225°)$.

6. Use the appropriate difference formula to find (without tables) $\sin 15° = \sin(60° - 45°)$.

7. Evaluate $\sin(\alpha - \beta)$ if $\sin\alpha = -\frac{4}{5}$ and $\cos\beta = \frac{1}{3}$, where α is in the third quadrant and β is in the first quadrant.

8. Evaluate $\cos(\alpha - \beta)$ if $\sin\alpha = \frac{8}{17}$ and $\cos\beta = -\frac{4}{5}$, where α is in the second quadrant and β is in the third quadrant.

9. Find $\sin 2\theta$, $\cos 2\theta$, and $\tan 2\theta$ if $\cos\theta = \frac{2}{3}$ and θ is in the fourth quadrant.

10. Find $\sin 2\theta$, $\cos 2\theta$, and $\tan 2\theta$ if $\sin\theta = \frac{3}{4}$ and θ is in the second quadrant.

11. Use a half-value formula to find (without tables) $\sin 67.5° = \sin(135°/2)$.

12. Use a half-value formula to find (without tables) $\cos 105° = \cos(210°/2)$.

In Exercises 13 through 18, solve for x where $0° \leq x < 360°$.

13. $\cos x(2\sin x - 1) = 0$

14. $2\cos 3x - 1 = 0$

15. $2\cos^2 x + 3\cos x + 1 = 0$

16. $3\tan 2x = \sqrt{3}$

17. $\sin x \cos x = \dfrac{1}{2}$

18. $\sin x - \cos x = 1$

In Exercises 19 through 24, evaluate each expression.

19. $\sin(\arcsin 0.7)$

20. $\cos(\arccos -0.2)$

21. $\sin(\arccos \frac{3}{5})$

22. $\cos\left(\arcsin -\frac{\sqrt{3}}{2}\right)$

23. $\cos(2 \arccos \frac{3}{5})$

24. $\sin(\frac{1}{2} \arcsin \frac{4}{5})$

25. Express $\cos(\arccos x)$ in terms of x.

26. Express $\tan(\arccos x)$ in terms of x.

In Exercises 27 through 32, prove each identity.

27. $\csc\theta - \sin\theta = \cos\theta \cot\theta$

28. $\dfrac{\sec^2\theta - \csc^2\theta}{\sec^2\theta + \csc^2\theta} = 1 - 2\cos^2\theta$

29. $\dfrac{\csc^2\theta - \cot^2\theta}{\sec^2\theta - 1} = \cot^2\theta$

30. $(\sin\frac{\theta}{2} + \cos\frac{\theta}{2})^2 = 1 + \sin\theta$

31. $\cos(90° - \theta) = \sin\theta$

32. $\sin(90° + 2\theta) = 1 - 2\sin^2\theta$

33. The phase angle between the voltage and current in a series circuit is given by

$$\theta = \arctan \frac{X}{R}$$

where X is the reactance and R is the resistance of the circuit.
(a) What is the phase angle if $X = 2000$ and $R = 5000$?
(b) What is the phase angle if $X = 2000$ and $R = 1000$?
(c) What is the phase angle if $X = 2000$ and $R = 50$?

CHAPTER 15

Analytic Geometry

SECTION 86

The Distance Formula; Symmetry

In the study of calculus we must often be able to recognize certain types of curves and identify their corresponding characteristics. This is generally done in one of two ways. If we are given the equation, we proceed to find the important geometric properties. If, however, we begin with a geometric description of a curve or its properties, we then determine the corresponding entity, the equation. The basis of analytic geometry is the relationship of geometry to algebra. In effect, we will be determining and studying geometric properties by using algebraic methods. In so doing, we hope to discover the bonds that exist between algebra and geometry and how one enriches the other. We will now develop those concepts that will be needed to establish the basic relationships between a curve and its equation.

We begin with the development of the formula that gives the distance between any two points in the coordinate plane. We know that the distance between any two points on a number line is the absolute value of the difference between the coordinates of the points (Figure 86.1):

$$\overline{AC} = |5 - (-3)| = |-3 - 5| = 8$$
$$\overline{BC} = |5 - 1| = |1 - 5| = 4$$

If two points in the coordinate plane lie on the same horizontal or vertical line, the distance between them is determined in the same way.

Figure 86.1

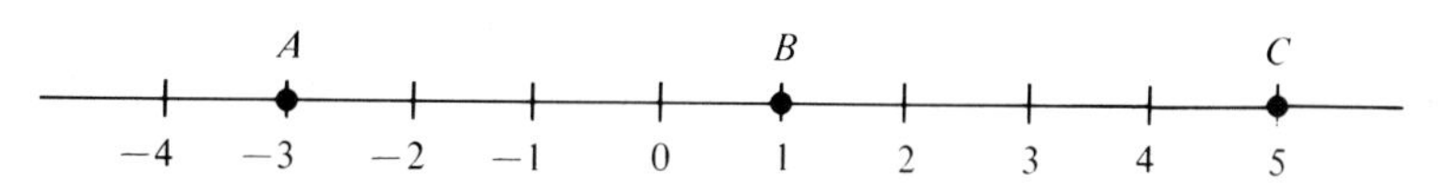

Let us consider the pairs of points in Figure 86.2. The distances are found as follows:

$$\overline{AB} = |3 - (-2)| = |3 + 2| = |5| = 5$$

$$\overline{CD} = |2 - (-1)| = |2 + 1| = |3| = 3$$

To find the distance between any two points in the coordinate plane, we make use of the Pythagorean theorem: The square of the length c of the hypotenuse of a right triangle equals the sum of the squares of the lengths of the other two sides, a and b. Thus, $c^2 = a^2 + b^2$.

Figure 86.2

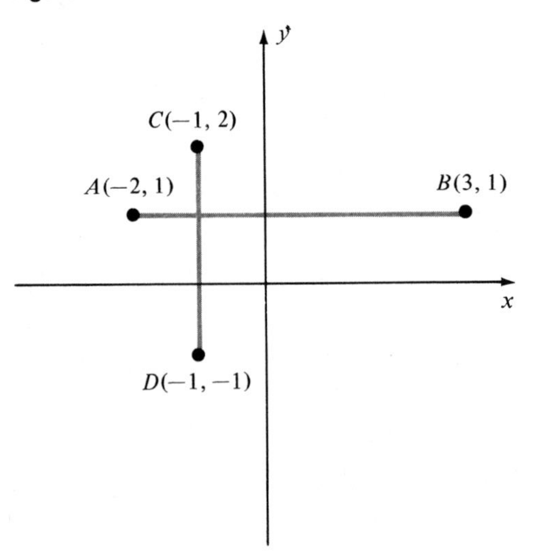

Theorem 86.1

Let $P_1(x_1, y_1)$ and $P_2(x_2, y_2)$ be any two points in the plane. The distance d between P_1 and P_2 is given by the formula

$$d = \overline{P_1P_2} = \sqrt{(x_2 - x_1)^2 + (y_2 - y_1)^2} \tag{86.1}$$

We can develop equation (86.1) in the following way. Let Q be the point of intersection between the line through P_1 parallel to the x axis and the line through P_2 parallel to the y axis (Figure 86.3). The coordinates of Q are (x_2, y_1). $\overline{P_1Q} = |x_2 - x_1|$ and $\overline{QP_2} = |y_2 - y_1|$. Since triangle P_1QP_2 is a right triangle, we can use the Pythagorean theorem. Thus, $(\overline{P_1P_2})^2 = (\overline{P_1Q})^2 + (\overline{QP_2})^2$. Since $(\overline{P_1Q})^2 = |x_2 - x_1|^2 = (x_2 - x_1)^2$ and $(\overline{QP_2})^2 = |y_2 - y_1|^2 = (y_2 - y_1)^2$, we have

$$(\overline{P_1P_2})^2 = (x_2 - x_1)^2 + (y_2 - y_1)^2$$

or

$$\overline{P_1P_2} = \sqrt{(x_2 - x_1)^2 + (y_2 - y_1)^2}$$

Figure 86.3

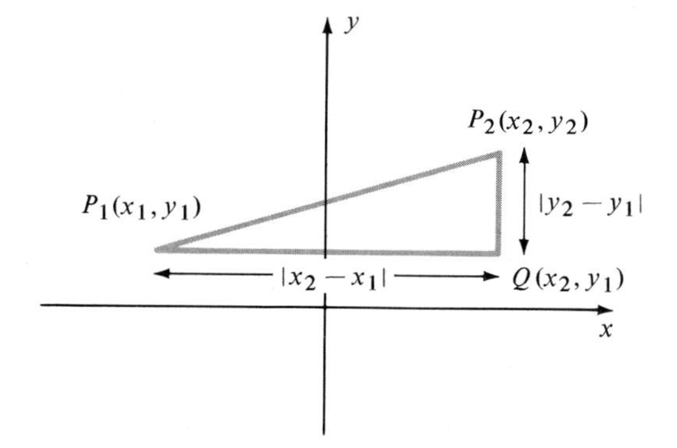

Example 86.1

Find the distance between $P_1(3, -2)$ and $P_2(1, 4)$.

Solution

$$\overline{P_1P_2} = \sqrt{(1-3)^2 + [4-(-2)]^2}$$
$$= \sqrt{(-2)^2 + (6)^2} = \sqrt{4+36} = \sqrt{40} = 2\sqrt{10}$$

Example 86.2

Find the distance r from the origin to any point (x, y).

Solution

We use (86.1) and let P_2 and P_1 correspond to (x, y) and $(0, 0)$ respectively. Thus,

$$r = \sqrt{(x-0)^2 + (y-0)^2}$$
$$r = \sqrt{x^2 + y^2}$$

Example 86.3

Use the distance formula to show that the points $P(3, 2)$, $Q(6, 3)$, and $R(12, 5)$ lie on a straight line.

Solution

$$\overline{PQ} = \sqrt{(6-3)^2 + (3-2)^2} = \sqrt{3^2 + 1^2} = \sqrt{10}$$
$$\overline{QR} = \sqrt{(12-6)^2 + (5-3)^2} = \sqrt{40} = 2\sqrt{10}$$
$$\overline{PR} = \sqrt{(12-3)^2 + (5-2)^2} = \sqrt{9^2 + 3^2} = \sqrt{90} = 3\sqrt{10}$$

Since $\overline{PQ} + \overline{QR} = \overline{PR}$, the points are collinear, i.e., they lie on a straight line.

Example 86.4

Find the equation of a circle whose center is at $(0, 0)$ and whose radius is equal to 1.

Solution

In Figure 86.4 we see that the radius r is equal to 1. We use the result of Example 86.2 to obtain $x^2 + y^2 = r^2 = 1$ or $x^2 + y^2 = 1$. This is the desired equation. ●

Figure 86.4

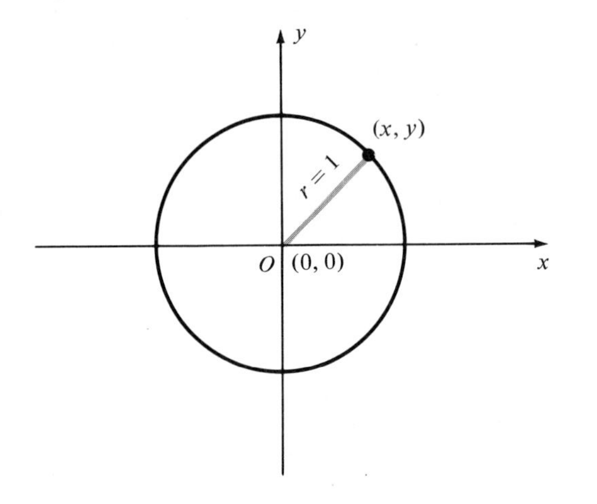

The graph of the circle in Figure 86.4 has an important geometric property. It is symmetric with respect to the x axis, the y axis, and the origin.

Symmetry with respect to the y axis means that the right half of the curve is the "mirror image" of the left half. Similarly, symmetry with respect to the x axis means that the upper half and lower half of the curve are mirror images. In Figure 86.4 the x and y axes are called **lines of symmetry**. In addition, the curve is symmetric with respect to the origin, since any line through the origin connecting two points on the curve will have O as its midpoint.

Algebraically, the following conditions are used to determine symmetry:

1. If we can replace x by $-x$ without changing the equation, the graph is symmetric about the y axis.
2. If we can replace y by $-y$ without changing the equation, the graph is symmetric about the x axis.
3. If we can replace x by $-x$ and y by $-y$ without changing the equation, the graph is symmetric about the origin.

Example 86.5

(a) Since $y = x^2$ is equivalent to $y = (-x)^2$, the graph of $y = x^2$ is symmetric about the y axis.

(b) The graph of $x^2/a^2 - y^2/b^2 = 1$ is symmetric about *each* axis, since replacing either x or y by its negative leaves the equation unchanged.

(c) Since $y = 1/x$ is equivalent to $-y = 1/-x$, the graph of $y = 1/x$ is symmetric about the origin. The graphs are shown in Figure 86.5.

Figure 86.5

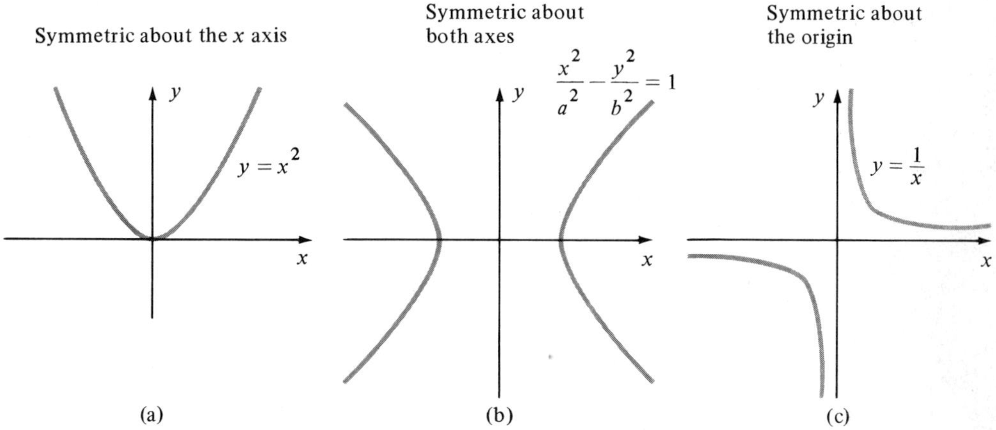

Symmetric about the x axis

$y = x^2$

(a)

Symmetric about both axes

$\frac{x^2}{a^2} - \frac{y^2}{b^2} = 1$

(b)

Symmetric about the origin

$y = \frac{1}{x}$

(c)

Example 86.5 shows that a curve symmetric about the origin is not necessarily symmetric about either axis. However, a curve symmetric about both axes *must* also be symmetric about the origin. Thus, the graph of the equation in Example 86.5(b) also has origin symmetry.

Exercises for Section 86

In Exercises 1 through 10, find the distances between each pair of points.

1. (4, 1) and (4, −6)

2. (2, 3) and (6, 3)

3. (2, 3) and (6, −3)

4. (1, 3) and (−2, 5)

5. (4, 1) and (7, 6)

6. (6, 4) and (−6, −1)

7. (5, 7) and (1, 4)

8. (−2, 3) and (7, 3)

9. (−1, −2) and (3, 6)

10. (4, 4) and (−2, 0)

11. Show that the points (−4, 3) and (0, −3) and (4, 4) represent the vertices of an isosceles triangle.

12. Show that the points (10, −7), (11, −8), (10, −9), and (9, −8) represent the vertices of a square.

13. Show that the points (1, 2), (9, 2), and (1, 7) represent the vertices of a right triangle.

14. The point $(x, 3)$ is 4 units from (2, 7). Find x.

15. Find the point on the y axis that is equidistant from the points (1, 3) and (5, 4).

In Exercises 16 through 18, use the distance formula to show that the points are collinear.

16. $(3, 0), (0, -4), (9, 8)$

17. $(1, 1), (0, -2), (2, 4)$

18. $(-2, 3), (1, 6), (3, 8)$

In Exercises 19 through 26, state what symmetry, if any, the equations have with respect to the axes and/or the origin.

19. $2x^2 + y = 0$

20. $x^2 + xy + y^2 = 0$

21. $x^3 + 2xy + y^2 = 0$

22. $y = 2x$

23. $xy = 4$

24. $9x^2 + 16y^2 = 144$

25. $y = x^2 + 1$

26. $4x^2 - 9y^2 = 36$

SECTION 87

The Straight Line

In this section we shall learn how to generate the equations of straight lines in the plane. Such equations will prove to be useful in our future work.

We begin our investigation with the following observation. Let L be any line not parallel to the axes. The inclination of L is the smallest angle (θ) between the positive direction of the x axis and L measured in a counterclockwise direction.

If L is parallel to or coincides with the x axis, its inclination is defined to be 0. If L is parallel to or coincides with the y axis, its inclination is $90°$. In any event, θ always has the property that $0 \leq \theta < 180°$ (Figure 87.1).

Figure 87.1

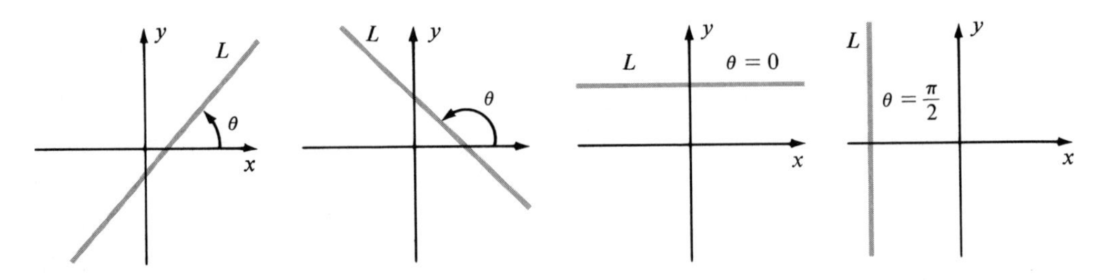

Another and sometimes more useful measure of the inclination of a line is its slope. If L is any line not parallel to the y axis, then the *slope* (m) is given by $\tan \theta$, where θ is the angle of the inclination of L. We note that the only lines for which the slope is not defined are those parallel to the y axis. In this case $\theta = 90°$ and $\tan 90°$ is undefined.

If L passes through the point $P_1(x_1, y_1)$ and $P_2(x_2, y_2)$,

$$m = \tan \theta = \frac{y_2 - y_1}{x_2 - x_1}$$

(Figure 87.2). In many instances, $y_2 - y_1$ is denoted by Δy (the change in y) and $x_2 - x_1$ is denoted by Δx (the change in x). Thus, we have

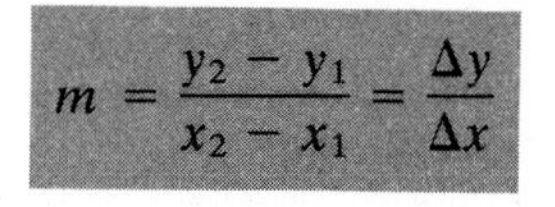

$$m = \frac{y_2 - y_1}{x_2 - x_1} = \frac{\Delta y}{\Delta x}$$

Figure 87.2

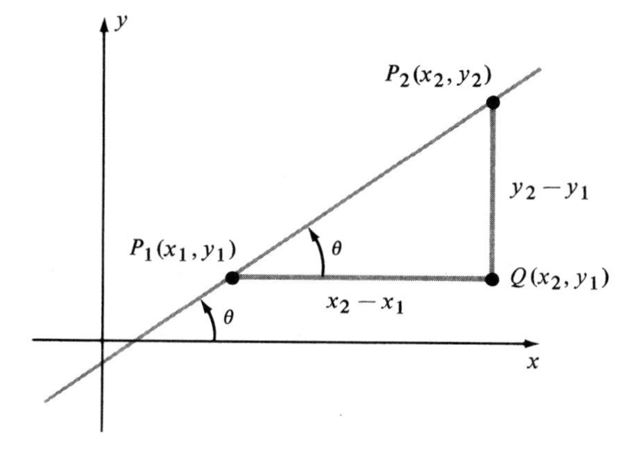

Example 87.1

Find the slope of the line passing through the points (1, 3) and (4, 5).

Solution

$$m = \frac{y_2 - y_1}{x_2 - x_1} = \frac{5 - 3}{4 - 1} = \frac{2}{3}$$

We obtain the same result if we change the order of subtraction. Thus,

$$m = \frac{3 - 5}{1 - 4} = \frac{-2}{-3} = \frac{2}{3}$$ ●

The change in y given by $(y_2 - y_1)$ is called the *rise* of the straight line, since it gives the vertical distance traveled in going from P_1 to P_2. Similarly, the change in x given by $(x_2 - x_1)$ is called the *run* of the straight line, since it gives the horizontal distance traveled in going from P_1 to P_2. Thus,

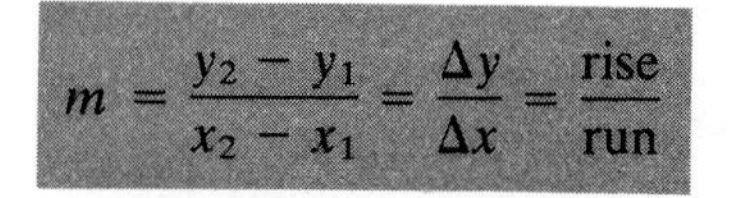

$$m = \frac{y_2 - y_1}{x_2 - x_1} = \frac{\Delta y}{\Delta x} = \frac{\text{rise}}{\text{run}}$$

We may interpret the rise over the run as the change in y per unit change in x. For example, a slope of 2 means a rise of 2 units for each unit of run. A slope of $\frac{2}{3}$ means a rise of $\frac{2}{3}$ unit for each unit of run. If the slope is negative, the line "falls" rather than rises as we move from left to right. A slope of -3 means a fall of 3 units for each unit of run (Figure 87.3).

Figure 87.3

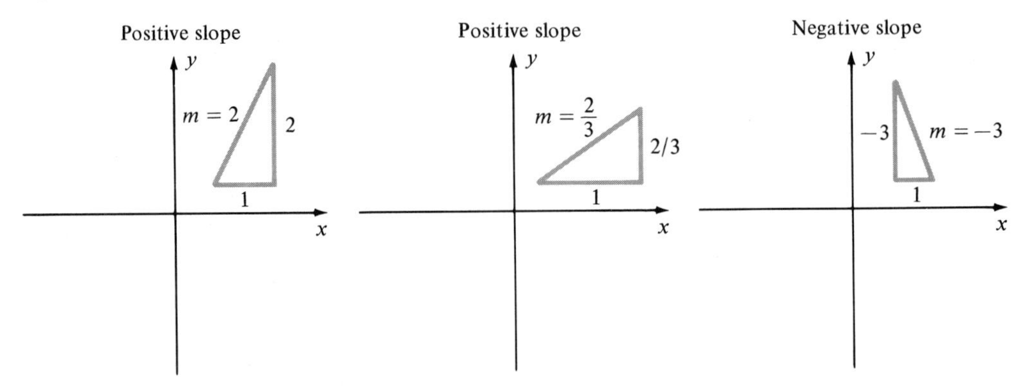

The value of the slope m is independent of the choice of P_1 and P_2. To show this, suppose that we choose two different points on L. We denote these points by $P_3(x_3, y_3)$ and $P_4(x_4, y_4)$ (see Figure 87.4). From P_1 and P_2 we have $m = (y_2 - y_1)/(x_2 - x_1)$ and from P_3 and P_4 we have $m' = (y_4 - y_3)/(x_4 - x_3)$. In Figure 87.4 we see that triangle P_1P_2Q is similar to triangle P_3P_4Q'. Since corresponding sides of similar triangles are proportional, we have

$$\frac{y_2 - y_1}{x_2 - x_1} = \frac{y_4 - y_3}{x_4 - x_3}$$

Thus, $m = m'$.

Figure 87.4

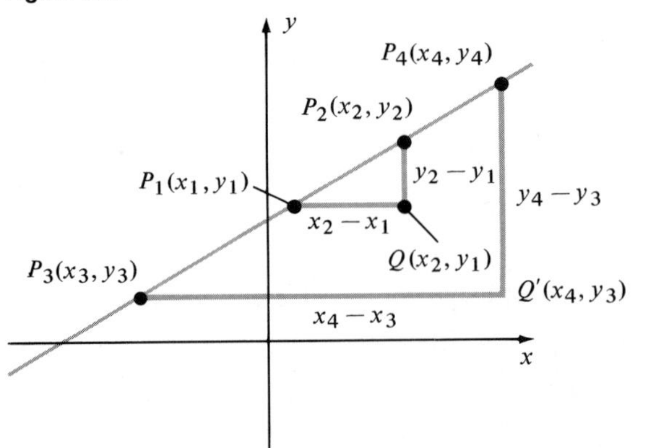

Example 87.2

Show that the points (0, 1), (2, 5), and (−2, −3) lie on the same straight line.

Solution

The slope of the line from (0, 1) to (2, 5) is

$$\frac{5-1}{2-0} = \frac{4}{2} = 2$$

The slope of the line from (2, 5) to (−2, −3) is

$$\frac{-3-5}{-2-2} = \frac{-8}{-4} = 2$$

Since the slopes are equal and there is a point in common, the points lie on the same line. ●

Any straight line is uniquely determined by two points. If the two points P_1 and P_2 are such that the line segment P_1P_2 is not parallel to the y axis, we may determine the equation of the straight line by choosing a third point on the line and equating the slopes of line segments. This is possible, since we have shown that the slope of a line is independent of the points selected. Since P_1, P_2, and P lie on L, the slope of $\overline{P_1P_2}$ is equal to the slope of $\overline{P_1P}$ (Figure 87.5). In terms of the coordinates of the points,

$$\frac{y-y_1}{x-x_1} = \frac{y_2-y_1}{x_2-x_1} \tag{87.1}$$

We call this the **two-point form** for the equation of a straight line.

Figure 87.5

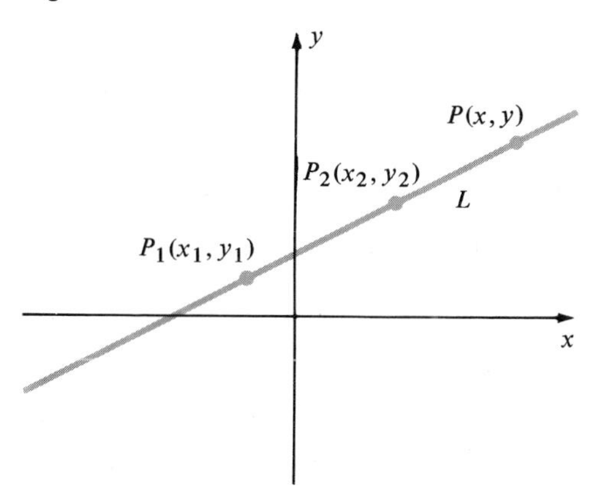

Example 87.3

Find the equation of the straight line passing through the points $(-1, 3)$ and $(2, 5)$.

Solution

Using the two-point form (87.1) for the equation of a straight line, we have $(y - 3)/(x + 1) = \frac{2}{3}$. Although this is the equation of the line, we may write it a bit more concisely by taking the cross multiples and collecting all terms on the left of the equal sign. Thus, $2x + 2 = 3y - 9$ or $2x - 3y + 11 = 0$. ***Note:*** In using the two-point form we assumed that P_1 was $(-1, 3)$ and P_2 was $(2, 5)$. If we reverse the P_1 and P_2, we have $(y - 5)/(x - 2) = -2/-3$ or $-2x + 4 = -3y + 15$. We may write this as $-2x + 3y - 11 = 0$ or $2x - 3y + 11 = 0$, as before. The graph is shown in Figure 87.6. ●

A line is also uniquely determined if the slope m and a point $P(x_1, y_1)$ are specified. Again, we generate the equation of such a line by choosing an arbitrary point P and equating the given slope m with the slope of $\overline{P_1P}$ (Figure 87.7). Thus,

$$m = \frac{y - y_1}{x - x_1} \quad \text{or} \quad y - y_1 = m(x - x_1) \tag{87.2}$$

This is called the **point-slope form** of the equation of a straight line.

Example 87.4

Find the equation of the straight line passing through the point $(2, 3)$ and whose slope is $-\frac{1}{3}$.

Solution

From our point-slope form (87.2), we have $y - 3 = -\frac{1}{3}(x - 2)$. Thus, $3y - 9 = -x + 2$ or $x + 3y - 11 = 0$. The graph is shown in Figure 87.8.

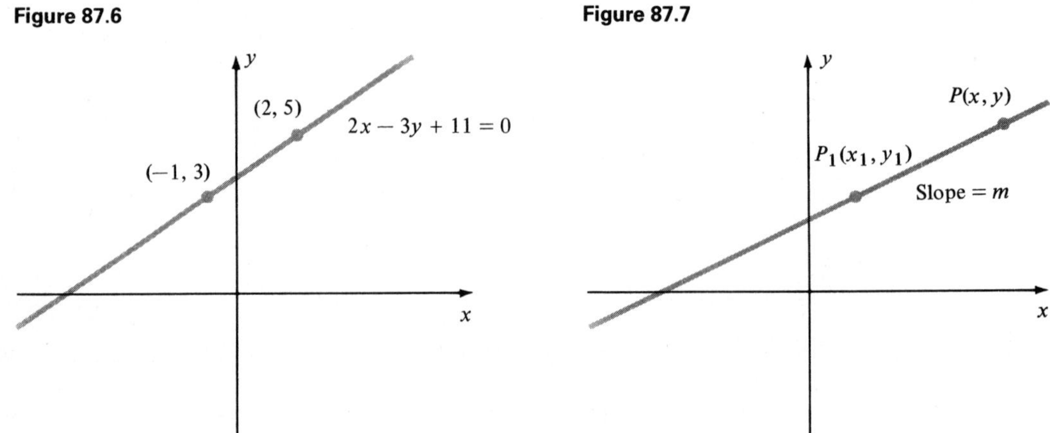

Figure 87.6

Figure 87.7

Figure 87.8

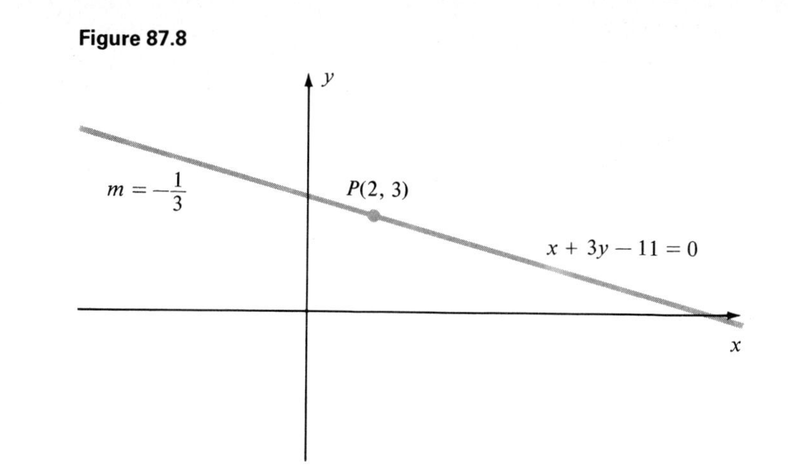

Now we consider a special case of the point-slope form. Suppose that L passes through the y axis. Then the y intercept of L will be of the form $(0, b)$. If L has a slope of m, we have $y - b = m(x - 0)$, from equation (87.2). Hence,

$$y = mx + b \tag{87.3}$$

This equation is called the **slope-intercept form** and has the following characteristics:

1. y is given as a function of x.
2. The coefficient of x is the slope of the line.
3. b is the y value of the intercept point on the y axis.

Example 87.5

Find the equation of a line with $m = 3$ and y intercept at $(0, 4)$.

Solution

From (87.3) we have $y = mx + b$, where $m = 3$ and $b = 4$. Hence, $y = 3x + 4$ is the equation we seek.

Example 87.6

Find the slope and y intercept of $2x + 3y - 5 = 0$.

Solution

We may write $2x + 3y - 5 = 0$ in slope-intercept form as $3y = -2x + 5$ or $y = -\frac{2}{3}x + \frac{5}{3}$. Hence, $m = -\frac{2}{3}$ and the y intercept is $(0, \frac{5}{3})$. ●

For any line parallel to the y axis, m will be undefined, since $m = \Delta y/\Delta x$, and in this case $\Delta x = 0$.

We can find the equation of such a line. The equation we seek depends upon any two points that lie on the line. If $P_1(x_1, y_1)$ lies on L, then P_2 must have coordinates of the form (x_1, y_2). Therefore, $P(x, y)$ is any point on the line if and only if $x = x_1$. Here we have an equation of a line parallel to the y axis which is independent of y. Geometrically, this is accounted for by the fact that the y value of P has no restriction (see Figure 87.9).

Figure 87.9

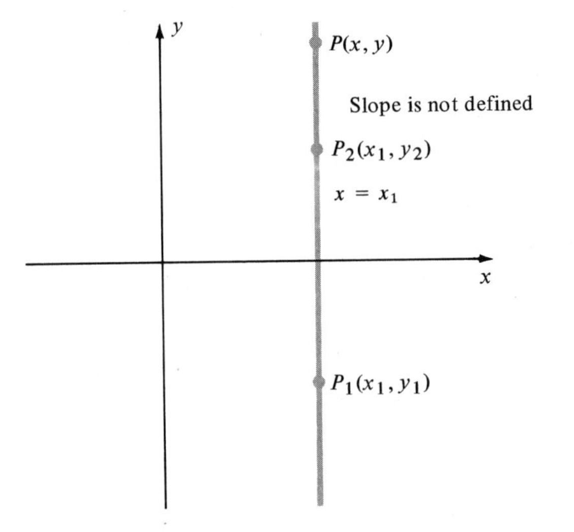

Algebraically, the relation $x = x_1$ is the abbreviated form of the equation $0y + x = x_1$. Since the coefficient of y is 0, y may take on any real value. Similarly, an equation for the horizontal line passing through $P_1(x_1, y_1)$ is $y = y_1$.

Example 87.7

Find the equations of the lines parallel to the y axis and the x axis, respectively, and passing through the point (1, 6).

Solution

For the line parallel to the y axis, the equation is of the form $x = x_1$. From the given point we have $x_1 = 1$, and thus $x = 1$ is the equation. For the line parallel to the x axis, the equation is of the form $y = y_1$. From the given point, $y_1 = 6$, and thus $y = 6$ is the desired equation. ●

We have used the concept of slope to study various forms of the equation of a straight line. We now extend the use to cover parallel and perpendicular lines.

Theorem 87.1

If L_1 is parallel to L_2, then $m_1 = m_2$ or m_1 and m_2 are undefined.

Example 87.8

Find the equation of the line parallel to $2x - 3y + 4 = 0$ and passing through point $(1, 5)$.

Solution

Since $2x - 3y + 4 = 0$ can be written as $y = \frac{2}{3}x + \frac{4}{3}$, the line we seek has a slope of $\frac{2}{3}$. From equation (87.2) we have $y - 5 = \frac{2}{3}(x - 1)$ or, equivalently, $2x - 3y + 13 = 0$.

Example 87.9

The lines $x - 2y + 4 = 0$ and $3x - 6y + 1 = 0$ are parallel, since they may be written as $y = \frac{1}{2}x + 2$ and $y = \frac{1}{2}x + \frac{1}{6}$ and have equal slopes.

Theorem 87.2

Suppose that L_1 and L_2 have slopes of m_1 and m_2, respectively. If L_1 is perpendicular to L_2 and the lines are not parallel to the axes, then $m_1 = -1/m_2$.

To demonstrate this, we suppose that the point of intersection of L_1 and L_2 lies on the x axis, as shown in Figure 87.10. From the figure we see that $\beta + \alpha_1 = 90°$; that is, α_1 and β are complementary angles. We have, therefore, $\tan \alpha_1 = \cot \beta = 1/\tan \beta$. But $\beta = (180° - \alpha_2)$ (Figure 87.10).

Figure 87.10

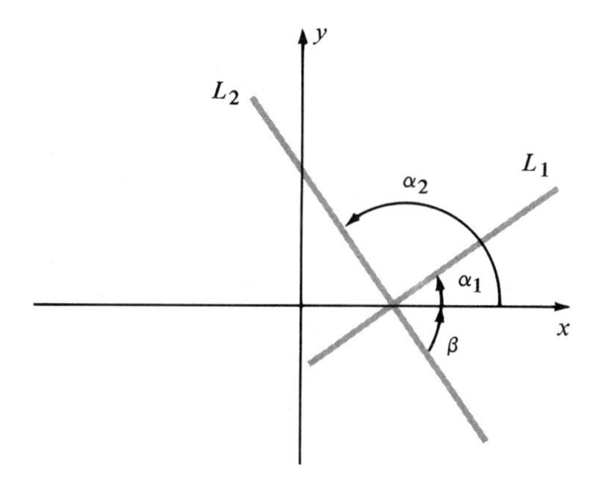

Hence,

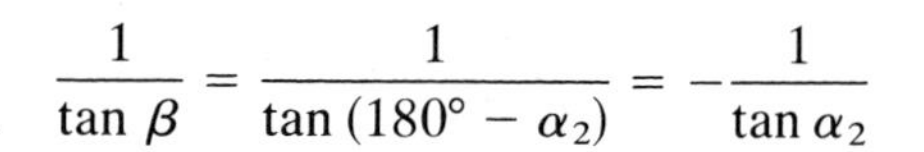

$$\frac{1}{\tan \beta} = \frac{1}{\tan (180° - \alpha_2)} = -\frac{1}{\tan \alpha_2}$$

Thus, we have $\tan \alpha_1 = -1/\tan \alpha_2$. Since $\tan \alpha_1 = m_1$ and $\tan \alpha_2 = m_2$, we have $m_1 = -1/m_2$ or $m_1 m_2 = -1$. We often state that "the slopes of perpendicular lines are negative reciprocals."

Example 87.10

Find the equation of the line perpendicular to $x - 2y + 3 = 0$ and passing through the point $(2, -1)$.

Solution

$x - 2y + 3 = 0$ is equivalent to $y = \frac{1}{2}x + \frac{3}{2}$. When we use Theorem 87.2, the slope of the line we seek is -2. Employing equation (87.2), we have $y + 1 = -2(x - 2)$ or $2x + y - 3 = 0$. The lines are shown in Figure 87.11.

Figure 87.11

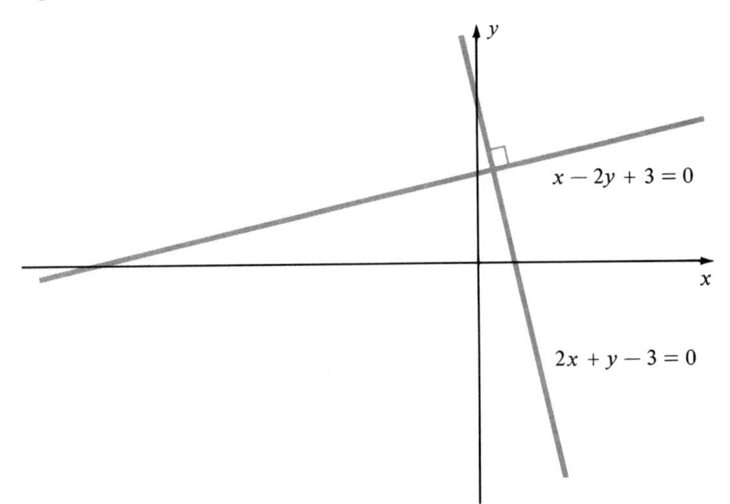

The equation $Ax + By + C = 0$, where A, B, and C are constants and both A and B are not equal to 0, is the **general form** of a first-degree equation in x and y. It is called a **linear equation** because its graph is always a straight line. This can be shown as follows. Suppose that $B \neq 0$. We solve $Ax + By + C = 0$ for y and obtain $y = -(A/B)x - C/B$. This is the equation of a straight line in slope-intercept form where $m = -A/B$ and $b = -C/B$. If $B = 0$ and $A \neq 0$, we have $Ax + C = 0$ or $x = -C/A$. We have seen that an equation of this form is the graph of a straight line parallel to the y axis.

In certain applications, the relationship that exists between the variables is linear. The following example illustrates such a situation.

Example 87.11

The pressure P of a fixed volume of gas varies linearly with the temperature T. If $P = 60$ when $T = 20$ and $P = 80$ when $T = 25$, find P as a function of T.

Solution
If we let P be the dependent variable and T the independent variable, the slope of the line is given by

$$m = \frac{P_2 - P_1}{T_2 - T_1}$$

From the values given we have

$$m = \frac{80 - 60}{25 - 20} = \frac{20}{5} = 4$$

We now use the point-slope form (87.2) to find $P - 60 = 4(T - 20)$ or $P = 4T - 20$.

Exercises for Section 87

In Exercises 1 through 8, find the slope (if it exists) of the line through the points. Draw the line.

1. $(2, 3), (1, 5)$

2. $(5, 2), (7, 2)$

3. $(0, 0), (4, 1)$

4. $(-2, 2), (3, 3)$

5. $(4, 2), (-5, -1)$

6. $(0, 0), (1, 1)$

7. $(1.3, 1.2), (2.5, 4.6)$

8. $(3, 5), (3, 7)$

In Exercises 9 through 18, find an equation of the line that satisfies the conditions.

9. Through the points given in Exercises 1 through 8.

10. Through the point $(1, 2)$ and parallel to the x axis.

11. Through the point $(1, 2)$ and perpendicular to the x axis.

12. Through the point $(2, 3)$ and having a slope of -1.

13. Through the point $(1, 4)$ and parallel to the line $x + 2y - 4 = 0$.

14. Through the point $(0, 0)$ and parallel to the line $y = 2x - 3$.

15. The x intercept is -2 and the y intercept is 5.

16. The x intercept is -2 and the slope is 4.

17. Through the point $(2, 5)$ and perpendicular to $x + y - 4 = 0$.

18. Through the point $(-2, -1)$ and perpendicular to $2x - 3y + 6 = 0$.

19. Find the equation of the three sides of the triangle with vertices $(1, 4)$, $(3, 0)$, and $(-2, 3)$.

20. Use slopes to show that the four points $(3, 1)$, $(4, 3)$, $(2, 4)$, and $(1, 2)$ are vertices of a rectangle.

21. Show that the points $(0, 2)$, $(2, 6)$, and $(-3, -4)$ are collinear by showing that the third point satisfies an equation for the line through the other two.

22. For which values of a are $ax - 2y + 8 = 0$ and $3ax + 6y + 4 = 0$ the equations of perpendicular lines?

23. If a ball is thrown directly upward with an initial velocity of 192 m/sec, the velocity at the end of t seconds is given by the equation $v = 192 - 32t$. Sketch the linear function for $0 \le t \le 6$. Find v when $t = 3$ sec.

24. Under certain conditions, the amount E that a spring is stretched by a force F is given by a linear function. If a spring of natural length 20 cm is stretched 5 cm by a weight of 100 g, find the equation relating E and F.

25. The voltage V in volts varies directly with the current I (in amperes). If $V = 15$ when $I = 0$ and $V = 10$ when $I = 150$, find the equation relating V and I. Graph the equation for $0 \le I \le 200$.

26. In a certain experiment it is found that the length of a metal bar is 100 cm at a temperature of 10°C, and 100.2 cm at a temperature of 100°C. If the length L of the bar varies directly as the temperature T, express L as a function of T.

SECTION 88

Conic Sections: The Circle

We continue our discussion of the relationship between algebra and geometry by studying types of curves called **conic sections**. Conic sections are the curves obtained by intersecting a plane with a cone. Figure 88.1 shows a cone of two **nappes**, an upper nappe and a lower nappe. The nappes meet at a point called the **vertex**.

There are exactly four ways in which a plane can intersect the cone if the plane does not contain the point of intersection. The curves thus obtained are called the four *conics*: the circle, the parabola, the ellipse, and the hyperbola. The curves are shown in Figure 88.2a, b, c, and d, respectively. The ellipse occurs when the intersecting plane cuts the cone obliquely and does not intersect the base. The circle occurs when the plane is perpendicular to the axis of the cone. Thus, the circle is sometimes said to be a special case of the ellipse. The parabola is generated when the intersecting plane cuts the cone obliquely and passes through its base. The hyperbola is generated by a plane that passes through both nappes of the cone but does not contain the axis of the cone.

Figure 88.1

In certain situations, the intersection of a cone and a plane may produce what we call "degenerate conics." If the plane passes through the vertex, the circle and the ellipse "degenerate" into a single point. The degenerate form of a parabola occurs when the intersecting plane is tangent to the surface of the cone as it passes through the vertex. In this case we have a single straight line. For the hyperbola, the degenerate case occurs when the plane contains the axis of the cone and yields two intersecting straight lines. We shall illustrate each of these situations in the following sections when we consider the conic sections individually.

We shall study the conic sections and their properties in detail, since they involve many scientific applications. We begin with the circle. A **circle** is the set of all points in a plane at a given distance from a fixed point. The fixed point is called the **center** of the circle and the measure of the given distance is called the **radius** of the circle.

If the center C is at the point (h, k) and the radius is r (Figure 88.3 on page 529), the equation is easily determined by means of the distance formula. The point $P(x, y)$ lies on the circle if and only if $\overline{CP} = r$, or, equivalently, if and only if $\sqrt{(x-h)^2 + (y-k)^2} = r$. Squaring both sides yields the more convenient form of the equation of the circle.

$$(x-h)^2 + (y-k)^2 = r^2 \qquad (88.1)$$

(Figure 88.4). If the center of the circle is at the origin, then $h = k = 0$ and (88.1) reduces to

$$x^2 + y^2 = r^2 \qquad (88.2)$$

Example 88.1

Find the equation of the circle with center at $(2, -1)$ and radius equal to 5.

Figure 88.2

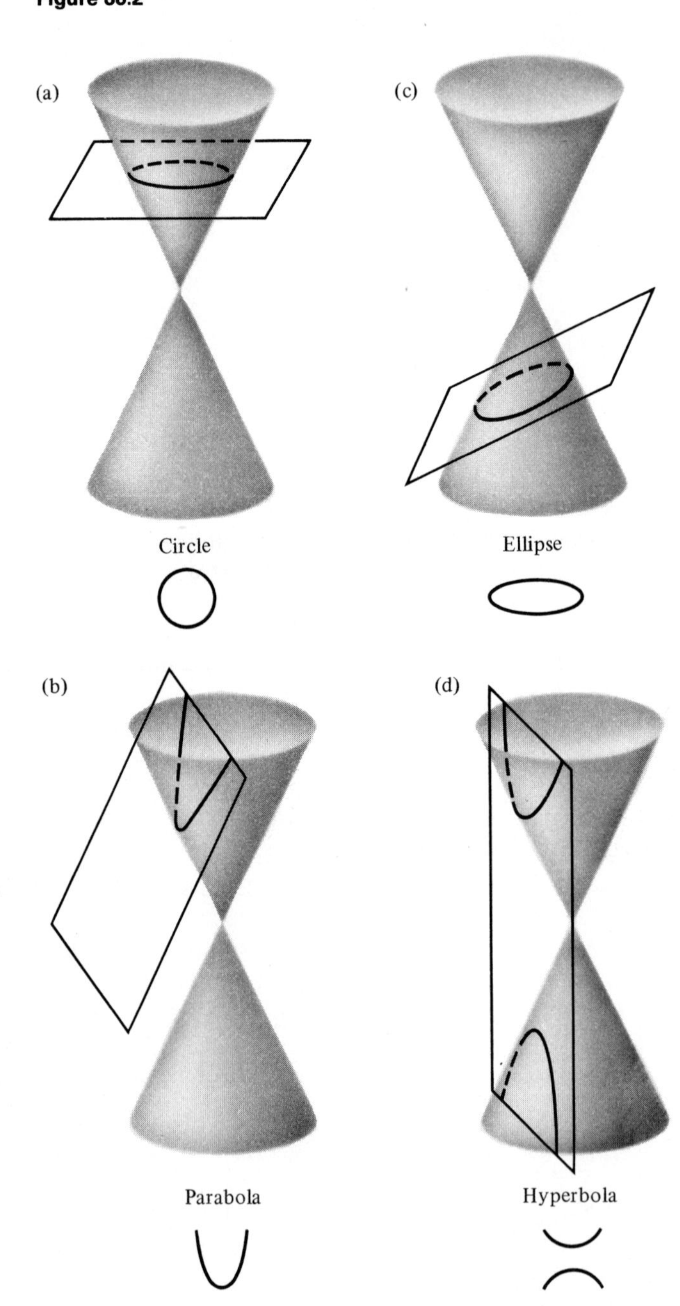

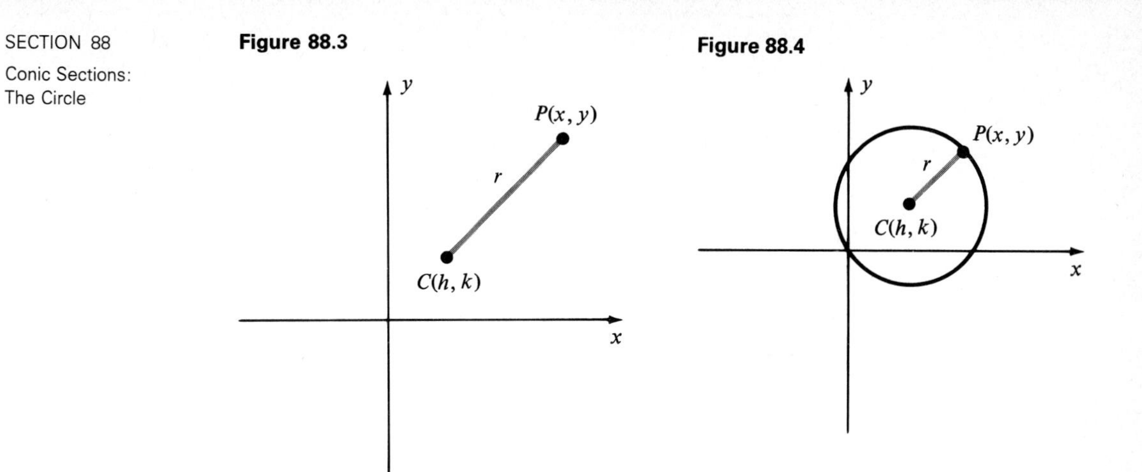

Figure 88.3

Figure 88.4

Solution

Here $h = 2$, $k = -1$, and $r = 5$. Substituting into (88.1), we have

$$(x - 2)^2 + [y - (-1)]^2 = 5^2 \quad \text{or} \quad (x - 2)^2 + (y + 1)^2 = 25$$

Squaring yields the expanded form

$$x^2 - 4x + 4 + y^2 + 2y + 1 = 25 \quad \text{or} \quad x^2 + y^2 - 4x + 2y - 20 = 0$$

The circle is shown in Figure 88.5. ●

Figure 88.5

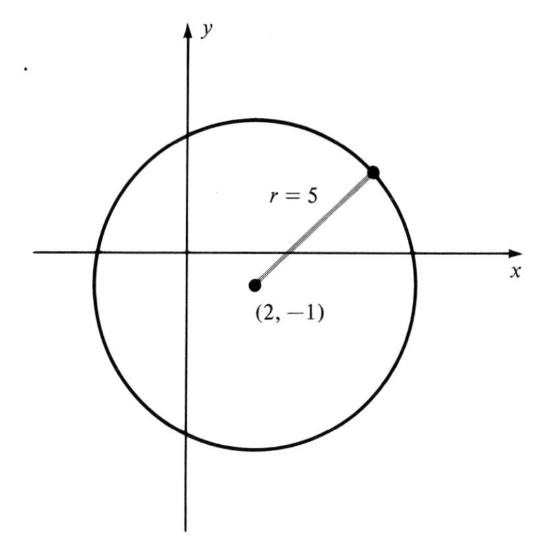

Example 88.2

Find the center and radius of the circle whose equation is $x^2 + y^2 + 4x - 6y - 3 = 0$.

Solution

If we are able to rewrite the equation in the form of equation (88.1), we will be able to read the coordinates of the center and the radius directly. To accomplish this, we must complete the square. We first rewrite

$$x^2 + y^2 + 4x - 6y - 3 = 0 \quad \text{as} \quad (x^2 + 4x) + (y^2 - 6y) = 3 \quad \text{(A)}$$

To complete the square for the x terms, we take half the x coefficient and square, getting $(\frac{4}{2})^2 = 4$. Similarly, for the y terms we have $(-\frac{6}{2})^2 = 9$. Adding these numbers to *both sides* of (A) yields $(x^2 + 4x + 4) + (y^2 - 6x + 9) = 3 + 4 + 9$. Writing the left-hand side in factored form, we have $(x + 2)^2 + (y - 3)^2 = 16$. Comparing this with equation (88.1), we see that the center is at $(-2, 3)$ and the radius is 4. This information enables us to sketch the circle in Figure 88.6.

Figure 88.6

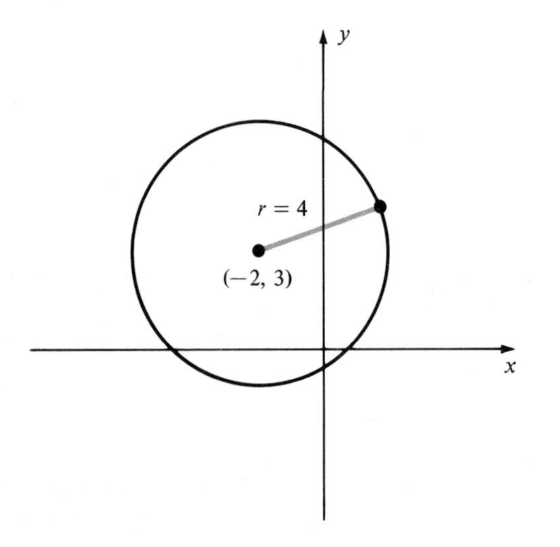

Example 88.3

Find the equation of the circle if the points $P(1, 2)$ and $Q(3, 8)$ are the endpoints of one of its diameters.

Solution

The center of the circle is the midpoint of $\overline{PQ}$ Hence, C is at $(2, 5)$. The radius is one half the diameter. Thus, $r = \frac{1}{2}PQ = \frac{1}{2}\sqrt{4 + 36} = \frac{1}{2}\sqrt{40} = \sqrt{10}$. Therefore, the equation of the circle is $(x - 2)^2 + (y - 5)^2 = 10$.

Example 88.4

Find the equation of the circle passing through the points $P(5, 3)$, $Q(6, 2)$, and $R(3, -1)$.

Solution

The equation of the circle is given by $(x - h)^2 + (y - k)^2 = r^2$, where we must determine the three constants, h, k, and r. Three conditions are necessary to determine these values. Since the circle must pass through the given points, h, k, and r may be determined by substituting the coordinates of the points for x and y and solving the three resulting equations simultaneously. Since P lies on the circle, we have

$$(5 - h)^2 + (3 - k)^2 = r^2 \quad \text{or} \quad 34 - 10h + h^2 - 6k + k^2 = r^2$$

Similarly, for Q we have

$$(6 - h)^2 + (2 - k)^2 = r^2 \quad \text{or} \quad 40 - 12h + h^2 - 4k + k^2 = r^2$$

and for R,

$$(3 - h)^2 + (-1 - k)^2 = r^2 \quad \text{or} \quad 10 - 6h + h^2 + 2k + k^2 = r^2$$

We now have three equations in three unknowns:

$$34 - 10h + h^2 - 6k + k^2 = r^2 \qquad \text{(a)}$$

$$40 - 12h + h^2 - 4k + k^2 = r^2 \qquad \text{(b)}$$

$$10 - 6h + h^2 + 2k + k^2 = r^2 \qquad \text{(c)}$$

Subtracting (c) from (b), we obtain $30 - 6h - 6k = 0$, and subtracting (b) from (a), we obtain $-6 + 2h - 2k = 0$. If we solve these simultaneously, we find that $h = 4$ and $k = 1$, $r = \sqrt{5}$, and the equation we seek is $(x - 4)^2 + (y - 1)^2 = 5$. ●

It is important for us to recognize that not all equations of the form we have shown represent a circle. Such equations may represent a point. For example, completing the square for $x^2 + y^2 + 6x - 2y + 10 = 0$ yields

$$(x^2 + 6x + 9) + (y^2 - 2y + 1) = -10 + 9 + 1 = 0$$

$$(x + 3)^2 + (y - 1)^2 = 0$$

This equation determines the single point $(-3, 1)$. (Why?)

Had we began with $x^2 + y^2 + 6x - 2y + 11 = 0$, we would have found that

$$(x^2 + 6x + 9) + (y^2 - 2y + 1) = -11 + 9 + 1 \quad \text{or} \quad (x + 3)^2 + (y - 1)^2 = -1$$

Since the left-hand side must be nonnegative for all (x, y) and the right-hand side is negative, the solution set for the equation is empty; that is, there are no points that satisfy the equation. The last two examples are often referred to as **degenerate cases of the circle**.

Exercises for Section 88

In Exercises 1 through 13, find the equation of the circle with the conditions given.

1. $r = 2$, $C(1, 3)$
2. $r = 5$, $C(-2, 4)$
3. $r = 1$, $C(0, 0)$
4. $r = 10$, $C(-1, -1)$
5. $r = 5$, $C(\sqrt{2}, \sqrt{3})$
6. $r = 3$, $C(0, 2)$
7. $r = 3$, $C(2, 0)$
8. $r = a$, $C(h, k)$
9. $C(1, 2)$, $P(3, 5)$, where P lies on the circle
10. $C(-2, 5)$, $P(-1, 3)$, where P lies on the circle
11. $C(2, 7)$, $P(2, 1)$, where P lies on the circle
12. $P(1, 3)$ and $Q(4, 7)$ are the endpoints of a diameter
13. $P(-2, -1)$ and $Q(-2, 6)$ are the endpoints of a diameter

In Exercises 14 through 25, if the equation is a circle, find the radius and the coordinates of the center, and sketch. If the equation is satisfied by a single point or no points, so state.

14. $x^2 + y^2 - 4x + 2y - 20 = 0$
15. $x^2 + y^2 - 6x - 8y - 10 = 0$
16. $x^2 + y^2 - 6x - 8y = 0$
17. $2x^2 + 2y^2 + 4x - 8y - 6 = 0$
18. $x^2 + y^2 + 8x - 4y + 4 = 0$
19. $3x^2 + 3y^2 - 4x + 2y + 6 = 0$
20. $x^2 + y^2 = 0$
21. $x^2 + y^2 + 4x - 12 = 0$
22. $x^2 + y^2 + 4y - 12 = 0$
23. $x^2 + y^2 + 4y + 12 = 0$
24. $x^2 + y^2 + 4x + 12 = 0$
25. $x^2 + y^2 + 4x + 4 = 0$
26. Find the equation of the circle passing through the following points:
    (a) $P(1, 3)$, $Q(-8, 0)$, and $R(0, 6)$
    (b) $P(7, 1)$, $Q(6, 2)$, and $R(-1, -5)$
    (c) $P(0, 0)$, $Q(4, 0)$, and $R(0, 4)$
27. Find the equation of the circle passing through $P(0, 0)$ and $Q(16, 0)$ with $r = 10$.
28. Find the equation of the circle passing through $P(1, 8)$ and $Q(-6, 7)$ with $r = 5$.
29. Find the equation of the circle concentric with $x^2 + y^2 - 2x - 4y - 4 = 0$ passing through $P(4, 7)$.
30. Find the equation of the circle concentric with $x^2 + y^2 + 4x + 6y - 21 = 0$ and passing through $(-2, 4)$.
31. Find the equation of all circles with:
    (a) Center on the x axis and $r = 5$.

(b) Center on the y axis and $r = 2$.
(c) Center at the origin and $r = a$.

32. Show that the circle $x^2 + y^2 - 9 = 0$ is symmetric with respect to the x axis, the y axis, and the origin.

33. Show that the circle $x^2 + y^2 + 2x - 3 = 0$ has symmetry only with respect to the x axis.

34. Show that the circle $x^2 + y^2 - 6y + 8 = 0$ has symmetry only with respect to the y axis.

35. Find the points of intersection of the circle $(x - 1)^2 + (y - 2)^2 = 9$ and the line $x + y - 6 = 0$.

SECTION 89

Translation of Axes

The shape of any curve is not affected by the position of the coordinate axes, although the equation of the curve is affected. In many cases the choice of a new coordinate axes may simplify the solution of a problem. Since the placement of the axes is arbitrary, we might prefer to move them. This type of movement is called a **transformation of axes**. As we shall see, a transformation enables us to change an equation or algebraic expression by substituting for the given variables their values in terms of another set of variables.

One of the most useful transformations is a *translation*. A translation of axes is simply a new set of coordinate axes $(x'y')$ chosen parallel to the xy axes. The geometric effect is the movement of the origin O to the point $O'(h, k)$ (see Figure 89.1).

Figure 89.1

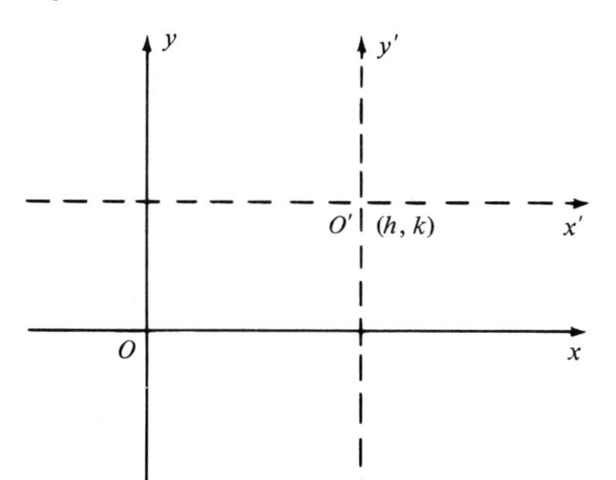

With the translation of axes, each point in the plane has two sets of coordinates: (x, y) and (x', y'). This relationship is shown in Figure 89.2. Suppose that P is any point in the plane. Then its coordinates are (x, y) in terms of the given axes with orgin at O, and (x', y') in terms of the new axes drawn with origin at (h, k). To determine x and y in terms of x', y', h, and k, we observe from Figure 89.2 that $x = \overline{QP} = \overline{QQ'} + \overline{Q'P} = h + x'$; similarly, $y = \overline{RP} = \overline{RR'} + \overline{R'P} = k + y'$. Thus,

$$\left.\begin{array}{lll} x = x' + h & \text{or} & x' = x - h \\ y = y' + k & \text{or} & y' = y - k \end{array}\right\} \tag{89.1}$$

These are called the **equations of translation.**

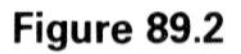

Figure 89.2

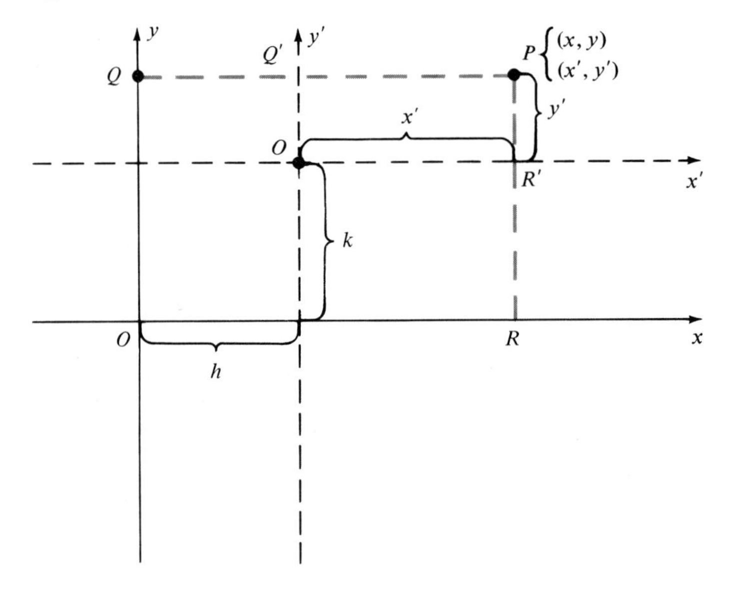

We can translate any equation in x and y to an equation in x' and y' by replacing x by $x' + h$ and y by $y' + h$. We must be careful to note that we have changed only the equation. The graph of the equation in x and y is exactly the same as the graph of the corresponding equation in x' and y'. Let us illustrate the use of the equation (89.1) by the following examples.

Example 89.1

The equation of a circle takes on its simplest form when the center is at the origin. Consider a circle of radius r and center at the point (h, k). The equation of such a circle is $(x - h)^2 + (y - k)^2 = r^2$. A translation of axes with O' at (h, k) will locate the center of the circle at O' (Figure

89.3). The equation of the circle in terms of x' and y' is obtained by replacing x by $x' + h$ and y by $y' + k$ or, equivalently, by replacing $x - h$ by x' and $y - k$ by y'. Thus, our new equation is $(x')^2 + (y')^2 = r^2$.

Figure 89.3

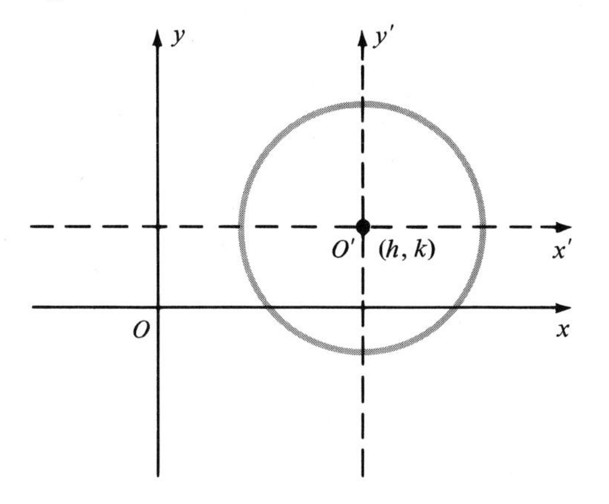

Example 89.2

Given the equation $x^2 + y^2 + 6x - 2y - 6 = 0$. Translate the axes so the equation in terms of x' and y' contains no first-degree terms.

Solution

We write the given equation as

$$(x^2 + 6x) + (y^2 - 2y) = 6$$

Completing the squares in x and y, we have

$$(x^2 + 6x + 9) + (y^2 - 2y + 1) = 6 + 9 + 1$$

or

$$(x + 3)^2 + (y - 1)^2 = 16$$

If we let $x' = x + 3$ and $y' = y - 1$, we obtain

$$(x')^2 + (y')^2 = 16$$

From equation (89.1) we note that the substitutions $x' = x + 3$ and $y' = y - 1$ translates the axes to a new origin at $(-3, 1)$. The graph of the equation is shown in Figure 89.4 on page 536.

Example 89.3

Given the equation $4x^2 + 9y^2 - 48x + 72y + 144 = 0$. Translate the axes so that the equation in terms of x' and y' contains no first-degree terms.

Figure 89.4

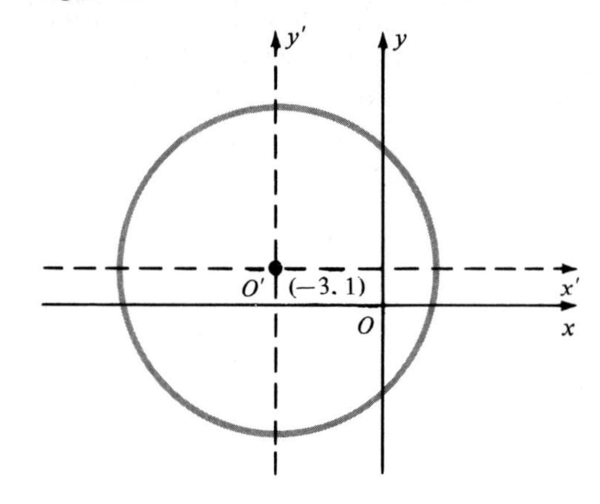

Solution

We rewrite the given equation in the form

$$4(x^2 - 12x) + 9(y^2 + 8y) = -144$$

Completing the square in x and y yields

$$4(x^2 - 12x + 36) + 9(y^2 + 8y + 16) = -144 + 4(36) + 9(16)$$

or

$$4(x - 6)^2 + 9(y + 4)^2 = 144$$

Dividing by 144, we obtain

$$\frac{(x-6)^2}{36} + \frac{(y+4)^2}{16} = 1 \qquad \text{(A)}$$

If we let $x' = x - 6$ and $y' = y + 4$, (A) becomes

$$\frac{(x')^2}{36} + \frac{(y')^2}{16} = 1$$

The substitutions $x' = x - 6$ and $y' = y + 4$ translate the axes to a new origin at $(6, -4)$. The graph of the equation, which is an ellipse, is shown in Figure 89.5. We shall study the ellipse in detail in Section 91.

The previous examples illustrate how an equation can be reduced to a simpler form by suitable translation of axes. In general, a second-degree equation that does not contain an xy term can be simplified by a suitable translation of axes. We can usually find the appropriate translation equations by first completing the square (if necessary) in x and y and then choosing x' and y' from the form of the equation.

In cases where completing the square is not possible, we can simplify the equation by the method outlined in the following example.

Figure 89.5

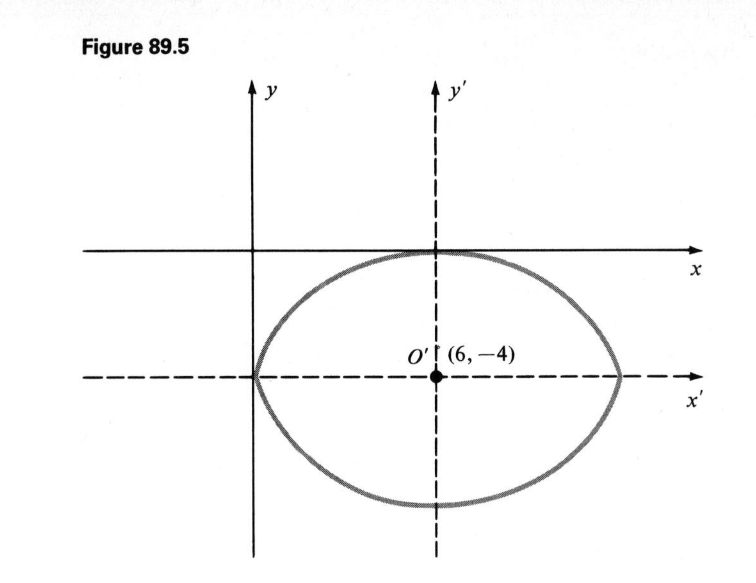

Example 89.4

Find translation equations that will simplify the equation $xy + 2x - 4y = 9$. Then graph the equation.

Solution

Since we do not know which translation works best, we let $x = x' + h$ and $y = y' + k$. Substituting in the equation yields

$$(x' + h)(y' + k) + 2(x' + h) - 4(y' + k) = 9$$

Expanding, we have

$$x'y' + kx' + hy' + hk + 2x' + 2h - 4y' - 4k = 9$$

Collecting terms and factoring gives

$$x'y' + (k + 2)x' + (h - 4)y' + 2h - 4k + hk = 9 \qquad \text{(A)}$$

Upon examining (A) we see that the equation readily simplifies if we choose $h = 4$ and $k = -2$. Substituting these values in (A) yields the form

$$x'y' + 0 \cdot x' + 0 \cdot y' + 8 + 8 - 8 = 9$$

or

$$x'y' = 1$$

Hence, the translation equations are

$$x = x' + 4 \quad \text{and} \quad y = y' - 2$$

Therefore, the origin of the new coordinate axes is translated to the point $(4, -2)$. The graph of the equation is shown in Figure 89.6 on page 538.

Figure 89.6

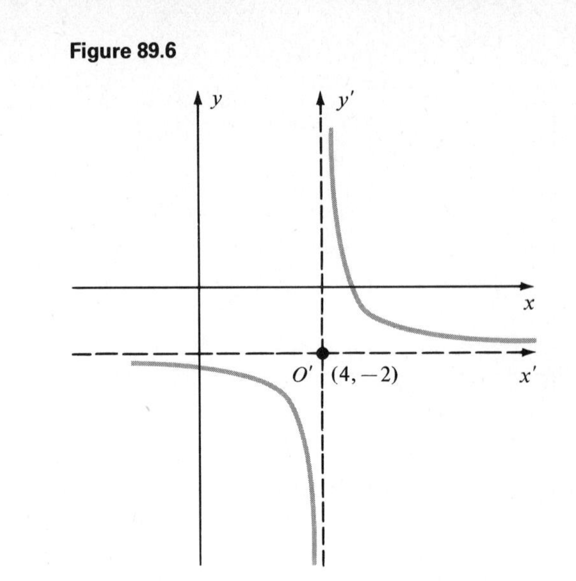

Exercises for Section 89

In Exercises 1 through 6, find the translation of axes that enables us to write each equation in a form that contains no first-degree terms. Write the translated equation in terms of x' and y'.

1. $x^2 + y^2 + 2x - 6y + 6 = 0$

2. $x^2 + y^2 - 3x + 4y - 2 = 0$

3. $2x^2 + 2y^2 + 4x - 4y - 1 = 0$

4. $x^2 + 2y^2 + 2x + 8y - 3 = 0$

5. $2x^2 + 3y^2 - 4x + 12y - 20 = 0$

6. $3x^2 - 4y^2 - 6x - 8y - 10 = 0$

7. What translation of axes enables us to write $(y - 2)^2 = 4(x + 1)$ as $(y')^2 = 4x'$?

8. What translation of axes enables us to write $y^2 - 6y - 4x + 5 = 0$ as $(y')^2 = 4x'$?

9. Simplify the equation $xy + x - 2y + 6 = 0$ using the method illustrated in Example 89.4.

10. Find the translation of axes that will enable us to write the following linear equations in the form $y' = mx'$.

(a) $\begin{cases} x + 2y - 4 = 0 \\ -2x - y + 5 = 0 \end{cases}$ (b) $\begin{cases} x - 3y + 2 = 0 \\ x + 3y - 4 = 0 \end{cases}$

In Exercises 11 through 18, graph each pair of equations on the same set of axes.

11. $y = x^2$ and $y - 3 = (x + 1)^2$

12. $x = y^2$ and $x - 1 = (y - 3)^2$

13. $y = 3^x$ and $y = 3^{x-1}$

14. $y = 2^x$ and $y + 1 = 2^{x-1}$

15. $y = x^3$ and $y = (x - 2)^3$

16. $y = x^3$ and $y + 1 = (x - 3)^3$

17. $y = \log x$ and $y = \log (x - 1)$

18. $y = \log x$ and $y + 2 = \log (x - 3)$

SECTION 90

The Parabola

The next conic section that we shall consider is the parabola. Like the ellipse and hyperbola, the parabola has many significant applications. The path of a projectile neglecting air resistance traces out a parabola. A hanging cable with uniform load will hang in the shape of a parabola. Many kinds of reflectors are parabolic in shape. Any ray or wave coming into a parabolic reflector parallel to the axis of symmetry will be reflected so that it passes through the focal point. Conversely, if a source of light rays is placed at the focus, those rays striking the reflector will be reflected into paths parallel to the axis of the parabola. This is the basic principle used for many types of lights, sound and radar detectors, and mirrors for telescopes and microscopes. In addition, the antenna of a radio telescope has a parabolic shape. We shall consider some of these applications in the exercises. Now let us study the algebraic and geometric properties of the parabola.

A **parabola** is the set of all points in a plane equidistant from a given fixed line and a given fixed point not on the line. The fixed line is called the **directrix**, and the fixed point is called the **focus**.

The equation of a parabola depends upon the location of its directrix and focal point. We simplify our initial discussion by choosing a point on the x axis as our focal point and a line perpendicular to the x axis and on the opposite side of the origin from the focus as our directrix. If the focal point F has coordinates $(p, 0)$, the equation of the directrix line is $x = -p$ (Figure 90.1 on page 546).

Suppose that the point $P(x, y)$ lies on the parabola shown in Figure 90.2. We note that P lies on the parabola if and only if $\overline{PR} = \overline{PF}$. The distance from P to F is given by the distance formula $\sqrt{(x - p)^2 + (y - 0)^2}$. The perpendicular distance from P to R is given by $|x + p|$.

Thus, according to our definition of a parabola,

$$\sqrt{(x - p)^2 + y^2} = |x + p|$$

Squaring both sides yields

$$(x - p)^2 + y^2 = (x + p)^2$$

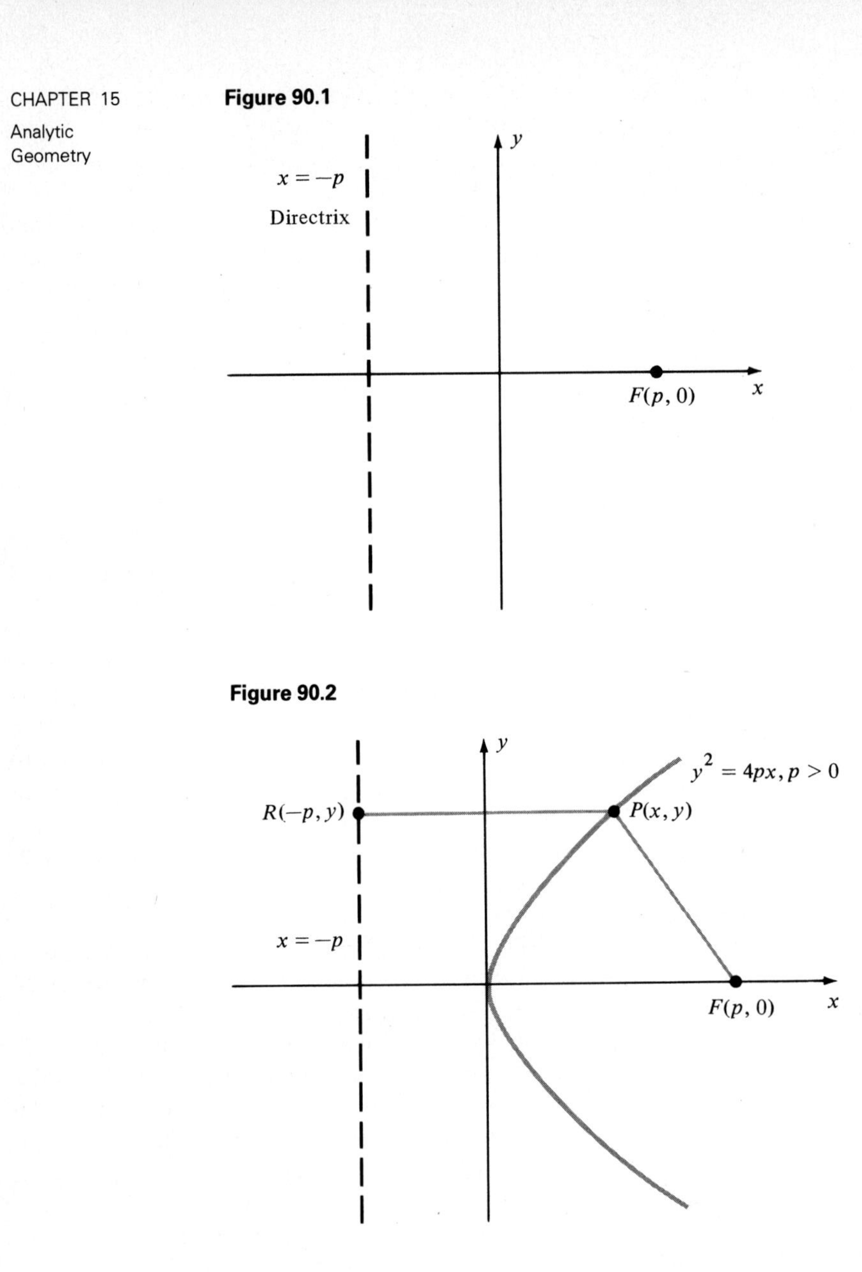

Expanding and simplifying we have

$$x^2 - 2px + p^2 + y^2 = x^2 + 2px + p^2$$

or

$$y^2 = 4px \qquad (90.1)$$

The line through the focal point drawn perpendicular to the directrix is called the **axis of symmetry** or the **axis of the parabola**. The point of intersection of this line and the curve is called the **vertex**. In this case the vertex is at (0, 0). Note that the graph is symmetric with respect to the x axis and opens to the right if $p > 0$. For $p < 0$, the graph is shown in Figure 90.3. The vertex is at (0, 0), the axis of symmetry is the x axis, and the curve opens to the left.

The analysis for the cases where the focal point lies on the y axis and the directrix line is perpendicular to the y axis does not differ significantly from what we have done and is left to the reader. These cases and their corresponding equations are shown in Figures 90.4 and 90.5.

Figure 90.3

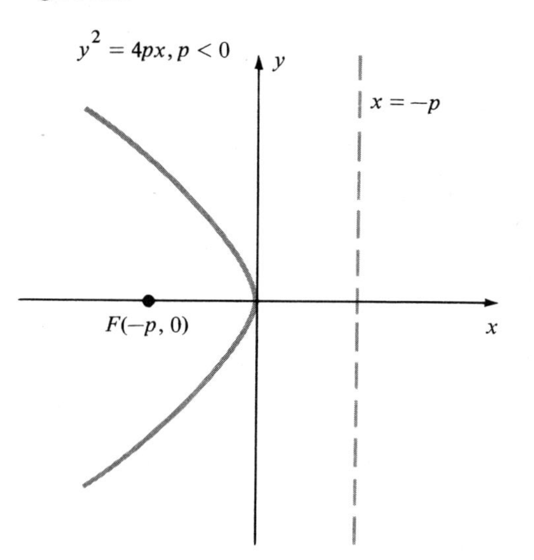

Figure 90.4

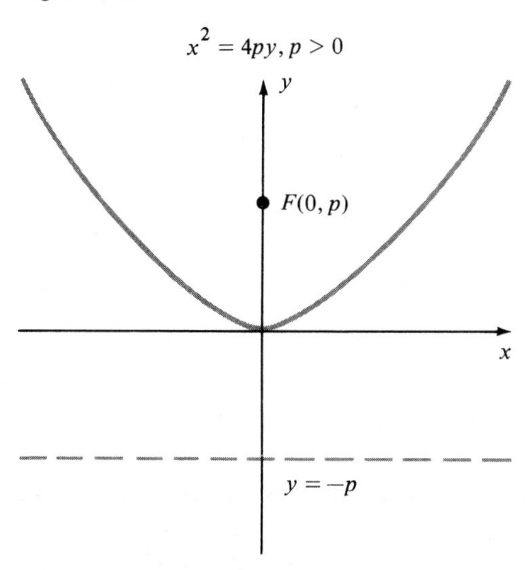

Figure 90.5

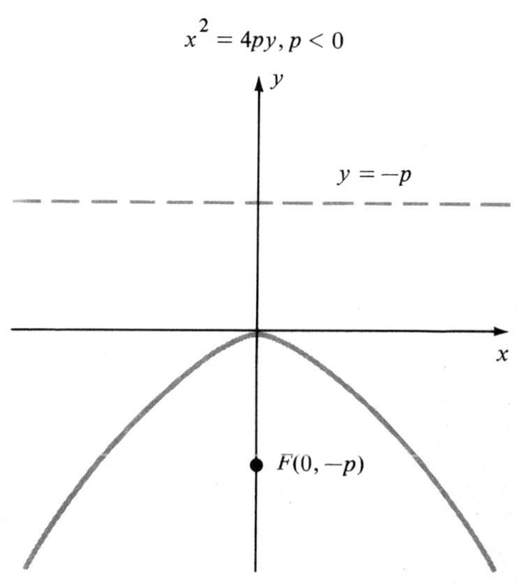

Example 90.1
Discuss and sketch $y^2 = 8x$.

Solution
The equation is of the form $y^2 = 4px$. If we take $4p = 8$, we have $p = 2$. The vertex is at $(0, 0)$ and the focal point is on the axis of symmetry (the x axis), p units from the vertex, that is, at $(2, 0)$. The directrix line is given by $x = -2$ and the curve opens to the right. (Why?) The graph is shown in Figure 90.6.

Example 90.2
Discuss and sketch $x^2 = -16y$.

Solution
The equation is of the form $x^2 = 4py$. Taking $4p = -16$, we have $p = -4$. Thus, the vertex is at $(0, 0)$, and the focal point lies on the y axis (the axis of symmetry) 4 units below the vertex at $(0, -4)$. The equation of the directrix line is given by $y = 4$ and the curve opens downward. The graph is shown in Figure 90.7.

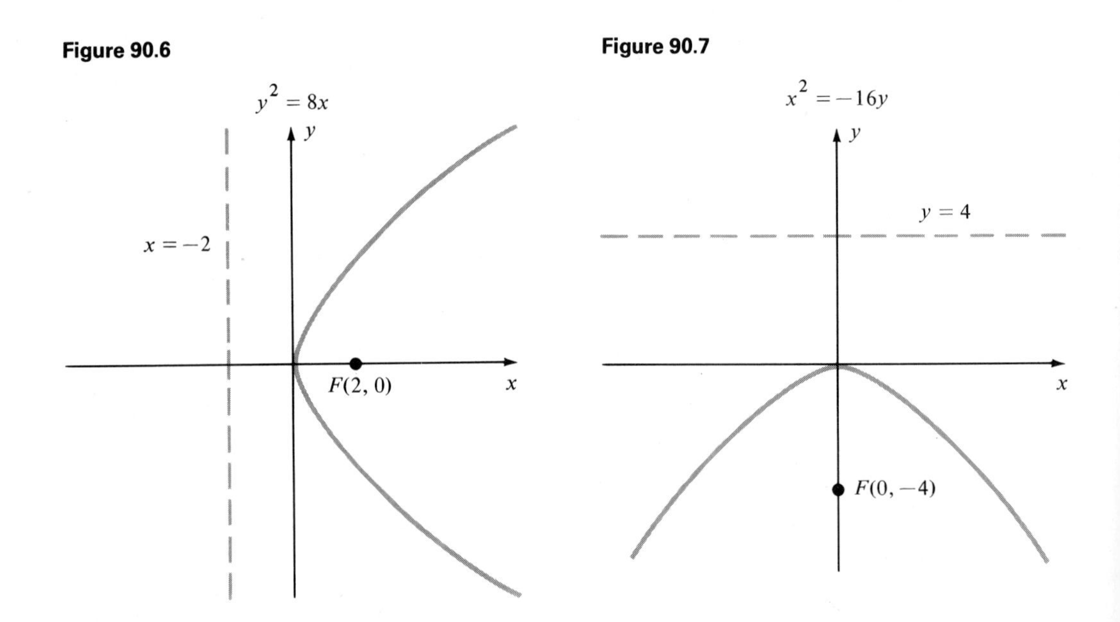

If the vertex of the parabola is at the point (h, k), the equations previously obtained no longer apply. However, we can make use of our translation equations to generate various forms of the equation of such a parabola.

Suppose that the vertex of our parabola opening to the right is at (h, k) and its axis is the line $y = k$ (see Figure 90.8). Let us introduce the new coordinate axis x' and y' parallel to the original axes and having their origin

Figure 90.8

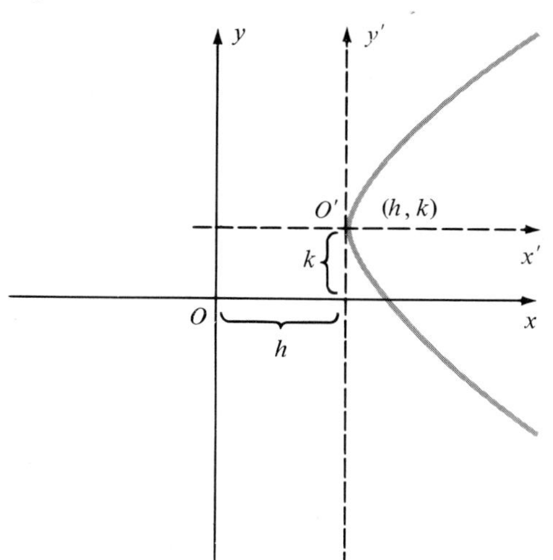

at the point (h, k). Since the vertex of the parabola is now at the origin of the $x'y'$ system, its equation is $y'^2 = 4px'$, where p is the distance from the vertex to the focus. The relationship between the old and new coordinate systems is

$$x = x' + h \qquad y = y' + k$$

or

$$x' = x - h \qquad y' = y - k$$

Substituting yields the desired equation in the original coordinate system,

$$(y - k)^2 = 4p(x - h) \qquad p > 0 \tag{90.2}$$

This equation has the following characteristics:

1. Vertex at (h, k).
2. Axis parallel to the x axis.
3. Focus p units to the right of vertex at $(h + p, k)$.
4. Directrix p units to the left of the vertex and perpendicular to the axis of symmetry. Its equation is $x = h - p$.
5. Parabola opens to the right.

Other forms of the equation of the parabola are

$$(y - k)^2 = 4p(x - h) \qquad p < 0 \tag{90.3}$$

$$(x - h)^2 = 4p(y - k) \qquad p > 0 \tag{90.4}$$

$$(x - h)^2 = 4p(y - k) \qquad p < 0 \tag{90.5}$$

These forms are shown respectively in Figure 90.9 on page 544.

Figure 90.9

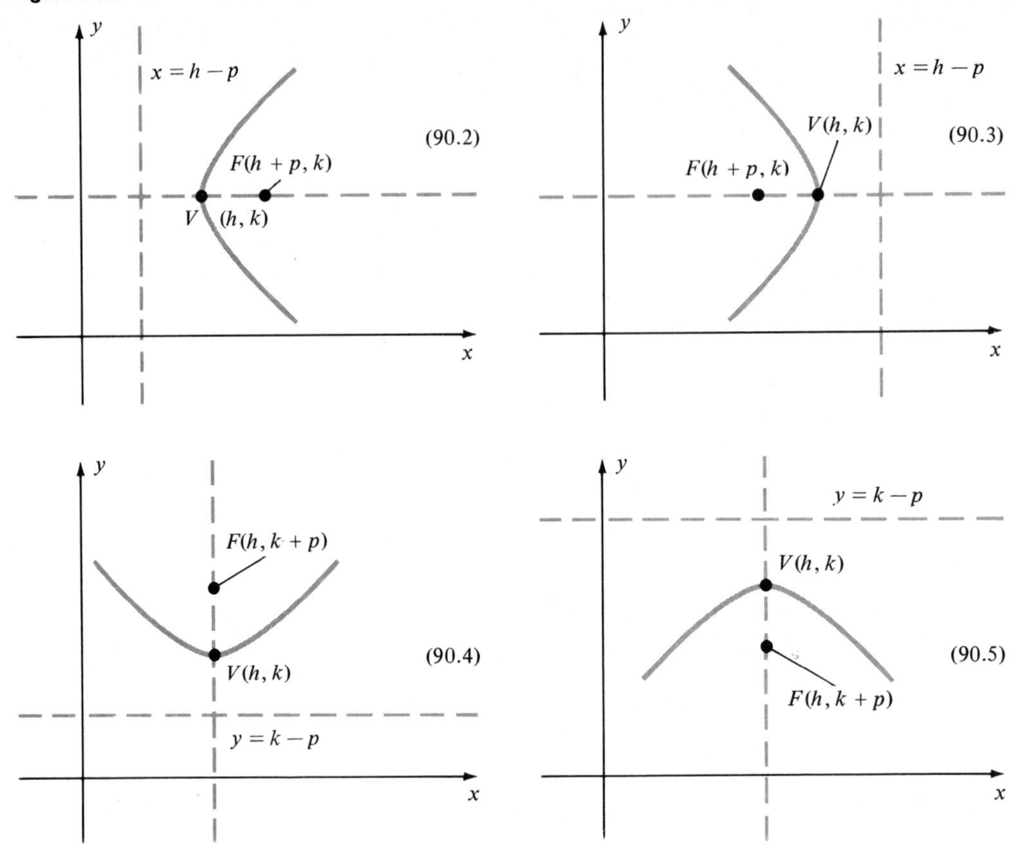

Example 90.3

Discuss and sketch $(x - 2)^2 = 12(y + 4)$.

Solution

This equation is of the form $(x - h)^2 = 4p(y - k)$. Thus, the parabola has vertex at $(2, -4)$ and opens upward. Setting $4p = 12$, we have $p = 3$. The focus is at the point $(2, -4 + 3)$ or $(2, -1)$. The equation of the directrix line is $y = k - p = -4 - 3 = -7$. The sketch is shown in Figure 90.10.

Example 90.4

Discuss and sketch $(y - 2)^2 = -8(x + 1)$.

Solution

This equation is of the form $(y - k)^2 = 4p(x - h)$. Therefore, the vertex is at $(-1, 2)$ and the parabola opens to the left. (Why?) Setting $4p = -8$,

Figure 90.10

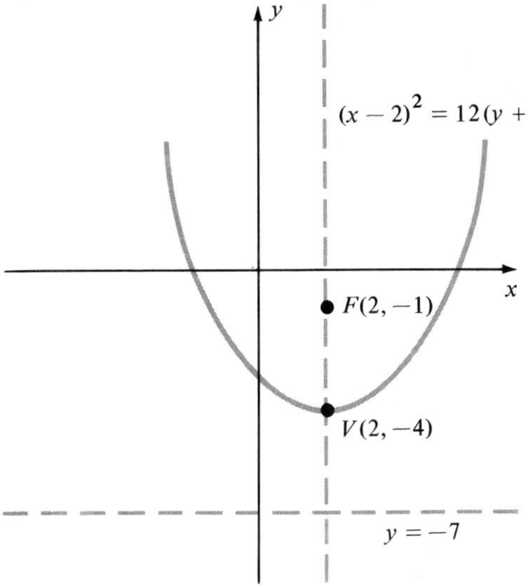

Figure 90.11

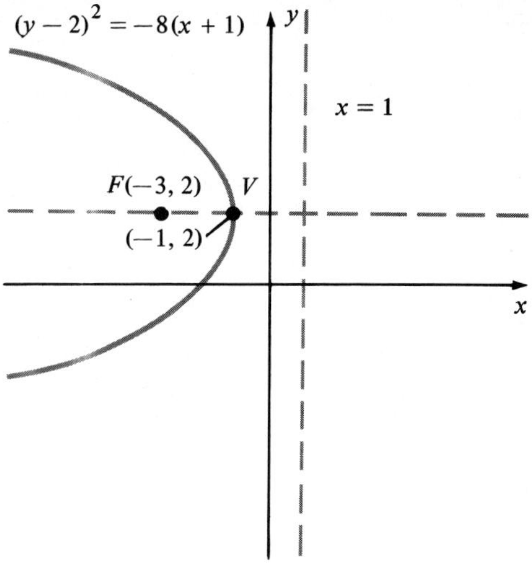

$p = -2$, the focus is at $(-3, 2)$ and the equation of the directrix line is $x = 1$. The sketch is shown in Figure 90.11.

Example 90.5

Find an equation of the parabola whose directrix is the line $y = -2$ and whose focus is the point $(3, 6)$.

Solution

Since the directrix is parallel to the x axis, our equation will have the form $(x - h)^2 = 4p(y - k)$. The vertex is halfway between the directrix and the focus. Thus, the coordinates of the vertex are $(3, 2)$ (Figure 90.12 on page 546). The directed distance from V to F is p, and thus $p = 6 - 2 = 4$. Therefore, the equation is $(x - 3)^2 = 16(y - 2)$.

If the equation of a parabola is not given in one of the standard forms, we may recognize the equation by noting that it is quadratic in one variable and linear in the other. Whenever we have such a situation, we complete the square in the variable that appears quadratically and then put the linear terms in the form $4p(x - h)$ or $4p(y - k)$. The standard form of the equation then yields the information regarding vertex, focus, axis of symmetry, directrix, and direction of opening.

Example 90.6

Discuss and sketch $x^2 + 2x - 4y - 3 = 0$.

Figure 90.12

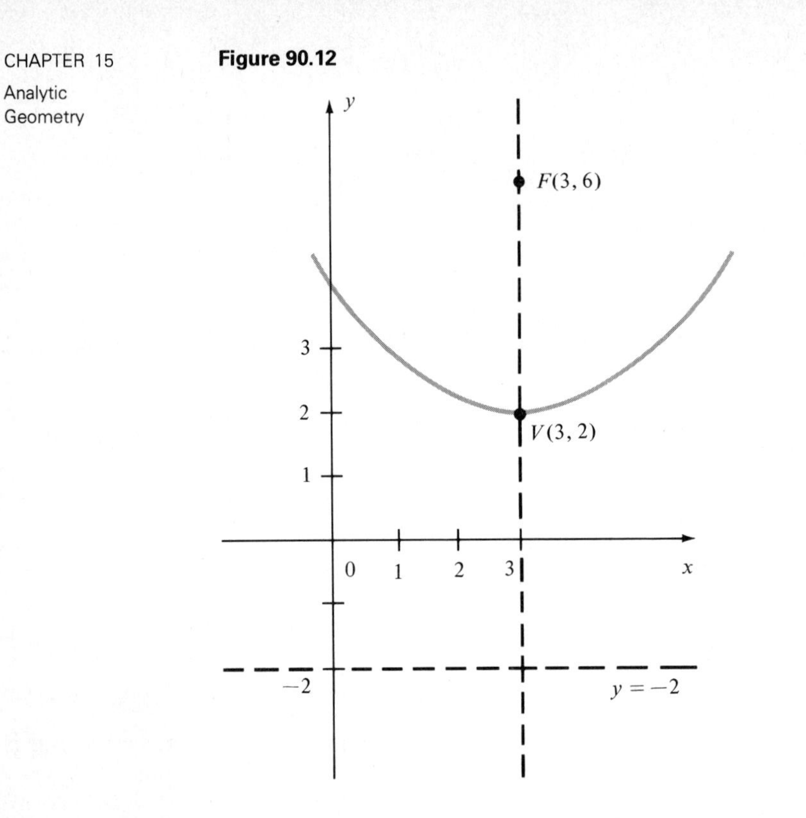

Solution

We proceed as we did for the equation of a circle, by completing the square.

$$x^2 + 2x = 4y + 3$$
$$x^2 + 2x + 1 = 4y + 3 + 1$$
$$(x + 1)^2 = 4y + 4$$
$$(x + 1)^2 = 4(y + 1)$$

Our equation is now of the form

$$(x - h)^2 = 4p(y - k)$$

The vertex is at $(-1, -1)$ and the parabola opens upward. Since $4p = 4$, $p = 1$, and the focus is at $(-1, 0)$. The equation of the directrix is $y = -2$. The sketch is shown in Figure 90.13.

If the linear term is missing, the equation is no longer a parabola. In this case we have either a pair of parallel lines, a single line, or no graph. These are sometimes referred to as the **degenerate cases of a parabola.**

Example 90.7

Discuss and sketch $x^2 + 2x - 3 = 0$.

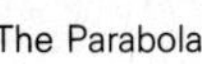

Figure 90.13

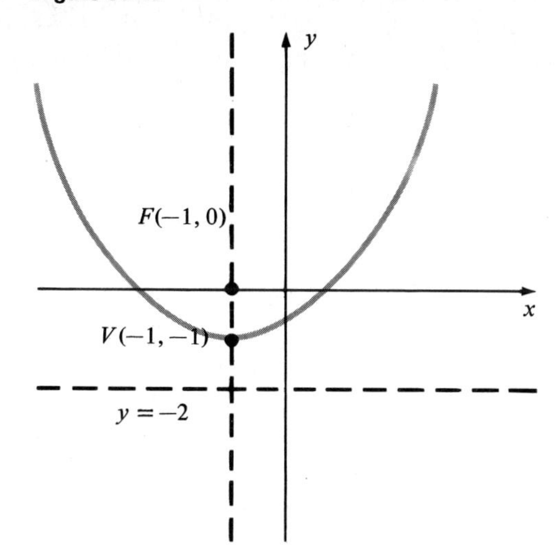

Solution

Since the equation does not involve both x and y, it cannot represent a parabola. Factoring yields

$$(x + 3)(x - 1) = 0 \quad \text{or} \quad x = -3, x = 1$$

We have a pair of parallel lines (Figure 90.14).

Figure 90.14

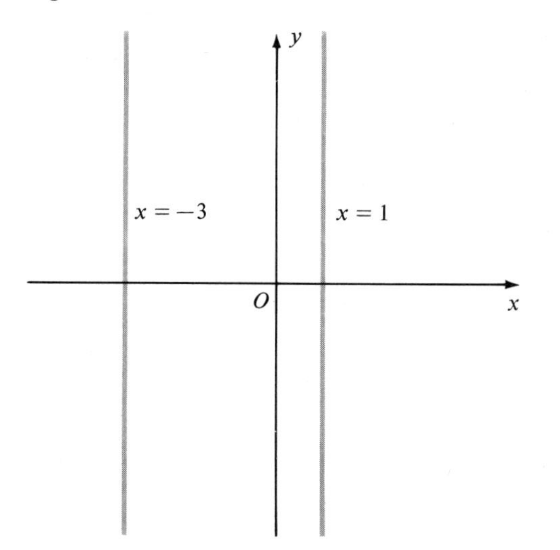

Our study of parabolas is not new. In Sections 24 and 25 we encountered the quadratic function and its applications. The results obtained in this section are consistent with our previous results. Let us demonstrate this consistency by reworking Example 24.6 using the methods of this section. Refer to Example 24.6 and then study the following procedure.

Example 90.8

An object is thrown vertically upward from the ground with an initial velocity of 80 ft/sec. Its height s above the ground is a function of time t and is given by the equation $s = 80t - 16t^2$. How high does the object rise above the ground?

Solution

We note that the function is quadratic in t and linear in s. Hence, the equation represents a parabola and is put into standard form as follows:

$$s = 80t - 16t^2$$

$$= -16(t^2 - 5t)$$

$$s - 100 = -16\left(t^2 - 5t + \frac{25}{4}\right)$$

$$s - 100 = -16\left(t - \frac{5}{2}\right)^2$$

or

$$\left(t - \frac{5}{2}\right)^2 = -\frac{1}{16}(s - 100) \qquad \text{(A)}$$

The form of (A) tells us that the parabola opens downward and has vertex at $(\frac{5}{2}, 100)$. The graph is shown in Figure 90.15 for the interval $0 \le t \le 5$. We see that the maximum height the object reaches is 100 ft. As an added

Figure 90.15

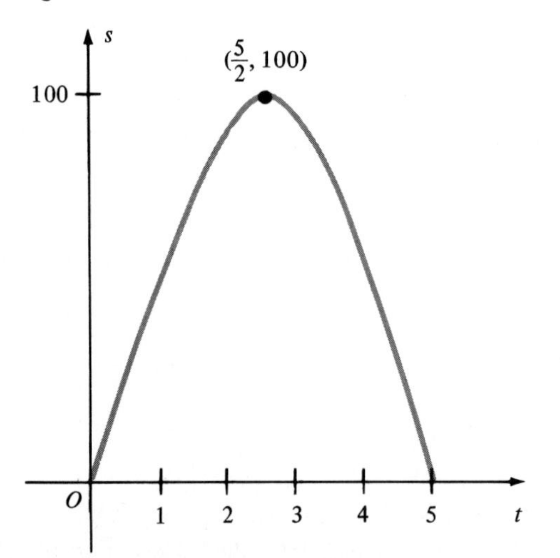

bit of information, we note that the maximum height occurs when $t = \frac{5}{2}$ sec.

Exercises for Section 90

In Exercises 1 through 20, discuss and sketch each equation

1. $y^2 = -4x$ **2.** $x^2 = 12y$

3. $x^2 = 4y$ **4.** $y^2 = 4x$

5. $(y - 2)^2 = 4(x - 3)$ **6.** $(x + 3)^2 = 6(y + 2)$

7. $(y + 1)^2 = -4(x + 1)$ **8.** $(y + 1)^2 = 8x$

9. $x^2 = 4y - 8$ **10.** $(y - 5)^2 = 2x + 3$

11. $x^2 + 6x - 4y - 3 = 0$ **12.** $y^2 + 4y - 8x + 36 = 0$

13. $y^2 - 2y + 8x - 39 = 0$ **14.** $y^2 + 8y - 6x + 4 = 0$

15. $4y^2 - 4y - 4x - 3 = 0$ **16.** $y^2 + 2y + 2x - 1 = 0$

17. $y^2 + 4y - 4 = 0$ **18.** $x^2 - x + 5 = 0$

19. $(x + y)^2 = 0$ **20.** $x^2 + y^2 = 4$

In Exercises 21 through 30, find an equation of the parabola described.

21. Focus $(0, 2)$, vertex $(0, 0)$ **22.** Focus $(-2, 4)$, vertex $(-2, 3)$

23. Focus $(0, 1)$, vertex $(-5, 1)$ **24.** Focus $(4, 4)$, vertex $(4, 5)$

25. Focus $(3, 5)$, vertex $(3, 9)$ **26.** Vertex $(3, 5)$, directrix $x = -1$

27. Vertex $(4, 0)$, directrix $x = 0$ **28.** Vertex $(0, 1)$, directrix $x = -1$

29. Vertex $(-2, 4)$, directrix $y = 12$

30. Vertex $(1, -2)$ and passing through $(5, 2)$

31. How high is a parabolic arch of span 12 ft and height 36 ft at a distance of 4 ft from the center of the span? At a distance of 3 ft from the center of the span?

32. When the load along a hanging cable is uniformly distributed, the cable hangs in the shape of an arc of a parabola. Suppose that we have such a cable with endpoints 100 ft apart horizontally and 40 ft above the lowest point.

(a) Select a coordinate axes and then find the equation of the parabola.

(b) Find the height of the cable above the low point at a distance of 20 ft from the end measured horizontally.

33. When an object is thrown straight upward with an initial velocity of 128 ft/sec, the height s (in feet) as a function of the time t (in seconds) is given by $s = 128t - 16t^2$. At what time does the object reach its maximum height and what is the height?

34. The power P produced by a varying electric current I is given by the equation $P = I^2R$, where the resistance R is constant. If $R = 8\ \Omega$, graph the function for the interval $0 \le I \le 4$.

35. A variable voltage is given by the formula $e = t^2 - 6t + 10$, t in seconds. Graph the formula for $1 \le t \le 6$ and state the minimum value for e in this interval.

36. Solve the following system of equations graphically.

$$\begin{cases} y = x - 5 \\ y^2 = 2x - 2 \end{cases}$$

SECTION 91

The Ellipse

Let us now consider the ellipse. One reason for studying this curve is its many physical applications. Many arches have elliptical shapes. The planets travel in elliptic orbits with the sun as one focal point. Certain machines have elliptic gears. The ellipse also has the focal property that a ray emanating from one focus is reflected to the other. We shall consider some of these applications in the exercises. For the moment we turn our interest to the development of the algebraic and geometric properties of the ellipse.

An **ellipse** is the set of points $P(x, y)$, the sum of whose distance from two fixed points (called the **foci**) is constant. The constant must be greater than the distance between the two given points. Let us choose the points $(-c, 0)$ and $(c, 0)$ to be the foci and $2a$ for the constant sum of the distance (Figure 91.1).

The distance from P to $(-c, 0)$ is $\sqrt{(x + c)^2 + (y - 0)^2}$, and the distance from P to $(c, 0)$ is $\sqrt{(x - c)^2 + (y - 0)^2}$. If $P(x, y)$ represents a point on the ellipse, the sum of these two distances is equal to $2a$. Thus, we have

$$\sqrt{(x - c)^2 + y^2} + \sqrt{(x + c)^2 + y^2} = 2a$$

or

$$\sqrt{(x - c)^2 + y^2} = 2a - \sqrt{(x + c)^2 + y^2}$$

Simpler equivalent equations are found by squaring both sides of the equation and simplifying:

$$x^2 - 2cx + c^2 + y^2 = 4a^2 - 4a\sqrt{(x + c)^2 + y^2} + x^2 + 2cx + c^2 + y^2$$

or

$$4a\sqrt{(x + c)^2 + y^2} = 4a^2 + 4cx$$

Figure 91.1

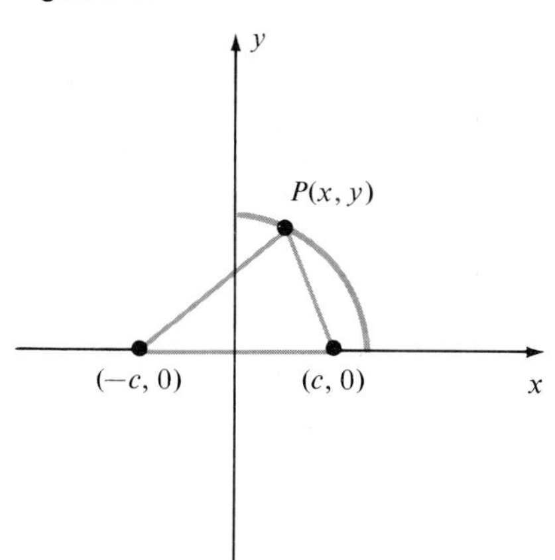

Dividing by 4 and squaring again yields

$$a^2(x^2 + 2cx + c^2 + y^2) = a^4 + 2a^2cx + c^2x^2$$

Simplifying again, we have

$$x^2(a^2 - c^2) + a^2y^2 = a^2(a^2 - c^2)$$

Dividing by $a^2(a^2 - c^2)$, we have

$$\frac{x^2}{a^2} + \frac{y^2}{a^2 - c^2} = 1$$

The triangle in Figure 91.1 has one side of length $2c$; the sum of the lengths of the other two sides is $2a$. Thus, $2a > 2c$, $a > c$, $a^2 > c^2$, and $a^2 - c^2 > 0$. Since $a^2 - c^2$ is positive, we replace it by another positive number, b^2. Thus,

$$\frac{x^2}{a^2} + \frac{y^2}{b^2} = 1 \qquad \text{where } b^2 = a^2 - c^2 \tag{91.1}$$

We note that in squaring both sides of the equation twice we introduced no extraneous roots, since in both steps both sides of the equation were nonnegative.

By setting y equal to 0 in equation (91.1), we obtain $x = \pm a$. The graph therefore has x intercepts at $(-a, 0)$ and $(a, 0)$. The line segment between $(-a, 0)$ and $(a, 0)$ is called the **major axis** of the ellipse, since it is the longer axis. Similarly, the graph has y intercepts at $(0, -b)$ and $(0, b)$. The line segment between $(0, -b)$ and $(0, b)$ is called the **minor axis** of the ellipse. The endpoints of the major axis $(\pm a, 0)$ are called the **vertices** of the ellipse and the endpoints of the minor axis $(0, \pm b)$ are called **covertices** of the ellipse. The point of intersection of the major and minor axis is called the **center** of the ellipse and in this case is at $(0, 0)$ (See Figure 91.2 on page 552).

Figure 91.2

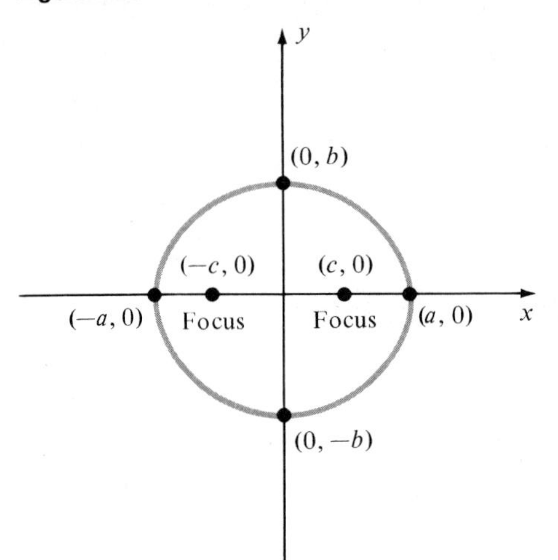

Note that the graph is symmetric with respect to the x axis, y axis, and origin.

Example 91.1

Discuss and sketch $9x^2 + 25y^2 = 225$.

Solution

Dividing by 225, we have $x^2/25 + y^2/9 = 1$, which is of the form of Equation (91.1). It is apparent that the equation represents an ellipse with $a^2 = 25$, $b^2 = 9$, and $c^2 = a^2 - b^2 = 16$. Thus, the ellipse has vertices at $(\pm 5, 0)$, covertices at $(0, \pm 3)$ and foci at $(\pm 4, 0)$. The major axis is of length 10, minor axis of length 6, and the center is at $(0, 0)$. The sketch is shown in Figure 91.3. ●

If the foci are located on the y axis at $(0, c)$ and $(0, -c)$, an analogous derivation would yield the equation

$$\frac{x^2}{a^2 - c^2} + \frac{y^2}{a^2} = 1$$

or

$$\frac{x^2}{b^2} + \frac{y^2}{a^2} = 1 \qquad \text{where } b^2 = a^2 - c^2 \tag{91.2}$$

One question is immediately apparent. How can we tell if we have

$$\frac{x^2}{a^2} + \frac{y^2}{b^2} = 1 \quad \text{or} \quad \frac{x^2}{b^2} + \frac{y^2}{a^2} = 1?$$

Figure 91.2

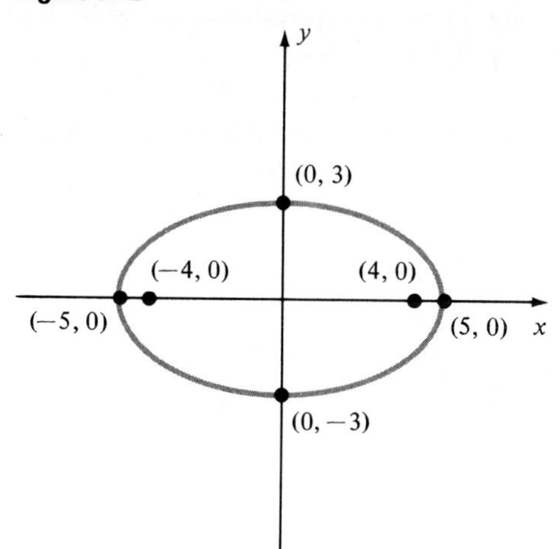

The answer is *size*. In both cases $a > b$. Thus, the larger denominator is always a^2 and the smaller is always b^2.

The equation $x^2/b^2 + y^2/a^2 = 1$, $a^2 > b^2$, has the following properties: vertices at $(0, \pm a)$, covertices at $(\pm b, 0)$, foci at $(0, \pm c)$. The major axis is of length $2a$, minor axis of length $2b$ and center is at $(0, 0)$. The sketch is shown in Figure 91.4.

Figure 91.4

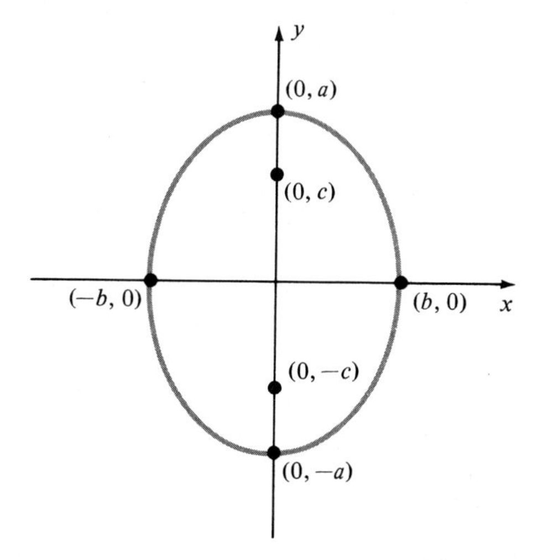

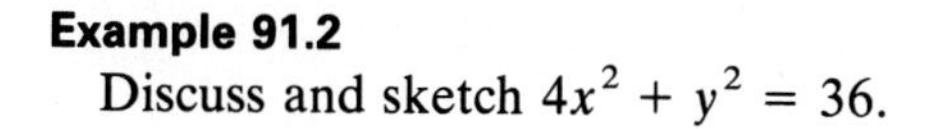

Example 91.2

Discuss and sketch $4x^2 + y^2 = 36$.

Solution
We may write the equation equivalently as $x^2/9 + y^2/36 = 1$. Now our equation is of form (91.2), with $a^2 = 36$, $b^2 = 9$, and $c^2 = a^2 - b^2 = 27$. Thus, the ellipse has vertices at $(0, \pm 6)$, covertices at $(\pm 3, 0)$, and foci at $(0, \pm 3\sqrt{3})$. The major axis is of length 12, minor axis of length 6, and the center is at $(0, 0)$. The sketch is shown in Figure 91.5.

Figure 91.5

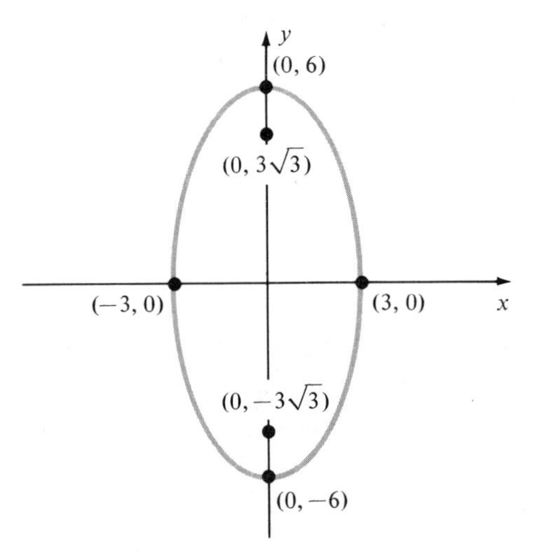

Example 91.3
Find an equation of the ellipse with vertices $(0, \pm 5)$ and foci $(0, \pm 3)$.

Solution
Since the vertices are on the y axis, the equation must be of the form $x^2/b^2 + y^2/a^2 = 1$. In addition, $a = 5$, $c = 3$, and $b^2 = a^2 - c^2 = 25 - 9 = 16$. Our result is $x^2/16 + y^2/25 = 1$.

If the center of the ellipse is not at the origin, and the major and minor axes are parallel to the x and y axes, we may use the method of Section 89 to write an equation for the ellipse. Suppose that we have an ellipse with center at (h, k) and foci at $(h - c, k)$ and $(h + c, k)$. The translation of axis $x' = x - h$ and $y' = y - k$ will readily show than an equation for this ellipse in terms of these new coordinates is $(x')^2/a^2 + (y')^2/b^2 = 1$. In terms of the original coordinates, we have

$$\frac{(x-h)^2}{a^2} + \frac{(y-k)^2}{b^2} = 1 \qquad \text{where } b^2 = a^2 - c^2 \tag{91.3}$$

Similarly, if the center is at (h, k) and the foci at $(h, k - c)$ and $(h, k + c)$ an equation for the ellipse is

$$\frac{(x-h)^2}{b^2} + \frac{(y-k)^2}{a^2} = 1 \qquad \text{where } b^2 = a^2 - c^2 \tag{91.4}$$

Example 91.4

Discuss and sketch the graph of

$$\frac{(x-2)^2}{9} + \frac{(y-1)^2}{25} = 1$$

Solution

The equation represents an ellipse with center at (2, 1) [compare with equation (91.4)]; $a^2 = 25$, $b^2 = 9$, and $c^2 = a^2 - b^2 = 16$. Both the vertices and the foci are above and below the center. The vertices are (2, 6) and (2, −4) and the foci are (2, 5) and (2, −3). The covertices are to the right and left of the center at a distance of 3 units and are at (−1, 1) and (5, 1). The major axis is of length 10 and the minor axis of length 6. The graph is drawn in Figure 91.6.

Figure 91.6

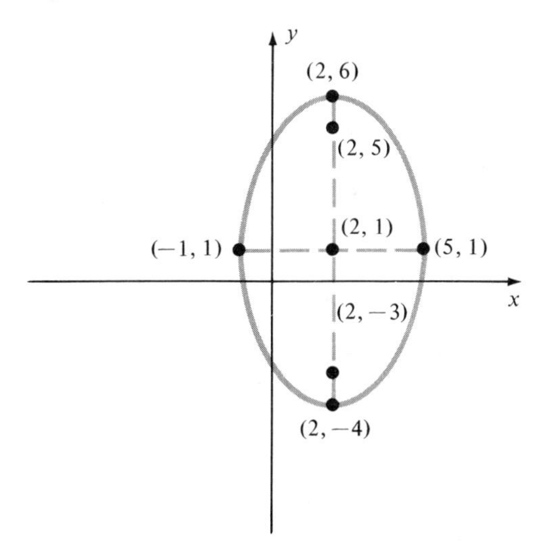

Example 91.5

Discuss and sketch $4x^2 + 9y^2 - 48x + 72y + 144 = 0$.

Solution

The equation can be put in a form that we recognize by completing the square.

$$4(x^2 - 12x) + 9(y^2 + 8y) = -144$$

$$4(x^2 - 12x + 36) + 9(y^2 + 8y + 16) = -144 + 144 + 144$$

$$4(x - 6)^2 + 9(y + 4)^2 = 144$$

$$\frac{(x - 6)^2}{36} + \frac{(y + 4)^2}{16} = 1 \qquad \text{(A)}$$

We now recognize (A) as the equation of an ellipse with center at $(6, -4)$; $a = 6$, $b = 4$, and $c = 2\sqrt{5}$. The vertices are $(0, -4)$ and $(12, -4)$, and the foci are $(6 + 2\sqrt{5}, -4)$ and $(6 - 2\sqrt{5}, -4)$. The covertices are $(6, 0)$ and $(6, -8)$. The sketch is shown in Figure 91.7. ●

Figure 91.7

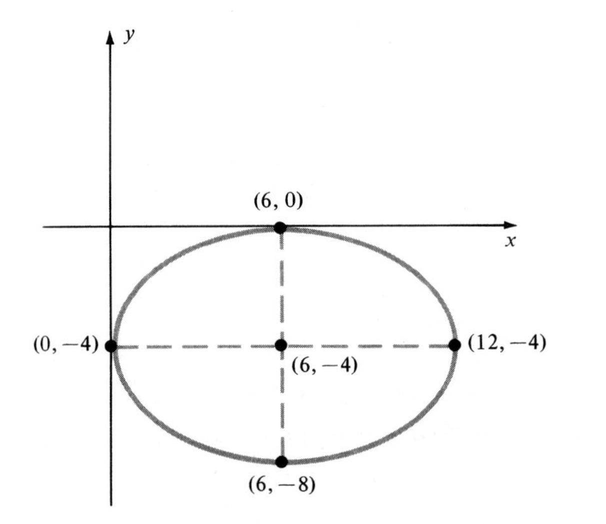

When completing the square, we must be aware of the possibility of degenerate cases. These include a single point or no points.

Example 91.6

Discuss the equation $x^2 + 2y^2 + 2x + 4y + 3 = 0$.

Solution

We put the given equation in standard form by completing the square in x and y as follows:

$$x^2 + 2x + 2(y^2 + 2y) = -3$$

$$(x^2 + 2x + 1) + 2(y^2 + 2y + 1) = -3 + 1 + 2$$

or

$$(x + 1)^2 + 2(y + 1)^2 = 0 \qquad \text{(A)}$$

Equation (A) is a degenerate ellipse, since the equation is only satisfied by the single point $(-1, -1)$.

Exercises for Section 91

In Exercises 1 through 20, discuss and sketch each ellipse.

1. $\dfrac{x^2}{144} + \dfrac{y^2}{81} = 1$

2. $\dfrac{x^2}{25} + \dfrac{y^2}{4} = 1$

3. $\dfrac{x^2}{4} + \dfrac{y^2}{25} = 1$

4. $\dfrac{x^2}{16} + \dfrac{y^2}{49} = 1$

5. $9x^2 + 4y^2 = 36$

6. $4x^2 + 9y^2 = 36$

7. $x^2 + y^2 = 1$

8. $4x^2 + y^2 = 1$

9. $9x^2 + y^2 = 9$

10. $4x^2 + 16y^2 = 64$

11. $\dfrac{(x-1)^2}{16} + \dfrac{(y+2)^2}{4} = 1$

12. $x^2 + 4y^2 - 2x - 3 = 0$

13. $4x^2 + y^2 + 8x + 10y + 13 = 0$

14. $\dfrac{(x-2)^2}{25} + \dfrac{(y-4)^2}{9} = 1$

15. $\dfrac{(x+5)^2}{169} + \dfrac{(y+2)^2}{144} = 1$

16. $25x^2 + 9(y-2)^2 = 225$

17. $9x^2 + 4y^2 + 36x - 24y + 36 = 0$

18. $x^2 + 4y^2 - 4x - 8y - 92 = 0$

19. $x^2 + 5y^2 - 2x + 20y + 16 = 0$

20. $x^2 + 2y^2 - 2x + 8y + 9 = 0$

In Exercises 21 through 30, find an equation of the ellipse described. Sketch the graph.

21. Center $(0, 0)$, focus $(0, 2)$, $a = 4$

22. Center $(0, 0)$, vertex $(0, 13)$, focus $(0, -12)$

23. Center $(0, 0)$, vertex $(0, 13)$, focus $(0, -5)$

24. Center $(0, 0)$, focus $(-3, 0)$, $a = 5$

25. Center $(0, 2)$, focus $(0, 0)$, $a = 6$

26. Center $(0, 0)$, covertex $(0, 5)$, focus $(12, 0)$

27. Center $(0, 0)$, vertex $(5, 0)$, passes through $(\sqrt{15}, 2)$

28. Vertices, $(1, 0)$ and $(1, 8)$, covertices $(2, 4)$ and $(0, 4)$

29. Vertices $(3, -4)$ and $(9, -4)$, covertices $(6, -6)$ and $(6, -2)$

30. Foci, $(0, \pm 4)$ and passes through $(\frac{12}{5}, 3)$

31. An ornamental arch in the form of a semiellipse has a span of 30 ft and a greatest height of 9 ft. A walkway is to be built between two vertical supports that are equidistance from the ends of the arch and from each other. Find the height of the supports.

32. The orbit of the earth around the sun is an ellipse with the sun at one of the foci. If the ellipse has a semimajor axis of 93 million miles and an eccentricity of approximately 0.016 (the eccentricity e of an ellipse is given the ratio c/a), find:
(a) How close the earth gets to the sun.
(b) The greatest distance between the earth and the sun.

33. A man-made satellite orbits the earth in an elliptical path whose center is at the center of the earth. If the altitude of the satellite ranges from 1000 to 2000 miles, find the equation of its path. The radius of the earth is approximately 4000 miles.

SECTION 92

The Hyperbola

The last conic section that we shall consider is the hyperbola. A **hyperbola** is the set of all points $P(x, y)$ in a plane such that the absolute value of the difference between the distance from P to two fixed points (called the **foci**) is constant.

Let us again choose the foci to be $(c, 0)$ and $(-c, 0)$ and the constant to be $2a$ (Figure 92.1). If $P(x, y)$ is a point on the hyperbola, we have,

Figure 92.1

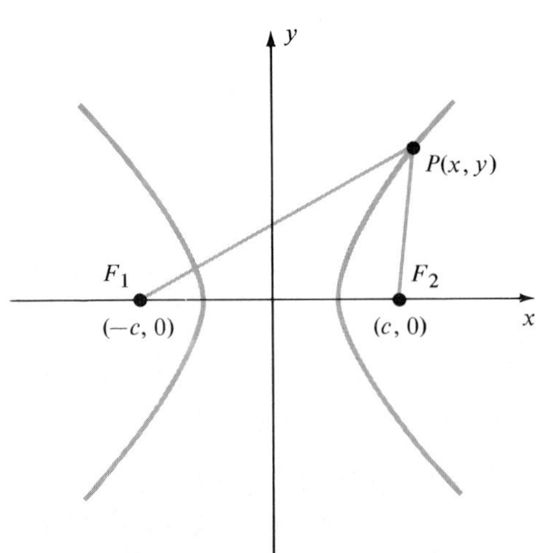

according to our definition, $\overline{PF_1} - \overline{PF_2} = \pm 2a$. Algebraically,

$$\sqrt{(x+c)^2+y^2} - \sqrt{(x-c)^2+y^2} = \pm 2a$$

We now follow the pattern used in the derivation of the equation of the ellipse; we isolate radicals and square as follows:

$$\sqrt{(x+c)^2+y^2} = \pm 2a + \sqrt{(x-c)^2+y^2}$$

$$x^2+2cx+c^2+y^2 = 4a^2 \pm 4a\sqrt{(x-c)^2+y^2} + x^2 - 2cx + c^2 + y^2$$

$$4cx - 4a^2 = \pm 4a\sqrt{(x-c)^2+y^2}$$

$$cx - a^2 = \pm a\sqrt{(x-c)^2+y^2}$$

$$c^2x^2 - 2a^2cx + a^4 = a^2(x^2 - 2cx + c^2 + y^2)$$

$$(c^2-a^2)x^2 - a^2y^2 = a^2(c^2-a^2)$$

$$\frac{x^2}{a^2} - \frac{y^2}{c^2-a^2} = 1 \qquad \text{(A)}$$

From Figure 92.1 we note that in triangle PF_1F_2, $PF_1 < PF_2 + F_1F_2$. Thus

$$PF_1 - PF_2 < F_1F_2$$

or

$$2a < 2c$$

$$a < c$$

Hence $a^2 < c^2$ and

$$c^2 - a^2 > 0$$

Since $c^2 - a^2$ is positive, we replace it by the positive number b^2 in equation (A). Thus, the equation of the hyperbola is

$$\frac{x^2}{a^2} - \frac{y^2}{b^2} = 1 \quad \text{where} \quad b^2 = c^2 - a^2 \qquad (92.1)$$

Now, let us use equation (92.1) to discuss some of the properties of the hyperbola. First, we note that the curve is symmetric with respect to the x axis, the y axis, and the origin. The hyperbola crosses the x axis, for if we substitute $y = 0$, we have $x^2/a^2 = 1$ *or* $x = \pm a$. The points $(a, 0)$ and $(-a, 0)$ are called the **vertices** and the line segment joining these points is called the **transverse axis** of the hyperbola and has a length of $2a$. The center of the hyperbola is the center of the transverse axis. The hyperbola does not cross the y axis, for if we substitute $x = 0$, we have $-y^2/b^2 = 1$, which has no real solutions. Further, if we write equation (92.1) in the form $y^2 = (b^2/a^2)(x^2 - a^2)$, y will not be real when $x^2 - a^2 < 0$ or equivalently, when $-a < x < a$.

The line segment joining the points $(0, b)$ and $(0, -b)$ is called the **conjugate axis**. Every hyperbola also has a pair of asymptotes. For $x^2/a^2 - y^2/b^2 = 1$, the asymptotes are given by $y = \pm(b/a)x$.

A formal proof of this statement is beyond the scope of this text, but let us give some indication of its validity. We begin with $x^2/a^2 - y^2/b^2 = 1$. Dividing by x^2,

$$\frac{1}{a^2} - \frac{y^2}{x^2b^2} = \frac{1}{x^2} \quad \text{or} \quad \frac{y^2}{x^2b^2} = \frac{1}{a^2} - \frac{1}{x^2}$$

Multiplying by b^2,

$$\frac{y^2}{x^2} = \frac{b^2}{a^2} - \frac{b^2}{x^2}$$

Taking the square root of both sides, we have

$$\left|\frac{y}{x}\right| = \sqrt{\frac{b^2}{a^2} - \frac{b^2}{x^2}}$$

For large values of x, y/x is very close to b/a, since b^2/x^2 tends to 0 as x grows "very large." Thus, the asymptotes for the hyperbola are given by $y = \pm(b/a)x$. [***Note:*** The asymptotes are not a part of the curve, they are only an aid in sketching. To sketch the asymptotes quickly, we plot $(\pm a, 0)$ and $(0, \pm b)$ and then construct the rectangle determined by them (Figure 92.2). The diagonals of the rectangle are the asymptotes.]

Figure 92.2

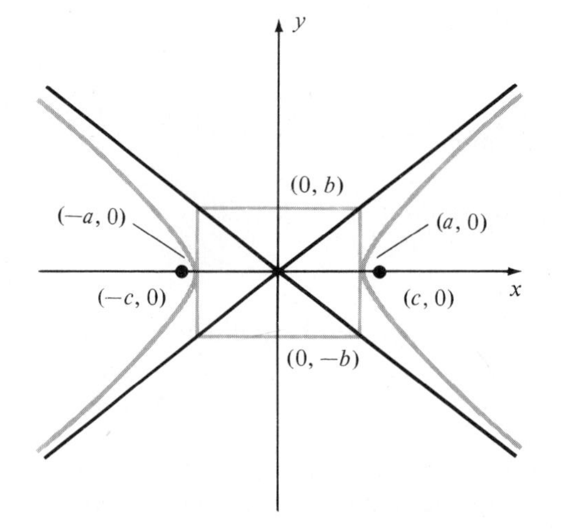

If the foci of the hyperbola are on the y axis at $(0, -c)$ and $(0, c)$, the equation will be

$$\frac{y^2}{a^2} - \frac{x^2}{b^2} = 1 \quad \text{where} \quad b^2 = c^2 - a^2 \tag{92.2}$$

The graph of (92.2) will then have branches opening upward and downward rather than right and left. In this case, the asymptotes are given by $y = \pm(a/b)x$.

Unlike the ellipse, the size of the numbers in the denominator have nothing to do with the determination of a and b. *a^2 is always the denominator of the positive term and b^2 is always the denominator of the negative term.* We carefully note that:

The asymptotes for

$$\frac{x^2}{a^2} - \frac{y^2}{b^2} = 1 \quad \text{are} \quad y = \pm\frac{b}{a}x$$

The asymptotes for

$$\frac{y^2}{a^2} - \frac{x^2}{b^2} = 1 \quad \text{are} \quad y = \pm\frac{a}{b}x$$

To avoid confusion, use the following trick. Replace 1 by 0 in the standard form of the equation and solve for y.

Example 92.1

Find the asymptotes for

$$\frac{x^2}{9} - \frac{y^2}{25} = 1.$$

Solution

If we replace 1 by 0 in the equation, we can solve for y as follows:

$$\frac{x^2}{9} - \frac{y^2}{25} = 0$$

$$\frac{y^2}{25} = \frac{x^2}{9}$$

$$y^2 = \frac{25}{9}x^2 \quad \text{or} \quad y = \pm\frac{5}{3}x$$

Thus, the lines $y = \frac{5}{3}x$ and $y = -\frac{5}{3}x$ are the desired asymptotes. We note again that for graphing purposes its more efficient to first plot the points $(\pm 3, 0)$ and $(0, \pm 5)$ and then construct the rectangle determined by them. The diagonals of this rectangle are the desired asymptotic lines.

Example 92.2

Discuss and sketch

$$\frac{x^2}{16} - \frac{y^2}{9} = 1.$$

Solution
Comparing the given equation to (92.1) we note that $a^2 = 16$, $b^2 = 9$, and $c^2 = a^2 + b^2 = 25$. Thus, the hyperbola has vertices at $(\pm 4, 0)$, foci at $(\pm 5, 0)$, and center at $(0, 0)$. Its asymptotes are $y = \pm\frac{3}{4}x$. The graph is shown in Figure 92.3.

Figure 92.3

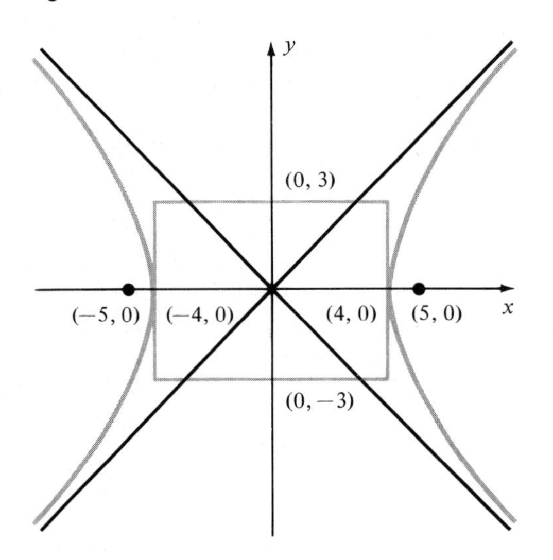

Example 92.3
Discuss and sketch

$$\frac{y^2}{9} - \frac{x^2}{16} = 1$$

Solution
This equation fits form (92.2); $a^2 = 9$, $b^2 = 16$, and $c^2 = 25$. The vertices are on the y axis at $(0, \pm 3)$ and the foci are at $(0, \pm 5)$. The center is at $(0, 0)$ and its asymptotes are $y = \pm\frac{3}{4}x$. The graph is shown in Figure 92.4. ●

Note that the left-hand sides of the equation given in Examples 92.2 and 92.3 differ only by a minus sign. They are called **conjugate hyperbolas**, since the transverse and conjugate axes of one, are respectively, the conjugate and transverse axes of the other.

Example 92.4
Find the equation of the hyperbola with foci at $(\pm 3, 0)$ and vertex at $(2, 0)$.

Figure 92.4

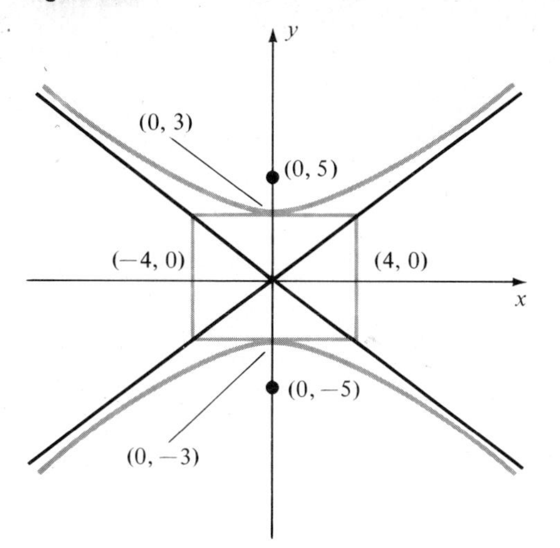

Solution
Since the foci are on the x axis, our equation must be of the form

$$\frac{x^2}{a^2} - \frac{y^2}{b^2} = 1$$

We are given that $a = 2$ and $c = 3$. Therefore, $b^2 = c^2 - a^2 = 9 - 4 = 5$. The equation is $x^2/4 - y^2/5 = 1$. ●

The foci of a hyperbola may be anywhere in the plane. If they are on a line parallel to one of the coordinate axis, an equation similar to that of the ellipse may be written. If the foci are at $(h - c, k)$ and $(h + c, k)$, the center must be at (h, k). We use the methods of Section 89 to introduce the translation $x' = x - h$ and $y' = y - k$ with origin O' at (h, k). In terms of these new coordinates, the equation of the hyperbola is

$$\frac{(x')^2}{a^2} - \frac{(y')^2}{b^2} = 1$$

In terms of the original coordinates, we have

$$\frac{(x - h)^2}{a^2} - \frac{(y - k)^2}{b^2} = 1 \quad \text{where} \quad b^2 = c^2 - a^2 \tag{92.3}$$

When the focal points are on a line parallel to the y axis and the center is at (h, k), our equation is

$$\frac{(y - k)^2}{a^2} - \frac{(x - h)^2}{b^2} = 1 \quad \text{where} \quad b^2 = c^2 - a^2 \tag{92.4}$$

Example 92.5
Discuss and sketch

$$\frac{(x-1)^2}{9} - \frac{(y-2)^2}{16} = 1.$$

Solution
The center of the hyperbola is $(1, 2)$ and its branches open to the right and left; $a^2 = 9$, $b^2 = 16$, and $c^2 = a^2 + b^2 = 25$. Thus, the vertices are $(-2, 2)$ and $(4, 2)$ and the foci are $(-4, 2)$ and $(6, 2)$. The equations of the asymptotes are $y - 2 = \pm\frac{4}{3}(x - 1)$, obtained by replacing 1 by 0 in the original equation. The sketch is shown in Figure 92.5.

Figure 92.5

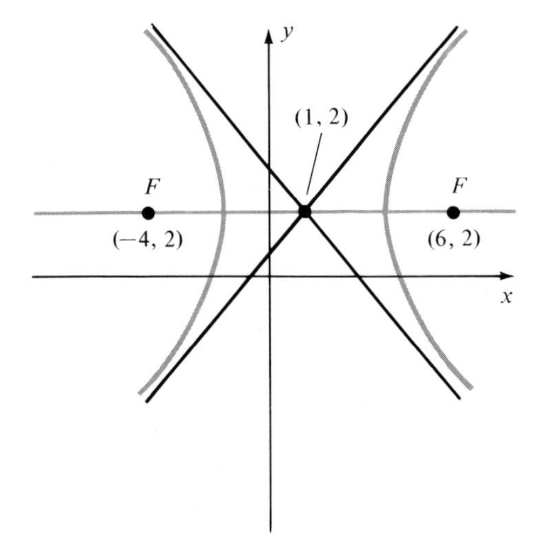

Example 92.6
Discuss and sketch $y^2 - 4x^2 + 2y + 8x - 7 = 0$.

Solution
We complete the square in x and y in order to reduce the equation to standard form.

$$(y^2 + 2y) - 4(x^2 - 2x) = 7$$

$$(y^2 + 2y + 1) - 4(x^2 - 2x + 1) = 7 + 1 - 4$$

$$(y + 1)^2 - 4(x - 1)^2 = 4$$

or

$$\frac{(y+1)^2}{4} - \frac{(x-1)^2}{1} = 1$$

The center of the hyperbola is $(1, -1)$ and its branches open upward and downward; $a^2 = 4$, $b^2 = 1$, and $c^2 = 5$. The vertices are $(1, -3)$ and $(1, 1)$ and the foci are $(1, -1 - \sqrt{5})$ and $(1, -1 + \sqrt{5})$. The equations of the asymptotes are $y + 1 = \pm 2(x - 1)$ or $2x + y - 1 = 0$ and $2x - y - 3 = 0$. The sketch is shown in Figure 92.6.

Figure 92.6

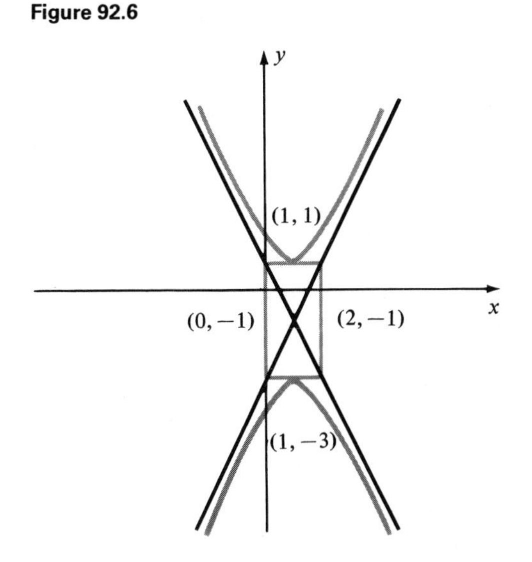

We note that in the general forms for the equation of a hyperbola [(92.1) through (92.4)], the denominators a^2 and b^2 determine in part the equation of the asymptotes. We especially note that if $a = b$, the rectangle used to draw the asymptotes is a square. In this case, the asymptotes are mutually perpendicular and the curve is called an **equilateral hyperbola** (see Exercise 4). The equation

$$xy = c \qquad c \neq 0 \tag{92.5}$$

is a special form of an equation of an equilateral hyperbola whose asymptotes are the coordinate axes. If $c > 0$, the vertices and foci lie on the line $y = x$; if $c < 0$, the vertices and foci lie on the line $y = -x$. If $c = 0$, the equation reduces to $xy = 0$, a degenerate form of the hyperbola consisting of the two straight lines $x = 0$ (y axis) and $y = 0$ (x axis). The graphs of $xy = c$ for $c > 0$ and $c < 0$ are shown in Figure 92.7 on page 566.

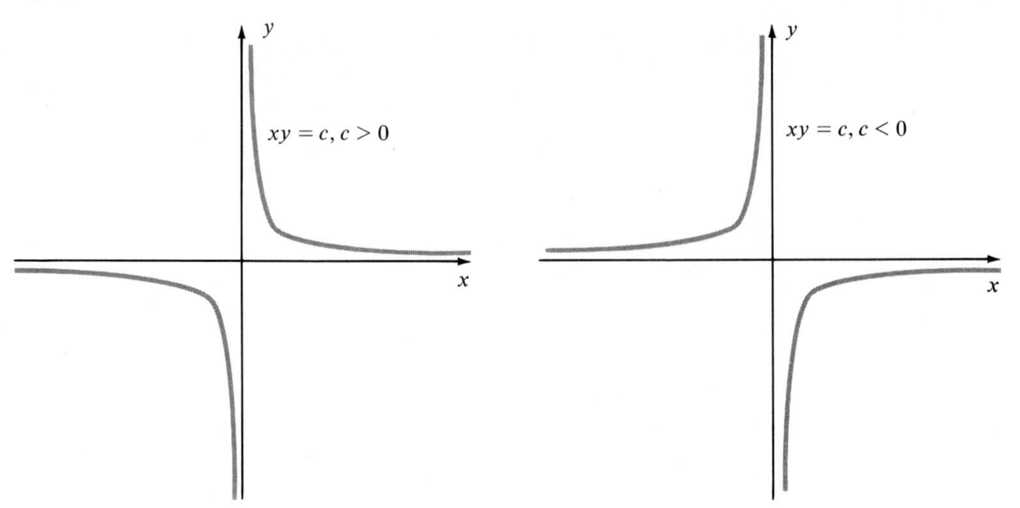

Exercises for Section 92

In Exercises 1 through 20, discuss and sketch each equation. Label the vertices and foci and sketch the asymptotes.

1. $\dfrac{x^2}{16} - \dfrac{y^2}{9} = 1$

2. $\dfrac{y^2}{25} - \dfrac{x^2}{16} = 1$

3. $\dfrac{x^2}{225} - \dfrac{y^2}{49} = 1$

4. $x^2 - y^2 = 1$ (equilateral hyperbola)

5. $\dfrac{y^2}{4} - \dfrac{x^2}{1} = 1$

6. $x^2 - 9y^2 = -36$

7. $9y^2 - 4x^2 = 9$

8. $9x^2 - 16y^2 - 9 = 0$

9. $\dfrac{(y-4)^2}{9} - \dfrac{(x+2)^2}{4} = 1$

10. $\dfrac{(x-3)^2}{16} - \dfrac{(y-1)^2}{25} = 1$

11. $\dfrac{(x-1)^2}{16} - \dfrac{(y+3)^2}{9} = 1$

12. $x^2 - y^2 - 2x + 2y - 2 = 0$

13. $y^2 - 4x^2 + 4y - 16x - 28 = 0$

14. $5x^2 - 4y^2 + 20x + 8y = 4$

15. $x^2 - 4y - 9y^2 = 0$

16. $36x^2 - 36y^2 + 24x + 36y + 31 = 0$

17. $4y^2 - x^2 - 2y - x - 16 = 0$

18. $4x^2 - y^2 - 2x - y - 16 = 0$

19. $4x^2 = y^2 - 4y + 8$

20. $9x^2 - 16y^2 - 18x + 32y - 151 = 0$

In Exercises 21 through 30, find an equation of the hyperbola described.

21. Center $(0, 0)$, focus $(8, 0)$, vertex $(6, 0)$

22. Center $(0, 0)$, vertex $(3, 0)$, asymptotes $y = \pm\frac{2}{3}x$

23. Vertices $(\pm 2, 0)$, focus $(4, 0)$

24. Asymptotes $y = \pm\frac{2}{3}x$, vertex $(9, 0)$

25. Vertices $(0, \pm 3)$, foci $(0, \pm 5)$

26. Vertices $(\pm 3, 0)$, foci $(\pm 5, 0)$

27. Center $(6, 1)$ and $(0, 1)$, focus $(8, 1)$

28. Asymptotes $4x - 3y + 13 = 0$ and $4x + 3y - 5 = 0$, focus $(-1, -2)$

29. Transverse axis of length 8, foci $(0, \pm 5)$

30. Center $(2, 5)$, vertex $(2, 7)$, focus $(2, 0)$

In Exercises 31 through 34, solve each system of equations algebraically.

31. $\begin{cases} x^2 - y^2 + 8x + 2y + 9 = 0 \\ x^2 - y^2 - 4x + 2y - 15 = 0 \end{cases}$

32. $\begin{cases} x^2 + y^2 = 16 \\ x^2 = 6y \end{cases}$

33. $\begin{cases} x^2 + y^2 = 16 \\ x - y = 4 \end{cases}$

34. $\begin{cases} y^2 - 4x^2 + 2y + 8x - 7 = 0 \\ \dfrac{(x - 1)^2}{1} + \dfrac{(y + 1)^2}{4} = 1 \end{cases}$

In Exercises 35 and 36, solve each system of equations graphically. Estimate the solutions to the nearest tenth.

35. $\begin{cases} 4x^2 + 9y^2 = 36 \\ x^2 - y^2 = 1 \end{cases}$

36. $\begin{cases} x^2 + y^2 = 4 \\ x - 2y - 1 = 0 \end{cases}$

37. The current in an electrical circuit is inversely proportional to the resistance. If the current is 1.4 A when the resistance is 20 Ω, evaluate the proportionality constant and sketch a graph of current versus resistance.

38. For an ideal gas at constant temperature, we have $pv = K$, where p represents pressure and v volume. For a certain gas the pressure is 3 atm when the volume is 4 liters. Sketch a graph of pressure versus volume.

39. Two listening posts, A and B, are 5000 m apart. An explosion is heard at these posts and from the difference in times it is determined that the site of the explosion is 800 m closer to A than to B. The site of the explosion lies on a hyperbolic curve. Find the equation of the hyperbola.

SECTION 93

General Second-Degree Equations: $Ax^2 + Bxy + Cy^2 + Dx + Ey + F = 0$

In Sections 89, 90, 91, and 92, we demonstrated that the circle, parabola, ellipse, and hyperbola could be described by second-degree equations in two variables. Now let us show that the graph of such an equation is always one of the conics (or a degenerate form).

We can write the general second-degree equation in x and y as

$$Ax^2 + Bxy + Cy^2 + Dx + Ey + F = 0 \tag{93.1}$$

where A, B, and C are not all zero.

In the case where $B = 0$, equation (93.1) reduces to $Ax^2 + Cy^2 + Dx + Ey + F = 0$. Now it follows from the preceding sections that after completing the square, we must obtain the equation of one of the conics. We summarize these results with the following theorem. Excluding degenerate cases, we have:

Theorem 93.1
The graph of

$$Ax^2 + Bxy + Cy^2 + Dx + Ey + F = 0$$

will be a

1. *Circle if* $A = C \neq 0$ *and* $B = 0$.
2. *Ellipse if* $A \neq C$ *but* A *and* C *have the same sign, and* $B = 0$.
3. *Parabola if* $A = 0$ *or* $C = 0$ *(but not both) and* $B = 0$.
4. *Hyperbola if* A *and* C *have opposite signs and* $B = 0$, *or if* $A = C = 0$, *and* $B \neq 0$.

In those cases where $B \neq 0$, equation (93.1) represents a conic section that has been rotated. One such example is shown in Figure 92.7.

Example 93.1
Identify the conic whose equation is

$$4x^2 - 5y^2 - 8x - 10y - 21 = 0$$

Solution
We note from the equation that $A = 4$, $C = -5$, and $B = 0$. Thus, by Theorem 93.1, the equation must represent a hyperbola. We shall verify our observation by completing the squares in x and y and writing the equation in standard form.

$$4(x^2 - 2x) - 5(y^2 + 2y) = 21$$

$$4(x^2 - 2x + 1) - 5(y^2 + 2y + 1) = 21 + 4 - 5$$

$$4(x - 1)^2 - 5(y + 1)^2 = 20$$

Dividing by 20 yields

$$\frac{(x - 1)^2}{5} - \frac{(y + 1)^2}{4} = 1$$

which is the equation of a hyperbola with center at $(1, -1)$ and branches opening left and right.

Example 93.2

(a) The equation $x^2 + y^2 - 4x + 6y - 7 = 0$ is the equation of a circle, since the coefficients of x^2 and y^2 are equal. ($A = C = 1$.)

(b) The equation $x^2 + 2y^2 - 4x + 8y - 11 = 0$ is the equation of an ellipse since the coefficients of x^2 and y^2 are unequal but have the same sign. ($A = 1, C = 2, A \neq C$.)

(c) The equation $3x^2 + 3y - 4x - 6 = 0$ is the equation of a parabola, since there is an x^2 term but no y^2 term. ($A = 3, C = 0$.)

(d) The equation $3x^2 - 4y^2 + 5x - 6y + 7 = 0$ is the equation of a hyperbola, since the coefficients of x^2 and y^2 have opposite signs ($A = 3$, $C = -4$.)

We note that in (a) through (d), $B = 0$.

Example 93.3

Identify the equation $x^2 - 4y^2 - 4x - 8y = 0$.

Solution

Since the coefficients of x^2 and y^2 have opposite signs and $B = 0$. Theorem 93.1 tells us that the equation must represent a hyperbola or a degenerate form. To see that we do, in fact, have a degenerate form of the hyperbola, we complete the square in x and y as follows:

$$(x^2 - 4x + 4) - 4(y^2 + 2y + 1) = 0 + 4 - 4$$

Thus,

$$(x - 2)^2 - 4(y + 1)^2 = 0 \tag{A}$$

Since the left side of (A) is the difference of two squares, it factors into

$$[(x - 2) + 2(y + 1)] \cdot [(x - 2) - 2(y + 1)] = 0$$

Setting each factor equal to zero yields

$$(x - 2) + 2(y + 1) = 0$$

and

$$(x - 2) - 2(y + 1) = 0$$

Since the equations are linear, they represent two straight lines. The graph of the lines is shown in Figure 93.1.

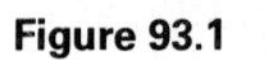

Figure 93.1

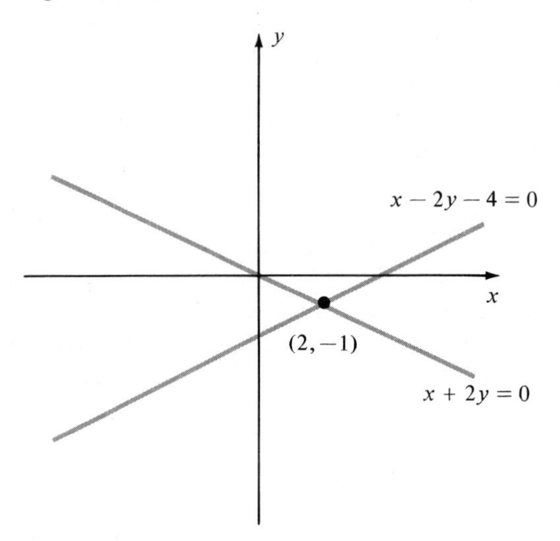

Exercises for Section 93

In Exercises 1 through 20, identify each equation as a circle, parabola, ellipse, or hyperbola.

1. $x^2 + y^2 + 6x + 4y = 12$

2. $x^2 + 4y^2 + 6x + 4y = 12$

3. $x^2 - 4y^2 + 6x - 7 = 0$

4. $x^2 + 4x = y^2 + 6y - 3$

5. $y^2 = 4x + 8$

6. $x^2 + y^2 + 8x = 6$

7. $3x^2 + 3y^2 + 6x + 9y + 100 = 0$

8. $4x^2 - y^2 + 4x - 6y - 11 = 0$

9. $x^2 + y^2 - 2y = 10$

10. $x^2 + 3y^2 - 4x = 5$

11. $x^2 - 3y^2 = 2$

12. $x^2 = y - 5y^2$

13. $x^2 = 5y - y^2$

14. $(x + 2)^2 = x^2 + y^2 + 4$

15. $x(1 - x) = 2(x^2 + y^2)$

16. $x(3 - x) = y(2 - y)$

17. $-x^2 - y^2 + 6x - 4y = 9$

18. $x^2 - 4 = 3y - 2x$

19. $x^2 = 4x + y^2$

20. $3(x^2 + 2y^2) = 4(x - y)$

SECTION 94
Graphs of Polar Equations

We have studied the straight line, the conic sections, and their corresponding graphs. Now let us consider curves in the plane defined by an equation in polar coordinates.

If the given equation is of the form $F(r, \theta) = 0$, the set of all points $P(r, \theta)$ that satisfy the equation will be called the **polar graph**. Keep in mind that the point $P(r, \theta)$ has many pairs of coordinates. However, P is on the graph if only one pair satisfies the equation. If the given equation is of the form $r = f(\theta)$, the previous remarks still hold. In this instance we usually think of θ as the independent variable and of r as the dependent variable.

For the sake of reference and convenience, we again list the relationships between rectangular and polar coordinates (Figure 94.1):

$$\left.\begin{aligned} x &= r\cos\theta \\ y &= r\sin\theta \end{aligned}\right\} \tag{94.1}$$

$$\left.\begin{aligned} &r^2 = x^2 + y^2 \\ &\tan\theta = \frac{y}{x} \quad \text{or} \quad \theta = \arctan\frac{y}{x} \qquad x \neq 0 \end{aligned}\right\} \tag{94.2}$$

Figure 94.1

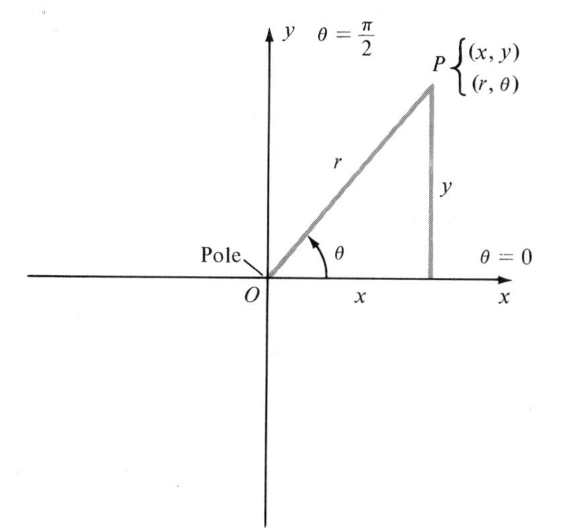

These equations enable us to change from one coordinate system to the other. If we hold r fixed at $r = r_1$ and let θ vary, we have a circle with center at the pole and radius of r_1. If we hold θ fixed at $\theta = \theta_1$ and let r vary, we have a line through the pole making an angle of θ_1 with the polar axis. Thus, we describe the horizontal or polar axis in terms of the equation $\theta = 0$ and the line through the pole and perpendicular to the polar axis (the y axis in rectangular coordinates) in terms of the equation $\theta = \pi/2$. See Figure 94.1.

If we begin with an equation that gives r explicitly in terms of θ, $r = f(\theta)$, we can arrive at a sketch by plotting points. We may obtain as many points as we wish simply by substituting values of θ and then computing the corresponding values for r. This method may be cumbersome and ineffective at times, so we attempt to simplify and order our work by using the following procedures in dealing with the properties of polar graphs.

These properties include tests for symmetry, extent (analogous to evaluating x and y rectangular coordinates), points at the pole, and a table of values.

I. Test for Symmetry

(a) The curve is *symmetric with respect to the pole* if the equation is unaltered when r is replaced by $-r$ (Figure 94.2).

(b) The curve is *symmetric with respect to the polar axis* if the equation is unaltered when θ is replaced by $-\theta$ (Figure 94.3).

(c) The curve is *symmetric with respect to the line* $\theta = \pi/2$ if the equation is unaltered when θ is replaced by $\pi - \theta$ (Figure 94.4).

If our given expression is $F(r, \theta) = 0$, we may formulate our symmetry tests as follows:

(a) If $F(r, \theta) = F(-r, \theta)$, the curve is symmetric with respect to the pole.

(b) If $F(r, \theta) = F(r, -\theta)$, the curve is symmetric with respect to the polar axis.

(c) If $F(r, \theta) = F(r, \pi - \theta)$, the curve is symmetric with respect to the line $\theta = \pi/2$.

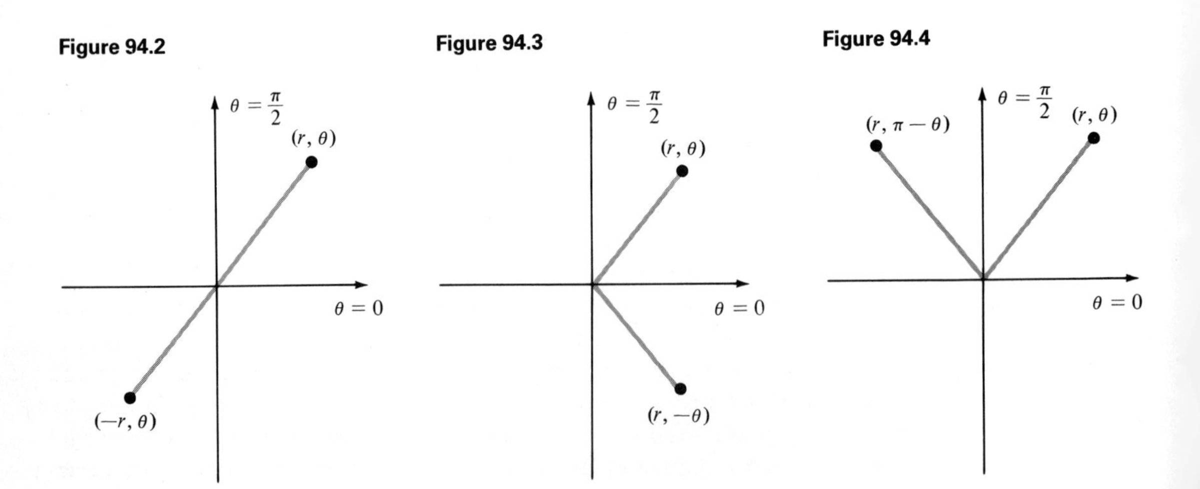

Our future discussion dealing with possible symmetries will be restricted to line symmetry with respect to the polar axis and the line $\theta = \pi/2$ and point symmetry with respect to the pole. It should be noted that these tests *are not unique* (see Exercises 29 and 30).

Tests (b) and (c) can often be used quickly and simply. Since $\cos(-\theta) = \cos\theta$ and $\sin(\pi - \theta) = \sin\theta$, it follows that any equation in which θ appears only as a function of $\cos\theta$ or only as a function of $\sin\theta$ must be symmetric with respect to the polar axis and the line $\theta = \pi/2$, respectively.

II. Determine the **extent** of the function for θ and r, noting especially the maximum and minimum values of r.

III. Find the values of θ for which the curve passes through the pole (i.e., set $r = 0$ and solve for θ).

IV. Construct a table of values.

The procedure outlined in steps I through IV will normally enable us to graph the given equation.

Example 94.1

Discuss and sketch the graph of $r = 4\cos\theta$.

Solution

Symmetry: We observe that the graph is symmetric with respect to the polar axis, since the equation is unaltered when θ is replaced by $-\theta$.
Extent: There is no restriction on θ, but since $\cos\theta$ has a period of 2π, we need only consider the interval $0 \le \theta \le 2\pi$. In this interval $-1 \le \cos\theta \le 1$ (Table 94.1). Hence, r varies from -4 to 4. The maximum value, 4, occurs when $\theta = 0$, and the minimum value, -4, occurs when $\theta = 180°$.

Table 94.1

θ	0°	30°	45°	60°	90°	180°
r	4	3.48	2.84	2	0	−4

Points at the Pole: The curve passes through the pole since $r = 0$ when $\theta = 90°$.

Graphing Comments

As θ varies from 0° to 90°, $\cos\theta$ decreases from 1 to 0, and r decreases from 4 to 0. The point $P(r, \theta)$ thus traces a continuous arc of the curve that lies above the polar axis and connects the point (4, 0) with the origin. We make use of our symmetry to graph the corresponding arc of the curve that lies below the axis.

For $90° < \theta \le 180°$, r becomes negative, achieving a minimum value of -4 when $\theta = 90°$. But the use of our symmetry has already given us the points in this interval, and we have no need to plot them a second time. For this reason, Table 94.1 ends at $\theta = 90°$.

As θ varies from 180° to 360°, the curve is traced a *second time*. The graph is a circle, shown in Figure 94.5. We verify this by converting our given equation to rectangular form as follows: Since $\cos\theta = x/r$, $r = 4\cos\theta$ is equivalent to $r = 4x/r$. Multiplying by r, we have $r^2 = 4x$. But $r^2 = x^2 + y^2$, and thus $x^2 + y^2 = 4x$. Completing the square of the terms in x, we have $(x - 2)^2 + y^2 = 4$, which is the equation of a circle with center (2, 0) and radius 2.

Figure 94.5

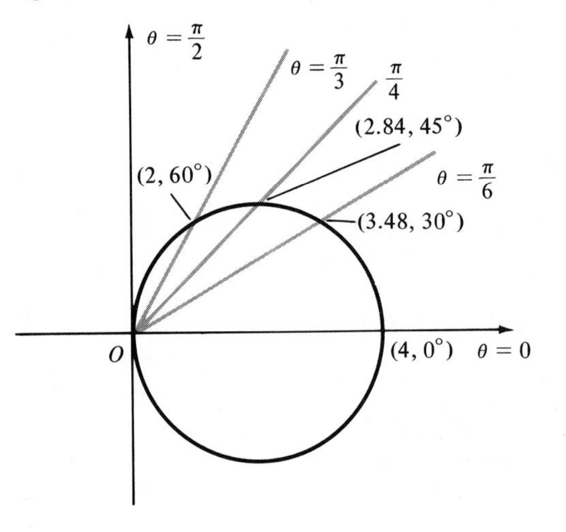

Remark 94.1. *The graph of* $r = a\cos\theta$, $a > 0$, *will always be a circle with center* $(a/2, 0)$, *radius* $a/2$, *and passing through the pole. The graph of* $r = a\sin\theta$, $a > 0$, *is the same figure rotated through an angle of* $\pi/2$ *or* 90°.

Example 94.2

Discuss and sketch the graph of $r = 2(1 + \cos\theta)$.

Solution

Symmetry: The graph is symmetric with respect to the polar axis, since the equation is unaltered when θ is replaced by $-\theta$.

Extent: Here, again, there is no restriction on θ. As before, we consider θ only in the interval $0 \le \theta \le 2\pi$, owing to the periodicity of cosine. Since $-1 \le \cos\theta \le 1$, the values for r vary from 0 to 4 (see Table 94.2). The maximum value, 4, occurs when $\theta = 0$, and the minimum value, 0, occurs when $\theta = 180°$ or π.

Points at the Pole: The curve passes through the pole since $r = 0$ when $\theta = \pi$.

Table 94.2

θ	0°	30°	45°	60°	90°	120°	135°	150°	180°
r	4	3.74	3.42	3	2	1	0.58	0.26	0

Graphing Comments

As θ varies from 0 to 180°, $\cos\theta$ decreases from 1 to -1 and r decreases from 4 to 0. Since the curve is symmetric with respect to the polar axis, Table 94.2 ends at $\theta = 180°$. We obtain the lower half of the curve by a reflection in the polar axis (Figure 94.6). The curve is called a **cardiod** because of its heart-shaped appearance.

Figure 94.6

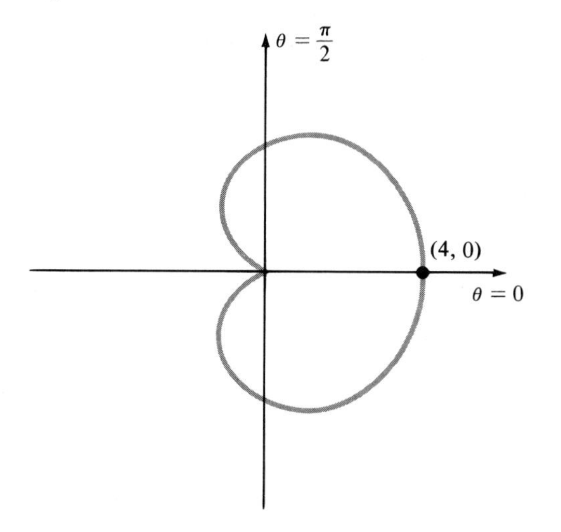

The rectangular form of the equation that is satisfied by the points on this curve can be found as follows. Since $\cos\theta = x/r$, we have $r = 2(1 + x/r)$. Multiplying by r yields $r^2 = 2(r + x)$ or $r^2 - 2x = 2r$. Squaring both sides, we have

$$(r^2 - 2x)^2 = 4r^2$$

Thus,

$$(x^2 + y^2 - 2x)^2 = 4(x^2 + y^2)$$

The last expression gives us some idea of why we might choose to work with the polar form rather than with the rectangular form of the curve.

Example 94.3

Discuss and sketch $r = 2 + \sin\theta$.

Solution

Symmetry: If we replace θ by $\pi - \theta$, the equation is unchanged and thus the graph is symmetric with respect to the line $\theta = \pi/2$.

Extent: There is no restriction on θ, so we choose the interval $0 \leq \theta \leq 2\pi$. In this interval $-1 \leq \sin\theta \leq 1$, and thus r varies from 1 to 3 (see Table 94.3). The maximum value, 3, occurs when $\theta = 90°$, and the mimimum value, 1, occurs when $\theta = 270°$, or, if we wish, when $\theta = -90°$.

Table 94.3

θ	0°	30°	45°	60°	90°	−30°	−45°	−60°	−90°
r	2	2.5	2.71	2.87	3	1.5	1.29	1.13	1

Points at the Pole: The curve does not pass through the pole because $\sin\theta$ is never equal to -2.

Graphing Comments

As θ varies from 0 to 90°, $\sin\theta$ increases from 0 to 1 and r increases continuously from 2 to 3. Since the curve is symmetric with respect to the line $\theta = \pi/2$, we consider some special values for θ in the interval $-90° \leq \theta < 0$ rather than in the interval $90° < \theta \leq 180°$. Once we have a sketch of the curve to the right of $\theta = 90°$, the remaining part of the curve is obtained by a reflection in the line $\theta = \pi/2$ (Figure 94.7). This curve is called a **limacon**. The rectangular form of the equation is $(x^2 + y^2 - y)^2 = 4(x^2 + y^2)$.

Figure 94.7

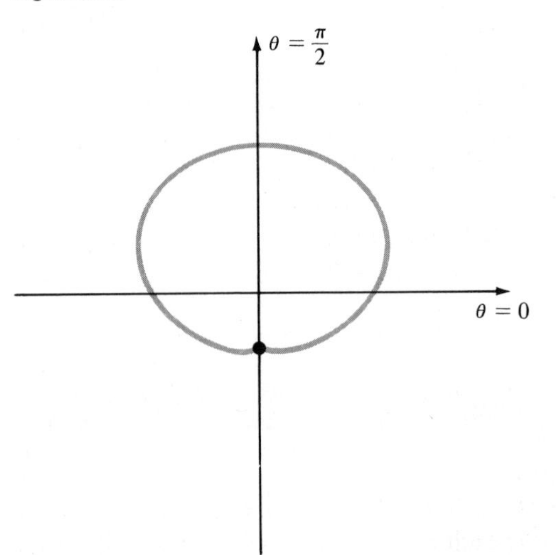

Example 94.4

Discuss and sketch $r = 1 - 2\cos\theta$.

Solution

Symmetry: The curve is symmetric with respect to the polar axis since the equation is unaltered when θ is replaced by $-\theta$.

Extent: There is no restriction on θ, so we consider θ only in the interval $0 \leq \theta \leq 2\pi$. Since $-1 \leq \cos\theta \leq 1$, r varies from -1 to 3. The maximum value 3 occurs when $\theta = 180$, and the minimum value -1 occurs when $\theta = 0$ (Table 94.4).

Points at the Pole: r will be 0 when $1 - 2\cos\theta = 0$. Solving this equation, we have $\cos\theta = \frac{1}{2}$ and thus $\theta = 60°$ and $\theta = 300°$. The curve will pass through the pole twice, once when $\theta = 60°$ and again when $\theta = 300°$.

Table 94.4

θ	0	30°	45°	60°	90°	120°	135°	150°	180°
r	-1	-0.74	-0.42	0	1	2	2.42	2.74	3

In plotting these points we must be careful to note that the quadrant in which the points lie is *not* determined by the signs of the polar coordinates (as is the case with rectangular coordinates) but by the *size* of θ and the *sign* of r.

Graphing Comments

As θ varies from 0 to 180°, $\cos\theta$ decreases from 1 to -1 and r increases from -1 to 3. We plot the points from Table 94.4 and carefully connect them with a smooth curve. The curve for the interval $180° < \theta \leq 360°$ is obtained by making use of our symmetry (Figure 94.8 on page 578). The curve is a limaçon.

The rectangular form of the equation is $(x^2 + y^2 + 2x)^2 = x^2 + y^2$.

The graphs of the equation shown in Examples 94.2 through 94.4 are specific examples of the graphs of equations of the form $r = a \pm b\cos\theta$ or $r = a \pm b\sin\theta$, $a > 0$. Such graphs are, in general, limaçons. In the case where $a = b$, we have a special type of limaçon, the cardiod. If $a > b$, we have a limaçon that does not pass through the pole (Figure 94.7). If $a < b$, the limaçon passes through the pole twice and has an interior loop, as in Example 94.4.

Example 94.5

Discuss and sketch the graph of $r^2 = 4\cos 2\theta$.

Solution

Symmetry: We observe that the equation is unchanged when r is replaced by $-r$, and therefore we have symmetry with respect to the pole. In

Figure 94.8

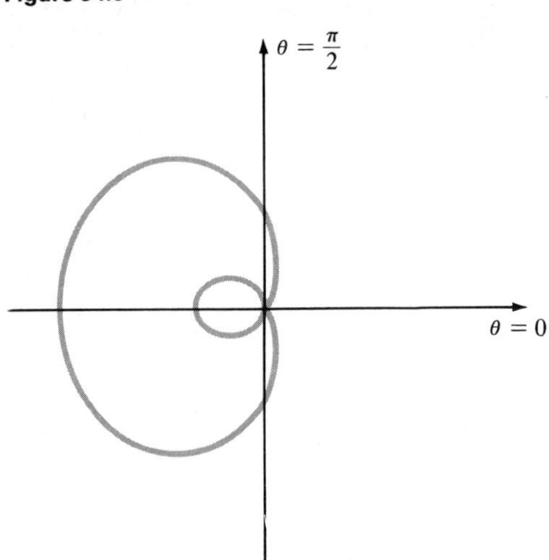

addition, since $\cos 2(-\theta) = \cos 2\theta$, we have symmetry with respect to the polar axis.

Extent: Since 2θ varies from 0 to 360° as θ varies from 0 to 180°, it appears that the entire curve can be obtained by examining values of θ in the interval $0 \leq \theta \leq 180°$. But r^2 can never be negative, and $\cos 2\theta$ is negative for $45° < \theta < 135°$. Thus, we consider only those values for θ in the interval $0 \leq \theta \leq 45°$ (see Table 94.5). For values of θ in this interval, r has a maximum value of 2 when $\theta = 0$ and a minimum value of 0 when $\theta = 45°$.

Table 94.5

θ	0	15°	$22\frac{1}{2}°$	30°	45°
2θ	0	30°	45°	60°	90°
r	2	1.86	1.68	1.42	0

Points at the Pole: $r = 0$ when $4 \cos 2\theta = 0$ or $\cos 2\theta = 0$. Thus, $2\theta =$ 90° or 270° and $\theta = 45°$ and 135°. The curve passes through the pole twice, once when $\theta = 45°$ and again when $\theta = 135°$.

Graphing Comments

As θ varies from 0 to $\pi/4$, $\cos 2\theta$ decreases from 1 to 0 and r decreases from 2 to 0. With these considerations and the points from Table 94.5, we have the quarter of the curve that lies in the first quadrant. The remaining curve is obtained through the use of symmetry (Figure 94.9). This curve is

Figure 94.9

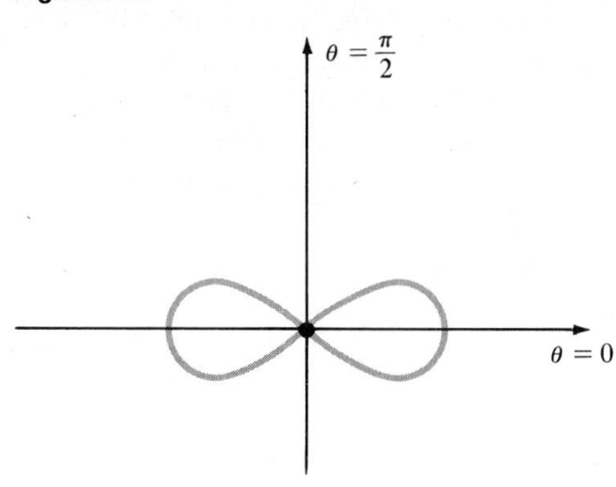

called a **lemniscate**. The rectangular form of the given equation is $(x^2 + y^2)^2 = 4(x^2 - y^2)$.

Graphs of equations of the form $r^2 = a^2 \cos 2\theta$ or $r^2 = a^2 \sin 2\theta$, $a > 0$, are lemniscates. The graph of $r^2 = a^2 \sin 2\theta$ is a rotation through an angle of 45° of the graph of $r^2 = a^2 \cos 2\theta$.

Example 94.6

Discuss and sketch $r = 2 \sin 3\theta$.

Solution

Symmetry: The equation is unaltered when θ is replaced by $\pi - \theta$, and thus the curve is symmetric with respect to the line $\theta = \pi/2$.

Extent: There is no restriction on θ, so we might choose the interval $0 \le \theta \le 360°$. But $\sin 3\theta$ has a period of 120°. Thus, we need only consider θ in the interval $0 \le \theta \le 120°$. We further note from the period of $\sin 3\theta$ that whatever figure appears for the interval $0 \le \theta \le 120°$ is repeated in the intervals $120° \le \theta \le 240°$ and $240° \le \theta \le 360°$. Thus, the remaining parts of the curve are obtained by a rotation through the angle of 120°.

In the interval $0 \le \theta \le 120°$, the maximum value for r is 2 and occurs when $3\theta = 90°$ or $\theta = 30°$. The minimum value is -2 and occurs when $\theta = 90°$.

Points at the Pole: $r = 0$ when $\sin 3\theta = 0$. Thus, $\theta = 0$; 60°, 120°, and 180°. The curve passes through the pole four times.

Table 94.6

θ	0°	10°	15°	20°	30°	40°	45°	50°	60°
3θ	0°	30°	45°	60°	90°	120°	135°	150°	180°
r	0	1	1.42	1.74	2	1.74	1.42	1	0

Graphing Comments

The points from Table 94.6 give us only the first loop of the curve shown in Figure 94.10, but this is sufficient since the remaining two loops may be obtained by a rotation of this initial loop through an angle of 120°. If our table of values had included special values of θ for $0 \leq \theta \leq 120°$, we would have obtained the first and third loops of the curve. The second loop could be graphed by then using the symmetry with respect to the line $\theta = \pi/2$. This type of curve is called a *petal curve*. In Example 94.6 we have a curve of three petals. Its equation in rectangular form is given by $(x^2 + y^2)^2 = 2y(3x^2 - y^2)$. The graph of $r = 3 \sin 2\theta$ is shown in Figure 94.11. The details are left until Exercise 9.

Figure 94.10

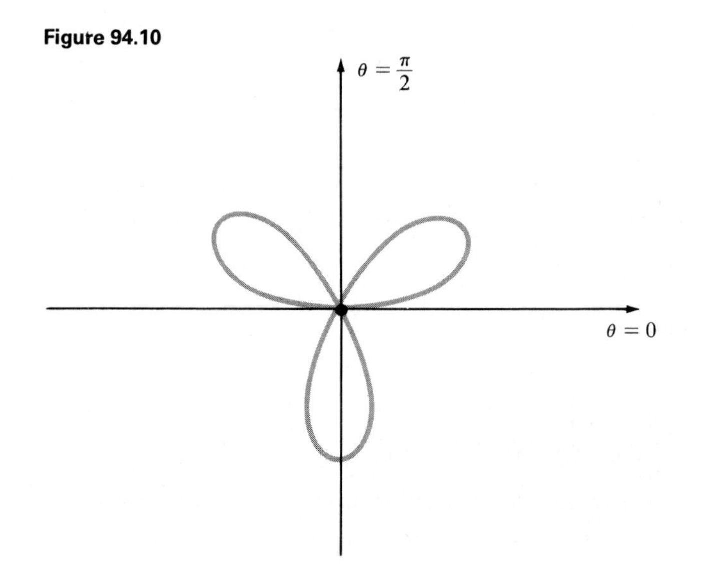

Figure 94.11

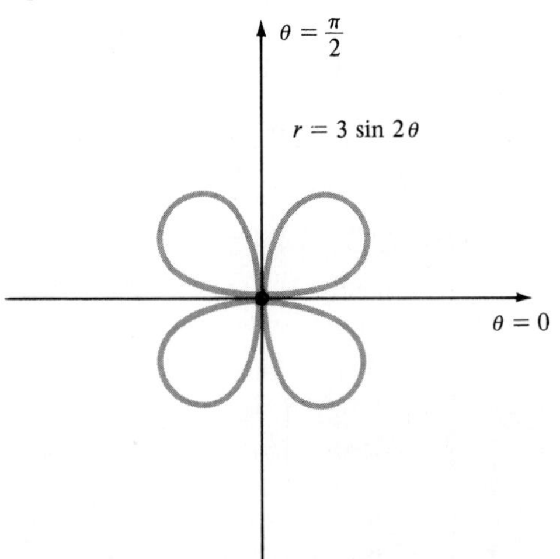

Curves of equations of the form $r = a \sin n\theta$ or $r = a \cos n\theta$, $a > 0$, n a positive integer, are generally called petal curves. If n is odd, the curve has n petals; if n is even, the curve has $2n$ petals. In the special case where $n = 1$, we have a circle passing through the pole.

Suppose that we wish to find the points of intersection of two curves whose equations are in polar form. We solve the equations simultaneously to obtain pairs of numbers that satisfy both equations. Although these points must be points of intersection of the curves, some points of intersection cannot be found in this way. The following example illustrates the problem.

Example 94.7

Find the points of intersection of the circle $r = \cos\theta$ and the cardiod $r = 1 - \cos\theta$.

Solution

Equating r in the pair of equations yields $\cos\theta = 1 - \cos\theta$ or $\cos\theta = \frac{1}{2}$. Thus, $\theta = 60°, 300°$ for $0 \leq \theta \leq 360°$. This gives us the points $(\frac{1}{2}, 60°)$ and $(\frac{1}{2}, 300°)$. The graphs of the two curves are shown in Figure 94.12.

As we see from Figure 94.12, there are three points of intersection. Both curves pass through the pole, but this point was not obtained by algebraic means. The reason for this lies in the fact that the pole has many different representations. On the curve $r = \cos\theta$, it is represented by $(0, 90°)$, and on the curve $r = 1 - \cos\theta$, it is represented by $(0, 0)$. Thus, although the pole does lie on both curves, it does not have a common representation that satisfies both equations. For this reason the intersections of polar equations are generally determined not only by algebraic methods, but also by graphing the curves. In this way we can be sure that we have found all points of intersection. ●

Figure 94.12

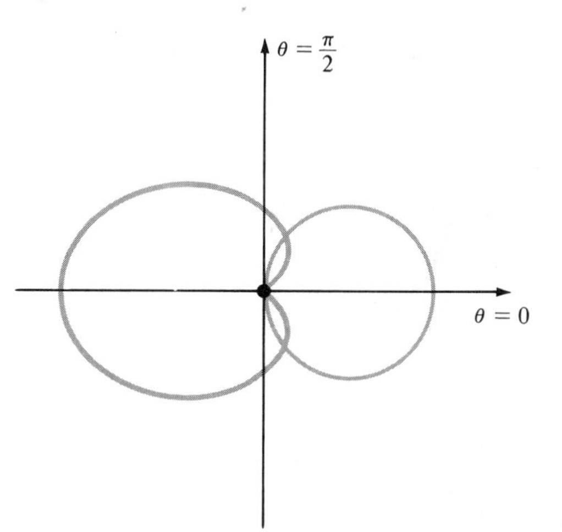

Figure 94.13

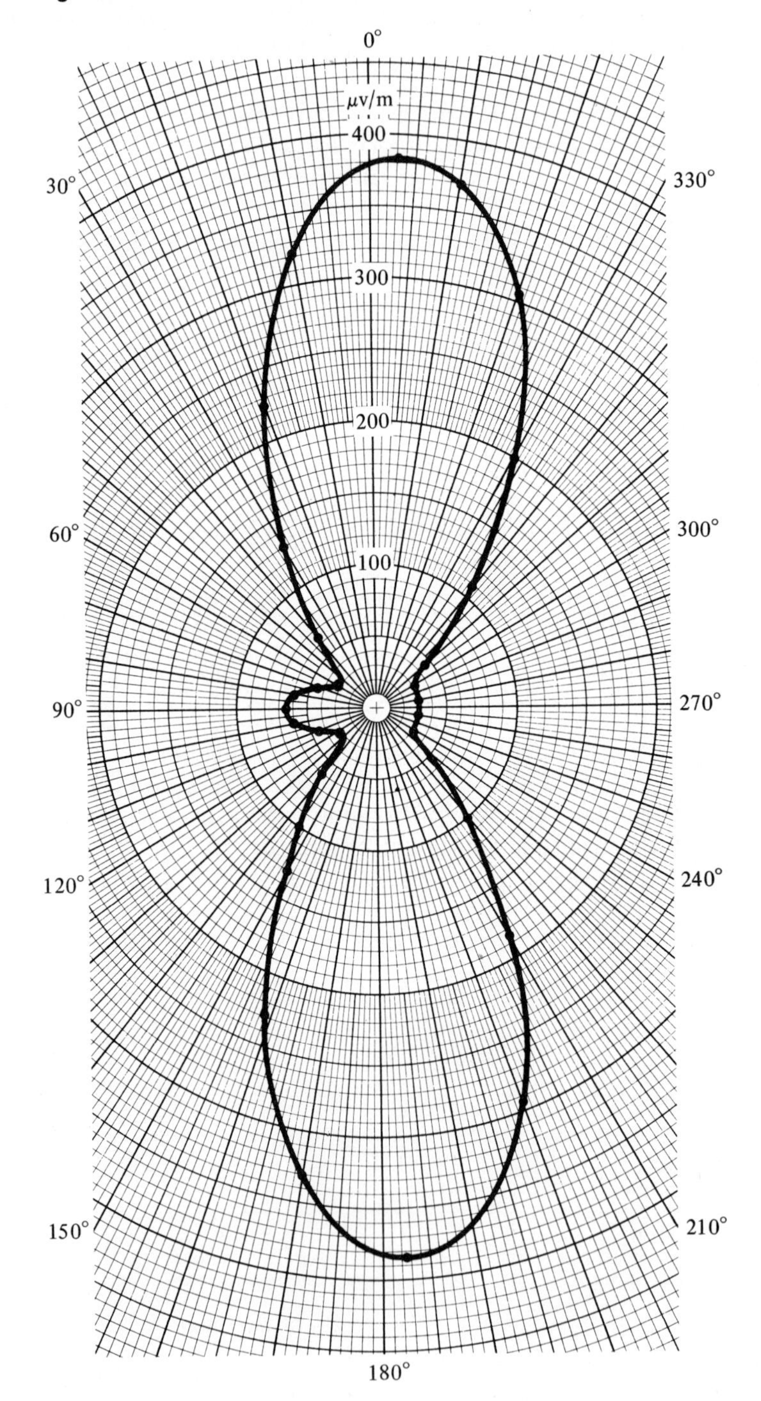

Polar coordinates are particularly useful in applications where energy is received or radiated in all directions. Such an application is found in the output of an antenna. As an illustration of this, consider the following example.

Example 94.8

The following data show the field strength of a broadcast station 1 mile from the antenna as a function of direction. Plot a graph of the antenna pattern.

Direction	*Field Strength* (μ V/m)	*Direction*	*Field Strength* (μ V/m)
10°	320	200°	295
20°	225	210°	185
30°	130	220°	100
40°	62	240°	30
60°	30	260°	30
70°	45	290°	30
80°	60	300°	30
90°	65	310°	45
100°	60	320°	110
110°	45	330°	200
130°	30	340°	305
140°	60	350°	370
150°	130	357°	380
160°	230		
170°	330		
183°	385		

Solution

Each data point is plotted on Figure 94.13 and connected to show the antenna pattern.

Exercises for Section 94

In Exercises 1 through 25, discuss and sketch each equation.

1. $r = 2$

2. $r = \dfrac{\pi}{2}$

3. $\theta = \dfrac{\pi}{6}$

4. $\theta = 2$

5. $r = \cos\theta$

6. $r = \sin\theta$

7. $r = 1 + \sin\theta$

8. $r = \cos 2\theta$

9. $r = 3\sin 2\theta$

10. $r = 2\cos 3\theta$

11. $r = 4 \cos 4\theta$

12. $r = 2(1 - \cos \theta)$

13. $r = 2(1 + \cos \theta)$

14. $r = 1 - 2 \sin \theta$

15. $r = 2 - \sin \theta$

16. $r = 2 - 2 \sin \theta$

17. $r = 1 + 3 \cos \theta$

18. $r = 2 - \cos \theta$

19. $r^2 = 4 \cos 2\theta$

20. $r^2 = 6 \sin 2\theta$

21. $r^2 = a^2 \cos 2\theta, a > 0$

22. $r^2 = a^2 \sin 2\theta, a > 0$

23. $r = \dfrac{1}{1 - \cos \theta}$ (parabola)

24. $r = a\theta, a > 0$ (spiral of Archimedes)

25. $r = \dfrac{1}{2 - \cos \theta}$ (ellipse)

26. Determine the polar equation of each of the following, and sketch.
(a) $x^2 + y^2 - 2ay = 0$
(b) $(x^2 + y^2)^2 = x^2 - y^2$
(c) $(x^2 + y^2)^2 = 2a^2xy$
(d) $(x^2 + y^2)^2 = 4a^2x^2y^2$

27. Determine the Cartesian equation of each of the following, and sketch.
(a) $r = \sin \theta + \cos \theta$
(b) $r = 4 \cos \theta$
(c) $\theta = \dfrac{\pi}{4}$
(d) $r^2(16 \cos^2 \theta + 9 \sin^2 \theta) = 144$
(e) $r = \dfrac{9}{5 - 4 \cos \theta}$

28. Sketch the following pairs of equations and find all points of intersection. Use the results of Exercises 1 through 25 where appropriate.
(a) $r = \cos 2\theta$ and $r = \sin \theta$
(b) $r = 1 - \sin \theta$ and $r = 1 - \cos \theta$
(c) $r = \cos \theta$ and $r = 1 + \sin \theta$
(d) $r = 1 + 2 \cos \theta$ and $r = 1 - 2 \cos \theta$
(e) $r = 2 \cos \theta$ and $r = 1$
(f) $r = 2(1 + \cos \theta)$ and $r = 6 \cos \theta$

29. Show that the graph of the equation $F(r, \theta) = 0$ is symmetric with respect to:
(a) The pole, if the point $(r, \pi + \theta)$ satisfies the equation whenever the point (r, θ) does.
(b) The polar axis, if the point $(-r, \pi - \theta)$ satisfies the equation whenever the point (r, θ) does.
(c) The line $\theta = \pi/2$, if the point $(-r, \theta)$ satisfies the equation whenever the point (r, θ) does.

30. Using symmetry tests (a), (b), and (c) and the results of Exercise 29, show that the graphs of the following equations are symmetric with respect to the pole, the polar axis, and the line $\theta = \pi/2$.
(a) $r^2 = \sin\theta$ (b) $r^2 = \cos\theta$

31. Show that $r^2 \cos 2\theta = A^2$ is an equilateral hyperbola.

32. Refer to Figure 94.13. What is the field strength at angles of 0°, 15°, 165°, 175°, and 335°?

33. In Figure 94.13, at what angles is the field strengh equal to 200 V/m?

34. When a standard broadcast antenna consists of two towers spaced $\frac{1}{4}$ wavelength apart and fed with equal power with a 90° phase difference, the horizontal pattern that results is given by the following sets of points:

(0°, 300)	(100°, 150)	(260°, 150)
(10°, 300)	(110°, 110)	(270°, 180)
(20°, 295)	(120°, 80)	(280°, 210)
(30°, 295)	(130°, 55)	(290°, 240)
(40°, 280)	(150°, 30)	(300°, 260)
(50°, 275)	(180°, 0)	(310°, 275)
(60°, 260)	(210°, 30)	(320°, 280)
(70°, 240)	(230°, 55)	(330°, 295)
(80°, 210)	(240°, 80)	(340°, 295)
(90°, 180)	(250°, 110)	(350°, 300)
		(360°, 300)

(a) Plot the graph of the data in polar coordinates.
(b) Identify the curve drawn in part (a).

35. A certain microphone has a directional response characteristic as given by the following data. Plot the graph of the data in polar coordinates.

Direction	*Relative Response (dB)*	*Direction*	*Relative Response (dB)*
0°	0	$202\frac{1}{2}$°	−21
$22\frac{1}{2}$°	−0.5	225°	−20
45°	−2.5	$247\frac{1}{2}$°	−17.5
$67\frac{1}{2}$°	−7.5	270°	−12.5
90°	−12.5	$292\frac{1}{2}$°	−7.5
$112\frac{1}{2}$°	−17.5	315°	−2.5
135°	−20	$337\frac{1}{2}$°	−0.5
$157\frac{1}{2}$°	−21	360°	0
180°	−22		

SECTION 95

Parametric Equations

It is often convenient to express the variables in an ordered pair, usually (x, y), in terms of a third variable, say t. When this is done, we generate a pair of equations of the form

$$x = f(t)$$
$$y = g(t) \qquad (95.1)$$

Equations such as (95.1) are called **parametric equations** and t is called the **parameter**.

Parametric equations are very useful in describing the complex paths traversed by particles, since they allow us to express x and y coordinates in terms of another variable.

The graph of the relation between x and y can usually be obtained by arbitrarily choosing values for t, and then obtaining the corresponding values for x and y. Alternately, if we are able to eliminate the parameter t, we obtain the Cartesian equation of the curve and then graph this equation in the usual way.

Example 95.1

Graph the function defined by the parametric equation $x = t - 1$, $y = t^2 - 4$, by assigning arbitrary values to t to obtain corresponding values for x and y.

Solution

By assigning values to the parameter t, we find the corresponding values of x and y shown in the table.

t	−3	−2	−1	0	1	2	3
x	−4	−3	−2	−1	0	1	2
y	5	0	−3	−4	−3	0	5

The graph is shown in Figure 95.1. We note that the parameter t does not appear in the graph. We further note that the curve appears to be a parabola (see Example 95.2).

Example 95.2

For the parametric equations given in Example 95.1, eliminate t, find the Cartesian equation, and identify the equation.

Figure 95.1

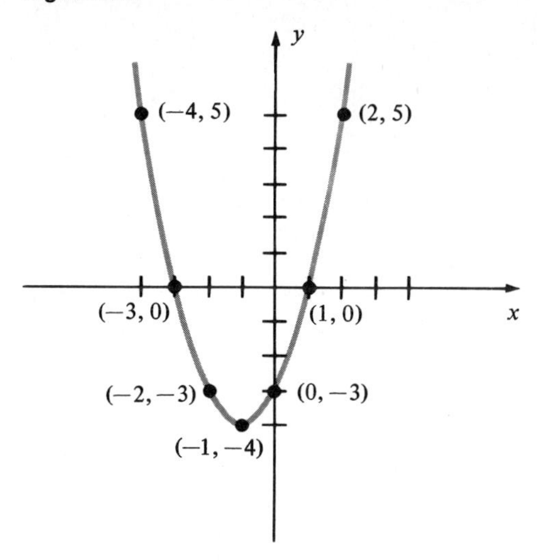

Solution

The equation $x = t - 1$ is equivalent to $t = x + 1$. Hence, we may write $y = t^2 - 4$ as $y = (x + 1)^2 - 4$ or $y = x^2 + 2x - 3$. The latter form is the Cartesian equation of a parabola that opens up, has vertex at $(-1, -4)$, y intercept at $(0, -3)$, and x intercepts at $(-3, 0)$ and $(1, 0)$. These observations are in complete agreement with the values shown on the graph of Figure 95.1.

In some cases a set of parametric equations will represent only a portion of a curve. This situation is illustrated in the following example.

Example 95.3

Let us consider the following sets of parametric equations:

$$\left.\begin{aligned} x &= t \\ y &= t \end{aligned}\right\} \qquad \text{(A)}$$

$$\left.\begin{aligned} x &= t^2 \\ y &= t^2 \end{aligned}\right\} \qquad \text{(B)}$$

If we eliminate the parameters, both (A) and (B) yield the Cartesian equation $y = x$. But the graphs are not the same. For (A) there are no restrictions on x or y. However, we note that for (B) $x \geq 0$ and $y \geq 0$, since t^2 is always nonnegative. Thus, the graph of (B) is only a portion of the line $y = x$, that portion for which $x \geq 0$ and $y \geq 0$. The graphs of (A) and (B) are shown in Figure 95.2 on page 588.

Example 95.4

Show that the equations $x = \cos 2t$ and $y = \sin t$ represent a portion of the parabola $x = 2y^2 - 1$.

Figure 95.2

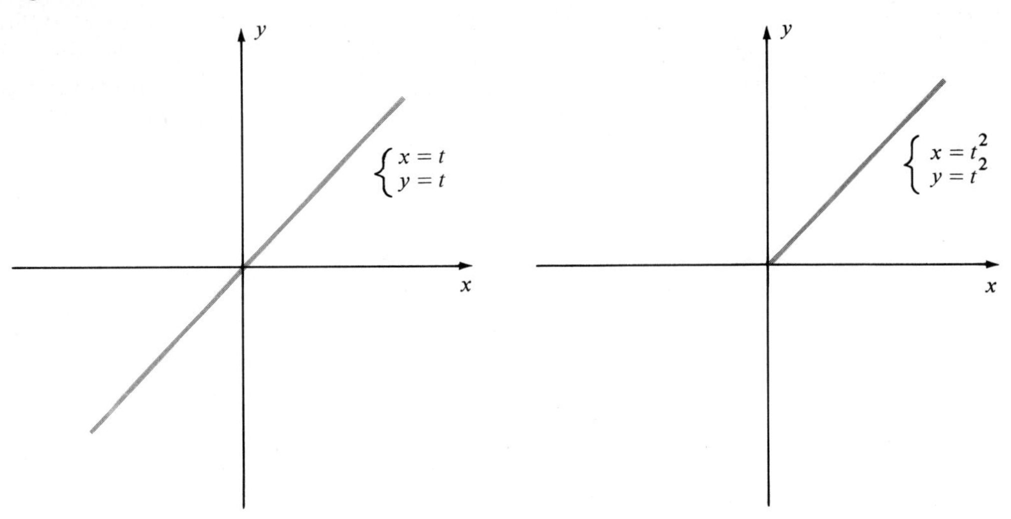

Solution

We begin by eliminating the parameter and thus finding the Cartesian equation. To do so we recall that the double-angle identity for cosine is given by $\cos 2t = 1 - 2\sin^2 t$. Thus, $x = \cos 2t = 1 - 2\sin^2 t$. Since $y = \sin t$, we have $x = 1 - 2y^2$. Since $-1 \le \cos 2t \le 1$ and $-1 \le \sin t \le 1$, we have the following restrictions for x and y:

$$-1 \le x \le 1 \quad \text{and} \quad -1 \le y \le 1$$

The graph of $x = 1 - 2y^2$ with these noted restrictions is shown in Figure 95.3.

Figure 95.3

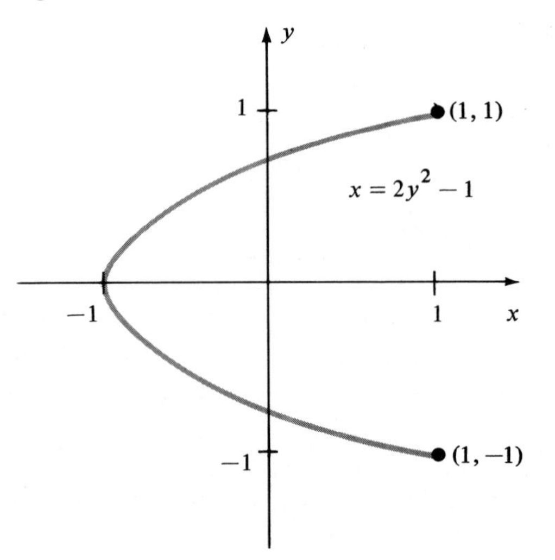

In the previous examples we were given the parametric equations and then proceeded to discuss the Cartesian equation and the graph of the curve. In many cases the development of these parametric equations can be a difficult process. In fact, the origin of some important parametric equations depends upon the methods of calculus. In the following example we illustrate an algebraic method that enables us to find parametric equations from a given Cartesian equation.

Example 95.5

Find a set of parametric equations for $x^2 + y^3 = 2xy$.

Solution

To introduce the parametric t, we let $y = tx$. Then we have

$$x^2 + t^3x^3 = 2x(tx)$$

or

$$x^2 + t^3x^3 = 2tx^2$$

Dividing by x^2 yields

$$1 + t^3x = 2t$$

Hence,

$$x = \frac{2t - 1}{t^3} \tag{A}$$

Since $y = tx$, we have

$$y = t\left(\frac{2t - 1}{t^3}\right)$$

or

$$y = \frac{2t - 1}{t^2} \tag{B}$$

The parametric equations we seek are given by (A) and (B).

Exercises for Section 95

In Exercises 1 through 10, graph the curve of each parametric equation. Eliminate the parameter and find the Cartesian equation.

1. $x = 2t,\ y = t - 1$

2. $x = 3t + 2,\ y = 2t - 3$

3. $x = 2t + 3,\ y = 3t - 5$

4. $x = t^2,\ y = t - 1$

5. $x = t + 1,\ y = t^2 - 3$

6. $x = \cos t,\ y = \sin t$
Hint: $\sin^2 t + \cos^2 t = 1$.

7. $x = 3 \cos t, y = 3 \sin t$

8. $x = 3 \sin t, y = \cos^2 t$

9. $x = 2 \cos t, y = \sin^2 t$

10. $x = \cos 2t, y = \cos t$

In Exercises 11 through 14, follow the procedure outlined in Example 95.5 to find a parametric representations for each equation.

11. $x + y = xy$

12. $x^2 + y^2 = 4x$

13. $x^2 + y^2 = 4$

14. $2x^2 - y^2 = 4x$

15. Graph the curve represented by the parametric equations $x = 2(\theta - \sin \theta)$ and $y = 2(1 - \cos \theta)$ for $0 \leq \theta \leq 4\pi$. We note that the graph is called a **cycloid**. It is the curve traced by a fixed point on the circumference of a circle rolling on a fixed straight line.

16. The parametric equations that give the motion of a projectile are

$$x = (v_0 \cos \alpha)t$$

$$y = (v_0 \sin \alpha)t - \tfrac{1}{2}gt^2$$

where v_0 is the initial velocity, α is the angle of inclination from the horizontal, and g is the acceleration due to gravity (approximately 32 ft/sec^2). Eliminate the parameter to show that the graph of the path of the projectile is a parabola.

17. For the projectile motion of Exercise 16, find a general expression for the range x. ***Hint:*** Find t when $y = 0$ and substitute in the equation for x.

18. For what value of t will the projectile of Exercise 16 reach its maximum height?

19. Suppose a projectile is fired with an initial velocity of 180 ft/sec and $\sin \alpha = \frac{3}{5}$ and $\cos \alpha = \frac{4}{5}$.
(a) Find the time when the projectile hits the ground.
(b) Find the distance the projectile travels.
(c) Find the maximum height reached.

20. Graph the parametric equations $x = 2t$ and $y = t^2 - 1$ in the interval $-2 \leq t \leq 1$. Eliminate the parameter and show that the curve is a parabola.

REVIEW QUESTIONS FOR CHAPTER 15

1. State the distance formula.

2. Explain symmetry with respect to the x axis; the y axis; the origin.

3. Explain the conditions used to determine symmetry.

4. What is the inclination of a line L?

5. What is the slope of a straight line?

6. What do we say about the slope of a line parallel to the y axis?

7. State the two-point form for the equation of a straight line.

8. State the point-slope form for the equation of a straight line.

9. State the slope-intercept form for the equation of a straight line.

10. State three characteristics of the slope-intercept form of the equation of a straight line.

11. How are the slopes of parallel lines related?

12. How are the slopes of perpendicular lines related?

13. Write the general form of a first-degree equation in x and y.

14. What are conic sections?

15. In how many ways can a plane intersect the cone if the plane does not contain the point of intersection of the two nappes?

16. What is a circle?

17. Write the general form for the equation of a circle with center at (h, k) and radius r.

18. What is a parabola?

19. What is the axis of symmetry of a parabola?

20. What is the vertex of a parabola?

21. What is an ellipse?

22. Given the ellipse $x^2/a^2 + y^2/b^2 = 1$, state the coordinates of the center, vertices, covertices, and foci.

23. Describe the major and minor axes of the ellipse given in Exercise 22.

24. Given the ellipse $(x - h)^2/a^2 + (y - k)^2/b^2 = 1$. Answer the questions asked in Exercise 22.

25. Repeat Exercise 24 for the equa ion

$$\frac{(x - h)^2}{b^2} + \frac{(y - k)^2}{a^2} = 1$$

26. What is a hyperbola?

27. Given the equation $x^2/a^2 - y^2/b^2 = 1$. State the coordinates of the center, the vertices, the foci, and the equation of the asymptotes.

28. Repeat Exercise 27 for the equations

$$\frac{(x - h)^2}{a^2} - \frac{(y - k)^2}{b^2} = 1$$

and

$$\frac{(y-k)^2}{a^2} - \frac{(x-h)^2}{b^2} = 1$$

29. What is a polar graph?

30. State the relationship between rectangular and polar coordinates.

31. State the tests for symmetry with respect to the pole, the polar axis, and the line $\theta = \pi/2$.

32. What is a parameter?

33. Why are parametric equations useful?

REVIEW EXERCISES FOR CHAPTER 15

1. Find the distance between the points.
(a) $(2, 3)$ and $(5, 7)$ (b) $(1, -3)$ and $(-4, 5)$

2. Show that the points $(0, -2)$, $(1, 1)$, and $(3, 7)$ are collinear.

3. Find the equation of the straight line through the points $(1, 2)$ and $(5, 4)$.

4. Find the equation of the straight line passing through the point $(1, 3)$ whose slope is 4.

5. Find the equation of the line whose slope is $\frac{1}{3}$ and whose y intercept is at $(0, -5)$.

6. Find the equation of the line parallel to $2x - 3y - 1 = 0$ and passing through the point $(4, 3)$.

7. Find the equation of the line perpendicular to $x + 2y - 3 = 0$ and passing through the point $(5, 2)$.

8. Find the equation of the circle with center at $(2, -4)$ and radius equal to 3.

9. Find the center and radius of the circle whose equation is $x^2 + y^2 + 4x + 6y - 12 = 0$.

10. Find the equation of the circle if the points $(2, 5)$ and $(6, -3)$ are the endpoints of a diameter.

11. Discuss and sketch the parabola whose equation is $(x - 2)^2 = 8(y + 3)$.

12. Discuss and sketch the parabola whose equation is $(y + 1)^2 = -4(x - 1)$.

13. Discuss and sketch the ellipse whose equation is $x^2 + 4y^2 = 36$.

14. Discuss and sketch the graph of the ellipse whose equation is

$$\frac{(x-1)^2}{25} + \frac{(y+2)^2}{9} = 1$$

15. Find an equation of the ellipse with center at $(0, 0)$, focus at $(4, 0)$, and $a = 5$.

16. Write the equations of the asymptotes of the hyperbola

$$\frac{x^2}{4} - \frac{y^2}{9} = 1$$

17. Write the equations of the asymptotes of the hyperbola

$$\frac{(y-1)^2}{4} - \frac{(x+2)^2}{16} = 1$$

18. Discuss and sketch the hyperbola whose equation is

$$\frac{x^2}{4} - \frac{y^2}{9} = 1$$

19. Discuss and sketch the hyperbola whose equation is

$$\frac{(x-2)^2}{16} - \frac{(y+1)^2}{25} = 1$$

20. Write an equation of the hyperbola with asymptotes $y = \pm\frac{2}{3}x$ and vertex at $(6, 0)$.

In Exercises 21 through 30, identify and write each equation in standard form.

21. $x^2 + y^2 - 4x + 2y - 4 = 0$

22. $x^2 + 2x - 4y - 7 = 0$

23. $4x^2 + 9y^2 + 18y - 135 = 0$

24. $x^2 - 4y^2 + 2x + 8y - 7 = 0$

25. $x^2 + y^2 - 6x - 8y - 11 = 0$

26. $9x^2 + 16y^2 - 36x - 96y + 36 = 0$

27. $y^2 + 4y - x + 5 = 0$

28. $x^2 + 2y^2 + 4x - 4y + 2 = 0$

29. $2x^2 - y^2 - 16x + 4y + 24 = 0$

30. $y^2 - 6x^2 + 2y + 36x - 59 = 0$

In Exercises 31 through 38, discuss and sketch the graph of each equation.

31. $r = 5\cos\theta$

32. $r = 10(1 + \cos\theta)$

33. $r = 3 + \cos\theta$

34. $r = 1 - 2\sin\theta$

35. $r^2 = 8\cos 2\theta$

36. $r = 2\sin 2\theta$

37. $r = 3\sin 3\theta$

38. $r = 4 - \cos\theta$

39. Sketch the following pairs of equations and find all the points of intersection.
(a) $r = \sin\theta$ and $r = \cos\theta$
(b) $r = 2\cos 2\theta$ and $r = 1$.

40. Graph the curve given by the parametric equations $x = 2t - 3$, $y = t - 1$. Eliminate the parametric and find the Cartesian equation.

41. Repeat Exercise 40 for $x = t - 1$, $y = t^2 - 1$.

42. Repeat Exercise 40 for $x = \sin^2 t$, $y = \cos^2 t$.

APPENDIX A

Basic Geometry

In this appendix we shall discuss some basic geometric facts and the terms that are used to illustrate these facts. The most basic concepts of geometry are the point, the line, and the plane. We accept these terms without definition as our starting point.

A.1 Lines

A *straight line* is a plane figure that extends infinitely far in two directions (Figure A.1). We note that the line has length but no width or thickness. A **line segment** is a part of a line between two points on a line (Figure A.2). The two points are called the **endpoints** of the segment. A **ray** consists of a point on a line and that part of the line extending infinitely in one direction from the point. We also say that a **ray** is a half-line with an initial point but no endpoint (Figure A.3).

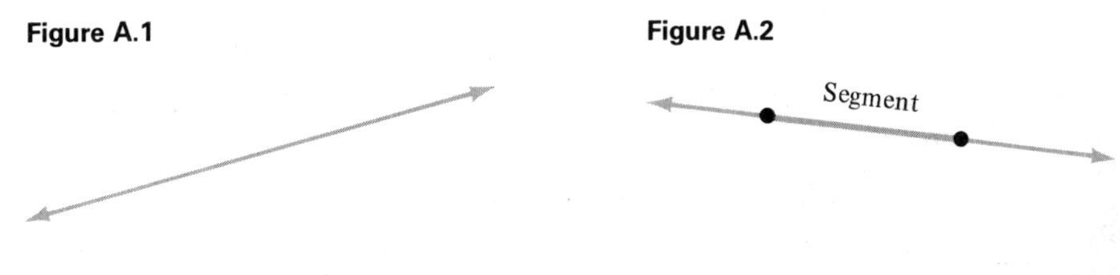

Figure A.1

Figure A.2

Figure A.3

Ray

Any two lines that **intersect** have a common point called the **point of intersection**. Two lines that do not intersect no matter how far they are extended are called **parallel lines**. Two lines are said to be coincident if one line lies entirely in the other line. Alternatively, if two lines have no points in common, they are **parallel**; if they have one point in common, they **intersect**; and if they have all points in common, they are coincident (Figure A.4).

Figure A.4

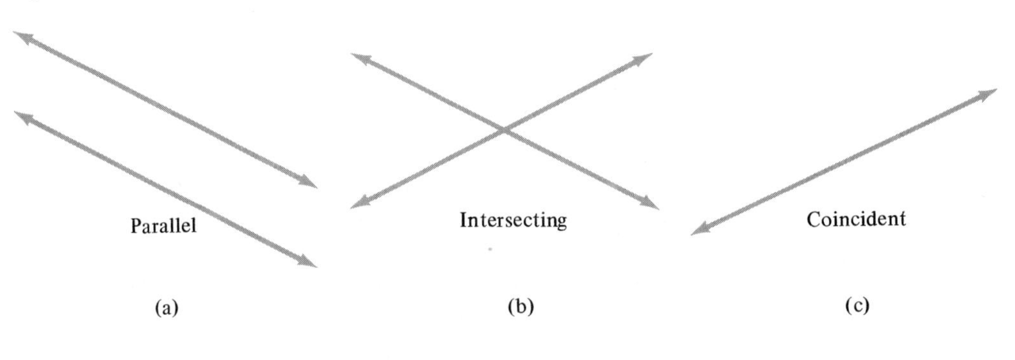

A.2 Angles

An *angle* is a figure formed by two rays emmanating from a common point. The common point is called the **vertex** of the angle. The measure of the angle is the amount of rotation necessary to rotate one ray about its endpoint in order to make the two sides coincide.

A **right angle** is an angle whose measure is 90°. The rays that form the sides of a right angle are said to be perpendicular. A **straight angle** is an angle whose measure is 180°. The sides of a straight angle form a straight line. An **acute angle** is an angle whose measure is greater than 0° and less than 90°. An **obtuse angle** is an angle whose measure is greater than 90° and less than 180°. A **reflex angle** is an angle whose measure is greater than 180° and less than 360°. These types of angles are shown in Figure A.5.

Two angles are **complementary** if their sum is 90°. Two angles are **supplementary** if their sum is 180°. Two intersecting lines form four angles. The nonadjacent pairs of angles (angles that do not have a common side) are called **vertical angles** and have equal measures. In Figure A.6 we see that $\angle a = \angle c$ and $\angle b = \angle d$.

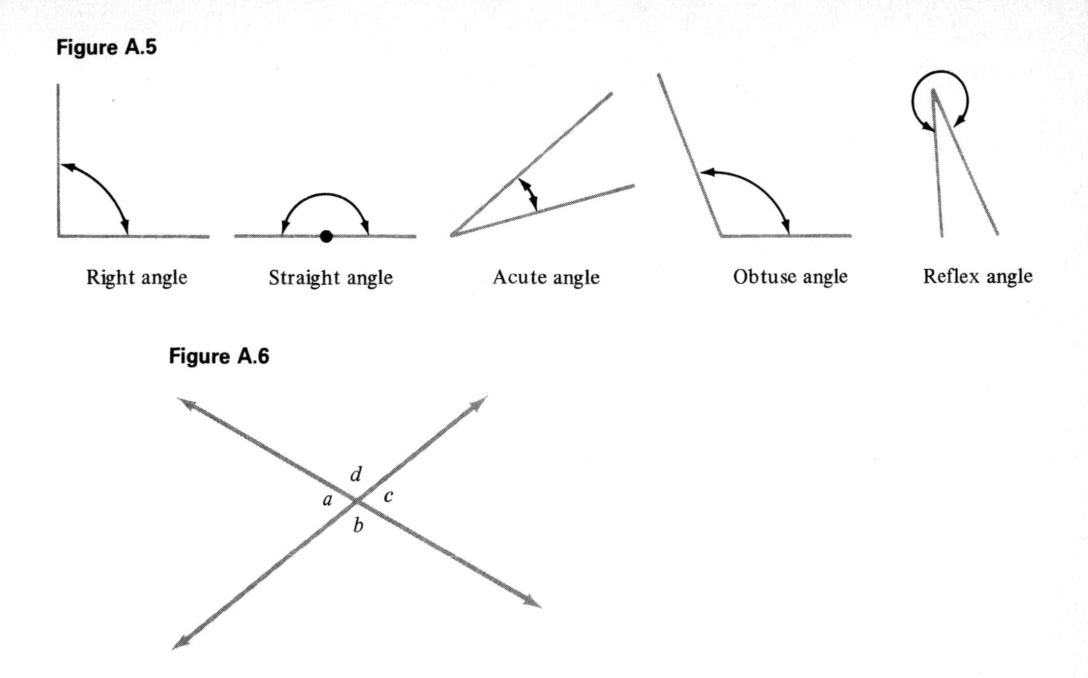

Another figure that yields a number of equal angles is formed by two parallel lines cut by a third line called a **transversal**. In this type of figure, **corresponding angles** are equal; **alternate interior angles** are equal; and **alternate exterior angles** are equal. Corresponding angles are pairs of nonadjacent angles on the same side of the transversal with one angle between the parallel lines and the other not between the parallel lines. In Figure A.7 the corresponding angles that are equal are given as follows: $\angle 1 = \angle 5$, $\angle 2 = \angle 6$, $\angle 3 = \angle 7$, and $\angle 4 = \angle 8$. Alternate interior angles are two nonadjacent angles between parallel lines and on opposite sides of the transversal. Thus, in Figure A.7 $\angle 3 = \angle 6$ and $\angle 4 = \angle 5$.

Alternate exterior angles are two nonadjacent angles on opposite sides of the transversal and not between the parallel lines. Thus, in Figure A.7 $\angle 1 = \angle 8$ and $\angle 2 = \angle 7$.

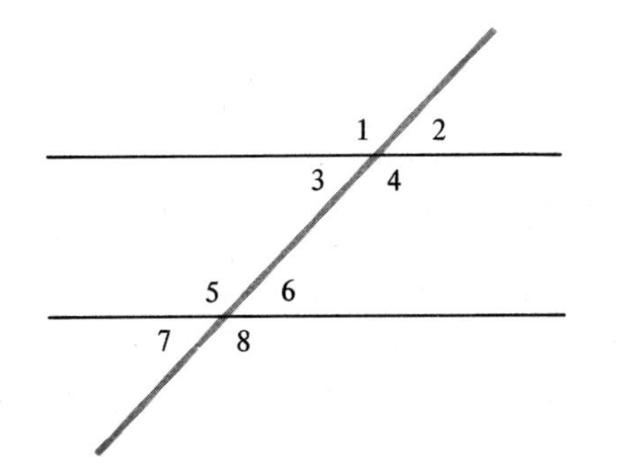

A.3 Triangles

A **triangle** is a plane figure formed by three connecting line segments. The line segments bounding the triangle are the sides of the triangle, and the endpoints of the line segments are the **vertices** of the triangle. The sum of the measures of the angles of any triangle is 180°.

Triangles are classified according to either the angles or the sides as follows:

Classification by Angles

1. An **acute triangle** is a triangle that has three acute angles.
2. An **equiangular triangle** is a triangle whose angles all have equal measure. All angles are 60°.
3. A **right triangle** is a triangle that has one right angle.
4. An **obtuse triangle** is a triangle that has one obtuse angle.

These triangles are shown in Figure A.8.

Figure A.8

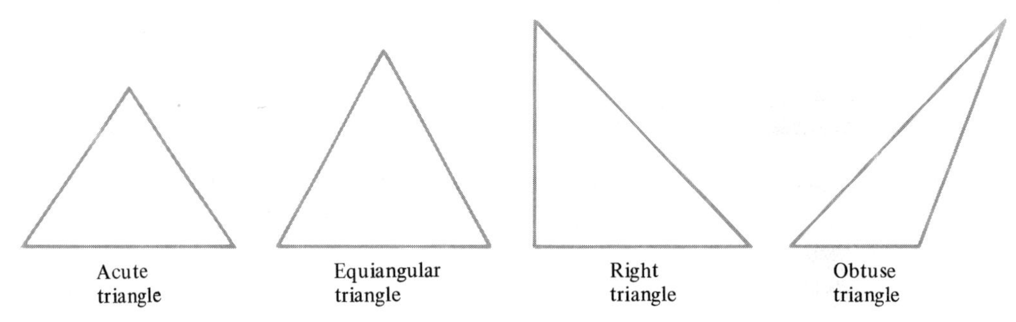

Classification by Sides

1. An **equilateral triangle** is a triangle that has all three sides equal in length. We note that an equilateral triangle is also equiangular.
2. An **isosceles triangle** is a triangle that has two sides equal in length. The third side of an isosceles triangle is called the *base* and the angle opposite the base is the **vertex angle.** The angles opposite the equal sides are the **base angles** and their measures are equal. We note that a equilateral triangle is isosceles. We further note that there are obtuse isosceles triangles, right isosceles triangles, and acute isosceles triangles.
3. A **scalene triangle** is a triangle that has no sides equal in length. In a scalene triangle, no angles have equal measure.

In Figure A.9 we show an equilateral triangle, an isosceles triangle, and a scalene triangle.

An **altitude** of a triangle is the perpendicular line segment from a vertex to the opposite side or the sides extended. The altitude is also the height of the triangle. We note that a triangle has three altitudes.

Figure A.9

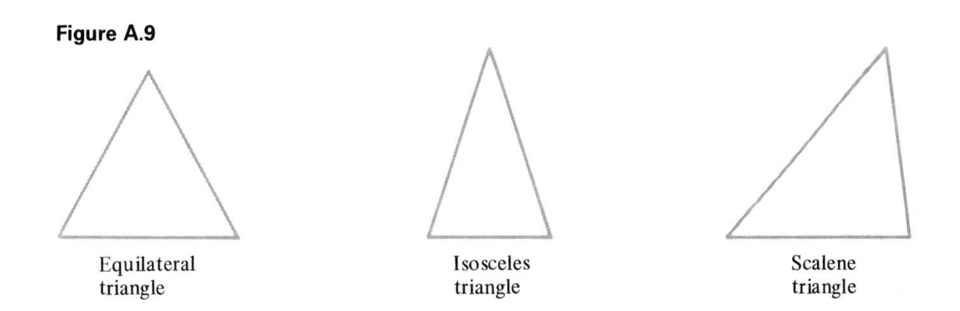

The **median** of a triangle is the line segment drawn from any vertex to the midpoint of the opposite side.

Two general relationships are of particular importance in working with triangles. The first involves similar triangles. Two triangles are said to be similar if their corresponding angles are equal. If we have similar triangles, the corresponding sides of these similar triangles are **proportional**. This fact is of great importance is the study of mathematics.

The second important relationship involves the sides of a right triangle. As we have noted, a right triangle is any triangle that contains a right angle. The two sides that are perpendicular to each other are the *legs* of the right triangle. The third side, the longest side, is the **hypotenuse** of the right triangle.

The general relationship among the sides of any right triangle is called the **Pythagorean theorem,** which states that *in any right triangle the sum of the square of the lengths of the legs is equal to the square of the length of the hypotenuse*. Thus, in Figure A.10, where a and b are the lengths of the legs and c is the length of the hypotenuse, we have

$$c^2 = a^2 + b^2$$

Equivalently, $c^2 - a^2 = b^2$ and $c^2 - b^2 = a^2$.

Figure. A.10

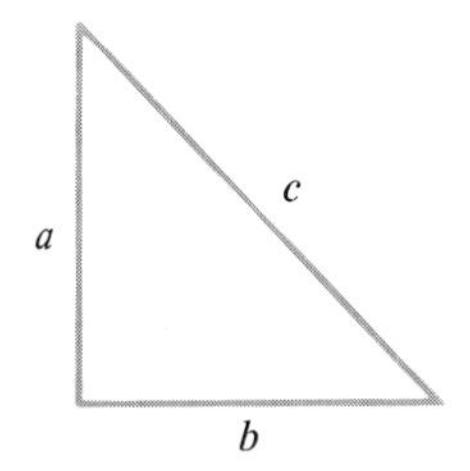

A.4 Quadrilaterals

A *quadrilateral* is a plane figure formed by four connecting line segments. Following are the most frequently encountered figures:

1. A **parallelogram** is a quadrilateral with two pairs of parallel sides.
2. A **rectangle** is a parallelogram with four right angles.
3. A **rhombus** is a parallelogram with four equal sides.
4. A **square** is a rectangle with four equal sides.
5. A **trapezoid** is a quadrilateral with one one pair of parallel opposite sides.

These figures are shown in Figure A.11.

Figure A.11

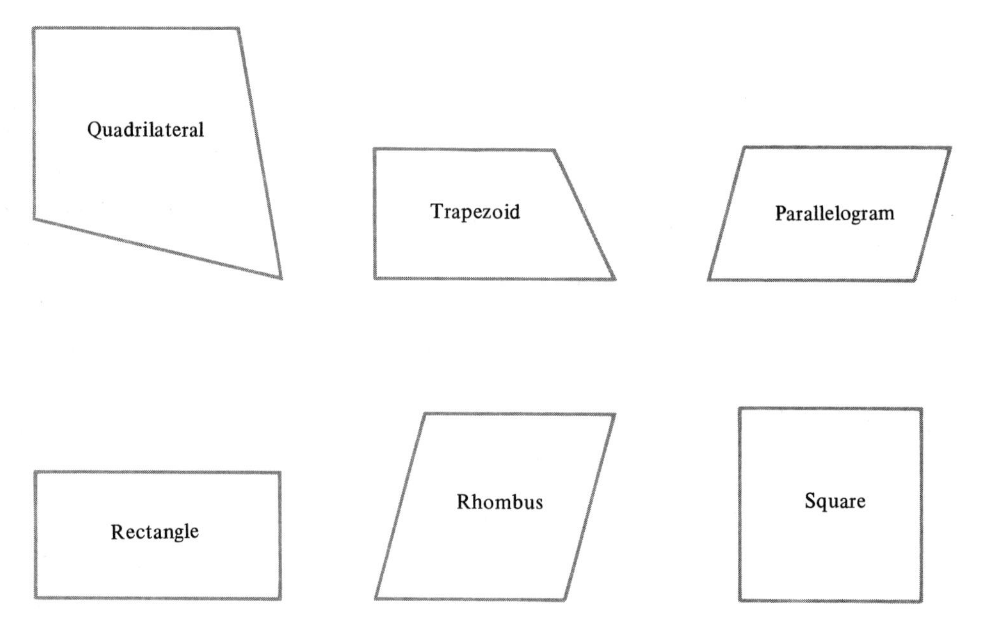

The following facts concerning quadrilaterals can be shown to be true:

1. The sum of the interior angles of a quadrilateral is 360°.
2. Any two opposite sides of a parallelogram are equal.
3. If the opposite sides of a quadrilateral are equal, the figure is a parallelogram.
4. The opposite angles of a parallelogram are equal.
5. If the opposite angles of a quadrilateral are equal, the figure is a parallelogram.
6. The diagonals of a rectangle (including a square) are equal.
7. The diagonals of a rhombus (including a square) are equal.

A.5 The Circle

A **circle** is a plane curve all of whose points are equidistant from a fixed point. A **radius** of a circle is a line segment joining the center of the circle with any point on the circle. A **secant** is any line passing through two points on the circle. A **tangent** is a line that touches a circle at only one point. A **chord** is a line segment joining any two points on the circle. An **arc** is a part of a circle between two points on the circle. A **sector** of a circle is the figure bounded by a circular arc and two radii drawn to the endpoints of the arc.

A.6 Basic Geometric Formulas

Plane Figure	*Perimeter*	*Area*
Triangle (Figure A.12)	$P = a + b + c$	$A = \frac{1}{2}bh$
Rectangle (Figure A.13)	$P = 2(a + b)$	$A = ab$
Square (Figure A.14)	$P = 4s$	$A = s^2$
Parallelogram (Figure A.15)	$P = 2(a + b)$	$A = bh$
Rhombus (Figure A.16)	$P = 4b$	$A = bh$
Trapezoid (Figure A.17)	$P = a + b + c + d$	$A = \frac{h}{2}(a + b)$

Figure A.12

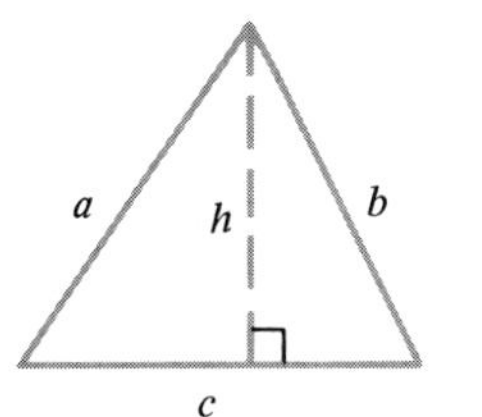

Figure A.13

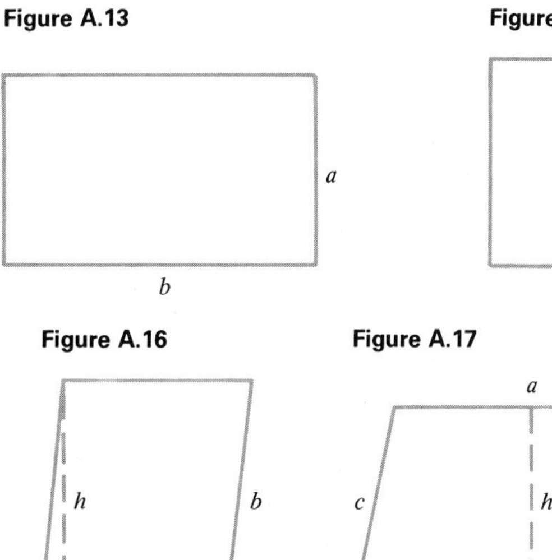

Figure A.14

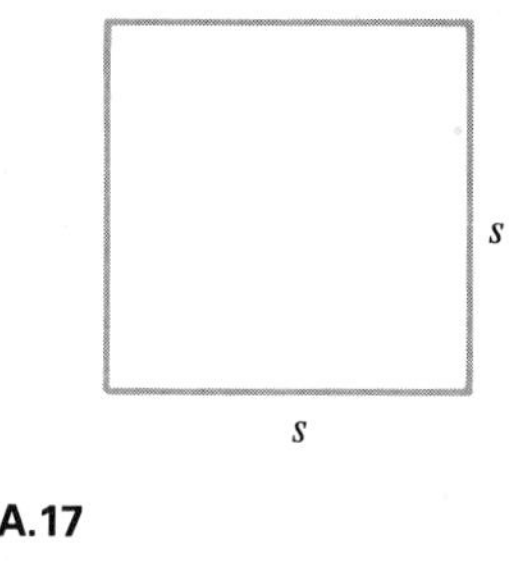

Figure A.15

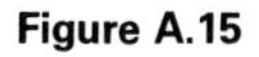

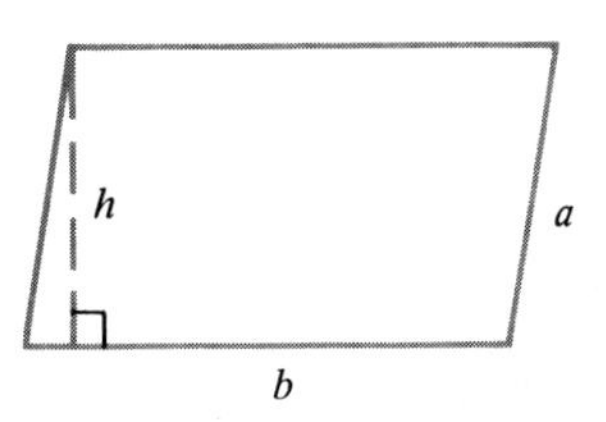

Figure A.16

Figure A.17

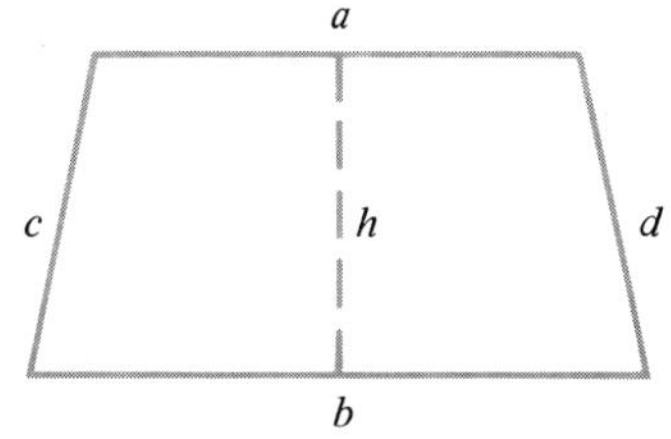

For a circle, the circumference is given by $C = 2\pi r$; the area, by $A = \pi r^2$ (Figure A.18).

Figure A.18

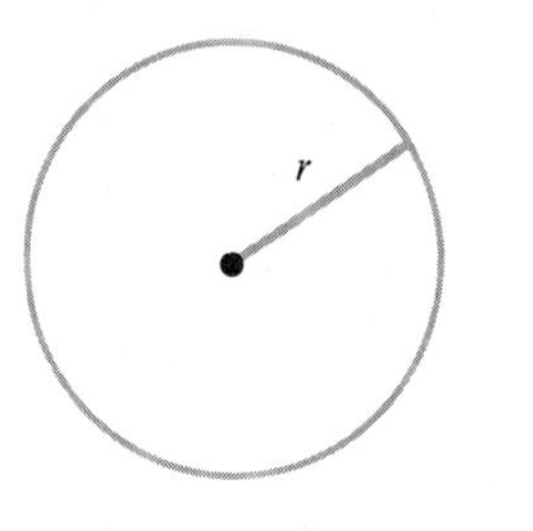

Solid	*Surface Area*	*Volume*
Cube (Figure A.19)	$A = 6e^2$	$V = e^3$
Rectangular solid (Figure A.20)	$A = 2(lw + lh + wh)$	$V = lwh$
Right circular cylinder (Figure A.21)	$A = 2\pi rh + 2\pi r^2$	$V = \pi r^2 h$
Right circular cone (Figure A.22)	$A = \pi r\sqrt{r^2 + h^2} + \pi r^2$ $= \pi rs + \pi r^2$	$V = \frac{1}{3}\pi r^2 h$
Sphere (Figure A.23)	$A = 4\pi r^2$	$V = \frac{4}{3}\pi r^3$

Figure A.19

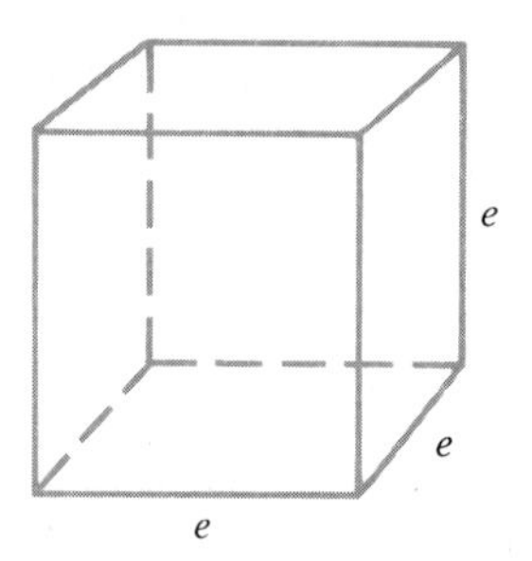

Figure A.20

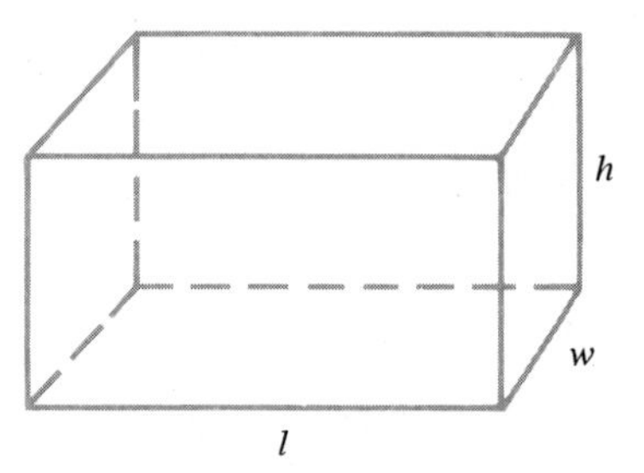

Figure A.21

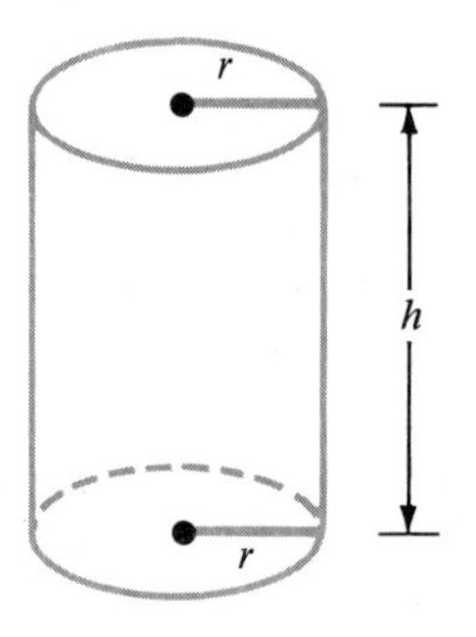

Figure A.22

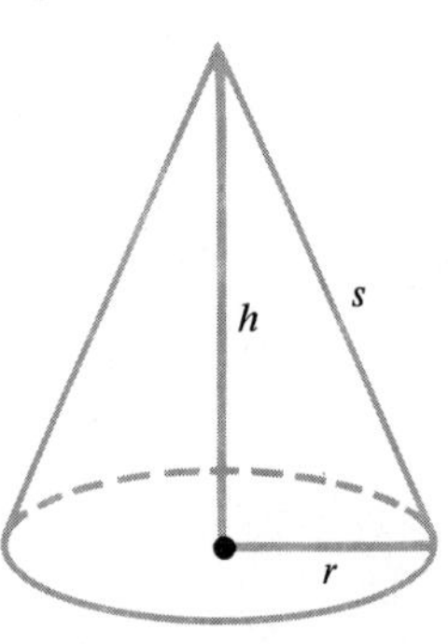

Figure A.23

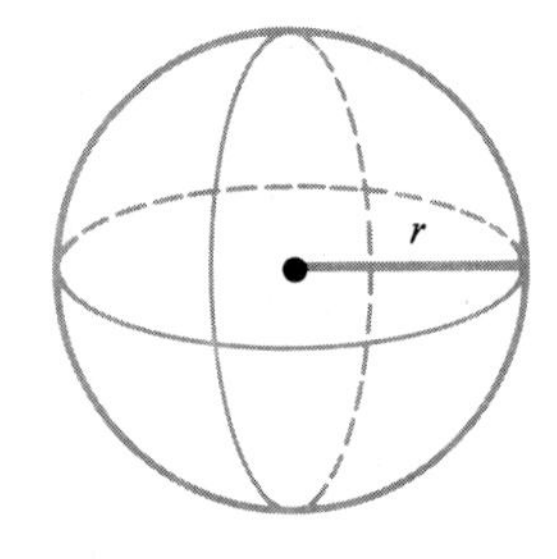

If B is the area of the base and L is the sum of the areas of the lateral faces, we have the following formulas:

Solid	*Surface Area*	*Formula*
Prism (Figure A.24)	$A = 2B + L$	$V = Bh$
Pyramid (Figure A.25)	$A = B + L$	$V = \frac{1}{3}Bh$

Figure A.24 **Figure A.25**

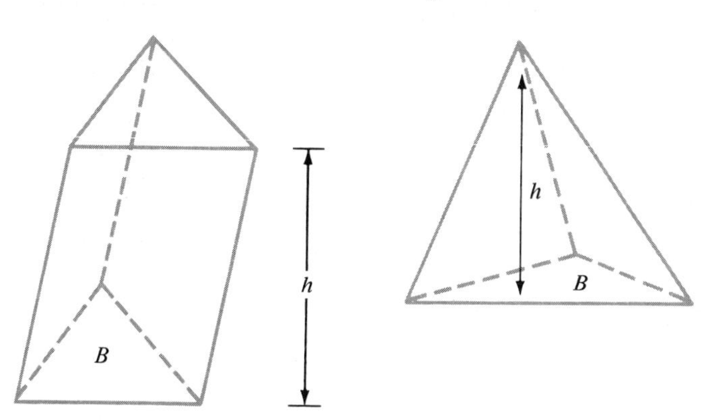

Exercises for Appendix A

In Exercises 1 through 15, answer true or false.

1. Adjacent angles are equal.
2. All acute angles are equal.
3. Two lines in a plane that have no points in common are parallel.
4. If two lines in a plane have two points in common, they coincide.
5. Each angle in an obtuse triangle is obtuse.
6. The acute angles in a right triangle are complementary.
7. An isosceles triangle has three equal sides.
8. If two angles of a triangle are equal, the triangle is isosceles.
9. A triangle can have three acute angles.
10. If two angles of a triangle are acute, the third angle must be acute.
11. A rhombus with one right angle is a square.
12. All squares are rectangles.

13. All rectangles are rhombuses.

14. The diagonals of a parallelogram are always equal in length.

15. The diagonals of a parallelogram are perpendicular.

16. Find the complement of $\angle A$ if $\angle A$ equals
(a) 10° (b) 20° (c) 38° (d) 75° (e) 90°

17. Find the supplement of $\angle B$ if $\angle B$ equals
(a) 10° (b) 36° (c) 90° (d) 118° (e) 150°

18. In Figure A.26 three lines intersect such that the x and y axes are perpendicular and $\angle a = 25°$. Find the measure of:
(a) $\angle b$ (b) $\angle d$
(c) $\angle f$ (d) The complement of $\angle a$
(e) The supplement of $\angle b$

Figure A.26

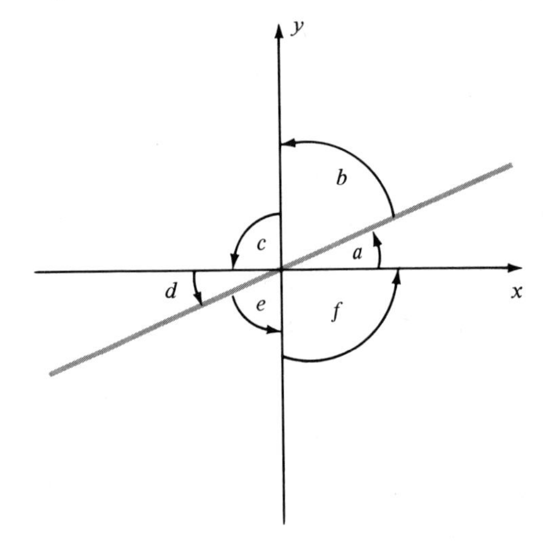

19. Two triangles are similar. The sides of the smaller triangle are 4, 10, and 12. The smallest side of the larger triangle is 6. Find the lengths of the other two sides.

20. Two triangles are similar. The sides of the larger triangle are 8, 14, and 20. The largest side of the smaller triangle is 16. Find the length of the other two sides.

21. The shadow cast by a flagpole is 40 ft long. At the same time, the shadow cast by a vertical yardstick is found to be $3\frac{1}{3}$ ft long. Find the height of the flagpole.

In Exercises 22 through 40, find the value indicated.

22. Triangle: altitude 11.3 in., base 14.4 in. Find the area.

23. Right triangle: legs have lengths of 8 and 12 in. Find the perimeter and the area.

24. Right triangle: legs have lengths of $\frac{3}{5}$ and $\frac{4}{5}$ m. Find the perimeter and the area.

25. Quadrilateral: sides 4, 6, 7, and 8 cm. Find the perimeter.

26. Parallelogram: base 13 cm, altitude 6 cm. Find the area.

27. Rectangle: length 14 in., width 6 in. Find the perimeter and the area.

28. Rectangle length is 13 in., width 5 in. Find the length of a diagonal.

29. Square: side 22.3 cm. Find the area and the perimeter.

30. Square: side $2\sqrt{2}$ m. Find the area and the length of a diagonal.

31. Rhombus: side 3.6 cm. Find the perimeter.

32. Square : area 4356 ft^2. Find the perimeter.

33. Trapezoid: bases 8.6 and 5.3 in., sides 6.4 and 2.3 in. Find the perimeter.

34. Trapezoid: bases 4.3 and 8.2 m; height 3.5 m. Find the area.

35. Trapezoid: bases 16.4 and 11.2 cm.; height 6.5 cm. Find the area.

36. Equilateral triangle: side 4 m. Find the perimeter and the area.

37. Parallelogram: base 3.16 cm, height 2.47 cm. Find the area.

38. Circle: radius 8 m. Find the circumference and the area.

39. Circle: diameter 5.6 cm. Find the circumference and the area.

40. Semicircle: diameter 5.6 cm. Find the circumference and the area.

41. The altitude of an equilateral triangle bisects the base. Find the area of an equilateral triangle whose perimeter is 12 cm. (Leave your answer in simplified radical form.)

42. A rug 9 by 12 ft is laid in a room 13 by $14\frac{1}{2}$ ft. How many square feet are not covered?

43. The altitude to the base of an isosceles triangle bisects the base. Find the altitude and the area of an isosceles triangle whose base is 8 cm long and whose equal sides are 10 cm long.

44. The perimeter of an equilateral triangle is equal to the perimeter of a square whose area is 64 m^2. Find the area of the triangle (Leave your answer in simplified radical form.)

In Exercises 45 through 55, find the volume of each figure.

45. A cube whose edge is 4.3 cm.

46. A cube whose edge is 2.16 ft.

47. A rectangular solid of length 5.0 m, width 3.0 m, and height 1.5 m.

48. A right circular cylinder of radius 5.14 cm and height 7.36 cm.

49. A right circular cone of radius 3.28 in. and height 4.10 in.

50. A sphere of radius 6.0 m.

51. A hemisphere (half a sphere) of diameter 4.2 ft.

52. A prism whose base is a square of side 11 cm and whose height is 7 cm.

53. A prism whose base is an equilateral triangle of side 6 cm and height 4 cm.

54. A pyramid with a rectangular base 8.5 by 4.6 in. and height 10.2 in.

55. A pyramid whose base is an equilateral triangle of side 6 in. and height 4 in.

In Exercises 56 through 60, find the total surface area of each figure.

56. Cube with edge 16.0 ft.

57. Rectangular solid of length 8 in., width 4 in., and height 3 in.

58. Circular cylinder of radius 8.6 cm and height 14.7 cm.

59. Sphere of radius 4.27 in.

60. Prism whose base is a parallelogram 2.4 by 3.8 in. and whose height is 4.9 in.

61. A certain brand of aluminum paint covers 250 ft^2 of surface per gallon. How many gallons will be needed to paint a spherical refinery tank 30 ft in diameter?

62. A circular building is to have a roof in the shape of a cone. The diameter of the building is 100 ft and the roof overhang is negligible. The apex of the conical roof is 30 ft above the top of the circular building wall. Find the area of the roof.

63. How many square feet of steel plate is needed to build a spherical refinery tank if the tank is to hold 50,000 gal (1 ft^3 = 7.48 gal).

64. Find the number of square feet of steel plate needed to construct a right cylindrical tank 20 ft high and 10 ft in diameter. Include the top and bottom of the tank.

65. A sewage-treatment plant has a water aerator system that uses circular tanks 100 ft in diameter. Find the surface area of the water exposed to the air in each tank.

APPENDIX B

The Metric System of Measurement

The metric system of measurement is a decimal system. Thus, like our number system, the decimal system is based on the number 10. A major advantage of such a system lies in the fact that a change in the size of the units can be accomplished by a move of a decimal point to the right or to the left.

Table B.1 illustrates the connection between the decimal numeral system and the metric system. From the table we notice the meaning of these prefixes:

deci means $\frac{1}{10}$ deca means 10

centi means $\frac{1}{100}$ hecto means 100

milli means $\frac{1}{1000}$ kilo means 1000

Table B.1

Thousandths	0.001	milli-
Hundredths	0.01	centi-
Tenths	0.1	deci-
Unit	1.	the unit of measure
Ten	10.	deca-
Hundreds	100.	hecto-
Thousands	1000.	kilo-

In the metric system, all measurements of length, area, volume, and weight are based upon the primary unit, the meter (m). Historically, the French are credited with the origin of this unit of measure. They decided to base this system of measure on a unit of length equal to 1 ten-millionth of the distance from the equator to the north pole. This distance was called 1 meter, meaning **measure**. The meter is now defined to be 1,650,763.73 wavelengths in vacuum of the orange-red line of the spectrum of krypton 86.

As we have noted, the metric system is based on the number 10. Thus, it is natural to divide the meter into 10 parts. Each part is $\frac{1}{10}$ of a meter and is called a **decimeter** (dm). The decimeter is further divided into 10 parts, each of which is called a **centimeter** (cm). One centimeter is thus $\frac{1}{100}$ of a meter. The centimeter is further divided into 10 parts, each called a **millimeter** (mm). Since the prefix "milli" means one one-thousandth, the millimeter is $\frac{1}{1000}$ of a meter.

For measures larger than 1 meter, we follow the procedure of using 10. A distance of 10 meters is called a **decameter** (dkm). A distance of 10 decameters is called a **hectometer** (hkm). A hectometer is 100 meters. Ten hectometers is called a **kilometer** (km). Table B.2 is a summary of the relations between the units of length in the metric system.

Table B.2

10 mm = 1 centimeter (cm)	$1 \text{ cm} = \frac{1}{100} \text{ m}$
10 cm = 1 decimeter (dm) 10 dm = 1 meter (m)	$1 \text{ mm} = \frac{1}{1000} \text{ m}$
10 m = 1 decameter (dkm)	100 cm = 1 m
10 dkm = 1 hectometer (hkm) 10 hkm = 1 kilometer (km)	1000 mm = 1 m

The metric system of measure is used in most countries of the world but not yet in the United States. Hence, it is often necessary to convert from the English system to the metric system, and vice versa. For measures of length, Table B.3 is useful.

Table B.3. Linear Measure

English to Metric	*Metric to English*
1 inch (in.) = 2.54 cm	1 mm = 0.04 in.
1 foot (ft) = 30.48 cm	1 cm = 0.39 in.
1 yard (yd) = 0.91 m	1 m = 39.37 in.
1 mile = 1.61 km	1 km = 0.62 mile

With the equivalents from Table B.3 it is an easy matter to develop a table of equivalents for square measure. For example, since 1 in. = 2.54 cm, 1 in.^{2} = $(2.54)^2$ cm^2 = 6.4516 cm^2. Similarly, since 1 m = 39.37 in., 1 m^2 = $(39.37)^2$ in.^{2} or 1 m^2 = 1550.1 in.^{2}. Other equivalents can be generated in the same fashion. The results are summarized in Table B.4.

Table B.4. Square Measure

English to Metric	*Metric to English*
1 in.^{2} = 6.45 cm^2	1 mm^2 = 0.002 in.^{2}
1 ft^2 = 0.09 m^2	1 cm^2 = 0.16 in.^{2}
1 yd^2 = 0.84 m^2	1 m^2 = 1549 in.^{2}
1 acre = 4047.9 m^2 = 0.4 hectare (ha)	1 km^2 = 0.39 mile^2
1 mile^2 = 2.59 km^2	1 km^2 = 247.10 acres^2
	1 ha = 10,000 m^2 = 2.5 acres

The relations between measures of volume are found in a manner similar to finding the relations between measures of area. In this instance, however, we take the cubes of linear units. For example, 1 in. = 2.54 cm. Thus, 1 in.^{3} = $(2.54)^3$ cm^3 or 1 in.^{3} = 16.39 cm^3. Additional relations are given in Table B.5.

Table B.5. Cubic Measure

English to Metric	*Metric to English*
1 in.^{3} = 16.39 cm^3	1 cm^3 = 0.06 in.^{3}
1 ft^3 = 0.03 m^3	1 m^3 = 35.32 ft^3
1 yd^3 = 0.76 m^3	1 km^3 = 0.24 mile^3

In the metric system, the liter is the basic unit of liquid capacity. A liter is by definition the volume of a cubic decimeter, that is, the capacity of a cube whose length, width, and height are all 1 decimeter or, equivalently, 10 centimeters. Basic equivalents of liquid measure are given in Table B.6.

Table B.6. Liquid Measure

English to Metric	*Metric to English*
1 fluid ounce (oz) = 30 milliliters (ml)	1 ml = 0.03 fluid oz
1 cup = 0.24 liter (ℓ)	1 ℓ = 2.11 pints
1 pint = 0.47 ℓ	1 ℓ = 1.06 quarts
1 quart = 0.95 ℓ	1 ℓ = 0.26 gal
1 gallon (gal) = 3.79 ℓ	1 ℓ = 1000 cm^3

In the English system the primary unit of weight is the pound. In the metric system the primary unit of weight is the gram. One gram of weight in the metric system is defined to be the weight of 1 cubic centimeter of pure water at a temperature of 4 degrees Celsius. The gram is changed into other units in the metric system by multiplying or dividing by 10 or powers of 10. The more important units and their equivalents are given in Table B.7.

Table B.7. Weights

English to Metric	*Metric to English*
1 oz = 28.35 grams (g)	1 g = 0.035 oz
1 pound (lb) = 453.59 g	1 kg = 2.20 lb
1 lb = 0.45 kilogram (kg)	1 ton (metric) = 1000 kg
	= 1.102 tons = 2205 lb
1 ton = 0.91 ton (metric)	

Two systems of measure are used for dealing with temperature. One system, the **Fahrenheit scale** (named after Gabriel Fahrenheit), has 212° as the boiling point of water and 32° as the freezing point of water.

The second system, the **Celsius scale** (developed by Anders Celsius), has a somewhat different scale. With the Celsius (or centigrade) thermometer, the boiling point of water is 100° and the freezing point of water is 0°. The term "centigrade" is an outgrowth of the fact that there are 100 graded intervals or graduations between 0° and 100° centigrade and, as we have previously noted, "centi" means $\frac{1}{100}$.

In working with these two systems of temperature measure, we must be sure to label degree readings. For example, a degree reading of 25 is meaningless and confusing. If we mean 25°F, we have quite a cold temperature (below freezing). If, however, we mean 25°C, we have a comfortably warm temperature.

The following two formulas enable us to convert from Fahrenheit to Celsius, and vice versa. When we know the Fahrenheit temperature and wish to convert to Celsius, we use the formula

$$C = \frac{5}{9}(F - 32)$$

When we know the Celsius temperature and wish to convert to Fahrenheit, we use the formula

$$F = \frac{9}{5}C + 32$$

For example, 20°C converts to 68°F as follows:

$$F = \frac{9}{5}(20) + 32$$

$$= \frac{180}{5} + 32$$

$$= 36 + 32 = 68 \quad \text{and hence} \quad 20°\text{C} = 68°\text{F}$$

To convert 95°F to Celsius, we use the equation

$$C = \frac{5}{9}(F - 32) \quad \text{with } F = 95$$

Hence,

$$C = \frac{5}{9}(95 - 32)$$

$$= \frac{5}{9}(63)$$

$$= 5 \cdot 7$$

$$= 35$$

Thus, 95°F = 35°C.

Exercises for Appendix B

In Exercises 1 through 30, use the tables in this appendix to change each measurement to the form indicated.

1. 24 in. to cm

2. 348 m to ft

3. 42.4 in. to mm

4. 5000 m to yd

5. 54 yd to m

6. 80 km to miles

7. 60 miles to km

8. 128 cm to in.

9. 3 m^2 to yd^2

10. 120 cm^2 to $in.^2$

11. 2.6 m^2 to $in.^2$

12. 32.4 km^2 to $miles^2$

13. 132 $in.^3$ to cm^3

14. 256 cm^3 to $in.^3$

15. 3.4 m^3 to ft^3

16. 12.3 gal to liters

17. 3.4 pints to liters

18. 14.6 l to quarts

19. 46 kg to lb

20. 4,285 g to lb

21. 16 lb to kg

22. 150 lb to kg

23. 500 g to oz

24. 120 kg to lb

25. 104°F to °C

26. 77°F to °C

27. 5°F to °C

28. 50°C to °F

29. 5°C to °F

30. −40°C to °F

31. The Eiffel Tower is 300 m high. Find the height in feet.

32. Express 80 km/hr as miles/hr (mph).

33. The speed of light is approximately 186,000 miles/sec. Express this as m/sec.

34. On a certain day the barometric pressure is 29.94 in. Express the pressure in milliliters.

35. A lot is 130 ft long and 85 ft wide. Find its area in square meters.

36. The speed limit in New York State is 55 mph. What is the speed limit in km/h?

37. How many miles is it from New York City to Los Angeles if the distance is 4690 km?

38. How much water (in liters) will an aquariam hold if it is 1 m long, 60 cm wide, and 60 cm high?

39. In Montana, the temperature once dropped from 44°F to −56°F in a period of 24 hr. This is a change in temperature of 100°F. Approximately how great a change is it on the Celsius scale?

40. The highest temperature ever recorded was 58°C, recorded in Africa. Find the equivalent temperature (to the nearest degree) on the Fahrenheit scale.

41. Sound travels at the rate of 1080 ft/sec. How many m/sec is this?

APPENDIX C

Dimensional Analysis

In the scientific and technical worlds we must develop the ability to make precise measurements. Measuring techniques often enable the scientist and technician to observe and describe what will happen under a given set of conditions, what changes will occur, and where and when these changes take place. Such information might then be used to develop new devices capable of performing complex operations.

Many kinds of units are used to measure the same quantity. For example, length is measured in feet, yards, meters, furlongs, fathoms, and many other units. Weight is measured in terms of pounds, grams, kilograms, and tons, just to name a few. The kind of unit we use is called the **dimension**. In our work with dimensional units and conversions we shall use the fundamental dimensions of length (L), mass (M), and time (T), since the dimensions of most quantities are just multiples of L, M, and T. In Table C.1 the basic units of length, mass, and time are illustrated for both the English and metric systems of measurement.

Table C.1. Units for Fundamental Dimensions

Physical Quantity	*Symbol*	*English Units*	*Metric Units*
length	L	foot	meter
mass	M	pound	gram
time	T	second	second

If we treat the units in Table C.1 as algebraic quantities that can be multiplied and divided, we can generate additional physical dimensions that are often encountered. Some of these derived units are shown in Table C.2.

Table C.2

Physical quantity	*Dimension*	*English Unit*	*Metric Unit*
Area	$L \times L = L^2$	ft^2	m^2
Volume	$L \times L \times L = L^3$	ft^3	m^3
Density $= \dfrac{\text{weight}}{\text{volume}}$	$\dfrac{M}{L^3}$	lb/ft^3	kg/m^3
Average velocity $= \dfrac{\text{distance}}{\text{time}}$	$\dfrac{L}{T}$	ft/sec	m/sec
Average acceleration $= \dfrac{\text{change in velocity}}{\text{time}}$	$\dfrac{L}{T} \div T = \dfrac{L}{T^2}$	ft/sec^2	m/sec^2
Pressure $= \dfrac{\text{force}}{\text{area}}$	$\dfrac{M}{L^2}$	lb/ft^2	kg/m^2
Work = distance × force	$L \times M$	ft · lb	kg · m
Power $= \dfrac{\text{work}}{\text{time}}$	$\dfrac{L \times M}{T}$	$\dfrac{\text{ft} \cdot \text{lb}}{\text{sec}}$	$\dfrac{\text{kg} \cdot \text{m}}{\text{sec}}$

It is often necessary to change units in order to be consistent with our calculations. Here again, we treat the units algebraically and multiply, divide, and cancel when possible. In so doing we must always be careful to multiply the original units by fractions that are equal to 1. For example, $\dfrac{12\text{ in.}}{1\text{ ft}}$, $\dfrac{60\text{ sec}}{1\text{ min}}$, $\dfrac{5280\text{ ft}}{1\text{ mile}}$, and $\dfrac{1\text{ m}}{100\text{ cm}}$ are all examples of conversion factors that are equal to 1.

The following examples illustrate the use of these conversion factors.

Example C.1

Convert 45 miles/hr (mph) to ft/sec.

Solution

To perform the conversion indicated, we need the following identities: 5280 ft = 1 mile; 60 min = 1 hr; 60 sec = 1 min. Thus, our conversion factors are

$$\frac{5280\text{ ft}}{1\text{ mile}} \qquad \frac{1\text{ hr}}{60\text{ min}} \qquad \frac{1\text{ min}}{60\text{ sec}}$$

We use these factors in the following multiplication, to obtain

$$\frac{45\ \text{miles}}{1\ \text{hr}} \times \frac{5280\ \text{ft}}{1\ \text{mile}} \times \frac{1\ \text{hr}}{60\ \text{min}} \times \frac{1\ \text{min}}{60\ \text{sec}} = \frac{45 \times 5280\ \text{ft}}{60 \times 60\ \text{sec}} = 66\ \text{ft/sec}$$

We note that in Example C.1 we did not change the physical quantity. We began with velocity and ended with velocity. *Only the units used to express the quantity were changed.*

Example C.2

Convert 440 yards to millimeters.

Solution

Here we need the following conversion factors:

$$\frac{3\ \text{ft}}{1\ \text{yd}} \qquad \frac{12\ \text{in.}}{1\ \text{ft}} \qquad \frac{2.54\ \text{cm}}{1\ \text{in.}} \qquad \frac{10\ \text{mm}}{1\ \text{cm}}$$

Thus,

$$\begin{aligned} 440\ \text{yd} &= 440\ \text{yd} \cdot \frac{3\ \text{ft}}{1\ \text{yd}} \times \frac{12\ \text{in.}}{1\ \text{ft}} \times \frac{2.54\ \text{cm}}{1\ \text{in.}} \times \frac{10\ \text{mm}}{1\ \text{cm}} \\ &= 402{,}336\ \text{mm} \end{aligned}$$

Note that we began with a measure of length and ended with a measure of length. *Only the units used to express the length have been changed.*

Example C.3

The density of zinc is $7.15\ \text{g/cm}^3$. Express the density in lb/in.^3.

Solution

Since $1\ \text{lb} = 454\ \text{g}$ and $1\ \text{in.} = 2.54\ \text{cm}$, our conversion factors are $\dfrac{1\ \text{lb}}{454\ \text{g}}$ and $\left(\dfrac{2.54\ \text{cm}}{1\ \text{in.}}\right)^3$. Thus,

$$\frac{7.15\ \text{g}}{\text{cm}^3} = \frac{7.15\ \text{g}}{\text{cm}^3} \times \frac{1\ \text{lb}}{454\ \text{g}} \times \frac{(2.54)^3\ \text{cm}^3}{1\ \text{in.}^3} = 0.04\ \text{lb/in.}^3$$

Note that it was necessary to cube the unit $\dfrac{2.54\ \text{cm}}{1\ \text{in.}}$ so that we could cancel the units cm^3.

Exercises for Appendix C

In Exercises 1 through 10, perform the conversion indicated. If necessary, use the tables in Appendix B.

1. Convert 4 ft 2 in. to meters.

2. Convert 2 yd to centimeters.

3. Convert 100 mm to inches.

4. Convert 1600 $in.^2$ to cm^2.

5. Convert 14 $lb/in.^2$ to kg/m^2.

6. Convert 75 mph to ft/sec.

7. Convert 75 mph to m/sec.

8. Convert 2.5 m^2 to $in.^2$.

9. Convert 32 ft/sec^2 to cm/sec^2.

10. Convert 550 ft · lb to kg · m.

11. The speedometer of an automobile reads 80 km/hr. How fast is the car going in mph? In ft/sec?

12. The density of bronze is 548 lb/ft^3. Express the density in terms of lb/m^3 and kg/m^3.

13. By definition, 1 (ℓ) of water weighs 1 kg. How many kilograms are there in 1 gal of water?

14. A gasoline tank holds 23 gal. How many liters does it hold?

15. The speed of light is approximately 1.86×10^5 miles/sec. Find the speed of light in m/sec.

16. The distance from the earth to the sun is about 93,000,000 miles. What is the distance in kilometers?

17. The focal length of a camera lens is 50 mm. What is the focal length in inches?

18. The speed of sound in air is about 1129 ft/sec. What is the speed in m/sec?

APPENDIX D

Approximation, Significant Digits, and Rounding Procedures

In technology, most of the numbers encountered are measures of physical objects. These measures are obtained by such instruments as meter sticks, calipers, balances, thermometers, and many more. If we measure anything other than a countable number of objects, the numeral used to indicate the measure is always subject to some error. The amount of error involved is due primarily to the precision of the instrument used and, to some degree, to the person making the measurement.

Numbers used in measurement will be either exact or approximate. For example, suppose that the number of slides in a certain tray is twenty-four. In this case the numeral 24 is exact, since it can be arrived at by a counting process. However, the length and width of each slide cannot be determined exactly. Any number used to represent either the length or the width will only be an approximate value that shows the limits within which the value is known. If we say that the length is 3 cm, we mean by convention that the value could actually be anywhere between 2.5 and 3.5; that is, $2.5 \leq 3 < 3.5$. However, if we say that the length is 3.0 cm, we mean that the actual value is between 2.95 and 3.05; that is, $2.95 \leq 3.0 < 3.05$.

To distinguish properly between the meaning or "significance" of numbers such as 3, 3.0, 3.00, and so on, we must understand the concept of **significant digits**. In effect, significant digits are those digits used to represent the accuracy of a number, that is, those digits that are known to be reliable.

We determine significant digits according to the following rules.

Rules for Determining Significant Digits

1. All nonzero digits are significant.
2. All zeros between significant digits are significant.
3. All zeros that lie to the right of the decimal point and the last nonzero digit from the right are significant.
4. All zeros between the first nonzero digit from the right and the zero with the tilde (~) over it inclusive are significant.

Example D.1

(a) 324.6 has four significant digits.
(b) 2004 has four significant digits.
(c) 260 has two significant digits.
(d) 260. has three significant digits.
(e) 260.0 has four significant digits.
(f) 0.0013 has two significant digits.
(g) 0.00130 has three significant digits.
(h) $45{,}00\tilde{0}$ has five significant digits.
(i) $45{,}0\tilde{0}0$ has four significant digits.
(j) 45,000 has two significant digits.

From Example D.1 we note that zeros at the beginning of a decimal fraction are not significant. We also note that those zeros at the end of a whole number are not significant if they are used only to locate the decimal point. Thus, 13, 130, 1300, 13,000, and 130,000 all have two significant digits.

In general, zero may or may not be significant, depending on its position in a numeral. This is illustrated in the following example.

Example D.2

Consider the number 0.003100. The three zeros preceding the 3 are not significant, but the two zeros following the 3 are significant. Thus, the numeral 0.003100 contains four significant figures.

The accuracy of a number is the position occupied by its rightmost significant digit.

Example D.3

(a) 2.16 is accurate to the *hundredths* place.
(b) 12.9 is accurate to the *tenths* place.
(c) 16,000 is accurate to the *thousands* place.
(d) 300 is accurate to the *hundreds* place.
(e) 640 is accurate to the *tens* place.
(f) 640. is accurate to the *units* place.
(g) 1.046 is accurate to the *thousandths* place.

We note that the annexation of zeros to the right of a numeral increases the degree of accuracy. The accuracy of 4.2 is to the *tenths* position, while the accuracy of 4.200 is to the *thousandths* position. Thus, the addition of zeros should be avoided unless we are actually indicating the accuracy of a number.

In some cases we may not be interested in a high degree of accuracy. For example, at a certain instant the distance from the earth to sun is 91,637,146 miles. Rather than work with this number, which contains eight significant digits, we may find it advantageous to "round off" the number. When we round off a number, we imply that although the rounded number is an approximation, it is accurate enough for practical purposes. Thus, we might round 91,637,146 to 92,000,000.

Example D.4

(a) 3125.57 rounded off to five significant digits is 3125.6
(b) 3125.57 rounded off to four significant digits is 3126.
(c) 3125.57 rounded off to three significant digits is 3130.
(d) 3125.57 rounded off to two significant digits is 3100.

In general, we observe the following rules when rounding off numbers.

Rules for Rounding

1. If the first significant digit dropped is less than 5, the last digit kept remains unchanged. In this case we have rounded the number downward.
2. If the first significant digit dropped is greater than 5, the last digit kept is increased by one. In this case we have rounded the number upward.
3. If the first significant digit dropped is equal to 5, the last digit kept remains unchanged if it is **even** and is increased by one if it is **odd**. In either case, the last digit kept is **always** an **even** number.

In performing algebraic operations with approximate numbers, we must be careful to express the result with the correct degree of accuracy. For example, suppose that the following numbers, representing measure in the same units, are to be added:

$$\begin{array}{r@{.}l} 13&214 \\ 234&6 \\ 7&0350 \\ 6&38 \end{array}$$

We observe that the second number in the list is the least accurate. Since 234.6 is accurate to the *tenths* position, the sum should also be rounded to the *tenths* position. Adding,

$$\begin{array}{r@{.}l} 13&214 \\ 234&6 \\ 7&0350 \\ 6&38 \\ \hline 261&2290 \end{array}$$

Rounding off to the tenths position yields 261.2.

We can simplify arithmetic operations and save time by using the smallest number of significant figures that will give the proper result.

For problems that involve addition or subtraction or both, we calculate the result as follows:

1. Round each number to one position more accurate than the least accurate number.
2. Perform the operations.
3. Round the result to the accuracy of the least accurate number.

Thus, we would write the previous set of numbers as follows:

$$\begin{array}{r} 13.21 \\ 234.6 \\ 7.04 \\ 6.38 \\ \hline 261.23 \end{array}$$

We now round 261.23 to the tenths position, to obtain the previous result, 261.2.

Example D.5

Add 1246.75, 134.5, 450, and 78.

Solution

The least accurate number is 450, which is accurate to the *tens* position. Therefore, before performing the addition, we shall round the remaining numbers to the *ones* position. Thus, 1246.75 is rounded to 1247; 134.5 is rounded to 134; and 78 is left unchanged. Adding, we have

$$\begin{array}{r} 1247 \\ 134 \\ 450 \\ 78 \\ \hline 1909 \end{array}$$

We now round 1909 to the *tens* place, to obtain 1910.

The product or quotient of approximate values is calculated according to the following rules:

1. Round off each factor to one more significant digit than the factor with the least number of significant digits.
2. Multiply or divide the factors as indicated.
3. Round the product to the same number of significant digits as the factor with the least number of significant digits.

Example D.6

Find the product of 50.35 and 0.025.

Solution

The factor 0.025 has the least number of significant digits, two. Therefore, we round the factor 50.35 to three significant digits. Thus, 50.35 is rounded to 50.4. The product $50.4 \times 0.025 = 1.2600$. We now round this result to two significant digits and obtain 1.3. Thus, $50.35 \times 0.025 = 1.3$.

Example D.7

Find the product of 16.235, 0.217, and 5.

Solution

The factor 5 has the least number of significant digits, one. Therefore, we round the other factors to two significant digits. Thus, 16.235 is rounded to 16 and 0.217 is rounded to 0.22. The product $16 \times 0.22 \times 5 = 17.60$. We now round this result to one significant digit and obtain 20. Thus, $16.235 \times 0.217 \times 5 = 20$.

Example D.8

Divide 435.7 by 3.6.

Solution

Since the divisor 3.6 has two significant digits, we round the dividend to three significant digits and divide until we have three significant digits in the quotient. We then round the result to two significant digits, as follows:

```
     121
 36)4360
    36
    --
     76
     72
     --
      40
      36
      --
```

We now round off 121 to obtain the final result, 120.

Exercises for Appendix D

In Exercises 1 through 16, indicate the number of significant figures and state the accuracy of each.

1. 24.04 cm

2. 3280 m^2

3. 40g

4. 0.02 $in.^3$

5. 0.020 g

6. 18,000 g

7. 4.213 m

8. $235,\tilde{0}00$ cm

9. 0.0005 g

10. 0.0396 m

11. 1.0396 l

12. 10,000 ft

13. $1\tilde{0},000$ in.

14. $10,00\tilde{0}$ kg

15. 0.0120 yd

16. 4221.0 mm

In Exercises 17 through 30, round off the given numbers to (a) four, (b) three, and (c) two significant digits.

17. 742,396

18. 82.1435

19. 0.07284

20. 5.26134

21. 231.45

22. 23,467

23. 10,265

24. 0.03124

25. 0.012450

26. 2315.2

27. 83,956

28. 1000.25

29. 23,850

30. 0.21475

In Exercises 31 through 50, perform the operation indicated on the numbers, using the method illustrated in Examples D.5 and D.8.

31. Add 21.5, 6.286, and 200.16.

32. Add 2.8, 3.45, and 1.125

33. Add 0.0213, 1.1234, 0.023, and 0.54

34. Add 9.52, 41.38, and 23.165

35. Add 0.2, 2, 6.2 and 12.4

36. $2.003 - 1.5$

37. $13 - 0.234$

38. 124×2.8

39. 7.34×2.3

40. 71.3×0.21

41. 0.946×23

42. 0.00435×4.6

43. 650.42×4.12

44. $12 \times 2.34 \times 5.678$

45. $0.1352 \times 1200 \times 6.25$

46. $6.256 \div 2.8$

47. $205 \div 8.5$

48. $4.424 \div 0.56$

49. $7.4 \div 0.13$

50. $40.6586 \div 2.46$

APPENDIX E

Hand Calculators

Hand calculators come with a great variety of features. The preference for one calculator over another is usually due to the needs of the person, budget limitations, or taste. For our purposes we shall assume that the machine in use can calculate trigonometric and logarithmic functions and display at least eight digits. Most of these machines will also have square root, reciprocal functions, a memory, and some internal constants (such as π). In addition, trigonometric operations can be performed in degrees and radians, and results converted to scientific notation.

These machines generally employ one of two logical systems. The first of these is parenthetical, the **reverse-Polish notation**, named in honor of Lukasciewicz. One advantage of this notation is that it enables the user to employ complex formulas and achieve results without writing intermediate steps. Calculators using reverse-Polish logic also have the feature of displaying intermediate results. This system is used primarily by Hewlett–Packard in the construction of their sophisticated pocket calculators.

The second system is **algebraic logic**. Calculations done with algebraic logic are carried out in the sequence in which we have learned algebra. Thus, the routine used in a calculator with algebraic logic is easy to learn. This advantage is somewhat negated by the fact that the number of entries needed to solve a problem using algebraic logic is frequently more than is required when using reverse-Polish notation.

Let us illustrate the differences between these types of logic with some examples.

Example E.1

Find (2 × 3) + (4 × 5), using (a) reverse-Polish logic and (b) algebraic logic.

Solution

The following steps yield the appropriate result.

(a)

Sequence	1	2	3	4	5	6	7	8	9
Entry	[2]	[↑]	[3]	[×]	[4]	[↑]	[5]	[×]	[+]

(b)

Sequence	1	2	3	4	5	6	7	8
Entry	[2]	[×]	[3]	[+]	[4]	[×]	[5]	[=]

Example E.2

Find (2 + 3) × (4 + 5), using (a) reverse-Polish logic and (b) algebraic logic.

Solution

(a)

Sequence	1	2	3	4	5	6	7	8	9
Entry	[2]	[↑]	[3]	[+]	[4]	[↑]	[5]	[+]	[×]

(b)

Sequence	1	2	3	4	5	6	7	8	9	10	11	12
Entry	[2]	[+]	[3]	[=]	[STO]	[4]	[+]	[5]	[=]	[×]	[RCL]	[=]

Example E.3

Find (2 + 3) ÷ (4 + 5), using (a) reverse-Polish logic and (b) algebraic logic.

Solution

(a)

Sequence	1	2	3	4	5	6	7	8	9
Entry	[2]	[↑]	[3]	[+]	[4]	[↑]	[5]	[+]	[÷]

(b)

Sequence	1	2	3	4	5	6	7	8	9	10	11	12	13
Entry	[2]	[+]	[3]	[=]	[STO]	[4]	[+]	[5]	[=]	[÷]	[RCL]	[$x \lessgtr y$]	[=]

Although it is not possible to cover all facets of all calculators, the following examples serve to illustrate how they might be used to solve some of the problems encountered in the text. The sequence shown in part (a) is reverse-Polish logic and the sequence shown in part (b) is algebraic logic.

These examples assume that the reader is familiar with the machine being used, has read the explanation in the owner's manual, and has worked out the sample examples shown.

Example E.4

Evaluate $\sqrt{5.64} \times \sin 25°$.

Solution

(a) Sequence 1 2 3 4 5 6

Entry [5.64] [↑] [$\sqrt{x}$] [25]* [sin] [×]

(b) Sequence 1 2 3 4 5 6

Entry [5.64] [$\sqrt{x}$] [×] [25]* [sin] [=]

1.00366276

Example E.5

Evaluate cos 30° + log 1.23.

Solution

(a) Sequence 1 2 3 4 5 6

Entry [30]* [↑] [cos] [1.23] [log] [+]

(b) Sequence 1 2 3 4 5 6

Entry [30][1] [cos] [+] [1.23] [log] [=]

0.95593052

Example E.6

Evaluate $\dfrac{\sin 1.40 + \ln 23.56}{\cos^{-1}(0.25)}$.

Solution

(a) Sequence 1 2 3 4 5 6 7 8 9

Entry [1.40]† [↑] [sin] [23.56] [ln] [+] [0.25]† [$\cos^{-1}$]‡ [÷]

(b) Sequence 1 2 3 4 5 6 7 8 9 10 11

Entry [1.40]† [sin] [+] [23.56] [ln] [=] [÷] [0.25] [arc] [cos] [=]

The result is 3.14463967.

Example E.7

Solve for x if $2^x = 5.28$.

Solution

We first simplify the expression by taking the common logarithm of both sides of the equation. This yields

$$\log 2^x = \log 5.28$$

$$x \log 2 = \log 5.28$$

or

$$x = \frac{\log 5.28}{\log 2}$$

* Calculations done in the degree mode. For calculators with D/R, switch to D.
† Calculations done in the radian mode. For calculators with D/R, switch to R.
‡ For calculators that [arc], the entry is [arc] followed by the appropriate function.

To find x we have

(a) Sequence 1 2 3 4 5 6

Entry [5.28] [↑] [log] [2] [log] [÷]

(b) Sequence 1 2 3 4 5 6

Entry [5.28] [log] [÷] [2] [log] [=]

} 2.40053793

(***Note***: The use of the ln function will yield the **same result**.)

Example E.8

Use the law of sines to solve for angle B if $a = 35$, $b = 48$, and $\angle A = 40°$.

Solution

The law of sines states that $\dfrac{a}{\sin A} = \dfrac{b}{\sin B} = \dfrac{c}{\sin C}$. Since we are solving for $\angle B$, we write

$$\sin B = \frac{b \sin A}{a}$$

or

$$B = \arcsin \frac{b \sin A}{a} = \arcsin \frac{48 \sin 40°}{35}$$

To find B we use the following steps:

(a) Sequence 1 2 3 4 5 6 7 8

Entry [48] [↑] [40]* [sin] [×] [35] [÷] [$\sin^{-1}$]‡

(b) Sequence 1 2 3 4 5 6 7 8 9

Entry [48] [×] [40]* [sin] [÷] [68] [=] [arc] [sin]

From these calculations, angle $B = 61.82836583°$. The reader is reminded that there is a second value for $\angle B$. Find this value and show that it is, in fact, a second solution for the problem.

For the reader who wishes to study the hand calculator in greater detail, the following books will be helpful:

The Complete Pocket Calculator Handbook, by Henry Millish (New York: Macmillan, 1976).

The Slide Rule, Electronic Hand Calculator, and Metrification in Problem Solving, by George C. Beakley and H. W. Leach (New York: Macmillan, 1975).

* Calculations done in the degree mode. For calculators with D/R, switch to D.

‡ For calculators that have [arc], the entry is [arc] followed by the appropriate function.

Exercises for Appendix E

Use a calculator to find solutions for the following sets of exercises taken from the sections indicated.

Section	*Exercise Numbers*
40	16–30
41	21–32
42	1–12
43	1–20
53	21–40
54	1–30
55	1–26
56	1–40
57	1–37

APPENDIX F

Tables

Table I. Four-Place Values of Trigonometric Functions

Angle θ in:									
Degrees	*Radians*	sin θ	csc θ	tan θ	cot θ	sec θ	cos θ		
0° 00′	0.0000	0.0000	No value	0.0000	No value	1.000	1.0000	1.5708	00° 00′
10	029	029	343.8	029	343.8	000	000	679	50
20	058	058	171.9	058	171.9	000	000	650	40
30	087	087	114.6	087	114.6	000	1.0000	621	30
40	116	116	85.95	116	85.94	000	.9999	592	20
50	145	145	68.76	145	68.75	000	999	563	10
1° 00′	0.0175	0.0175	57.30	0.0175	57.29	1.000	0.9998	1.5533	89° 00′
10	204	204	49.11	204	49.10	000	998	504	50
20	233	233	42.98	233	42.96	000	997	475	40
30	262	262	38.20	262	38.19	000	997	446	30
40	291	291	34.38	291	34.37	000	996	417	20
50	320	320	31.26	320	31.24	001	995	388	10
2° 00′	0.0349	0.0349	28.65	0.0349	28.64	1.001	0.9994	1.5359	88° 00′
10	378	378	26.43	378	26.43	001	993	330	50
20	407	407	24.56	407	24.54	001	992	301	40
30	436	436	22.93	437	22.90	001	990	272	30
40	465	465	21.49	466	21.47	001	989	243	20
50	495	494	20.23	495	20.21	001	988	213	10
3° 00′	0.0524	0.0523	19.11	0.0524	19.08	1.001	0.9986	1.5184	87° 00′
10	553	552	18.10	553	18.07	002	985	155	50
20	582	581	17.20	582	17.17	002	983	126	40
30	611	610	16.38	612	16.35	002	981	097	30
40	640	640	15.64	641	15.60	002	980	068	20
50	669	669	14.96	670	14.92	002	978	039	10
4° 00′	0.0698	0.0698	14.34	0.0699	14.30	1.002	0.9976	1.5010	86° 00′
10	727	727	13.76	729	13.73	003	974	981	50
20	756	765	13.23	758	13.20	003	971	952	40
30	785	785	12.75	787	12.71	003	969	923	30
40	814	814	12.29	816	12.25	003	967	893	20
50	844	843	11.87	846	11.83	004	964	864	10
5° 00′	0.0873	0.0872	11.47	0.0875	11.43	1.004	0.9962	1.4835	85° 00′
10	902	901	11.10	904	11.06	004	959	806	50
20	931	929	10.76	934	10.71	004	957	777	40
30	960	958	10.43	963	10.39	005	954	748	30
40	.0989	.0987	10.13	.0992	10.08	005	951	719	20
50	.1018	.1016	9.839	.1022	9.788	005	948	690	10
6° 00′	0.1047	0.1045	9.567	0.1051	9.514	1.006	0.9945	1.4661	84° 00′
10	076	074	9.309	080	9.255	006	942	632	50
20	105	103	9.065	110	9.010	006	939	603	40
30	134	132	8.834	139	8.777	006	936	573	30
40	164	161	8.614	169	8.556	007	932	544	20
50	193	190	8.405	198	8.345	007	929	515	10
7° 00′	0.1222	0.1219	8.206	0.1228	8.141	1.008	0.9925	1.4486	83° 00′
10	251	248	8.016	257	7.953	008	922	457	50
20	280	276	7.834	287	7.770	008	918	428	40
30	309	305	7.661	317	7.596	009	914	399	30
40	338	334	7.496	346	7.429	009	911	370	20
50	367	363	7.337	376	7.269	009	907	341	10
8° 00′	0.1396	0.1392	7.185	0.1405	7.115	1.010	0.9903	1.4312	82° 00′
								Radians	*Degrees*
		cos θ	sec θ	cot θ	tan θ	csc θ	sin θ	*Angle θ*	

Table I. Four-Place Values of Trigonometric Functions (continued)

Angle θ in: Degrees	Radians	sin θ	csc θ	tan θ	cot θ	sec θ	cos θ		
8° 00′	0.1396	0.1392	7.185	0.1405	7.115	1.010	0.9903	1.4312	82° 00′
10	425	421	7.040	435	6.968	010	899	283	50
20	454	449	6.900	465	827	011	894	254	40
30	484	478	765	495	691	011	890	224	30
40	513	507	636	524	561	012	886	195	20
50	542	536	512	554	435	012	881	166	10
9° 00′	0.1571	0.1564	6.392	0.1584	6.314	1.012	0.9877	1.4137	81° 00′
10	600	593	277	614	197	013	872	108	50
20	629	622	166	644	6.084	013	868	079	40
30	658	650	6.059	673	5.976	014	863	050	30
40	687	679	5.955	703	871	014	858	1.4021	20
50	716	708	855	733	769	015	853	1.3992	10
10° 00′	0.1745	0.1736	5.759	0.1763	5.671	1.015	0.9848	1.3963	80° 00′
10	774	765	665	793	576	016	843	934	50
20	804	794	575	823	485	016	838	904	40
30	833	822	487	853	396	017	833	875	30
40	862	851	403	883	309	018	827	846	20
50	891	880	320	914	226	018	822	817	10
11° 00′	0.1920	0.1908	5.241	0.1944	5.145	1.019	0.9816	1.3788	79° 00′
10	949	937	164	.1974	5.066	019	811	759	50
20	.1978	965	089	.2004	4.989	020	805	730	40
30	.2007	.1994	5.016	035	915	020	799	701	30
40	036	.2022	4.945	065	843	021	793	672	20
50	065	051	876	095	773	022	787	643	10
12° 00′	0.2094	0.2079	4.810	0.2126	4.705	1.022	0.9781	1.3614	78° 00′
10	123	108	745	156	638	023	775	584	50
20	153	136	682	186	574	024	769	555	40
30	182	164	620	217	511	024	763	526	30
40	211	193	560	247	449	025	757	497	20
50	240	221	502	278	390	026	750	468	10
13° 00′	0.2269	0.2250	4.445	0.2309	4.331	1.026	0.9744	1.3439	77° 00′
10	298	278	390	339	275	027	737	410	50
20	327	306	336	370	219	028	730	381	40
30	356	334	284	401	165	028	724	352	30
40	385	363	232	432	113	029	717	323	20
50	414	391	182	462	061	030	710	294	10
14° 00′	0.2443	0.2419	4.134	0.2493	4.011	1.031	0.9703	1.3265	76° 00′
10	473	447	086	524	3.962	031	696	235	50
20	502	476	4.039	555	914	032	689	206	40
30	531	504	3.994	586	867	033	681	177	30
40	560	532	950	617	821	034	674	148	20
50	589	560	906	648	776	034	667	119	10
15° 00′	0.2618	0.2588	3.864	0.2679	3.732	1.035	0.9659	1.3090	75° 00′
10	647	616	822	711	689	036	652	061	50
20	676	644	782	742	647	037	644	032	40
30	705	672	742	773	606	038	636	1.3003	30
40	734	700	703	805	566	039	628	1.2974	20
50	763	728	665	836	526	039	621	945	10
16 00′	0.2793	0.2756	3.628	0.2867	3.487	1.040	0.9613	1.2915	74° 00′
		cos θ	sec θ	cot θ	tan θ	csc θ	sin θ	*Radians*	*Degrees*
								Angle θ	

Table I. Four-Place Values of Trigonometric Functions (continued)

Angle θ in:									
Degrees	*Radians*	sin θ	csc θ	tan θ	cot θ	sec θ	cos θ		
16° 00′	0.2793	0.2756	3.628	0.2867	3.487	1.040	0.9613	1.2915	74° 00′
10	822	784	592	899	450	041	605	886	50
20	851	812	556	931	412	042	596	857	40
30	880	840	521	962	376	043	588	828	30
40	909	868	487	.2944	340	044	580	799	20
50	938	896	453	.3026	305	045	572	770	10
17° 00′	0.2967	0.2924	3.420	0.3057	3.271	1.046	0.9563	1.2741	73° 00′
10	.2996	952	388	089	237	047	555	712	50
20	.3025	.2979	357	121	204	048	546	683	40
30	054	.3007	326	153	172	048	537	654	30
40	083	035	295	185	140	049	528	625	20
50	113	062	265	217	108	050	520	595	10
18° 00′	0.3142	0.3090	3.236	0.3249	3.078	1.051	0.9511	1.2566	72° 00′
10	171	118	207	281	047	052	502	537	50
20	200	145	179	314	3.018	053	492	508	40
30	229	173	152	346	2.989	054	483	479	30
40	258	201	124	378	960	056	474	450	20
50	287	228	098	411	932	057	465	421	10
19° 00′	0.3316	0.3256	3.072	0.3443	2.904	1.058	0.9455	1.2392	71° 00′
10	345	283	046	476	877	059	446	363	50
20	374	311	3.021	508	850	060	436	334	40
30	403	338	2.996	541	824	061	426	305	30
40	432	365	971	574	798	062	417	275	20
50	462	393	947	607	773	063	407	246	10
20° 00′	0.3491	0.3420	2.924	0.3640	2.747	1.064	0.9397	1.2217	70° 00′
10	520	448	901	673	723	065	387	188	50
20	549	475	878	706	699	066	377	159	40
30	578	502	855	739	675	068	367	130	30
40	607	529	833	772	651	069	356	101	20
50	636	557	812	805	628	070	346	072	10
21° 00′	03665	0.3584	2.790	0.3839	2.605	1.071	0.9336	1.2043	69° 00′
10	694	611	769	872	583	072	325	1.2014	50
20	723	638	749	906	560	074	315	985	40
30	752	665	729	939	539	075	304	956	30
40	782	692	709	.3973	517	076	293	926	20
50	811	719	689	.4006	496	077	283	897	10
22° 00′	0.3840	0.3746	2.669	0.4040	2.475	1.079	0.9272	1.1868	68° 00′
10	869	773	650	074	455	080	261	839	50
20	898	800	632	108	434	081	250	810	40
30	927	827	613	142	414	082	239	781	30
40	956	854	595	176	394	084	228	752	20
50	985	881	577	210	375	085	216	723	10
23° 00′	0.4014	0.3907	2.559	0.4245	2.356	1.086	0.9205	1.1694	67° 00′
10	043	934	542	279	337	088	194	665	50
20	072	961	525	314	318	089	182	636	40
30	102	.3987	508	348	300	090	171	606	30
40	131	.4014	491	383	282	092	159	577	20
50	160	041	475	417	264	093	147	548	10
24° 00′	0.4189	0.4067	2.459	0.4452	2.246	1.095	0.9135	1.1519	66° 00′
								Radians	*Degrees*
		cos θ	sec θ	cot θ	tan θ	csc θ	sin θ	*Angle θ*	

Table I. Four-Place Values of Trigonometric Functions (continued)

Angle θ in:									
Degrees	*Radians*	sin θ	csc θ	tan θ	cot θ	sec θ	cos θ		
24° 00′	0.4189	0.4067	2.459	0.4452	2.246	1.095	0.9135	1.1519	66° 00′
10	218	094	443	487	229	096	124	490	50
20	247	120	427	522	211	097	112	461	40
30	276	147	411	557	194	099	100	432	30
40	305	173	396	592	177	100	088	403	20
50	334	200	381	628	161	102	075	374	10
25° 00′	0.4363	0.4226	2.366	0.4663	2.145	1.103	0.9063	1.1345	65° 00′
10	392	253	352	699	128	105	051	316	50
20	422	279	337	734	112	106	038	286	40
30	451	305	323	770	097	108	026	257	30
40	480	331	309	806	081	109	013	228	20
50	509	358	295	841	066	111	.9001	199	10
26° 00′	0.4538	0.4384	2.281	0.4877	2.050	1.113	0.8988	1.1170	64° 00′
10	567	410	268	913	035	114	975	141	50
20	596	436	254	950	020	116	962	112	40
30	625	462	241	.4986	2.006	117	949	083	30
40	654	488	228	.5022	1.991	119	936	054	20
50	683	514	215	059	977	121	923	1.1025	10
27° 00′	0.4712	0.4540	2.203	0.5095	1.963	1.122	0.8910	1.0996	63° 00′
10	741	566	190	132	949	124	897	966	50
20	771	592	178	169	935	126	884	937	40
30	800	617	166	206	921	127	870	908	30
40	829	643	154	243	907	129	857	879	20
50	858	669	142	280	894	131	843	850	10
28° 00′	0.4887	0.4695	2.130	0.5317	1.881	1.133	0.8829	1.0821	62° 00′
10	916	720	118	354	868	134	816	792	50
20	945	746	107	392	855	136	802	763	40
30	.4974	772	096	430	842	138	788	734	30
40	.5003	797	085	467	829	140	774	705	20
50	032	823	074	505	816	142	760	676	10
29° 00′	0.5061	0.4848	2.063	0.5543	1.804	1.143	0.8746	1.0647	61° 00′
10	091	874	052	581	792	145	732	617	50
20	120	899	041	619	780	147	718	588	40
30	149	924	031	658	767	149	704	559	30
40	178	950	020	696	756	151	689	530	20
50	207	.4975	010	735	744	153	675	501	10
30° 00′	0.5236	0.5000	2.000	0.5774	1.732	1.155	0.8660	1.0472	60° 00′
10	265	025	1.990	812	720	157	646	443	50
20	294	050	980	851	709	159	631	414	40
30	323	075	970	890	698	161	616	385	30
40	352	100	961	930	686	163	601	356	20
50	381	125	951	.5969	675	165	587	327	10
31° 00′	0.5411	0.5150	1.942	0.6009	1.664	1.167	0.8572	1.0297	59° 00′
10	440	175	932	048	653	169	557	268	50
20	469	200	923	088	643	171	542	239	40
30	498	225	914	128	632	173	526	210	30
40	527	250	905	168	621	175	511	181	20
50	556	275	896	208	611	177	496	152	10
32° 00′	0.5585	0.5299	1.887	0.6249	1.600	1.179	0.8480	1.0123	58° 00′
								Radians	*Degrees*
		cos θ	sec θ	cot θ	tan θ	csc θ	sin θ	*Angle θ*	

Table I. Four-Place Values of Trigonometric Functions (continued)

Angle θ in:									
Degrees	*Radians*	sin θ	csc θ	tan θ	cot θ	sec θ	cos θ		
32° 00′	0.5585	0.5299	1.887	0.6249	1.600	1.179	0.8480	1.0123	58° 00′
10	614	324	878	289	590	181	465	094	50
20	643	348	870	330	580	184	450	065	40
30	672	373	861	371	570	186	434	036	30
40	701	398	853	412	560	188	418	1.0007	20
50	730	422	844	453	550	190	403	.9977	10
33° 00′	0.5760	0.5446	1.836	0.6494	1.540	1.192	0.8387	0.9948	57° 00′
10	789	471	828	536	530	195	371	919	50
20	818	495	820	577	520	197	355	890	40
30	847	519	812	619	511	199	339	861	30
40	876	544	804	661	501	202	323	832	20
50	905	568	796	703	492	204	307	803	10
34° 00′	0.5934	0.5592	1.788	0.6745	1.483	1.206	0.8290	0.9774	56° 00′
10	963	616	781	787	473	209	274	745	50
20	.5992	640	773	830	464	211	258	716	40
30	.6021	664	766	873	455	213	241	687	30
40	050	688	758	916	446	216	225	657	20
50	080	712	751	.6959	437	218	208	628	10
35° 00′	0.6109	0.5736	1.743	0.7002	1.428	1.221	0.8192	0.9599	55° 00′
10	138	760	736	046	419	223	175	570	50
20	167	783	729	089	411	226	158	541	40
30	196	807	722	133	402	228	141	512	30
40	225	831	715	177	393	231	124	483	20
50	254	854	708	221	385	233	107	454	10
36° 00′	0.6283	0.5878	1.701	0.7265	1.376	1.236	0.8090	0.9425	54° 00′
10	312	901	695	310	368	239	073	396	50
20	341	925	688	355	360	241	056	367	40
30	370	948	681	400	351	244	039	338	30
40	400	972	675	445	343	247	021	308	20
50	429	.5995	668	490	335	249	.8004	279	10
37° 00′	0.6458	0.6018	1.662	0.7536	1.327	1.252	0.7986	0.9250	53° 00′
10	487	041	655	581	319	255	696	221	50
20	516	065	649	627	311	258	951	192	40
30	545	088	643	673	303	260	934	163	30
40	574	111	636	720	295	263	916	134	20
50	603	134	630	766	288	266	898	105	10
38° 00′	0.6632	0.6157	1.624	0.7813	1.280	1.269	0.7880	0.9076	52° 00′
10	661	180	618	860	272	272	862	047	50
20	690	202	612	907	265	275	844	.9018	40
30	720	225	606	.7954	257	278	826	.8988	30
40	749	248	601	.8002	250	281	808	959	20
50	778	271	595	050	242	284	790	930	10
39° 00′	0.6807	0.6293	1.589	0.8098	1.235	1.287	0.7771	0.8901	51° 00′
10	836	316	583	146	228	290	753	872	50
20	865	338	578	195	220	293	735	843	40
30	894	361	572	243	213	296	716	814	30
40	923	383	567	292	206	299	698	785	20
50	952	406	561	342	199	302	679	756	10
40° 00′	0.6981	0.6428	1.556	0.8391	1.192	1.305	0.7660	0.8727	50° 00′
								Radians	*Degrees*
		cos θ	sec θ	cot θ	tan θ	csc θ	sin θ	*Angle θ*	

Table I. Four-Place Values of Trigonometric Functions (continued)

Angle θ in:									
Degrees	*Radians*	sin θ	csc θ	tan θ	cot θ	sec θ	cos θ		
40° 00′	0.6981	0.6428	1.556	0.8391	1.192	1.305	0.7660	0.8727	50° 00′
10	.7010	450	550	441	185	309	642	698	50
20	039	472	545	491	178	312	623	668	40
30	069	494	540	541	171	315	604	639	30
40	098	517	535	591	164	318	585	610	20
50	127	539	529	642	157	322	566	581	10
41° 00′	0.7156	0.6561	1.524	0.8693	1.150	1.325	0.7547	0.8552	49° 00′
10	185	583	519	744	144	328	528	523	50
20	214	604	514	796	137	332	509	494	40
30	243	626	509	847	130	335	490	465	30
40	272	648	504	899	124	339	470	436	20
50	301	670	499	.8952	117	342	451	407	10
42° 00′	0.7330	0.6691	1.494	0.9004	1.111	1.346	0.7431	0.8378	48° 00′
10	359	713	490	057	104	349	412	348	50
20	389	734	485	110	098	353	392	319	40
30	418	756	480	163	091	356	373	290	30
40	447	777	476	217	085	360	353	261	20
50	476	799	471	271	079	364	333	232	10
43° 00′	0.7505	0.6820	1.466	0.9325	1.072	1.367	0.7314	0.8203	47° 00′
10	534	841	462	380	066	371	294	174	50
20	563	862	457	435	060	375	274	145	40
30	592	884	453	490	054	379	254	116	30
40	621	905	448	545	048	382	234	087	20
50	650	926	444	601	042	386	214	058	10
44° 00′	0.7679	0.6947	1.440	0.9657	1.036	1.390	0.7193	0.8029	46° 00′
10	709	967	435	713	030	394	173	.7999	50
20	738	.6988	431	770	024	398	153	970	40
30	767	.7009	427	827	018	402	133	941	30
40	796	030	423	884	012	406	112	912	20
50	825	050	418	.9942	006	410	092	883	10
45° 00′	0.7854	0.7071	1.414	1.000	1.000	1.414	0.7071	0.7854	45° 00′
								Radians	*Degrees*
		cos θ	sec θ	cot θ	tan θ	csc θ	sin θ	*Angle θ*	

Table II. Four-Place Common Logarithms

											Proportional Parts								
No.	0	1	2	3	4	5	6	7	8	9	1	2	3	4	5	6	7	8	9
10	0000	0043	0086	0128	0170	0212	0253	0294	0334	0374	4	8	12	17	21	25	29	33	37
11	0414	0453	0492	0531	0569	0607	0645	0682	0719	0755	4	8	11	15	19	23	26	30	34
12	0792	0828	0864	0899	0934	0969	1004	1038	1072	1106	3	7	10	14	17	21	24	28	31
13	1139	1173	1206	1239	1271	1303	1335	1367	1399	1430	3	6	10	13	16	19	23	26	29
14	1461	1492	1523	1553	1584	1614	1644	1673	1703	1732	3	6	9	12	15	18	21	24	27
15	1761	1790	1818	1847	1875	1903	1931	1959	1987	2014	3	6	8	11	14	17	20	22	25
16	2041	2068	2095	2122	2148	2175	2201	2227	2253	2279	3	5	8	11	13	16	18	21	24
17	2304	2330	2355	2380	2405	2430	2455	2480	2504	2529	2	5	7	10	12	15	17	20	22
18	2553	2577	2601	2625	2648	2672	2695	2718	2742	2765	2	5	7	9	12	14	16	19	21
19	2788	2810	2833	2856	2878	2900	2923	2945	2967	2989	2	4	7	9	11	13	16	18	20
20	3010	3032	3054	3075	3096	3118	3139	3160	3181	3201	2	4	6	8	11	13	15	17	19
21	3222	3243	3263	3284	3304	3324	3345	3365	3385	3404	2	4	6	8	10	12	14	16	18
22	3424	3444	3464	3483	3502	3522	3541	3560	3579	3598	2	4	6	8	10	12	14	15	17
23	3617	3636	3655	3674	3692	3711	3729	3747	3766	3784	2	4	6	7	9	11	13	15	17
24	3802	3820	3838	3856	3874	3892	3909	3927	3945	3962	2	4	5	7	9	11	12	14	16
25	3979	3997	4014	4031	4048	4065	4082	4099	4116	4133	2	3	5	7	9	10	12	14	15
26	4150	4166	4183	4200	4216	4232	4249	4265	4281	4298	2	3	5	7	8	10	11	13	15
27	4314	4330	4346	4362	4378	4393	4409	4425	4440	4456	2	3	5	6	8	9	11	13	14
28	4472	4487	4502	4518	4533	4548	4564	4579	4594	4609	2	3	5	6	8	9	11	12	14
29	4624	4639	4654	4669	4683	4698	4713	4728	4742	4757	1	3	4	6	7	9	10	12	13
30	4771	4786	4800	4814	4829	4843	4857	4871	4886	4900	1	3	4	6	7	9	10	11	13
31	4914	4928	4942	4955	4969	4983	4997	5011	5024	5038	1	3	4	6	7	8	10	11	12
32	5051	5065	5079	5092	5105	5119	5132	5145	5159	5172	1	3	4	5	7	8	9	11	12
33	5185	5198	5211	5224	5237	5250	5263	5276	5289	5302	1	3	4	5	6	8	9	10	12
34	5315	5328	5340	5353	5366	5378	5391	5403	5416	5428	1	3	4	5	6	8	9	10	11
35	5441	5453	5465	5478	5490	5502	5514	5527	5539	5551	1	2	4	5	6	7	9	10	11
36	5563	5575	5587	5599	5611	5623	5635	5647	5658	5670	1	2	4	5	6	7	8	10	11
37	5682	5694	5705	5717	5729	5740	5752	5763	5775	5786	1	2	3	5	6	7	8	9	10
38	5798	5809	5821	5832	5843	5855	5866	5877	5888	5899	1	2	3	5	6	7	8	9	10
39	5911	5922	5933	5944	5955	5966	5977	5988	5999	6010	1	2	3	4	5	7	8	9	10
40	6021	6031	6042	6053	6064	6075	6085	6096	6107	6117	1	2	3	4	5	6	8	9	10
41	6128	6138	6149	6160	6170	6180	6191	6201	6212	6222	1	2	3	4	5	6	7	8	9
42	6232	6243	6253	6263	6274	6284	6294	6304	6314	6325	1	2	3	4	5	6	7	8	9
43	6335	6345	6355	6365	6375	6385	6395	6405	6415	6425	1	2	3	4	5	6	7	8	9
44	6435	6444	6454	6464	6474	6484	6493	6503	6513	6522	1	2	3	4	5	6	7	8	9
45	6532	6542	6551	6561	6571	6580	6590	6599	6609	6618	1	2	3	4	5	6	7	8	9
46	6628	6637	6646	6656	6665	6675	6684	6693	6702	6712	1	2	3	4	5	6	7	7	8
47	6721	6730	6739	6749	6758	6767	6776	6785	6794	6803	1	2	3	4	5	5	6	7	8
48	6812	6821	6830	6839	6848	6857	6866	6875	6884	6893	1	2	3	4	4	5	6	7	8
49	6902	6911	6920	6928	6937	6946	6955	6964	6972	6981	1	2	3	4	4	5	6	7	8
50	6990	6998	7007	7016	7024	7033	7042	7050	7059	7067	1	2	3	3	4	5	6	7	8
51	7076	7084	7093	7101	7110	7118	7126	7135	7143	7152	1	2	3	3	4	5	6	7	8
52	7160	7168	7177	7185	7193	7202	7210	7218	7226	7235	1	2	2	3	4	5	6	7	7
53	7243	7251	7259	7267	7275	7284	7292	7300	7308	7316	1	2	2	3	4	5	6	6	7
54	7324	7332	7340	7348	7356	7364	7372	7380	7388	7396	1	2	2	3	4	5	6	6	7

Table II. Four-Place Common Logarithms (continued)

No.	0	1	2	3	4	5	6	7	8	9	Proportional Parts 1	2	3	4	5	6	7	8	9
55	7404	7412	7419	7427	7435	7443	7451	7459	7466	7474	1	2	2	3	4	5	5	6	7
56	7482	7490	7497	7505	7513	7520	7528	7536	7543	7551	1	2	2	3	4	5	5	6	7
57	7559	7566	7574	7582	7589	7597	7604	7612	7619	7627	1	2	2	3	4	5	5	6	7
58	7634	7642	7649	7657	7664	7672	7679	7686	7694	7701	1	1	2	3	4	4	5	6	7
59	7709	7716	7723	7731	7738	7745	7752	7760	7767	7774	1	1	2	3	4	4	5	6	7
60	7782	7789	7796	7803	7810	7818	7825	7832	7839	7846	1	1	2	3	4	4	5	6	6
61	7853	7860	7868	7875	7882	7889	7896	7903	7910	7917	1	1	2	3	4	4	5	6	6
62	7924	7931	7938	7945	7952	7959	7966	7973	7980	7987	1	1	2	3	3	4	5	6	6
63	7993	8000	8007	8014	8021	8028	8035	8041	8048	8055	1	1	2	3	3	4	5	5	6
64	8062	8069	8075	8082	8089	8096	8102	8109	8116	8122	1	1	2	3	3	4	5	5	6
65	8129	8136	8142	8149	8156	8162	8169	8176	8182	8189	1	1	2	3	3	4	5	5	6
66	8195	8202	8209	8215	8222	8228	8235	8241	8248	8254	1	1	2	3	3	4	5	5	6
67	8261	8267	8274	8280	8287	8293	8299	8306	8312	8319	1	1	2	3	3	4	5	5	6
68	8325	8331	8338	8344	8351	8357	8363	8370	8376	8382	1	1	2	3	3	4	4	5	6
69	8388	8395	8401	8407	8414	8420	8426	8432	8439	8445	1	1	2	2	3	4	4	5	6
70	8451	8457	8463	8470	8476	8482	8488	8494	8500	8506	1	1	2	2	3	4	4	5	6
71	8513	8519	8525	8531	8537	8543	8549	8555	8561	8567	1	1	2	2	3	4	4	5	5
72	8573	8579	8585	8591	8597	8603	8609	8615	8621	8627	1	1	2	2	3	4	4	5	5
73	8633	8639	8645	8651	8657	8663	8669	8675	8681	8686	1	1	2	2	3	4	4	5	5
74	8692	8698	8704	8710	8716	8722	8727	8733	8739	8745	1	1	2	2	3	4	4	5	5
75	8751	8756	8762	8768	8774	8779	8785	8791	8797	8802	1	1	2	2	3	3	4	5	5
76	8808	8814	8820	8825	8831	8837	8842	8848	8854	8859	1	1	2	2	3	3	4	5	5
77	8865	8871	8876	8882	8887	8893	8899	8904	8910	8915	1	1	2	2	3	3	4	4	5
78	8921	8927	8932	8938	8943	8949	8954	8960	8965	8971	1	1	2	2	3	3	4	4	5
79	8976	8982	8987	8993	8998	9004	9009	9015	9020	9025	1	1	2	2	3	3	4	4	5
80	9031	9036	9042	9047	9053	9058	9063	9069	9074	9079	1	1	2	2	3	3	4	4	5
81	9085	9090	9096	9101	9106	9112	9117	9122	9128	9133	1	1	2	2	3	3	4	4	5
82	9138	9143	9149	9154	9159	9165	9170	9175	9180	9186	1	1	2	2	3	3	4	4	5
83	9191	9196	9201	9206	9212	9217	9222	9227	9232	9238	1	1	2	2	3	3	4	4	5
84	9243	9248	9253	9258	9263	9269	9274	9279	9284	9289	1	1	2	2	3	3	4	4	5
85	9294	9299	9304	9309	9315	9320	9325	9330	9335	9340	1	1	2	2	3	3	4	4	5
86	9345	9350	9355	9360	9365	9370	9375	9380	9385	9390	1	1	2	2	3	3	4	4	5
87	9395	9400	9405	9410	9415	9420	9425	9430	9435	9440	0	1	1	2	2	3	3	4	4
88	9445	9450	9455	9460	9465	9469	9474	9479	9484	9489	0	1	1	2	2	3	3	4	4
89	9494	9499	9504	9509	9513	9518	9523	9528	9533	9538	0	1	1	2	2	3	3	4	4
90	9542	9547	9552	9557	9562	9566	9571	9576	9581	9586	0	1	1	2	2	3	3	4	4
91	9590	9595	9600	9605	9609	9614	9619	9624	9628	9633	0	1	1	2	2	3	3	4	4
92	9638	9643	9647	9652	9657	9661	9666	9671	9675	9680	0	1	1	2	2	3	3	4	4
93	9685	9689	9694	9699	9703	9708	9713	9717	9722	9727	0	1	1	2	2	3	3	4	4
94	9731	9736	9741	9745	9750	9754	9759	9763	9768	9773	0	1	1	2	2	3	3	4	4
95	9777	9782	9786	9791	9795	9800	9805	9809	9814	9818	0	1	1	2	2	3	3	4	4
96	9823	9827	9832	9836	9841	9845	9850	9854	9859	9863	0	1	1	2	2	3	3	4	4
97	9868	9872	9877	9881	9886	9890	9894	9899	9903	9908	0	1	1	2	2	3	3	4	4
98	9912	9917	9921	9926	9930	9934	9939	9943	9948	9952	0	1	1	2	2	3	3	4	4
99	9956	9961	9965	9969	9974	9978	9983	9987	9991	9996	0	1	1	2	2	3	3	3	4

Table III. Four-Place Natural Logarithms (Base e; $\log_e 10 = \ln 10 = 2.3026$)

	0.00	0.01	0.02	0.03	0.04	0.05	0.06	0.07	0.08	0.09
1.0	0.0000	0.0100	0.0198	0.0296	0.0392	0.0488	0.0583	0.0677	0.0770	0.0862
1.1	0.0953	0.1044	0.1133	0.1222	0.1310	0.1398	0.1484	0.1570	0.1655	0.1740
1.2	0.1823	0.1906	0,1989	0.2070	0.2151	0.2231	0.2311	0,2390	0.2469	0.2546
1.3	0.2624	0.2700	0.2776	0.2852	0.2927	0.3001	0.3075	0.3148	0.3221	0.3293
1.4	0.3365	0.3436	0.3507	0.3577	0.3646	0.3716	0.3784	0.3853	0.3920	0.3988
1.5	0.4055	0.4121	0.4187	0.4253	0.4318	0.4383	0.4447	0.4511	0.4574	0.4637
1.6	0.4700	0.4762	0.4824	0.4886	0.4947	0.5008	0.5068	0.5128	0.5188	0.5247
1.7	0.5306	0.5365	0.5423	0.5481	0.5539	0.5596	0.5653	0.5710	0.5766	0.5822
1.8	0.5878	0.5933	0.5988	0.6043	0.6098	0.6152	0.6206	0.6259	0.6313	0.6366
1.9	0.6419	0.6471	0.6523	0.6575	0.6627	0.6678	0.6729	0.6780	0.6831	0.6881
2.0	0.6932	0.6981	0.7031	0.7080	0.7129	0.7178	0.7227	0.7275	0.7324	0.7372
2.1	0.7419	0.7467	0.7514	0.7561	0.7608	0.7655	0.7701	0.7747	0.7793	0.7839
2.2	0.7885	0.7930	0.7975	0.8020	0.8065	0.8109	0.8154	0.8198	0.8242	0.8286
2.3	0.8329	0.8373	0.8416	0.8459	0.8502	0.8544	0.8587	0.8629	0.8671	0.8713
2.4	0.8755	0.8796	0.8838	0.8879	0.8920	0.8961	0.9002	0.9042	0.9083	0.9123
2.5	0.9163	0.9203	0.9243	0.9282	0.9322	0.9361	0.9400	0.9439	0.9478	0.9517
2.6	0.9555	0.9594	0.9632	0.9670	0.9708	0.9746	0.9783	0.9821	0.9858	0.9895
2.7	0.9933	0.9969	1.0006	1.0043	1.0080	1.0116	1.0152	1.0188	1.0225	1.0260
2.8	1.0296	1.0332	1.0367	1.0403	1.0438	1.0473	1.0508	1.0543	1.0578	1.0613
2.9	1.0647	1.0682	1.0716	1.0750	1.0784	1.0818	1.0852	1.0886	1.0919	1.0953
3.0	1.0986	1.1019	1.1053	1.1086	1.1119	1.1151	1.1184	1.1217	1.1249	1.1282
3.1	1.1314	1.1346	1.1378	1.1410	1.1442	1.1474	1.1506	1.1537	1.1569	1.1600
3.2	1.1632	1.1663	1.1694	1.1725	1.1756	1.1787	1.1817	1.1848	1.1878	1.1909
3.3	1.1939	1.1969	1.2000	1.2030	1.2060	1.2090	1.2119	1.2149	1.2179	1.2208
3.4	1.2238	1.2267	1.2296	1.2326	1.2355	1.2384	1.2413	1.2442	1.2470	1.2499
3.5	1.2528	1.2556	1.2585	1.2613	1.2641	1.2669	1.2698	1.2726	1.2754	1.2782
3.6	1.2809	1.2837	1.2865	1.2892	1.2920	1.2947	1.2975	1.3002	1.3029	1.3056
3.7	1.3083	1.3110	1.3137	1.3164	1.3191	1.3218	1.3244	1.3271	1.3297	1.3324
3.8	1.3350	1.3376	1.3403	1.3429	1.3455	1.3481	1.3507	1.3533	1.3558	1.3584
3.9	1.3610	1.3635	1.3661	1.3686	1.3712	1.3737	1.3762	1.3788	1.3813	1.3838
4.0	1.3863	1.3888	1.3913	1.3938	1.3962	1.3987	1.4012	1.4036	1.4061	1.4085
4.1	1.4110	1.4134	1.4159	1.4183	1.4207	1.4231	1.4255	1.4279	1.4303	1.4327
4.2	1.4351	1.4375	1.4398	1.4422	1.4446	1.4469	1.4493	1.4516	1.4540	1.4563
4.3	1.4586	1.4609	1.4633	1.4656	1.4679	1.4702	1.4725	1.4748	1.4771	1.4793
4.4	1.4816	1.4839	1.4861	1.4884	1.4907	1.4929	1.4951	1.4974	1.4996	1.5019
4.5	1.5041	1.5063	1.5085	1.5107	1.5129	1.5151	1.5173	1.5195	1.5217	1.5239
4.6	1.5261	1.5282	1.5304	1.5326	1.5347	1.5369	1.5390	1.5412	1.5433	1.5454
4.7	1.5476	1.5497	1.5518	1.5539	1.5560	1.5581	1.5602	1.5623	1.5644	1.5665
4.8	1.5686	1.5707	1.5728	1.5748	1.5769	1.5790	1.5810	1.5831	1.5851	1.5872
4.9	1.5892	1.5913	1.5933	1.5953	1.5974	1.5994	1.6014	1.6034	1.6054	1.6074
5.0	1.6094	1.6114	1.6134	1.6154	1.6174	1.6194	1.6214	1.6233	1.6253	1.6273
5.1	1.6292	1.6312	1.6332	1.6351	1.6371	1.6390	1.6409	1.6429	1.6448	1.6467
5.2	1.6487	1.6506	1.6525	1.6544	1.6563	1 6582	1.6601	1.6620	1.6639	1.6658
5.3	1.6677	1.6696	1.6715	1.6734	1.6752	1.6771	1.6790	1.6808	1.6827	1.6845
5.4	1.6864	1.6882	1.6901	1.6919	1.6938	1.6956	1.6974	1.6993	1.7011	1.7029

Table III. Four-Place Natural Logarithms (continued)

	0.00	0.01	0.02	0.03	0.04	0.05	0.06	0.07	0.08	0.09
5.5	1.7047	1.7066	1.7084	1.7102	1.7120	1.7138	1.7156	1.7174	1.7192	1.7210
5.6	1.7228	1.7246	1.7263	1.7281	1.7299	1.7317	1.7334	1.7352	1.7370	1.7387
5.7	1.7405	1.7422	1.7440	1.7457	1.7475	1.7492	1.7509	1.7527	1.7544	1.7561
5.8	1.7579	1.7596	1.7613	1.7630	1.7647	1.7664	1.7681	1.7699	1.7716	1.7733
5.9	1.7750	1.7766	1.7783	1.7800	1.7817	1.7834	1.7851	1.7868	1.7884	1.7901
6.0	1.7918	1.7934	1.7951	1.7967	1.7984	1.8001	1.8017	1.8034	1.8050	1.8066
6.1	1.8083	1.8099	1.8116	1.8132	1.8148	1.8165	1.8181	1.8197	1.8213	1.8229
6.2	1.8245	1.8262	1.8278	1.8294	1.8310	1.8326	1.8342	1.8358	1.8374	1.8390
6.3	1.8405	1.8421	1.8437	1.8453	1.8469	1.8485	1.8500	1.8516	1.8532	1.8547
6.4	1.8563	1.8579	1.8594	1.8610	1.8625	1.8641	1.8656	1.8672	1.8687	1.8703
6.5	1.8718	1.8733	1.8749	1.8764	1.8779	1.8795	1.8810	1.8825	1.8840	1.8856
6.6	1.8871	1.8886	1.8901	1.8916	1.8931	1.8946	1.8961	1.8976	1.8991	1.9006
6.7	1.9021	1.9036	1.9051	1.9066	1.9081	1.9095	1.9110	1.9125	1.9140	1.9155
6.8	1.9169	1.9184	1.9199	1.9213	1.9228	1.9242	1.9257	1.9272	1.9286	1.9301
6.9	1.9315	1.9330	1.9344	1.9359	1.9373	1.9387	1.9402	1.9416	1.9430	1.9445
7.0	1.9459	1.9473	1.9488	1.9502	1.9516	1.9530	1.9544	1.9559	1.9573	1.9587
7.1	1.9601	1.9615	1.9629	1.9643	1.9657	1.9671	1.9685	1.9699	1.9713	1.9727
7.2	1.9741	1.9755	1.9769	1.9782	1.9796	1.9810	1.9824	1.9838	1.9851	1.9865
7.3	1.9879	1.9892	1.9906	1.9920	1.9933	1.9947	1.9961	1.9974	1.9988	2.0001
7.4	2.0015	2.0028	2.0042	2.0055	2.0069	2.0082	2.0096	2.0109	2.0122	2.0136
7.5	2.0149	2.0162	2.0176	2.0189	2.0202	2.0215	2.0229	2.0242	2.0255	2.0268
7.6	2.0281	2.0295	2.0308	2.0321	2.0334	2.0347	2.0360	2.0373	2.0386	2.0399
7.7	2.0412	2.0425	2.0438	2.0451	2.0464	2.0477	2.0490	2.0503	2.0516	2.0528
7.8	2.0541	2.0554	2.0567	2.0580	2.0592	2.0605	2.0618	2.0631	2.0643	2.0656
7.9	2.0669	2.0681	2.0694	2.0707	2.0719	2.0732	2.0744	2.0757	2.0769	2.0782
8.0	2.0794	2.0807	2.0819	2.0832	2.0844	2.0857	2.0869	2.0882	2.0894	2.0906
8.1	2.0919	2.0931	2.0943	2.0956	2.0968	2.0980	2.0992	2.1005	2.1017	2.1029
8.2	2.1041	2.1054	2.1066	2.1078	2.1090	2.1102	2.1114	2.1126	2.1138	2.1150
8.3	2.1163	2.1175	2.1187	2.1199	2.1211	2.1223	2.1235	2.1247	2.1259	2.1270
8.4	2.1282	2.1294	2.1306	2.1318	2.1330	2.1342	2.1353	2.1365	2.1377	2.1389
8.5	2.1401	2.1412	2.1424	2.1436	2.1448	2.1459	2.1471	2.1483	2.1494	2.1506
8.6	2.1518	2.1529	2.1541	2.1552	2.1564	2.1576	2.1587	2.1599	2.1610	2.1622
8.7	2.1633	2.1645	2.1656	2.1668	2.1679	2.1691	2.1702	2.1713	2.1725	2.1736
8.8	2.1748	2.1759	2.1770	2.1782	2.1793	2.1804	2.1815	2.1827	2.1838	2.1849
8.9	2.1861	2.1872	2.1883	2.1894	2.1905	2.1917	2.1928	2.1939	2.1950	2.1961
9.0	2.1972	2.1983	2.1994	2.2006	2.2017	2.2028	2.2039	2.2050	2.2061	2.2072
9.1	2.2083	2.2094	2.2105	2.2116	2.2127	2.2138	2.2148	2.2159	2.2170	2.2181
9.2	2.2192	2.2203	2.2214	2.2225	2.2235	2.2246	2.2257	2.2268	2.2279	2.2289
9.3	2.2300	2.2311	2.2322	2.2332	2.2343	2.2354	2.2364	2.2375	2.2386	2.2396
9.4	2.2407	2.2418	2.2428	2.2439	2.2450	2.2460	2.2471	2.2481	2.2492	2.2502
9.5	2.2513	2.2523	2.2534	2.2544	2.2555	2.2565	2.2576	2.2586	2.2597	2.2607
9.6	2.2618	2.2628	2.2638	2.2649	2.2659	2.2670	2.2680	2.2690	2.2701	2.2711
9.7	2.2721	2.2732	2.2742	2.2752	2.2762	2.2773	2.2783	2.2793	2.2803	2.2814
9.8	2.2824	2.2834	2.2844	2.2854	2.2865	2.2875	2.2885	2.2895	2.2905	2.2915
9.9	2.2925	2.2935	2.2946	2.2956	2.2966	2.2976	2.2986	2.2996	2.3006	2.3016

Table IV. Four-Place Logarithms of the Trigonometric Functions
Subtract 10 from each entry in this table.

Angle θ	*L* sin *θ*	*L* csc *θ*	*L* tan *θ*	*L* cot *θ*	*L* sec *θ*	*L* cos *θ*	
0° 00′	No value	No value	No value	No value	10.0000	10.0000	90° 00′
10′	7.4637	12.5363	7.4637	12.5363	.0000	.0000	50′
20′	.7648	.2352	.7648	.2352	.0000	.0000	40′
30′	7.9408	12.0592	7.9409	12.0591	.0000	.0000	30′
40′	8.0658	11.9342	8.0658	11.9342	.0000	.0000	20′
50′	.1627	.8373	.1627	.8373	.0000	10.0000	10′
1° 00′	8.2419	11.7581	8.2419	11.7581	10.0001	9.9999	89° 00′
10′	.3088	.6912	.3089	.6911	.0001	.9999	50′
20′	.3668	.6332	.3669	.6331	.0001	.9999	40′
30′	.4179	.5821	.4181	.5819	.0001	.9999	30′
40′	.4637	.5363	.4638	.5362	.0002	.9998	20′
50′	.5050	.4950	.5053	.4947	.0002	.9998	10′
2° 00′	8.5428	11.4572	8.5431	11.4569	10.0003	9.9997	88° 00′
10′	.5776	.4224	.5779	.4221	.0003	.9997	50′
20′	.6097	.3903	.6101	.3899	.0004	.9996	40′
30′	.6397	.3603	.6401	.3599	.0004	.9996	30′
40′	.6677	.3323	.6682	.3318	.0005	.9995	20′
50′	.6940	.3060	.6945	.3055	.0005	.9995	10′
3° 00′	8.7188	11.2812	8.7194	11.2806	10.0006	9.9994	87° 00′
10′	.7423	.2577	.7429	.2571	.0007	.9993	50′
20′	.7645	.2355	.7652	.2348	.0007	.9993	40′
30′	.7857	.2143	.7865	.2135	.0008	.9992	30′
40′	.8059	.1941	.8067	.1933	.0009	.9991	20′
50′	.8251	.1749	.8261	.1739	.0010	.9990	10′
4° 00′	8.8436	11.1564	8.8446	11.1554	10.0011	9.9989	86° 00′
10′	.8613	.1387	.8624	.1376	.0011	.9989	50′
20′	.8783	.1217	.8795	.1205	.0012	.9988	40′
30′	.8946	.1054	.8960	.1040	.0013	.9987	30′
40′	.9104	.0896	.9118	.0882	.0014	.9986	20′
50′	.9256	.0744	.9272	.0728	.0015	.9985	10′
5° 00′	8.9403	11.0597	8.9420	11.0580	10.0017	9.9983	85° 00′
10′	.9545	.0455	.9563	.0437	.0018	.9982	50′
20′	.9682	.0318	.9701	.0299	.0019	.9981	40′
30′	.9816	.0184	.9836	.0164	.0020	.9980	30′
40′	8.9945	11.0055	8.9966	11.0034	.0021	.9979	20′
50′	9.0070	10.9930	9.0093	10.9907	.0023	.9977	10′
6° 00′	9.0192	10.9808	9.0216	10.9784	10.0024	9.9976	84° 00′
	L cos *θ*	*L* sec *θ*	*L* cot *θ*	*L* tan *θ*	*L* csc *θ*	*L* sin *θ*	*Angle θ*

Table IV. Four-Place Logarithms of the Trigonometric Functions (continued)

Angle θ	L sin θ	L csc θ	L tan θ	L cot θ	L sec θ	L cos θ	
6° 00′	9.0192	10.9808	9.0216	10.9784	10.0024	9.9976	84° 00′
10′	.0311	.9689	.0336	.9664	.0025	.9975	50′
20′	.0426	.9574	.0453	.9547	.0027	.9973	40′
30′	.0539	.9461	.0567	.9433	.0028	.9972	30′
40′	.0648	.9352	.0678	.9322	.0029	.9971	20′
50′	.0755	.9245	.0786	.9214	.0031	.9969	10′
7° 00′	9.0859	10.9141	9.0891	10.9109	10.0032	9.9968	83° 00′
10′	.0961	.9039	.0995	.9005	.0034	.9966	50′
20′	.1060	.8940	.1096	.8904	.0036	.9964	40′
30′	.1157	.8843	.1194	.8806	.0037	.9963	30′
40′	.1252	.8748	.1291	.8709	.0039	.9961	20′
50′	.1345	.8655	.1385	.8615	.0041	.9959	10′
8° 00′	9.1436	10.8564	9.1478	10.8522	10.0042	9.9958	82° 00′
10′	.1525	.8475	.1569	.8431	.0044	.9956	50′
20′	.1612	.8388	.1658	.8342	.0046	.9954	40′
30′	.1697	.8303	.1745	.8255	.0048	.9952	30′
40′	.1781	.8219	.1831	.8169	.0050	.9950	20′
50′	.1863	.8137	.1915	.8085	.0052	.9948	10′
9° 00′	9.1943	10.8057	9.1997	10.8003	10.0054	9.9946	81° 00′
10′	.2022	.7978	.2078	.7922	.0056	.9944	50′
20′	.2100	.7900	.2158	.7842	.0058	.9942	40′
30′	.2176	.7824	.2236	.7764	.0060	.9940	30′
40′	.2251	.7749	.2313	.7687	.0062	.9938	20′
50′	.2324	.7676	.2389	.7611	.0064	.9936	10′
10° 00′	9.2397	10.7603	9.2463	10.7537	10.0066	9.9934	80° 00′
10′	.2468	.7532	.2536	.7464	.0069	.9931	50′
20′	.2538	.7462	.2609	.7391	.0071	.9929	40′
30′	.2606	.7394	.2680	.7320	.0073	.9927	30′
40′	.2674	.7326	.2750	.7250	.0076	.9924	20′
50′	.2740	.7260	.2819	.7181	.0078	.9922	10′
11° 00′	9.2806	10.7194	9.2887	10.7113	10.0081	9.9919	79° 00′
10′	.2870	.7130	.2953	.7047	.0083	.9917	50′
20′	.2934	.7066	.3020	.6980	.0086	.9914	40′
30′	.2997	.7003	.3085	.6915	.0088	.9912	30′
40′	.3058	.6942	.3149	.6851	.0091	.9909	20′
50′	.3119	.6881	.3212	.6788	.0093	.9907	10′
12° 00′	9.3179	10.6821	9.3275	10.6725	10.0096	9.9904	78° 00′
10′	.3238	.6762	.3336	.6664	.0099	.9901	50′
20′	.3296	.6704	.3397	.6603	.0101	.9899	40′
30′	.3353	.6647	.3458	.6542	.0104	.9896	30′
40′	.3410	.6590	.3517	.6483	.0107	.9893	20′
50′	.3466	.6534	.3576	.6424	.0110	.9890	10′
13° 00′	9.3521	10.6479	9.3634	10.6366	10.0113	9.9887	77° 00′
	L cos θ	L sec θ	L cot θ	L tan θ	L csc θ	L sin θ	*Angle θ*

Table IV. Four-Place Logarithms of the Trigonometric Functions (continued)

Angle θ	*L* sin θ	*L* csc θ	*L* tan θ	*L* cot θ	*L* sec θ	*L* cos θ	
13° 00′	9.3521	10.6479	9.3634	10.6366	10.0113	9.9887	77° 00′
10′	.3575	.6425	.3691	.6309	.0116	.9884	50′
20′	.3629	.6371	.3748	.6252	.0119	.9881	40′
30′	.3682	.6318	.3804	.6196	.0122	.9878	30′
40′	.3734	.6266	.3859	.6141	.0125	.9875	20′
50′	.3786	.6214	.3914	.6086	.0128	.9872	10′
14° 00′	9.3837	10.6163	9.3968	10.6032	10.0131	9.9869	76° 00′
10′	.3887	.6113	.4021	.5979	.0134	.9866	50′
20′	.3937	.6063	.4074	.5926	.0137	.9863	40′
30′	.3986	.6014	.4127	.5873	.0141	.9859	30′
40′	.4035	.5965	.4178	.5822	.0144	.9856	20′
50′	.4083	.5917	.4230	.5770	.0147	.9853	10′
15° 00′	9.4130	10.5870	9.4281	10.5719	10.0151	9.9849	75° 00′
10′	.4177	.5823	.4331	.5669	.0154	.9846	50′
20′	.4223	.5777	.4381	.5619	.0157	.9843	40′
30′	.4269	.5731	.4430	.5570	.0161	.9839	30′
40′	.4314	.5686	.4479	.5521	.0164	.9836	20′
50′	.4359	.5641	.4527	.5473	.0168	.9832	10′
16° 00′	9.4403	10.5597	9.4575	10.5425	10.0172	9.9828	74° 00′
10′	.4447	.5553	.4622	.5378	.0175	.9825	50′
20′	.4491	.5509	.4669	.5331	.0179	.9821	40′
30′	.4533	.5467	.4716	.5284	.0183	.9817	30′
40′	.4576	.5424	.4762	.5238	.0186	.9814	20′
50′	.4618	.5382	.4808	.5192	.0190	.9810	10′
17° 00′	9.4659	10.5341	9.4853	10.5147	10.0194	9.9806	73° 00′
10′	.4700	.5300	.4898	.5102	.0198	.9802	50′
20′	.4741	.5259	.4943	.5057	.0202	.9798	40′
30′	.4781	.5219	.4987	.5013	.0206	.9794	30′
40′	.4821	.5179	.5031	.4969	.0210	.9790	20′
50′	.4861	.5139	.5075	.4925	.0214	.9786	10′
18° 00′	9.4900	10.5100	9.5118	10.4882	10.0218	9.9782	72° 00′
10′	.4939	.5061	.5161	.4839	.0222	.9778	50′
20′	.4977	.5023	.5203	.4797	.0226	.9774	40′
30′	.5015	.4985	.5245	.4755	.0230	.9770	30′
40′	.5052	.4948	.5287	.4713	.0235	.9765	20′
50′	.5090	.4910	.5329	.4671	.0239	.9761	10′
19° 00′	9.5126	10.4874	9.5370	10.4630	10.0243	9.9757	71° 00′
10′	.5163	.4837	.5411	.4589	.0248	.9752	50′
20′	.5199	.4801	.5451	.4549	.0252	.9748	40′
30′	.5235	.4765	.5491	.4509	.0257	.9743	30′
40′	.5270	.4730	.5531	.4469	.0261	.9739	20′
50′	.5306	.4694	.5571	.4429	.0266	.9734	10′
20° 00′	9.5341	10.4659	9.5611	10.4389	10.0270	9.9730	70° 00′
	L cos θ	*L* sec θ	*L* cot θ	*L* tan θ	*L* csc θ	*L* sin θ	*Angle θ*

Table IV. Four-Place Logarithms of the Trigonometric Functions (continued)

Angle θ	*L* sin *θ*	*L* csc *θ*	*L* tan *θ*	*L* cot *θ*	*L* sec *θ*	*L* cos *θ*	
20° 00′	9.5341	10.4659	9.5611	10.4389	10.0270	9.9730	70° 00′
10′	.5375	.4625	.5650	.4350	.0275	.9725	50′
20′	.5409	.4591	.5689	.4311	.0279	.9721	40′
30′	.5443	.4557	.5727	.4273	.0284	.9716	30′
40′	.5477	.4523	.5766	.4234	.0289	.9711	20′
50′	.5510	.4490	.5804	.4196	.0294	.9706	10′
21° 00′	9.5543	10.4457	9.5842	10.4158	10.0298	9.9702	69° 00′
10′	.5576	.4424	.5879	.4121	.0303	.9797	50′
20′	.5609	.4391	.5917	.4083	.0308	.9692	40′
30′	.5641	.4359	.5954	.4046	.0313	.9687	30′
40′	.5673	.4327	.5991	.4009	.0318	.9682	20′
50′	.5704	.4296	.6028	.3972	.0323	.9677	10′
22° 00′	9.5736	10.4264	9.6064	10.3936	10.0328	9.9672	68° 00′
10′	.5767	.4233	.6100	.3900	.0333	.9667	50′
20′	.5798	.4202	.6136	.3864	.0339	.9661	40′
30′	.5828	.4172	.6172	.3828	.0344	.9656	30′
40′	.5859	.4141	.6208	.3792	.0349	.9651	20′
50′	.5889	.4111	.6243	.3757	.0354	.9646	10′
23° 00′	9.5919	10.4081	9.6279	10.3721	10.0360	9.9640	67° 00′
10′	.5948	.4052	.6314	.3686	.0365	.9635	50′
20′	.5978	.4022	.6348	.3652	.0371	.9629	40′
30′	.6007	.3993	.6383	.3617	.0376	.9624	30′
40′	.6036	.3964	.6417	.3583	.0382	.9618	20′
50′	.6065	.3935	.6452	.3548	.0387	.9613	10′
24° 00′	9.6093	10.3907	9.6486	10.3514	10.0393	9.9607	66° 00′
10′	.6121	.3879	.6520	.3480	.0398	.9602	50′
20′	.6149	.3851	.6553	.3447	.0404	.9596	40′
30′	.6177	.3823	.6587	.3413	.0410	.9590	30′
40′	.6205	.3795	.6620	.3380	.0416	.9584	20′
50′	.6232	.3768	.6654	.3346	.0421	.9579	10′
25° 00′	9.6259	10.3741	9.6687	10.3313	10.0427	9.9573	65° 00′
10′	.6286	.3714	.6720	.3280	.0433	.9567	50′
20′	.6313	.3687	.6752	.3248	.0439	.9561	40′
30′	.6340	.3660	.6785	.3215	.0445	.9555	30′
40′	.6366	.3634	.6817	.3183	.0451	.9549	20′
50′	.6392	.3608	.6850	.3150	.0457	.9543	10′
26° 00′	9.6418	10.3582	9.6882	10.3118	10.0463	9.9537	64° 00′
10′	.6444	.3556	.6914	.3086	.0470	.9530	50′
20′	.6470	.3530	.6946	.3054	.0476	.9524	40′
30′	.6495	.3505	.6977	.3023	.0482	.9518	30′
40′	.6521	.3479	.7009	.2991	.0488	.9512	20′
50′	.6546	.3454	.7040	.2960	.0495	.9505	10′
27° 00′	9.6570	10.3430	9.7072	10.2928	10.0501	9.9499	63° 00′
	L cos *θ*	*L* sec *θ*	*L* cot *θ*	*L* tan *θ*	*L* csc *θ*	*L* sin *θ*	*Angle θ*

Table IV. Four-Place Logarithms of the Trigonometric Functions (continued)

Angle θ	*L* sin *θ*	*L* csc *θ*	*L* tan *θ*	*L* cot *θ*	*L* sec *θ*	*L* cos *θ*	
27° 00′	9.6570	10.3430	9.7072	10.2928	10.0501	9.9499	63° 00′
10′	.6595	.3405	.7103	.2897	.0508	.9492	50′
20′	.6620	.3380	.7134	.2866	.0514	.9486	40′
30′	.6644	.3356	.7165	.2835	.0521	.9479	30′
40′	.6668	.3332	.7196	.2804	.0527	.9473	20′
50′	.6692	.3308	.7226	.2774	.0534	.9466	10′
28° 00′	9.6716	10.3284	9.7257	10.2743	10.0541	9.9459	62° 00′
10′	.6740	.3260	.7287	.2713	.0547	.9453	50′
20′	.6763	.3237	.7317	.2683	.0554	.9446	40′
30′	.6787	.3213	.7348	.2652	.0561	.9439	30′
40′	.6810	.3190	.7378	.2622	.0568	.9432	20′
50′	.6833	.3167	.7408	.2592	.0575	.9425	10′
29° 00′	9.6856	10.3144	9.7438	10.2562	10.0582	9.9418	61° 00′
10′	.6878	.3122	.7467	.2533	.0589	.9411	50′
20′	.6901	.3099	.7497	.2503	.0596	.9404	40′
30′	.6923	.3077	.7526	.2474	.0603	.9397	30′
40′	.6946	.3054	.7556	.2444	.0610	.9390	20′
50′	.6968	.3032	.7585	.2415	.0617	.9383	10′
30° 00′	9.6990	10.3010	9.7614	10.2386	10.0625	9.9375	60° 00′
10′	.7012	.2988	.7644	.2356	.0632	.9368	50′
20′	.7033	.2967	.7673	.2327	.0639	.9361	40′
30′	.7055	.2945	.7701	.2299	.0647	.9353	30′
40′	.7076	.2924	.7730	.2270	.0654	.9346	20′
50′	.7097	.2903	.7759	.2241	.0662	.9338	10′
31° 00′	9.7118	10.2882	9.7788	10.2212	10.0669	9.9331	59° 00′
10′	.7139	.2861	.7816	.2184	.0677	.9323	50′
20′	.7160	.2840	.7845	.2155	.0685	.9315	40′
30′	.7181	.2819	.7873	.2127	.0692	.9308	30′
40′	.7201	.2799	.7902	.2098	.0700	.9300	20′
50′	.7222	.2778	.7930	.2070	.0708	.9292	10′
32° 00′	9.7242	10.2758	9.7958	10.2042	10.0716	9.9284	58° 00′
10′	.7262	.2738	.7986	.2014	.0724	.9276	50′
20′	.7282	.2718	.8014	.1986	.0732	.9268	40′
30′	.7302	.2698	.8042	.1958	.0740	.9260	30′
40′	.7322	.2678	.8070	.1930	.0748	.9252	20′
50′	.7342	.2658	.8097	.1903	.0756	.9244	10′
33° 00′	9.7361	10.2639	9.8125	10.1875	10.0764	9.9236	57° 00′
10′	.7380	.2620	.8153	.1847	.0772	.9228	50′
20′	.7400	.2600	.8180	.1820	.0781	.9219	40′
30′	.7419	.2581	.8208	.1792	.0789	.9211	30′
40′	.7438	.2562	.8235	.1765	.0797	.9203	20′
50′	.7457	.2543	.8263	.1737	.0806	.9194	10′
34° 00′	9.7476	10.2524	9.8290	10.1710	10.0814	9.9186	56° 00′
	L cos *θ*	*L* sec *θ*	*L* cot *θ*	*L* tan *θ*	*L* csc *θ*	*L* sin *θ*	*Angle θ*

Table IV. Four-Place Logarithms of the Trigonometric Functions (continued)

Angle θ	*L* sin *θ*	*L* csc *θ*	*L* tan *θ*	*L* cot *θ*	*L* sec *θ*	*L* cos *θ*	
34° 00′	9.7476	10.2524	9.8290	10.1710	10.0814	9.9186	56° 00′
10′	.7494	.2506	.8317	.1683	.0823	.9177	50′
20′	.7513	.2487	.8344	.1656	.0831	.9169	40′
30′	.7531	.2469	.8371	.1629	.0840	.9160	30′
40′	.7550	.2450	.8398	.1602	.0849	.9151	20′
50′	.7568	.2432	.8425	.1575	.0858	.9142	10′
35° 00′	9.7586	10.2414	9.8452	10.1548	10.0866	9.9134	55° 00′
10′	.7604	.2396	.8479	.1521	.0875	.9125	50′
20′	.7622	.2378	.8506	.1494	.0884	.9116	40′
30′	.7640	.2360	.8533	.1467	.0893	.9107	30′
40′	.7657	.2343	.8559	.1441	.0902	.9098	20′
50′	.7675	.2325	.8586	.1414	.0911	.9089	10′
36° 00′	9.7692	10.2308	9.8613	10.1387	10.0920	9.9080	54°00′
10′	.7710	.2290	.8639	.1361	.0930	.9070	50′
20′	.7727	.2273	.8666	.1334	.0939	.9061	40′
30′	.7744	.2256	.8692	.1308	.0948	.9052	30′
40′	.7761	.2239	.8718	.1282	.0958	.9042	20′
50′	.7778	.2222	.8745	.1255	.0967	.9033	10′
37° 00′	9.7795	10.2205	9.8771	10.1229	10.0977	9.9023	53° 00′
10′	.7811	.2189	.8797	.1203	.0986	.9014	50′
20′	.7828	.2172	.8824	.1176	.0996	.9004	40′
30′	.7844	.2156	.8850	.1150	.1005	.8995	30′
40′	.7861	.2139	.8876	.1124	.1015	.8985	20′
50′	.7877	.2123	.8902	.1098	.1025	.8975	10′
38° 00′	9.7893	10.2107	9.8928	10.1072	10.1035	9.8965	52° 00′
10′	.7910	.2090	.8954	.1046	.1045	.8955	50′
20′	.7926	.2074	.8980	.1020	.1055	.8945	40′
30′	.7941	.2059	.9006	.0994	.1065	.8935	30′
40′	.7957	.2043	.9032	.9068	.1075	.8925	20′
50′	.7973	.2027	.9058	.0942	.1085	.8915	10′
39° 00′	9.7989	10.2011	9.9084	10.0916	10.1095	9.8905	51° 00′
10′	.8004	.1996	.9110	.0890	.1105	.8895	50′
20′	.8020	.1980	.9135	.0865	.1116	.8884	40′
30′	.8035	.1965	.9161	.0839	.1126	.8874	30′
40′	.8050	.1950	.9187	.0813	.1136	.8864	20′
50′	.8066	.1934	.9212	.0788	.1147	.8853	10′
40° 00′	9.8081	10.1919	9.9238	10.0762	10.1157	9.8843	50° 00′
10′	.8096	.1904	.9264	.0736	.1168	.8832	50′
20′	.8111	.1889	.9289	.0711	.1179	.8821	40′
30′	.8125	.1875	.9315	.0685	.1190	.8810	30′
40′	.8140	.1860	.9341	.0659	.1200	8800	20′
50′	.8155	.1845	.9366	.0634	.1211	.8789	10′
41° 00′	9.8169	10.1831	9.9392	10.0608	10.1222	9.8778	49° 00′
	L cos *θ*	*L* sec *θ*	*L* cot *θ*	*L* tan *θ*	*L* csc *θ*	*L* sin *θ*	*Angle θ*

Table IV. Four-Place Logarithms of the Trigonometric Functions (continued)

Angle θ	L sin θ	L csc θ	L tan θ	L cot θ	L sec θ	L cos θ	
41° 00′	9.8169	10.1831	9.9392	10.0608	10.1222	9.8778	49° 00′
10′	.8184	.1816	.9417	.0583	.1233	.8767	50′
20′	.8198	.1802	.9443	.0557	.1244	.8756	40′
30′	.8213	.1787	.9468	.0532	.1255	.8745	30′
40′	.8227	.1773	.9494	.0506	.1267	.8733	20′
50′	.8241	.1759	.9519	.0481	.1278	.8722	10′
42° 00′	9.8255	10.1745	9.9544	10.0456	10.1289	9.8711	48° 00′
10′	.8269	.1731	.9570	.0430	.1301	.8699	50′
20′	.8283	.1717	.9595	.0405	.1312	.8688	40′
30′	.8297	.1703	.9621	.0379	.1324	.8676	30′
40′	.8311	.1689	.9646	.0354	.1335	.8665	20′
50′	.8324	.1676	.9671	.0329	.1347	.8653	10′
43° 00′	9.8338	10.1662	9.9697	10.0303	10.1359	9.8641	47° 00′
10′	.8351	.1649	.9722	.0278	.1371	.8629	50′
20′	.8365	.1635	.9747	.0253	.1382	.8618	40′
30′	.8378	.1622	.9772	.0228	.1394	.8606	30′
40′	.8391	.1609	.9798	.0202	.1406	.8594	20′
50′	.8405	.1595	.9823	0177	.1418	.8582	10′
44° 00′	9.8418	10.1582	9.9848	10.0152	10.1431	9.8569	46° 00′
10′	.8431	.1569	.9874	.0126	.1443	.8557	50′
20′	.8444	.1556	.9899	.0101	.1455	.8545	40′
30′	.8457	.1543	.9924	.0076	.1468	.8532	30′
40′	.8469	.1531	.9949	.0051	.1480	.8520	20′
50′	.8482	.1518	9.9975	.0025	.1493	.8507	10′
45° 00′	9.8495	10.1505	10.0000	10.0000	10.1505	9.8495	45° 00′
	L cos θ	L sec θ	L cot θ	L tan θ	L csc θ	L sin θ	*Angle θ*

Table V. Values of the Exponential Function

t	e^t	e^{-t}	t	e^t	e^{-t}
0.00	1.0000	1.0000	2.5	12.182	0.0821
0.05	1.0513	0.9512	2.6	13.464	0.0743
0.10	1.1052	0.9048	2.7	14.880	0.0672
0.15	1.1618	0.8607	2.8	16.445	0.0608
0.20	1.2214	0.8187	2.9	18.174	0.0550
0.25	1.2840	0.7788	3.0	20.086	0.0498
0.30	1.3499	0.7408	3.1	22.198	0.0450
0.35	1.4191	0.7047	3.2	24.533	0.0408
0.40	1.4918	0.6703	3.3	27.113	0.0369
0.45	1.5683	0.6376	3.4	29.964	0.0334
0.50	1.6487	0.6065	3.5	33.115	0.0302
0.55	1.7333	0.5769	3.6	36.598	0.0273
0.60	1.8221	0.5488	3.7	40.447	0.0247
0.65	1.9155	0.5220	3.8	44.701	0.0224
0.70	2.0138	0.4966	3.9	49.402	0.0202
0.75	2.1170	0.4724	4.0	54.598	0.0183
0.80	2.2255	0.4493	4.1	60.340	0.0166
0.85	2.3396	0.4274	4.2	66.686	0.0150
0.90	2.4596	0.4066	4.3	73.700	0.0136
0.95	2.5857	0.3867	4.4	81.451	0.0123
1.0	2.7183	0.3679	4.5	90.017	0.0111
1.1	3.0042	0.3329	4.6	99.484	0.0101
1.2	3.3201	0.3012	4.7	109.95	0.0091
1.3	3.6693	0.2725	4.8	121.51	0.0082
1.4	4.0552	0.2466	4.9	134.29	0.0074
1.5	4.4817	0.2231	5.0	148.41	0.0067
1.6	4.9530	0.2019	5.5	244.69	0.0041
1.7	5.4739	0.1827	6.0	403.43	0.0025
1.8	6.0496	0.1653	6.5	665.14	0.0015
1.9	6.6859	0.1496	7.0	1,096.6	0.0009
2.0	7.3891	0.1353	7.5	1,808.0	0.0006
2.1	8.1662	0.1225	8.0	2,981.0	0.0003
2.2	9.0250	0.1108	8.5	4,914.8	0.0002
2.3	9.9742	0.1003	9.0	8,103.1	0.0001
2.4	11.023	0.0907	10.0	22,026	0.00005

Table VI. Powers and Roots

No.	Sq.	Sq. Root	Cube	Cube Root	No.	Sq.	Sq. Root	Cube	Cube Root
1	1	1.000	1	1.000	51	2,601	7.141	132,651	3.708
2	4	1.414	8	1.260	52	2,704	7.211	140,608	3.733
3	9	1.732	27	1.442	53	2,809	7.280	148,877	3.756
4	16	2.000	64	1.587	54	2,916	7.348	157,464	3.780
5	25	2.236	125	1.710	55	3,025	7.416	166,375	3.803
6	36	2.449	216	1.817	56	3,136	7.483	175,616	3.826
7	49	2.646	343	1.913	57	3,249	7.550	185,193	3.849
8	64	2.828	512	2.000	58	3,364	7.616	195,112	3.871
9	81	3.000	729	2.080	59	3,481	7.681	205,379	3.893
10	100	3.162	1,000	2.154	60	3,600	7.746	216,000	3.915
11	121	3.317	1,331	2.224	61	3,721	7.810	226,981	3.936
12	144	3.464	1,728	2.289	62	3,844	7.874	238,328	3.958
13	169	3.606	2,197	2.351	63	3,969	7.937	250,047	3.979
14	196	3.742	2,744	2.410	64	4,096	8.000	262,144	4.000
15	225	3.873	3,375	2.466	65	4,225	8.062	274,625	4.021
16	256	4.000	4,096	2.520	66	4,356	8.124	287,496	4.041
17	289	4.123	4,913	2.571	67	4,489	8.185	300,763	4.062
18	324	4.243	5,832	2.621	68	4,624	8.246	314,432	4.082
19	361	4.359	6,859	2.668	69	4,761	8.307	328,509	4.102
20	400	4.472	8,000	2.714	70	4,900	8.367	343,000	4.121
21	441	4.583	9,261	2.759	71	5,041	8,426	357,911	4.141
22	484	4.690	10,648	2.802	72	5,184	8.485	373,248	4.160
23	529	4.796	12,167	2.844	73	5,329	8.544	389,017	4.179
24	576	4.899	13,824	2.884	74	5,476	8.602	405,224	4.198
25	625	5.000	15,625	2.924	75	5,625	8.660	421,875	4.217
26	676	5.099	17,576	2.962	76	5,776	8.718	438,976	4.236
27	729	5.196	19,683	3.000	77	5,929	8.775	456,533	4.254
28	784	5.292	21,952	3.037	78	6,084	8.832	474,552	4.273
29	841	5.385	24,389	3.072	79	6,241	8.888	493,039	4.291
30	900	5.477	27,000	3.107	80	6,400	8.944	512,000	4.309
31	961	5.568	29,791	3.141	81	6,561	9.000	531,441	4.327
32	1,024	5.657	32,768	3.175	82	6,724	9.055	551,368	4.344
33	1,089	5.745	35,937	3.208	83	6,889	9.110	571,787	4.362
34	1,156	5.831	39,304	3.240	84	7,056	9.165	592,704	4.380
35	1,225	5.916	42,875	3.271	85	7,225	9.220	614,125	4.397
36	1,296	6.000	46,656	3.302	86	7,396	9.274	636,056	4.414
37	1,369	6.083	50,653	3.332	87	7,569	9.327	658,503	4.431
38	1,444	6.164	54,872	3.362	88	7,744	9.381	681,472	4.448
39	1,521	6.245	59,319	3.391	89	7,921	9.434	704,969	4.465
40	1,600	6.325	64,000	3.420	90	8,100	9.487	729,000	4.481
41	1,681	6.403	68,921	3.448	91	8,281	9.539	753,571	4.498
42	1,764	6.481	74,088	3.476	92	8,464	9.592	778,688	4.514
43	1,849	6.557	79,507	3.503	93	8,649	9.644	804,357	4.531
44	1,936	6.633	85,184	3.530	94	8,836	9.695	830,584	4.547
45	2,025	6.708	91,125	3.557	95	9,025	9.747	857,375	4.563
46	2,116	6.782	97,336	3.583	96	9,216	9.798	884,736	4.579
47	2,209	6.856	103,823	3.609	97	9,409	9.849	912,673	4.595
48	2,304	6.928	110,592	3.634	98	9,604	9.899	941,192	4.610
49	2,401	7.000	117,649	3.659	99	9,801	9.950	970,299	4.626
50	2,500	7.071	125,000	3.684	100	10,000	10.000	1,000,000	4.642

ANSWERS TO SELECTED EXERCISES

Section 1, p. 2

1. negative integer, rational, real **3.** rational, real **5.** rational, real
7. positive integer, rational, real **9.** positive integer, rational, real
11. non-real **13.** rational, real **15.** irrational, real

Section 2, p. 5

1. 19 **3.** -10 **5.** 18 **7.** -2 **9.** 2 **11.** 9
13. -12 **15.** 12 **17.** 0 **19.** -66 **21.** 0 **23.** 30
25. -6 **27.** -9 **29.** negative; positive.

Section 3, p. 7

1. (a) $<$ (b) $>$ (c) $>$ (d) $<$ (e) $>$ (f) $<$ (g) $>$ (h) $<$ (i) $<$ (j) $>$
3. (a) 12 (b) 12 (c) 0 (d) 4 (e) 4 (f) 3 (g) 0 (h) 14 (i) 10

Section 4, p. 10

1. 8 **3.** 64 **5.** $\frac{8}{27}$ **7.** 27 **9.** 64 **11.** non-real
13. 5 **15.** non-real **17.** -11 **19.** $\frac{4}{3}$ **21.** $\frac{5}{2}$ **23.** 4
25. 36 **27.** $\frac{1}{32}$ **29.** 9 **31.** $4\sqrt{2}$ **33.** $2\sqrt{6}$ **35.** 5
37. $3^{1/4}$ **39.** 2 **41.** 2 **43.** 2 **45.** 3 **47.** 1

Section 5, p. 13

1. 3 **3.** 2 **5.** 6 **7.** $3\sqrt[3]{3}$ **9.** $3\sqrt[3]{4}$ **11.** $5\sqrt{3}$
13. $6\sqrt{5}$ **15.** $2^{17/12}$ **17.** $2\sqrt[3]{3} - 2\sqrt[3]{4}$ **19.** $5\sqrt{2}$ **21.** $-11\sqrt{2}$
23. $\frac{\sqrt{2}}{2}$ **25.** $\frac{\sqrt{15}}{3}$ **27.** $\sqrt{2}$ **29.** $\frac{\sqrt[3]{10}}{2}$ **31.** $\sqrt[6]{54}$ **33.** $\sqrt[6]{7}$
35. $2\sqrt[6]{\frac{4}{27}}$

Review Exercises for Chapter 1, p. 15

1. 16 **3.** 1 **5.** 0 **7.** 1 **9.** 39 **11.** 5 **13.** 3
15. 10 **17.** 0 **19.** $\frac{18}{5}$ **21.** 0 **23.** undefined
25. indeterminant **27.** indeterminant **29.** 0 **31.** 16
33. 16 **35.** 1 **37.** $\frac{3}{4}$ **39.** $\frac{1}{125}$ **41.** $3^{5/2}$ **43.** 4
45. $4\sqrt{3}$ **47.** 3 **49.** non-real **51.** $2\sqrt{6}$ **53.** $5\sqrt{2}$
55. 2 **57.** $7\sqrt{3}$ **59.** $2 + 20\sqrt{2}$

Section 6, p. 18

1. 3 **3.** 3 **5.** 4 **7.** C^2 **9.** -3 **11.** -1
13. a, b, e **15.** -4 **17.** $-\frac{5}{3}$ **19.** 0 **21.** -18
23. **(a)** $3 + x + y + z$
(b) $y - x$
(c) $xy + (x - y)$
(d) $\dfrac{x + y}{xy}$
(e) $3x + 5$
(f) $x - (x + 7)$
25. $12(x + y)$ **27.** $\dfrac{p}{2} + 5$

Section 7, p. 21

1. $7x + 3$ **3.** $2x^2 - 2y^2 - x - y$ **5.** $-5xy + 10xz$
7. $2x^3 + 3x^2 + 9x - 10$ **9.** $2y - 5x - 6$ **11.** $-7 - 2y$ **13.** $-4b$
15. $3 - 3x + 2y$ **17.** $-2x + 3y + 2$ **19.** $4ab + a + b$
21. -38 **23.** EI **25.** $15 - 4x$ **27.** $5x^2 - 5y^2 - 3x$
29. $5a^3 + 2a^2 - ab + 5$ **31.** $2l + 2w + 2$
33. (a) 6 (b) $2\sqrt{6}$ (c) $5\sqrt{2}$

Section 8, p. 24

1. $-6a^3b^5$ **3.** $10r^2s + 5rs^2$ **5.** $6m^2 - m - 2$ **7.** $4x^2 + 4x + 1$
9. $3a^4b + 6a^3b^2 + 3a^2b^3$ **11.** $x^3 - 3x^2 - 7x + 6$
13. $x^7 + x^5 + x^4 + 3x^3 + x^2 + 3$ **15.** $x^4 + x^2y^2 + y^4$
17. $6x^4 + 14x^3 + 10x^2 + 2x$ **19.** $b^2 + 8b + 16$ **21.** $a^4 - b^4$
23. $9 - 6y^2 + y^4$ **25.** $9x^4y^2 - 3x^2y + \frac{1}{4}$ **27.** $lw + 3l + 2w + 6$
29. $a^2 + 2ab + b^2 + 2ac + 2bc + c^2$ **31.** 1488 **33.** 361 **35.** 525

Section 9, p. 28

1. $3x^4$ **3.** $-\dfrac{5ac}{b}$ **5.** $4xy^2 - \dfrac{3}{y} + x^2$ **7.** $x^3 - 3x + 1 + \dfrac{6}{x} + \dfrac{1}{x^2}$
9. $x^4 - 3x^3 + x^2 + 6x + 1$ **11.** $x + 1$ **13.** $x + 4$ **15.** $x + 4$
17. $x + 4, R = 1$ **19.** $x + 9, R = 36$ **21.** $2x^2 + 7x - 4, R = -4$
23. $x^4 - 2x^2 + x + 2, R = -2x - 1$ **25.** $a^2 + 2a + 4$
27. $v^2 - v + 1$ **29.** $x^2 + x + 1$

Section 10, p. 32

1. $2x(2 - 5x)$ **3.** $4xy(xy - 3)$ **5.** $(x - 3)(x + 3)$
7. $(4 - 5x)(4 + 5x)$ **9.** $(x - 5)(x + 3)$ **11.** $(2r - s)(2r + 5s)$
13. $(3x - 1)^2$ **15.** $(x - 6)^2$ **17.** $(x + 1)^2$ **19.** $(b - a)(y + x)$
21. $(x + b)(x + y)$ **23.** $(x - 1)(x + 1)(3x - 2)$
25. $(x + y - 1)(x + y - 4)$ **27.** $(2x + 1)(4x^2 - 2x + 1)$

29. $(2x - 1)(4x^2 + 2x + 1)$ **31.** $(v - t)(v^2 + vt + t^2)$
33. $(4x^2 + 1)(2x + 1)(2x - 1)$ **35.** $(x + y - 1)(x^2 + y^2 - xy - 2x + y + 1)$
37. $(x^4 + 1)(x^2 + 1)(x + 1)(x - 1)$
39. $(x - 2)(x^2 + 2x + 4)(x + 2)(x^2 - 2x + 4)$
41. $(x^2 + 3)(x^2 - 5)$ **43.** $6xy(x^4 - 6y^4)$ **45.** Prime
47. $(x + y)(y - x + 1)$ **49.** $pq^2(p^2 + 9)(p + 3)(p - 3)$
51. (a) $(4x - 1)(2x + 1)$ (b) $(x + 1)(x - 22)$ (c) $(5x + 2)(2x - 1)$
(d) $(6x - 5)(2x + 3)$

Section 11, p. 35

1. $\frac{2}{3}$ **3.** $-2x$ **5.** $\dfrac{3b}{b + c}$ **7.** $5x$ **9.** $\dfrac{2 + x}{3 - x}$ **11.** $\dfrac{3(x + y)}{4}$

13. $\dfrac{1}{p - q}$ **15.** $\frac{1}{4}$ **17.** $\dfrac{a}{x - y}$ **19.** $\dfrac{(x^2 + 5x + 6)}{(x + 1)}$ **21.** 1

23. In lowest terms **25.** $\dfrac{x - 3}{x - 5}$ **27.** $\dfrac{3(a + b)^2 - x}{(a - b)}$

29. $\dfrac{(x + y)}{x^2 + xy + y^2}$ **31.** $\dfrac{(x^2 + 1)(x + 1)}{(x^2 + x + 1)}$ **33.** $\dfrac{-2(x + 2y)}{5(4x^2 + 2xy + x^2)}$

Section 12, p. 38

1. $\frac{8}{3}$ **3.** $\frac{3}{2}$ **5.** $\dfrac{z^2}{xy}$ **7.** $\dfrac{16}{z}$ **9.** 1 **11.** 1

13. $\dfrac{-(1 + 3x)}{(x + 2)}$ **15.** $\dfrac{(x - 5)(x + 4)}{(x - 2)(x + 2)}$ **17.** $\dfrac{(x^2 + xy + y^2)}{(3x + 2y)}$

19. $\dfrac{(x - 2)(x - 8)}{(x - 1)(x + 3)}$ **21.** $(p - 3)(p - 9)$ **23.** $\dfrac{(x + 2)^2(2 - x)}{(3x - 2)}$

25. $\dfrac{3(2x - 1)(x + 7)}{2(x + 3)^2}$

Section 13, p. 43

1. $\frac{7}{5}$ **3.** $-\frac{1}{9}$ **5.** $\dfrac{a + b}{y}$ **7.** $\dfrac{r + s^2}{st}$ **9.** $\dfrac{16 - xy}{10x}$

11. $\dfrac{bc + ac + ab}{abc}$ **13.** $\dfrac{(t + 1)}{(t - 2)}$ **15.** $\dfrac{p^2 - 6p + 10}{(p - 2)(p - 3)(p - 4)}$

17. $\dfrac{4xy}{x^2 - y^2}$ **19.** $\dfrac{2a^2b - 2ab^2 + 1}{2ab}$ **21.** $\dfrac{2x - 1}{(2x + 1)^2}$ **23.** $\dfrac{5x + 21}{x^2 - 9}$

25. $\dfrac{s + rs - r - r^2 - 2}{rs}$ **27.** $\dfrac{x^3 - 3}{x^2 - 1}$ **29.** $\dfrac{(4p^2 + 3p + 1)}{(p + 2)(p + 3)(p - 3)}$

31. $\dfrac{ac + b^2}{(a - b)(c - b)}$ **33.** $\dfrac{3y + 3x}{5y + 5x}$ **35.** $\dfrac{a}{b}$ **37.** $\dfrac{y - 4}{y + 2}$

39. $\dfrac{2r + 2s + 1}{s}$ **41.** $\dfrac{x^2 + x + 2}{x^2 + 2}$

Review Exercises for Chapter 2, p. 45

1. ab **3.** $4\sqrt{7}$ **5.** $2xy - 4yz + 2xz$ **7.** $2b$ **9.** $2x^2 - 4x + 11$

11. $2x^6y^7$ **13.** $a^2 - 4a + 4$ **15.** $x^3 - x^2y + xy^2 - y^3$ **17.** $2x^4 - 10x^3 + 8x^2$ **19.** $\frac{4y}{x}$ **21.** $x^3 - 2x^2 + 6x - 8$ **23.** $x - 4$

25. $(x^2 + 1)(x - 1)$ **27.** $\frac{12a^2}{bc}$ **29.** 1 **31.** $\frac{1}{x + 1}$ **33.** $x(x - 3)$

35. $\frac{(5x + 7)}{(x^2 - 1)}$ **37.** $\frac{2x^2 + 9x + 19}{(x + 1)(x + 2)(x + 3)}$ **39.** $\frac{3x - 1}{(x + 1)(x + 3)(x - 3)}$

41. xy

Section 14, p. 50

1. 4 **3.** 4 **5.** 0 **7.** 17 **9.** 0 **11.** $\frac{5b}{4a - b}$

13. $\frac{a}{a + 1}$ **15.** $\frac{l - a}{n - 1}$ **17.** $\frac{K}{v}$ **19.** $\frac{3v}{\pi r^2}$ **21.** $\frac{P - 2l}{2}$

23. $180 - A - C$ **25.** $\frac{2A}{h} - C$ **27.** $\frac{Fr^2}{Gm_2}$ **29.** $\frac{x}{x + k}$ **31.** $\frac{E}{R}$

33. (a) $\frac{5}{9}(F - 32)$ (b) $\frac{9C}{5} + 32$ (c) 32, 50, 107.6, 212 (d) -40

35. 22, 22, 30 **37.** 6x6

Section 15, p. 54

1. Quadratic **3.** Quadratic **5.** Linear **7.** Quadratic

9. ± 3 **11.** ± 4 **13.** $0, \frac{1}{3}$ **15.** $0, \frac{2}{3}$ **17.** $0, \frac{1}{3}$ **19.** $\frac{1}{4}, \frac{1}{3}$

21. $-6, -1$ **23.** $-\frac{1}{3}, \frac{1}{3}$ **25.** $\frac{3}{2}, \frac{3}{2}$ **27.** 3, 5 **29.** ± 5

31. $\pm\sqrt{\frac{v}{\pi h}}$ **33.** $\pm\sqrt{\frac{2K}{\theta}}$ **35.** (a) 3 sec (b) 4 sec (c) $3\sqrt{2}$ sec

37. 3 in. **39.** 196 in.^{2}

Section 16, p. 60

1. $-5, -1$ **3.** $\frac{3 \pm \sqrt{2}}{2}$ **5.** $3 \pm \sqrt{5}$ **7.** non-real **9.** non-real

11. ± 3 **13.** ± 4 **15.** $0, \frac{1}{3}$ **17.** $0, \frac{2}{3}$ **19.** $0, \frac{1}{3}$ **21.** $\frac{1}{4}, \frac{1}{3}$

23. $\frac{-5 + \sqrt{129}}{4} = 12.95$ amps

Section 17, p. 62

1. $\frac{13}{6}$ **3.** $-\frac{45}{2}$ **5.** 4 **7.** 12 **9.** 3 **11.** no solution

13. no solution **15.** 2 **17.** 60 **19.** $\frac{CC_1C_3}{C_1C_3 - CC_3 - CC_1}$

Section 18, p. 66

1. 49 **3.** 27 **5.** $\frac{7 + \sqrt{13}}{2}$ **7.** 4 **9.** no solution

11. no solution **13.** 0 **15.** $1 \pm \sqrt{10}$ or $-2.16, 4.16$ **17.** 5

19. 3 **21.** $\frac{T^2g}{4\pi^2}$

Section 19, p. 69

1. $\pm\sqrt{2}$ **3.** $-1, -\sqrt[3]{5}$ **5.** 9 **7.** ± 1 **9.** $-\frac{1}{64}, \frac{27}{8}$
11. 1, 16 **13.** $\pm\dfrac{\sqrt{2}}{2}$ **15.** $-\sqrt[3]{2}, 1$ **17.** 576 **19.** $3 \pm \sqrt{2}$
21. $\frac{1}{3}, \dfrac{2 - \sqrt[3]{5}}{3}$ **23.** 1, 2, 3, 4 **25.** $\frac{1}{8}, 1$ **27.** $\pm 1, \pm 2^{3/4}$

Section 20, p. 73

1. 7 cm by 14 cm **3.** 25 in.^{2} **5.** 25 liters **7.** $\frac{120}{47}$ min or 2 min 33 sec
9. 400 mph **11.** \$3000 at 5.5% and \$5000 at 8% **13.** 6 hr and 12 hr
15. 18 and 20 **17.** 13, 15 **19.** $5\frac{1}{2}$ hr after the faster train starts; 412.5 miles
21. 2 ft **23.** $\dfrac{-3 + \sqrt{29}}{2}$ or 1.2 sec **25.** 3.5 sec

Review Exercises for Chapter 3, p. 75

1. 12 **3.** $-\frac{18}{11}$ **5.** 4 **7.** 5 **9.** 3 **11.** ± 1 **13.** 3, 8
15. $-1, -3$ **17.** 2 **19.** $-1, -3$ **21.** $-1 \pm 2\sqrt{3}$
23. $\dfrac{-3 \pm \sqrt{5}}{2}$ **25.** 1, 4 **27.** $1 \pm \sqrt{6}$ **29.** $\frac{1}{2}, 3$ **31.** 1
33. 3 **35.** 7 **37.** 2 **39.** $\pm 1, \pm 3$ **41.** $\sqrt[3]{\dfrac{2}{3}}, -\sqrt[3]{3}$
43. 4 cm **45.** $\dfrac{1}{2\pi\sqrt{LC}}$ **47.** 4 **49.** 42 and 54 **51.** 4 oz
53. 2 ft

Section 21, p. 82

1. $A = \dfrac{\pi d^2}{4}$ **3.** $r = \sqrt{\dfrac{A}{\pi}}$ **5.** $A = s^2$ **7.** $A = \dfrac{d^2}{2}$
9. $A = \dfrac{\sqrt{3}s^2}{4}$ **11.** $A = 3b$ **13.** $V = 4l^3 - 6l^2$
15. $A = \dfrac{p^2}{16} + \dfrac{(80 - p)^2}{4\pi}$ **17.** $h = \sqrt{169 - x^2}$ **19.** 200, 224, 0
21. 2, 2.5, 10.1 **23.** $2x^2 + 4xh + h^2 + x + h + 1, 2h^2 + h + 1$
25. $-(1 + 2x + h)$ **27.** $w \neq \pm 1$ **29.** $y < 2$ **31.** none
33. $x \neq 0$ **35.** non-real for all s

Section 22, p. 87

1. **3.**

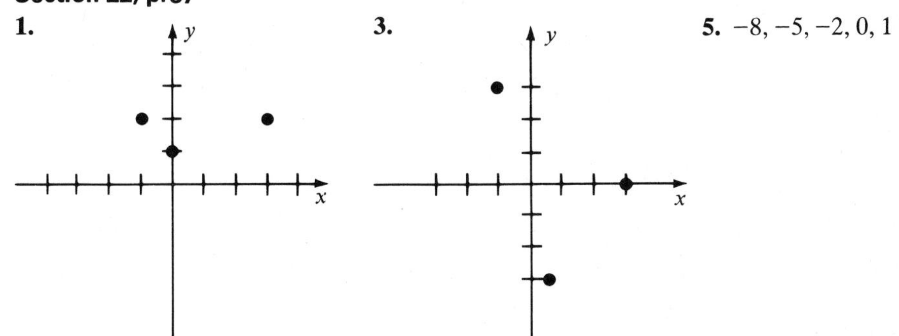

5. $-8, -5, -2, 0, 1$

7. 2, 7, 2, 23 **9.** 1st and 4th quadrant **11.** 2nd and 3rd quadrant
13. 2 units to the left of the y axis **15.** $C(4, y), y \neq 2$

Section 23, p. 92

1. $(-\frac{1}{2}, 0), (0, 1)$ **3.** $(2, 0), (0, 6)$ **5.** $(3, 0), (0, 2)$
7. $(-1, 0), (0, 1)$ **9.** $(\frac{4}{3}, 0), (0, 4)$ **11.** $(-\frac{1}{2}, 0), (0, 2)$

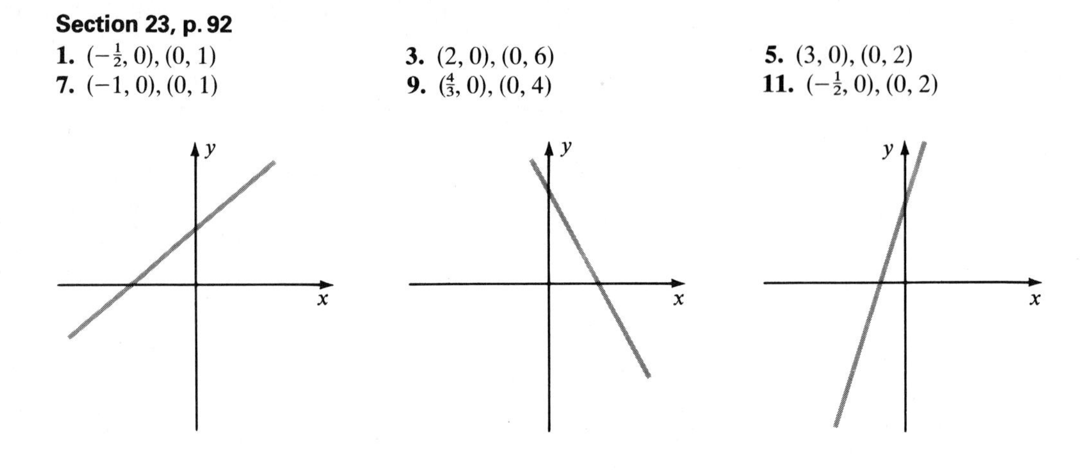

13. $(\frac{3}{2}, 0), (0, -3)$ **15.** $(0, -3)$ **17.** $-\frac{7}{4}$ **19.** 2

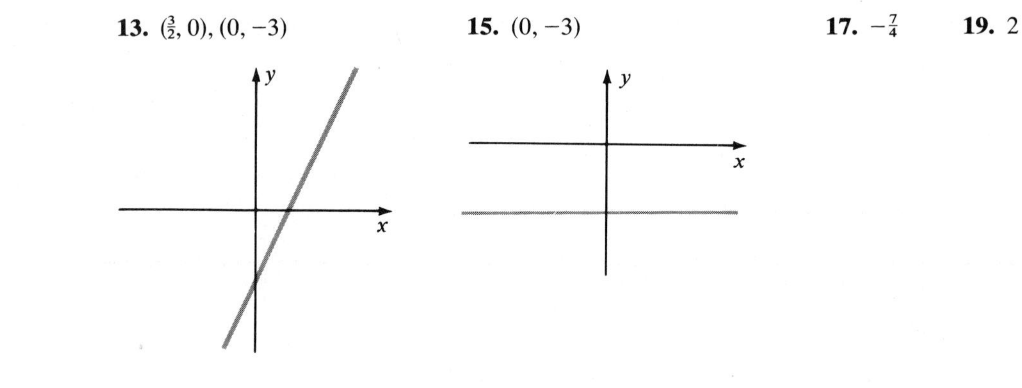

21. **23.** **25.**

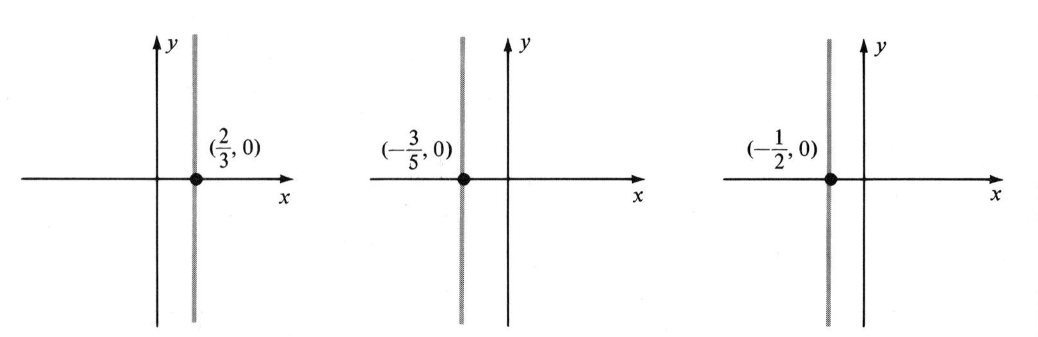

27. (a) 3 sec (b)

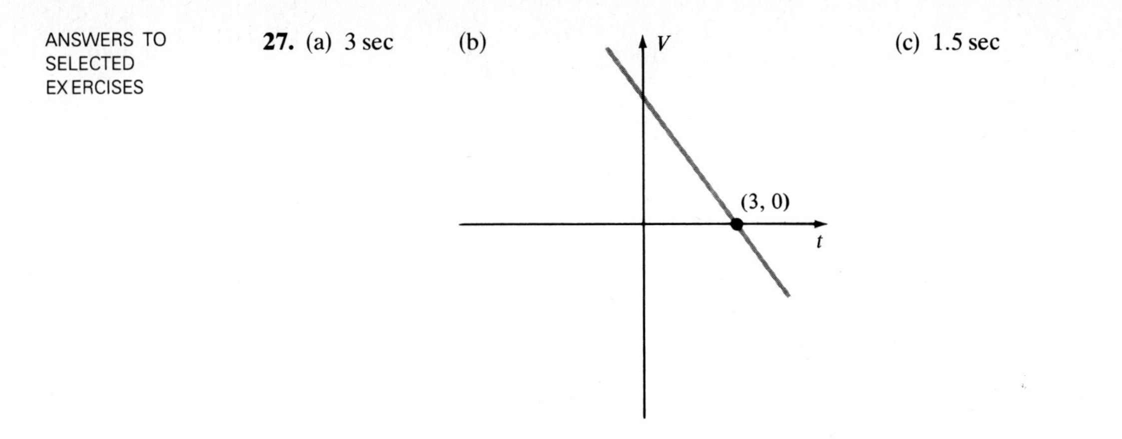

(c) 1.5 sec

29. (a) 128 (b) 64 (c) 0 (d)

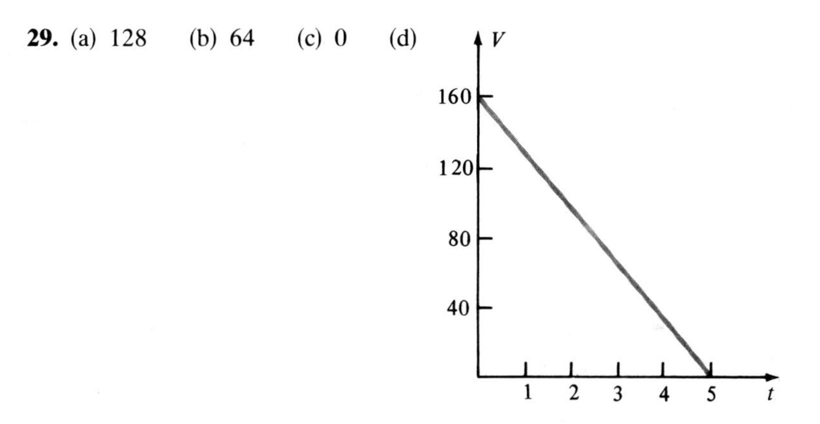

31. (a) $R = \dfrac{T + 120}{400}$ (b)

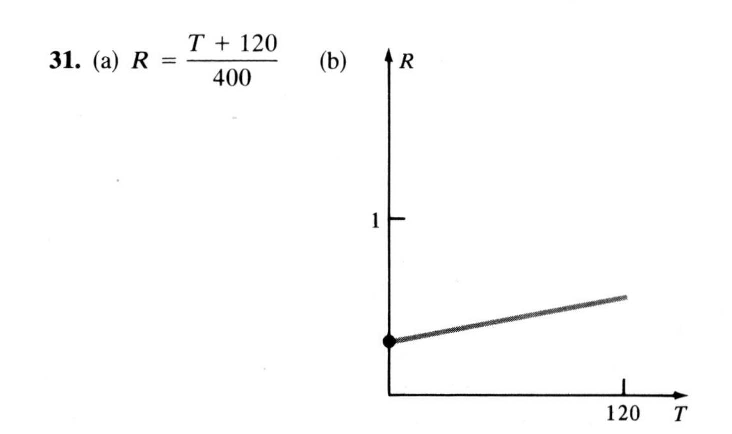

33. (a) $t = 200 - \dfrac{R}{2}$ (b) (c) $(0, 400)$

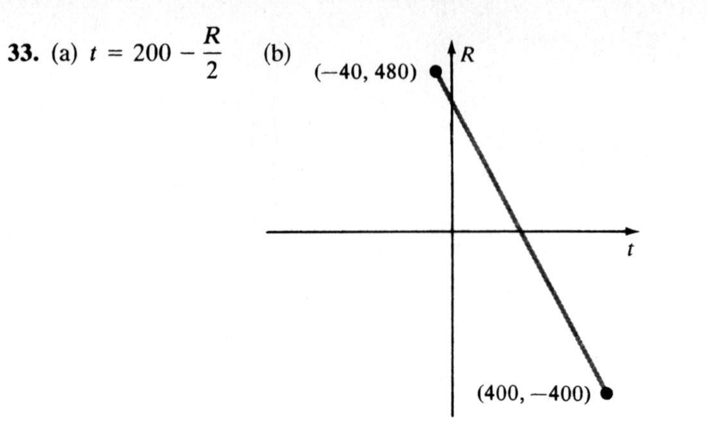

Section 24, p. 102

1. $\left(\dfrac{-3 \pm \sqrt{17}}{2}, 0\right)$

$(0, 2)$

Vertex $(-\frac{3}{2}, \frac{17}{4})$

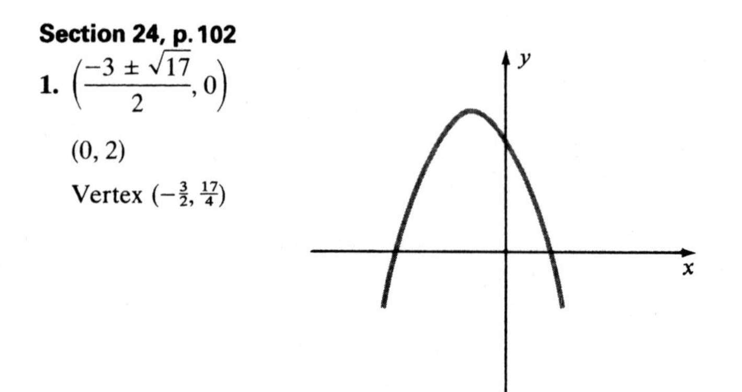

3. $\left(\dfrac{-1 \pm \sqrt{17}}{2}, 0\right)$

$(0, -4)$

Vertex $(-\frac{1}{2}, -\frac{17}{4})$

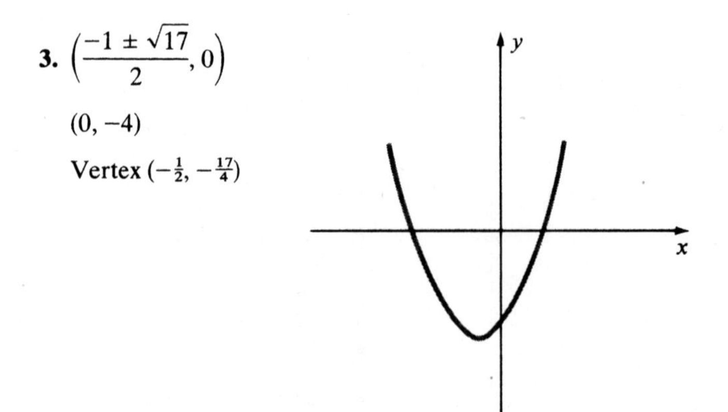

5. $(-\frac{1}{2}, 0), (2, 0)$

$(0, 2)$

Vertex $(\frac{3}{4}, \frac{25}{8})$

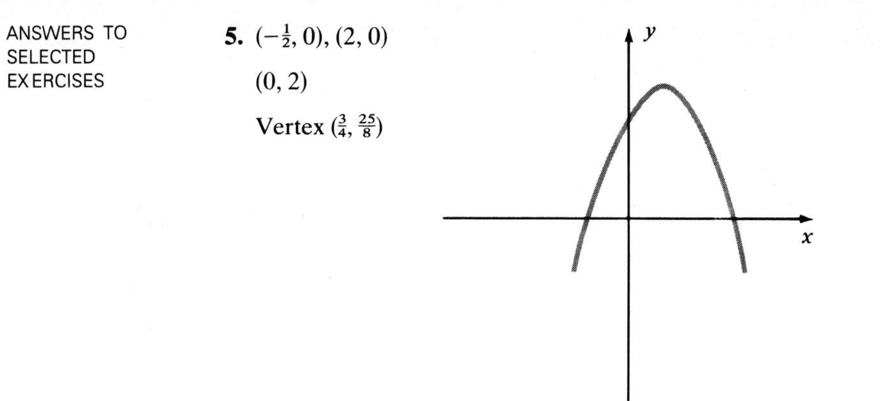

7. $(-\frac{5}{2}, 0), (\frac{1}{3}, 0)$

$(0, -5)$

Vertex $(-\frac{13}{12}, -12\frac{1}{24})$

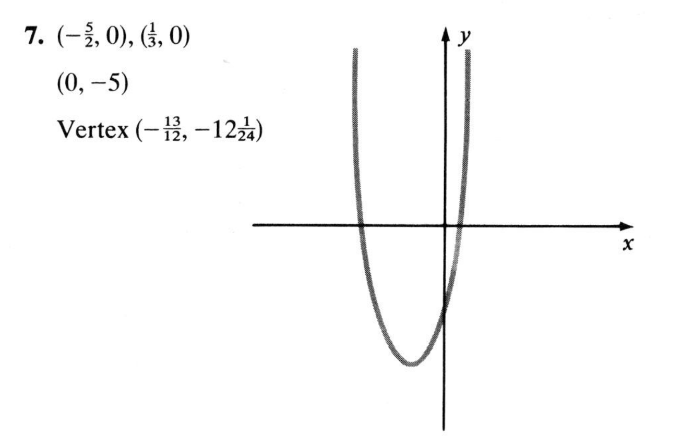

9. $\left(\dfrac{1 \pm \sqrt{3}}{2}, 0\right)$

$(0, -2)$

Vertex $(\frac{1}{2}, -3)$

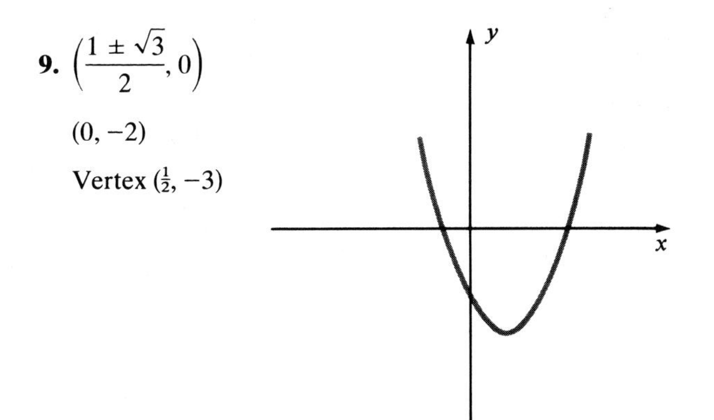

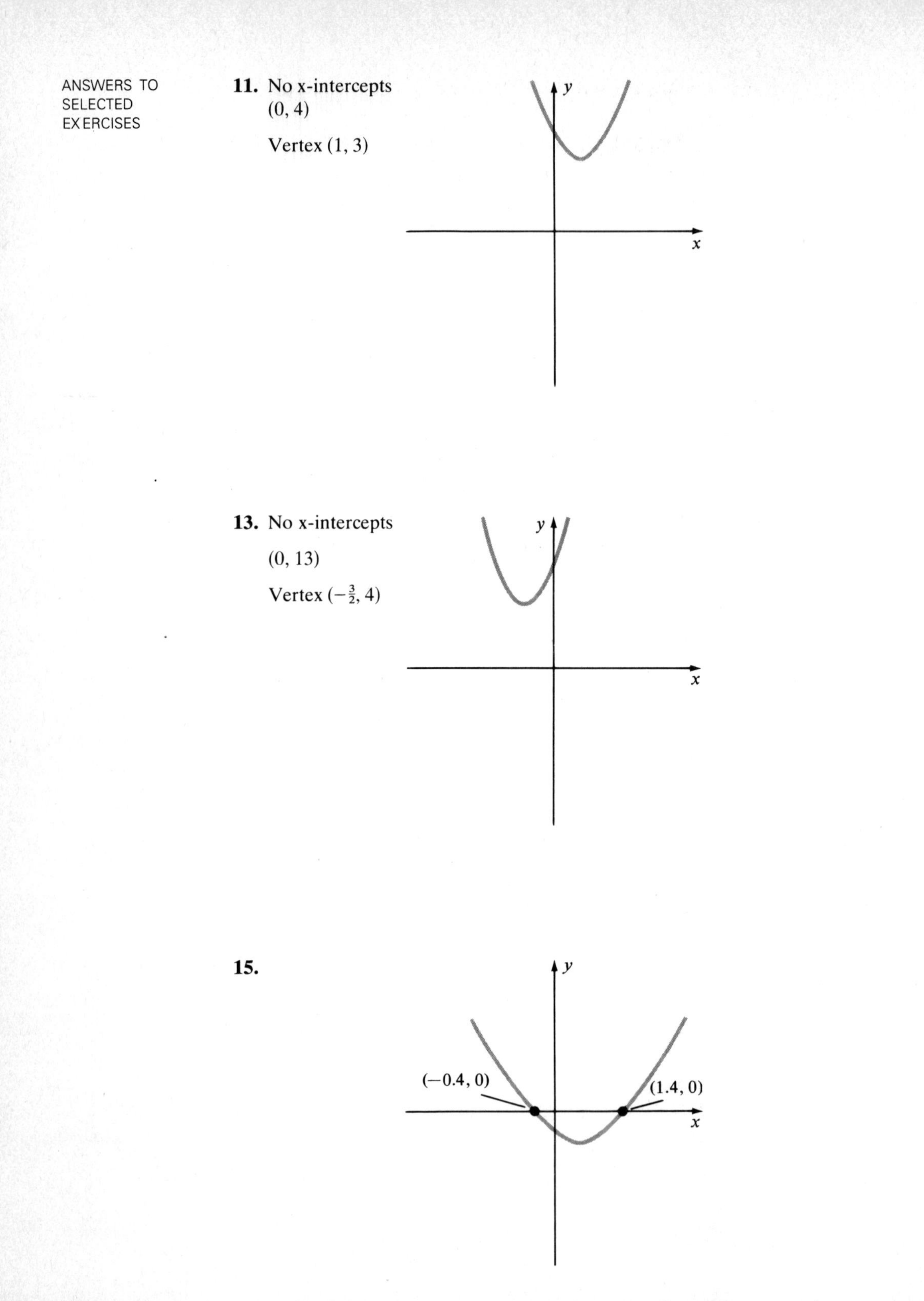

11. No x-intercepts
(0, 4)
Vertex (1, 3)

13. No x-intercepts
(0, 13)
Vertex $(-\frac{3}{2}, 4)$

15.

17.

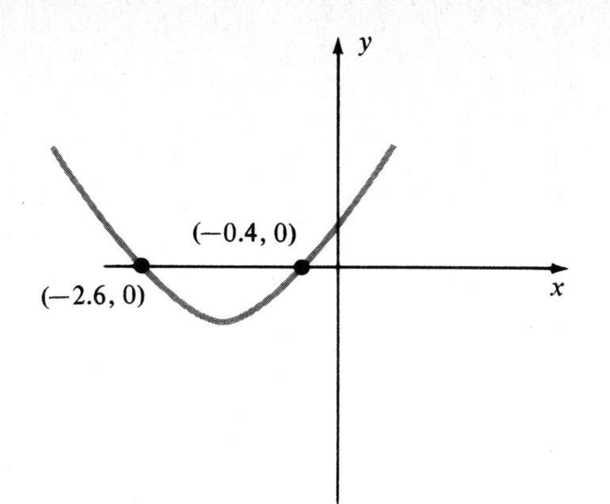

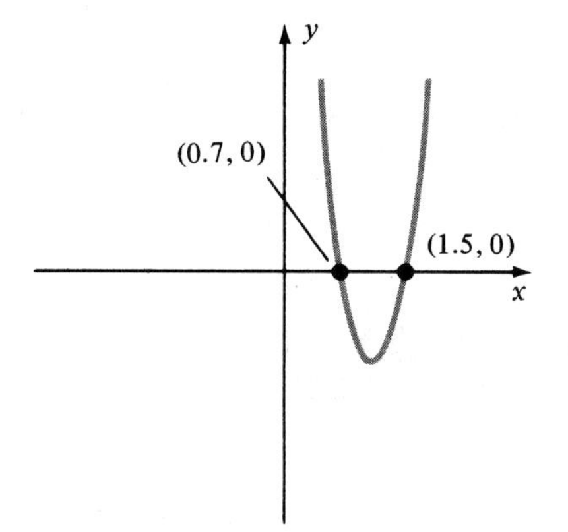

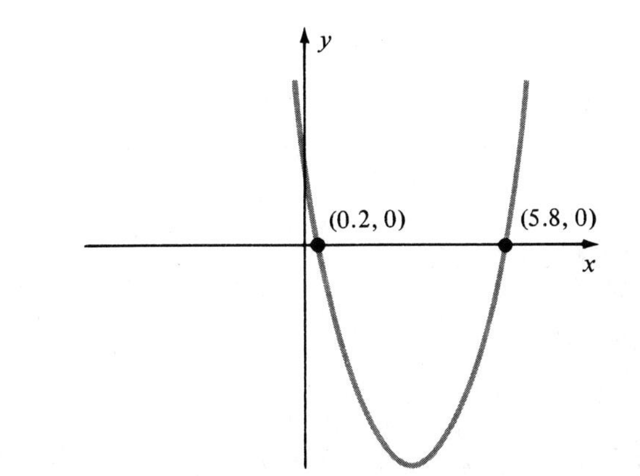

23.

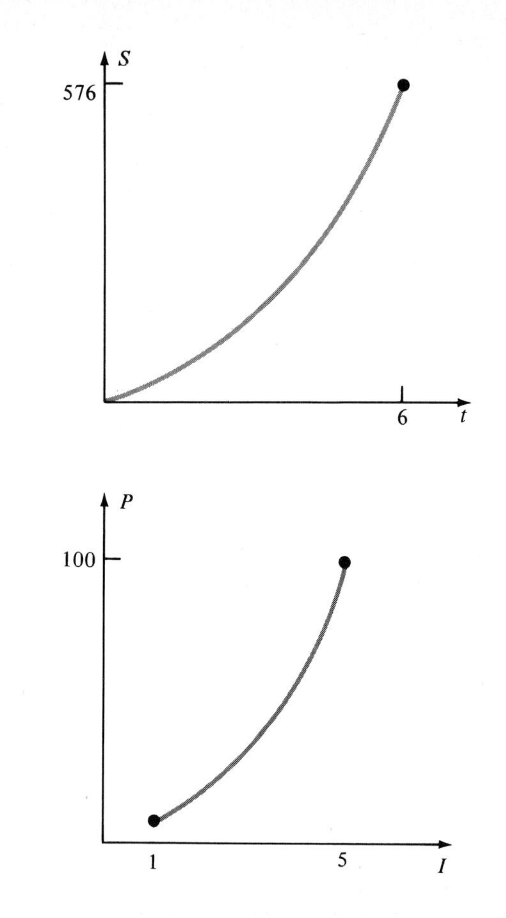

25.

Section 25, p. 107

1. (2, 2); min **3.** $(-\frac{1}{2}, \frac{19}{4})$; max **5.** (15, 225); max
7. (2, −8); min. **9.** 14, 14 **11.** 6 ft by 6 ft; 36 ft^2
13. 30 ft by 30 ft **15.** 50 **17.** $12.50
19. Base $= \dfrac{40}{\pi + 4}$; Vertical side $= \dfrac{20}{\pi + 4}$ **21.** 5 sec
23. Allow $\dfrac{40\sqrt{3}\,\pi}{9 + \sqrt{3}\,\pi} = 15.07$ in. for the circle and the remaining length for the triangle.
25. Max. occurs when entire length is used for the circle
27. 40 sec; 2.56×10^4 ft

Review Exercises for Chapter 4, p. 109

1. $A = \pi(r + 2)^2$ **3.** $P = 4s^2$ **5.** $A = \dfrac{b(b + 2)}{2}$ **7.** 3, 4, 0
9. $1, -1, a^2 - 3a + 1$ **11.** 3, 1, 3 **13.** All $x, x \neq 2$
15. All $x, x \neq \pm 2$ **17.** $x \geq -\frac{3}{2}$ **19.** $x > -1$

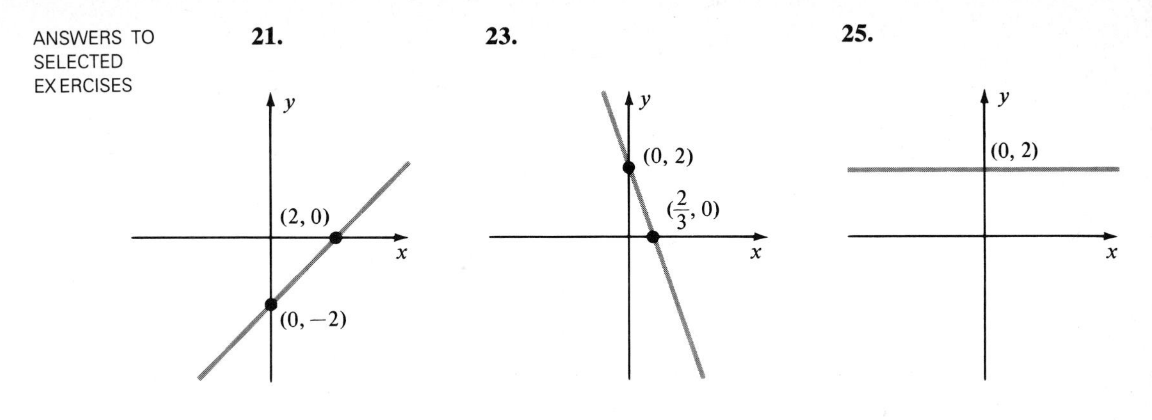

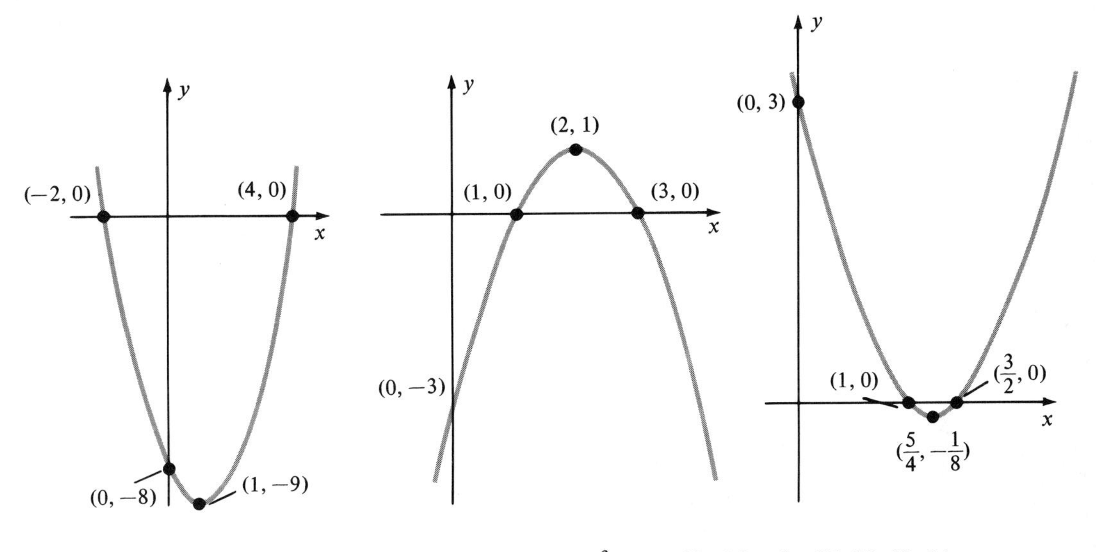

33. $-2, 4$ **35.** $1, 3$ **37.** $1, \frac{3}{2}$ **39.** Max. for 27, 28, 30, 31
Min. for 29 and 32

41. \$1.50

Section 26, p.113

1. $3 < 9$ **3.** $-3 \leq -1$ **5.** $r + 3 > -4$ **7.** $1 \leq t \leq 4$
9. $x - 3 \geq 0$ **11.** $3x - 4 \leq 0$ **13.** $-3 \leq 4r \leq -1$
15. $x + 1 > 3$ or $x + 1 < -2$ **17.** $y \leq 0$ **19.** $x \geq 0$
21. $-2 < y < 3$

Section 27, p.118

1. $x > -2$ **3.** $x > -\frac{3}{4}$ **5.** $x \geq -4$ **7.** $x > -3$ **9.** $x > -1$
11. $x \leq -\frac{1}{2}$ **13.** $x > 4$ **15.** $0 < x < \frac{9}{2}$
17. $-\frac{9}{2} < x < -\frac{1}{2}$ **19.** $0 < t < \frac{9}{4}$ **21.** $50 < F < 86$
23. \$20,000 **25.** \$3,000 **27.** $20 < R_2 < 60$

Section 28, p. 122

1. $x < -3$ or $x > 1$ **3.** $2 \leq x \leq \frac{7}{2}$ **5.** $-5 < x < 1$
7. $x \leq 1$ or $x \geq 5$ **9.** $x < -\frac{1}{2}$ or $x > 3$ **11.** All x **13.** No x
15. $x \leq 0$ or $x \geq \frac{1}{2}$ **17.** All x **19.** $-\frac{5}{2} < x < \frac{5}{2}$ **21.** $x \leq 1$ or $x \geq 5$
23. $x \leq -3$ or $x \geq 3$ **27.** $\frac{1}{2} < t < 3$

Section 29, p. 128

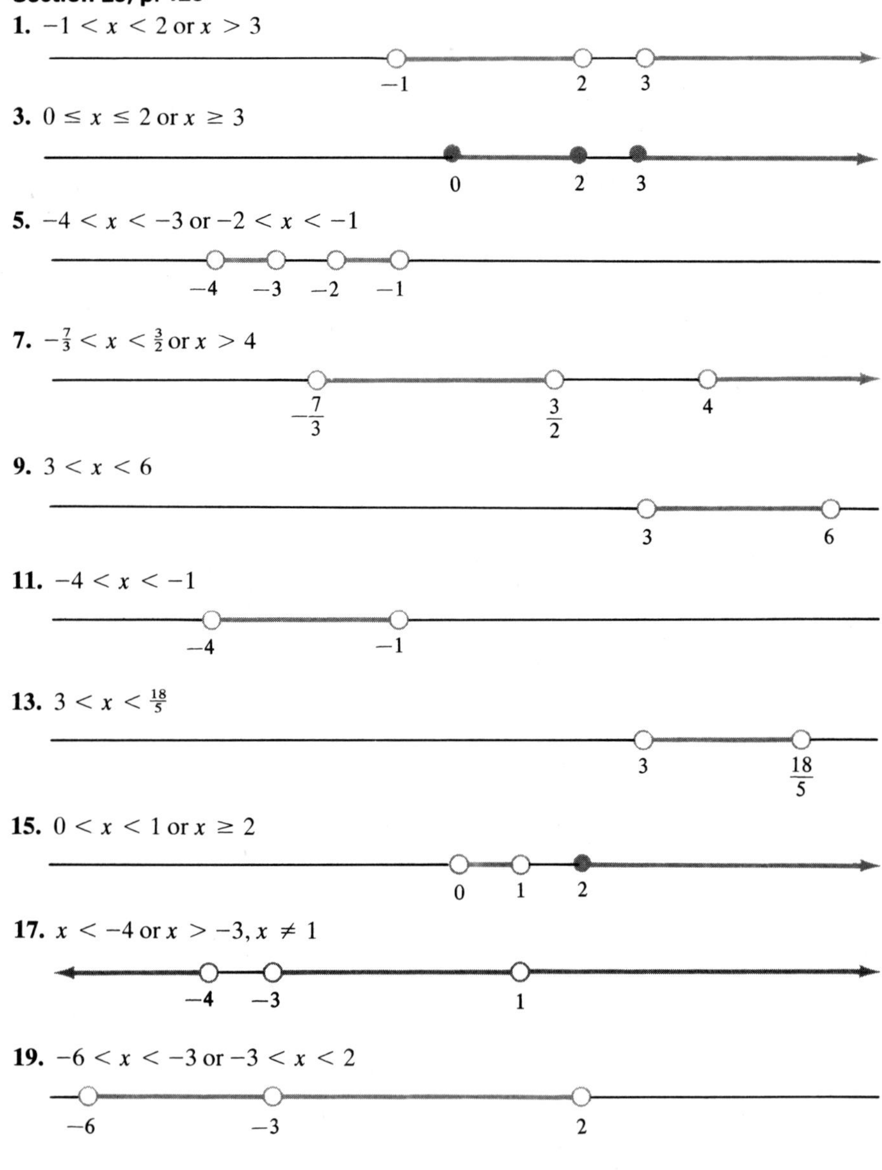

1. $-1 < x < 2$ or $x > 3$

3. $0 \leq x \leq 2$ or $x \geq 3$

5. $-4 < x < -3$ or $-2 < x < -1$

7. $-\frac{7}{3} < x < \frac{3}{2}$ or $x > 4$

9. $3 < x < 6$

11. $-4 < x < -1$

13. $3 < x < \frac{18}{5}$

15. $0 < x < 1$ or $x \geq 2$

17. $x < -4$ or $x > -3$, $x \neq 1$

19. $-6 < x < -3$ or $-3 < x < 2$

Section 30, p. 131

1. $-8 \leq x \leq 8$ **3.** $x \leq -3$ or $x \geqq 3$ **5.** $-3 < x < 3$
7. $5 < x < 7$ **9.** $x \leq -5$ or $x \geq 1$ **11.** $\frac{1}{3} < x < 1$
13. $x \leq -\frac{2}{5}x$ or $x \geq 2$ **15.** $0 < x < 6$

17. $-5 < x < -1 - \sqrt{2}$ or $-1 + \sqrt{2} < x < 3$
19. $-2 < x < 4$ **21.** $1 < t < 5$

Review Exercises for Chapter 5, p. 132

1. $4 > 2$ **3.** $x + 1 < 4$ **5.** $r + 1 < r - 1$ **7.** $x^2 + 1 > 0$
9. $0 < x^2 + 1 < 5$ **11.** $x < 1$ **13.** $x > 2$ **15.** $x \leq 2$ **17.** $x \geq \frac{2}{7}$
19. $2 \leq x \leq 5$ **21.** $x < 2$ or $x > 5$ **23.** $-1 \leq x \leq 5$
25. All x **27.** $-5 \leq x \leq 1$ or $x \geq 2$ **29.** $x < -2$ or $\frac{3}{2} < x < 3$
31. $-1 < x < 5$ **33.** $x \leq -1$ or $x \geq 4$ **35.** $x < \frac{1}{3}$ or $x > 1$
37. (a) $\frac{2}{3} < t < 6$ (b) $0 < t < \frac{2}{3}$ **39.** $0 < t < \frac{5}{2}$

Section 31, p. 139

1.

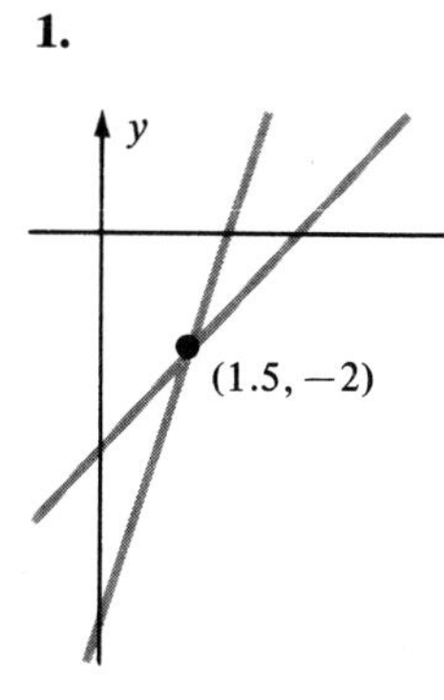

3.

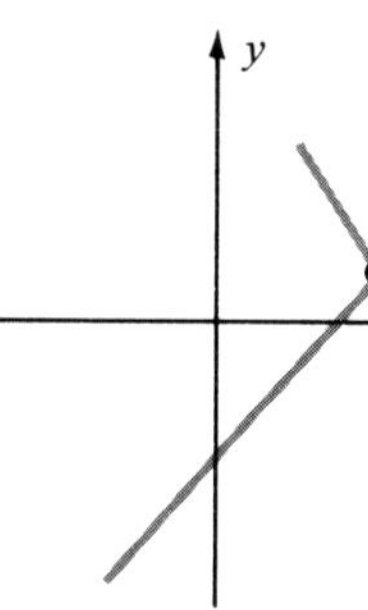

5.

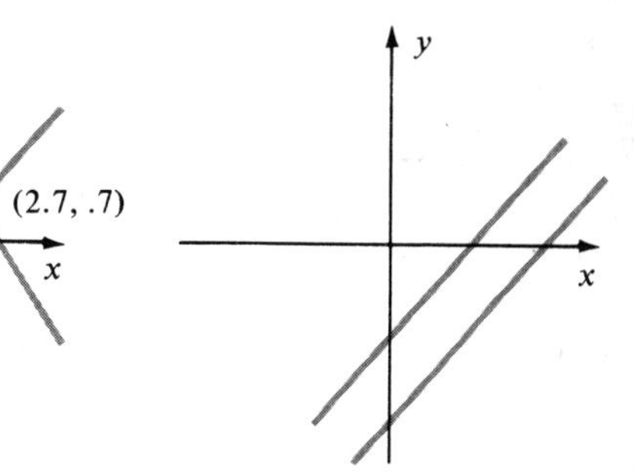

7.

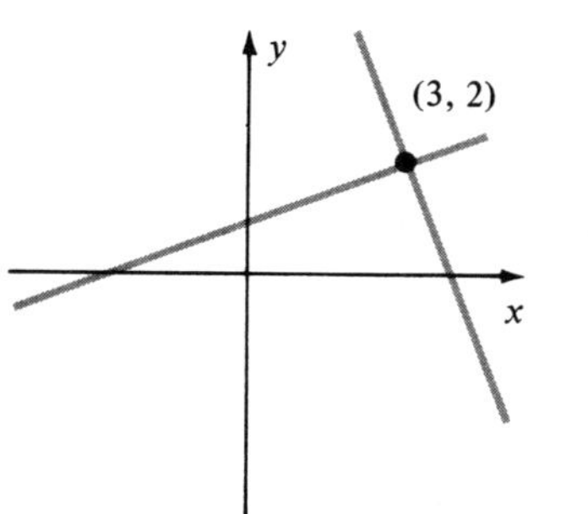

9.

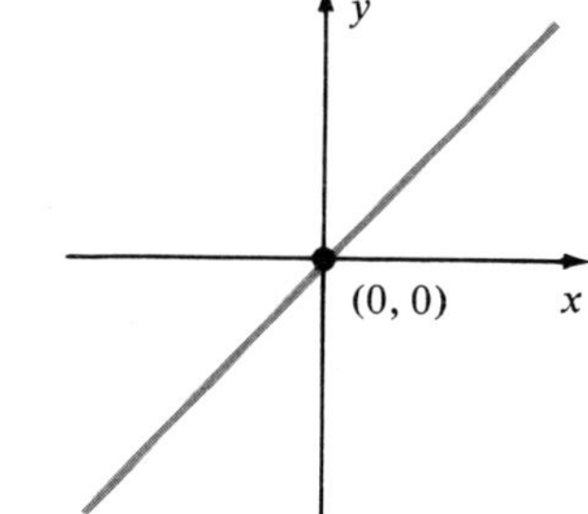

11

y
(−1, 3)
x

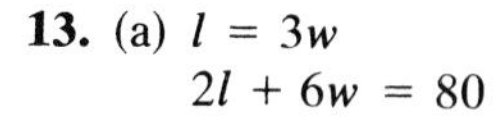

13. (a) $l = 3w$
$2l + 6w = 80$

(b)

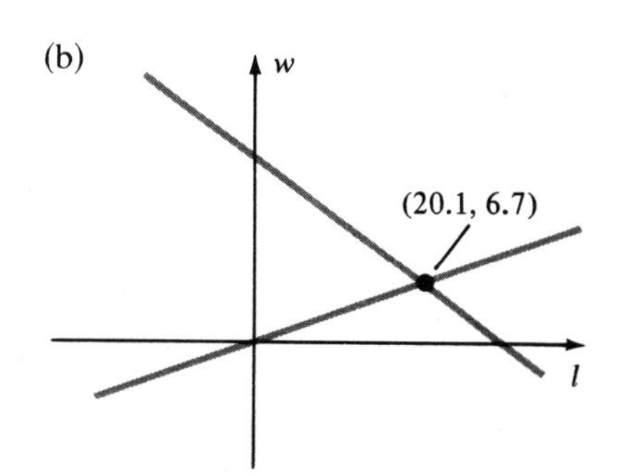

Section 32, p.146

1. $x = 1.5, y = -2$ **3.** $x = \frac{8}{3}, y = \frac{2}{3}$ **5.** Inconsistent **7.** $x = 3, y = 2$
9. $x = 0, y = 0$ **11.** $x = -1, y = 3$ **13.** $x = 2, y = 2$
15. $x = -14.5, y = -6.5$ **17.** $x = \frac{1}{4}, y = \frac{15}{16}$ **19.** $x = 1, y = \frac{5}{3}$
21. $x = \frac{1}{2}, y = -1$ **23.** $x = \frac{1}{3}, y = \frac{1}{2}$ **25.** 8 and 12
27. 30 mph; 300 mph **29.** $\frac{25}{3}$ gal **31.** 17 **33.** $y = 3x + 2$
35. $F = \frac{9}{5}C + 32$ **37.** $R_0 = 24.55$ ohms, $\alpha = 0.0045$
39. 35 ft/sec and 20 ft/sec **41.** 10 and 12 pounds

Section 33, p.152

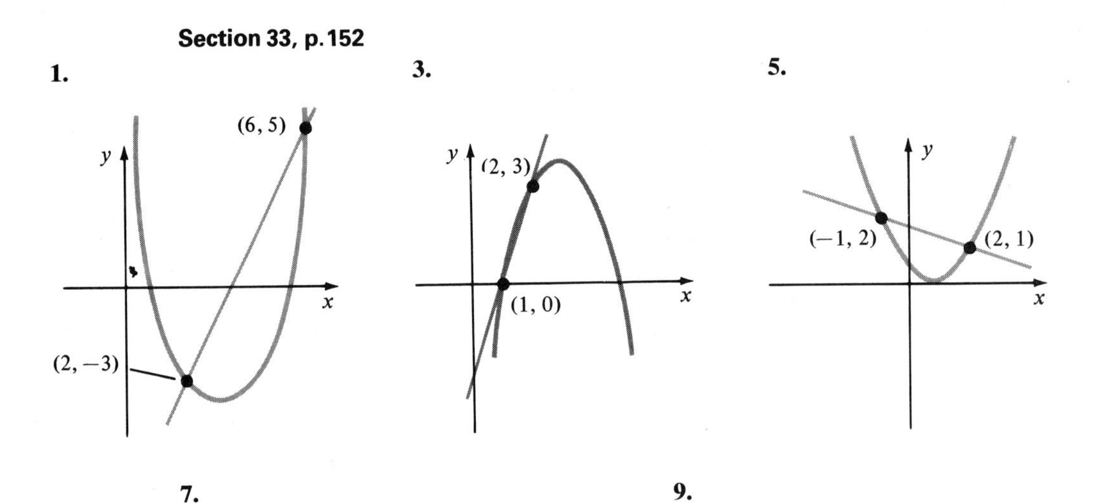

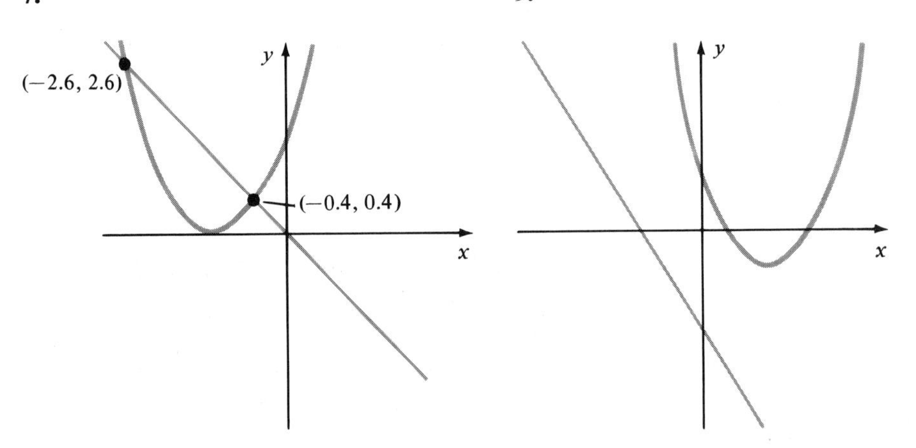

Section 34, p.159

1. $(2, -3), (6, 5)$ **3.** $(1, 0), (2, 3)$ **5.** $(-1, 4), (2, 1)$
7. $\left(\frac{-3+\sqrt{5}}{2}, \frac{3-\sqrt{5}}{2}\right), \left(\frac{-3-\sqrt{5}}{2}, \frac{3+\sqrt{5}}{2}\right)$ **9.** no solution
11. $(-4, -7), \left(\frac{5}{2}, \frac{-15}{4}\right)$ **13.** $\left(\frac{\pm 3\sqrt{2}}{2}, \frac{9}{2}\right)$ **15.** no solution
17. $\left(\frac{\sqrt{2}}{2}, \frac{\sqrt{2}}{2}\right), \left(-\frac{\sqrt{2}}{2}, -\frac{\sqrt{2}}{2}\right)$ **19.** $(-5, -6), (3, 2)$ **21.** $(\pm 3, 0)$

23. $(\pm 3, 2), (\pm 3, -2)$ **25.** $(-3, 4), (4, -3)$ **27.** $\left(-\frac{1}{3}, \frac{5}{3}\right), (2, 4)$
29. $\left(4\sqrt{\frac{21}{13}}, -\sqrt{\frac{21}{13}}\right), \left(-4\sqrt{\frac{21}{13}}, \sqrt{\frac{21}{13}}\right), (2\sqrt{7}, \sqrt{7}), (-2\sqrt{7}, -\sqrt{7})$
31. $(2, 1), (-2, -1)$ **33.** 7 and 9 **35.** 8 in. $\times$ 12 in.
37. Initial dimensions are 40×75 **39.** \$10,000 at 6% and \$11,000 at 5%

Section 35, p.164

1. $x = 3, y = 0, z = -5$ **3.** $U = 2, V = -2, W = 3$
5. $x = -3, y = 1, z = -3$ **7.** $s = 1, r = 2, t = -1$
9. $x = -1, y = 2, z = -2$ **11.** $a = -1, b = \frac{2}{3}, c = 2$
13. $y = -x^2 + 2x - 3$ **15.** 5, 17, 12
17. \$4000 at 5%, \$6000 at 6% and \$10,000 at 9%
19. 25 lbs of A, 75 lbs, of B, and 150 lbs, of C
21. $a = 25, b = 80, c = 60.$
23. $R_1 = 1400$ ohms, $R_2 = 275$ ohms, $R_3 = 725$ ohms
25. $I_1 = \frac{10}{11}, I_2 = \frac{8}{11}, I_3 = -\frac{2}{11}$ **27.** $I_1 = \frac{3}{19}, I_2 = \frac{9}{19}, I_3 = \frac{6}{19}$

Section 36, p.172

1. 11 **3.** 0 **5.** -22 **7.** 2 **9.** -45 **11.** $x = -1, y = 3$
13. $x = -\frac{1}{2}, y = \frac{5}{2}$ **15.** dependent **17.** $x = 9, y = 3$
19. $x = -\frac{3}{2}, y = 3$ **21.** $x = 2, y = 0$ **23.** $x = \frac{16}{13}, y = -\frac{2}{13}$
25. Inconsistent **27.** $x = -\frac{17}{4}, y = -\frac{19}{6}$ **29.** $x = 1, y = 2$
31. 26 mph and 39 mph **33.** \$6000 **35.** $\frac{1}{3}$ and $\frac{1}{5}$

Section 37, p.179

1. $x = 3, y = 0, z = -5$ **3.** $U = 2, V = -2, W = 3$
5. $x = -3, y = 1, z = -3$ **7.** $s = 1, r = 2, t = -1$
9. $x = -1, y = 2, z = -2$ **11.** $a = -1, b = \frac{2}{3}, c = 2$
13. $y = 6x^2 + 3x + 1$ **15.** 2 A's, 3 B's, 4 C's
17. first 60, second 40, third 30

Review Exercises for Chapter 6, p. 181

1. $x = 3, y = -2$ **3.** $x = -4, y = -5$ **5.** $x = 3, y = -1$
7. $x = 2, y = 1$ **9.** $x = -2, y = 3$ **11.** $x = 2, y = 0$
13. $x = -1, y = 2$ **15.** $(-2, -2), (1, 1)$ **17.** $(-2, 5), (4, -7)$
19. $x = -\frac{3}{2}, y = \frac{13}{2}$ **21.** $x = 4, y = 1$ **23.** $x = 1, y = 1$
25. dependent **27.** $(1, 2), (6, 17)$ **29.** $(\pm 2, 0)$
31. $(-2, 5), (4, -7)$ **33.** $(2, 4), (3, 15)$
35. $x = \frac{111}{49}, y = -\frac{17}{49}, z = -\frac{24}{49}$ **37.** $(1, 1)$ **39.** $(-1, \frac{3}{2})$
41. dependent **43.** $(2, \frac{1}{2}, -1)$ **45.** $(\frac{1}{2}, \frac{2}{9}, \frac{1}{3})$ **47.** 12 and 42
49. $f(x) = 3$ **51.** \$15,000 and \$10,000
53. $\frac{4}{3}$ liters of 80% solution and $\frac{8}{3}$ liters of 50% solution

Section 38, p.191

1. $\frac{\pi}{4}$ **3.** $\frac{\pi}{12}$ **5.** $\frac{2\pi}{3}$ **7.** 0.43 **9.** $\frac{\pi}{6}$ **11.** 30° **13.** 80°

15. 210° **17.** 180° **19.** 133°41′24″ **21.** $\frac{35\pi}{6} = 18.33$ cm

23. 10.7 cm^2 **25.** 1.8 mph **21.** 2.1 ft **29.** 3°49′11″; 0.067

31. 18.53 miles per sec **33.** 31.4 in., 41.9 in.

Section 39, p.196

1. 1st or 4th **3.** 4th **5.** 1st or 4th **7.** plus **9.** plus

11. plus **13.** minus **15.** minus **17.** plus

19. $\sin\theta = \frac{4}{5}$
$\cos\theta = \frac{3}{5}$
$\tan\theta = \frac{4}{3}$
$\csc\theta = \frac{5}{4}$
$\sec\theta = \frac{5}{3}$
$\cot\theta = \frac{3}{4}$

21. $\sin\theta = \frac{4}{5}$
$\cos\theta = -\frac{3}{5}$
$\tan\theta = -\frac{4}{3}$
$\csc\theta = \frac{5}{4}$
$\sec\theta = -\frac{5}{3}$
$\cot\theta = -\frac{3}{4}$

23. $\sin\theta = -\frac{5}{13}$
$\cos\theta = -\frac{12}{13}$
$\tan\theta = \frac{5}{12}$
$\csc\theta = -\frac{13}{5}$
$\sec\theta = -\frac{13}{12}$
$\cot\theta = \frac{12}{5}$

25. $\sin\theta = 0$
$\cos\theta = 1$
$\tan\theta = 0$
$\csc\theta$ is undefined
$\sec\theta = 1$
$\cot\theta$ is undefined

27. See # 25

29. $\cos\theta = \frac{12}{13}$
$\tan\theta = \frac{5}{12}$
$\csc\theta = \frac{13}{5}$
$\sec\theta = \frac{13}{12}$
$\cot\theta = \frac{12}{5}$

31. $\sin\theta = -\frac{4}{5}$
$\tan\theta = \frac{4}{3}$
$\sec\theta = -\frac{5}{3}$
$\csc\theta = -\frac{5}{4}$
$\cot\theta = \frac{3}{4}$

33. $\cos\theta = \pm\frac{15}{17}$
$\tan\theta = \pm\frac{8}{15}$
$\csc\theta = \frac{17}{8}$
$\sec\theta = \pm\frac{17}{15}$
$\cot\theta = \pm\frac{15}{8}$

35. $\sin\theta = \pm\frac{3}{5}$
$\cos\theta = \pm\frac{4}{5}$
$\csc\theta = \pm\frac{5}{3}$
$\sec\theta = \pm\frac{5}{4}$
$\cot\theta = \frac{4}{3}$

37. $\sin\theta = \frac{3}{5}$
$\tan\theta = -\frac{3}{4}$
$\sec\theta = -\frac{5}{4}$
$\csc\theta = \frac{5}{3}$
$\cot\theta = -\frac{4}{3}$

39. $\sin\theta = \pm\frac{\sqrt{11}}{6}$
$\cos\theta = \frac{5}{6}$
$\tan\theta = \pm\frac{\sqrt{11}}{5}$
$\csc\theta = \pm\frac{6}{\sqrt{11}}$
$\cot\theta = \pm\frac{5}{\sqrt{11}}$

41.

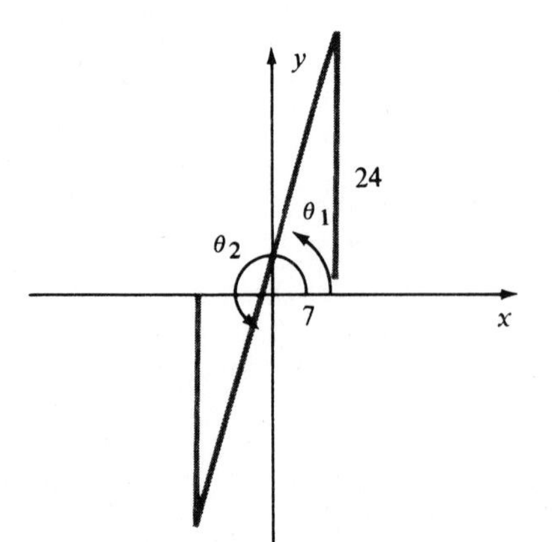

43.

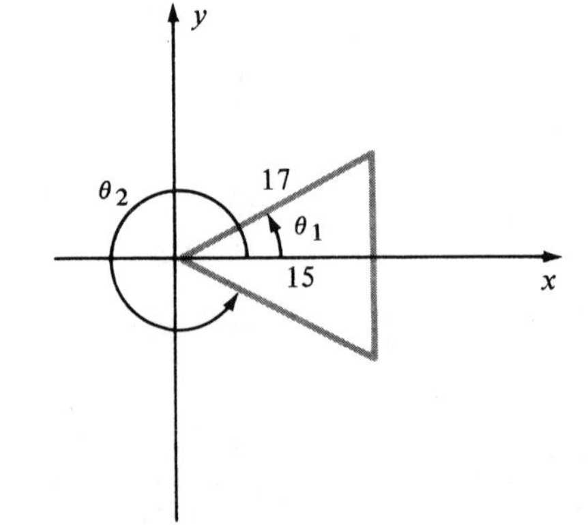

Section 40, p. 202

1. $\sin 0° = 0$
$\cos 0° = 1$
$\tan 0° = 0$
$\csc 0°$ is undefined
$\sec 0° = 1$
$\cot 0°$ is undefined

3. $\sin 180° = 0$
$\cos 180° = -1$
$\tan 180° = 0$
$\csc 180°$ is undefined
$\sec 180° = -1$
$\cot 180°$ is undefined

5. $\sin 135° = \frac{\sqrt{2}}{2}$
$\cos 135° = -\frac{\sqrt{2}}{2}$
$\tan 135° = -1$
$\csc 135° = \sqrt{2}$
$\sec 135° = -\sqrt{2}$
$\cot 135° = -1$

7. $\sin 225° = -\frac{\sqrt{2}}{2}$
$\cos 225° = -\frac{\sqrt{2}}{2}$
$\tan 225° = 1$
$\csc 225° = -\sqrt{2}$
$\sec 225° = -\sqrt{2}$
$\cot 225° = 1$

9. $\sin 300° = -\frac{\sqrt{3}}{2}$
$\cos 300° = \frac{1}{2}$
$\tan 300° = -\sqrt{3}$
$\csc 300° = -\frac{2\sqrt{3}}{3}$
$\sec 300° = 2$
$\cot 300° = -\frac{\sqrt{3}}{3}$

11. $\sin 330° = -\frac{1}{2}$
$\cos 330° = \frac{\sqrt{3}}{2}$
$\tan 330° = -\frac{\sqrt{3}}{3}$
$\csc 330° = -2$
$\sec 330° = \frac{2\sqrt{3}}{3}$
$\cot 330° = -\sqrt{3}$

13. $\sin -30° = -\frac{1}{2}$
$\cos -30° = \frac{\sqrt{3}}{2}$
$\tan -30° = -\frac{\sqrt{3}}{3}$
$\csc -30° = -2$
$\sec -30° = \frac{2\sqrt{3}}{3}$
$\cot -30° = -\sqrt{3}$

15. $\sin -150° = -\frac{1}{2}$
$\cos -150° = -\frac{\sqrt{3}}{2}$
$\tan -150° = \frac{\sqrt{3}}{3}$
$\csc -150° = -2$
$\sec -150° = -\frac{2\sqrt{3}}{3}$
$\cot -150° = \sqrt{3}$

17. True **19.** False **21.** False **23.** True **25.** True
27. False **29.** True

Section 41, p. 211

1. 82° **3.** 46° **5.** 46° **7.** 45° **9.** 0.4436 **11.** 3.0777
13. 0.0029 **15.** 0.5185 **17.** 0.5911 **19.** 2.7653 **21.** 16°19′54″
23. 31°11′33″ **25.** 41°40′27″ **27.** 78°1′45″ **29.** 24°24′21″
31. 54°58′10″ **33.** 67°50′; 0.9261 **35.** 38°; 0.7813
37. 36°50′; −1.3351 **39.** 78°; −0.9781 **41.** 57°; −1.8361
43. 12°; 1.9563 **45.** θ = 4°10′09″, 175°49′51″
47. θ = 63°26′06″, 243°26′06″ **49.** θ = 184°10′09″, 355°49′51″
51. $V = 119.8355$

Section 42, p. 216

1. $B = 58°; a = 4, b = 7$ **3.** $c = 28, A = 32°; B = 58°$
5. $A = 27°, a = 7, b = 14$ **7.** $B = 44°, b = 52, c = 75$
9. $A = 26°40', b = 13.4, c = 29.9$ **11.** $A = 40°, B = 50°, a = 10$
13. 93 ft **15.** 100 in. **17.** 275 ft **19.** 7.2 ft
21. 20°33′ **23.** 100 ft **25.** 31° **27.** 32°15′

Section 43, p. 225

1. $b = 142, c = 100, C = 43°$ **3.** no solution
5. $A = 36°52', B = 53°08', C = 90°$
7. $B_1 = 42°27', C_1 = 107°33', c_1 = 38; B_2 = 137°33', C_2 = 12°27', c_2 = 13$
9. $B_1 = 62°07', C_1 = 72°53', c_1 = 27$ or $B_2 = 117°53', C_2 = 17°07', c_2 = 8$
11. $A = 26°21', B = 47°42'; C = 105°58'$
13. $C = 37°20', a = 980, b = 929$ **15.** $A = 26°, B = 119°, C = 35°$
17. 1st solution $B = 139°22', C = 23°18', b = 26.7$ 2nd solution $B = 5°58'$, $C = 156°42', b = 4.3$
19. $b = 5.52, A = 26°19', C = 49°46'$ **21.** Area $= \frac{1}{2}bc \sin A$
25. 1000 ft **27.** 13,952 yd^2; 2.88 acres **29.** 335 miles

Section 44, p. 234

1. Scalar **3.** Scalar **5.** Scalar

7.

y

x

9.

$\vec{B}$

$\vec{A} + \vec{B}$

$\vec{A}$

11.

$\vec{B}$

$-\vec{A}$

$\vec{B} - \vec{A}$

13. **15.**

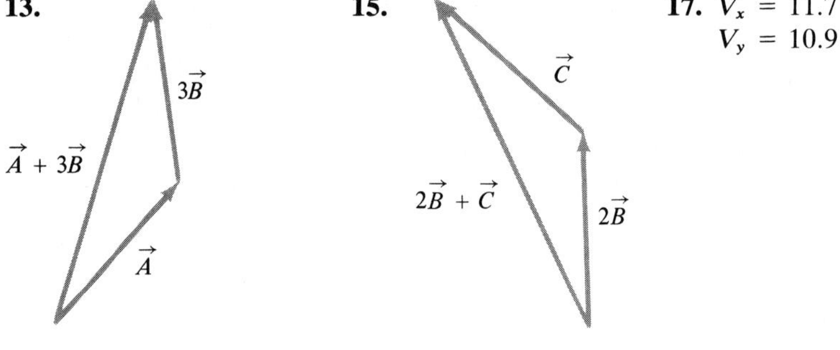

17. $V_x = 11.7$
$V_y = 10.9$

19. $V_x = -99.5$
$V_y = -67.1$
21. $V_x = V_y = 0.7$ **23.** (22.7, 16.5) **25.** 22.9, 45°
27. 1 in the direction of A **29.** 18.6 in the direction 29°23′ measured from A
31. 310 in the direction 12°17′ measured from A **33.** 18.44, 48°32′
35. 0.66, 143° **37.** 2468, −77°43′ **39.** 18.55, 37°04′

Section 45, p. 242

1. 120 lb, 41°38′ **3.** 544 ft-lb **5.** 76.16 lb, 66°48′ south of east
7. 11.4, 127°53′ **9.** $F_1 = 71.26, F_2 = 79.86$
11. 605 mph 7°08′ north of east **13.** 16.16 mph, 68°12′
15. 14°30′ south of east **17.** 77, 76° west of north **19.** 162.7 lbs

Review Exercises for Chapter 7, p. 244

1. (a) $\frac{\pi}{6}$ (b) $\frac{\pi}{3}$ (c) π (d) $\frac{4\pi}{3}$ (e) $\frac{11\pi}{12}$

3. $\sin\theta = \frac{3}{5}$
$\cos\theta = \frac{4}{5}$
$\tan\theta = \frac{3}{4}$
$\csc\theta = \frac{5}{3}$
$\sec\theta = \frac{5}{4}$
$\cot\theta = \frac{4}{3}$

5. $\sin\theta = \frac{12}{13}$
$\cos\theta = -\frac{5}{13}$
$\tan\theta = -\frac{12}{5}$
$\csc\theta = \frac{13}{12}$
$\sec\theta = -\frac{13}{5}$
$\cot\theta = -\frac{5}{12}$

7. $\sin\theta = 0$
$\cos\theta = -1$
$\tan\theta = 0$
$\csc\theta$ is undefined
$\sec\theta = -1$
$\cot\theta$ is undefined

9. $\sin\theta = \frac{5}{\sqrt{29}}$
$\cos\theta = -\frac{2}{\sqrt{29}}$
$\tan\theta = -\frac{5}{2}$
$\csc\theta = \frac{\sqrt{29}}{5}$
$\sec\theta = -\frac{\sqrt{29}}{2}$
$\cot\theta = -\frac{2}{5}$

11. $\cos\theta = -\frac{4}{5}$
$\tan\theta = \frac{3}{4}$

13. $\sin\theta = \frac{4}{5}$
$\csc\theta = \frac{5}{4}$

15. $\sin\theta = \pm\frac{8}{17}$
$\tan\theta = \pm\frac{8}{15}$

17. $\cos\theta = \pm\frac{3}{5}$
$\sec\theta = \pm\frac{5}{3}$

19. $\frac{1}{2}, -\frac{\sqrt{3}}{2}$

21. 0.6428

23. −0.6561 **25.** −0.6561 **27.** 0.2456 **29.** −1.2091
31. 37° **33.** 61°50′ **35.** 71°10′ **37.** 5°50′ **39.** 37°10′
41. 23°, 157° **43.** 152°30′, 207°30′
45. (a) $A = 14°, B = 46°, c = 36$ (b) $b = 9.8, c = 9.4, C = 70°$
(c) 1st solution $B = 38°41', C = 111°19', c = 7.45$; 2nd solution $B = 141°19'$, $C = 8°41', c = 1.21$
47. 2.74, 7.52 **49.** 48°11′ **51.** 148.6 lb

Section 46, p. 251

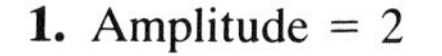
1. Amplitude = 2

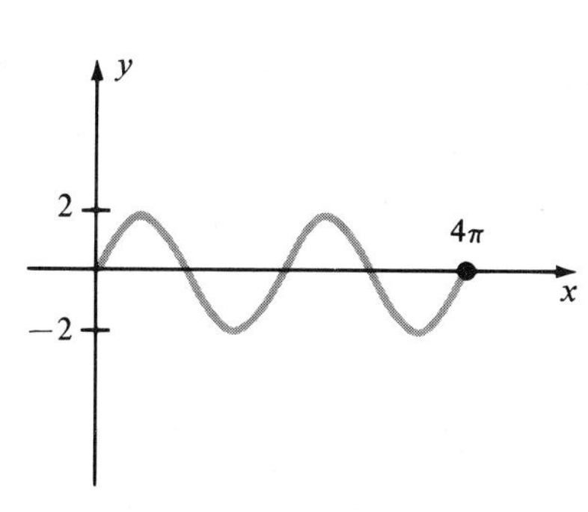

3. Amplitude = 4

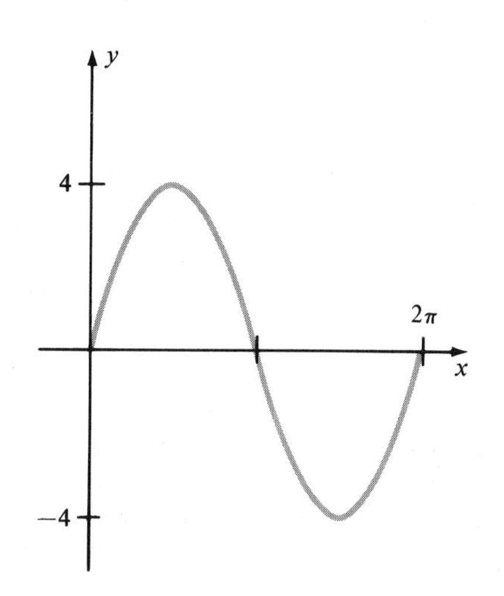

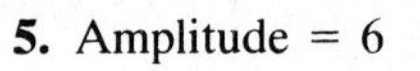

5. Amplitude = 6

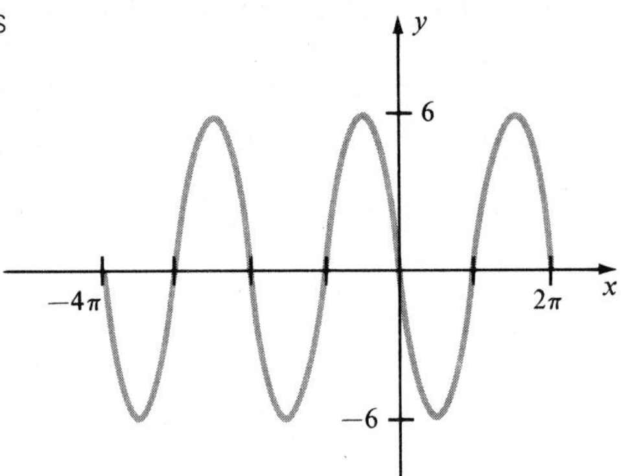

7. Amplitude = 3

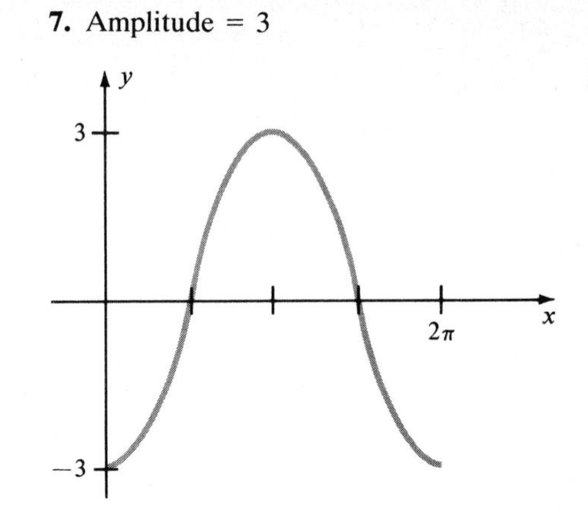

9. Amplitude = $\frac{7}{2}$

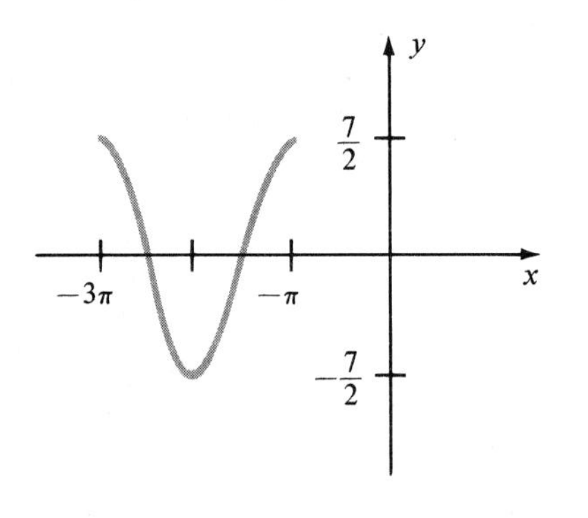

11. Amplitude = 6

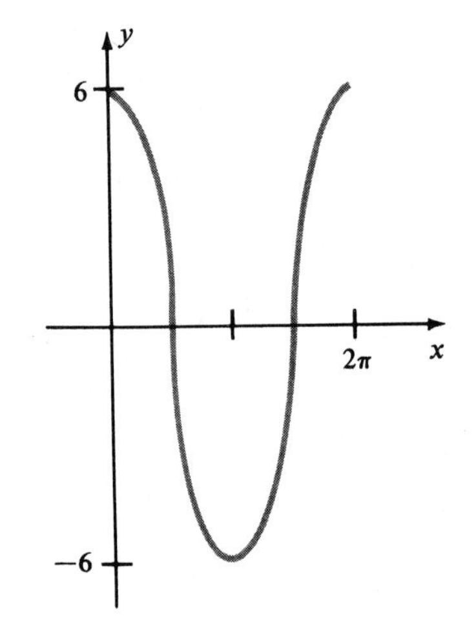

13. Amplitude = 0.9

15. Amplitude = 0.5

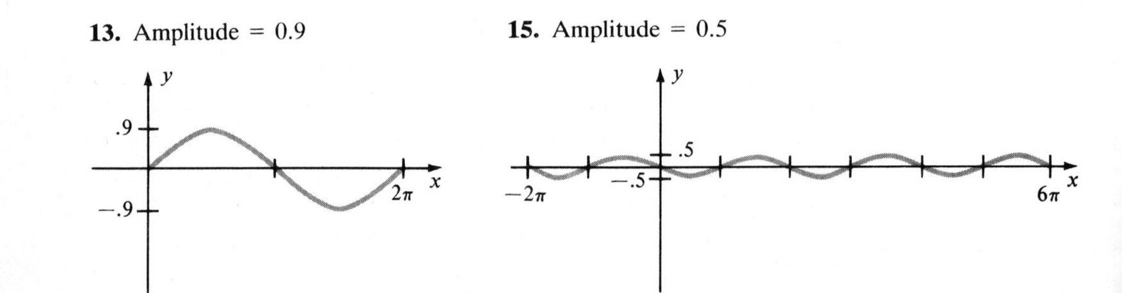

Section 47, p. 258

1. Amplitude = 2
Period = 2
Phase shift = $-\dfrac{3}{\pi}$

3. Amplitude = 4
Period = 2π
Phase shift = 2π

5. Amplitude = 3
Period = 2
Phase shift = 2

7. Amplitude = 1
Period = 2π
Phase shift = -1

9. $6, 12, \pi, \dfrac{\pi}{2}$

11. $y = 3 \sin (2x - 2\pi)$

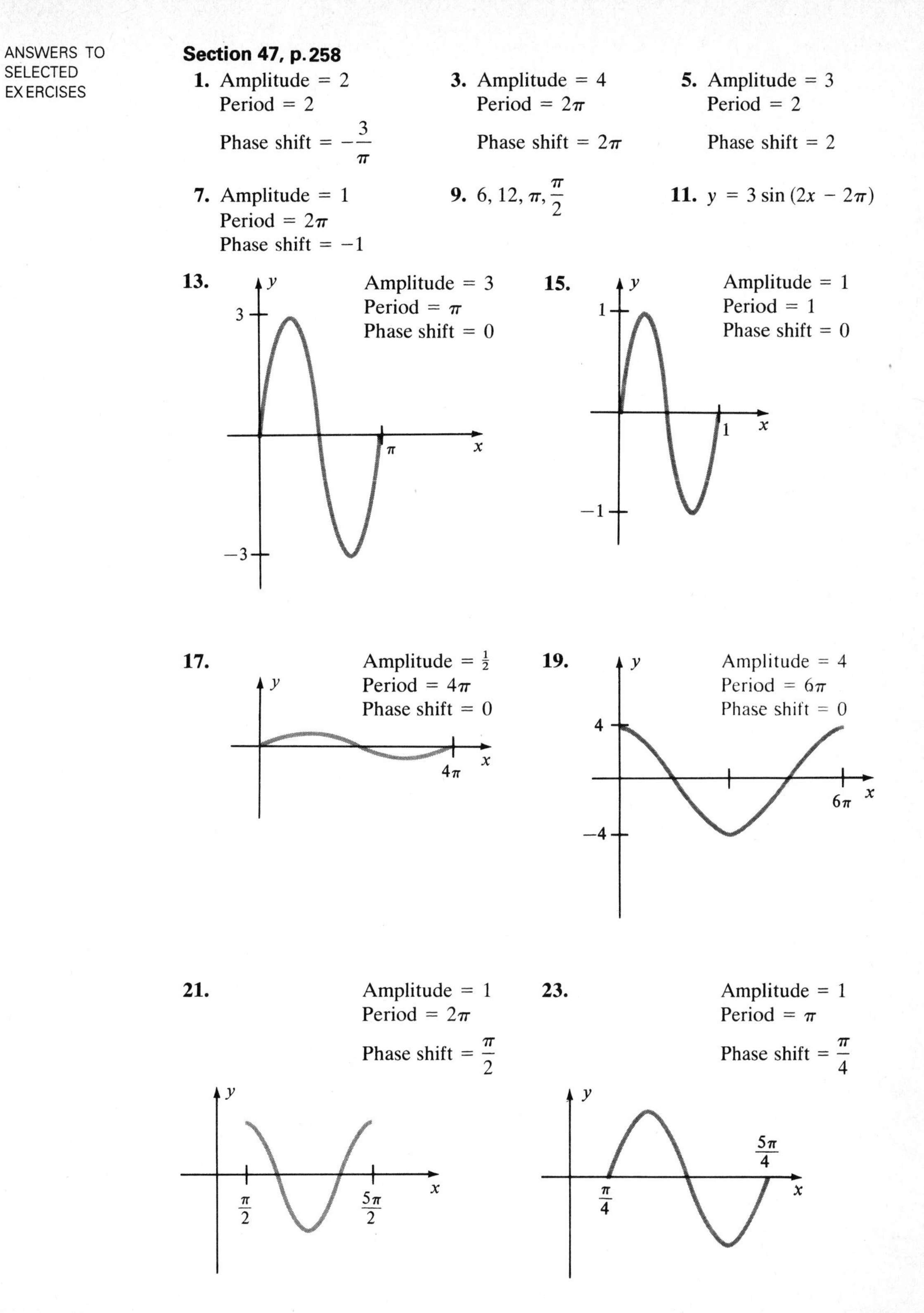

13. Amplitude = 3
Period = π
Phase shift = 0

15. Amplitude = 1
Period = 1
Phase shift = 0

17. Amplitude = $\frac{1}{2}$
Period = 4π
Phase shift = 0

19. Amplitude = 4
Period = 6π
Phase shift = 0

21. Amplitude = 1
Period = 2π
Phase shift = $\dfrac{\pi}{2}$

23. Amplitude = 1
Period = π
Phase shift = $\dfrac{\pi}{4}$

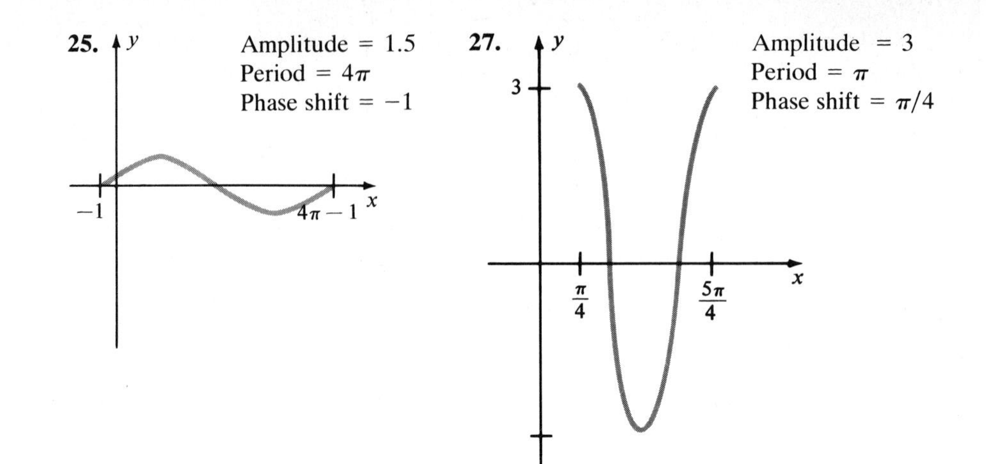

25. Amplitude = 1.5
Period = 4π
Phase shift = −1

27. Amplitude = 3
Period = π
Phase shift = $\pi/4$

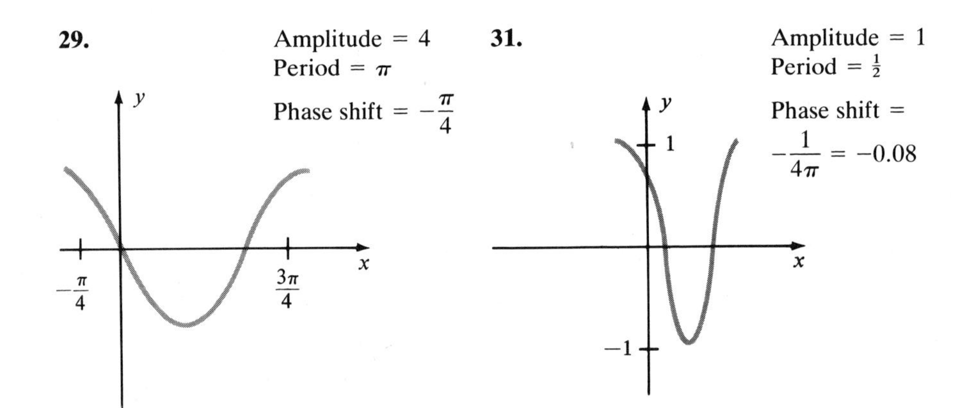

29. Amplitude = 4
Period = π
Phase shift = $-\frac{\pi}{4}$

31. Amplitude = 1
Period = $\frac{1}{2}$
Phase shift = $-\frac{1}{4\pi} = -0.08$

33.

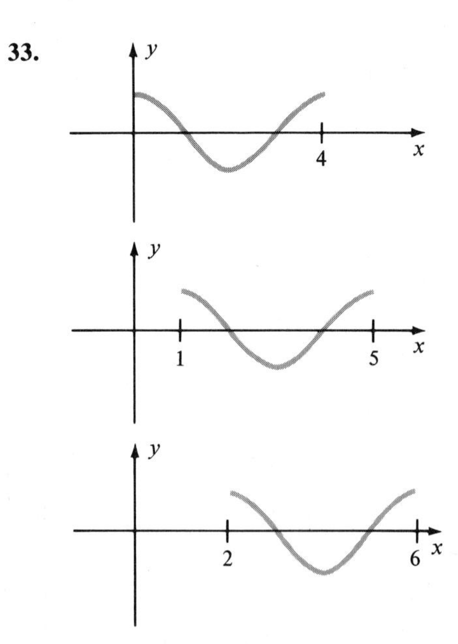

For $t = 0$, $y = 2\cos(.5\pi x)$

For $t = 0.0025$, $y = 2\cos\left[\frac{\pi}{2}(x - 1)\right]$

For $t = 0.005$, $y = 2\cos[\pi(0.5x - 1)]$

35.

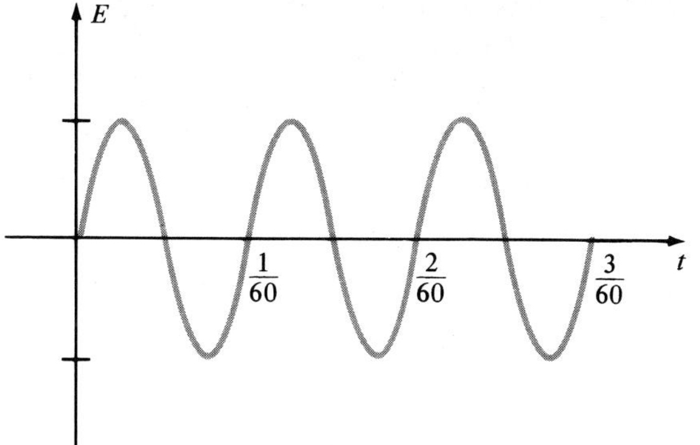

37. Graphs are the same. *Conclusion:* $\sin\left(\frac{\pi}{2} + x\right) = \cos x$

39. **41.**

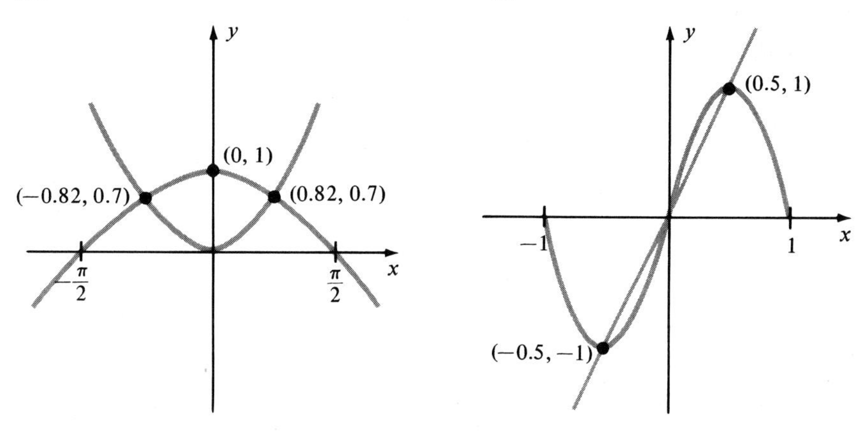

43. $x = -\frac{5\pi}{6}, -\frac{\pi}{2}, -\frac{\pi}{6}, \frac{\pi}{6}, \frac{\pi}{2}, \frac{5\pi}{6}, \frac{7\pi}{6}, \frac{3\pi}{2}, \frac{11\pi}{6}, \frac{13\pi}{6}, \frac{5\pi}{2}, \frac{17\pi}{6}$

Section 48, p. 226

1. **3.** **5.**

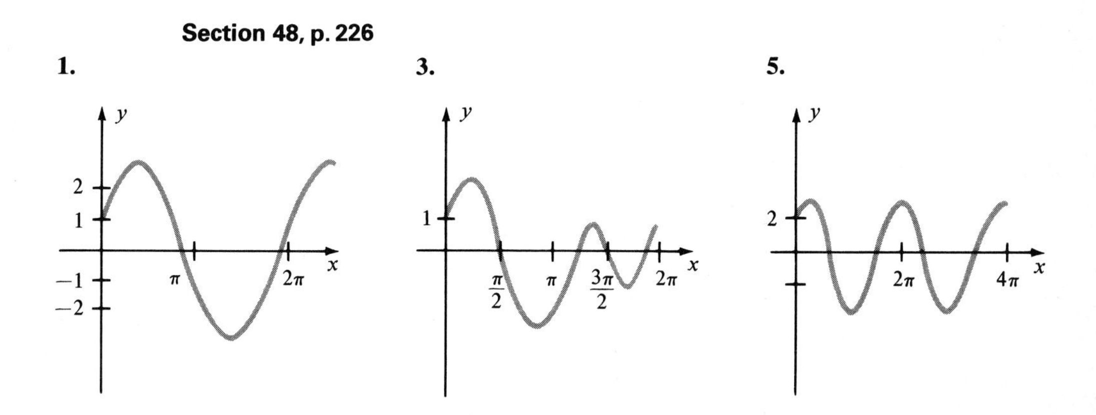

7. **9.** **11.**

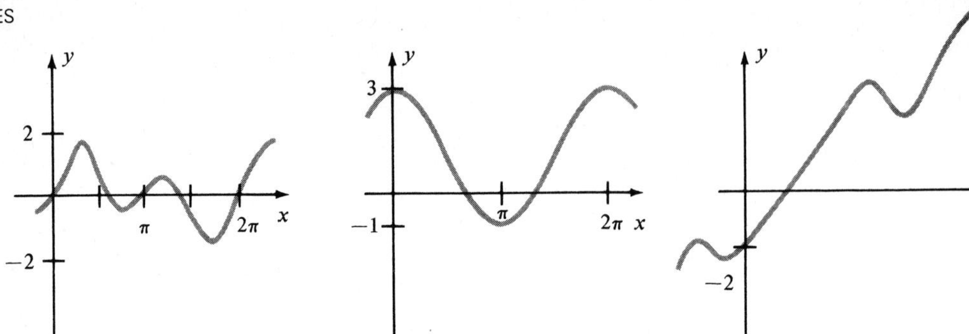

13.

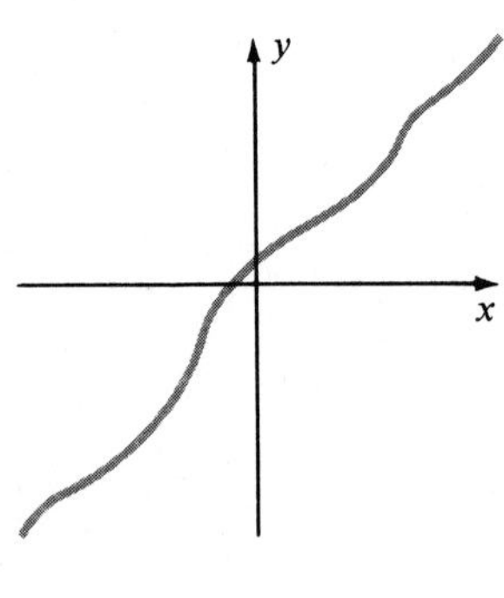

15.

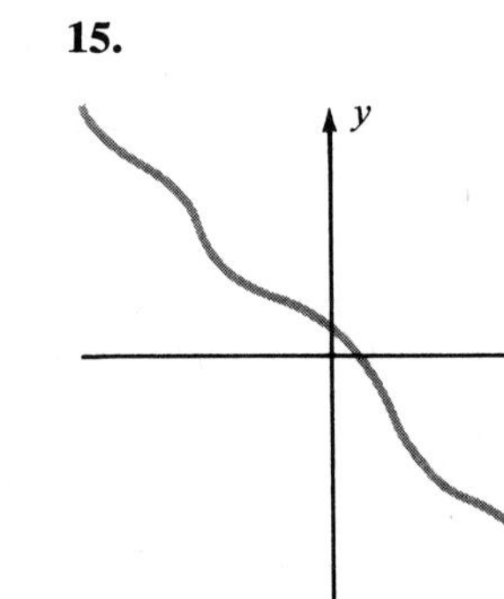

17.

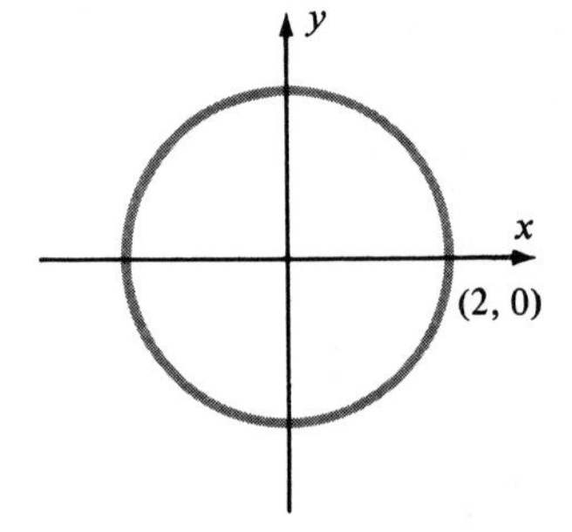

19.

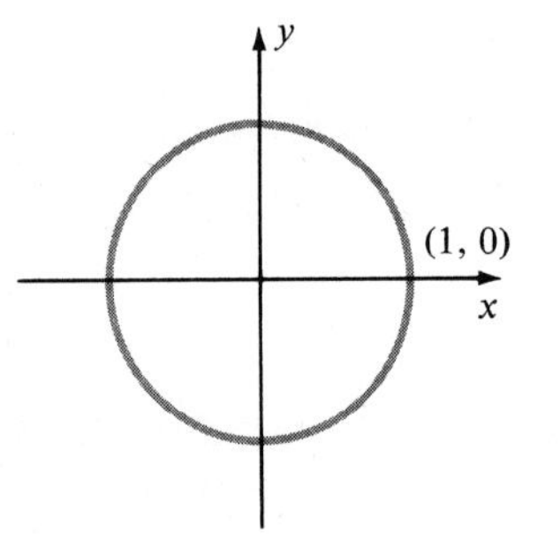

21.

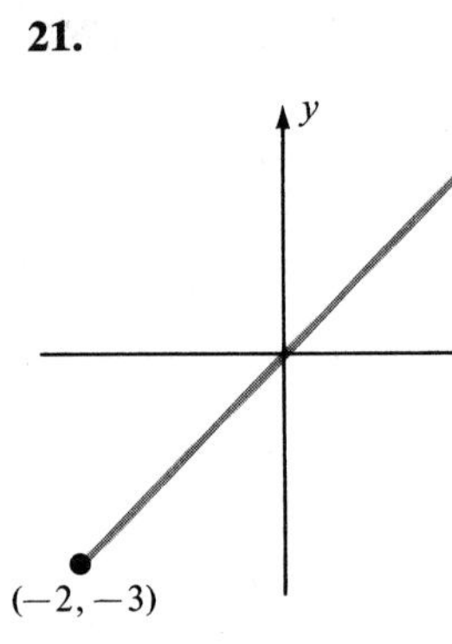

23.

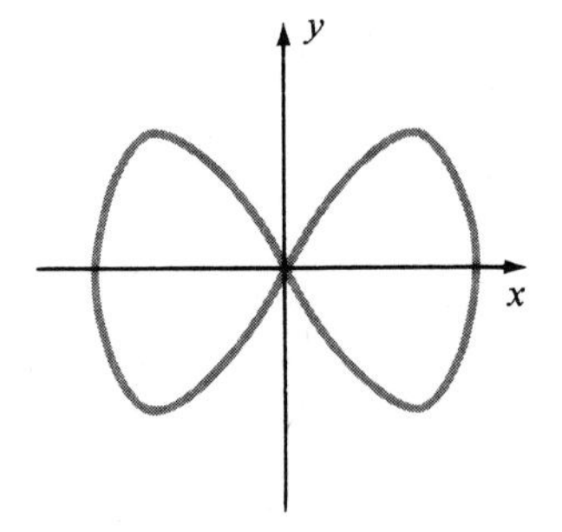

25. Amplitude = 6

Period = $\dfrac{\pi}{60w}$

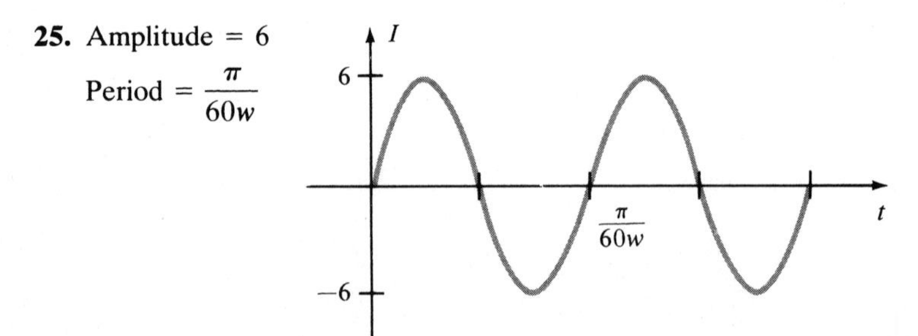

27. $x = \cos 2t$, $y = \sin 2t$, amplitude is 1, period is π, and the frequency is $\dfrac{1}{\pi}$

29. $x = 4\cos\frac{1}{2}t$, $y = 4\sin\frac{1}{2}t$, amplitude is 4, period is 4π, and the frequency is $\dfrac{1}{4\pi}$

Section 49, p. 272

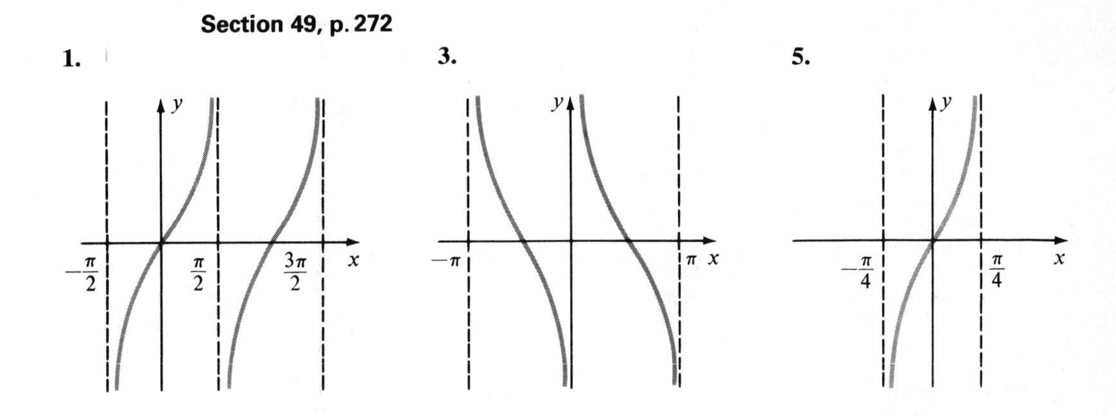

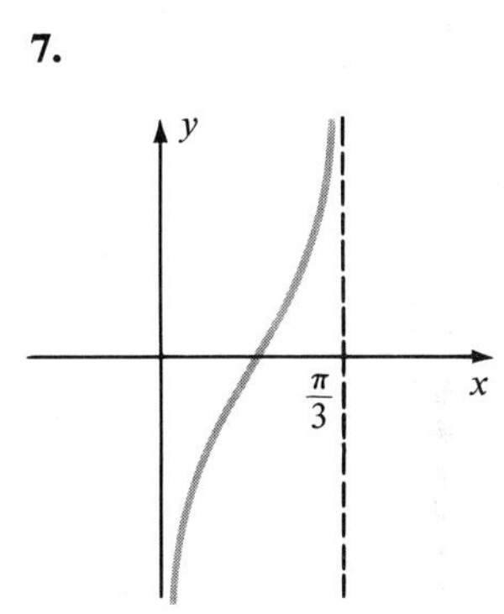

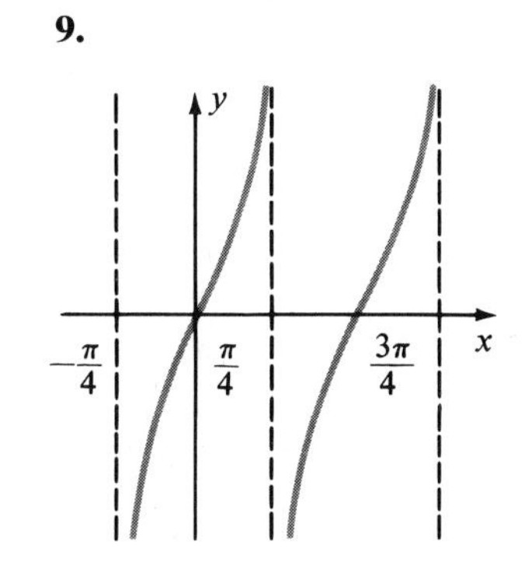

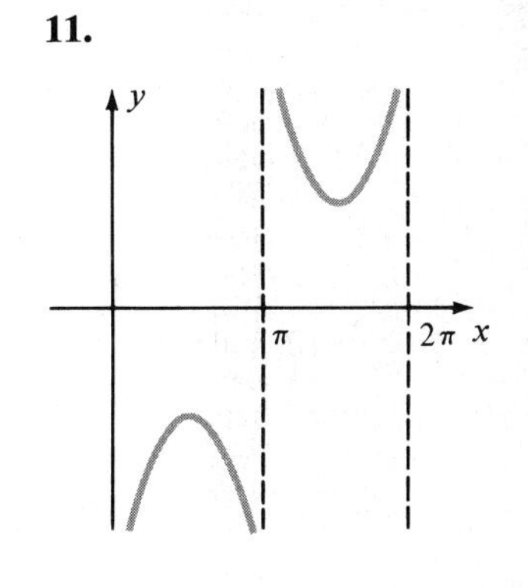

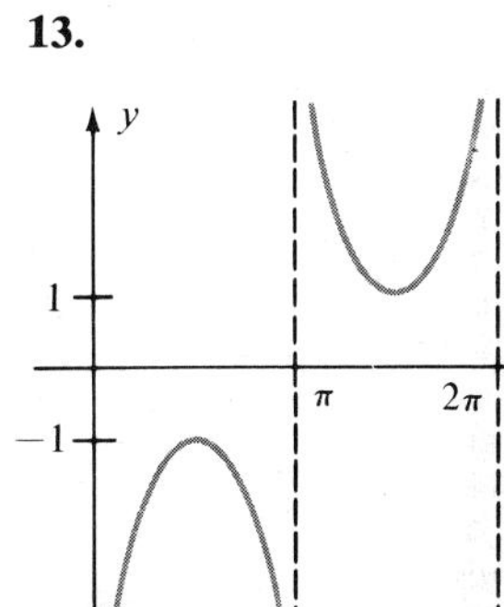

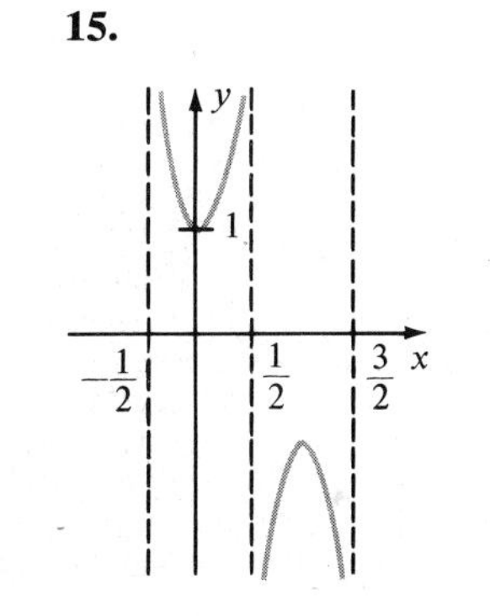

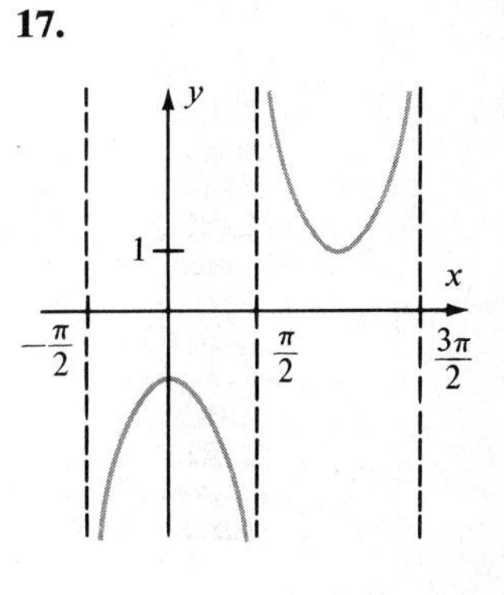

19. **21.** **23.** No

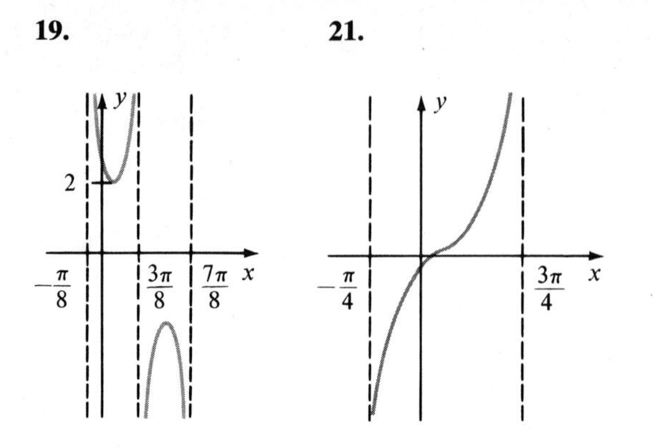

Review Exercises for Chapter 8, p. 273

1. Amplitude = 2
Period = 2π
Phase shift = 0

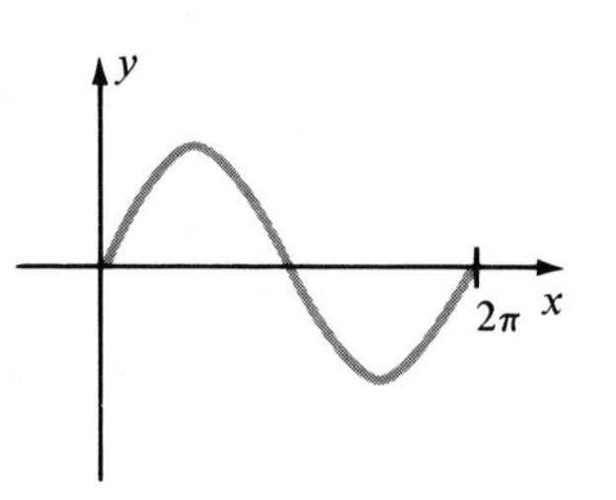

3. Amplitude = 3
Period = 2π
Phase shift = 0

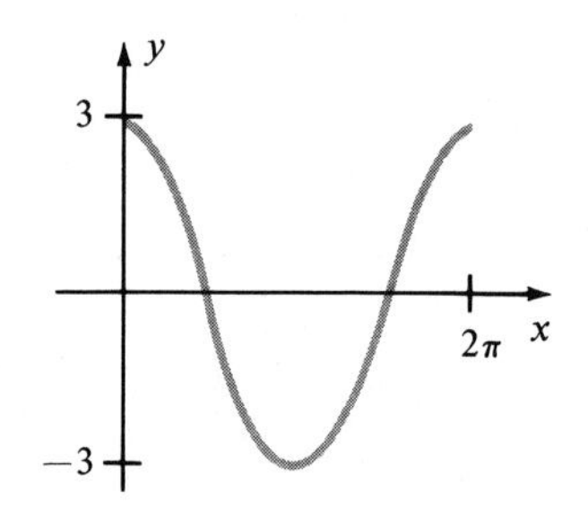

5. Amplitude = 2
Period = $\frac{2\pi}{3}$
Phase shift = 0

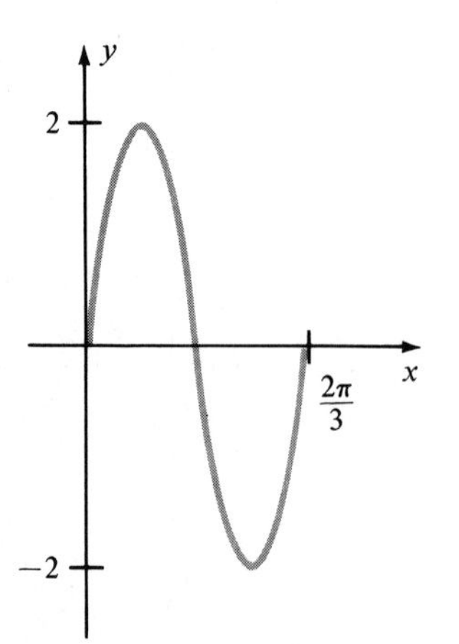

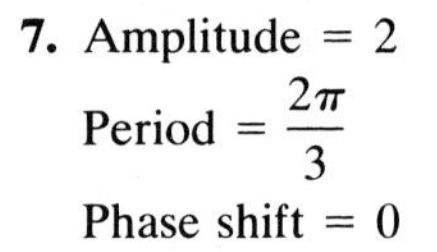
7. Amplitude = 2
Period = $\frac{2\pi}{3}$
Phase shift = 0

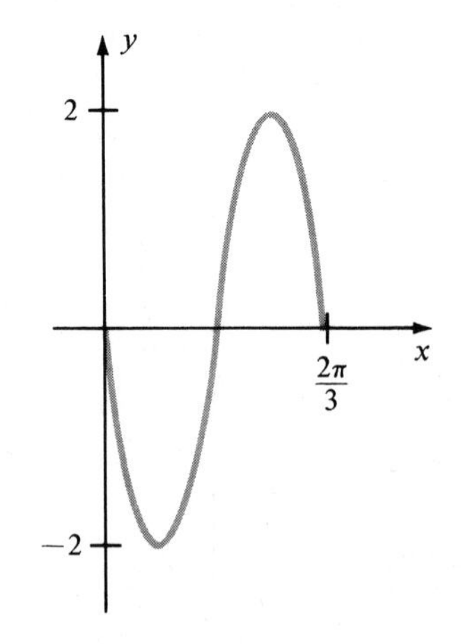

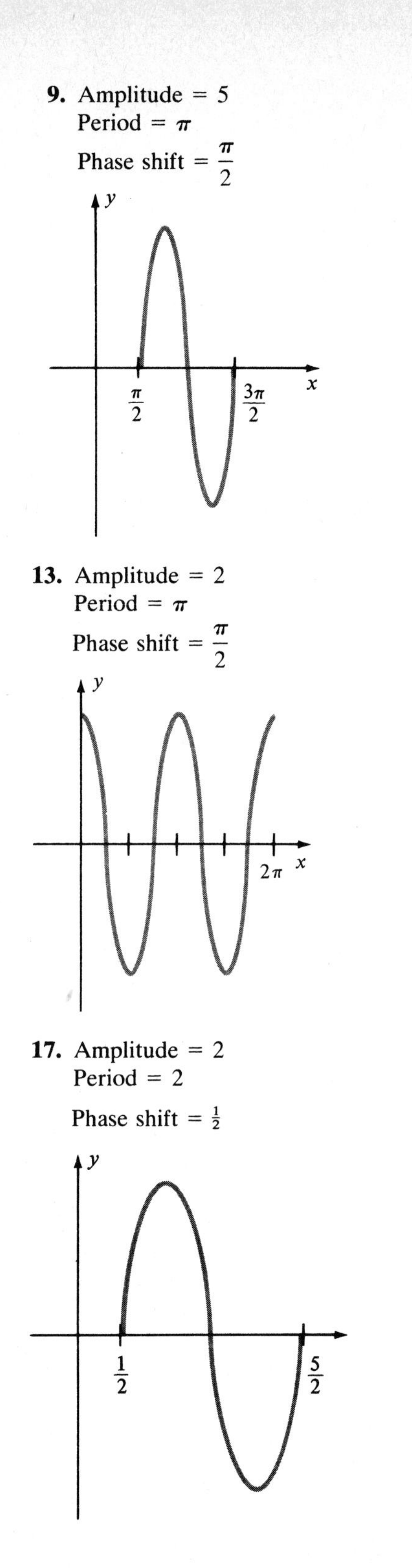

9. Amplitude = 5
Period = π
Phase shift = $\frac{\pi}{2}$

13. Amplitude = 2
Period = π
Phase shift = $\frac{\pi}{2}$

17. Amplitude = 2
Period = 2
Phase shift = $\frac{1}{2}$

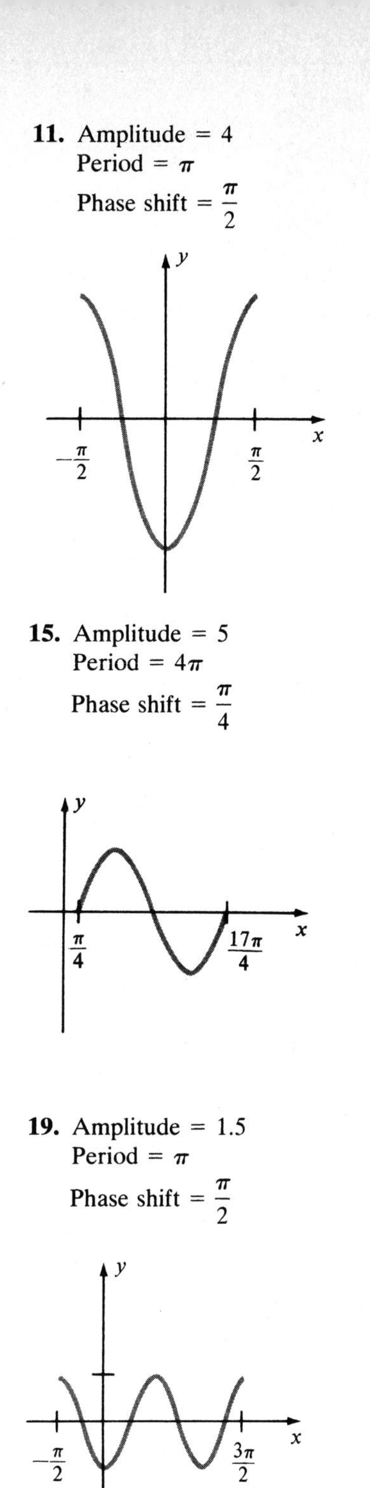

11. Amplitude = 4
Period = π
Phase shift = $\frac{\pi}{2}$

15. Amplitude = 5
Period = 4π
Phase shift = $\frac{\pi}{4}$

19. Amplitude = 1.5
Period = π
Phase shift = $\frac{\pi}{2}$

21. Period $= \pi$

Phase shift $= 0$

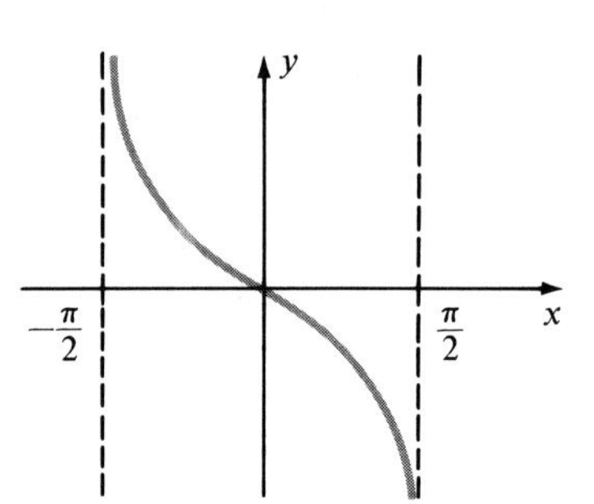

23. Period $= \frac{\pi}{2}$

Phase shift $= -\frac{\pi}{2}$

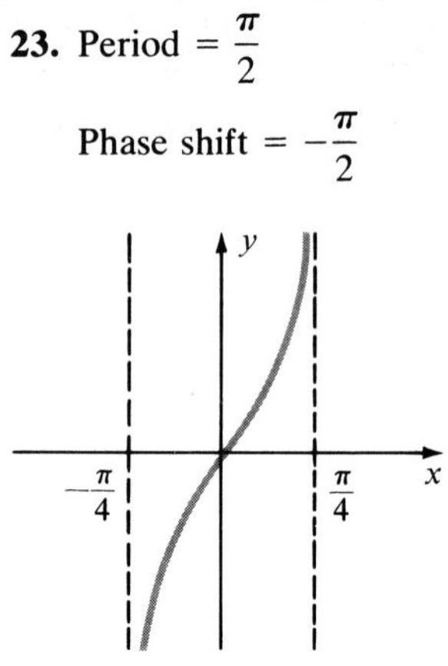

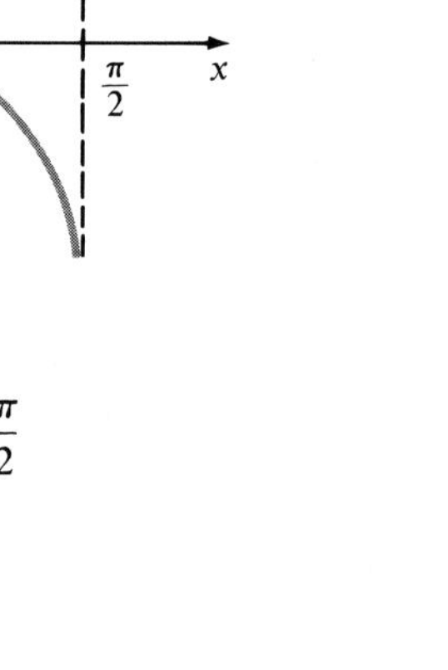

25. Period $= \pi$

Phase shift $= \frac{\pi}{2}$

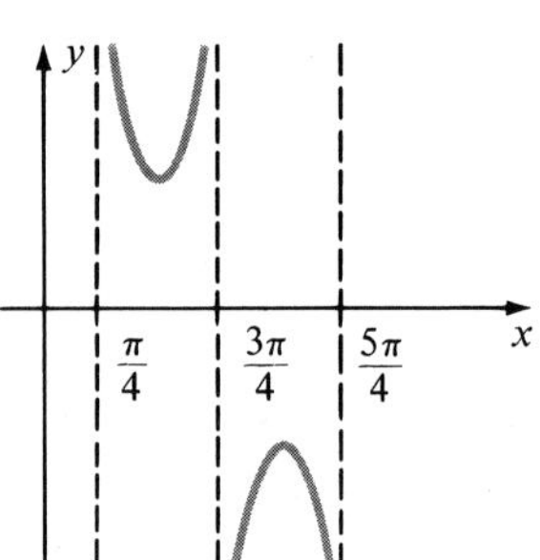

27. Period $= 2\pi$

Phase shift $= \pi$

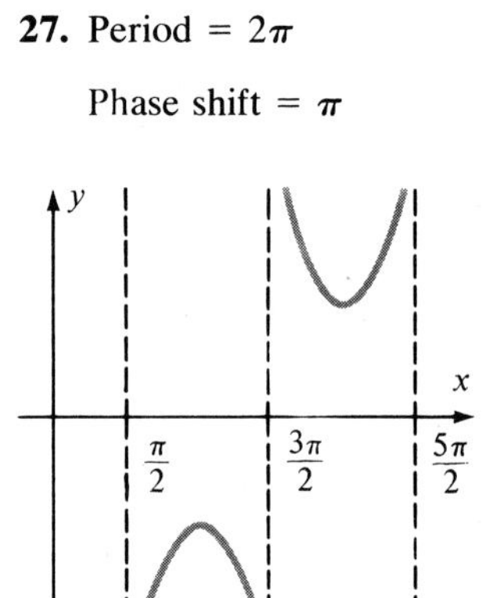

29. Period $= \frac{2\pi}{3}$

Phase shift $= \frac{\pi}{3}$

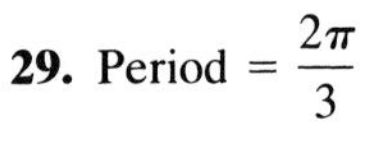

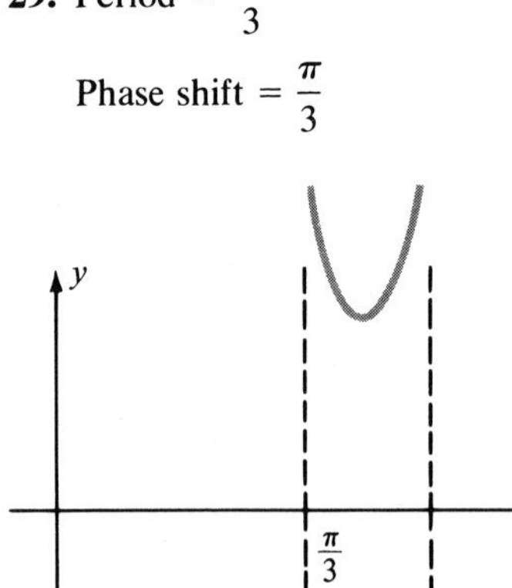

31. (a)

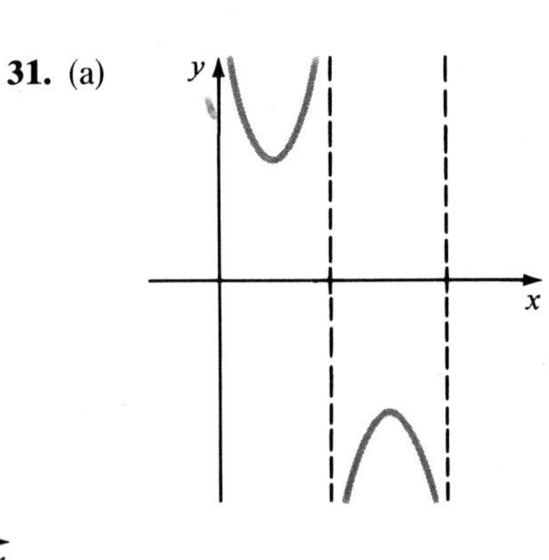

31. (b)

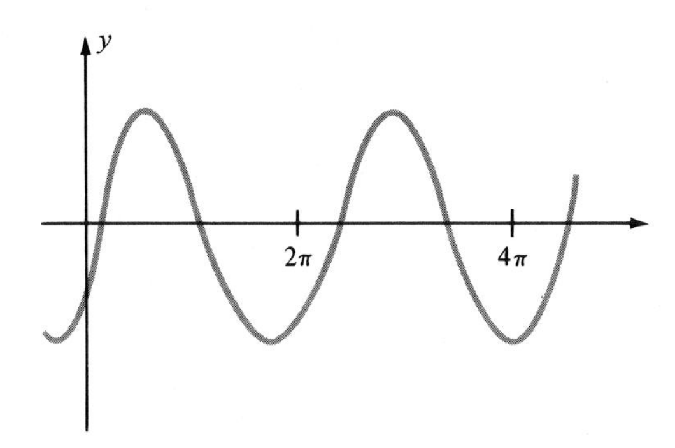

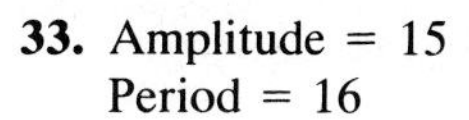

33. Amplitude = 15
Period = 16

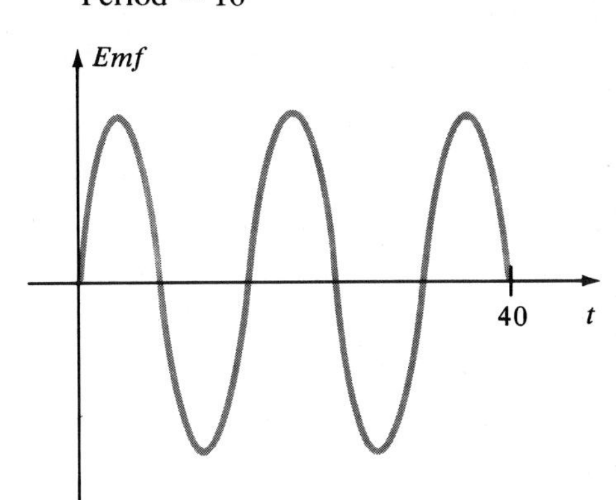

Section 50, p. 279

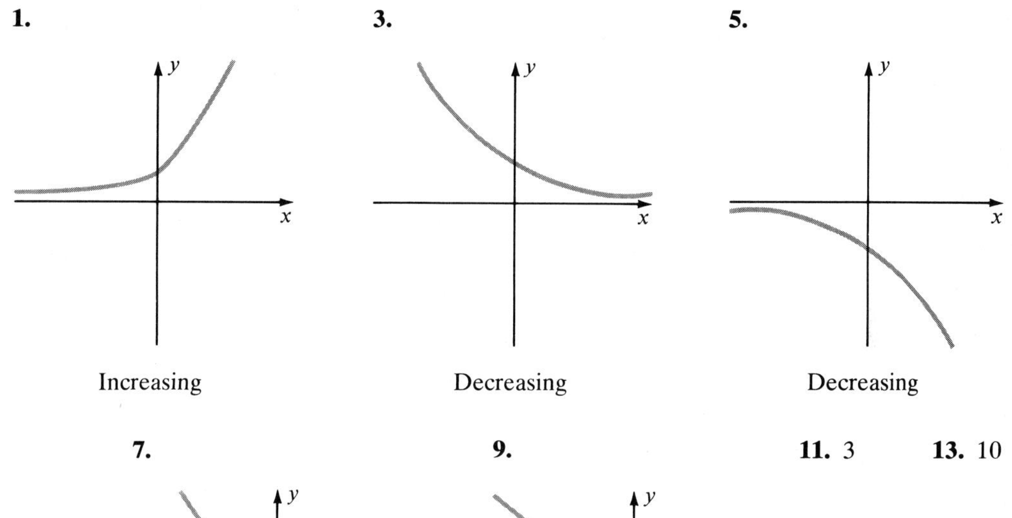

1. Increasing

3. Decreasing

5. Decreasing

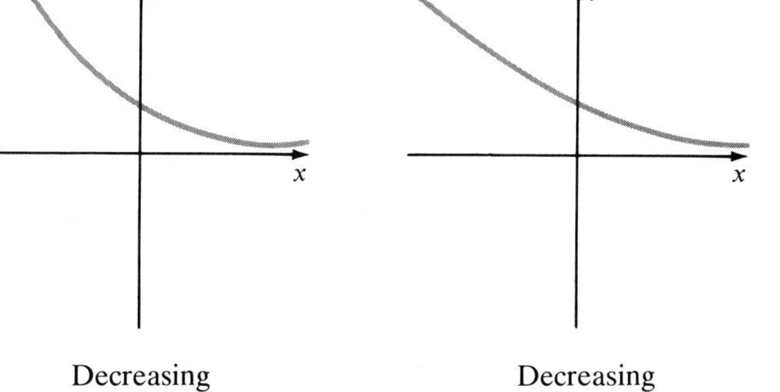

7. Decreasing

9. Decreasing

11. 3 **13.** 10

15. $\frac{1}{2}$ **17.** 16 **19.** $81\sqrt{3}$ **21.** $x = 4$ **23.** -3 **25.** 0
27. 2 **29.** 3 **31.** -1 **33.** 1 **35.** (a) 400 (b) $\frac{2}{5}$ (c) 3600

37. (a)

x	-2	-1	0	1	2	3	4
x^2	4	1	0	1	4	9	16
2^x	$\frac{1}{4}$	$\frac{1}{2}$	1	2	4	8	16

37. (b)

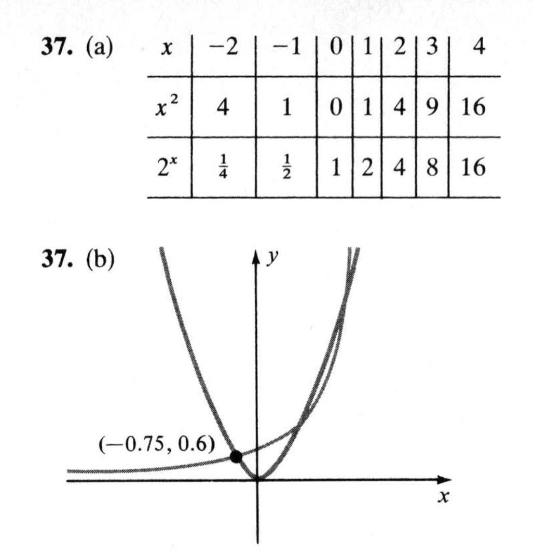

Section 51, p. 285

1. **3.** **5.**

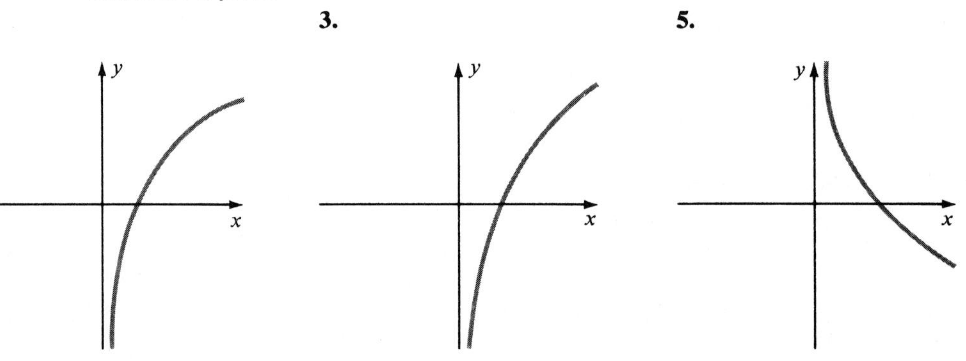

7. $\log_3 9 = 2$ **9.** $\log_{10} 1 = 0$ **11.** $\log_4 2 = \frac{1}{2}$ **13.** $\log_{10} \frac{1}{1000} = -3$
15. $\log_{64} \frac{1}{4} = \frac{1}{3}$ **17.** $\log_{10} 10 = 1$ **19.** $\log_e 1 = 0$ **21.** $6^2 = 36$
23. $10^2 = 100$ **25.** $10^{\circ} = 1$ **27.** $27^{4/3} = 81$ **29.** $(\frac{1}{3})^{-2} = 9$
31. $8^{-2} = \frac{1}{64}$ **33.** $10^5 = 10^5$ **35.** $e^{-6} = e^{-6}$ **37.** 4
39. 10 **41.** 10 **43.** 5 **45.** 16 **47.** 6 **49.** ± 1

Section 52, p. 290

1. $\log_{10} a + \log_{10} b$ **3.** $2 \log_{10} x - \log_{10} y$ **5.** $3 \log_{10} x + 6 \log_{10} y$
7. $2 \log_{10} x + 3 \log_{10} y - 4 \log_{10} z$ **9.** $-(\log_{10} x + \frac{1}{2} \log_{10} y)$
11. $\log_2 35$ **13.** $\log_3 2$ **15.** $\log_5 3$ **17.** $\log_2 2^{1/2} = \frac{1}{2}$
19. $\log_{10} 6$ **21.** 10 **23.** 7 **25.** 9 **27.** $-1, 4$ **29.** 10
31. $y = \dfrac{x^2}{10}$ **33.** $y = \dfrac{6}{x}$ **35.** $y = \dfrac{x}{6}$ **37.** $y = 8x^{2/3}$
39. $y = \dfrac{x}{9}$ **41.** Let $b = 2$ and $x = 4$ **43.** Let $b = 2, n = 3, x = 4$
45. Let $b = 2, x = 4, y = 8$ **47.** 0.9542 **49.** 3.9542
51. -0.9542

Section 53, p. 300

1. 3.25×10^2 **3.** 1.0×10^4 **5.** 1.0×10^{-3} **7.** 1.24×10^{-1}
9. 1.0×10^6 **11.** 2 **13.** 4 **15.** −3 **17.** −1 **19.** 6
21. 0.5105 **23.** −0.4895 **25.** 1.7193 **27.** −2.8539
29. −3.4597 **31.** 0.5047 **33.** 3.7876 **35.** −2.5095
37. 6.4181 **39.** 3.9032 **41.** 1.9542 **43.** −0.0458
45. 1.9084 **47.** 1.4313 **49.** −0.4771 **51.** 1.3222
53. 1.2552 **55.** 1.9912 **57.** 0.4225 **59.** 0.7235

Section 54, p. 304

1. 8.03 **3.** 6560 **5.** 0.0245 **7.** 0.00466 **9.** 0.00643
11. 1.6346 **13.** 5163 **15.** 0.1406 **17.** 0.01706
19. 0.04843 **21.** 152 **23.** 6310 **25.** 0.00768
27. 0.00746 **29.** 0.04457

Section 55, p. 310

1. 559 **3.** 18.18 **5.** 1562 **7.** 1520 **9.** 8010
11. 4.313 **13.** 11.27 **15.** 0.0449 **17.** 2.89 **19.** 0.763
21. 4.644 **23.** 0.3010 **25.** 4.1918 **27.** 5.439 sec
29. 996 cm/sec^2
31. (a) 1.1×10^{-6} (b) 1.0×10^4 (c) 5.71×10^{-4} (d) 2.22×10^5
(e) 2.77×10^0 **33.** $19,672 **35.** $64,420 **37.** 5.96%
39. −123 **41.** (a) 1.4427 (b) 8.4978 **43.** 7.74 km/sec
45. 90.6 min **47.** 71.1

Section 56, p. 318

1. 9.6990 **3.** 0.0 **5.** 10.0396 **7.** 9,8696 **9.** 11.1787
11. 9.6990 **13.** not defined **15.** not defined **17.** 0.0
19. not defined **21.** 23° **23.** 68°30′ **25.** 20°10′ **27.** 35°27′
29. 63°15′ **31.** $B = 48°40'$, $b = 19$, $c = 26$
33. $A = 55°56'$, $B = 34°04'$, $c = 29.1$ **35.** $C = 20°$, $a = 50.09$, $c = 19.78$
37. $A = 56°45'$, $a = 6.74$, $c = 6.65$
39. $A = 36°11'$, $B = 43°32'$, $C = 100°17'$
41. 198 newtons, 24° **43.** 37 lb (30° chord), 26 lb (45° chord)
45. 50° with the 120 lb force and 26° with the 210 lb force

Section 57, p. 326

1. 0.8544 **3.** −1.4482 **5.** 4.8363 **7.** 1.7253 **9.** 4.8331
11. 4.5 **13.** 5.08 **15.** 134 **17.** 0.413 **19.** 0.0709
21. e^3 **23.** $e^{3/2}$ **25.** e^{-2} **27.** $2 + e^{4/3}$ **29.** $e^{1/2}$
31. (a) 2.9958 (b) 5.2984 (c) 1.3864 (d) 3.4660 (e) 0.3466
(f) 1.6094 (g) 3.2188 (h) −0.6932 (i) −3.9120 (j) 0.2230
33. (a) 0.4939 (b) 3.2552 (c) 10.8671 (d) 0.1644 (e) −1.1059
(f) −4.8062
35. (a) 2.1827 (b) 2.1911 (c) 6.2288 (d) 0.3562 (e) −1.6959
(f) 1.8453
37. (a) 1.5261 (b) 1.0759 (c) −27.2846 (d) 2.3026 (e) 0.3507
(f) 6.4094

41. For $t = 0, y = 1$
For $y = 0$, $\cos 2\pi t = 0$ and $t = \frac{1}{4}, \frac{3}{4}, \frac{5}{4}, \frac{7}{4}, \frac{9}{4}, \frac{11}{4}, \frac{13}{4}, \frac{15}{4}$.
$\text{Cos}\, 2\pi t = 1$, for $t = 0, 1, 2, 3, 4$ and $\cos 2\pi t = -1$ for $t = \frac{1}{2}, \frac{3}{2}, \frac{5}{2}, \frac{7}{2}$.
We use these values to construct the following tables.

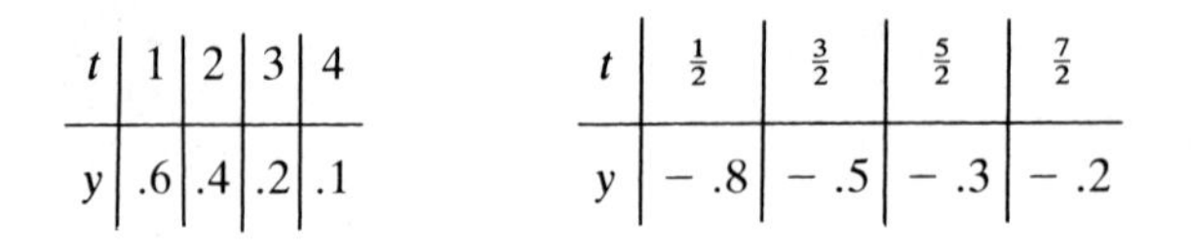

t	1	2	3	4
y	.6	.4	.2	.1

t	$\frac{1}{2}$	$\frac{3}{2}$	$\frac{5}{2}$	$\frac{7}{2}$
y	− .8	− .5	− .3	− .2

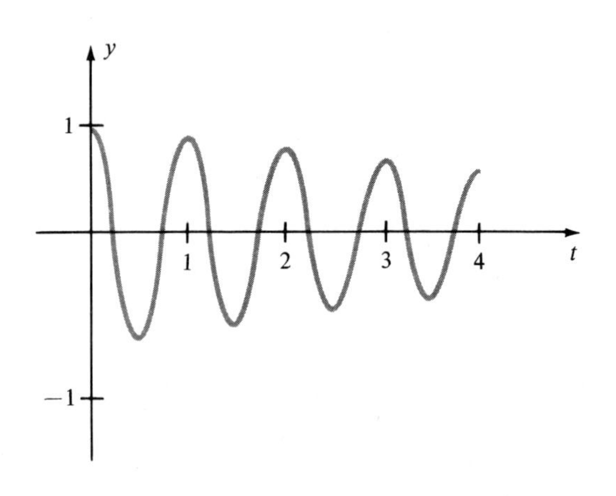

43. 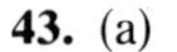 (a) 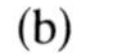 (b)

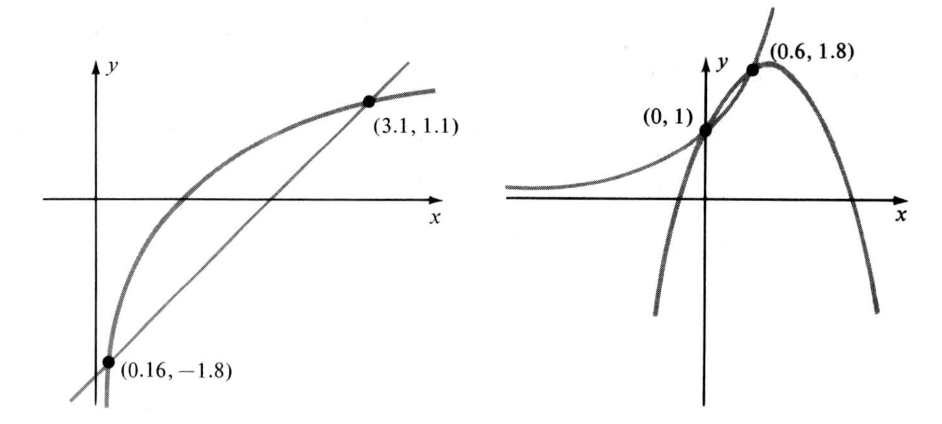

Section 58, p. 333

1. 1.442×10^5 **3.** 17.66 gm; 12.48 gm **5.** 0.3010 yr **7.** 125%
9. 1.3863 yr **11.** 51.56° **13.** 2.3 sec **15.** 3.73 lb
17. $K = 0.045$

Section 59, p. 347

1. log–log **3.** log–log **5.** log–log **7.** semilog **9.** log–log

11.

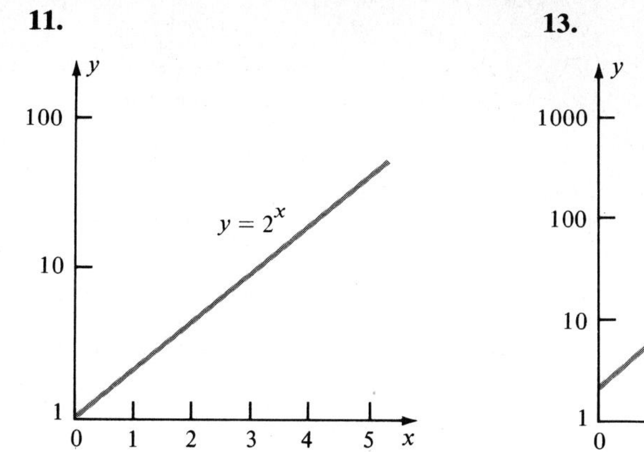

13.

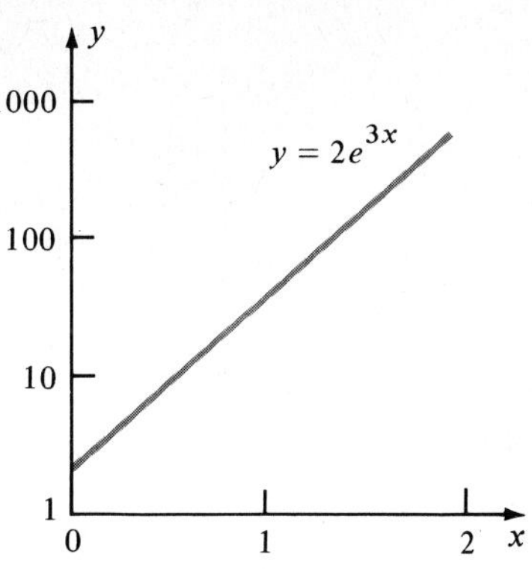

15.

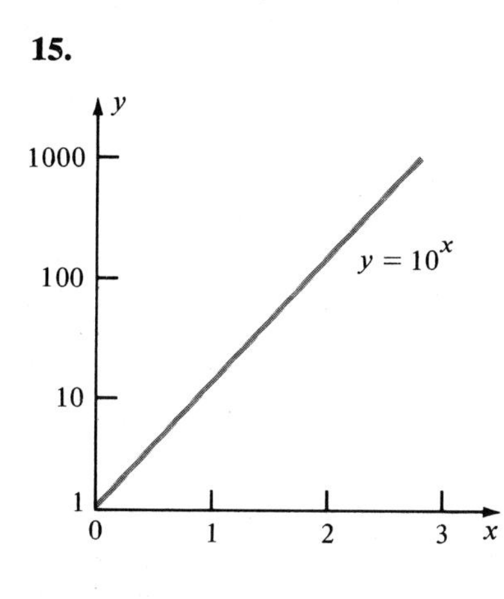

17.

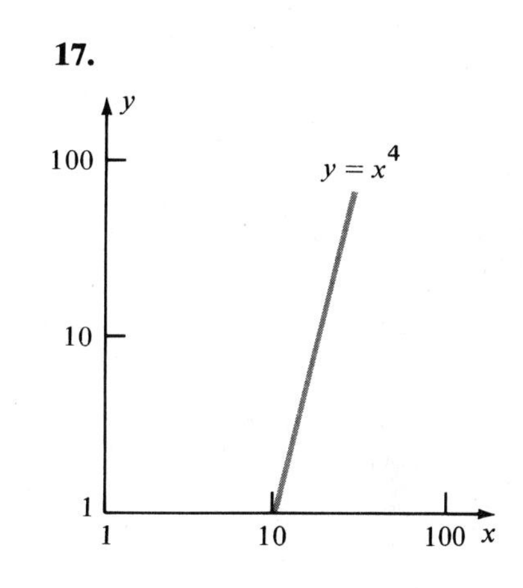

19.

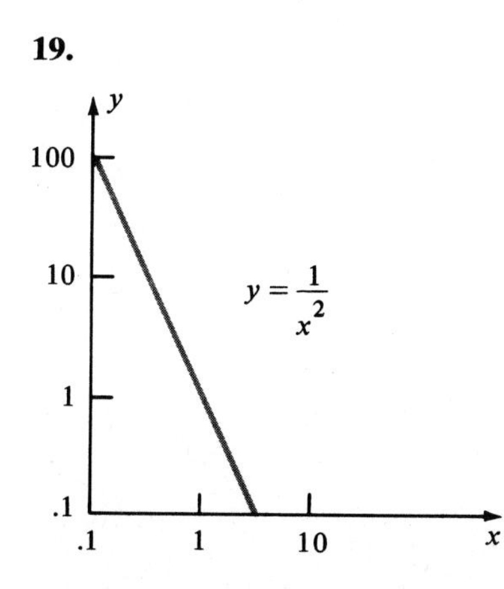

21.

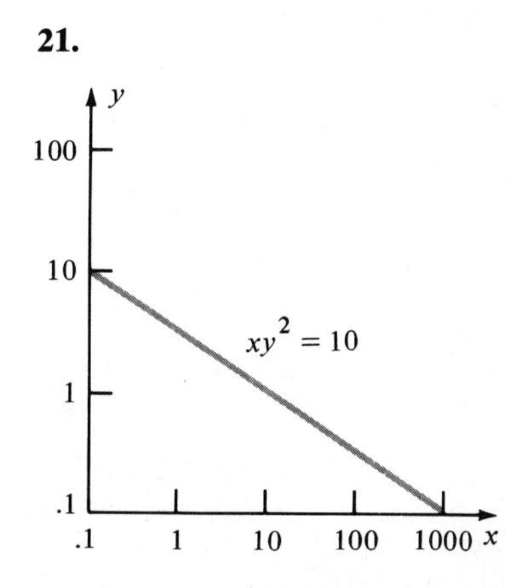

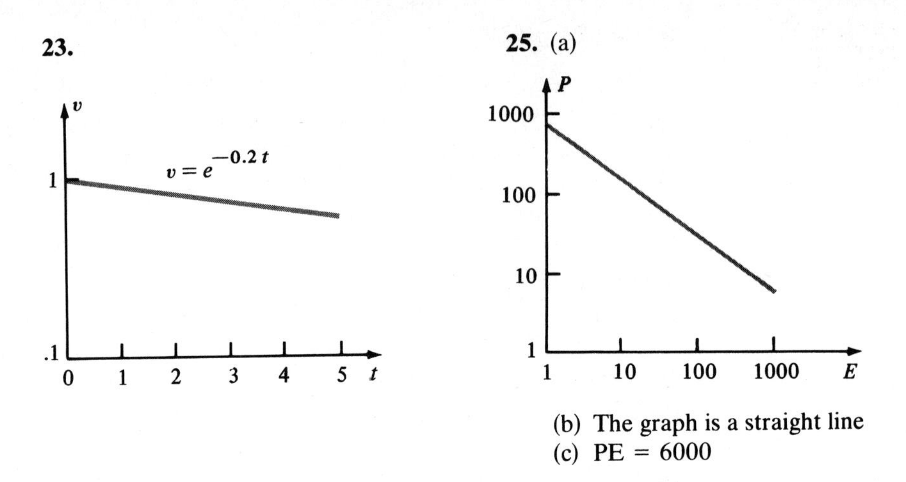

(b) The graph is a straight line
(c) PE = 6000

Review Exercises for Chapter 9, p. 349

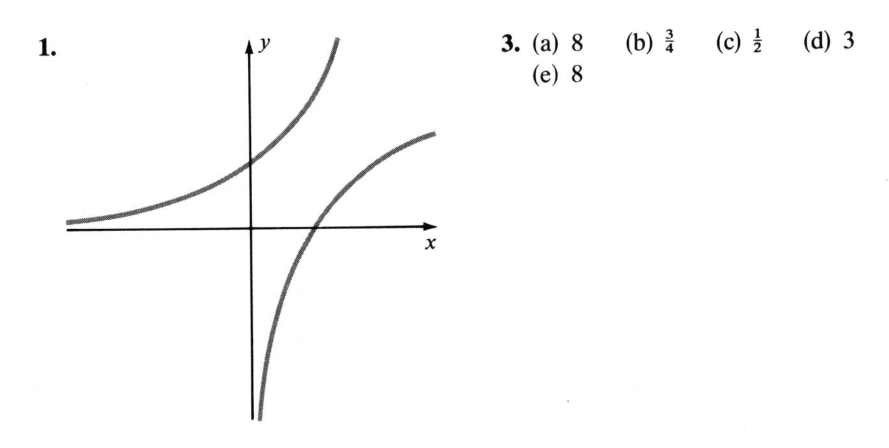

3. (a) 8 (b) $\frac{3}{4}$ (c) $\frac{1}{2}$ (d) 3 (e) 8

5. $\log_4 64$; 3 **7.** 9 **9.** $y = \dfrac{16}{15x}$ **11.** 2.3692 **13.** −0.8069
15. −3.6680 **17.** 3.8501 **19.** 5.8551 **21.** 5.7730
23. 386 **25.** 0.0437 **27.** 0.354 **29.** 2.97 **31.** 54.8
33. 0.247 **35.** 513 **37.** 0.746 **39.** 5.08 **41.** 0.0709
43. 0.4771 **45.** 0.7608 **47.** (a) 168.4 (b) 24.2 (c) 2.98
49. (a) 46°15′ (b) 32°20′ (c) 30°13′ (d) 42°04′ (e) 20°13′
51.

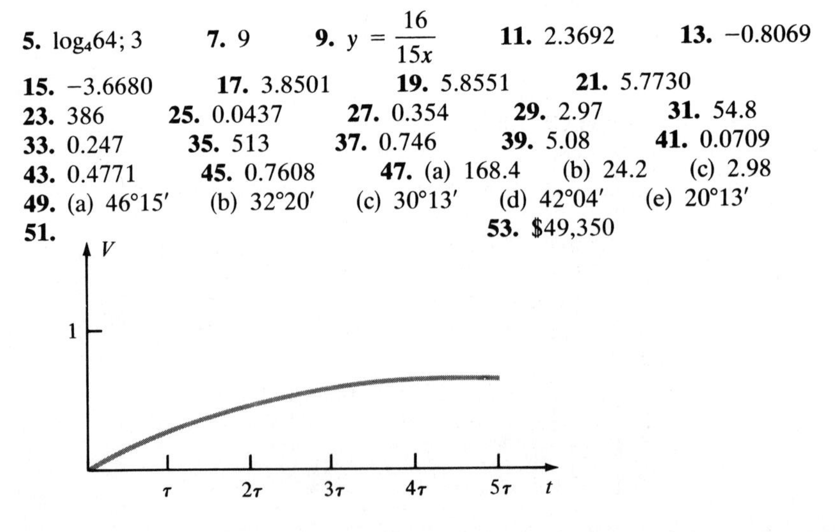

53. $49,350

Section 60, p. 355

1. $2j$ **3.** $6j$ **5.** $\frac{j}{2}$ **7.** $\frac{j}{10}$ **9.** $\pi^2 j$ **11.** $5\sqrt{3}j$ **13.** j
15. $-j$ **17.** $-j$ **19.** -1 **21.** 1 **23.** -2 **25.** $1 - 2j$
27. $5 - 7j$ **29.** $1 + 2j$ **31.** $-3 + 2j$ **33.** 0 **35.** $-2j$
37. $-6 - 2j$ **39.** $3 - j$ **41.** 9 **43.** $-j$ **45.** $4j$
47. -3 **49.** $x = 3, y = -2$ **51.** $x = 3, y = 2$
53. $x = 1, y = -2$ **55.** $x = 3, y = 2$ **57.** $x = -5, y = -\frac{5}{2}$

Section 61, p. 360

1. $2 + 3j$ **3.** $14 - 9j$ **5.** $1 + j$ **7.** $2 - 5j$ **9.** $-2 + j$
11. $2 + 2j$ **13.** $-72j$ **15.** $-8 + 4j$ **17.** $-5j$ **19.** $14 - 2j$
21. $-4\sqrt{3}j$ **23.** $-5 + 12j$ **25.** $9 - 40j$ **27.** $-(4\sqrt{2} + 2\sqrt{3})j$
29. $-2 + 2j$ **31.** $\frac{9 + 6j}{13}$ **33.** $2 + j$ **35.** $\frac{7 - 6j}{17}$ **37.** j
39. $\frac{5 - 12j}{13}$ **41.** $\frac{-2 + 23j}{13}$ **43.** $\frac{-2j}{3}$ **45.** $\frac{4j}{5}$ **47.** $\frac{3 + 4j}{25}$
49. -1 **51.** 5; real and unequal **53.** 0; real and equal
55. $6\sqrt{3}$; real and unequal **57.** $2\sqrt{6}$; real and unequal
59. -4; imaginary **61.** ± 2 **63.** 2 **65.** $\frac{-1 \pm \sqrt{3}j}{2}$
67. $-1 \pm j$ **69.** $\frac{-3 \pm 5j}{2}$ **71.** $2a$ **73.** $22 + 6j$

Section 62, p. 365

1. $3 + 2j$ **3.** $-4j$ **5.** 2 **7.** $1 - 3j$ **9.** $1 - j$

11.

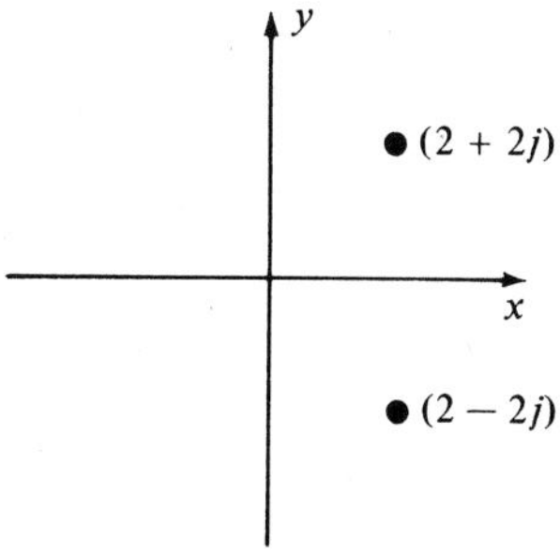

13.

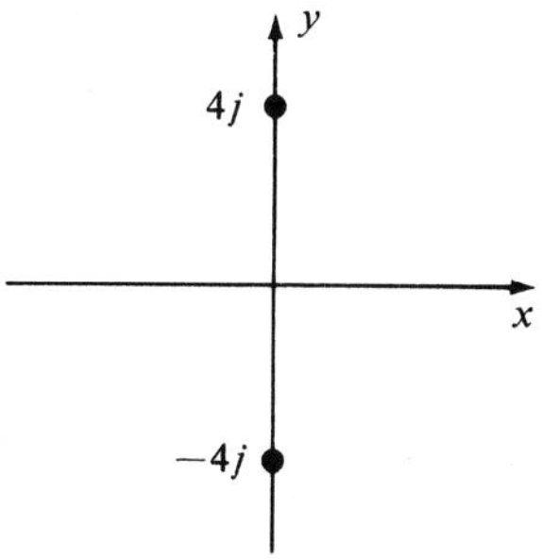

15.

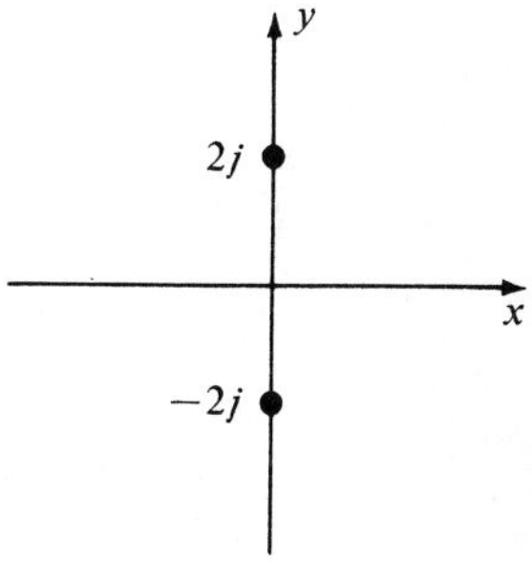

Section 63, p. 371

1. $2\sqrt{2}(\cos 45° + j \sin 45°)$ **3.** $\cos 135° + j \sin 135°$
5. $3(\cos 30° + j \sin 30°)$ **7.** $2(\cos 60° + j \sin 60°)$
9. $8(\cos 90° + j \sin 90°)$ **11.** $27(\cos 0° + j \sin 0°)$
13. $5.83[\cos 59°02' + j \sin 59°02']$ **15.** $4.47(\cos 116°34' + j \sin 116°34')$
17. $\sqrt{3} + j$ **19.** 3 **21.** $2j$ **23.** $-2\sqrt{3} + 2j$
25. $-1.2856 + 1.5321j$ **27.** $-2 - 2\sqrt{3}j$ **29.** $-j$ **31.** $\sqrt{2} - \sqrt{2}j$
33. $1.5321 - 1.2856j$ **35.** 26.9 **37.** $2e^{0.5236j}$ **39.** $2e^{1.0472j}$
41. $5e^{-2.2143j}$
43. $3(\cos 57° + j \sin 57°)$; $1.6209 + 2.5244j$
45. $4(\cos 103° + j \sin 103°)$; $-0.9088 + 1.9477j$
47. $2(\cos 210° + j \sin 210°)$; $-1.7321 - j$
49. $6(\cos 286° + j \sin 286°)$; $1.7020 - 5.7535j$

Section 64, p. 377

1. $4 + 4\sqrt{3}j$ **3.** $-5 + 5\sqrt{3}j$ **5.** $-\sqrt{2} - \sqrt{2}j$ **7.** $-12j$

9. $1.1491 + 0.9642j$ **11.** $-2 + 2\sqrt{3}j$ **13.** $\frac{-3\sqrt{3}}{2} + \frac{3}{2}j$ **15.** -4

17. $\frac{\sqrt{3}}{4} - \frac{1}{4}j$ **19.** -2 **21.** $\frac{\sqrt{2}}{2} + \frac{\sqrt{2}}{2}j$ **23.** -5

25. $-8 + 8\sqrt{3}j$ **27.** $32(\cos 90° + j \sin 90°)$ **29.** $3(\cos 90° + j \sin 90°)$

Section 65, p. 384

1. $32 + 32\sqrt{3}j$ **3.** $16 - 16\sqrt{3}j$ **5.** $-\frac{\sqrt{3}}{2} + \frac{1}{2}j$ **7.** 32

9. $-1 + \sqrt{3}j$ **11.** $-4 - 4j$ **13.** 1 **15.** -1

17. $512 - 512\sqrt{3}j$ **19.** $\sqrt{3} + j$

21. $\cos 0° + j \sin 0° = 1$

$\cos 120° + j \sin 120° = -\frac{1}{2} + \frac{\sqrt{3}}{2}j$

$\cos 240° + j \sin 240° = -\frac{1}{2} - \frac{\sqrt{3}}{2}j$

23. $2(\cos 45° + j \sin 45°) = \sqrt{2} + \sqrt{2}j$

$2(\cos 135° + j \sin 135°) = -\sqrt{2} + \sqrt{2}j$

$2(\cos 225° + j \sin 225°) = -\sqrt{2} - \sqrt{2}j$

$2(\cos 315° + j \sin 315°) = \sqrt{2} - \sqrt{2}j$

25. $\cos 135° + j \sin 135° = -\frac{\sqrt{2}}{2} + \frac{\sqrt{2}}{2}j$

$\cos 315° + j \sin 315° = \frac{\sqrt{2}}{2} - \frac{\sqrt{2}}{2}j$

27. $\cos 45° + j \sin 45° = \frac{\sqrt{2}}{2} + \frac{\sqrt{2}}{2}j$

$\cos 225° + j \sin 225° = -\frac{\sqrt{2}}{2} - \frac{\sqrt{2}}{2}j$

29. $\cos 30° + j\sin 30° = \frac{\sqrt{3}}{2} + \frac{1}{2}j$

$\cos 120° + j \sin 120° = -\frac{1}{2} + \frac{\sqrt{3}}{2}j$

$\cos 210° + j \sin 210° = -\frac{\sqrt{3}}{2} - \frac{1}{2}j$

$\cos 300° + j \sin 300° = \frac{1}{2} - \frac{\sqrt{3}}{2}j$

31. $2, -1 + \sqrt{3}j$ **33.** $\pm 1, \frac{1}{2} \pm \frac{\sqrt{3}}{2}j, -\frac{1}{2} \pm \frac{\sqrt{3}}{2}j$

35. $\cos 45° + j \sin 45°$

$\cos 117° + j \sin 117°$

$\cos 189° + j \sin 189°$

$\cos 261° + j \sin 261°$

$\cos 333° + j \sin 333°$

37. $1, \frac{-1 \pm \sqrt{3}j}{2}$

Review Exercises for Chapter 10, p. 396

1. $5 + 2j$ **3.** $-\frac{1}{2} + \frac{1}{3}j$ **5.** $-21 - j$ **7.** $1 - j$ **9.** $\frac{3 + 89j}{122}$
11. $\frac{-5 + 12j}{169}$ **13.** $1 - j$ **15.** (a) $2 + 4j$ (b) $-2j$ (c) $3 + 4j$
17. $x = 3, y = 1$ **19.** $x = 2, y = -3$ **21.** $\pm 6j$ **23.** $2 \pm 2j$
25. $\frac{-2 \pm 2\sqrt{5}j}{3}$ **27.** $6\sqrt{2}(\cos 225° + j \sin 225°)$
29. $8(\cos 240° + j \sin 240°)$ **31.** $3(\cos 132° + j \sin 132°)$
33. $-4\sqrt{2} - 4\sqrt{2}j$ **35.** $\frac{-2\sqrt{3}}{3} + \frac{2}{3}j$ **37.** -2
39. $-\frac{1}{2} + \frac{\sqrt{3}}{2}j$ **41.** $-\frac{1}{2}$ **43.** $-\frac{1}{2} - \frac{\sqrt{3}}{2}j$ **45.** $-1 - \sqrt{3}j$
47. $-1.5321 + 1.2856j$ **49.** $4, -2 \pm 2\sqrt{3}j$ **51.** $\pm 2j, \sqrt{3} \pm j, -\sqrt{3} \pm j$
53. (a) $90 + 188.4j$ (b) $100 + 188.4j$ (c) $1000 - 1214j$

Section 66, p. 391

1. $\frac{2}{9}$ **3.** $\frac{2}{3}$ **5.** $\frac{2}{3}$ **7.** $\frac{1}{2.54}$ **9.** $\frac{3}{2}$ **11.** $\frac{3}{4}, \frac{3}{4}, \frac{9}{16}$ **13.** 6
15. 3.6 **17.** ± 10 **19.** 3 **21.** ± 16 **23.** ± 6 **25.** $\pm\frac{3}{4}$
27. $\frac{448}{15}$ lb **29.** 1020 km **31.** $5666\frac{2}{3}$ ft^3/min **33.** 88 ft/sec
35. 105 ft

Section 67, p. 395

1. $C = Kr$ **3.** $V = Kt$ **5.** $W = Kr^3$ **7.** $S = Kt$ **9.** $H = KI^2$
11. 15 **13.** 20 **15.** 6 **17.** $\frac{25}{4}$ **19.** 32 **21.** 576 ft
23. 1250 gal/min **25.** 36 lb **27.** $8V; 27V$ **29.** 144π ft^3
31. 1.8 yr

Section 68, p. 401

1. $y = kxz$ **3.** $y = k\sqrt{x}\sqrt[3]{z}$ **5.** $s = \frac{k}{\sqrt{2t}}$ **7.** $v = \frac{k}{\sqrt[3]{5}}$ **9.** $y = \frac{k}{x^4}$
11. 6 **13.** $8\sqrt{2}$ **15.** $\frac{32\sqrt{3}}{3}$ **17.** 0.25 **19.** 36
21. 2 hr **23.** 187.5 watts **25.** $16\frac{2}{3}$ hr **27.** 1.62×10^7
29. 80 units **31.** 25 lb/in.^{2} **33.** 190 lb; 128 lb; 50 lb

Section 69, p. 405

1. $\frac{75}{2}$ **3.** $\frac{400}{27}$ **5.** $\frac{2}{\sqrt[3]{36}}$ **7.** Increased 32 times **9.** (a) Halved
(b) Increase of 18% **11.** 6 **13.** Increased 8 times

Review Exercises for Chapter 11, p. 406

1. 6 **3.** $\frac{10}{3}$ **5.** ± 6 **7.** $\pm 6\sqrt{2}$ **9.** 5 **11.** $\frac{128}{27}$ **13.** $\frac{15}{2}$
15. $\frac{2\sqrt{3}}{3}$ **17.** $\frac{45}{4}$ **19.** Increased 8 times **21.** 49.92 cm

23. 4π **25.** 150 sq. units **27.** $\frac{160\pi}{3}$ cubic units **29.** $\frac{5}{4\pi}$

31. $F = \frac{KQ_1Q_2}{r}$ **33.** $T = \frac{1500°}{\sqrt{x^2 + y^2}}$; 300°

Section 70, p. 413

1. $x^2 + x + 5, R = 11$ **3.** $x^3 + 4x^2 + 5x + 6, R = 4$
5. $x^2 + 3x + 13, R = 32$ **7.** $2x^2 - x, R = 4$
9. $x^3 - 4x^2 - x - 7, R = -3$ **11.** $x^3 + x^2 + x + 1, R = 0$
13. $2x^3 + 4x^2 + 7x + 17, R = 29$ **15.** $3x^2 + 3x - 6, R = 0$
17. $x^3 + 6x^2 + 11x + 6, R = 0$ **19.** $3x^2 + 6, R = 0$
21. $x^2 - x + 2, R = 8$ **23.** $4x^3 + 3x^2 - 2x - 1, R = 0$
25. $2x^3 + 2x^2 + x - 2, R = 0$

Section 71, p. 417

1. -2 **3.** 4 **5.** 10 **7.** 0 **9.** 0 **11.** Yes; $(x - 5)(x^2 + 7x + 10)$
13. No **15.** No **17.** Yes, $(x + \frac{3}{2})(4x^2 - 10x + 4)$
19. Yes, $(x - 2)(x^3 + 6x^2 + 11x + 6)$ **21.** True **23.** True
25. True **27.** True **29.** True **31.** $K = 2$ **33.** $p = 24$

Section 72, p. 423

1. 1, 2 is a double root, 3 is a triple root
3. 0 is a triple root, -1 is a root of multiplicity four, 2 is a root of multiplicity five
5. 1, 3 is a triple root **7.** $(x - 1)(x - 2)^2(x - 4) = 0$
9. $(x - 2)(x^2 - 2x + 2) = 0$ **11.** $-1, 2, 3$ **13.** $-1, -2, -3$
15. $\frac{1}{2}, \frac{-1 \pm \sqrt{5}}{2}$ **17.** $1, \pm 2j$ **19.** $-3, -\frac{1}{2}, 2, 3$ **21.** $-1, 1, 2, 2$
23. $-1, 2, 2, 4$ **25.** $\pm j, \frac{1}{2}, \frac{1}{2}, 3$

Section 73, p. 431

1. $\frac{1}{2}, \pm\sqrt{2}$ **3.** $\frac{3}{2}, 1 \pm j$ **5.** $-2, 1, 3$ **7.** $-\frac{1}{2}, \frac{3}{4}, 1$ **9.** $-4, -2, 3$
11. $-3, -2, \frac{1}{2}, 2$ **13.** $-3, \frac{1}{2}, \pm j$ **15.** $-1, -1, -1, 3$ **17.** $4, 1 \pm \sqrt{5}$
19. $-1, -1, 2$ **21.** $x \leq -1$ or $1 \leq x \leq 2$ **23.** $-\frac{1}{2} \leq x \leq \frac{3}{4}$ or $x \geq 1$
25. $-1 \leq x \leq 3$
27. 1 negative and 2 imaginary roots
29. Two or zero real positive roots; two or zero real negative roots; four, two or zero imaginary roots
31. $t = \frac{3}{2}, 2, 3$ **33.** 8 cm, 9 cm, 10 cm **35.** 5 cm

Section 74, p. 437

1. 1.5 **3.** 1.9 **5.** 1.6 **7.** 0.4 **9.** 2.2 **11.** 2.17
13. -0.62 **15.** 1.62 **17.** -1.24 **19.** 1.73 **21.** 1.82
23. 0.5 m

Review Exercises for Chapter 12, p. 439

1. $2x^2 + x - 2, R = -10$ **3.** $3x^3 - x^2 + 2x - 13, R = 13$
5. $x^4 - x^3 - 2x^2 + 8x - 1, R = 6$ **19.** $-\frac{3}{2}, -1, 2$
21. $P(x) = (2x - 5)(x^2 - 3x + 1)$ **23.** 2

25. $(x-1)(2x-1)(3x+2)=0$ **27.** $1, 1 \pm j$ **29.** $-1, \frac{1}{2}, 1$
31. $-3, -2, 1, 2$ **33.** $x \le 1$ **35.** -0.4
39. (a) $t = 1, 2, 6$ (b) $1 < t < 2$ or $t > 6$ **41.** 12

Section 75, p. 444

1. 3, 6, 9, 12; 30 **3.** $\frac{1}{2}, \frac{1}{3}, \frac{1}{4}, \frac{1}{5}; \frac{1}{11}$ **5.** $\frac{1}{2}, \frac{1}{3}, \frac{1}{4}, \frac{1}{5}; \frac{1}{11}$ **7.** $\frac{1}{2}, \frac{1}{2}, \frac{3}{8}, \frac{1}{4}; \frac{5}{512}$
9. $\frac{3}{2}, \frac{9}{5}, \frac{27}{10}, \frac{81}{17}; \frac{3^{10}}{101}$ **11.** $-\frac{1}{2}, \frac{1}{3}, -\frac{1}{4}, \frac{1}{5}; \frac{1}{11}$ **13.** $e, \frac{e^2}{2}, \frac{e^3}{3}, \frac{e^4}{4}; \frac{e^{10}}{10}$
15. $1, 4, 27, 256; 10^{10}$ **17.** $3n+1$ **19.** $\frac{(-1)^n}{n}$ **21.** $3n+2$
23. $(-1)^{n+1}$ **25.** $\frac{n}{n+1}$ **27.** $\frac{1}{3^n}$ **29.** $\sqrt{n}$ **31.** 2, 3, 4, 5
33. 1, 2, 6, 24 **35.** 5, 19, 75, 299 **37.** $\frac{5}{8}$ ft; $\frac{10}{2^{10}}$ ft **39.** 7.29×10^6

Section 76, p. 448

1. $\sum_{k=1}^{10} k$ **3.** $\sum_{k=1}^{n-1} k$ **5.** $\sum_{k=1}^{30} k^2$ **7.** $\sum_{k=1}^{100} 3k$ **9.** $\sum_{k=1}^{7} 2^k$
11. $\sum_{k=1}^{6} k^2$ **13.** $\sum_{k=1}^{9} (-1)^{k-1}$ **15.** $\sum_{k=1}^{6} \frac{2k}{2^k}$ **17.** 21 **19.** $\frac{77}{60}$
21. $\frac{20}{3}$ **23.** $\frac{31}{32}$ **25.** $2+3+4+5+6$
27. $2 \cdot 1 + 3 \cdot 2 + 4 \cdot 3 + \cdots + 100 \cdot 99$
29. $2 \cdot 1^2 + 2 \cdot 2^2 + 2 \cdot 3^2 + \cdots + 2n^2$ **31.** $-\frac{1}{2} + \frac{1}{4} - \frac{1}{8} + \frac{1}{16}$
33. $\frac{1}{1 \cdot 2} + \frac{1}{2 \cdot 3} + \frac{1}{3 \cdot 4} + \frac{1}{4 \cdot 5} + \frac{1}{5 \cdot 6}$ **35.** $\frac{3}{1} + \frac{3^2}{2} + \frac{3^3}{3} + \cdots + \frac{3^{10}}{10}$

Section 77, p. 452

1. 29, 155 **3.** $-15, -42$ **5.** 7, 12 **7.** 81, 860 **9.** 46, 200
11. $a = 7, d = 4$ **13.** $a = -33, S_n = -33$ **15.** $n = 12$
17. $a = -4, d = 2$ **19.** 56 **21.** $d = \frac{10}{13}, S_n = 520$
23. $a = -5, d = 2$ **25.** 5050 **27.** 36 months
29. (a) 8 ft (b) 81 ft (c) 41 **31.** (a) \$550 (b) \$3250

Section 78, p. 457

1. $\frac{1}{32}, S_7 = \frac{127}{32}$ **3.** $\frac{625}{2}, \frac{521}{4}$ **5.** $\frac{1}{243}, S_9 = \frac{19679}{486}$
7. $\frac{7776}{78125}, S_8 = \frac{6^8 - 5^8}{6^2 \cdot 5^7}$ **11.** $l = \frac{24}{125}, n = 4$ **13.** $l = 243, S_8 = \frac{3280}{9}$
15. $r = 2, S_4 = 30$ **17.** $l = \frac{-\sqrt{3}}{27}, S_8 = \frac{80}{9(3-\sqrt{3})}$
19. $l = \frac{16}{243}, S_6 = \frac{697}{486}$ **21.** $\frac{3^9}{2^8}$ **23.** 2 **25.** $a = 81, l = 1$
27. 243 **29.** \$3132 **31.** 14,800 **33.** 23.73%
35. $40 - \frac{5}{2^8}$ cm

Section 79, p. 461

1. 2 **3.** 8 **5.** 9 **7.** $\frac{20}{9}$ **9.** $\frac{50}{27}$ **11.** $-\frac{8}{3}$
13. Does not exist; $|r| = |-2| > 1$ **15.** Does not exist; $|r| = |3| > 1$
17. 2 **19.** 20 **21.** $\frac{4}{9}$ **23.** $\frac{7}{9}$ **25.** $\frac{14}{99}$ **27.** $\frac{41}{333}$
29. $2\frac{13}{99}$ **31.** 150 ft **33.** 216 cm **35.** $-\frac{1}{2}$

Section 80, p. 466

1. (a) 15 (b) 15 (c) 1 (d) 1 (e) 100
3. $x^6 - 6x^5y + 15x^4y^2 - 20x^3y^3 + 15x^2y^4 - 6xy^5 + y^6$
5. $x^6 - 12x^5y + 60x^4y^2 - 160x^3y^3 + 240x^2y^4 - 192xy^5 + 64y^6$
7. $16x^4 - 96x^3 + 216x^2 - 216x + 81$
9. $16x^4 - 96x^3y + 216x^2y^2 - 216xy^3 + 81y^4$
11. $x^6y^6 - 12x^6y^5 + 60x^6y^4 - 160x^6y^3 + 240x^6y^2 - 192x^6y + 64x^6$
13. $\left(\frac{x}{y}\right)^4 + 4\left(\frac{x}{y}\right)^2 + 6 + 4\left(\frac{y}{x}\right)^2 + \left(\frac{y}{x}\right)^4$
15. $x^{15} - 5x^{12}y^2 + 10x^9y^4 - 10x^6y^6 + 5x^3y^8 - y^{10}$
17. $5376a^3b^6$ **19.** $101{,}376x^{15}y^7$ **21.** $30{,}030x^{10}y^5$
23. $\frac{28}{81}x^4y^6$ **25.** $-252x^{10}y^{10}$ **27.** $59{,}136x^{12}y^6$
29. $50{,}319{,}360x^6y^{12}$ **31.** 1.062 **33.** 0.851 **35.** 7.071

Review Exercises for Chapter 13, p. 468

1. 1, 4, 7; 22 **3.** 1, 1, $\frac{4}{3}$; 16 **5.** 1, −3, 9; -3^7 **7.** 5, 11, 23
9. (a) 200 (b) 35 (c) 287 **11.** Geometric, $\frac{1}{15{,}552}, \frac{(6^7 - 1)}{10 \cdot 6^5}$
13. Geometric, $-\frac{3}{64}, \frac{255}{64}$ **15.** Geometric, $-\frac{1}{4^{10}}, \frac{4^{12} - 1}{5 \cdot 4^{10}}$
17. Arithmetic, 25, 117 **19.** 72 **21.** $16 \cdot (\frac{4}{3})^5$
23. (a) 18 (b) 4 (c) −4 (d) 3 (e) $\frac{4}{5}$ **25.** $\frac{1}{8}$
27. (a) 5.099 (b) 3.072 **29.** (a) 20 ft (b) $\frac{5}{54}$ ft (c) 960 ft
31. 67,000 **33.** 0.49, 0.24, $(0.7)^n$

Section 81, p. 476

Note: *Answers are intermediate steps used in proving the given identities.*

1. $\tan\theta \cdot \frac{1}{\tan\theta} \cdot \frac{1}{\cos\theta} \cdot \cos\theta = 1$ **3.** $\sin\theta \cdot \frac{1}{\cos\theta} = \tan\theta$

5. $\cos\theta \cdot \frac{\sin\theta}{\cos\theta} = \sin\theta$ **7.** $\frac{\cos^2\theta}{\sin^2\theta} = \cot^2\theta$ **9.** $\sin^2\theta \cdot \csc^2\theta = 1$

11. $\frac{\sec^2\theta}{\csc\theta} = \frac{\sin\theta}{\cos^2\theta} = \frac{\sin\theta}{1 - \sin^2\theta}$ **13.** $\frac{\sec^2\theta}{\csc^2\theta} = \frac{\sin^2\theta}{\cos^2\theta} = \tan^2\theta$

15. $\frac{\sin^2\theta}{\cos^2\theta} - \frac{\sin^2\theta\cos^2\theta}{\cos^2\theta} = \frac{\sin^2(1 - \cos^2\theta)}{\cos^2\theta} = \sin^2\theta \cdot \frac{\sin^2\theta}{\cos^2\theta}$

17. $1 - 2\sin^2\theta = \cos 2\theta$ **19.** $\frac{1}{\cos x} + \frac{\cos x}{\sin x} + \frac{\sin x}{\cos x} = \frac{\sin x + \cos^2 x + \sin^2 x}{\cos x \sin x}$

21. $\frac{1 + \frac{1}{\csc x}}{1 - \frac{1}{\csc x}} = \frac{\csc x + 1}{\csc x - 1}$ **23.** $\frac{\sin x(1 + \cos x)}{\sin^2 x} = \frac{\sin x(1 + \cos x)}{1 - \cos^2 x} = \frac{\sin x}{1 - \cos x}$

25. $\frac{\cos x(1 - \sin x)}{2(1 - \sin^2 x)} + \frac{\cos x(1 + \sin x)}{2(1 - \sin^2 x)} = \frac{2\cos x}{2\cos^2 x} = \sec x$

27. $4 \cdot \frac{1}{\csc x} + \frac{\frac{1}{\csc x}}{\frac{1}{\sec x}} = \frac{4}{\csc x} + \frac{\sec x}{\csc x}$

29. $\dfrac{\sin x - 1 - \sin x - 1}{\sec^2 x - \tan^2 x} = \dfrac{-2}{1 + \tan^2 x - \tan^2 x}$

31. $\cos x = \dfrac{\sqrt{5}}{3}$

$\tan x = \dfrac{2}{\sqrt{5}}$

$\csc x = \dfrac{3}{2}$

$\sec x = \dfrac{3}{\sqrt{5}}$

$\cot x = \dfrac{\sqrt{5}}{2}$

33. $\sec x = \dfrac{\sqrt{5}}{2}$

$\cos x = \dfrac{2}{\sqrt{5}}$

$\sin x = -\dfrac{1}{\sqrt{5}}$

$\csc x = -\sqrt{5}$

$\cot x = -2$

35. $\cos x = \pm\sqrt{1 - \sin^2 x}$

$\sec x = \pm\dfrac{1}{\sqrt{1 - \sin^2 x}}$

$\tan x = \pm\dfrac{\sin x}{\sqrt{1 - \sin^2 x}}$

$\cot x = \pm\dfrac{\sqrt{1 - \sin^2 x}}{\sin x}$

$\csc x = \dfrac{1}{\sin x}$

37. (a) $\cos^2 x = 1 - \sin^2 x$

(b) $\sec^2 x = \dfrac{1}{1 - \sin^2 x}$

(c) $\dfrac{\sec x + 1}{\sin x + \tan x} = \dfrac{1}{\sin x}$

Section 82, p. 484

1. $\dfrac{\sqrt{6} - \sqrt{2}}{4}$ **3.** $\dfrac{\sqrt{6} + \sqrt{2}}{4}$ **5.** $\dfrac{\sqrt{2} - \sqrt{6}}{4}$ **7.** $\dfrac{3 - \sqrt{3}}{3 + \sqrt{3}}$

9. $-\left(\dfrac{\sqrt{6} + \sqrt{2}}{4}\right)$ **11.** $-\left(\dfrac{4}{\sqrt{2} + \sqrt{6}}\right)$ **13.** $\frac{56}{65}$ **15.** $\frac{36}{85}$

17. $-\frac{77}{85}$ **19.** $\sin 90° \cos x + \cos 90° \sin x = \cos x$

21. $\dfrac{\tan 180° + \tan x}{1 + \tan 180° \tan x} = \tan x$ **23.** $\sin 270° \cos x + \cos 270° \sin x = -\cos x$

25. $\cos x \cos 45° + \sin x \sin 45° = \dfrac{\sqrt{2}}{2}(\cos x + \sin x)$

27. $y = 2[\cos x \cos \pi - \sin x \sin \pi] = -2 \cos x$

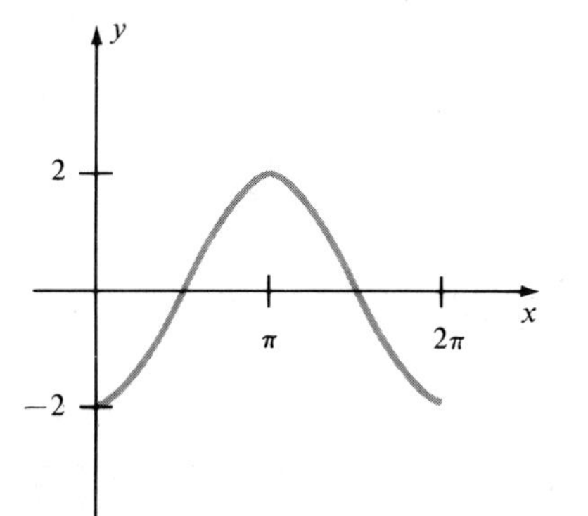

29. $y = 2\left[\sin x \cos\frac{\pi}{2} + \cos x \sin\frac{\pi}{2}\right] = 2\cos x$

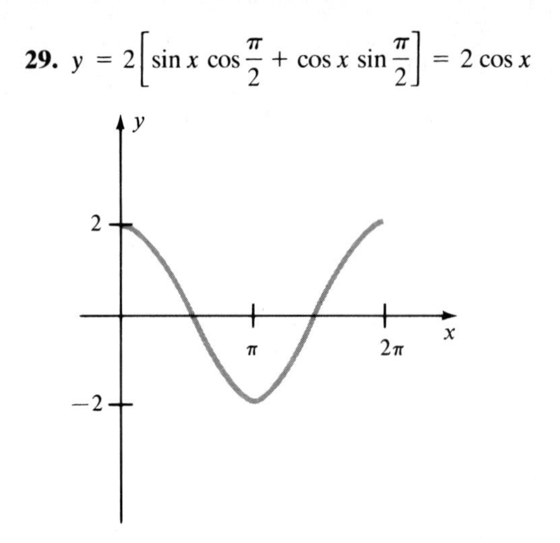

31. $(\cos x \cos y - \sin x \sin y)(\cos x \cos y + \sin x \sin y) = \cos^2 x \cos^2 y - \sin^2 x \sin^2 y$
$= \cos^2 x(1 - \sin^2 y) - \sin^2 y(1 - \cos^2 x) = \cos^2 x - \sin^2 y$

33. $\frac{1}{2}\sin x \cos y + \frac{1}{2}\cos x \sin y + \frac{1}{2}\sin x \cos y - \frac{1}{2}\sin x \cos y = \sin x \cos y$

35. $\frac{1}{2}\cos x \cos y - \frac{1}{2}\sin x \sin y + \frac{1}{2}\cos x \cos y + \frac{1}{2}\sin x \sin y = \cos x \cos y$

37. $n = \pm\sqrt{\frac{2 + \sin\alpha}{2} \quad \frac{\sqrt{3}\cos\alpha}{\sqrt{3}}}$

Section 83, p. 492

1. $\frac{\sqrt{3}}{2}$ **3.** $-\frac{1}{2}$ **5.** $-\sqrt{3}$ **7.** $\frac{\sqrt{2+\sqrt{3}}}{2}$ **9.** $\frac{\sqrt{2+\sqrt{3}}}{2}$ **11.** $\frac{\sqrt{2-\sqrt{2}}}{2}$

13. $\frac{24}{25}$ **15.** $\frac{119}{169}$ **17.** $\sin 2x = \frac{24}{25}$, $\cos 2x = \frac{7}{25}$ **19.** $-\frac{\sqrt{5}}{5}$ **21.** -2

23. $\frac{1 - \cos 2x}{2}$ **25.** $\frac{2}{1 - \cos^2 x + \sin^2 x} = \frac{2}{2\sin^2 x} = \csc^2 x$

27. $\frac{1}{\left(\frac{\sin x}{\cos x} + \frac{\cos 2x}{\sin 2x}\right)} = \frac{1}{\left(\frac{2\sin^2 x + \cos 2x}{2\sin x \cos x}\right)} = \frac{2\sin x \cos x}{2\sin^2 x + \cos^2 x - \sin^2 x} = \sin 2x$

29. $\frac{1}{2}\sin\left(\frac{x}{2}\right) = 2\left[\frac{1}{2}\sin\left(\frac{x}{4}\right)\cos\left(\frac{x}{4}\right)\right] = \sin\left(\frac{x}{4}\right)\cos\left(\frac{x}{4}\right)$

31. $\sqrt{\frac{1 - \cos x}{1 + \cos x}} = \sqrt{\frac{(1 - \cos x)(1 + \cos x)}{(1 + \cos x)(1 + \cos x)}} = \sqrt{\frac{1 - \cos^2 x}{(1 + \cos x)^2}} = \frac{\sin x}{1 + \cos x}$

33. $\sin^2 x + 2\sin x \cos x + \cos^2 x = 1 + 2\sin x \cos x = 1 + \sin 2x$

35. $t = 2\pi\sqrt{\frac{50}{980}\left(\frac{18 - \sqrt{3}}{16}\right)}\ \text{sec} = 1.44\ \text{sec}$

Section 84, p. 497

1. $\frac{7\pi}{6}, \frac{11\pi}{6}$ **3.** 0 **5.** $\frac{7\pi}{6}, \frac{3\pi}{2}, \frac{11\pi}{6}$ **7.** $\frac{7\pi}{6}, \frac{3\pi}{2}, \frac{11\pi}{6}$ **9.** 0

11. $0, \frac{2\pi}{3}, \pi, \frac{4\pi}{3}$ **13.** $\frac{\pi}{6}, \frac{5\pi}{6}, \frac{3\pi}{2}$ **15.** $\frac{\pi}{4}, \frac{3\pi}{4}, \frac{5\pi}{4}, \frac{7\pi}{4}$

17. $\frac{\pi}{3}, \frac{5\pi}{6}, \frac{4\pi}{3}, \frac{11\pi}{6}$ **19.** $\frac{\pi}{3}, \frac{5\pi}{3}$ **21.** $0, \frac{\pi}{3}, \pi, \frac{5\pi}{3}$ **23.** $\frac{2\pi}{3}, \pi$

25. $\frac{\pi}{2}, \frac{3\pi}{2}$ **27.**

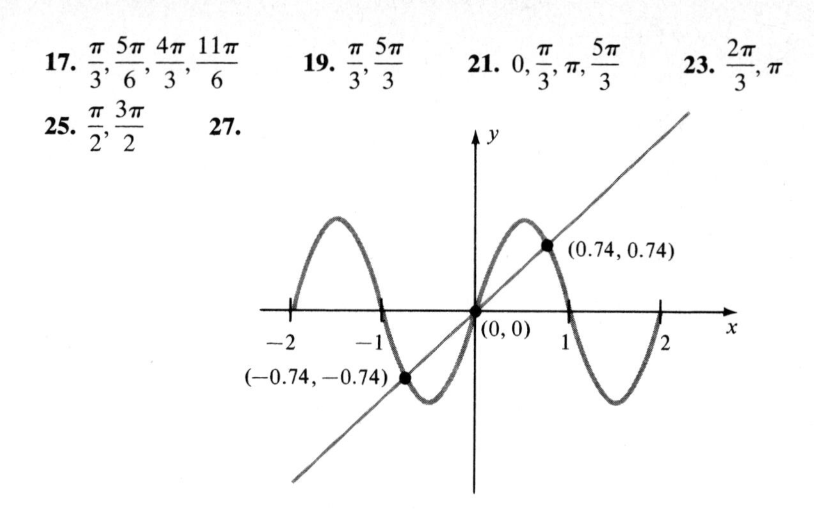

29. 60°, 120°, 420°, 480°

31. $-\frac{15\pi}{4}, -\frac{13\pi}{4}, -\frac{11\pi}{4}, -\frac{9\pi}{4}, -\frac{7\pi}{4}, -\frac{5\pi}{4}, -\frac{3\pi}{4}, -\frac{\pi}{4}, \frac{\pi}{4}, \frac{3\pi}{4}$

33. 10°, 50°, 130°, 170°, 250°, 290°, 370°, 410°, 490°, 530°, 610°, 650°, 730°, 770°, 850°, 890°, 970°, 1010°

Section 85, p. 506

1. $\frac{\pi}{3}$ **3.** $\frac{\pi}{6}$ **5.** 0 **7.** $\frac{3\pi}{4}$ **9.** $\frac{\pi}{3}$ **11.** 1.107

13. 0.305 **15.** 1.266 **17.** $\frac{\pi}{6}$ **19.** $\frac{3}{5}$ **21.** $\frac{\sqrt{3}}{2}$ **23.** 0

25. −0.2 **27.** $\frac{\pi}{3}$ **29.** 0 **31.** $\sqrt{1-x^2}$ **33.** x

35. $2x\sqrt{1-x^2}$ **37.** $\frac{x}{\sqrt{1-x^2}}$ **39.** $\frac{2x}{1-x^2}$ **41.** $\frac{4\sqrt{5}+6}{15}$

Review Exercises for Chapter 14, p. 508

1. $\sec x = \frac{3}{2}$

$\sin x = \frac{\sqrt{5}}{3}$

$\csc x = \frac{3}{\sqrt{5}}$

$\tan x = \frac{\sqrt{5}}{2}$

$\cot x = \frac{2}{\sqrt{5}}$

3. $\cos x = \frac{1}{5}$

$\sin x = -\frac{2\sqrt{6}}{5}$

$\csc x = \frac{-5}{2\sqrt{6}}$

$\tan x = -2\sqrt{6}$

$\cot x = -\frac{1}{2\sqrt{6}}$

5. $\frac{\sqrt{2}-\sqrt{6}}{4}$

7. $\frac{6\sqrt{2}-4}{15}$

9. $\sin 2\theta = \frac{-4\sqrt{5}}{9}$

$\cos 2\theta = -\frac{1}{9}$

$\tan 2\theta = 4\sqrt{5}$

11. $\dfrac{\sqrt{2+\sqrt{2}}}{2}$ **13.** 30°, 90°, 150°, 270° **15.** 120°, 180°, 240°

17. 45°, 225° **19.** 0.7 **21.** $\frac{4}{5}$ **23.** $-\frac{7}{25}$ **25.** x

27. $\dfrac{1}{\sin\theta} - \sin\theta = \dfrac{1-\sin^2\theta}{\sin\theta} = \dfrac{\cos^2\theta}{\sin\theta} = \cos\theta\cot\theta$

29. $\dfrac{\dfrac{1}{\sin^2\theta} - \dfrac{\cos^2\theta}{\sin^2\theta}}{\tan^2\theta} = \dfrac{\dfrac{1-\cos^2\theta}{\sin^2\theta}}{\tan^2\theta} = \dfrac{1}{\tan^2\theta} = \cot^2\theta$

31. $\cos 90^\circ \cos\theta + \sin 90^\circ \sin\theta = \sin\theta$ **33.** (a) 21°48′ (b) 63°26′ (c) 88°34′

Section 86, p. 515

1. 7 **3.** $2\sqrt{13}$ **5.** $\sqrt{34}$ **7.** 5 **9.** $4\sqrt{5}$ **15.** $(0, \frac{31}{2})$

19. y-axis **21.** none **23.** origin **25.** y-axis

Section 87, p. 525

1. -2 **3.** $\frac{1}{4}$ **5.** $\frac{1}{3}$ **7.** $\frac{17}{6}$

9. (1) $2x + y - 7 = 0$ (3) $x - 4y = 0$ (5) $x - 3y + 2 = 0$
(7) $17x - 6y - 14.9 = 0$

11. $x = 1$ **13.** $x + 2y - 9 = 0$ **15.** $5x - 2y + 10 = 0$

17. $x - y + 3 = 0$

19. $2x + y - 6 = 0$
$3x + 5y - 9 = 0$
$x - 3y + 11 = 0$

23.

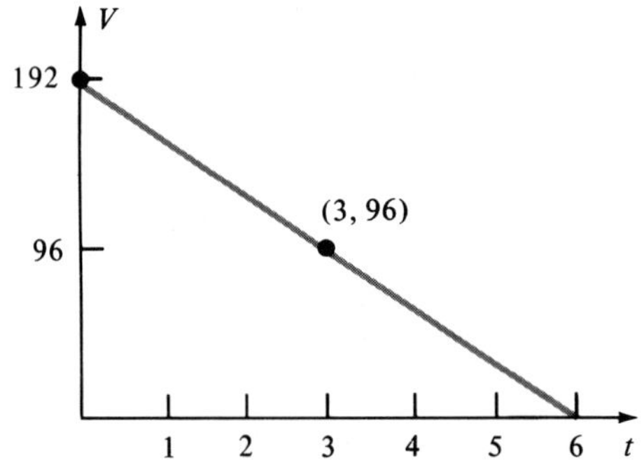

25. $I + 30V - 450 = 0$

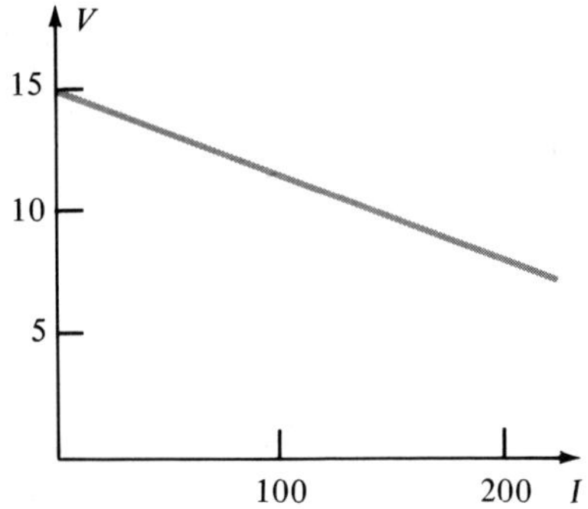

Section 88, p. 532

1. $(x-1)^2 + (y-3)^2 = 4$ **3.** $x^2 + y^2 = 1$

5. $(x-\sqrt{2})^2 + (y-\sqrt{3})^2 = 25$ **7.** $(x-2)^2 + y^2 = 9$

9. $(x-1)^2 + (y-2)^2 = 13$ **11.** $(x-2)^2 + (y-7)^2 = 36$

13. $(x+2)^2 + (y-\frac{5}{2})^2 = \frac{49}{4}$ **15.** $r = \sqrt{35}, C(3, 4)$

17. $r = \sqrt{8}, C(-1, 2)$ **19.** no points **21.** $r = 4, C(-2, 0)$

23. no points **25.** $(-2, 0)$ **27.** $(x-8)^2 + (y-6)^2 = 100$

29. $(x-1)^2 + (y-2)^2 = 34$

31. (a) $(x-h)^2 + y^2 = 25$ (b) $x^2 + (y-k)^2 = 4$ (c) $x^2 + y^2 = a^2$

35. (1, 5), (4, 2)

Section 89, p.538

1. $x' = x + 1, y' = y - 3, (x')^2 + (y')^2 = 4$
3. $x' = x + 1, y' = y - 1, 2(x')^2 + 2(y')^2 = 5$
5. $x' = x - 1, y' = y + 2, 2(x')^2 + 3(y')^2 = 34$
7. $x' = x + 1, y' = y - 2$
9. For $x = x' + 2$ and $y = y' - 1, x'y' + 8 = 0$

11.

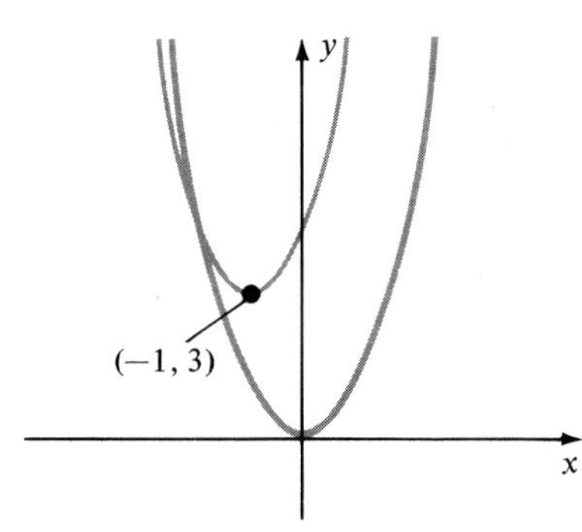

13.

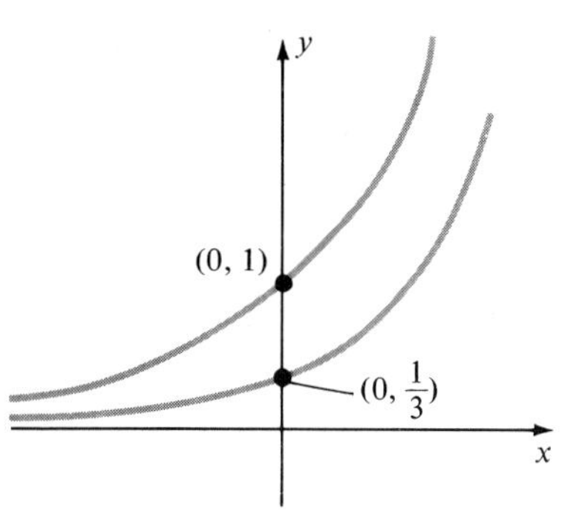

15.

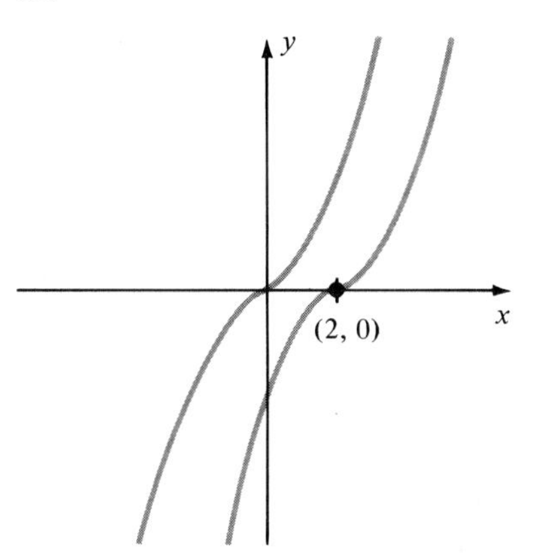

17.

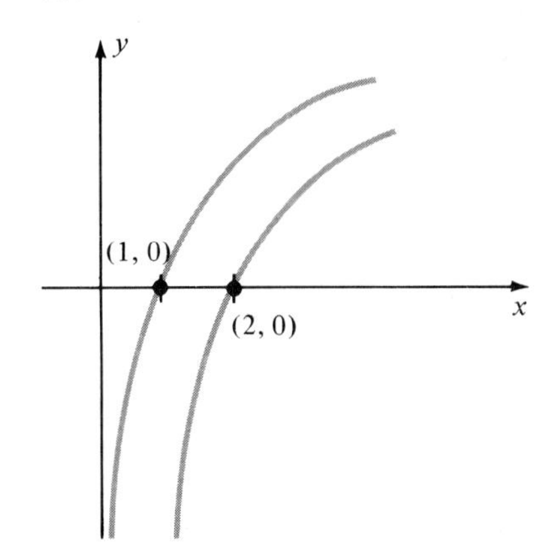

Section 90, p.549

	Vertices	Focus	Eq. of Directrix
1.	(0, 0)	(−1, 0)	$x = 1$
3.	(0, 0)	(0, 1)	$y = -1$
5.	(3, 2)	(4, 2)	$x = 2$
7.	(−1, −1)	(−2, −1)	$x = 0$
9.	(0, 2)	(0, 3)	$y = 1$
11.	(−3, −3)	(−3, −2)	$y = -4$
13.	(5, 1)	(3, 1)	$x = 7$
15.	$(-1, \frac{1}{2})$	$(-\frac{3}{4}, \frac{1}{2})$	$x = -\frac{5}{4}$

17. *Two straight lines:* $y = -2 \pm 2\sqrt{2}$ **19.** *One straight line:* $x + y = 0$
21. $x^2 = 8y$ **23.** $(y - 1)^2 = 20(x + 5)$ **25.** $(x - 3)^2 = -16(y - 9)$
27. $y^2 = 16(x - 4)$ **29.** $(x + 2)^2 = 32(y - 4)$ **31.** 16 ft, 9 ft
33. 4 sec, 256 ft **35.** Minimum value is $e = 1$

Section 91, p. 557

	Center	Vertices	Covertices	Foci
1.	$(0, 0)$	$(\pm 12, 0)$	$(0, \pm 9)$	$(\pm 3\sqrt{7}, 0)$
3.	$(0, 0)$	$(0, \pm 5)$	$(\pm 2, 0)$	$(0, \pm\sqrt{21})$
5.	$(0, 0)$	$(0, \pm 3)$	$(\pm 2, 0)$	$(0, \pm\sqrt{5})$
7.	$(0, 0)$	Circle, $r = 1$		
9.	$(0, 0)$	$(0, \pm 3)$	$(\pm 1, 0)$	$(0, \pm 2\sqrt{2})$
11.	$(1, -2)$	$(-3, -2), (5, -2)$	$(1, -4), (1, 0)$	$(1 \pm 2\sqrt{3}, -2)$
13.	$(-1, -5)$	$(-1, -9), (-1, -1)$	$(-3, -5), (1, -5)$	$(-1, -5 \pm 2\sqrt{3})$
15.	$(-5, -2)$	$(-18, -2), (8, -2)$	$(-5, -14), (-5, 10)$	$(-10, -2), (0, -2)$
17.	$(-2, 3)$	$(-2, 0), (-2, 6)$	$(-4, 3), (0, 3)$	$(-2, 3 \pm \sqrt{5})$
19.	$(1, -2)$	$(1 \pm \sqrt{5}, -2)$	$(1, -3), (1, -1)$	$(3, -2), (-1, -2)$

21. $\dfrac{x^2}{12} + \dfrac{y^2}{16} = 1$ **23.** $\dfrac{x^2}{144} + \dfrac{y^2}{169} = 1$ **25.** $\dfrac{x^2}{32} + \dfrac{(y-2)^2}{36} = 1$

27. $\dfrac{x^2}{25} + \dfrac{y^2}{10} = 1$ **29.** $\dfrac{(x-6)^2}{9} + \dfrac{(y+4)^2}{4} = 1$ **31.** $6\sqrt{2}$ ft

33. $\dfrac{x^2}{3.6(10^7)} + \dfrac{y^2}{2.5(10^7)} = 1$

Section 92, p. 566

	Vertices	Foci	Asymptotes
1.	$(\pm 4, 0)$	$(\pm 5, 0)$	$y = \pm\frac{3}{4}x$
3.	$(\pm 15, 0)$	$(\pm\sqrt{274}, 0)$	$y = \pm\frac{7}{15}x$
5.	$(0, \pm 2)$	$(0, \pm\sqrt{5})$	$y = \pm 2x$
7.	$(0, \pm 1)$	$\left(0, \pm\dfrac{\sqrt{13}}{2}\right)$	$y = \pm\frac{2}{3}x$
9.	$(-2, 1), (-2, 7)$	$(-2, 4 \pm \sqrt{13})$	$y - 4 = \pm\frac{3}{2}(x + 2)$
11.	$(5, -3), (-3, -3)$	$(6, -3), (-4, -3)$	$y + 3 = \pm\frac{3}{4}(x - 1)$
13.	$(-2, 2), (-2, -6)$	$(-2, -2 \pm 2\sqrt{5})$	$y + 2 = \pm 2(x + 2)$
15.	$(0, 0), (0, -\frac{4}{9})$	$\left(0, \dfrac{-2 \pm 2\sqrt{10}}{9}\right)$	$y + \dfrac{2}{9} = \pm\dfrac{x}{3}$
17.	$(-\frac{1}{2}, -\frac{7}{4}), (-\frac{1}{2}, \frac{9}{4})$	$(-\frac{1}{2}, \frac{1}{4} \pm 2\sqrt{5})$	$y - \frac{1}{4} = \pm\frac{1}{2}(x + \frac{1}{2})$
19.	$(-1, 2), (1, 2)$	$(\pm\sqrt{5}, 2)$	$y - 2 = \pm 2x$

21. $\dfrac{x^2}{36} - \dfrac{y^2}{28} = 1$ **23.** $\dfrac{x^2}{4} - \dfrac{y^2}{12} = 1$ **25.** $\dfrac{y^2}{9} - \dfrac{x^2}{16} = 1$

27. $\dfrac{(x-3)^2}{9} - \dfrac{(y-1)^2}{16} = 1$ **29.** $\dfrac{y^2}{16} - \dfrac{x^2}{9} = 1$ **31.** no solution

33. $(4, 0)$ and $(0, -4)$ **35.**

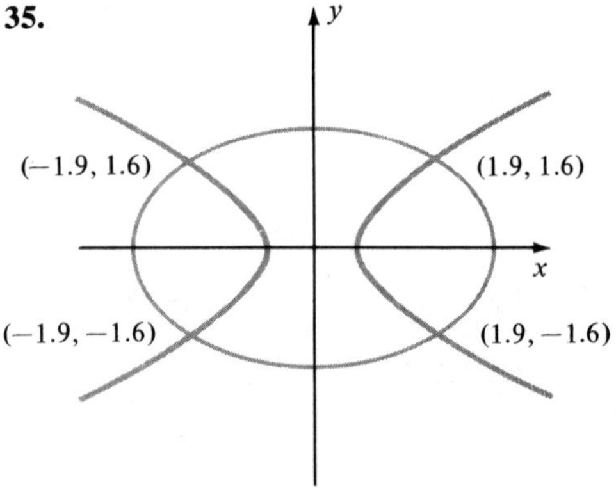

y
(−1.9, 1.6)
(1.9, 1.6)
x
(−1.9, −1.6)
(1.9, −1.6)

37. $K = 28$

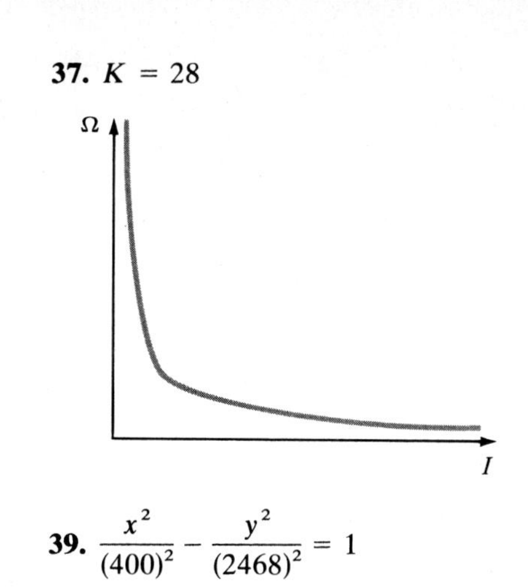

39. $\dfrac{x^2}{(400)^2} - \dfrac{y^2}{(2468)^2} = 1$

Section 93, p. 570

1. Circle **3.** Hyperbola **5.** Parabola **7.** Circle **9.** Circle **11.** Hyperbola **13.** Circle **15.** Ellipse **17.** Circle **19.** Hyperbola

Section 94, p. 583

1.

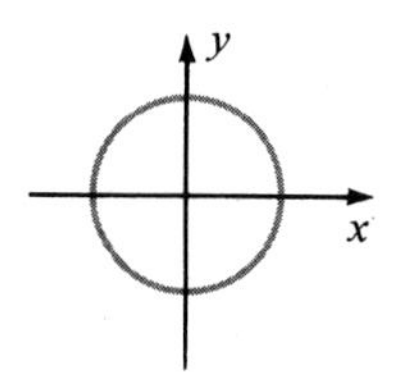

3.

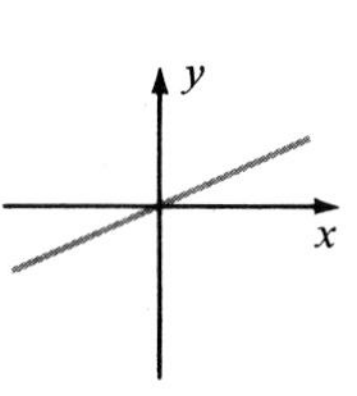

5.

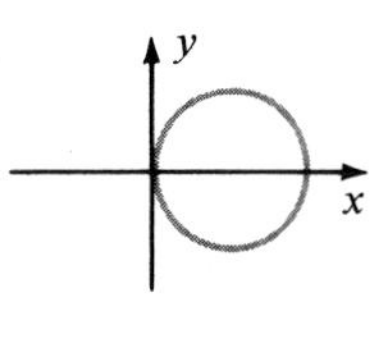

7.

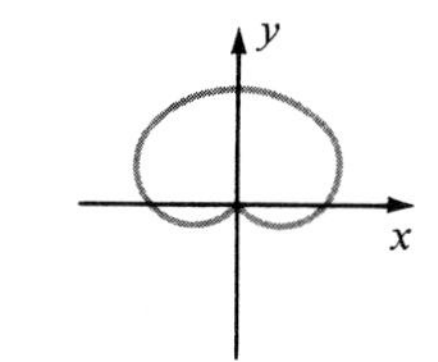

9.

11.

13.

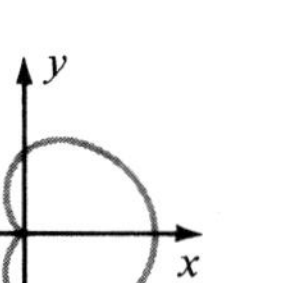

15.

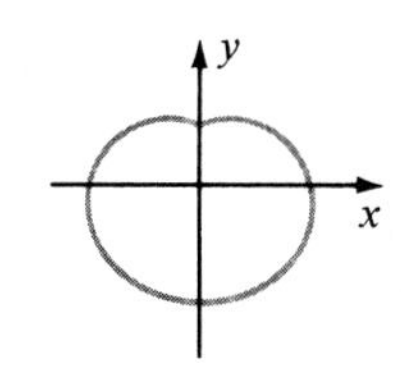

17.

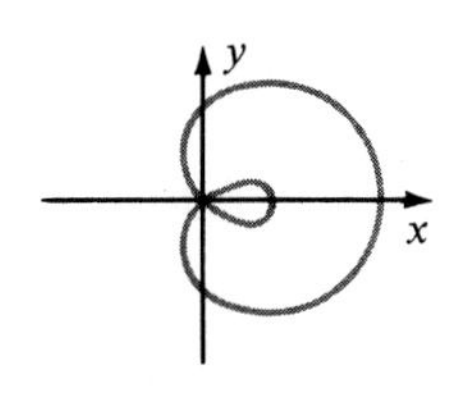

19.

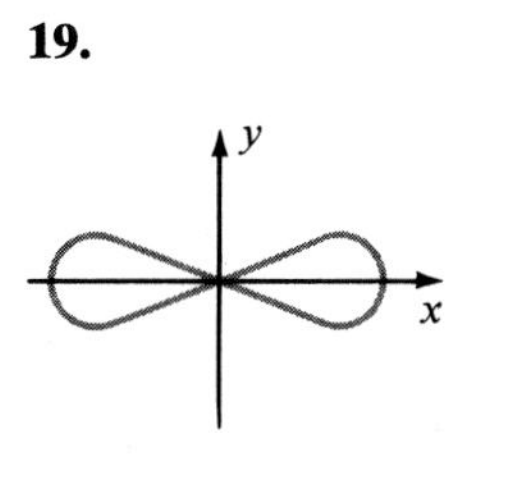

21.

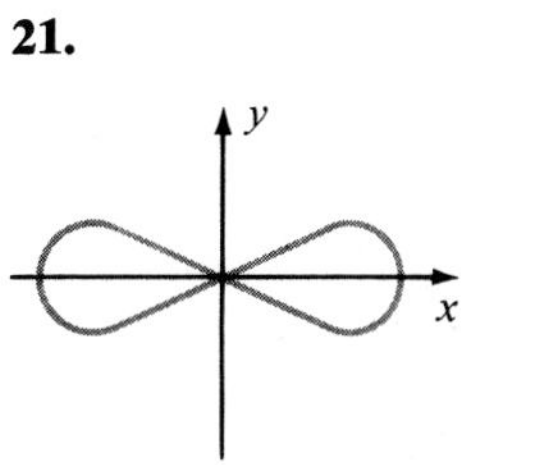

23.

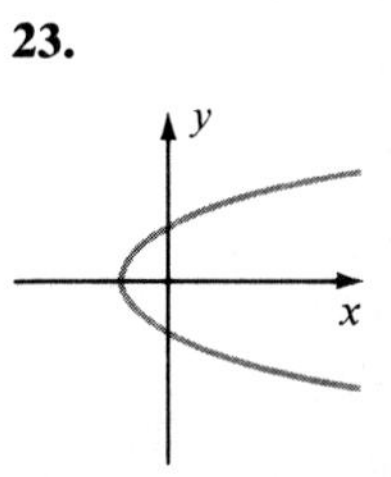

25.

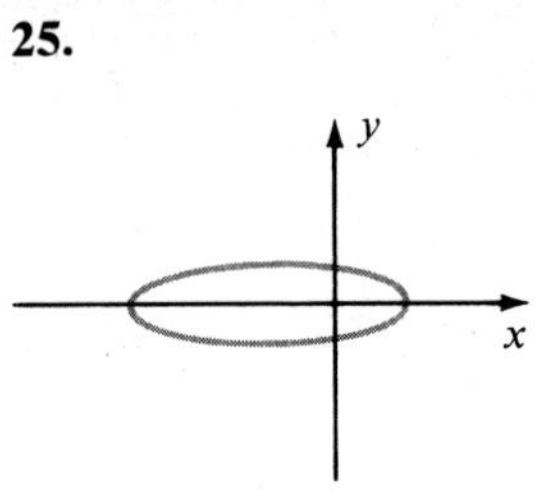

27. (a) $(x - \frac{1}{2})^2 + (y - \frac{1}{2})^2 = \frac{1}{2}$

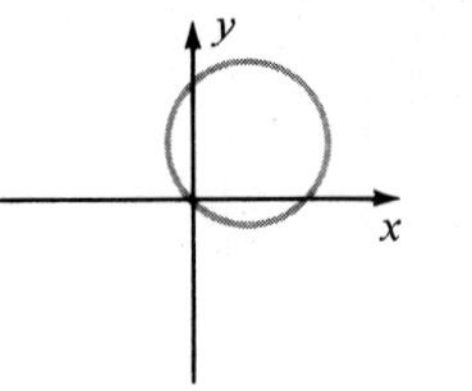

27. (b) $(x - 2)^2 + y^2 = 4$

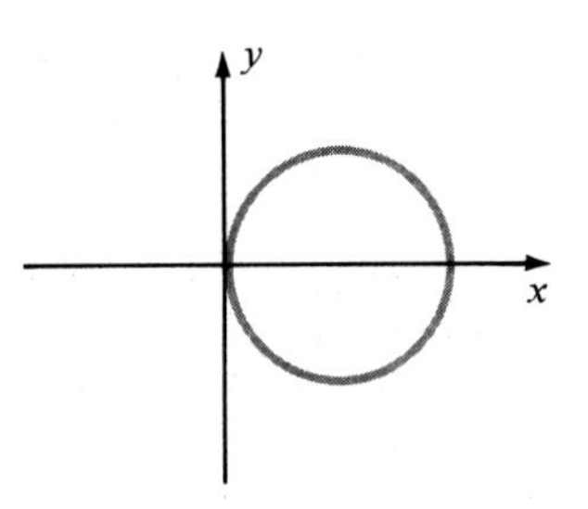

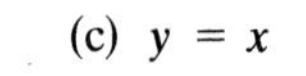
(c) $y = x$

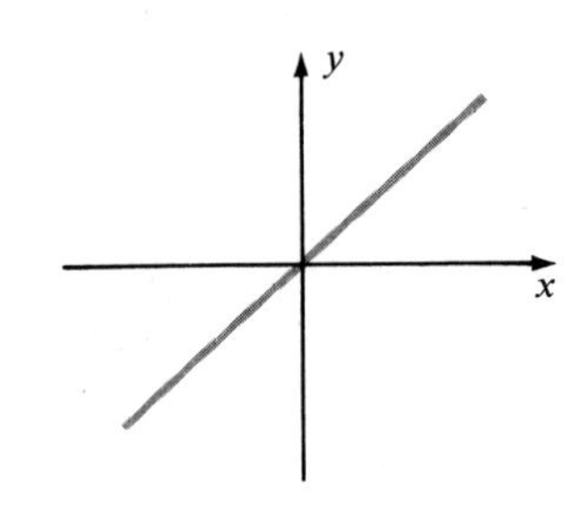

(d) $\frac{x^2}{9} + \frac{y^2}{16} = 1$

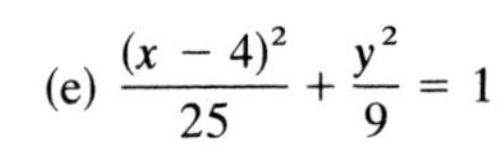
(e) $\frac{(x - 4)^2}{25} + \frac{y^2}{9} = 1$

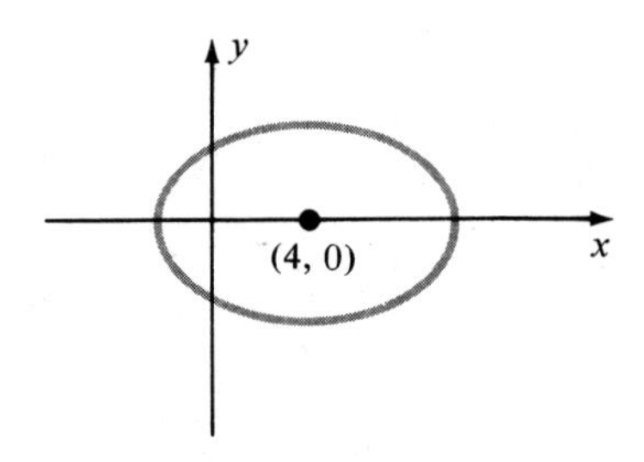

33. $22\frac{1}{2}°$, 157°, 208°, 330°

35.

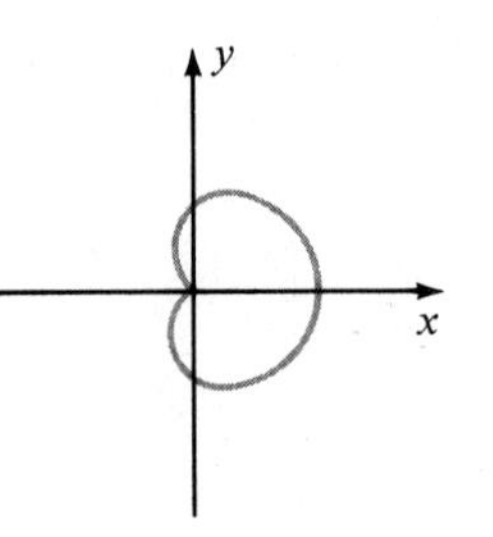

Section 95, p. 589

1. $x - 2y - 1 = 0$ **3.** $3x - 2y - 19 = 0$ **5.** $y = x^2 - 2x - 2$

7. $x^2 + y^2 = 9$ **9.** $y = 1 - \frac{x^2}{4}$ **11.** $x = \frac{t+1}{t}, y = t + 1$

13. $x = \frac{2}{\sqrt{t^2 + 1}}, y = \frac{2t}{\sqrt{t^2 + 1}}$

15.

17. Range $= \frac{v_0^2}{g} \sin 2\alpha$

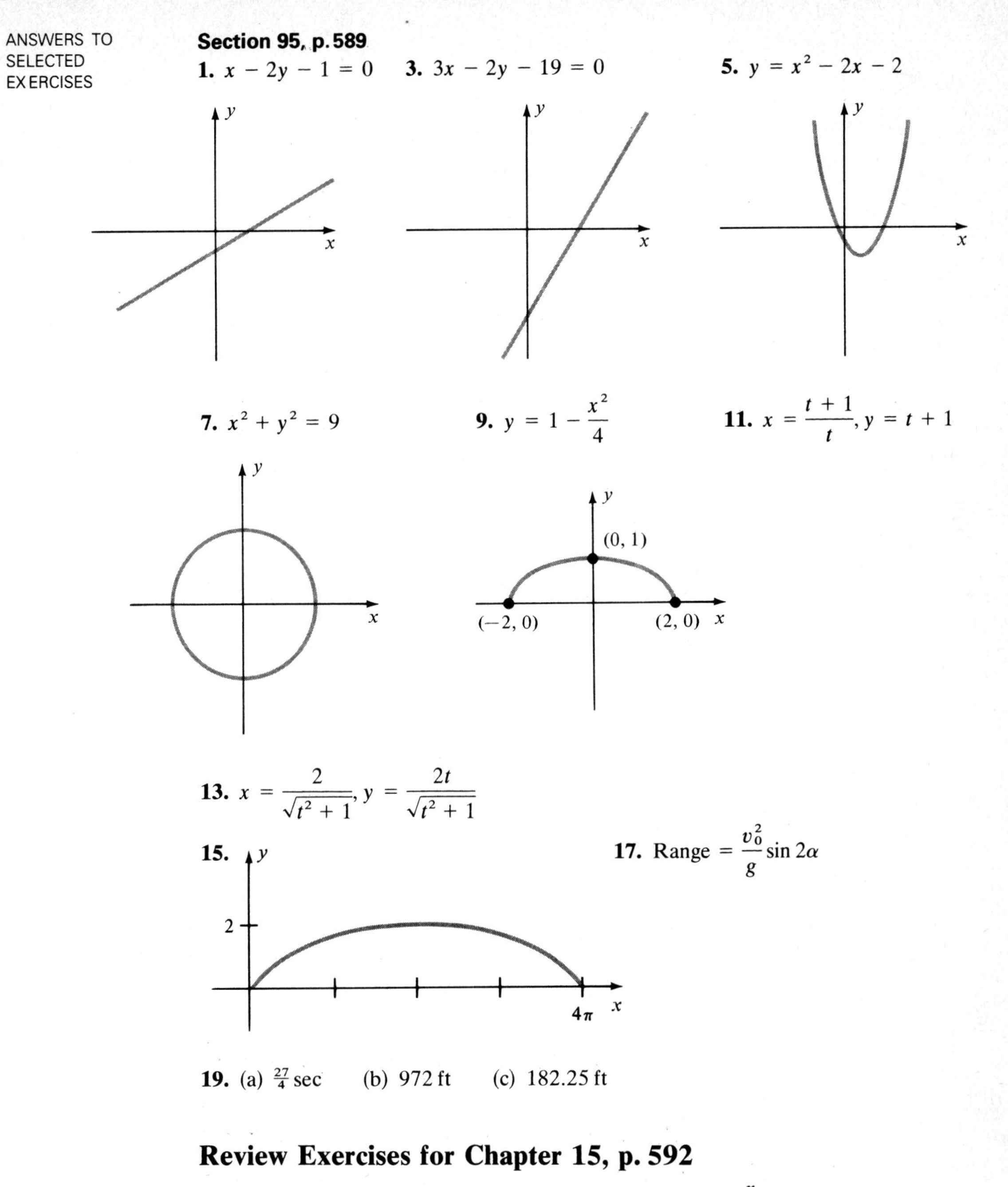

19. (a) $\frac{27}{4}$ sec (b) 972 ft (c) 182.25 ft

Review Exercises for Chapter 15, p. 592

1. (a) 5 (b) $\sqrt{89}$ **3.** $x - 2y + 3 = 0$ **5.** $y = \frac{x}{3} - 5$

7. $2x - y - 8 = 0$ **9.** $C(-2, -3), r = 5$

11. Vertex $(2, -3)$
Focus $(2, -1)$
Directrix $y = -5$

13. Center $(0, 0)$
Vertices $(\pm 6, 0)$
Foci $(\pm 3\sqrt{3}, 0)$
Covertices $(0, \pm 3)$

15. $\dfrac{x^2}{25} + \dfrac{y^2}{9} = 1$

17. $y - 1 = \pm\frac{1}{2}(x + 2)$

19. Center $(2, -1)$
Vertices $(-2, -1), (6, -1)$
Foci $(2 \pm \sqrt{41}, -1)$
Asymptotes: $y + 1 = \pm\frac{5}{4}(x - 2)$

21. Circle: $(x - 2)^2 + (y + 1)^2 = 9$

23. Ellipse: $\dfrac{x^2}{36} + \dfrac{(y + 1)^2}{16} = 1$

25. Circle: $(x - 3)^2 + (y - 4)^2 = 36$

27. Parabola: $(y + 2)^2 = x - 1$

29. Hyperbola: $\dfrac{(x - 4)^2}{2} - \dfrac{(y = 2)^2}{4} = 1$

31. **33.**

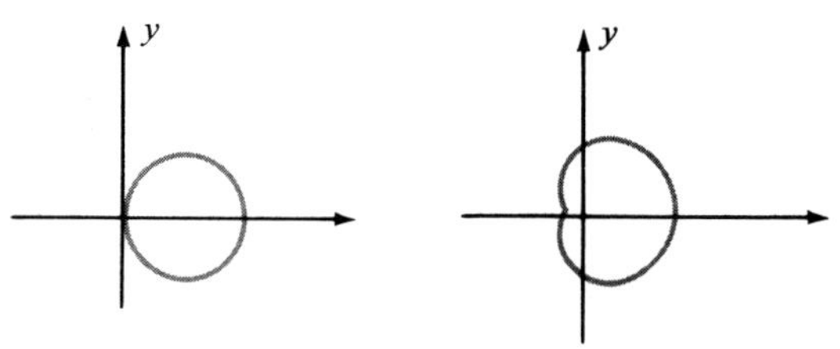

35.

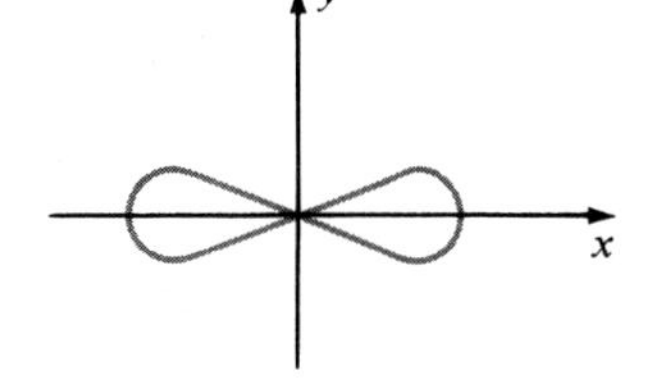

37.

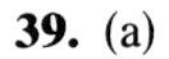
39. (a)

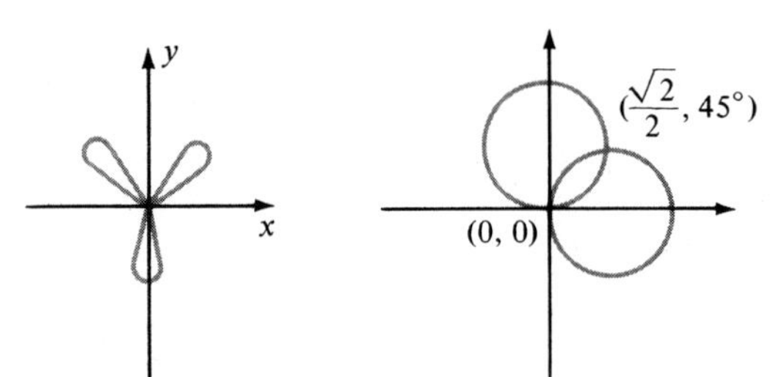

(b) $r = 1$ for $\theta = \pm\dfrac{\pi}{6}, \pm\dfrac{\pi}{3}, \pm\dfrac{2\pi}{3}, \pm\dfrac{5\pi}{6}$

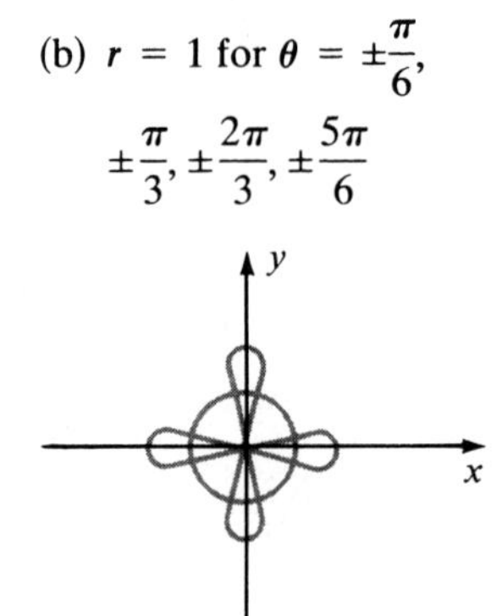

41. $y = x^2 + 2x$

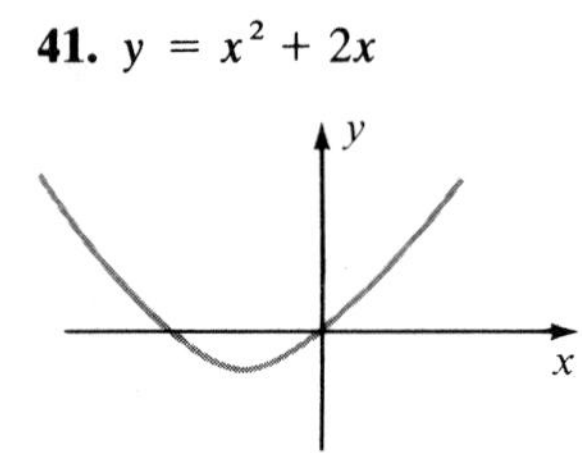

Appendix A, 603

1. False **3.** True **5.** False **7.** False **9.** True
11. True **13.** False **15.** False
17. (a) 170° (b) 144° (c) 90° (d) 62° (e) 30° **19.** 15,18
21. 36 ft **23.** 34.4 in.; 48 in.^{2} **25.** 25 cm **27.** 40 in.; 84 in.^{2}
29. 497.3 cm^2; 89.2 cm **31.** 14.4 cm **33.** 22.6 in. **35.** 89.7 cm^2

37. 7.81 cm^2 **39.** 17.6 cm; 24.6 cm^2 **41.** $2\sqrt{3}$ cm^2
43. $2\sqrt{21}$ cm; $8\sqrt{21}$ cm^2 **45.** 79.5 cm^3 **47.** 22.5 m^3
49. 46.19 in.^{3} **51.** 19.4 ft^3 **53.** $36\sqrt{3}$ cm^3 **55.** $12\sqrt{3}$ cm^3
57. 136 in.^{2} **59.** 229.12 in.^{2}

Appendix B, 611

1. 60.96 **3.** 1076.96 **5.** 49.14 **7.** 96.6 **9.** 3.57
11. 4027.4 **13.** 2163.48 **15.** 120.09 **17.** 1.6 **19.** 101.41
21. 7.26 **23.** 17.5 **25.** 40° **27.** −15° **29.** 41°
31. 328.08 **33.** 2.9946×10^8 **35.** 994.5 **37.** 2907.8
39. 56° **41.** 329.4

Appendix C, 615

1. 1.27 **3.** 4 **5.** 9836.6 **7.** 3.35 **9.** −975.36
11. 49.6; 73 **13.** 3.79 **15.** 2.99×10^7 **17.** 2

Appendix D, 621

1. 4, hundredths place **3.** 1, tens place **5.** 2, thousandths place
7. 4, thousandths place **9.** 1, ten-thousandths place
11. 5, ten-thousandths place **13.** 2, thousands place
15. 3, ten-thousandths place

17.	**19.**	**21.**	**23.**	**25.**
742,400	0.07284	231.4	10,260	0.02145
742,000	0.0728	231	10,300	0.0214
740,000	0.073	230	$1\tilde{0}$,000	0.021

27.	**29.**	**31.**	**33.**	**35.**
83,960	23,850	227.0	1.71	21
$84,\tilde{0}00$	23,800			
84,000	24,000			

37. 13 **39.** 16.9 **41.** 22 **43.** 2680 **45.** 1000
47. 24 **49.** 57

INDEX